## 李国杰院士简介

李国杰，1943 年出生于湖南，1985 年在美国普渡大学获得博士学位。主要从事高性能计算机、并行算法、互联网、人工智能、大数据等领域的研究和产业化工作，发表了 150 多篇学术论文，出版了《创新求索录》《创新求索录（第二集）》个人文集。主持研制了“曙光一号”并行计算机和“曙光 1000”大规模并行计算机等高性能计算机，获得了国家科学技术进步奖一等奖和首届何梁何利基金科学与技术进步奖等奖励。1995 年创建的曙光公司已成为我国高性能计算机的骨干企业。1990—2000 年担任国家智能计算机研究开发中心主任，2000—2011 年担任中国科学院计算技术研究所所长，现任该所首席科学家，兼任中国计算机学会名誉理事长等职。1995 年当选中国工程院院士，2001 年当选发展中国家科学院（TWAS）院士。

我独立从事科研工作是从这里（中国科学院计算技术研究所图书馆阅览室）开始的。1980 年我从中国科学技术大学转到中国科学院计算技术研究所代培读硕士，经常在这个阅览室里查阅资料。当时中国科学院计算技术研究所图书馆拥有的计算机类外文专著、期刊和国际会议论文集的数量为全国之最，吸引了全国各地的学者

我走上科研道路的引路人是我的硕士导师夏培肃教授。她推荐我去美国读博士，也是在她的召唤和安排下，1987 年我回国到中国科学院计算技术研究所工作。上图为 1988 年在北京大学参加一次关于并行计算的国际研讨会后，我陪同夏培肃教授与李政道教授交流

1981 年 9 月至 1985 年 5 月，我在美国普渡大学攻读博士学位。我的导师华云生（Benjamin W. Wah）教授比我还年轻，是当时美国计算机界的新起之秀。华教授现在是香港中文大学常务副校长。此照片摄于 1981 年华教授家中

1986 年，IEEE Computer Society 出版社出版了华云生教授与我编著的《Computers for artificial intelligence applications》，此书连续 3 年都是其社畅销书

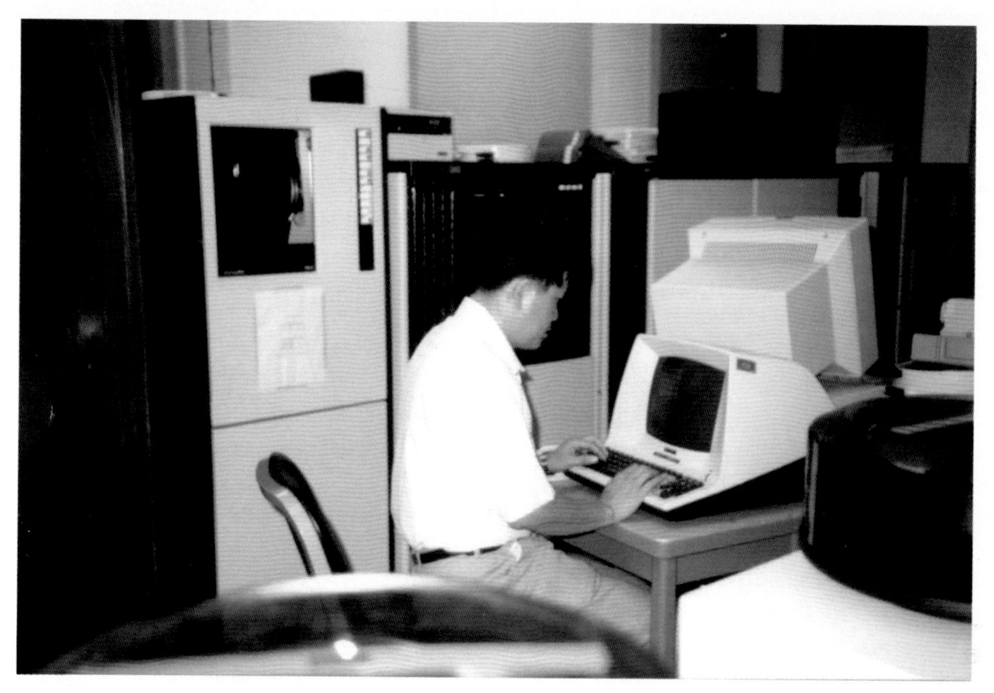

我在普渡大学的 4 年中，除了上课，几乎天天泡在办公室和实验室中。当时还没有个人计算机（PC），我使用的是 VAX 小型计算机，终端只显示命令行，没有图形界面

1984 年，我在美国人工智能协会（AAAI）年会上做学术报告，当时用的是胶片投影仪

1992 年，我邀请知识工程之父爱德华 · 阿尔伯特 · 费根鲍姆（Edward Albert Feigenbaum）教授（于 1994 年获得图灵奖）来国家智能计算机研究开发中心讲课。坐在第一排听课的是（右起）王鼎兴、戴汝为、张钹、许卓群教授

1993 年，我邀请精简指令集计算机（RISC）发明人、2016 年图灵奖获得者大卫 · 帕特森（David Patterson）教授来国家智能计算机研究开发中心讲课。上图是我向他介绍国家智能计算机研究开发中心正在研制的曙光并行计算机

1992 年，我与“曙光一号”并行计算机研制组开现场工作会，分析引进的 Encore 并行计算机主板布局

1995 年，在北京友谊宾馆召开“曙光 1000”大规模并行计算机成果发布会，我在会上做研制工作报告

1980 年，普渡大学黄铠教授来国内讲学时，接收我去普渡大学读博士，这改变了我的人生历程。徐志伟（右）是黄教授后来招收的博士，1996 年回国后也到中国科学院计算技术研究所工作，他是我科研工作的主要合作者，我们联名发表了几篇论文（上图是我们 2007 年相聚时的合影）

这张笑容灿烂的照片是我与部分学生的合影，本书收集的许多论文的合作者都在场，包括（第一排）孙凝晖（左 5）、程学旗（右 4）、熊劲（右 2），（第二排）陈明宇（右 2）、马捷（右 4），（第三排）张国清（右 1）等

20 世纪 90 年代初期，国家智能计算机研究开发中心的研发环境

以“曙光一号”并行计算机知识产权为基础建立的曙光公司，20 多年来经历了许多波折，现在市值已接近 600 亿元，成为我国发展高性能计算机的龙头企业。此照片的背景是耸立在曙光公司门前的硅立方超级计算机——地球模拟器

2001 年，我获得普渡大学颁给杰出校友的 Outstanding Engineer 奖

2001 年，我被选为发展中国家科学院（TWAS）院士，2002 年在印度接受院士证书

# 李国杰院士 学术论文选集

李国杰 著

人民邮电出版社
北京

**图书在版编目（CIP）数据**

李国杰院士学术论文选集 / 李国杰著. -- 北京 : 人民邮电出版社, 2020.9
ISBN 978-7-115-54691-3

Ⅰ. ①李… Ⅱ. ①李… Ⅲ. ①计算机科学－文集 Ⅳ. ①TP3-53

中国版本图书馆CIP数据核字(2020)第156925号

## 内 容 提 要

本书汇集了李国杰院士1981—2019年发表（包括联名发表）的150多篇学术论文中的51篇影响较大的论文，内容包括数据流计算机和Systolic阵列、并行组合搜索、智能计算机、人工智能与计算理论、并行算法、高性能计算机、微处理器、下一代网络与通信、大数据系统等。本书还讲述了相关研究背后的历史故事和感悟，清晰展示了李国杰院士的科研历程及学术成就，可供从事计算机科学技术研究、对近40年计算机技术发展历程有兴趣的读者参考。

◆ 著　　　　李国杰
责任编辑　吴娜达
责任印制　彭志环

◆ 人民邮电出版社出版发行　　北京市丰台区成寿寺路11号
邮编　100164　　电子邮件　315@ptpress.com.cn
网址　http://www.ptpress.com.cn
北京捷迅佳彩印刷有限公司印刷

◆ 开本：787×1092　1/16　　彩插：6
印张：44.5　　2020年9月第1版
字数：1000千字　　2020年9月北京第1次印刷

定价：320.00元

**读者服务热线：(010)81055493　印装质量热线：(010)81055316**
**反盗版热线：(010)81055315**

# 序 言

1978 年我考上了研究生，1980 年初从中国科学技术大学转到中国科学院计算技术研究所（以下简称计算所）代培，师从夏培肃和韩承德老师，才真正开始独立从事科研工作。1981 年，我在《电子计算机动态》（后改名《计算机研究与发展》）期刊上发表了我的处女作《一种新的体系结构——数据流计算机》。从那以后，我先后在各种期刊和国际会议上发表了 150 多篇学术论文。这些论文涉及计算机系统结构、组合搜索、并行算法、人工智能、智能计算机、高性能计算、微处理器设计、网络与通信、大数据系统等诸多领域。除了少数文章发表在 *Nature* 子刊、*Science*、*Physical Review* 和 *Nucleic Acids Research* 之外，大多数文章发表在计算机领域的重要期刊和会议上，如 *Proceeding of the IEEE*、*Communication of the ACM*、*IEEE Computer*、*IEEE Micro*、*IEEE Transaction on Computer*、*IEEE Transaction on Software Engineering*、*IEEE Transaction on System,Man and Cybernetics*、《计算机学报》《软件学报》《通信学报》《模式识别与人工智能》《中国科学院院刊》等期刊和 ISCA、AAAI、ICPP 等著名国际会议。最近的一篇文章发表在 2019 年 6 月的《中国科学院院刊》上。从 1981 年到 2019 年，时间跨度近 40 年。我从这 150 多篇学术论文中挑选了 51 篇，出版了这本学术论文选集，这本选集基本上反映了我一生的科研工作。

我攻读博士学位、做博士后研究和回国工作的前几年，发表的论文多数是以本人为第一作者。担任国家智能计算机研究开发中心主任以后，特别是 1999 年年底担任中国科学院计算技术研究所所长以后，我的主要精力花在组织和发展科研队伍、确定团队的科研方向、筹措科研经费、参与制定国家科技计划等方面，亲力亲为做学术研究和指导研究生已不是我的主要工作，近十几年我的学生由较年轻的导师与我共同培养。40 年来，我培养的硕士生、博士生已毕业 70 多人，他们在学习期间发表的论文有 200 多篇，只有我做了明显贡献的论文我才会署名。本选集收集的与学生合作的论文大多是 1998 年以前入学的学生写的，1998 年以后入学的学生有 40 多人，他们写的论文我几乎都没有署名。

改革开放以来，计算所在高性能计算机、CPU 芯片、未来网络和通信、人工智能、大数据等领域做出了出色的成绩，从本选集的论文中可以看出计算所技术研究的脉络。技术的发展是波浪式的，往往是二三十年一个周期。年轻的学者一般只看近 20 年以内的论文，殊不知当前的一些热点 30 年前就曾经发表过许多相关的研究成果。重新阅读这些过去的论文，脑海中油然而生李白的诗句：“今人不见古时月，今月曾经照古人”。本选集有不少综述性质的长文，这些论文大多是出版机构的约稿。据我的体会，写一篇有分量的综述论文比写一篇原创性论文还难，而好的综述论文往往比一般的研究性论文历史价值更悠久。

Systolic Array 现在被翻译成“脉动阵列”，1979 年我在计算所读硕士时以此为研究方向，当时还没有合适的中文翻译，我的硕士论文题目是《阵列流水算法和流水式阵列处理机》。本选集第 1 章收录了我在《计算机学报》上最早发表的论文《用参数确定法设计阵列流水算法》，这是我硕士论文的一部分。我到美国学习以后，将这篇论文修改为《设计最优脉动阵列的方法（The design of optimal systolic arrays）》，发表在 *IEEE Transactions on Computers* 1985 年第 1 期。从修改这篇论文开始，我学会了做基础研究的一个基本方法：面对一个科学问题，先要问“此问题有没有最优算法”，如果有，就要找到它并证明其是最优算法。做到这一点等于为此问题的理论研究画上句号。如果没有，就要证明此问题是 NP 难问题，再找尽可能好的启发式算法。一个问题可能用许多不同的 Systolic 算法求解，用我的论文提出的参数法可以给出最优的 Systolic 算法，也就是说，不用再找其他 Systolic 算法了，这应该是一个很漂亮的科研成果。此论文已被引用 311 次，至今仍是我执笔写的引用率最高的学术论文。由于 Google 公司的 TPU 芯片采用了 Systolic 技术，Systolic 阵列再次成为热点，重新引起人们关注。2019 年全球几万人参加的超级计算机大会（Supercomputing Conference）上，一篇关于 GPU 设计的论文引用了我这篇论文。一篇论文 34 年后仍被人引用，说明寻求最优算法的研究思路有较长的生命力。

本选集第 2 章反映了我在美国普渡大学 4 年的博士生涯。我的博士论文研究方向是“并行组合搜索”。受日本第五代计算机的影响，20 世纪 80 年代学术界普遍认为并行处理是解决人工智能问题的主要途径。我的研究结果表明：并行处理不是“万能药”。并行处理只能提高串行计算机可求解问题的计算效率，不能用来扩大求解指数复杂性问题的求解规模。而且，在并行组合搜索中，可能出现“并行不如串行”等异常现象。我分析了出现各种异常现象的原因，给出了保证性能的充分条件和获得超线性加速的必要条件。1984 年我在美国人工智能协会（AAAI）年会上发表了有关这方面的研究成果。现在 AAAI 是人工智能界的顶级会议，近几年发表的论文半数以上出自华人作者，但当时我去开会时没有遇到国内来的学者。1985 年 6 月在 *IEEE Computer* 上发表的长文 *Multiprocessing of combinatorial search problems* 较全面地阐述了我的博士论文研究成果，这篇论文至今已有 140 次引用，最近的引用是在 2017 年。回国以后我继续指导我的学生做并行组合搜索方向的研究，孙凝晖是我回国后招收的第一个硕士生，他 1992 年发表的关于组合搜索任务的负载平衡方法是其硕士论文的一部分，论文名为《采用有效切分的负载平衡》。现在他已经是计算所所长，基于他对高性能计算的突出贡献，2019 年被选为中国工程院院士。

20 世纪 80 年代，美国也跟随日本开展了智能计算机研究，但美国学者的研究方向较为广泛，不限于 Prolog 计算机。1984 年我和华云生教授在计算机体系结构国际研讨会（ISCA）上发表了一篇求解组合搜索问题的多处理机（取名 MANIP 计算机）的论文。这是试图用并行处理对付组合爆炸的一次尝试，后来我转做并行组合搜索的异常性分析，MANIP 计算机的研究没有继续做下去。由于我在 IEEE Computer Society 出版社出版的教辅（Tutorial）*Computers for artificial intelligence applications* 的影响很大，好几家期刊约华云生教授和我写智能计算机方面的综述，我在伊利诺伊大学厄巴纳 - 香槟分校做博士后期间，写了几篇有关智能计算机的综述论文，分别发表在 *SIGART Newsletter*、*Proceedings of the IEEE* 和 *IEEE Transactions on Systems, Man, and Cybernetics* 期刊上。最长的论文 *Computers for symbolic processing* 在本选集长达 58 页。这些综述论文对现在的学者了解上一波人工智能浪潮颇有帮助，论文提出的一些关于智能计算机的观点对于今天的研究仍然有借鉴意义。特别是这些综述论文中都附有数百篇参考文献，今天的学者要溯源智能计算机，了解 30 年以前这一领域的研究成果，这些参考文献可以提供难得的索引。近几年深度学习和神经网络火热，各种神经网络加速器不断推出。由于半导体工艺技术的进步，今天的神经网络加速芯片的性能已比 30 年前高出几个数量级，但神经网络计算机面临的基本科学问题并没有改变。1990 年，我在国内的《模式识别与人工智能》期刊上发表了一篇论文

《神经网络计算机的体系结构》，这篇论文重点讨论了设计神经网络计算机应考虑的几个技术难点，包括映射理论、虚拟处理器、并行粒度、符号处理与神经网络计算相结合等，这些问题今天仍然需要重视。

20 世纪 80 年代中期，计算所的科研条件很差。我回国到计算所工作，计算所没有上机条件，只能在我自己从美国带回来的 8 位 IBM PC 上做研究。令人难以置信的是，我竟然在如此简陋的条件下做模拟实验，写出了一篇论文《基于熵的人工神经网络系统理论初探》，发表在 1990 年的《计算机学报》上。这个研究方向可能很有价值，可惜后来没有继续做下去。刚回国时我协助夏培肃先生指导了她的博士生唐志敏，第 4 章收录了他的一篇论文。唐志敏现在是在服务器 CPU 正面战场上能与 Intel 公司叫板的本土企业成都海光集成电路设计有限公司的技术负责人。20 世纪 90 年代初，我指导了一位中国科学技术大学转到我这里做论文研究的博士生姚新，第 4 章中的论文 *General simulated annealing* 是他博士论文的一部分。他后来成为国际上演化计算的领军人物，目前在南方科技大学担任计算机系主任。他指导的博士陈天石现在是智能加速芯片的"独角兽"——北京中科寒武纪科技有限公司的总裁。国家智能计算机研究开发中心成立以后，不但开展了并行计算机的研制，也进行了人工智能理论的研究，而且与中国自动化学会一起创办了《模式识别与人工智能》期刊，我长期担任此刊的副主编。20 世纪 90 年代初我在此期刊上发表了几篇论文，其中一篇《人工智能的计算复杂性研究》今天读来仍有新鲜感，对于目前的人工智能研究可能有启发价值。关于 SAT 问题的综述性论文是与顾钧教授合写的。顾钧是一位很有独创性的学者，他是我国信息领域第一个主持 973 计划项目的首席科学家，可惜几年前突然人间蒸发，杳无音讯。国家智能计算机研究开发中心早期还做过一件有意思的事，将国内 3 家（包括计算所）印刷体文字识别技术集成在一起，识别率提高一个百分点，误识率降低一个数量级，达到千分之一。论文《技术综合集成在模式识别中的应用》反映了这个成果，其主要贡献者刘昌平现在是汉王科技股份有限公司的总裁。

算法研究一直是我的主要兴趣。我做算法研究不是对别人提出的算法做一些增量性的改进，而是试图从全局的角度更深入地理解一些典型的算法。第 5 章选入的第一篇论文 *Parallel processing of serial dynamic programming problems* 是这类研究的代表。根据动态规划（DP）的不同递归表示形式，我将 DP 分成 Monoadic-Serial、Polyadic-Serial、Monoadic-Nonserial 和 Polyadic-Nonserial 4 类，并给出了适合不同类型 DP 的并行处理方法。后来这一分类方法被几位著名学者写入《Introduction to parallel computing》等流行的教科书，现在这种动态规划的分类已成为普遍接受的知识，没有人关心它的出处了。这篇论文发表在当时刚创办的国际会议 Computer Software and Application Conference（COMPSAC）上。如今COMPSAC 已成为软件工程的顶级会议，但当时还并没有多少人关注。我回国后早期指导的学生大多也是做算法分析和复杂性研究，第 5 章选入的几篇论文多数是我指导的学生写的。值得一提的是我的学生卜东波牵头写的关于蛋白质结构分析的论文发表在生物领域著名期刊 *Nucleic Acids Research* 上，已被引用 732 次。网格技术是 21 世纪初热门的研究方向，后来推动了云计算的发展。徐志伟研究员是我在普渡大学读博士时的同学，他到普渡大学时是我去机场接他的，他回国后到计算所工作时，也是我去接他的，这是我们的缘分。我与他合作写过几篇文章，第 5 章选入的关于网格系统软件的论文是我们合作研究的成果。

高性能计算是我回国后最主要的研究方向，选入本选集的论文达 11 篇之多。我回国后被李政道先生邀请加入他创办的中国高等科学技术中心，与物理学家有过一段合作的经历，所以我有一篇在《计算物理》期刊上的论文。我回国后 30 多年最牵肠挂肚的是高性能计算机产业。曙光计算机现在已成为中国发展高技术实现产业化的一个样板，曙光系列高性能计算机连续研制了 8 代产品，2019 年鉴定的“曙光 7000”的性能已经达到几百 PFlops 水平，曙光人的基因在研制“曙光一号”并行计算机时已经形成。第 6 章收录了 3 篇介绍“曙光一号”并行计算机等早期产品研发工作的论文。陈鸿安和樊建平是国家智能计算机研究开发中心派往美国研发“曙光一号”并行计算机的骨干人员，他们分别介绍了“曙光一号”并行计算机的系统结构和操作系统。樊建平后来担任过计算所副所长，近十几年任中国科学院深圳先进技术研究院院长。1994 年我作为大会特邀报告人，在日本召开的一次国际会议上介绍了国家智能计算机研究开发中心的并行处理研究进展，重点介绍了当时刚开始研制的东方 1 号 MPP 计算机的设计思路（后改名为“曙光 1000”大规模并行计算机），此报告也收录在第 6 章中。这一章的论文多数是以我的学生为第一作者，其中与我合写论文较多的是熊劲。她是我 1990 年招收的研究生，一直在计算所做存储系统方面的研究。曙光高性能计算机的应用很广，水稻基因组测试是具有国际影响的一个范例。我将 2002 年在 *Science* 杂志发表的论文 *A draft sequence of the rice genome* 的前 3 页也摘录在高性能计算这一章中，因为这是一篇特殊的论文，迄今已被引用 3 715 次，作者有 100 位。我也被列为论文作者，我认为这是作为“曙光 3000”超级计算机研制者的代表。“曙光 3000”在水稻基因测试中发挥了不可替代的作用，2002 年 *Science* 杂志出版了一期水稻基因组专辑，“曙光 3000”的照片印在期刊封面上。《发展高性能计算需要思考的几个战略性问题》是我 2019 年发表的新论文，对我国超级计算机的发展提出了一些值得重视的建议。

CPU 是全国上下都感到揪心的核心技术，从 2000 年开始计算所率先发力，先后研制成功几款 CPU 芯片，其中影响最大的是龙芯 CPU。第 7 章选录了 4 篇与 CPU 研发有关的论文：胡伟武牵头写的论文有一篇，范东睿牵头写的论文有两篇，最后一篇是我 2012 年在 IPDPS 国际会议的特邀报告的内容摘要。有趣的是，他们写的论文都有一篇发表在 *IEEE Micro* 上，一篇介绍 Godson-3，另一篇介绍 Godson-T。*IEEE Micro* 是芯片领域的顶级期刊，一个单位 3 年之内在这种期刊上发表两篇介绍 CPU 研制的文章，实属不易。龙芯 CPU 现在已成为国防和党政办公应用的主流国产芯片，本书收录的还是 10 年前的论文，今天的技术水平已大大提高。Godson-T 是我担任首席科学家的 973 计划项目的成果。此项目指南原定的目标是发展半导体工艺，捅破摩尔定律的天花板，可能是我在项目申请答辩会上提出要在 973 计划项目的支持下，研制出每秒万亿次（TFlops）计算能力的 CPU 芯片的交账目标打动了评委，才选择了我们做这个项目。十几年前就想做万亿次级的 CPU，确实面临巨大的挑战。好在课题组不负众望，真做出了一款 64 核的众核芯片，为国内其他芯片研制单位趟出了一条可行之路。计算所做芯片研制的不止这两支团队，但这几篇论文足以反映计算所在微处理器设计方向的努力。特别是 *New methodologies for parallel architecture* 这篇论文，比较了 Godson-3 和 Godson-T 两款芯片，总结了 Manycore 和 Multicore 两种体系结构，提出设计可扩展和可重构 CPU 的新思路，值得一读。

计算所不仅做高性能计算和 CPU 研究，在网络和通信领域也有所作为。1994 年 5 月，我的学生陈明宇牵头，在国家智能计算机研究开发中心开通了曙光 BBS 站。第 8 章第一篇论文是他执笔写的关于大型 E-MAIL 服务系统的论文。21 世纪初我主持了中国科学院的 IPv6 网络研究重点项目，在重庆建立了中国第一个 IPv6 示范区。本章收录了一篇关于 IPv6 业务技术的论文。这篇论文是与中国联合网络通信集团有限公司合作的成果，论文的第一作者张云勇是我指导的博士后，现在是中国联合网络通信集团有限公司产品中心总经理。另一位作者刘韵洁院士后来与我在申请和承接国家未来网络实验平台的过程中有更密切的合作。第 8 章收录了一篇有趣的论文，是我指导的在职博士生张国清执笔的。他通过分析和实验指出，不是要增加连接链路，而是删去一些他称为“害群之马（Black Sheep）”的连接链路，可以提高网络通信的传输效率。这篇论文发表在 2007 年 *Physical Review* 上，被引用 182 次。为了促进计算机界和通信界的交流，我推动举办“计算机与通信技术融合”香山科学研讨会，每年一次，已经开了 6 次，在学术界产生了一定的影响。周一青研究员在我的一次报告的基础上增加了许多她的观点，写了一篇论文《未来移动通信系统中的通信与计算融合》，此文 2018 年发表在《电信科学》期刊上，是本选集收录的较新的论文。

大数据是近年来又一个热点技术。2012 年中国计算机学会成立大数据专家委员会，我担任首任主任，对推动我国大数据技术的发展做了一点贡献。我在《中国科学院院刊》等期刊上发表的关于大数据的论文，已收入我的报告文集《创新求索录（第二集）》，本书不再转载。第 9 章选入的 5 篇论文都是与我的学生或同事的合作研究成果。前两篇论文完成在 21 世纪初，当时还没有“大数据”的说法，论文讨论的都是海量数据处理的策略和系统。这些论文发表在国内期刊上，但引用率很高。何清法写的工作流引擎的论文被引用 418 次，实为难得。他现在担任神州通用数据库公司的董事长。另两篇是与卜东波、白硕合写的，白硕当时是国家智能计算机研究开发中心理论组的负责人，他发起并组织的人工智能讨论班很有人气，吸引了北京大学、清华大学等高校的年轻学者。卜东波在算法方面有较深厚的功底，他多年在中国科学院大学上算法大课，几百人的教室座无虚席，还要另开一个同步的视频教室。徐志伟与我合写的 *Computing for the masses*，作为 Contributed Articles 发表在 *Communications of the ACM* 期刊上，阐述了中国科学院组织的中国信息领域 2050 年路线图研究的基本观点。在这次战略咨询研究中我是信息领域的牵头人，因为出版了咨询报告英文版，所以产生了较大的国际影响。我近几年指导的学生中，只有黄俊铭写的关于信息传播规律的论文被纳入本选集，这篇论文 2014 年发表在 *Nature* 的子刊 *Scientific Report* 上。我的学生程学旗是国内大数据研究的领军人物之一，他牵头写的《大数据系统和分析技术综述》受到同行重视，刊登在 2014 年《软件学报》上，已被引用 301 次，也收录在本选集中。

院士们纷纷出版学术论文集，已成为一种风气，我不算是积极跟随者。这是因为，与全力以赴在第一线做科研的科学家相比，我发表的论文数量比他们少，论文的学术影响力也比他们小，我常常为此感到遗憾。我在美国读博士和做博士后时，被同学们戏称为“论文机器”，发表论文的质量和速度超出周围的同学。如果保持当时的势头，之后 30 年应该发表更多更好的论文。可是，人在江湖，身不由己。我回国后不久就到了科研管理的岗位，管理国家智能计算机研究开发中心、曙光公司和计算所。近 30 年来，每天思考的问题主要是如何缩小我国在信息领域核心技术上与国外的差距，如何改变高性能计算机和 CPU 受制于人的局面，如何推动云计算、大数据和人工智能等产业的发展。因此，我写的文章、做的报告多半是一些与技术发展战略有关的宏观思考。我已将这些文章、报告整理成两本文集，分别于 2008 年和 2018 年出版了《创新求索录》和《创新求索录（第二集）》。鱼和熊掌不可兼得，有所舍才能有所得。宏观思考方面的文章也可能有较大的影响力。2012 年 6 月我和程学旗在《中国科学院院刊》上发表的文章“大数据的研究现状与科学思考”至今被引用超过 1 000 次，引用数超过我写的任何一篇学术论文，令人颇感欣慰。

我的本性更喜欢做理论研究。1968 年在北京大学临近毕业要分专业的时候，同学们纷纷报金属、半导体等专业，我毅然报了理论物理专业。但是，留学回国以后，再也没有走“论文机器”的理论研究道路，被大的潮流卷入“发展高科技，实现产业化”的道路。当时挑选我做国家智能计算机开发中心主任的国家科学技术委员会（1998 年后更名为科学技术部）领导常常拿邓稼先的事迹鼓励我，说国家最缺的不是写论文的学者，而是邓稼先这样的能为国分忧的领军人才。说来也巧，我与邓稼先都是普渡大学毕业的博士，算是校友。但我深知自己没有他那样的才能，在接受国家科学技术委员会聘任时我还提出一个条件：要同意我每年有 3 个月以上的时间出国继续做理论研究。1991 年我去伊利诺伊大学厄巴纳 - 香槟分校做了 3 个月访问学者，以后工作一忙，就再也没有机会出国做研究了。1995 年我出任曙光公司总裁时，我的老朋友汪成为曾戏谑过：“国家少了一个科学家，多了一个二流的企业家。”我走的这条路究竟是历史的误会还是正确的道路，恐怕是“仁者见仁，智者见智”了。

我只是一个天资一般的书生。我清醒地知道，在专门的学术分支上我个人难以做出影响后世的重大发现与发明，是时代把我铸造成有一点宏观思维能力的所谓“战略型科学家”。2005 年，*IEEE Spectrum* 期刊派人在中国做了较长时间调查后，在刊物上发表了中国科技十杰（TEN TO WATCH）的人物介绍，其中对我的介绍是：“李国杰希望中国引领世界的信息技术，为了实现这个目标，他的研究工作是设计从微处理器、刀片服务器直到超级计算机。”真是“当局者迷，旁观者清”，美国人不关心我写过什么论文，只关注我做过什么事，他们认为我就是一个想改变中国技术落后局面的挑战者。

我的学术论文基本上折射了计算所近 30 年主要的研究方向，许多论文背后隐藏着令人回味的故事。对于信息领域而言，选择做什么往往比知道怎么做还重要。在本选集每一章之前，我都附加了一篇短文，说明当时我为什么要选择这个研究方向，同时追忆一些留下深刻印象的趣人趣事，为后人理解当时的历史背景添一点旁证。

本选集收集的论文是我走过的科研之路的标签。这些论文背后有许许多多看不见的同事和学生在做实验、写软件、调试计算机。趁此机会，我不但要感谢这些论文的合作者，还要感谢多年来支持、帮助我的所有同事和朋友，更要感谢这个伟大的时代。

李国杰

2020 年 6 月

# 目 录

## 第 1 章 数据流计算机和 Systolic 阵列 / 001



## 第 2 章 并行组合搜索 / 063

## 第 6 章 高性能计算机 / 455



## 第 7 章 微处理器 / 549



## 第 8 章 下一代网络与通信 / 593

## 第 9 章 大数据系统与算法 / 629



备注：因收录的论文发表的时间不同，文中技术术语和计量单位的表述可能与现在规范不完全一致，本选集遵从当时的表述。同样参考文献标示方式也都不尽相同，但为了便于读者查阅相关资料，本选集采用 GB/T 7714—2015《信息与文献 参考文献著录规则》进行了统一标示修改。

## 第1章

# 数据流计算机和Systolic阵列

# 柳暗花明又一村

1974—1976 年，作为湖南省邵阳市计算机厂的派出人员，我在清华大学参加了第四机械工业部组织的 DJS-140 计算机联合设计，接触了一些科研工作，实现了我从物理专业到计算机专业的转行，但直到 1980 年我转到中国科学院计算技术研究所攻读硕士学位后，才真正开始独立从事科研工作。

1980 年，美国普渡大学黄铠教授来北京讲学，在中国科学院研究生院开大课讲计算机体系结构。在讲课中他介绍了美国麻省理工学院（MIT）Dennis 教授发明的数据流计算机（Data Flow Computer），其工作原理与传统的指令控制计算机完全不同，引起我的好奇和兴趣。Dennis 教授没有在期刊和会议上发表过关于数据流计算机的论文，他长达几十页的论文只发布在 MIT 的技术报告 Memo 上，我只能向 MIT 索取。我认真阅读了他的论文后，写了一篇介绍数据流计算机的论文，发表在《电子计算机动态》（即现在的《计算机研究与发展》）刊物上。这是我生平第一次发表科技论文，也是国内第一篇介绍数据流计算机的论文。从这篇论文开始，我与数据流计算机就结下了缘分。

1984 年我去密西根大学参加计算机体系结构国际会议（ISCA），会上有一个高规格的颁奖仪式，给 Dennis 教授颁发 Eckert-Mauchly 奖，这是 ACM/IEEE 的大奖。当年的 Dennis 教授英姿焕发，颇有“一览众山小”的气势。他在答谢演说中的一段话我至今记忆犹新。他说：“再过 20 年，我们还会用 Fortran 编程序吗？不会！20 年后的编程语言一定是数据流语言！”他的豪壮预言并没有实现，后来的 30 年数据流计算机逐渐被人遗忘，只有他的关门弟子、中国留学生高光荣等少数学者在苦苦坚持。高光荣与我一样，都是改革开放后第一批出国的留学生，30 年来我们一直有密切合作。数据流计算机的另一位领军人物是 MIT 的 Arvind 教授，现在是 MIT 计算机学院院长。曾任 IBM 中国研究院院长的沈晓卫是他指导的博士，沈晓卫也是夏培肃先生的硕士生，出国前上过我讲的计算机课。通过他的联系，我与 Arvind 教授有较密切的接触。1996 年，我邀请 Arvind 教授访问我的工作场所国家智能计算机研究开发中心，陪同他看了一场茶座式的中央民族学院演出，演出期间演员们下台给他献哈达，他感到受宠若惊，十分得意。以后每年我访问 MIT 时，计算机系的教授排着队约时间向我介绍他们的科研成果，我开始觉得奇怪，后来才明白是 Arvind 教授在系里透露了他在中国受到的热情招待，教授们希望我邀请他们访问中国。

数据流计算机虽然长期在国际上受到冷遇，但在我牵头的 973 计划项目中受到重视。在高光荣教授的指导下，中国科学院计算技术研究所范东睿研究员领导的课题组在 Godson-T 众核芯片研制中，充分运用了数据流执行模型，形成了高通量流式处理的睿芯系列芯片产品。该产品将数据流思想与人工智能应用有机结合，在语音识别、图像识别等 AI 领域实现了规模化应用。2017 年，在 IEEE MICRO 会议 50 周年之际，高光荣教授获得了罗摩克里希纳·劳奖（B. Ramakrishna Rau Award）。高教授是第一位获得该大奖的出身于中国大陆的华人科学家，该奖体现了国际学术界对他领导的数据流并行体系结构研究成果和应用前景的充分认可。高教授与他的导师 Dennis 最近合著了一本新书：《数据流模型与系统》，为了体现华人对计算机技术的贡献，高教授坚持先在国内出版中文版，后出版英文版，并希望我为这本新书写序言。

在我读硕士期间还有一项技术与数据流计算一样引起我极大兴趣，那就是脉动阵列（Systolic Array）。1980 年我刚转到中国科学院计算技术研究所攻读硕士学位时，夏培肃先生就要我先到河北涿州石油物探中心去实习，了解石油物探业务对并行计算的需求，在实习中我开始接触 FIR 滤波等信号处理技术。刚好黄铠教授在讲课中介绍了美国卡内基梅隆大学（CMU）孔祥重教授发明的脉动阵列，对 FIR 滤波这类应用特别有效，我的硕士论文就选定研究脉动算法设计。在夏先生的帮助下，我直接与孔教授取得联系，获得了一大批参考资料。人们通常以为，学数学出身的人转行研究计算机算法会得心应手，但学物理出身对研究计算机算法可能没有什么帮助，但我的硕士论文却得益于我学物理的背景。脉动阵列计算中数据像脉搏跳动一样有规则地移动，恰好到需要计算的时刻，数据就及时赶到，不能不令人赞叹韵律之美。我想，如果把数据块看成刚体，数据运动一定会符合刚体运动的时间空间关系。凭着物理学者的灵感，我找到了脉动阵列数据移动规律的“运动学”方程，从而得到了设计 Systolic 算法的通用方法。

国家智能计算机研究开发中心成立后不久，我率团访问美国。在访问 CMU 期间，孔祥重教授在他自己出资的餐馆里宴请我们，他建议中国大陆像中国台湾一样从鼠标、显示器、板卡做起，不要以高性能计算机和智能计算机为目标。从他的建议中我感受到，学术上有成就的人还要有“身土不二”的情怀，才能提出适合国情的发展战略。

技术往往是波浪式的发展。脉动阵列技术与数据流计算一样，热过一阵以后又冷落了二三十年。但随着大数据和人工智能的兴起，这两项技术再次受到重视，Google 公司的 TPU 芯片采用了脉动阵列，Intel 公司研制的超级计算机中用到数据流技术，应验了“30 年河东、30 年河西”的规律。做研究不能只追逐当时的热门，在大家不看好的低潮中坚持正确的方向往往能“柳暗花明又一村”。

发表于《电子计算机动态》1981年第11期

# 一种新的体系结构——数据流计算机

李国杰

中国科学技术大学

## 1 引言

电子计算机从诞生到现在经历了几代变革，但冯・诺依曼计算机一直占统治地位。它是基于控制流概念设计的，以指令指针（即程序计数器）的移动来激发指令执行。数据通过访问指令指定的存储单元而取得。众所周知，冯・诺依曼型计算机处理并行计算问题存在着本质上的困难。近年来各种新型的计算机体系，如阵列机、流水线机、多处理机等都采用了并行技术，但这些体系结构仍旧没有摆脱传统的以控制流为主的设计思想。由于指令相关、地址空间相关等的矛盾，并行性受到限制。这些计算机的并行是局部的。并行理论的研究导致了计算机体系结构的又一次重大革新，出现了以数据流概念为主的设计思想，即数据流计算机。数据流计算机由操作数的到达来驱动指令执行，取消了程序计数器，摆脱了传统的同步控制的束缚，采用异步控制，从而以一种很自然的方式表示算法中的并行性，提供了全局并行的可能性，其计算速度可以比通常的计算机速度高1～2个数量级。由于数据流计算机具有上述明显的优点，所以它引起了计算机界浓厚的兴趣。近年来美、英、法等国都在从事这方面的研制，1980年10月，在国际信息处理联合会（International Federation for Information Processing，IFIP）举办的第八届世界计算机大会上，数据流计算机受到重视。目前研制的数据流计算机大多是专用机，但长远的目标是研发出数据流通用机。数据流计算机采用异步并行系统，可以用Petri网作理论模型。一般是先研究数据流语言，然后在这种语言基础上考虑硬件实现。

## 2 数据流语言

数据流语言是一种用图表示的机器语言。数据流计算机是面向这种语言的计算机，它的硬件是数据流语言的解释器。数据流语言通过编码成为机器指令，并存于计算机中。

数据流语言表示为有向偶图，它有两种不同的节点，分别称为链（Link）和动作（Actor）。由偶图性质可知，两个动作之间必有一链，两个链之间必有一动作，如图1所示。其中，圆圈表示动作，小圆点表示链。图1表示快速傅里叶变换中的蝶式运算。粗略地看，每一个动作表示进行一步运算，其结果由链传送到下一动作。根据Petri网的规定，信息由记号（Token）运载，图1中用大黑圆点表示。在数据流语言中，指令的执行顺序由所谓点火规则（Firing Rule）规定。下面以J.B. Dennis 1973年提出的第一个数据流语言方案[1]为例加以说明：（1）一个节点（动作或链）的每一输入弧上都有记号，而任何输出弧上都没有记号，则称此节点就绪；

（2）任何就绪的节点都可以点火；（3）链的点火是从输入弧上移去记号放到它的每一条输出弧上；（4）一个动作点火是从它的每一输入弧上移去记号，用输入记号传送来的值确定一个结果，作为输出记号，放在它的输出弧上（注意：链只有一条输入弧，而动作只有一条输出弧）。链和动作的点火规则如图 2 所示。

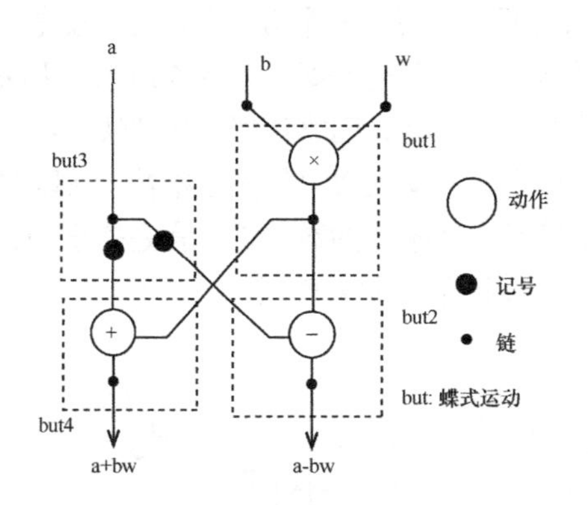


图 1　数据流语言的表示

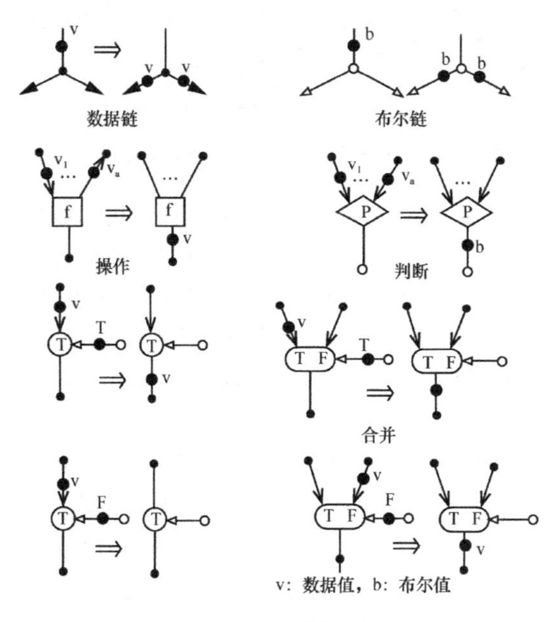


图 2　链和动作的点火规则

Dennis 方案中有两种链：数据链和布尔链。前者在数据弧上传送数据，记号的值可以是整数、实数、复数或字符串等。后者在控制弧上传送布尔值，记号的值只有真（T）和假（F）两种。动作分几种类型：操作 f 进行各种运算，输出数据值；判断 P 输出布尔值；控制型动作（门和合并）的点火规则与其他动作有区别。对于 T 门，如果控制记号为真，则让输入记号通过；如果控制记号为假，则吸收输入记号，不放到输出弧上，即阻止记号通过。F 门则反之，控制记号为假时让输入记号通过。合并（Merge）动作将 T 门与 F 门合在一起，点火规则与单独的 T、F 门一样。控制型动作在条件语句和循环语句中起开关作用。下面将一段类似 ALGOL 语言写的源程序用数据流语言表示出来。源程序如下：

```
input (W, X)
y: = X;    t: = 0;
While t≠W do
begin
if y＞1 then y: = y÷2
else y: =y × 3;
t: = t+1;
end
output y
```

对应的数据流语言如图 3 所示。图 3 中 *X*、*W* 是输入链，虚线小框内的部分表示条件语句 if-then-else，其余右边部分表示循环迭代语句 While-do。其中 3 个合并动作的控制弧用←表示其初始状态控制记号为假值，即允许 *X*、*W* 及 *t* 的初值 0 进入循环程序。除了这 3 个控制记号外，初始时其他弧上没有记号。

人们自然会提出一个问题：用图表示的数据流语言如何被计算机接受，成为硬件可执行的指令？解决的办法是将数据流程序图中一个或几个动作及其输出链直接编码成一条机器指令（注意：不是输入链）。图 1 的小框 but 1 可编码为一指令，占用一个指令单元，也有人称为模板（Template），如图 4 所示。一个指令单元包括以下几部分。一是操作码（C-mul 表示复数乘法）。二是操作数。放在接收器中，此处为 3 个，其中两个放乘数与被乘数，它们等待数据链把要计算的值送来，第三个 $N$ 表示空。三是目标地址。它指出运算结果送到那个单元。小黑点表示输出的数据记号。目标信息是指令必不可少的组成部分。

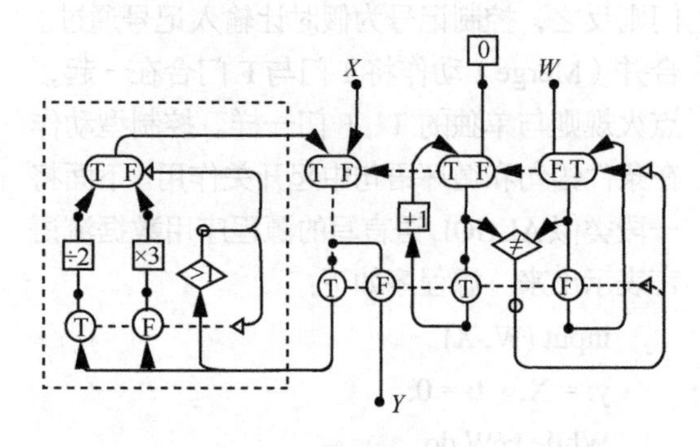


图 3　数据流语言

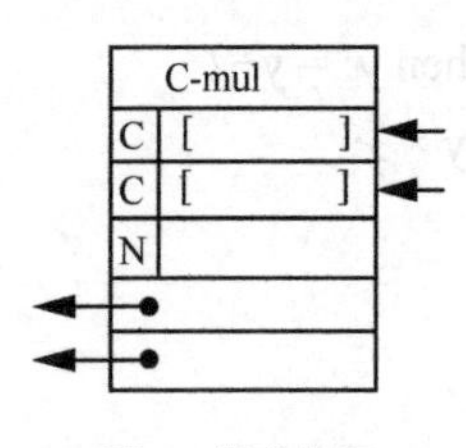


图 4　指令单元

值得注意的是，在数据流计算机的指令中，没有通常的存取操作数所需的地址码。操作数不是指令“取”来的，而是前一步已进行的操作“送”来的。常数则事先存在指令单元中。这是数据流计算机与一般控制流计算机的主要区别。正因为操作数是自动送来的，数据的相关性就由程序直接表示出来了。操作数未全部到达，指令就不会操作，一旦操作数到齐，指令就可以执行。通过目标地址把整个程序连接成一个整体。图 5 画出了如图 1 所示的程序的指令单元及其连接方式。图 5 中“*”表示回答信号。数据流计算机的操作与流水线类似。根据点火规则，一个操作要能进行，除了操作数到齐之外，还要求输出链上没有记号，即上一次运算的结果已被取走。图 1 中加、减操作能否进行要看它的下一步操作是否完成，如果完成了，发一个回答信号（*），所以要回答信号与操作数都到齐才算真正就绪。图 5 中指令操作码后的椭圆圈内的一对数字，分子表示需要的回答信号个数，分母表示预先假定已收到的回答信号个数。如图 5 所示，乘法结果送加、减操作，因乘法慢，加法快，不必回答。但乘法的输入数可能产生很快，所以要回答。满足上述条件的数据流语言是安全的，它保证计算机不会发生死锁。

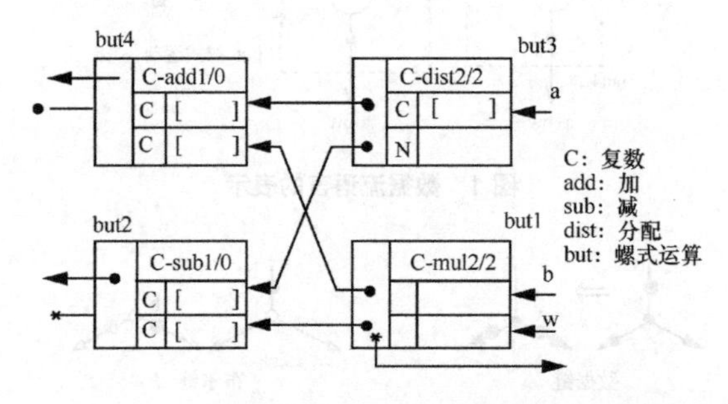


图 5　图 1 所示的程序的指令单元及其连接方式

在带有条件语句和循环迭代语句的数据流程序图中，其机器指令编码与上述情况类似，但判断动作输出的是布尔值。合并动作的输入值一个是数据值，另一个是布尔值，输出是数据值。图 3 的虚线框内的条件语句所对应的指令单元及其连接如图 6 所示。

在上述的数据流程序中，记号传送的值限制在整数、实数、复数或字符串范围内，称为基本的数据流程序。这种程序等价于 **While- do** 和 **if-then-else** 写的表达式功能。以后又提出了扩展的数据流语言[1-2]，它可以用 **Apply** 操作调用一个数据流过程，并具有较复杂的数据结构。限于篇幅，本文对扩展的数据流语言不进行详细讨论。

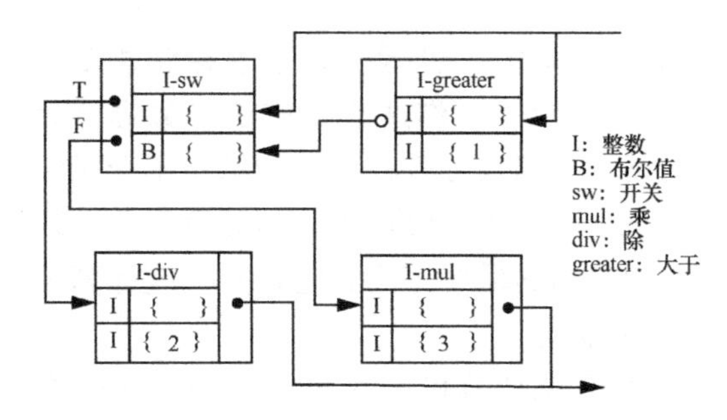


图 6　图 3 的虚线框内的条件语句所对应的指令单元及其连接（注：图中省略了回答信号）

在数据流语言中，没有 go to 语句，没有全程变量及其他有副作用的功能。每一操作只局限于它的输入输出链，不影响其他的操作。每个操作都是独立并行的，有多少操作就绪，原则上就可以有多少操作并行。执行数据流语言表示的程序时，控制流和数据流都由数据流程序图的点火规则确定，同时有许多条控制的路径，充分发挥了并行效率。

数据流计算机的用户用高级语言写源程序，这种高级语言用一般的句文表示，但要求它易于检查程序的数据相关性，并容易编译成数据流语言。目前正在研究各种有效的源语言，如美国麻省理工学院研制的 VAL 语言和法国 Plas 等人研究的单赋值语言[3]等。在单赋值语言中，各种语句都被看成赋值语句，每一变量只赋值一次，也就是说存储单元写进一个数，在程序执行过程中不再改变，执行这种语言写的程序可以不考虑指令的先后次序，几乎能直接转换成数据流语言。法国研制的 LAU 系统就是建立在单赋值语言的基础上的[3]。

## 3　数据流计算机的体系结构和实现方法

数据流计算机由 5 个主要部件组成，如图 7 所示。这五大部件是：存储部件、处理部件、仲裁网络、分配网络和控制网络。

当存储器中的指令单元从分配网和控制网收到全部需要的数据包和控制包时，指令就绪。就绪的指令与它的操作数一起形成一个操作包，经过仲裁网识别操作码，按先进先出次序送到相应的处理单元，执行完一个操作所得的结果又通过分配网和控制网送到存储器，作为它的目标指令单元的操作数。这就是执行一条数据流指令的大致循环过程。

在数据流计算机中没有通常的中央处理器（CPU），CPU 的功能分散在各个部件中实现。操作码规定的操作在处理部件中进行，但指令顺序的控制由存储器的指令单元实现，操作码的译码由仲裁网实现。各部件之间以固定尺寸的、离散的信息包的形式进行通信。信息流方向如图 7 所示，它构成一个环形网络，信息单方向流动，各部件之间的包通信完全是异步的。5 个部件可以独立操作，不需要统一的时钟信号。因此数据流计算机有时也称包通信计算机。

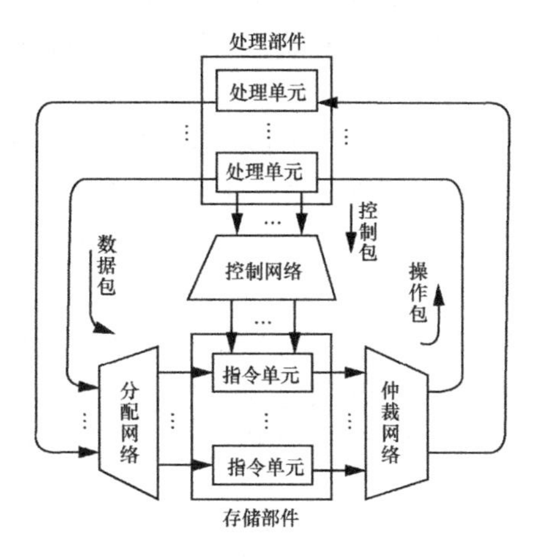


图 7　数据流计算机

下面以美国麻省理工学院 Dennis 领导的计算机体系结构组 1979 年提出的修改方案为

例[4-6]，进一步分析各个部件的功能和结构。

（1）存储部件

存储部件（即指令单元存储器）是数据流计算机的核心，它不只存储指令和数据，而且决定哪些指令可以执行，用数据到达来驱动指令执行，取代了通常的程序计数器。怎样判别一条指令的操作数是否到齐呢？可为每一指令设置一个就绪计数器，开始时置应输入的记号总数，每到一个记号，计数器减 1，计数器为 0 表示所要求的数据、控制记号及回答信号已到齐。也可以在指令单元的每一接收器中设一个标志位，操作数到达就改变标志位，所有的标志位都改过了，说明输入记号已到齐。

第 2 节已提到一个指令单元包括指令（由操作码和目标地址组成）和接收器。麻省理工学院设计的数据流计算机（以下简称“MIT 机”）的每一个指令单元规定有 3 个接收器，允许一条指令最多有 5 个目标地址。每一个指令单元有唯一的单元标识，每个接收器有接收器标识。一个目标地址包括单元标识和接收器标识两部分，即两个整数。一个指令单元的物理结构由输入接口模块、3 个接收器及输出接口模块组成。

数据流语言的程序在装入时，每个指令单元的接收器根据指令要求设置相应的接收器类型和接收器模式。接收器类型分为布尔型、整数、复数 3 种；模式分成变量和常数两种。运行时根据接收器类型和模式接收不同的信息。记号传送来的信息必须与对应的类型和模式相符，这样才可以达到校验的目的。

数据流计算机中采用的存储单元不是一般的磁心或 MOS 存储器，而是一种称为位流水线（Bit-Pipeline）的与速度无关电路[7]。这种电路信息的流入和流出可同时进行（相当于读和写同时进行），它的存储速度与其他逻辑元件工作速度一样快。数据流计算机是异步系统，各种操作的结果不受元件及连接线传输延迟的影响。与速度无关电路是 Pefri 网的重要应用，它是数据流计算机的电路基础。

数据流计算机的每一个指令占用一个指令单元（不止一个字），比传统的计算机占用的存储空间多得多。为了节省存储空间，可采取存储器分层的办法，由位流水线组成的指令单元只作高速缓存，存放最活跃的指令，其余大部分指令存于成本较低的大容量存储器中。为了减少指令单元的入口和出口，从而简化仲裁网和分配网，可将指令单元分块，每块只有一对输入输出口，同时把仲裁网分成几个子网，每个子网对应一组处理单元。

（2）处理部件

处理部件包括各种处理单元（即功能单元），如算术逻辑运算单元、判断及扇出控制单元等。进入处理部件的操作包中有目标信息，用来形成数据包和控制包的目标地址。处理部件也采取流水线方式工作，以获得最高处理效率。为提高整个系统的性能，多采用一些功能单元当然有好处，但功能单元的多少必须与指令单元的容量匹配，而且功能单元越多，路径选择网络就越复杂。下面以整数操作单元为例简单说明功能单元的工作情况。整数操作单元具有加、减、位测试和比较等功能（为减少仲裁网的出口数，操作单元的功能应强一些，不要分得太细），它分成为整数控制和整数运算两个子模块。整数控制子模块对接收到的操作包进行判断，看需要何种操作，然后发送命令包到整数运算子模块，运算后把结果包送回整数控制子模块形成数据包和控制包。处理部件中各个处理单元都是独立并行的，慢的操作不会影响快的操作，处理单元处理完一个操作包，马上由仲裁网供给下一个操作包。

（3）路径选择网络

数据流计算机是一个包通信系统。除了存储部件和处理部件外，其余 3 个部分都是通信网络。这些网络的基本元件是仲裁单元、开关单元和缓冲器。仲裁网入口多、出口少，主要由仲裁单元构成；分配网入口少、出口多，主要由开关单元构成。这两个网的结构如图 8 所示。

仲裁网分为三级。第一、二级的仲裁单元把来自许多指令单元的操作包汇集到较少数量的通道中；第二级的开关单元按指令所要求的处理能力把操作包流分裂为许多分枝；最后由第三级合併，使每一处理单元对应一个仲裁网的出口，只接收一个操作包流。当有几个相同类型的操作包要输出时，调度的原则是依次轮流，即先进先出。前一个操作包未处理完，后面的操作包在缓冲器中等待。所以仲裁网起到指令译码、分配处理单元和先进先出缓冲区的作用。仲裁网入口信号多，采用字节串行方式，减少与存储部件的联系。输出出口少，采用并行方式，提高传输速度使之与进口匹配。因此在仲裁网中需要有一个步骤实现串行到并行的转换。相反，在分配网中需要并行到串行转换。分配网与控制网的结构与仲裁网类似，不赘述。

数据流计算机的路径选择网既不同于一般的总线结构，也不同于多处理机系统的纵横开关。在数据流计算机中，信息流的方向是单向的，所以不要求像纵横开关和总线那样必须

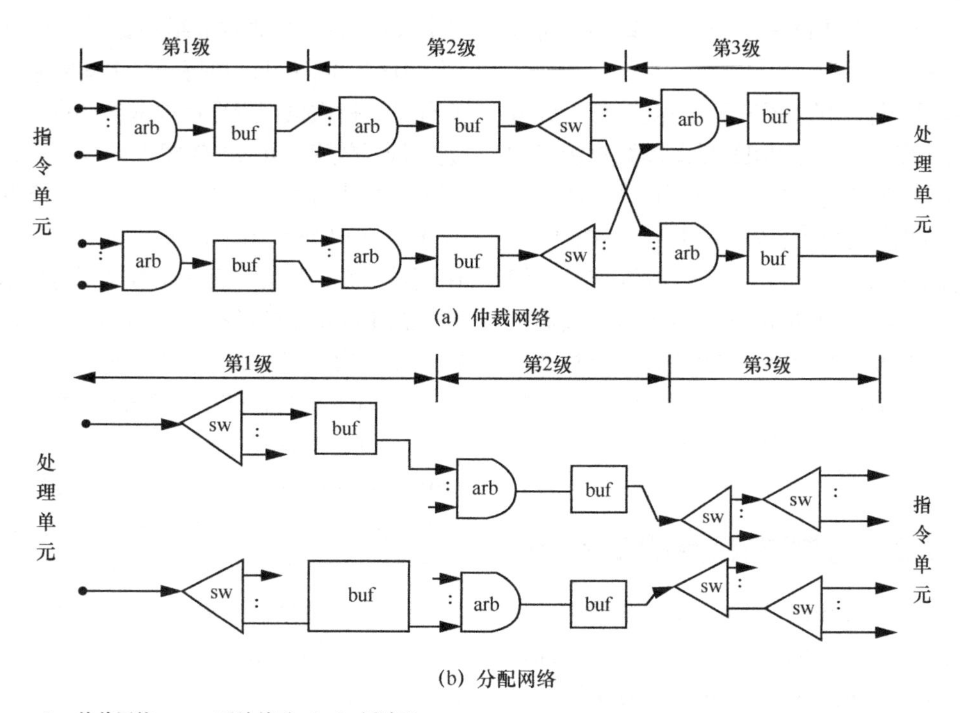

(a) 仲裁网络

(b) 分配网络

arb：仲裁网络，sw：开关单元，buf：缓冲器

**图 8 仲裁网络和分配网络的网络结构**

立即给回答信号，而且往往有几个指令单元就绪，准备利用同一处理单元，所以数据包和控制包的传输延迟无关紧要，在一定范围内它不会影响系统的处理速度。数据流计算机的路径选择网络保证了存储器中各指令之间的通信与它们在存储器中的位置无关，程序员可以随机地存放指令，而不必考虑复杂的内存分配。数据流计算机的路径选择网络的逻辑复杂性为 $O(N\log N)$，虽然比纵横开关的复杂性 $O(N^2)$ 好一些，但仍然大于线性关系。为了减少其复杂性，提出了许多新的想法，如动态分配资源、选择最佳级数等，这方面的研究是数据流计算机的重要课题。数据流计算机的一条指令的执行周期等于信息包通过两个网与相应的处理单元的时间。一般处理单元的延迟时间是固定的，而两个网络由于冲突，延迟时间将随机变化。这种系统的关键是仲裁网的出口和分配网的进口，这部分电路采用 ECL 代替 TTL 可以提高效率。

数据流计算机还没有真正实现，但设计方案已有许多种。除了 MIT 的基本数据流机和具有过程调用功能的数据流机外[4-5]，还有数据流多处理机方案[2]、法国的 LAU 系统[3]、英国曼彻斯特大学的带名字标号数据流机和纽卡斯尔大学设计的带可修改存储器的数据流计算机（JUMBO）[8]等。JUMBO 计算机的设计结合了控制流和数据流计算机两种不同的思想，独具特色。下面将专门介绍它们的设计原理。

## 4 数据流与控制流相结合的设计思想

在介绍 JUMBO 计算机之前，先比较一下数据流计算机和常规的控制流计算机。下面以一个赋值语句的几种不同的实现方法说明两种设计的区别。赋值语句为：

a: =(b + c) (b − c)

图 9 画出了 4 种不同的实现方法。图 9（a）是冯 • 诺依曼方法；图 9（b）是扩展的多控制流方法，图 9（c）是数据流计算机的实现方法。其中方框表示指令，$I_1$、$I_2$、$I_3$、a、b、c、$t_1$、$t_2$ 都是名字或地址，“○”表示存储单元，“?”表示数据记号，实线弧表示数据流，虚线弧表示控制流。从图 9 能明显地看出在常规的控制流方法中，通过指令共享存储单元来实现数据通信。产生数据的指令和使用这一数据的指令公用同一名字的存储单元，通过“访问”才能取得数据。而数据流计算机没有单独包含一个数的存储单元或嵌在指令中的数据名字的概念。指令不能指定到某单元去取数，而是由记号直接传送“值”，这样，一个常数如果要被不同的指令使用一千次，则要占一千个接收器，在循环过程中，常数也需要再生。数据流机实现数组运算也不方便，而控制流机可通过变址寻址很方便地对数组等数据结构进行加工。但控制流计算机的每一个中间结果都要送存储器，每次执行指令都要取数和存数，显然这降低了效率。JUMBO 机取两者之长，采用了一个折中方案，如图 9（d）所示。它保留了 a、b、c 作为地址名字，而去掉了 $t_1$ 和 $t_2$ 这些中间结果存储单元，直接用数据记号传送。因此 JUMBO 机中的数据分成两类。一类是常数或需要多次使用的数，称之为要存储的数据，存于所谓半永久存储器（类似于通常的存储器）；另一类是只用一次的数据，由数据记号传送到指令激活部件中的目标指令。前一类数据是有地址的，通过访问存取，可以修改内容，这是控制流方式；后一类数据没有名字，采用数据流方式。这样既节约了存储空间，便于加数据结构，又节省了读写中间结果的时间。

JUMBO 的总框架如图 10 所示，它分成三大模块，每一模块由一台微处理机实现相应的

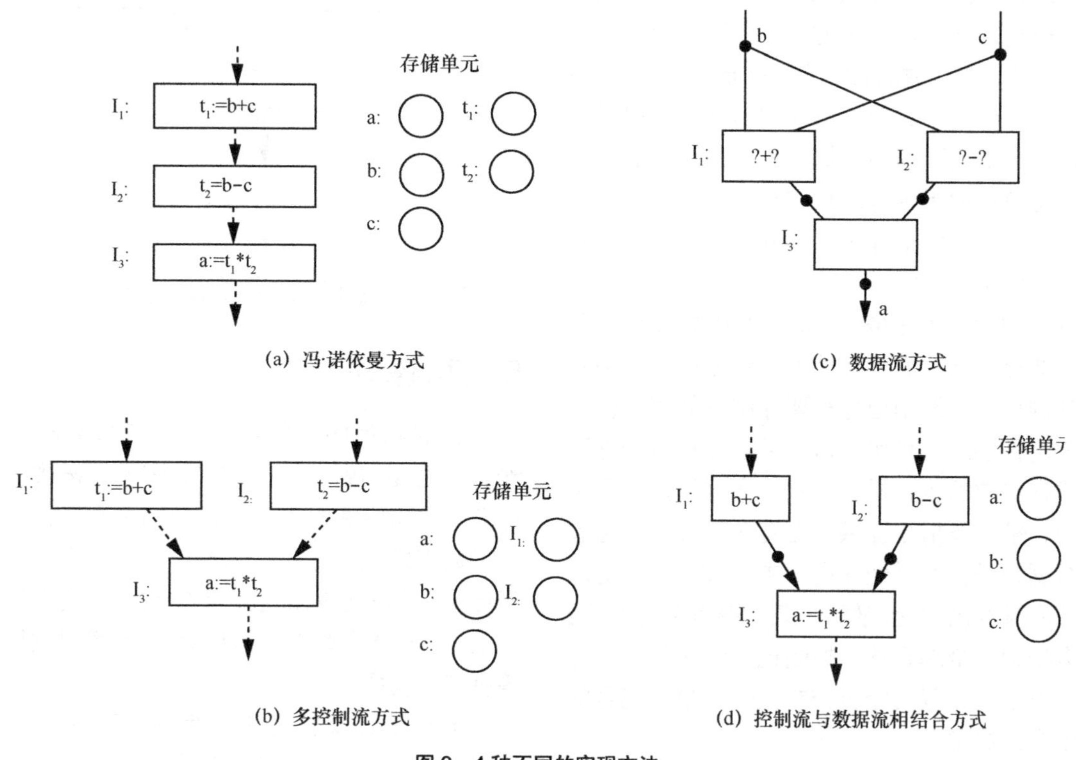


**图 9　4 种不同的实现方法**

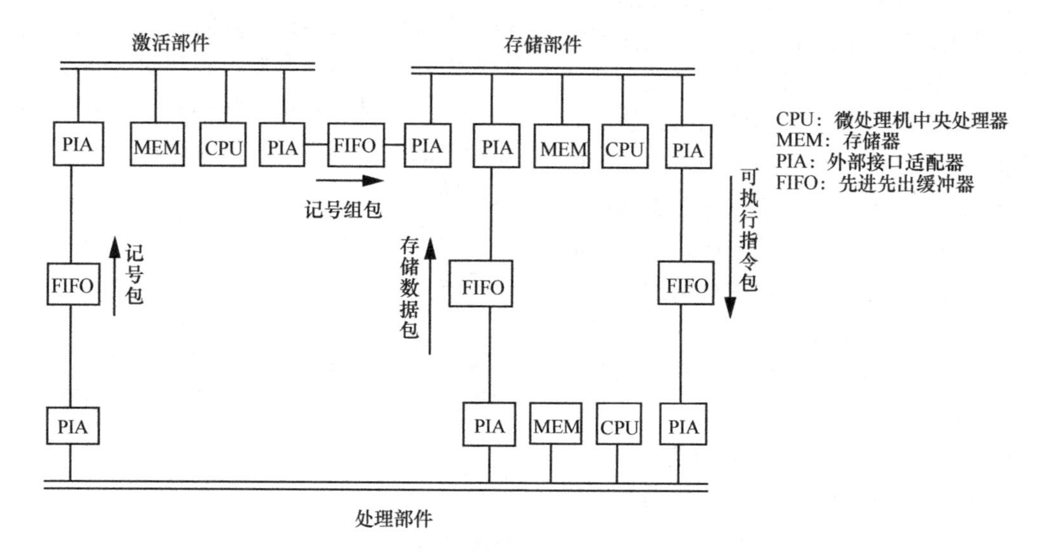


**图 10　JUMBO 的总框架**

功能。模块之间由 8 位总线通过外部接口适配器（PIA）互相连接。3 个模块独立异步地工作。第一个模块是指令激活单元，它收集数据记号，输出一个记号组包。这部分功能类似于

MIT机的指令存储部件，但JUMBO机激活单元输出的各个记号中的操作数不一定是值本身，还可以是数的名字（即地址）。第二个模块是存储器，其功能与通常的存储器相似，通过“访问”为指令提供真正的操作数，形成可执行的指令包。第3个模块是处理部件。这一部分还包括仲裁网和分配网的功能。处理部件由多个相同的处理单元组成，因而减少了仲裁网的任务。JUMBO机规模小、路径选择网络较简单。处理部件输出的结果有数据记号、控制记号和要存储的数据3种。前两种形成记号包，送激活部件；后一种形成存储数据包，送存储器。在这种折中方案中，给一条指令提供数据有3种方式：通过数据记号；通过嵌在指令中的数据；按地址访问某存储单元。激活一条指令有两种方式：数据记号全部到达；控制记号通知某些事件已发生。有些指令由控制记号和数据记号的组合来激发。

下面以一小段程序为例说明JUMBO机的信息形式，程序如图11所示。这段程序的内容是：指令1存值26于x单元，送一个控制记号到指令3，指令1的信息形式是存储数据包；指令2送一个数据记号（值为10）到指令3，信息形式是记号；指令3把记号的值10与x单元相加，其信息形式先是存储的指令，记号到达后变成可执行的指令；指令3的结果由数据记号（值为36）送到指令4，指令4将36与x单元相乘，其信息形式与指令3类似。

纽卡斯尔大学设计JUMBO机的目的是在组合的模型中研究控制流和数据流的相互影响。他们力求结构简单，只用了3台微处理机，使每一所大学都可以研制。这种计算机的程序既可以按纯数据流方式编写也可按多控制流方式编写，或两者组合编写。

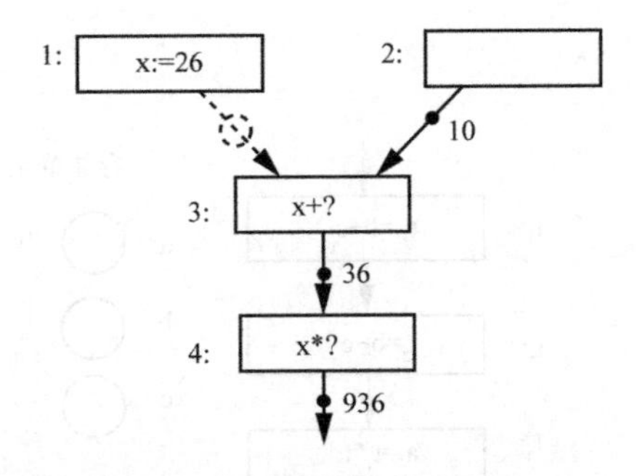


**图11　体现JUMBO机信息形式的程序示例**

## 5　几点看法

数据和程序的关系是计算机体系结构涉及的主要问题。数据流计算把冯·诺依曼计算机中指令和数据的关系颠倒过来了，突出了数据的主导作用，在理论上这一变革对于计算机的发展来说是一个突破。在实现技术上，它与流水线机、多机系统、计算机网络等有密切联系。

数据流计算机是否有生命力取决于它的成本。数据流计算机必须有大容量的存储器和大规模的路径选择网络，它所得到的高速度以耗费更多的存储空间和网络连接元件为代价。这是一种用空间换时间的策略。因此器件的发展程度对数据流计算机的实现有决定性的影响。

数据流计算机具有模块化的特点。与当前迅速发展的微处理机技术是协调的。采用微处理机做模块，按照多机系统的方式构成数据流计算机是一条重要的实现途径。将控制流计算机的设计思想和数据驱动原理结合起来，互相补充，将有利于新的计算机体系的实现。

## 参考文献：

[1] DENNIS J B. First version of data flow procedure language[J]. Lecture Notes in Computer Science, 1974, 19: 362-376.

[2] RUMBAUGH J. A data flow multiprocessor[J]. IEEE Transactions on Computers, 1977, C-26(2): 138-146.

[3] PLAS A. LAU system architecture: a parallel data-driven processor design based on single assignment[C]// International Conference on Parallel Processing. Piscataway: IEEE Press, 1976: 293-302.

[4] DENNIS J B. A preliminary architecture for a basic data flow processor[C]// The 2nd Annual Symposium on Computer Architecture. [S.l.:s.n.], 1975: 126-132.

[5] DENNIS J B. The varieties of data flow computer[C]// Advanced Computer Architecture and Processing Techniques. [S.l.:s.n.], 1986: 200-209.

[6] DENNIS J B. A highly parallel processor using a data flow machine language[C]// Advanced Computer Architecture and Processing Techniques. [S.l.:s.n.], 1986: 210-277.

[7] MISUNAS D. Petri net and speed indepent design[J]. Communications of the ACM, 1973, 16(8): 474-481.

[8] TRELEAVEN P C. Pricipal components of a data flow computer[J]. Euromicro, 1978.

发表于《计算机学报》1982年第2期

# 用参数确定法设计阵列流水算法

李国杰

中国科学技术大学

**摘　要：**阵列流水算法是与集成电路的迅速发展相适应的高速并行算法，它可在流水式阵列处理机中直接由硬件实现。本文通过分析阵列流水的时间—空间关系得出阵列流水原理，并以此原理为基础提出了设计阵列流水算法的一种统一的方法——阵列流水参数确定法。处理递归问题的各种阵列流水方案都可以由此方法推导出来。运用确定参数法设计了矩阵乘法、FIR滤波、DFT、三角形矩阵求逆等新的、更合理的阵列流水算法。

## 1　引言

微电子技术的迅速发展促使计算机体系结构发生了革命性变化，连接线已成为器件与系统的主要成本，因此建立在高度集成的器件基础上的新型计算机体系必须考虑局部而且规则的连接。集成电路的发展使阵列处理和流水线这两项最主要的并行技术密切结合了起来。目前的器件水平已达到一个封装内包含几万甚至几十万个门电路的水平，因此可以考虑在一片或若干片组成的流水式阵列中直接实现较复杂的科学计算。本文采用“流水式阵列处理机”（以下简称“PAP”）这一名称代表一类以大规模或超大规模集成电路为基础的新计算机体系结构。在 PAP 上实现的算法称为阵列流水算法，这种算法与通常的计算机算法很不一样。为了说明 PAP 及阵列流水算法的特点，本文先提出一个实现矩阵乘法的阵列流水方案（本文讨论的就是这一类算法的设计方法）。图 1 表示两个三阶矩阵相乘，即 $A \cdot B = C$。每一小方块代表一个可做乘加运算的处理单元，这些处理单元通过简单而规则的连接网络构成一个阵列。阵列中的数据流有 3 个方向：菱形数据块 A 向上平移；菱形数据块 B 向左平移；正方形数据块 C 向左上角平移。在统一的时钟控制下，A、B、C 以同步方式流动，每拍前进一个单元，在一个单元内相遇的操作数实现一次乘加运算 $c'=c+ab$。三阶矩阵经过 7 拍可以得到全部结果。

一般来说，PAP 以及阵列流水算法具有以下特点。

（1）PAP 由若干处理单元（可以成千上万个）构成，具有规则的阵列结构，它可以用统一的时钟同步，但不同于通常意义下的 SIMD 计算机，同一时刻不同的处理单元可执行不同的操作。

（2）PAP 的控制方式类似于一般的流水线计算机，但它可能有几十甚至几百条流水线，这些流水线可以沿几个不同方向流动数据。不同方向的流水线中或同一方向流水线中以不同速度流动的数据在某一处理单元相遇并进行操作。任意一步运算所需的操作数必须

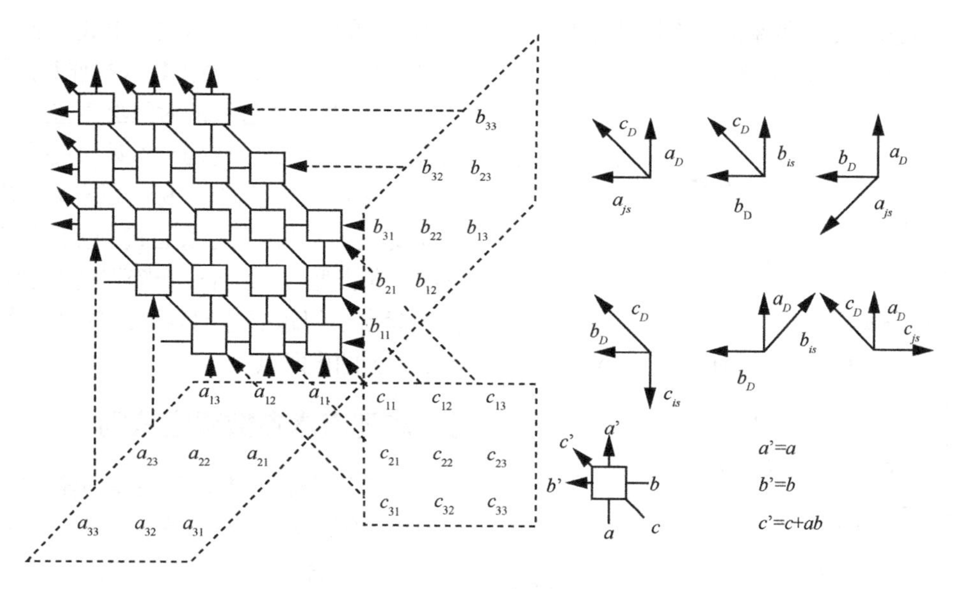


**图 1 矩阵乘法的阵列流水方案**

在同一拍时间内到达一个处理单元，称这种流水方式为阵列流水。

（3）阵列流水算法着眼于数据流，主要考虑输入数据迭代运算的中间结果在什么时刻达到什么位置，因此其算法常常可用几何方式表示。阵列流水算法主要用来解决递归计算问题，过去常采用循环语句实现递归。PAP 将循环语句向量化，直接用硬件实现各种递归运算，因此，这种算法设计必须与体系设计结合起来。本文讨论的阵列流水算法的设计方法也就是 PAP 的核心部分——流水式阵列结构的设计方法。

（4）PAP 是专用机，它特别适合处理大规模的科学计算问题。由于每一数据从存储器读出后要反复使用许多次，所以它可以达到很高的处理速度（如每秒几亿次操作），而对存储器没有过高的要求。

（5）PAP 最主要的特点是具有局部而规则的连接，处理单元类型很少，整个系统的数据流控制流十分规则。这种高度模块化的体系结构大大降低了设计和制造的成本，适于超大规模集成，也正由于数据流很规则，所以可以考虑采用统一的方法设计阵列流水算法。

近几年国外很注意研究与集成电路发展相适应的新型计算机，阵列流水方式受到重视[1-4]。但迄今见到的有关文献大都是针对某一问题设计一种方案，没有确定的原则可循。本文从分析递归公式出发，得到阵列流水原理，使阵列流水算法设计建立在一定的理论基础上。

## 2 递归运算的硬件实现

研究阵列流水算法可从分析递归运算入手，为了反映在流水线上实现递归的动态过程，本文采用迭代式。先考虑以下几个例子。

（1）矩阵乘向量

$$\begin{cases} y_i^{(0)} = 0 \\ y_i^{(k)} = y_i^{(k-1)} + a_{ik} x_k \end{cases} \tag{1}$$

（2）矩阵乘法

$$\begin{cases} z_{ij}^{(0)} = 0 \\ z_{ij}^{(k)} = z_{ij}^{(k-1)} + x_{ik} y_{kj} \end{cases} \tag{2}$$

（3）反褶积，即解系数矩阵是带形下三角 Toeplits 矩阵的线性方程组

$$\begin{cases} y_i^{(0)} = b_i \\ y_i^{(k)} = y_i^{(k-1)} - a_{m+1-k} x_{i-m+k}, 1 \leqslant k \leqslant m \\ x_i = y_i^{(m-1)} / a_1 \end{cases} \quad (3)$$

（4）矩阵 LU 分解

$$\begin{cases} a_{ij}^{(0)} = a_{ij} \\ a_{ij}^{(k)} = a_{ij}^{(k-1)} - l_{ik} u_{kj} \quad (1 \leqslant k < i, j) \\ u_{ij} = a_{ij}^{(i-1)} \quad (i \leqslant j) \\ l_{ij} = a_{ij}^{(j-1)} / u_{jj} \quad (i < j) \end{cases} \quad (4)$$

（5）两维比较

$$\begin{cases} z_{ij}^{(0)} = 1 \\ z_{ij}^{(k)} = z_{ij}^{(k-1)} \wedge (x_{ik} = y_{ik}) \end{cases} \quad (5)$$

这些递归计算各不相同。其中，式（1）、式（2）、式（5）是一阶递归，式（3）是 $m$ 阶递归，记为 $R\langle n,m \rangle$；式（4）是 $n-1$ 阶递归，即一般递归，记为 $R\langle n \rangle$。对于 $R\langle m,n \rangle$ 和 $R\langle n \rangle$ 递归，输出结果需要反馈作为输入。上述各迭代式中，各变量的下标也不一样，向量只有一个下标，矩阵有两个下标。式（3）的下标是表达式，各式中操作类型和次序也不一定一样，如式（5）表示逻辑操作。为了研究在流水式阵列中实现递归计算的一般规律，本文用统一的函数形式表示一类迭代式，具体如下。

$$z^{(k)}(i,j) = f\left[z^{(k-1)}(i,j), x(i,k), y(k,j)\right] \quad (6)$$

其中，$f$ 表示一个处理单元的功能，$x$、$y$、$z$ 看成下标 $i$、$j$、$k$ 的函数，向量是矩阵的特殊情形，变量 $z$ 被迭代处理，称为迭代变量，$x$ 和 $y$ 的值在流水中不改变，称为输入变量。如果 $x$ 或 $y$ 是向量，式（6）中 $x(i, k)$ 或 $y(k, j)$ 表示下标表达式中有下标变量 $i, k$ 或 $k, j$。递归表达式（6）中有两类角标变量：$i$ 和 $j$ 表示数据在阵列中的位置；$k$ 表示迭代步数，下标中的 $k$ 则指明第 $k$ 步迭代计算的输入操作数的位置，角标 $k$ 可以把流水过程的时间次序与空间次序联系起来。下面讨论的确定参数法只关心 $k$ 出现在行下标中还是在列下标中以及 $k$ 的符号，不需要考虑具体的下标表达式，也不考虑操作的次序和类型。凡设计一个阵列流水方案，总要先写出该问题的迭代表达式并以它作为参数法设计的出发点。

## 3　阵列流水参数确定法则

从上面介绍的矩阵乘法阵列流水方案可以看出3个数据块中各元素的位置相互之间有关系，为了确定这些关系需要规定一些参数。一般来说，PAP 有 3 类参数：一类是空间参数，即各数据项如何分布；一类是时间参数，如迭代运算周期；还有一类是速度参数，即数据流速度。只要把这3类参数的相互关系搞清楚了，这个 PAP 也就完全确定了。

现在定义几个阵列流水参数。首先说明一下本文所指的“距离”的含义。规定两相邻处理单元的距离为距离单位（由连接线直接相连的单元称为相邻单元）。如果把带一条对角线的正方形连接理解为与它拓扑同构且功能等价的正六边形连接，则所有相邻单元的距离一致。定义时间的单位为一拍，在一拍时间内可以完成一次迭代运算并将数据送到相邻的处理单元或缓冲器。

**定义 1**　数据流速度。数据 A 在它的流水线方向一拍时间内前进的距离称为数据流速度，记为 $A_D$。如果两个相邻处理单元之间有 $m-1$ 个附加的缓冲器，一个数据到下一个处理单元需 $m$ 拍，则它的数据流速度 $A_D=1/m$。

**定义 2**　数据分布。一个矩阵 $\boldsymbol{x}$ 的各元素 $x_{ij}$ 分布在阵列中，$x_{i,j}$ 与 $x_{i+1,j}$ 的距离称为 $\boldsymbol{x}$ 的行距，记为 $x_{is}$。$x_{i,j}$ 与 $x_{i,j+1}$ 距离称为 $\boldsymbol{x}$ 的列距，

记为 $x_{js}$。一个向量的两相邻数据项 $x_i$ 和 $x_{i+1}$ 的距离称为项距，记为 $x_s$。数据分布沿下标递增还是递减取决于迭代表达式中 $k$ 的符号。以 $x_{ks}$ 表示 $x(i,k)$ 与 $x(i,k+1)$ 的距离，如果 $k$ 属于行下标，则 $k+1$ 意味着 $i+1$，所以 $x_{ks}=x_{is}$；如果 $k$ 属于列下标，则 $x_{ks}=x_{js}$，对于向量，有 $x_{ks}=x_s$。

**定义 3** 时间因子。迭代运算的周期记为 $m_k$，即相继两次迭代 $z_{ij}^{(k-1)}$ 与 $z_{ij}^{(k)}$ 相隔 $m_k$ 拍。迭代计算 $z_{ij}^{(k)}$ 与 $z_{i+1,j}^{(k)}$ 的时间间隔，以拍为单位，称为行时间因子，记为 $m_i$。迭代计算 $z_{ij}^{(k)}$ 与 $z_{i,j+1}^{(k)}$ 的时间间隔称为列时间因子，记为 $m_j$。

以上共有 12 个参数，即 3 个数据流速度 $x_D$、$y_D$、$z_D$；6 个数据分布，其中 3 个行距为 $x_{is}$、$y_{is}$、$z_{is}$，3 个列距为 $x_{js}$、$y_{js}$、$z_{js}$；还有 3 个时间因子，即迭代周期 $m_k$、行时间因 $m_i$、列时间因子 $m_j$。对于一维运算，数据分布和时间因子个数相应减少。在本文讨论的问题中，数据流速度不能大于 1，而时间因子 $m_i$，$m_j \geqslant 1$。

阵列流水原理如下。

用阵列流水方法实现迭代计算：$z^{(k)}(i,j)=f[z^{(k-1)}(i,j), x(i,k), y(k,j)]$。各变量的数据流速度、数据分布及阵列流水时间因子必须满足以下关系。

$$
\begin{cases}
m_k x_D + x_{ks} = m_k z_D & (7)\\
m_k y_D + y_{ks} = m_k z_D & (8)\\
m_i x_D + x_{is} = m_i y_D & (9)\\
m_i z_D + z_{is} = m_i y_D & (10)\\
m_j y_D + y_{js} = m_j x_D & (11)\\
m_j z_D + z_{js} = m_j x_D & (12)
\end{cases}
$$

现对式（7）～式（12）加以说明。把阵列画成如图 2 所示的正方形连接并不失一般性，图 2 中 A、B、C、D 不一定是相邻的处理单元。图 2（a）表示某一时刻处理单元 C 在计算 $z^{(k)}(i,j)=f[z^{(k-1)}(i,j), x(i,k), y(k,j)]$，此刻，$x(i,k+1)$ 在 B，所以 $CB$ 表示 $x_{ks}$。$y(k+1,j)$ 在 D，$CD$ 表示 $y_{ks}$。$x_{ks}$、$y_{ks}$ 表示行距还是列距取决于 $k$ 属于什么下标。根据阵列流水要求，下一步要计算的操作数必须在同一拍时间内到达同一处理单元，因此经 $m_k$ 拍以后，$z^{(k)}(i,j)$、$x(i,k+1)$ 和 $y(k+1,j)$ 都到达 A 单元，在 A 中计算：$z^{(k+1)}(i,j)=f[z^{(k)}(i,j),x(i,k+1),y(k+1,j)]$，如图 2（b）所示。在体系设计时可以不考虑各操作数在一拍之内的延迟，把 $m_k$ 当作常数。由定义 1 可知，$z_D$ 是 $z$ 在一拍内前进的距离，所以 $CA = m_k z_D$。同理 $BA=m_k x_D$，$DA=m_k y_D$。由矢量合成法则，$BA + CB=CA$，$DA + CD = CA$，直接得到式（7）与式（8）。

假设某时刻在处理单元 D 进行某步计算 $z^{(k)}(i,j)=f[z^{(k-1)}(i,j),x(i,k),y(k,j)]$，$z(i+1,j)$ 在处理单元 C，如图 2（c）所示。阵列流水中同一变量的各数据项的相对位置在流水时不改变，即与迭代步数无关，所以 $DC=z_{is}$。此时刻 $x(i+1,k)$ 在 B，所以 $DB = x_{is}$ 根据阵列流水要求，经过 $m_i$ 拍，$z(i+1,j)x(i+1,k)$ 与 $y(k,j)$ 都到达 A 单元，并计算 $z^{(k)}(i+1,j)=f[z^{(k-1)}(i+1,\ j),x(i+1,k),y(k,j)]$。

由定义 1 可知，$BA=m_i x_D$，$CA = m_i z_D$，$DA= m_i y_D$。根据矢量合成原则，有：$BA + DB = DA, CA + DC = DA$。因此，直接得到式（9）和式（10）；同理可得式（11）与式（12）。

对于一维运算，式（11）与式（12）不存在。注意在式（9）～式（12）的推导过程中，我们利用了下述条件：递归式（6）中输入变量 $x$ 的下标不明显地包含 $j$，$y$ 的下标不明显地包含 $i$。本文最后一节将讨论当 $x$ 与 $y$ 同时包含下标 $i$ 或 $j$ 时如何运用阵列流水原理。式（7）～式（12）反映了阵列流水的基本时空关系。对于 $m_k=m_i=m_j$ 的情形，经过简单的代换，式（10）与式（12）分别变为 $z_{is}=-y_{ks}$，$z_{js}=-x_{ks}$，

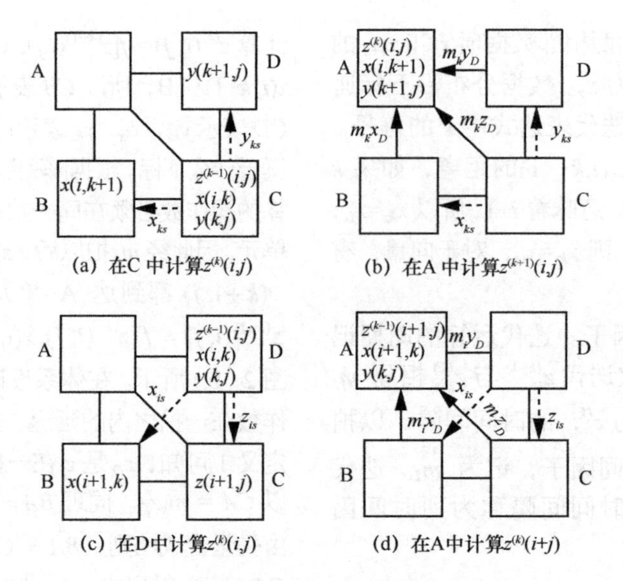


**图 2　阵列流水原理的说明**

则相互关系更加清楚。

图 1 用矩阵乘法验证了阵列流水原理。由图 1 很容易看出各参数满足式（7）～式（12）（在这个例子中，$m_k$=$m_i$=$m_j$=1）。这一方案是根据阵列流水原理设计的，它的处理速度比 H. T. Kung 提出的 Systolic 方案[1,3]及 K. Hwang 提出的方案[4]都快，而所用器件与 Systolic 方案一样多。

任意阵列流水方案由上述各参数完全确定，因此给定某些合理的已知条件，解参数方程（7）～（12），求出待定参数，这样就确定了一种阵列流水算法，称这种方法为阵列流水参数确定法（以下简称参数法）。它是一种设计算法的算法，一个问题往往有几组独立而且合理的已知条件，因而可能有几种 PAP 方案，可以进行比较与折中，从中选出最佳方案。这就使设计更灵活、更合理。如果已有一个 PAP，其数据流方向固定，要在它上面计算某一新问题，可将那些不能改变的限制作为已知条件，用参数法设计合适的算法，从而可扩大 PAP 的应用范围。目前已提出的各种阵列流水方案，如 H. T. Kung 提出的各种 Systolic[1-3] 方案都可由参数法方便地得出，而且运用参数法可以得出更好的方案。

## 4　阵列流水参数确定法的应用举例

（1）FIR 滤波

FIR 滤波是数字信号处理中特别常用的技术之一，它实质上是做矩阵乘向量运算。如下式所示。

$$\begin{bmatrix} y_1 \\ y_2 \\ \vdots \\ y_n \end{bmatrix} = \begin{bmatrix} a_1 & a_2 \cdots a_m & \cdots 0 \\ 0 & a_1\ a_2 \cdots a_m \cdots 0 \\ \vdots & & \vdots \\ 0 \cdots & a_1\ a_2 \cdots a_m \end{bmatrix} \begin{bmatrix} x_1 \\ x_2 \\ \vdots \\ x_n \end{bmatrix}, n > m$$

其中，矩阵 $A$ 是 Toeplits 矩阵，计算时不必重复输入同样的数据，可将 $a_1 \cdots a_m$ 分别存在一维连接的 $m$ 个处理单元中，运算时不移动。根据它的迭代式

$$\begin{cases} y_i^{(0)} = 0 \\ y_i^{(k)} = y_i^{(k-1)} + a_k x_{i+k-1}, 1 \leqslant k \leqslant m \end{cases}$$

和阵列流水原理可列出以下 4 个方程。

$$\begin{cases} m_k a_D + a_s = m_k y_D & (13) \\ m_k x_D + x_s = m_k y_D & (14) \\ m_i x_D + x_s = m_i a_D & (15) \\ m_i y_D + y_s = m_i a_D & (16) \end{cases}$$

这个问题所有的参数都在一条直线上，可以按标量处理。因为 $a$ 固定，所以已知条件是 $a_D= 0$, $a_s= 1$，再假设输入变量是密集的，即 $x_D= 1$，$x_s=-1$。求解方程（13）～（16），得到 $m_k = 2$，$m_i= 1$，$y_D=1/2$，$y_s=-1/2$。根据这些参数可画出实现 FIR 滤波的 PAP，方案如图 3 所示。

图 3 中两处理单元之间的短线表示一个缓冲器。用这一方案实现 $n$ 个数据 FIR 滤波（滤波因子 $m$ 个），流水线建立时间为 $2m-1$ 拍，阵列流水时间 $n-1$ 拍总时间为 $n+2m-2$ 拍。一般 FIR 滤波 $m=(1/10\sim1/5)n$。因此这一方案的并行效率很高 $(E=\dfrac{mn}{m(2m+n-2)}=0.7\sim0.8)$，这一方案的另一优点是控制简单。一个数据进入流水线后要使用 $m$ 次，所以对存储器和片子 I/O 引出线的要求不高，这是一个较理想的 FIR 滤波方案。

如果已知条件是 $a_D= 0$，$a_s= 1$，$y_D= 1$，$x_D=-1$，即输入 $x$ 与输出 $y$ 相对流动，用参数法可以得到另一种滤波方案，$m_k=1$，$m_i=2$, $x_s=2$，$y_s=-2$。该方案与 Systolic 方案一样[3]。由于数据项距大，需要两拍出一个结果，总时间为 $2(n+m-1)$，效率 $E<1/2$。

若假设已知条件是 $a_D= 0$，$a_s= 1$，$y_D=-1$，$y_s=1$（$y_D=1$，$y_s=-1$ 不合理），还可以求得一种 FIR 滤波方案，其速度与图 3 所示方案一样，但控制略为复杂，需要先送 $m/2$ 个 $x$ 数据到阵列中。

从这个例子可以看出 PAP 的性能是所有参数的多元函数，反映了参数的综合效果。采用参数法可设计多种方案，再进行折中。

（2）多项式乘法

设两个多项式相乘 $(\sum_{l=0}^{n-1}a_l x^l)(\sum_{j=0}^{n-1}b_j x^j)=\sum_{i=0}^{2n-2}c_i x^i$，式中 $c_i=\sum_{k=1}^{n}a_{k-1}b_{i-k+1}$，当 $l, j<0$ 或 $l, j\geqslant n$ 时，$a_l=b_j=0$。迭代计算式为：

$$\begin{cases} c_i^{(k)}=c_i^{(k-1)}+a_{k-1}b_{i-k+1} \\ c_i^{(0)}=0 \end{cases}$$

这个计算式与 FIR 滤波公式的区别只是 $b$ 的下标 $k$ 带负号。我们规定 $b_s$ 沿下标递增方向，则 $b_{ks}=-b_s$，假设已知条件为 $m_k=m_i=1$，$a_D=0$, $a_s=1$，采用参数法可求得 $c_D=1$，$c_s=-1$，$b_D=\dfrac{1}{2}$, $b_s=-\dfrac{1}{2}$。图 4 画出了多项式乘法的阵列流水方案，它与 FIR 滤波不同之处只是输入变量与迭代变量交换位置。

这种方案实现两个 $n$ 级多项式相乘需 $n$ 个处理单元 $3n$ 拍时间（假设 $a$ 预先输入占 $n+1$ 拍）。同样，如果给定另外的已知参数，如 $c_D=0$, $c_s=1$，或 $b_D=0$, $b_s=1$ 等可得到另外的方案。

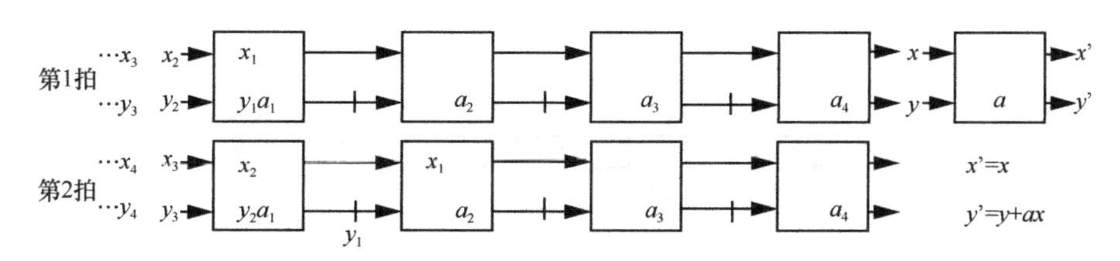


图 3　FIR 滤波的 PAP 方案（$m$=4）

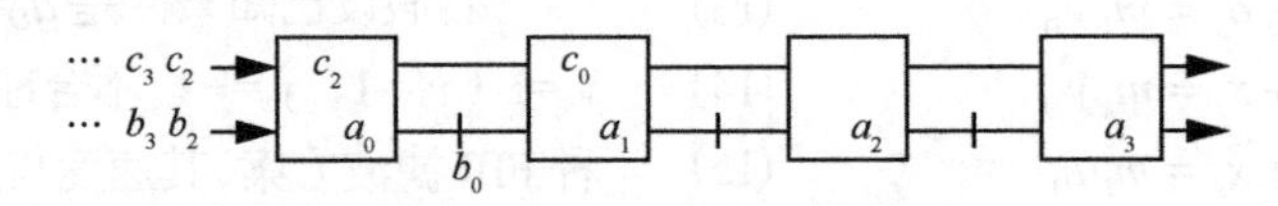


图4　多项式乘法的阵列流水方案

（3）离散傅里叶变换（DFT）

近十年来，快速傅里叶变换（FFT）得到了广泛应用，但用并行方式直接实现 FFT 有大量非局部而且不规则的连接线，比如 shuffle-exchange 网。为了适应高度集成，研究具有规则、局部连接的 DFT 阵列流水算法是有意义的。在一定范围内，采用总面积相同的硅片（减少连线面积，增加处理单元面积）DFT 可以取得与 FFT 同样的处理速度。DFT 的计算式是 $y_i=\sum_{k=0}^{n-1} x_k w^{ik}, i=0,1,\cdots,n-1$，其中$w=\mathrm{e}^{-\mathrm{j}2\pi/n}$ 。如果采用矩阵乘向量方式，每一个旋转因子 $w^{ik}$ 只用一次，输入输出量大，由于器件外引线限制，实现较困难。我们采用求多项式值的方式：$y_i=(\cdots(x_{n-1}w^i+x_{n-2})w^i+\cdots)w^i+x$。其迭代式为：

$$\begin{cases} y_i^{(0)}=0 \\ y_i^{(k)}=y_i^{(k-1)}\cdot w^i+x_{n-k}, 1\leqslant k\leqslant n \end{cases}$$

函数形式是 $y_i^{(k)}=f[y^{(k-1)}(i),w(i),x(k)]$。在这个等式中 $w(i)$ 中 $i$ 是 $w$ 的幂，而不是下标，$w^i$ 也有数据分布，参数法仍然适用，由于 $w$ 不包括下标 $k$，对应的方程式中 $w_{ks}$ 不出现。设 $m_k=m_i=1$，由参数法列出以下 4 个方程：

$$\begin{cases} w_D=y_D \\ x_D+x_s=y_D \\ w_D+w_s=x_D \\ y_D+y_s=x_D \end{cases}$$

如果预先把1, $w\cdots w^{n-1}$ 分别放在 $n$ 个一维连接的处理单元中，运算时不移动，即已知条件为 $w_D=0$, $w_s=1$。解上面的方程组得到 $x_D=1$, $y_D=0$, $x_s=-1$, $y_s=1$，注意 $x$ 的下标 $k$ 带负号，$x_s$ 的正方向是下标递减方向。根据这些参数画出 DFT 阵列流水方案，如图 5 所示。

计算结束后，保存在 $n$ 个处理单元的结果再串行输出。这一方案实现 $n$ 点 DFT 需 $3n-1$ 拍（不计预先送 $w^i$ 的时间，$w^i$ 可保留在处理单元中）。这一方案的最大优点是输入输出引线少。对于 VLSI 系统来说，外引线是系统的瓶子口，尽量减少引出线是设计 PAP 的一条重要原则。

如果已知条件是 $x_D=0$, $x_s=1$，可得到另一种 DFT 方案，计算速度稍快但多一倍引出线。

（4）三角形矩阵求逆

上面 3 个例子都是一阶递归，而且都是一维问题，不能充分反映阵列流水的特点。三角形矩阵求逆是二维 $R\langle n\rangle$ 递归问题，需要反馈输入。设有上三角形矩阵：

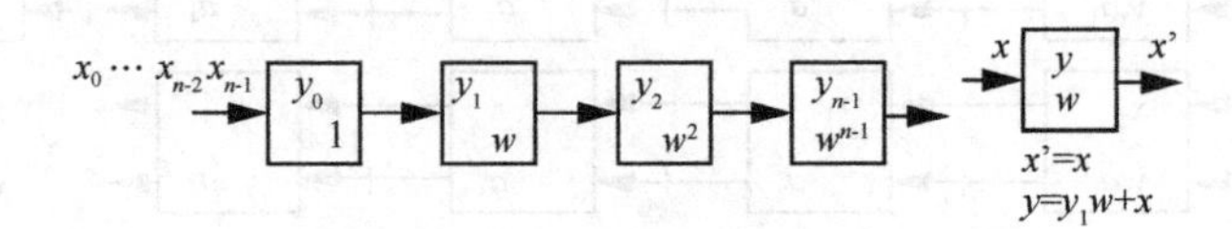


图5　DFT 的阵列流水方案

$$U=\begin{pmatrix} u_{11}u_{12}\cdots u_{1n} \\ u_{22}\cdots u_{2n} \\ \cdots \\ u_{nn} \end{pmatrix}$$

其逆矩阵为：

$$V=\begin{pmatrix} v_{11}v_{12}\cdots v_{1n} \\ v_{22}\cdots v_{2n} \\ \cdots \\ v_{nn} \end{pmatrix}$$

上三角形矩阵求逆由计算 $v_{nn}$ 开始，然后计算 $v_{n-1,n-1}, v_{n-1,n}$，直到 $v_{1n}$。它的迭代式是：

$$\begin{cases} w_{ij}^{j+1}=0 \quad (i\neq j) \\ w_{ij}^{(k)}=w_{ij}^{(k+1)}-u_{ik}v_{kj}, \qquad i<k\leqslant j \\ v_{ij}=w_{ij}^{(i+1)}/u_{ij},(i\neq j), v_{ii}=1/u_{ii} \end{cases}$$

因为迭代时下标 $k$ 和 $i$ 都是由大到小，所以我们可以确定已知条件 $m_k=-1$，$m_i=-1$，$m_j=1$。由参数法可以列出以下 6 个方程。

$$\begin{cases} -u_D+u_{is}=-w_D & (17) \\ -v_D+v_{is}=-w_D & (18) \\ -u_D+u_{is}=-v_D & (19) \\ -w_D+w_{is}=-v_D & (20) \\ v_D+v_{js}=u_D & (21) \\ w_D+w_{js}=u_D & (22) \end{cases}$$

我们可把 3 个数据流速度作为已知条件，例如规定 $w$ 向左流动、$v$ 向上流动、$u$ 向左上角流动，如图 6 所示，速度大小均为 1。解向量方程（17）～（22）可得到 6 个数据分布参数。实际过程是根据每一个方程式画一个矢量合成图，根据求出的参数可直接画出 $u_{ij}$、$v_{ij}$、$w_{ij}$ 在平面上的分布。再考虑数据反馈要求，容易画出这一问题的阵列流水方案，如图 6 所示。注意对于 $R\langle n\rangle$ 递归（$R\langle m,n\rangle$ 递归也一样），阵列流水中迭代变量的数据流方向要改变。在这个例子中，$w_{ij}$ 经 $j-i$ 步迭代（最后一步做除法）产生 $v_{ij}$，数据流方向由向左变为向上。$v_{ij}$ 是非独立的输入变量，这个方案采用两种处理单元，斜边小圆圈代表除法器，最下面的除法器输入 1 作为被除数，产生 $v_{ii}$。此方案实现了 $n$ 阶三角形矩阵求逆需 $n(n+1)/2$ 个处理单元、$2n-1$ 拍时间的需求。

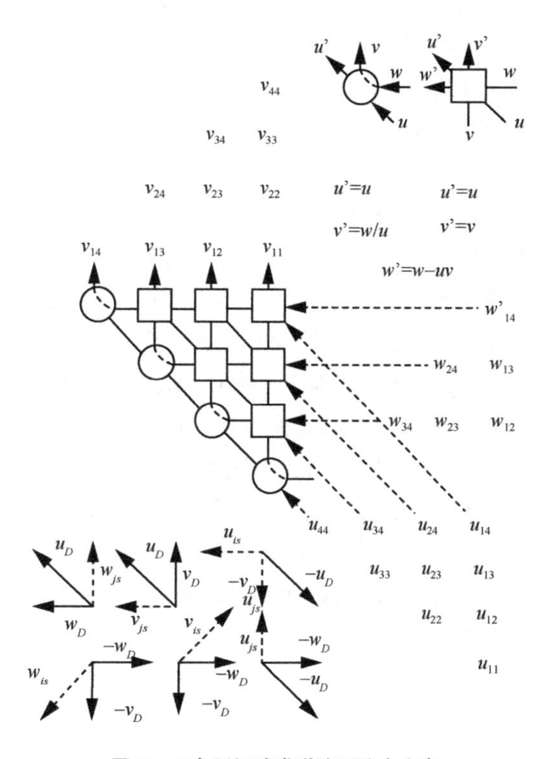


图 6　三角形矩阵求逆阵列流水方案

关于上面的问题，如果给定其他的数据流方向，还可以得到许多不同的方案，它们的数据分布不同，反馈方式也不同。这些方案要么与图 6 所示的方案同构，要么更复杂，或速度较慢。

（5）非数值计算问题

在推导式（7）～式（12）时，对迭代公式函数表示式中的 $f$ 未加限制，因此参数法同样适用于各种逻辑操作，如比较、排队、组合算法及数据库操作等。阵列流水结构取决于迭代公式的形式，关键是输入变量的下标。将阵列流水原理的几个等式稍加变换，由式（9）和式（10）得到：

$$m_i z_D + z_{is} = m_i x_D + x_{is} \tag{23}$$

由式（11）和式（12）可得到：

$$m_j z_D + z_{js} = m_j y_D + y_{js} \tag{24}$$

这两个等式表示了迭代变量对它的某个下标的数据分布与带此下标的输入变量的关系，这两个等式也适合于 $x$ 和 $y$ 同时带下标 $i$ 或 $j$ 的情形。

在关系数据库与其他许多逻辑判断问题中需要对两个集合的对应元素的属性进行比较，本文用 $i$ 表示元素在集合中的次序，$k$ 表示属性编号。逐个元素逐项性质比较的过程用迭代公式表示为 $E_i^{(k)} = E_i^{(k-1)} \wedge (x_{i,k}\ \rho\ y_{i,k}),\ E_i^{(0)} = 1$。式中，$\rho$ 代表比较操作，若 $x_{i,k}=y_{i,k}$，则 $\rho$ 操作结果为真，即布尔值为 1。对应元素的所有属性相同，则 $E_i = 1$，否则 $E_i = 0$。这一过程相当于用按位比较方式判断两个向量对应元素是否相等。用参数法设计时，注意 $x$ 和 $y$ 的下标都有 $i$，假定 $m_k = 1$，$m_i = 1$，可列出如下 4 个方程：

$$\begin{cases} x_D + x_{ks} = E_D \\ y_D + y_{ks} = E_D \\ E_D + E_s = x_D + x_{is} \\ E_D + E_s = y_D + y_{is} \end{cases}$$

我们仍采用与前面例子同样的带一条对角线的正方形连接方式。设已知条件为 $x_D$、$y_D$、$E_D$ 和 $E_s$，解上面这个方程组得到 $x_{ks}$、$y_{ks}$、$x_{is}$、$y_{is}$，可画出实现比较功能的阵列流水方案，如图 7 所示。

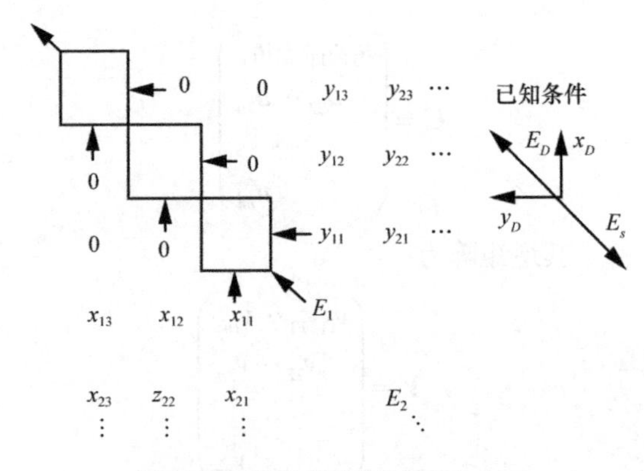


图 7　实现比较功能的阵列流水方案

如果要求两个集合中的元素两两相互比较，而不只是比较相同序号的元素，可以采用 $E_{ij}$ 表示 $x_i$ 与 $y_i$ 比较的结果（布尔值），这一比较过程的迭代表达式为：

$$E_{ij}^{(k)} = E_{ij}^{(k-1)} \wedge (x_{i,k} \rho y_{i,k}), E_{ij}^{(0)} = 1$$

其函数形式是 $E_{ij}^{(k)} = f[E^{(k-1)}(i,j), x(i,k), y(j,k)]$。注意这里 $i$ 和 $j$ 不表示行和列，而分别表示 $x$ 集合中第 $i$ 个元素与 $y$ 集合中第 $j$ 个元素。它们都是元素的序号，$k$ 表示属性的序号。这一迭代表达式的函数形式与本文引言一开始列举的矩阵乘法相同，运用参数法时 6 个方程式也就相同，因此可以用图 1 所示的流水式阵列来实现上述两维比较功能。但在这种情况下 $y_{ks}$ 和 $y_{js}$ 的含义与矩阵乘法不同，$y_{ks}$ 表示属性之间的分布距离，相当于矩阵的列距，而 $y_{js}$ 相当于矩阵的行距，因此进行两维比较时，$y$ 的数据项 $y_{jk}$ 要放在矩阵乘法方案 $b_{kj}$ 的位置，而且每个处理单元不是进行算术乘加而是进行逻辑操作。

## 5　关于 PAP 的实现

采用阵列流水结构的单片 VLSI 器件已经问世[2]，但在实现由许多片器件构成的 PAP 时还存在一些问题，例如怎样保证存储器及时地按算法要求提供必需的数据。对于许多条流水线同步工作的阵列流水，只要有一条流水线数据供应不上，整个阵列流水都会受影响。还有一个重要的问题是器件外引线限制，如果一个封装内包含若干处理单元，由图 1 和图 6 的连接方式可以看出，如果采用通常的位并行方式，外引出线数将大大超过一片器件所允许的引线数，因此克服输入输出外引线带宽限制是实现 PAP 必须解决的课题。限于篇幅，本文没有论述 PAP 的具体实现，这些问题将在另外的文章中专门讨论。本文认为，位串行联线流水技术可以提高数据的复用次数，降低对存储器的要求，从而使整个 PAP 系统取得很高的处理能力。PAP 将是一种有发展前途的计算机体系结构。

## 致谢

本文在夏培肃、韩承德两位导师具体指导下写成，并得到美国普渡大学黄铠教授与卡内基梅隆大学 H.T.Kung 教授的热情帮助，谨表示衷心感谢。

## 参考文献：

[1] KUNG H T. Let's design algorithms for VLSI system[C]// Conference on VLSI: Architecture, Design, Fabrication. [S.l.:s.n.], 1979: 65-90.
[2] FOSTER M J, KUNG H T. The design of special-purpose VLSI chip[J]. Computer,1980,13(1): 26-40.
[3] MEAD C A, CONWAGL A. Introduction to VLSI system[M]. Massachusetts: Addsion-Wesley, 1980.
[4] HWANG K. VLSI computing structure for solving large-scale linear system of equations[C]// The 1980 International Conference on Parallel Processing. Piscataway: IEEE Press, 1980.

发表于*IEEE Transactions on Computers*,1985年第1期

# The design of optimal systolic arrays

Guo-jie LI[1], Benjamin W. WAH[2]

1. Student member, IEEE

2. Member, IEEE

**Abstract:** Conventional design of systolic arrays is based on the mapping of an algorithm onto an interconnection of processing elements in a VLSI chip. This mapping is done in an ad hoc manner, and the resulting configuration usually represents a feasible but suboptimal design. In this paper, systolic arrays are characterized by three classes of parameters: the velocities of data flows, the spatial distributions of data, and the periods of computation. By relating these parameters in constraint equations that govern the correctness of the design, the design is formulated into an optimization problem. The size of the search space is a polynomial of the problem size, and a methodology to system-atically search and reduce this space and to obtain the optimal design is proposed. Some examples of applying the method, including matrix multiplication, finite impulse response filtering, deconvolution, and triangular-matrix inversion, are given.

**Key words:** data distribution, data flow, parameter method, period, recurrence equation, systolic array, velocity

## 1 Introduction

The evolution in Very Large Scale Integration (VLSI) technology has had a great impact on computer architecture. Specialized algorithms can be implemented on a VLSI chip using multiple, regularly connected Processing Elements (PE's) to exploit the great potential of pipelining and multiprocessing. This type of array processor has been referred to as a systolic array[1]. One of the many advantages of this approach is that each input data item can be used a number of times once it is accessed, and thus, a high computation throughput can be achieved with only modest bandwidth. Other advantages include modular expandability, simple and regular data and control flows, and use of simple and uniform cells.

Systolic arrays have been classified into (semi-) systolic arrays with global data communications and (pure) systolic arrays without global data communications[1]. In semi-systolic arrays,a data item accessed from memory is broadcast to and used by a number of possibly nonidentical cells concurrently. Although this approach is potentially faster than systolic arrays without data broadcast, providing (or collecting) a data item to (or from) all the cells in each cycle requires the use of a global bus that may eventually slow down the processing speed as the

This work was supported in part by a David Ross Grant from the Purdue Research Foundation and by the National Science Foundation under Grant ECS 80-16580.

number of cells increases. On the other hand, a pure systolic array eliminates the use of broadcast buses and implements the algorithm in pipelines extending in different directions. Several data items flowing along different pipes with the same or different rates may meet and interact. The PE's operate synchronously, that is, each data item must stay in a PE for one and only one clock cycle, and all the necessary operands to be processed by a PE in each computational step must arrive at this PE simultaneously. This mode of pipelining is referred to as systolic processing.

In general, systolic arrays are somewhat inflexible to implement because they must be algorithmically specialized. Studies have been made to design reconfigurable interconnections for the pure systolic array approach in order to provide the flexibility needed while retaining the benefits of uniformity and locality[2]. Another approach is based on the mapping of different algorithms onto a fixed computer architecture. This is exemplified in the mapping of algorithms onto wavefront array processors[3] and linear array of processors[4]. Special controls are included in order to tailor the algorithms to the architectures.

The efficient mapping of algorithms onto architectures is very important in a direct implementation, and also essential as a benchmark for comparison when algorithms are mapped onto a fixed architecture. Previously, the mapping of algorithms has been done in a rather ad hoc fashion. The proposed designs usually represent feasible but suboptimal solutions. Previous studies can be classified into four categories. In the first type, transformations are performed at the algorithm representation level, and a direct mapping is made from this level to the architecture[5-7]. In the second type, transformations are performed at the algorithm-model level, and there are procedures for deriving the model from the algorithm representation and for mapping the model into hardware[4,8-19]. In the third type, transformations are performed on a previously designed architecture to obtain a new architecture[20]. In the last type, transformations are performed to map the systolic architecture into the function implemented and to prove the correctness of the design[21]. A critical review of the methodologies can be found in [22].

For the above methods, very little can be said about the optimality of the resulting design. Most of the transformational methods are heuristic in nature. Although minimum execution-time algorithms can be systematically derived and heuristics for optimizing space were studied[10], no formal method for optimally mapping the transformed algorithms onto systolic architectures is proposed.

The objective of this paper is to provide a systematic methodology for the design of optimal pure planar systolic arrays for algorithms that are represent table as linear recurrence processes. Planar systolic arrays are those in which the interconnections can be laid out in a plane without crossing each other. Linear recurrence processes (to be defined later) with one- or two-dimensional inputs are suitable for implementation on these systolic arrays, and inputs with a larger number of dimensions have to be partitioned first. The merit of a design is measured by the completion time ($T$) or the product of the VLSI chip area and the completion time ($A \times T$) or the product of the VLSI chip area and the square of the completion time ($A\times T^2$). However, area is a complex function of the number of PE's, the number of buffers, the interconnection pattern, and the available technology, and has to be assessed for each configuration separately. The discussion on area measurement is outside the scope of this paper, and the number of PE's is used as an indicator for the area required.

Systolic designs are characterized by three classes of parameters: velocities of data flows, spatial distributions of data, and periods of computation. The relationships among these parameters are represented as constraint equations, and the completion time and hardware complexity of a design can be expressed in terms of these parameters (Section 3). A systematic methodology to solve the design problem as an optimization problem is discussed in Section 4. Examples illustrating the method are shown in Section 5.

## 2 Characteristics of recurrence processes

In this section, the characteristics of recurrence processes and their relationships to systolic processing are discussed. The following are examples of linear recurrences.

• Finite Impulse Response (FIR) filtering. FIR filtering, an important technique in signal processing[23-25], can be considered as a matrix-vector multiplication

$$\begin{bmatrix} y_1 \\ y_2 \\ \vdots \\ y_n \end{bmatrix} = \begin{bmatrix} a_1 \cdots a_m \cdots 0 \\ a_1 \cdots a_m \vdots \\ \vdots \quad\quad a_m \\ \quad\quad \vdots \\ 0 \quad \cdots \quad a_1 \end{bmatrix} \begin{bmatrix} x_1 \\ x_2 \\ \vdots \\ x_n \end{bmatrix} \quad n > m \tag{1}$$

where the matrix is an upper-triangular Toeplitz band matrix. $y_i$, $1 \leqslant i \leqslant n$, is a summation of product terms, the $k$th term of which is denoted as $y_i^k$. The operation can be written as a recurrence equation

$$\begin{aligned} & y_i^0 = 0 \quad 1 \leqslant i \leqslant n \\ & y_i^k = y_i^{k-1} + a_k x_{i+k-1} \quad 1 \leqslant i \leqslant n, 1 \leqslant k \leqslant m, \\ & x_j = 0 \quad \text{for } j > n \end{aligned} \tag{2}$$

• Two-dimensional matrix multiplication[1,24,26-27]. Each term in the multiplication of two two-dimensional matrices, $\boldsymbol{C} = \boldsymbol{A}\times\boldsymbol{B}$, is a summation of product terms. The $k$th term of $c_{i,j}$ can be computed as

$$\begin{aligned} & c_{i,j}^0 = 0 \quad\quad 1 \leqslant i,j \leqslant n \\ & c_{i,j}^k = c_{i,j}^{k-1} + a_{i,k} b_{k,j} \quad 1 \leqslant i,j,k \leqslant n \end{aligned} \tag{3}$$

• Discrete Fourier Transform (DFT). DFT is usually regarded as a matrix-vector multiplication, $\boldsymbol{Y} = \boldsymbol{\Omega}\times\boldsymbol{X}$, in which $\boldsymbol{\Omega}$ contains power terms of $\omega = \mathrm{e}^{2\pi\sqrt{-1}/n}$, the $n$th root of unity. Each term of $\boldsymbol{Y}$ can be expressed as a summation of product terms: $y_i = \sum_{k=0}^{n-1} x_k \omega^{ik}, 0 \leqslant i \leqslant n-1$. This form of representation requires additions, multiplications, and power operations. Rewriting it in a different way, the $k$ th term of $y_i$ is

$$\begin{aligned} & y_i^0 = 0 \quad\quad 0 \leqslant i \leqslant n-1 \\ & y_i^k = y_i^{k-1}\omega^i + x_{n-k} \quad 1 \leqslant k \leqslant n, 0 \leqslant i \leqslant n-1 \end{aligned} \tag{4}$$

This representation requires multiplications and additions only.

• Polynomial multiplication. The multiplication of two polynomials $\left(\sum_{i=0}^{n-1} a_i x^i\right) \times \left(\sum_{j=0}^{n-1} b_j x^j\right) = \left(\sum_{i=0}^{2n-2} c_i x^i\right)$ can be reduced to finding the coefficients of the resulting polynomial. The $k$th term of $c_i$ is expressed as

$$
\begin{aligned}
& c_i^0 = 0 && 0 \leqslant i \leqslant 2n-2 \\
& c_i^k = c_i^{k-1} + a_{k-1} b_{i-k+1} && 0 \leqslant i \leqslant 2n-2 \\
& && 1 \leqslant k \leqslant n, b_j = 0 \\
& && \text{for } j < 0 \text{ or } j \geqslant n
\end{aligned} \tag{5}
$$

• Deconvolution. This is the inverse of FIR filtering [see Eq. (1)] that solves for vector $\boldsymbol{X}$ given vector $\boldsymbol{Y}$ and the Toeplitz matrix. By equating the product on the left-hand side with vector $\boldsymbol{Y}$ on the right-hand side, $a_1 x_i = y_i - \sum_{k=1}^{m-1} a_{m-k+1} x_{i+m-k}$, $1 \leqslant i \leqslant n$. This can be expressed as a recurrence with a temporary variable $z_i$

$$
\begin{aligned}
& z_i^0 = y_i \qquad 1 \leqslant i \leqslant n \\
& z_i^k = z_i^{k-1} - a_{m-k+1} x_{i+m-k} \\
& \qquad 1 \leqslant k \leqslant m-1, 1 \leqslant i \leqslant n, x_j = 0 \text{ for } j > 0 \\
& x_i = \frac{z_i^{m-1}}{a_1} \qquad 1 \leqslant i \leqslant n
\end{aligned} \tag{6}
$$

• Triangular matrix inversion. Given an upper-triangular matrix $\boldsymbol{U}$ such that $u_{i,j}=0$ $i > j$, $\boldsymbol{V} = \boldsymbol{U}^{-1}$ (the inverse of $U$) can be expressed as a recurrence by using a temporary variable $w_{i,j}$[28]

$$
\begin{aligned}
& w_{i,j}^{j+1} = 0 && 1 \leqslant i < j \leqslant n \\
& w_{i,j}^{k} = w_{i,j}^{k+1} - u_{i,k} v_{k,j} && 1 \leqslant i < k \leqslant j \leqslant n \\
& v_{i,j} = \frac{w_{i,j}^{i+1}}{u_{i,i}} && 1 \leqslant i < j \leqslant n \\
& v_{i,i} = \frac{1}{u_{i,i}} && 1 \leqslant i \leqslant n
\end{aligned} \tag{7}
$$

• Two-dimensional tuple comparison. Given two two-dimensional matrices $\boldsymbol{A}$ and $\boldsymbol{B}$, the operation of finding whether the $i$th row of $\boldsymbol{A}$ is identical to the $j$th row of $\boldsymbol{B}$ can be expressed as a recurrence[27].

$$
\begin{aligned}
& c_{i,j}^0 = \text{TRUE} && 1 \leqslant i, j \leqslant n \\
& c_{i,j}^k = c_{i,j}^{k-1} \wedge (a_{i,k} = b_{j,k}) && 1 \leqslant i, j, k \leqslant n
\end{aligned} \tag{8}
$$

Eq. (2) and Eq's (4)~(6) above are one-dimensional recurrences, while others represent two-dimensional recurrences.

In general, linear recurrences for the computation of a two-dimensional result $\boldsymbol{Z}$ from two two-dimensional inputs $\boldsymbol{X}$ and $\boldsymbol{Y}$ can be expressed as

$$
z_{i,j}^k = f\left[ z_{i,j}^{k-\delta}, x(i,k), y(k,j) \right] \qquad \delta = 1 \text{ or } -1 \tag{9}
$$

where $f$ is a function to be executed by a PE, and $k$ is a positive integer bounded by a linear function of $i$, $j$, and the problem size. The size of a problem is characterized by a finite set of integers. For example, the size of the FIR filtering problem [see Eq. (2)] is characterized by $\{n,m\}$. $z_{i,j}^{k-\delta}$ is the intermediate result of the last step of iterative computation at position $(i, j)$. $x(i,k)$ and y $(k, j)$ are linear functions in linear recurrences to define the indexes of $\boldsymbol{X}$ and $\boldsymbol{Y}$ in the $k$th step of iteration. In the following discussion, the coefficients of $i$, $j$, and $k$ in $x(i, k)$ and $y(k, j)$are 1 or −1. This assumption will be extended in Section 5 so that the coefficients can be any integer. $\delta$ is defined within a recurrence to indicate the order of evaluation. When $\delta=-1$, the recurrence is called a forward recurrence, and $z^k$ is defined in terms of $z^{k+1}$. When $\delta=-1$, the recurrence is termed a backward recurrence, and $z^k$ is defined in terms of $z^{k-1}$. The triangular matrix inversion recurrence defined above is a forward recurrence, while others are backward recurrences.

There are four ways of representing a recurrence process when the operations performed in computing $z_{i,j}$ are commutative: forward recurrence, backward recurrence, and the corresponding recurrences in which the evaluation order of terms are reversed. As an example, the following three recurrences are equivalent ways of representing matrix multiplication in addition to Eq. (3).

$$c_{i,j}^0 = 0; \quad c_{i,j}^k = c_{i,j}^{k-1} + a_{i,n-k+1} b_{n-k+1,j} \quad k = 1, \cdots, n \tag{10}$$

$$c_{i,j}^{n+1} = 0; \quad c_{i,j}^k = c_{i,j}^{k+1} + a_{i,k} b_{k,j} \quad k = n, \cdots, 1 \tag{11}$$

$$c_{i,j}^{n+1} = 0; \quad c_{i,j}^k = c_{i,j}^{k+1} + a_{i,n-k+1} b_{n-k+1,j} \quad k = n, \cdots, 1 \tag{12}$$

These four formulations represent only two unique ways of computing $c_{i,j}$. Eq. (10) and Eq. (11) are equivalent as far as the evaluation order of terms is concerned. The same can be said about Eq. (3) and Eq. (12). In designing systolic algorithms, only the evaluation order is important, and it is insignificant whether a recurrence is written in the forward or backward form. In this paper, it is assumed that the two alternate ways of solving the same problem are expressed as backward recurrences. Since the complexities of the resulting design may depend on the order of evaluation, the two alternative recurrences must be studied.

The key to systolic processing is that the appropriate data must be in the appropriate place at the time they have to be processed. That is, both timing and data distributions are very important, and the relationship between them must be identified. The coupling between time and space is provided by index $k$ in the general recurrence formula. The superscript $k$ is concerned with time, that is, the number of steps of iterative operation, whereas the subscript $k$ is concerned with data distribution. For example, in matrix multiplication, $c_{i,j}^k$ can be computed provided that $c_{i,j}^{k-1}$, $a_{i,k}$, and $b_{k,j}$ arrive at the same PE simultaneously. In other words, the $k$th step of computing $c_{i,j}$ requires the intermediate result of the $(k-1)$ st step of computing $c_{i,j}$, an entry in row $i$ and column $k$ of matrix $\boldsymbol{A}$, and an entry in row $k$ and column $j$ of matrix $\boldsymbol{B}$.

## 3 Parameters for designing systolic arrays

In this section, a set of parameters characterizing the behavior and the correctness of systolic arrays are defined[29]. These parameters are illustrated with respect to the matrix multiplication problem that can be represented in a backward recurrence $c_{i,j}^k = f\left[c_{i,j}^{k-1}, a(i,k), b(k,j)\right]$ [see Eq. (3)]. A possible systolic design is depicted in Figure 1. The data flows of the three rhomboidal data blocks are in three directions: $\boldsymbol{A}$ moves towards the north, $\boldsymbol{B}$ moves towards −120° north, and $\boldsymbol{C}$ moves towards −60° north. During a clock cycle, each PE receives three data items from three different pipes and executes a multiply-add operation. These data items advance into neighboring PE's along their own pipes synchronously in the next clock cycle. This design is for illustration and requires the minimal completion time but not the minimal number of PE's. If hardware or #PE×$T$ or #PE×$T^2$ is to be optimized, one of the matrices has to be stationary; however, the concept of vector composition of zero vectors cannot be depicted clearly.

A systolic array consists of a mesh of interconnected PE's. The distance between two directly connected neighboring PE's is defined to be unity. If buffers exist between two adjacent

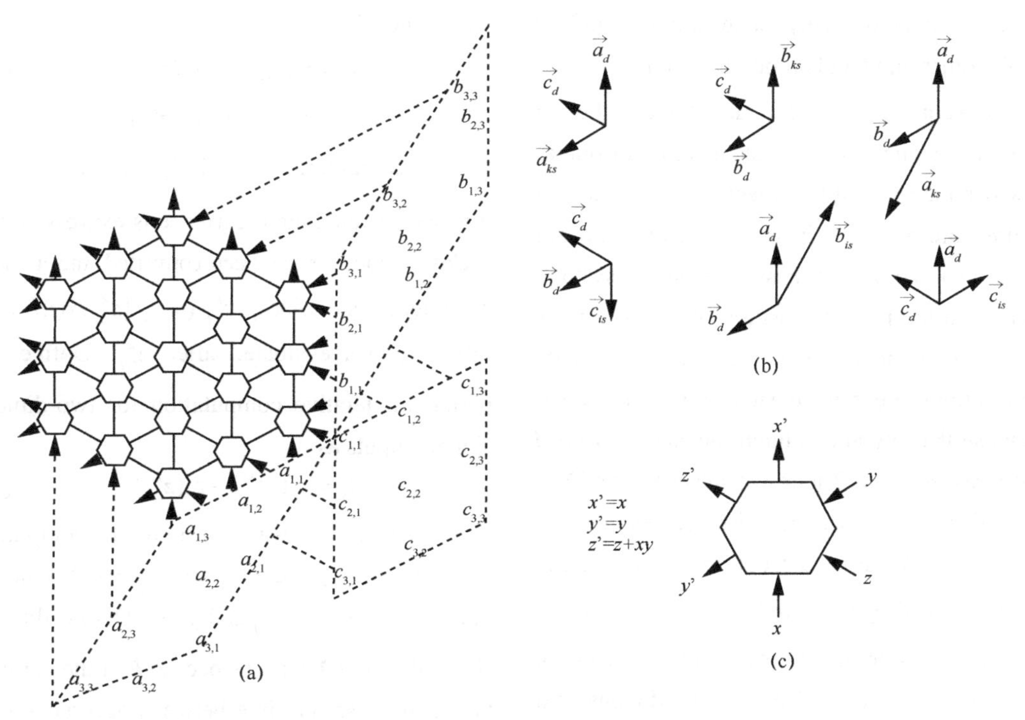

Figure 1. The systolic processor for two-dimensional matrix multiplication $[c_{i,j}^0 = 0; c_{i,j}^k = c_{i,j}^{k-1} + a_{i,k}b_{k,j}, 1 \leqslant i,j,k \leqslant n]$.
(a) ystolic processor. (b) Vector equations with ($t_i$=$t_j$=$t_k$=1). (c) Structure of cell

PE's, they are equally spaced along the link. A clock cycle is a unit of time during which one iterative operation is computed in a PE and data advance into neighboring PE's or buffers, or a datum advances from one buffer into the next stage.

The ways that data are fed into a systolic array are related and can be characterized by three parameters: velocity, space, and time. If these parameters are known, the systolic array can be completely determined. The first two parameters defined below are vectors[1]; the third parameter is a scalar.

[1] All vectors in this paper are indicated by symbols in bold letters with an arrow on top(e.g., $\vec{x}$ ). The magnitude(absolute value) of a vector is enclosed in vertical bars. A negative value indicates that the direction of the vector is reversed.

**Parameter 1**: Velocity of data flow.The velocity of a datum $x$ is defined as the directional distance passed by $x$ during a clock cycle and is denoted as $\vec{x}_d$. Since the distance between adjacent PE's is unity, and buffers, if they exist, are equally spaced between PE's, the magnitude of $\vec{x}_d$ must be a rational number of the form $i/j$ where $i$ and $j$ are integers and $i \leqslant j$. This means that in $j$ clock cycles, $x$ has propagated through $i$ PE's and $j-i$ buffers. If $i = 1$, then there are $j-1$ buffers between two neighboring PE's in the pipelining direction of $x$. Otherwise, there are $j-i$ buffers between $i$+1 PE's, and their positions have to be determined from the recurrence process. The locations of these buffers for one of the input directions can be chosen randomly, and these determine uniquely the locations of buffers for the other input and output directions.

**Parameter 2:** Data distribution. For a two-dimensional array ***X*** used as input or output of a systolic array, the elements along a row or a column are arranged in a straight line and are equally spaced as they pass through the systolic array, and the relative positions of the elements are iteration independent. Other forms of data distributions are not considered in this paper. Suppose the row and column indexes of ***X*** are $i$ and $j$, respectively. The row displacement of ***X*** is defined as the directional distance between $x_{i,j}$ and $x_{i+1,j}$ as ***X*** passes through the systolic array and is denoted by $\vec{x}_{is}$. Similarly, the column dis-placementof is defined as the directional distance between $x_{i,j}$ and $x_{i,j+1}$. If ***X*** is a one-dimensional array, the index in accessing ***X*** is implicit, and the item displacement of $X(\vec{x}_s)$ is defined as the directional distance between $x_i$ and $x_{i+1}$. Since data are equally spaced along rows and columns, the ro $X(\vec{x}_{js})$ w and column displacements are independent of the values of $i$ and $j$. Note that the direction of data distribution is defined along a subscript-increasing direction, and the magnitudes of all data-distribution vectors are nonzero rational numbers due to pipelining.

Referring to Figure 1, array ***A*** is referenced through indexes $i$ and $k$ in the recurrence formula; hence, the data distribution vectors are defined by $\vec{a}_{is}$ and $\vec{a}_{ks}$.

**Parameter 3**: Period.Suppose the time at which a computation is performed is defined by the function $\tau_c$, and the time at which an input is accessed for a particular computation is $\tau_a$. The periods of $i$ and $j$ for two-dimensional outputs are defined as

$$t_i = \tau_c(z_{i+1,j}^k) - \tau_c(z_{i,j}^k) \tag{13}$$

$$t_j = \tau_c(z_{i,j+1}^k) - \tau_c(z_{i,j}^k) \tag{14}$$

In computing $z_{i,j}^k (= f\left[z_{i,j}^{k-1}, x(i,k), y(k,j)\right])$, it is assumedthat the recurrence is expressed in a backward form or has been converted into a backward form, and hence, $z_{i,j}^{k+1}(= f[z_{i,j}^k, x(i,k+1), y(k+1,j)])$ is evaluated after $z_{i,j}^k$. Define the period of iterative computation for two-dimensional outputs as

$$t_k = \tau_c(z_{i,j}^{k+1}) - \tau_c(z_{i,j}^k) \tag{15}$$

Note that $t_k$ is always positive. In computing $z_{i,j}$, items $x_{i,k}$ and $x_{i,k+1}$ are accessed sequentially, and so are $y_{k,j}$ and $y_{k+1,j}$. Define the periods of ***X*** and ***Y*** with respect to $k$ in the computation of $z_{i,j}$ as the time between accessing successive elements of ***X*** and ***Y***. Formally,

$$t_{kx} = \tau_a(x_{i,k+1}) - \tau_a(x_{i,k}) \tag{16}$$

$$t_{ky} = \tau_a(y_{k+1,j}) - \tau_a(y_{k,j}) \tag{17}$$

$t_{kx}$ and $t_{ky}$ may be negative depending on the order of access defined in the subscript-access functions $x(i,k)$ and $y(k,j)$. Since data needed in the computation of $z_{i,j}^{k+1}$ after the computation of $z_{i,j}^k$ must be assembled in time $t_k$, it is true that

$$t_k = |t_{kx}| = |t_{ky}| \tag{18}$$

As an example, the systolic array in Figure 1 is implemented with $t_{ka} = t_{kb} = 1$ because $a_{i,k}$ is accessed one cycle before $a_{i,k+1}$,and so are $b_{k,j}$ and $b_{k+1,j}$. It is also seen that $t_i = t_j = 1$. As another example, elements of vector ***A*** are accessed in increasing order of indexes in the recurrence for polynomial multiplication [see Eq. (5)]. The period $t_{ka}$ is, therefore, positive. On the other hand, vector ***B*** is accessed in decreasing order of indexes, and hence, $t_{kb}$ is negative.

It is important to note that the computations in a systolic array are periodic, and hence, all the periods are independentof *i*, *j*,and *k*. Furthermore, periods represent time intervals between computations, and their absolute values must be rational numbers greater than or equal to 1 because directly connected PE's are separated at unit distances, and computations can only be performed in PE's. When absolute values of periods are less than 1, there must be a bus that can broadcast data, and this is outside the scope of our present discussion.

There is a total of 13 parameters for two-dimensional linear recurrences, of which three are for the velocities of data flow, $\vec{x}_d, \vec{y}_d, \vec{z}_d$ , six are for data distributions, $\vec{x}_{is}, \vec{x}_{js}, \vec{y}_{is}, \vec{y}_{js}, \vec{z}_{is}, \vec{z}_{js}$ ,and four are for the periods, $t_{kx}, t_{ky}, t_i, t_j$.For one-dimensional problems, only nine parameters $\vec{x}_d, \vec{y}_d, \vec{z}_d$, $\vec{x}_s, \vec{y}_s, \vec{z}_s, t_{kx}, t_{ky}$, and $t_i$ exist. These parameters can be used in constraint equations to govern the correctness of the design and in performance measures to define the number of PE's needed and the completion time. The following theorem states the relationships among these parameters.

**Theorem 1** (Theorem of Systolic Processing). Suppose a two-dimensional recurrence computation $z_{i,j}^k = f\left[z_{i,j}^{k-1}, x(i,k), y(k,j)\right]$ is implemented in a systolic array, then the velocities, data distributions, and periods must satisfy the following vector equations:

$t_{kx}\vec{x}_d + \vec{x}_{ks} = t_{kx}\vec{z}_d$ (data movement for ***X*** and ***Z*** between computing $z_{i,j}^{k-1}$ and $z_{i,j}^k$) (19)

$t_{ky}\vec{y}_d + \vec{y}_{ks} = t_{ky}\vec{z}_d$ (data movement for ***Y*** and ***Z*** between computing $z_{i,j}^{k-1}$ and $z_{i,j}^k$) (20)

$t_i\vec{x}_d + \vec{x}_{is} = t_i\vec{y}_d$ (data movement for ***X*** and ***Y*** between computing $z_{i,j}^k$ and $z_{i+1,j}^k$) (21)

$t_i\vec{z}_d + \vec{z}_{is} = t_i\vec{y}_d$ (data movement for ***Y*** and ***Z*** between computing $z_{i,j}^k$ and $z_{i+1,j}^k$) (22)

$t_j\vec{y}_d + \vec{y}_{js} = t_j\vec{x}_d$ (data movement for ***X*** and ***Y*** between computing $z_{i,j}^k$ and $z_{i,j+1}^k$) (23)

$t_j\vec{z}_d + \vec{z}_{js} = t_j\vec{x}_d$ (data movement for ***X*** and ***Z*** between computing $z_{i,j}^k$ and $z_{i,j+1}^k$) (24)

For one-dimensional problems, only Eq. (19)~(22) are necessary.

**Proof**: Without loss of generality, orthogonally connected PE's with diagonal connections are used in our proof. It is also assumed that $t_k = t_{kx} = t_{ky} > 0$. In Figure 2, A, B, *C* and D represent four PE's that do not have to be directly connected. While PE C is computing $z_{i,j}^k = f\left[z_{i,j}^{k-1}, x(i,k), y(k,j)\right], x_{x(i,k+1)}$ is in PE B, [2]and $y_{y(k+1,j)}$ is in PE D [Figure 2(a)]. Since $t_{kx} = t_{ky} > 0$ , $\overrightarrow{CB}$ represencts $\vec{x}_{ks}$ , and $\overrightarrow{CD}$ represencts $\vec{y}_{ks}$. According to the characteristics of systolic processing, the operands needed in the next iteration must arrive at the same PE simultaneously after $t_k$ units of time. Hence, $z_{i,j}^k, x_{x(i,k+1)}$, and $y_{y(k+1,j)}$ arrive at PE A simultaneously that computes $z_{i,j}^{k+1} = f\left[z_{i,j}^k, x(i,k+1), y(k+1,j)\right]$ [Figure 2(b)]. We have $\overrightarrow{CA} = t_k\vec{z}_d, \overrightarrow{BA} = t_k\vec{x}_d, \overrightarrow{DA} = t_k\vec{y}_d$. From the principle of vector composition, $\overrightarrow{BA} + \overrightarrow{CB} = \overrightarrow{CA}$ and $\overrightarrow{DA} + \overrightarrow{CD} = \overrightarrow{CA}$, so Eq. (19) and Eq. (20) are proved. The other cases in

[2] $X_{x(i,j)}$ represents the element of ***X*** defined by the subscript-access function $x(i,j)$

which $t_{kx}$ and $t_{ky}$ have different signs or are negative can be proved similarly.

To prove Eq. (21) and Eq. (22), suppose that while PE D is computing $z_{i,j}^{k}=f[z_{i,j}^{k-1},x(i,k),y(k,j)]$, PE C is computing $z_{i+1,j}^{p}, p<k$, and PE B has $x_{x(i+1,k)}$ [Figure 2(c)]. Therefore, $\overrightarrow{DC}=\vec{z}_{is}$, and $\overrightarrow{DB}=\vec{x}_{is}$. In accordance with the characteristics of systolic processing, $k$–$p$ steps of the iterative computation are performed after $t_i$ units of time, and $z_{i+1,j}^{k-1}$, $x_{x(i+1,k)}$, and $y_{y(k,j)}$ arrive at PE A for the computation of $z_{i+1,j}^{k}=f\left[z_{i+1,j}^{k-1},x(i+1,k),y(k,j)\right]$. Therefore, $\overrightarrow{BA}=t_i\vec{x}_d$, $\overrightarrow{CA}=t_i\vec{z}_d$, and $\overrightarrow{DA}=t_i\vec{y}_d$. From the principle of vector composition, $\overrightarrow{BA}+\overrightarrow{DB}=\overrightarrow{DA}$ and $\overrightarrow{CA}+\overrightarrow{DC}=\overrightarrow{DA}$; thus, Eq. (21) and Eq. (22) are proved. The cases for negative periods can be proved similarly.

Last, Eq. (23) and Eq. (24) can be proved similarly and will not be shown here.

Performance of a systolic design can be expressed in terms of the defined parameters. The number of PE's required (denoted by #PE) is studied here, and the completion time can be expressed as a function of the PE configuration and velocity.

The number of streams of data flow of a matrix $X$ in the direction of data flow is defined as the number of distinct lines that must be drawn in parallel to the direction of data flow so that each element of the matrix lies in exactly one line. For a one-dimensional vector with $n$ elements, the number of streams can be one (elements are fed serially) or $n$ (elements are fed in parallel). For a two-dimensional $n$-by-$n$ matrix, this number depends on the directions of $\vec{x}_{is}$ and $\vec{x}_{js}$.

If $\vec{x}_{is}$ and $\vec{x}_{js}$ are in the same or opposite directions, then the number of streams of data flow can be one (serial input) or $n^2$ (parallel input). If $\vec{x}_{is}$ and $\vec{x}_{js}$ are in different directions, then the number of streams is given by $n+(n-1)\ell$ where $\ell$ is an integer between 0 and $n$. To see that this is true, when $\ell$ is 0, each row or column of the matrix lies in a single stream, and the number of streams is $n$. The extreme case happens when each element of the matrix lies in a different stream, and the number of streams is $n^2$. As an example, each matrix in Figure 1 has five streams of data flow ($n$=3, $\ell$=1).

#PE depends on the directions in which the inputs are moving. There are four possible cases. First, one of the input matrices or the output matrix is stationary, and the others are moving. Assuming that all the elements of the stationary matrix are used in the computation, #PE is given by the size of the stationary matrix. Second, both input and output matrices are moving in the same or opposite directions. #PE is given by the product of the minimum number of streams of data flow and the distance traveled between the time that the first elements of the input matrices meet and the time that the last elements of the input matrices meet. Third, there are two independent directions of data flow. If the two input matrices are flowing in the same or opposite directions and the output matrix is flowing in a different direction, #PE is given by the number of streams of data flow of the output matrix. If the two input matrices are flowing in differentdirections and the output matrix is flowing in a direction of one of the inputs, and if each stream of data flow in an input matrix has to be interacted with every other stream of data flow in the other input matrix, #PE is given by the product of the numbers of streams of data flow of the input matrices. When the last assumption on the interaction of data flows is false, #PE reduced by a term that has to be determined from the recurrence. Lastly, there are three independent directions of data flow. In this case, #PE for the two input matrices can be computed as before. However, this number can be further reduced by the flow of the output matrix.

For example, matrices A, B, and C in Figure 1 are flowing in three different directions and have five streams of data flow each. #PE for A and B is 5×5=25. However, matrix C is flowing in a different direction which cuts off two corners of the PE configuration. #PE is reduced to 19.

Based on the last item of the result that will be computed, the time required for all the computations in a systolic array can be derived in terms of $t_k$,$t_i$,$t_j$, and the problem size to be solved. However, any load and drain times of the input-output matrices would depend on the directions and distributions of data flows on a given configuration of the PE's, and have to be analyzed for each case individually. Other design requirements such as the number of input-output pins can also be expressed in terms of the PE configuration and the defined parameters.

For example, the time needed for multiplying two $n$-by-$n$ matrices is $nt_k+(n-1)|t_i|+(n-1)|t_j|$ since, from Eq. (3), it takes $nt_k$ steps to compute $c_{1,1}$, $(n-1)|t_i|$ steps from computing $c_{1,1}$ to $c_{n,1}$, and $(n-1)|t_j|$ steps from computing $c_{n,1}$ to $c_{n,n}$. In the design in Figure 1, no load or drain time is necessary.

We have been less specific in the discussion on hardware and time complexities because

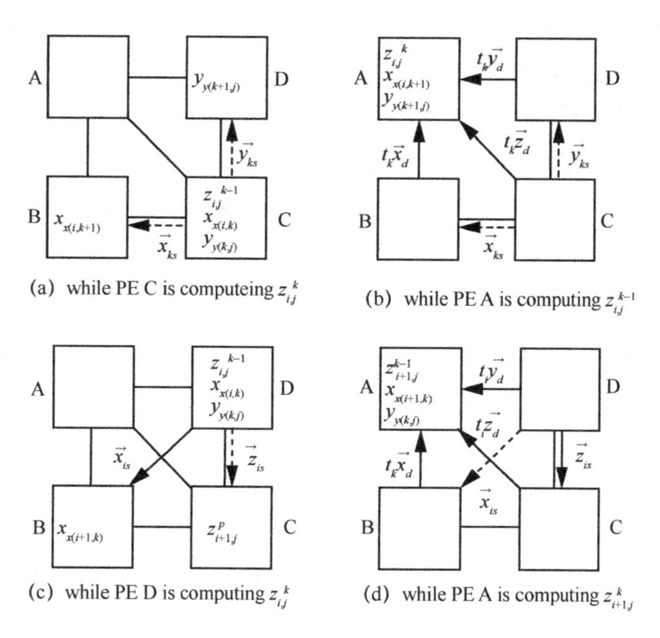


Figure 2. Proof of Theorem 1

they are dependent on the recurrence. In fact, closed-form expressions for these complexities under all possible situations are very complex, and it would be better to analyze the cases separately. This will be illustrated in the examples.

## 4 Design methodology for systolic arrays

The design of an optimal systolic array for a linear recurrence can be formulated into an optimization problem. The constraints of optimization are provided by Theorem 1, which shows the fundamental space-time relationships and governs the corrections in systolic processing. The objective function to be minimized can be expressed in terms of the parameters of systolic processing and the problem size. We first show the formulation of the optimization and then explain the various constraints. The design problem is formulated as

$$\text{minimize} \quad \#PE\times T^2 \text{ or } \#PE\times T \text{ or } T \quad (25)$$

subject to Eq. (19)~(24) and

$$\frac{1}{t_{j\max}} \leqslant |\vec{x}_d| \leqslant 1 \quad \text{or} \quad |\vec{x}_d| = 0 \quad (26)$$

$$\frac{1}{t_{i\max}} \leqslant |\vec{y}_d| \leqslant 1 \quad \text{or} \quad |\vec{y}_d| = 0 \quad (27)$$

$$\frac{1}{t_{k\max}} \leqslant |\vec{z}_d| \leqslant 1 \quad \text{or} \quad |\vec{z}_d| = 0 \quad (28)$$

$$1 \leqslant |t_k| \leqslant t_{k\max}; 1 \leqslant |t_i| \leqslant t_{i\max}; 1 \leqslant |t_j| \leqslant t_{j\max} \quad (29)$$

$$\begin{aligned} &|t_k||\vec{z}_d| = k_1 \leqslant t_{k\max}; |t_i||\vec{y}_d| = k_2 \leqslant t_{i\max} \\ &|t_j||\vec{x}_d| = k_3 \leqslant t_{j\max} \end{aligned} \quad (30)$$

$$|\vec{x}_{is}| \neq 0; \quad |\vec{x}_{ks}| \neq 0; \quad |\vec{y}_{ks}| \neq 0 \quad (31)$$

$$|\vec{y}_{js}| \neq 0; \quad |\vec{z}_{is}| \neq 0; \quad |\vec{z}_{js}| \neq 0 \quad (32)$$

$$t_k = |t_{kx}| = |t_{ky}| \quad \text{recurrence determines relative signs of } t_{kx} \text{ and } t_{ky} \quad (33)$$

$k_1, k_2, k_3, t_{k\max}, t_{i\max}$, and $t_{j\max}$ are integers. All other parameters are rational numbers, Moreover, $t_{k\max}, t_{i\max}$, and $t_{j\max}$ are functions of the problem size and $T_{\text{serial}}$, the number of times that function f in the recurrence has to be executed in order to compute all the required results.

The terms in Eq. (30) represent the distances traversed between computations. Since a computation must be performed in a PE, the distance traversed must coincide with the locations of PE's. The upper bounds of $k_1$, $k_2$, and $k_3$ are the maximum values of $t_k, t_i$, and $t_j$ because the maximum values of speeds are 1 [see Eq. (26)~(28)]. To derive these upper bounds, recall that the total computation time $T$ is a function of $t_k$, $|t_i|$, and $|t_j|$. In order for systolic processing to be more efficient than a serial computation, it is necessary for $T \leqslant T_{\text{serial}}$. By using the minimum values for two of the periods ($t_k = 1, |t_i| = 1$, $|t_j| = 1$) in the above inequality, the upper bound for the other period ($t_{k\max}, t_{i\max}$, or $t_{j\max}$) is obtained. As an example, in multiplying two 3-by-3 matrices, $T = 3t_k + 2|t_i| + 2|t_j| \leqslant T_{\text{serial}} = 27$. By setting $|t_i| = |t_j| = 1, t_{k\max} = [(27-4)/3] = 8$. Similarly, $t_{i\max} = t_{j\max} = 11$. Sometimes, $T$ is not a function of a period, say $t_i$. According to the definition of $t_i$, this is the time between computing $z_{i,j}^k$ and $z_{i+1,j}^k$. Suppose $i$ ranges from 1 to $n-1$; then $n|t_i|$ units of time are needed for the computation from $z_{1,j}^k$ to $z_{n,j}^k$. The inequality $n|t_i| \leqslant T_{\text{serial}}$ allows $t_{i\max}$ to be solved.

The constraints in Eq. (26)~(28) state that

data cannot travel more than one unit per unit of time because directly connected PE's are separated at unit distances, and no broadcasting is allowed. The lower bounds on velocities are derived from Eq. (30). Velocities smaller than the given lower bounds do not have to be considered because there exists a more efficient design of computing the recurrence in a single PE (with time $T_{serial}$). The constraints in Eq. (29) follow directly from the definitions of $t_k$,$t_i$, and $t_j$.

The set of distinct values of speeds and periods are related to the number of buffers among the PE's. Specifically, for $t_k, |\vec{z}_d|, k_1$, and $t_{k\max}$, $k_1$ represents the number of PE'straversed by a datum between two successive iterative computations, the maximum of which is $t_{k\max}$. Let $\eta$ be the maximum number of iterations required for computing a result ($\eta$ is usually a linear function of the problem size). For a given $k_1$, the maximum number of PE's in the pipeline is $(\eta-1)k_1+1$ since the first iteration is computed in a single PE, and the remaining $\eta$-1 iterations require $(\eta-1)k_1$ PE's. The total number of buffers in the pipeline, $b$, satisfies

$$0 \leqslant b \leqslant [(\eta-1)t_{k\max}+1]-[(\eta-1)k_1+1]= \eta(t_{k\max}-k_1) \quad (34)$$

Once $b$ is chosen, $|\vec{z}_d|$ and $|t_k|$ can be determined

$$|\vec{z}_d| = \frac{(\eta-1)k_1}{(\eta-1)k_1+b} \quad (35)$$

$$|t_k| = \frac{k_1}{|\vec{z}_d|} = k_1 + \frac{b}{\eta-1} \quad (36)$$

As a result, there are $O(\eta t_{k\max}^2)$ combinations of values of $t_k$ and $|\vec{z}_d|$. Similarly, for $t_i$ and $|\vec{y}_d|$, $t_j$ and $|\vec{x}_d|$, there are $O(\eta t_{i\max}^2)$ and $O(\eta t_{j\max}^2)$ combinations of value, respectively.

Data flow in the above optimization problem can be in one, two, or three independent directions. There are five possibilities: a) all inputs and outputs are flowing in the same or opposite directions; b) the inputs are flowing in the same or opposite directions, and the output is in a different direction; c), d) the inputs are flowing in two different directions, and the output is flowing in a direction of one of the inputs; and e) inputs and outputs are flowing in three different directions. The directions of vectors can be reversed, and velocities can be zero vectors. Due to the assumption of unit distance between directly connected PE's, when data are flowing in two independent directions, they must be orthogonal to each other; and when data are flowing in three independent directions, they must be in multiples of 120° to each other. Furthermore, the magnitudes of velocities and periods are chosen from a finite set. Therefore, the optimization of design of a systolic array for a given recurrence has a finite search space of complexity $O(\eta^3 t_{i\max}^2 t_{j\max}^2 t_{k\max}^2)$.

The worst case complexity shown above is very large. There are two ways to reduce this complexity. First, instead of requiring that $T \leqslant T_{serial}$, the requirement that $T \leqslant O(T_{serial}/\#PE)$ may be used. This reduces the values of $t_{k\max}$, $t_{i\max}$, and $t_{j\max}$, and in turn reduces the search complexity. Second, it is noted that systolic designs for linear recurrences are extendible because the systolic processing equations governing the correctness of the design (Theorem 1) are independent of the problem size. To reduce the search complexity, an optimal design for a smaller problem can be found and is used to extend to the systolic design for a larger version of the same problem. This method can also be applied when the maximum number of PE's that can be implemented in a single chip is smaller than that required by the optimal design. In general, this method will not lead to an optimal design for the original problem. This is due to the fact that the objective function is monotonically increasing with the problem size, and the fact that the objective function for one design is better at a given problem size does not imply that this design is better at a different problem size.

The optimal solution to the above design problem can be found by exhaustive enumeration. However, the search time for an arbitrarily large recurrence can be long, if not impossible. By recognizing that the completion time is a linear function of $t_k$,$t_i$, and $t_j$, their values can be ordered so that periods that do not lead to an optimal solution are eliminated from further consideration each time a feasible solution is found. This strategy is explained with respect to the minimization of completion time and #PE×$T^2$.

To minimize the completion time, the different directions of data flows are first determined (five possibilities). The maximum values of $t_k$,$t_i$, and $t_j$ are found. A set of $t_k$,$t_i$, and $t_j$ that minimize the completion time is selected from the set of possible values. The speeds of data flows are evaluated from Eq. (30) by using $k_1$=$k_2$=$k_3$=1. The six remaining unknowns on spatial distributions can be solved from the systolic processing Eq. (19)~(24). If no feasible solution is found, the procedure is repeated by finding another set of periods so that the completion time is increased by the least amount. The above steps are carried out for all the five combinations of data flow directions. If no feasible solution is found, one of the $k_1$, $k_2$, or $k_3$ is increased by 1, and the procedure repeats. It is obvious that the first feasible solution found is the optimal solution that minimizes the completion time.

To minimize #PE×$T^2$, it is necessary to know the lower bound on #PE. For linear recurrences with two-dimensional ($n$-by-$n$) inputs, the lower bound on #PE can be one (both inputs are serial), $n$ (one input is serial and the other has $n$ streams of data flow), or $n^2$ (both inputs have $n$ streams of data flow). It is easy to prove from the systolic processing equations that serial inputs usually do not lead to feasible solutions, and the lower bound is $n^2$. Repeating the procedure in minimizing completion time, a feasible design is first found. Suppose this design requires $P_1$ PE's and $T_1$ clock cycles to complete, then it is true that any design with $T_2 \geqslant \sqrt{P_1}T_1/n$ will not lead to a better solution (since $P_2T_2 \geqslant n^2T^2 \geqslant P_1T_1^2$) and can be eliminated from consideration. The search is continued to find better solutions with completion time between $T_1$ and $T_2$.

By systematically enumerating and reducing the search space that is a polynomial function of the problem size, the optimal design can be solved very efficiently. The method of designing optimal systolic arrays using the parameters defined in this paper is referred to as the parameter method. The steps in the design are sketched as follows.

**Step 1:** Write the recurrence formula for the problem to be solved.

**Step 2:** Write the corresponding systolic

processing equations (Theorem 1) and the constraints on the values of parameters.

**Step 3:** Write the objective function based on the design requirements in terms of the systolic processing parameters and the problem size.

**Step 4:** Find the parameter values that minimize the objective function by enumerating over the limited search space.

**Step 5:** Design a basic cell for the systolic array and find a possible interconnection of cells from the parameters obtained. Eliminate cells that do not perform any useful computation.

In the 3-by-3 matrix-multiplication problem, recall that the computation time needed is $3t_k+(3-1)|t_i|+(3-1)|t_j|$. The completion time is minimized when $t_k$,$|t_i|$ and $|t_j|$ are as small as possible. The search is started with $t_k=t_i=t_j=1$ on all the combinations of directions of data flows. If no feasible solution is found, the signs of $t_i$ or $t_j$ are negated, and the search repeats. In this example, when data are flowing in three different directions, $t_k=t_i=t_j=1$ results in a solution that satisfies the constraints of Eq. (19)~(24) and Eq. (26)~(33) and minimizes the completion time. The corresponding vectors are depicted in Figure 1(b). By using these vectors, the velocities and spatial distributions of data flows can be determined, and a basic cell design is shown in Figure 1(c). These cells are connected together into a mesh. Some of the cells are eliminated because no computation is performed there. The final VLSI structure is shown in Figure 1(a). This is the fastest matrix-multiplication scheme, which can be completed in 7 units of time with 19 cells.

On the other hand, if #PE $\times T^2$ is to be minimized, the search has to be continued to find out all the feasible designs with completion time less than $\sqrt{19\times 7^2/3^2}=10.2$. By assuming that the output matrix is stationary, a feasible design can be found with $t_k=t_i=t_j=1$,7 units of computation time, and 3 units of drain time. This reduces the search space further to finding all feasible designs with less than 10 units of completion time. In fact, this is the optimal solution that minimizes #PE$\times T^2$.

Note that the recurrence for two- dimensional tuple com-parison [see Eq. (8)] isidentical to the recurrence for matrix multiplication except for the operations performed in the PE's, and hence, the systolic array for matrix multiplication can be applied in this case.

## 5 Examples

(1) Direct applications of the methodology

- FIR filtering. The operation can be represented as a one-dimensional linear recurrence $y_i^k=f\left[y_i^{k-1},x(k,i),a(k)\right]$ [see Eq. (2)]. Another recurrence that evaluates the terms in a reverse order can be written as $y_i^k=y_i^{k-1}+a_{m-k+1}x_{m-k+i},1\leqslant i\leqslant n,1\leqslant k\leqslant m$. In both recurrences, the inputs areaccessed in the same order, and hence $t_{ka}=t_{kx}$. It takes $mt_k$ units of time to compute $y_1$ ($m$ is the window size of the FIR filter) and $(n-1)|t_i|$ units of time to compute the remaining $y_i$'s. The total computation time, disregarding possible load and drain times, is $mt_k+(n-1)|t_i|$. For a problem with $m$= 4 and $n$= 6, $T_{serial}$=24. From this, $t_{k\max}$ = $\lceil 19/4\rceil=5$, $t_{i\max}=\lceil 20/5\rceil=4$. The design problem can be formulated as minimize #PE $\times[4|t_k|+5|t_i|$ +load time + drain time$]^2$ (37)

subject to

$$t_{kx}\vec{x}_d+\vec{x}_s=t_{kx}\vec{y}_d \quad (38)$$

$$t_{ka}\vec{a}_d+\vec{a}_s=t_{ka}\vec{y}_d \quad (39)$$

$$t_i\vec{x}_d+\vec{x}_s=t_i\vec{a}_d \quad (40)$$

$$t_i\vec{y}_d+\vec{y}_s=t_i\vec{a}_d \quad (41)$$

$$\frac{1}{4}\leqslant|\vec{a}_d|\leqslant 1 \text{ or } |\vec{a}_d|=0$$

$$\frac{1}{5}\leqslant|\vec{y}_d|\leqslant 1 \text{ or } |\vec{y}_d|=0$$

$$1\leqslant t_k\leqslant 5 \quad 1\leqslant|t_i|\leqslant 4$$

$$|\vec{a}_s|\neq 0 \quad |\vec{x}_s|\neq 0 \quad |\vec{y}_s|\neq 0$$

$$|t_i||\vec{a}_d|=k_1\leqslant 4 \quad t_k|\vec{y}_d|=k_2\leqslant 5$$

$$t_k=|t_{kx}|=|t_{ka}| \quad t_{kx}=t_{ka}$$

where $k_1,k_2$ are integers; all the other magnitudes are rational numbers. It is not necessary to bound $|\vec{x}_d|$ because $\vec{x}_d$ is uniquely determined when $t_k,t_i,\vec{y}_d$ and $\vec{a}_d$ are set.

The search space in the above problem is quite reasonable. However, the problem can be solved very efficiently by using the defined search order. First, consider $t_{kx}=t_{ka}=-1$ and $t_i=1$. Substituting into Eq. (38)~(41) results in four equations with six unknowns. From FIR-filtering applications, it is known that the $a_i$'s are defined constants and can be fixed in the systolic array without preloading. Assume that the $a_i$'s are statically placed in the PE's, $|\vec{a}_d|=0$, and from Eq. (39), $\vec{a}_s=-\vec{y}_d$. Since $|\vec{a}_s|\neq 0$ and $|\vec{y}_d|\leqslant 1$, $\vec{y}_d$ can be set to 1 or $-1$. Solving both cases results in a systolic design that requires four PE's, $m+n-1$ units of computation time and the preloading of $x_1,\cdots,x_m$ into the pipe[30]. The completion time is, therefore, $2m+n-1$ units.

On the other hand, if $t_{kx}=t_{ka}=1, t_i=-1$, and $|\vec{a}_d|=0$, then $|\vec{y}_d|=|\vec{y}_s|=1, |\vec{x}_d|=|\vec{x}_s|=1/2$, and $|\vec{a}_s|=1$. This is a one-dimensional solution (all the vectors are pointing in the same direction). A feasible systolic design satisfying the above parameters is depicted in Figure 3. This design does not require elements of $X$ to be preloaded, and the completion time of the algorithm is 9 units. It should be noted that $t_{kx}=t_{ka}=1$ implies that the recurrence formula in Eq. (2) is used.

To see that the design in Figure 3 is optimal, the performance measure #PE $\times T^2$ for $n=6$, $m=4$ is 324. If the number of PE's is decreased to one, $T=T_{\text{serial}}=24$, and #PE $\times T^2$ = 576. The lower bound on #PE is $m$=4. The proposed design achieves this lower bound in the minimal completion time and, hence, is optimal.

It is of interest to note that discrete Fourier transform [see Eq. (4)], polynomial multiplication [see Eq. (5)], and many pattern-matching problems have the same form of recurrence equations as FIR filtering. The systolic design in Figure 3 can be applied except that different functions are performed in the PE's[30].

• Band-matrix multiplication. The systolic array for matrix multiplication (Figure 1) can be used for band-matrix multiplieation. However, the number of cells required is large considering that most of the terms in the result are zero. By recognizing that the result is also a band matrix, a different recurrence can be written to compute the elements in the band alone. The proposed methodology can be applied to obtain the optimal systolic design.

In this example, we illustrate the flexibility of our design method by showing that additional constraints (based on insights) can be included in the

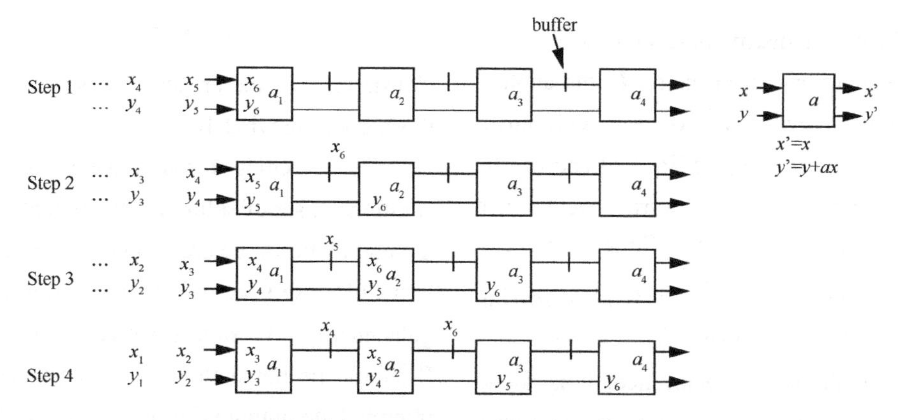


Figure 3. Systilic array for FIR filtering ($n$=6, $m$=4).

optimization. Given that the band width is $m$, the most efficient direction of sending the band matrix into a systolic array is along the diagonal. These required directions of data flows can be added as constraints to the original design problem

$$\vec{a}_{is} + \vec{a}_{js} = k_1\vec{a}_d \qquad k_1 \text{ is a rational number} \quad (42)$$

$$\vec{b}_{is} + \vec{b}_{js} = k_2\vec{b}_d \qquad k_2 \text{ is a rational number} \quad (43)$$

The details of solution are not shown here. The resulting design for $m$=3 has three independent directions of data flow (Figure 4). The design is optimal and requires $m^2$ PE's and $m+n-1$ units of completion time.

(2) Recurrences with feedback

In this section, the design of systolic arrays for recurrences with feedback are exemplified. When outputs are routed back into a systolic array, the direction of data flows has to be changed. The difficulties in designing such a systolic array are that feedbacks pose another direction of data flows in the inputs, and the correctness of feedbacks is hard to express in systolic processing equations because the format of feedbacks is governed by the way outputs are generated. One method is to treat the feedbacks as an independent input stream and to design the systolic array in the usual way. After an efficient design is obtained, the feedbacks have to be checked to determine whether they are generated before they are fed back into the systolic array. The process is repeated until an efficient and correct design is found.

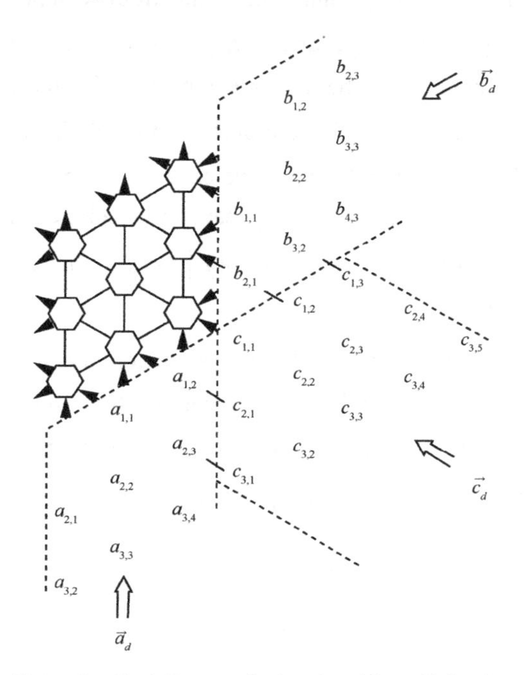

Figure 4. Systolic array for band-matrix multiplication (the cell used is the same as that of matrix multiplication)

• Triangular-matrix inversion: The inversion of an upper-triangular matrix **U** into another upper-triangular matrix **V** has been represented in a forward recurrence [see Eq. (7)] and can be rewritten into a backward recurrence $w_{i,j}^k = f\left[w_{i,j}^{k-1}, u(i,k), v(k,j)\right]$. Consecutive elements of **U** and **V** are accessed in decreasing order of subscripts, and hence, $t_{ku} = t_{kv} < 0$.

By examining the recurrence, it is found that the evaluation of $w_{i,j}$ requires one more term than that of $w_{i,j-1}$. Therefore, $w_{i,j-1}$ can be completed before $w_{i,j}$, which implies that $t_i > 0$. On the other hand, the evaluation of requires one less term than that of $w_{i-1,j}$, which implies that $t_i < 0$. From the recurrence, it is also seen that the last item to be computed is $v_{1,n}$ rather than $v_{1,1}$. It takes $(n-1)|t_i|$ units of time from computing $v_{n,n}$ to $v_{2,n}$; and $n|t_{ku}|$ units of time to compute $v_{1,n}$. The computation time is, thus, $n|t_{ku}| + (n-1)|t_i|$. For a problem with $n$=4, $T_{serial}$= 20. By solving $T \leqslant T_{serial}$, we have $t_{k\max} = 5$ and $t_{i\max} = 6$. Since $v_{1,1}, \cdots, v_{1,4}$ have to be computed, we have another equality $4|t_j| \leqslant T_{serial}$ that results in $t_{j\max} = 5$.

To minimize #PE×$T^2$, parameters are first chosen to minimize the computation time: $t_{ku} = t_{kv} = -1$, $t_i = -1$, and $t_j = 1$. From these, the following systolic processing equations are obtained:

$$-\vec{u}_d + \vec{u}_{ks} = -\vec{w}_d \tag{44}$$

$$-\vec{v}_d + \vec{v}_{ks} = -\vec{w}_d \tag{45}$$

$$-\vec{u}_d + \vec{u}_{is} = -\vec{v}_d \tag{46}$$

$$-\vec{w}_d + \vec{w}_{is} = -\vec{v}_d \tag{47}$$

$$\vec{v}_d + \vec{v}_{js} = \vec{u}_d \tag{48}$$

$$\vec{w}_d + \vec{w}_{js} = \vec{u}_d \tag{49}$$

Constraints on parameter values are similar to those stated previously.

By searching through the set of feasible solutions, a set of solution vectors that minimize the completion time are shown in Figure 5(a). In terms of these vectors, a systolic array for triangular-matrix inversion is depicted in Figure 5(b). It should be noted that $w_{i,j}$ is considered as an intermediate output with initial value of 0. When $w_{i,j}$ arrives at one of the dividers denoted by circles, the computation of $w_{i,j}$ is finished, and $v_{i,j}$ is generated. $v_{i,j}$ is then fed back and moves downward. Note that in Figure 5(b), the triangular block of **V** above the systolic processor represents a pseudo-input-matrix, whereas another triangular block of **V** to the right of the array shows the output sequence. In this scheme, a triangular array of $n(n+1)/2$ PE's can compute an $n$th-order triangular-matrix inversion in $2n-1$ units of time. Since there is no feasible design with one or $n$ PE's, the proposed design minimizes #PE×$T^2$.

• Deconvolution. In the previous examples, processing times in the PE's are assumed to be identical. This implies that the period of processing is independent of the PE along the direction of data flow.

In some applications, different operations may be performed in PE's along a direction of data flow. For example, the last operation in deconvolution is division, which may take more time than multiplication and may become the bottleneck of data flow. Referring to the recurrence for deconvolution $z_i^k = f\left[z_i^{k-1}, x(i,k), a(k)\right]$ [see Eq. (6)], the

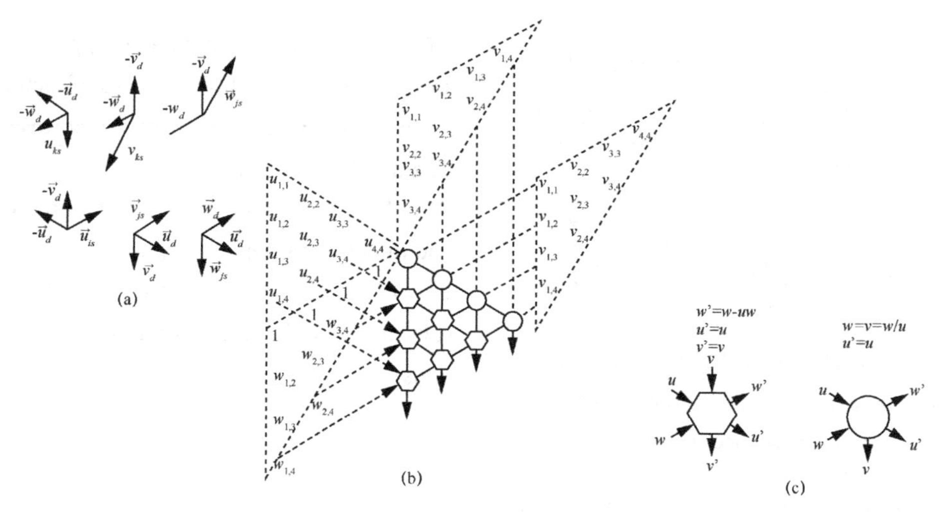

Figure 5. Systolic array for triangular-matrix inversion($n$=4)

$x_i$'s computed are fed back into the pipe for future computation. Let the delay of a division PE be $w$ and the delay of other PE's be 1. The last iteration of computing $z_i$ and the division of $z_i$ to obtain $x_i$ takes ($w$+1) units of time. This is the period for the $\boldsymbol{X}$ inputs.

$$|t_i| = w+1 \quad (50)$$

Due to the extra time involved in division, the design of feedback signals is more complicated than that of triangular matrix inversion. The fact that inputs and feedbacks are one-dimensional permits us to write down the relationships of data flows more formally in systolic processing equations [Figure 6(a)]. Suppose PE A is the PE in which $z_i^{m-1}$ is being computed using $x_{i+1}$ before being sent to the division unit (PE D). $z_{i-1}^{p}, p < m-1$ (for backward recurrences), exists in PE B. In (1 + $w$) units of time, $z_i^{m-1}$ is sent to PE D and is converted to $x_i$, which is then sent to PE C. At that time, $z_{i-1}^{m-1}$ will be computed in PE C using $x_i$. From the above discussion, we have another systolic processing equation

$$\vec{z}_d + \vec{x}_d = -\vec{z}_s + (1+w)\vec{z}_d \Rightarrow \vec{x}_d = w\vec{z}_d - \vec{z}_s \quad (51)$$

Eq. (50) and Eq. (51) must be included in the optimization. Note that $\boldsymbol{X}$ is a one-dimensional vector, and therc is one spatial-distribution parameter $\vec{x}_s$, although the subscript-access function is a function of $i$ and $k$. By letting $w$ = 2, $|\vec{a}_d| = 0, |\vec{a}_s| = -1$, and observing that $t_{kx} = t_{ka}$ the other parametes can be solved: $t_{ka} = -3/2, t_i = 3$, $|\vec{z}_d| = -2/3, |\vec{z}_s| = 2, |\vec{x}_d| = 2/3$ and $|\vec{x}_s| = -2$ . These parametes satisfy the systolic processing equations, but the feedback $\boldsymbol{X}$ cannot be generated at the right time from $\boldsymbol{Z}$. if $t_i = -3$, the values of the other parametes are $t_k = -3/2$ , $|\vec{z}_d| = 2/3$, $|\vec{z}_s| = 2, |\vec{x}_d| = -2/3$, $|\vec{x}_s| = -2$. For $m$=4, $n$=5, the completion time $T = (m-1+w)t_k + \ (n-1)|t_i|$ and #PE $\times T^2$ are minimized. It should be noted that the velocities of data flows are averaged over three clock cycles. By first inserting buffers for the $\boldsymbol{Z}$ inputs, the positions of buffers for the $\boldsymbol{X}$ inputs can be decided. A feasible assignment of buffers is shown in Figure 6(b).

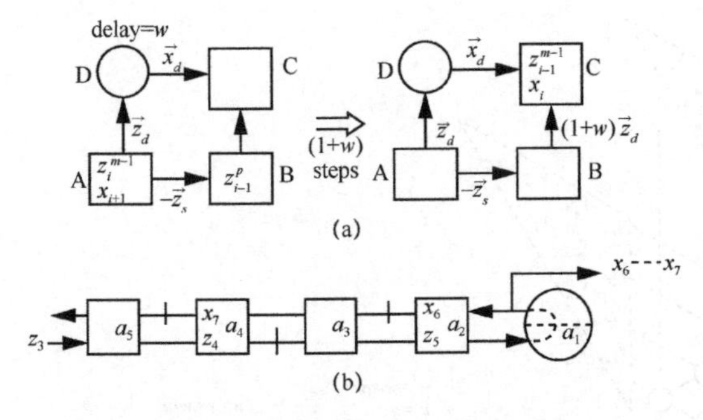


Figure 6. Systolic array for deconvolution. (a) Relationships of data flow. (b) Systolic array for $n$=5, $m$=4 after six clock cycles.

(3) Generalizations on the recurrence formula

In this section, extended forms of the general recurrence formula [see Eq. (9)] are discussed. The systolic processing equations presented in Theorem 1 have to be modified.

• Indexes in subscript-access functions. In the original recurrence formula (9), $x$ is a function of $i$ and $k$, and $y$ is a function of $k$ and $j$. In general, $x$ and $y$ are functions of $i$, $j$, and $k$. The general recurrence formula can be expressed as

$$z_{i,j}^{k}=f\left[z_{i,j}^{k-1},x(i,j,k),y(i,j,k)\right] \tag{52}$$

Systolic processing Eq. (21)-(24) have to be changed

$$t_i\vec{z}_d+\vec{z}_{is}=t_i\vec{x}_d+\vec{x}_{is} \tag{53}$$

$$t_i\vec{z}_d+\vec{z}_{is}=t_i\vec{y}_d+\vec{y}_{is} \tag{54}$$

$$t_j\vec{z}_d+\vec{z}_{js}=t_j\vec{x}_d+\vec{x}_{js} \tag{55}$$

$$t_j\vec{z}_d+\vec{z}_{js}=t_j\vec{y}_d+\vec{y}_{js} \tag{56}$$

The proof of Eq. (53) and Eq. (54) is similar to the proof shown in Theorem 1 except that an additional PE is needed where $Y_{y(i+1,j,k)}$ is stored(Figure 2). $Y_{y(i+1,j,k)}$ is moved to PE A in $t_i$ units of time. Similar arguments apply to Eq. (55) and Eq. (56). If a subscript-access function does not involve a particular index, the corresponding data distribution parameter is zero.

As an example, the recurrence for one-dimensional tuple comparison: $c_i^0=\text{TRUE}$, $c_i^k=c_i^{k-1}\wedge(a_{i,k}-b_{i,k}),1\leqslant i\leqslant n,1\leqslant k\leqslant m$, results in the following systolic processing equations:

$$t_{ka}\vec{a}_d+\vec{a}_{ks}=t_{ka}\vec{c}_d \tag{57}$$

$$t_{kb}\vec{b}_d+\vec{b}_{ks}=t_{kb}\vec{c}_d \tag{58}$$

$$t_i\vec{c}_d+\vec{c}_s=t_i\vec{a}_d+\vec{a}_{is} \tag{59}$$

$$t_i\vec{c}_d+\vec{c}_s=t_i\vec{b}_d+\vec{b}_{is} \tag{60}$$

• Coefficients of indexes. The coefficients of $i$, $j$, and $k$ used in the subscript-access functions $x$ and $y$ are 1 (regardless of sign). This means that all elements of $\boldsymbol{X}$ and $\boldsymbol{Y}$ are used in the course of computation of the recurrence. When these coefficients are greater than 1, some items in the inputs will be skipped. As an example, suppose we have the following recurrence for FIR filtering:

$$\begin{aligned}&y_i^0=0\quad 1\leqslant i\leqslant n\\&y_i^k=y_i^{k-1}+a_{2k}x_{i+3k-1}\\&\qquad 1\leqslant i\leqslant n,1\leqslant k\leqslant m,x_j=0\text{ for }j>n\end{aligned} \tag{61}$$

In this case, every other item of $\boldsymbol{A}$ and every other two items of $\boldsymbol{X}$ will be skipped.

Consider the case in which the coefficients of $k$ can be any integer (regardless of sign). The general form of the recurrence formula is

$$z_{i,j}^{k}=f\left[z_{i,j}^{k-1},x(i,uk),y(vk,j)\right]\ u,v\text{ are positive integers} \tag{62}$$

In this case, systolic processing Eq. (19)

and Eq. (20) have to be modified

$$t_{kx}\vec{x}_d + u\vec{x}_{ks} = t_{kx}\vec{z}_d \quad (63)$$

$$t_{ky}\vec{y}_d + v\vec{y}_{ks} = t_{ky}\vec{z}_d \quad (64)$$

The other four systolic processing equations remain changed. The proof for these is very similar to that of Theorem 1. The data distribution parameters are augmented in this case because nonadjacent data are used. Similar changes have to be made on the other four systolic processing equations when the coefficients of $i$ and $j$ in the subscript-access functions are integers not equal to 1.

For the recurrence given in Eq. (61), systolic processing Eq. (38) and Eq. (39) have to be changed to Eq. (63) and Eq. (64) with $u = 2, v = 3$. Solving these with Eq. (40) and Eq. (41) while assuming $t_k = t_i = 1$, we obtain $|\vec{y}_d| = 1$, $|\vec{y}_s| = -1$, $|\vec{a}_d| = 0$, $|\vec{x}_d| = -1/2$, and $|\vec{a}_s| = |\vec{x}_s| = 1/2$. The interestingpoint is that the separation between the $a_i$'s is 1 /2 unit, which implies that only half of the $a_i$'s reside in PE's. The other half do not have to be loaded since they are never used. The throughput of the system remains unchanged, that is, one output per clock cycle.

## 6 Conclusion

Conventional approaches in mapping algorithms onto systolic arrays are usually done in an ad hoc fashion. Theresulting design represents a feasible, but not necessarily optimal, solution. In this paper, we propose a systematic methodology for the optimal mapping of algorithms that are represented as linear recurrences onto systolic arrays. The characteristics of systolic arrays are parameterized by the velocities of data flows, spatial distributions and periods of computation. The design problem is formulated into an optimization problem with objective function constraint equations that are functions of the defined parameters. An efficient order of searching the solution space is also proposed.

Numerous examples illustrating the methodology are shown. Cases including feedbacks and general recurrence formulas are discussed. Although the systolic-processing equations defined does not exhaust all possible situations, the theory developed can guide the designers in obtaining the necessary equations. For example, the mapping of recurrences onto a fixed architecture, such as the wavefront array processors, essentially restricts the search space of the optimization problem. Design requirements, such as hardware and I/O constraints, can also be included in the optimization. The art of designing systolic arrays is, therefore, reduced to a systematic methodology.

## Acknowledgment

The authors are indebted to Prof. P. S. Xia and Prof. C. D. Han for their helpful Comments and discussions. Thanks are also due to Prof. K. Hwang and Prof. H. T. Kung for providing valuable information.

## References:

[1] KUNGH T. Why systolic architecture[J]. Computer, 1982: 37-46.
[2] SNYDER L. Introduction to the configurable, highly parallel computer[J]. Computer, 1982, 15(1): 47-56.
[3] KUNG S Y, ARUN K S, GAL-EZER R J, et al. Wavefront array processor: language, architecture, and applications[J]. IEEE Transactions on Computer, 1982, C-31: 1054-1066.
[4] RAMAKRISHNAN I V, FUSSELL D S, SLLBER-

SCHATZ A.On mapping homogeneous graphs on a linear array-processor model[C]// International Conferences on Parallel Processing. [S.l.:s.n], 1983: 440-447.

[5] JOHNSSON L, COHEN D. A mathematical approach to modelling the flow of data and control in computational networks[M]. VLSI Systems and Computations. Maryland: Computer Science Press, 1981: 213-225.

[6] WEISER V, DAVIS A. A wavefront notion tool for VLSI array design[M]. VLSI Systems and Computations. Maryland: Computer Science Press, 1981: 226-234.

[7] LAM M, MOSTOW J. A transformational model of VLSI systolic design[C]// IFIP WG 10.2 6th International Symposium on Computer Hardware Descnption Languages and Their Applications. [S.l.:s.n], 1983.

[8] CAPPELLO P R, STEIGLITZ K.Unifying VLSI array designs with geometric transformations[C]// International Conferences on Parallel Processing. [S.l.:s.n], 1983: 448-457.

[9] CHENG H D, LIN W C, FU K S. Space-time domain expansion approach to VLSI and its application to hierarchical scene matching[C]// International Conferences on Pattern Recognition. [S.l.:s.n], 1984.

[10] FORTES J A B. Algorithm transformations for parallel processing and VLSI architecture design[D]. Los Angeles: University of Southern California, 1983.

[11] MOLDOVAN D I. On the analysis and synthesis of VLSI algorithms[J]. IEEE Transactions on Computers, 1982, C-31: 1121-1126.

[12] GANNON D. Pipelining array computations for MIMD parallelism: a functional specification[C]// International Conferences on Parallel Processing. [S.l.:s.n], 1982: 284-286.

[13] JOVER J M, KAILATH T. Design framework for systolic-type arrays[C]// IEEE International Conference on Acoustics, Speech and Signal Processing. Piscataway: IEEE Press,1984.

[14] KUHN R H. Optimization and interconnection complexity for parallel processors, single stage networks and decision trees[D]. Champaign-Urbana: University of Illinois at Urbana-Champaign, 1980.

[15] KUNG H T, LIN W T. An algebra for VLSI algorithm design[C]// The Elliptic Problem Solvers Conference. The Netherland: Science Direct, 1983.

[16] KUNG S Y. On supercomputing with systolic/wavefront array processors[J]. Procedings of the IEEE, 1984, 72(7): 867-884.

[17] MIRANKER W L, WINKLER A. Space-time representations of computational structures[J]. Computing, 1984, 32: 93-114.

[18] QUINTON P. Automatic synthesis of systolic arrays from uniform recurrent equations[C]// The 11th Annual International Symposium on Computer Architecture. New York: ACM Press, 1984: 208-214.

[19] SCHWARTZ D A, BARNWELL III T P. A graph theoretic technique for the generation of systolic implementations for shift-invariant flow graphs[C]// International Conference on Acoustics, Speech and Signal Processing. [S.l.: s.n.], 1984.

[20] LEISERSON C E, ROSE F M, SAXE J B. Optimizing synchronous circuitry by retiming[C]// The 3rd Caltech Conference on Very Large Scale Integration. Maryland: Computer Science Press, 1983: 87-116.

[21] CHEN M C, MEAD C A. Concurrent algorithms as space-time recursion equations[C]// USC Workshop VLSI Modern Signal Processing. [S.l.:s.n], 1982: 31-52.

[22] FORTES J A B, FU K S, WAH B W. Systematic approaches to the design of algorithmically specified systolic arrays[D]. West Lafayette: Purdue University, 1984.

[23] FISHER A L. Systolic algorithms for running order statistics in signal and image processing[M]. VLSI Systems and Computations. Maryland: Computer Science Press, 1981:265-271.

[24] HOROWITZ E. VLSI architecture for matrix computations[C]// International Conferences on Parallel Processing. [S.l.:s.n], 1979: 124-127.

[25] KUNG H T, RUANCE L M, YEN D W L. A two-level pipelined systolic array for convolutions[M]. VLSI Systems and Computations. Maryland: Computer Science Press, 1981: 255-264.

[26] HWANG K, CHENG Y H. Partitioned matrix algorithms for VLSI arithmetic systems[J]. IEEE Transactions on Computer, 1982, C-31: 1215-1224.

[27] KUNG H T. Highly concurrent systems[M]. Introduction to VLSI System, New Jersey: Addison- Wesley, 1980.

[28] HWANG K, CHENG Y H. VLSI computing structure for solving large scale linear system of equations[C]// International Conferences on Parallel Processing. [S.l.:s.n], 1980: 217-227.

[29] LI G J. Array pipelining algorithms and pipelined array processors[D]. Beijing: Institute of Computing Technology, Chinese Academy of Science, 1981.

[30] LI G J, WAH B W. The design of optimal systolic algorithms[C]// Computer Software and Applications Conference. [S.l.:s.n], 1983: 310-319.

发表于International Conference on Parallel Processing论文集，1985年

# Systolic processing for dynamic programming problems

Guo-jie LI, Benjamin W. WAH

School of Electrical Engineering Purdue University,West Lafayette, IN 47907

**Abstract:** In this paper, we investigate systolic processing for problems formulated in dynamic programming. These problems are classified as monadic-serial, polyadic-serial, monadic-nonserial, and polyadic-nonserial. Problems in serial formulations can be implemented easily in systolic arrays; however, nonserial problems may have to be transformed into a serial one before an efficient implementation can be found. A monadic-serial dynamic programming problem can be solved as the search of an optimal path in a multistage graph and can be computed as a string of matrix multiplications. An efficient systolic array is presented. For solving a polyadic-serial problem, the optimal architecture consists of an interconnection of $N/\log_2 N$ systolic arrays, where $N$ is the number of stages in the problem. A second method to solve a polyadic-serial dynamic programming problem is to view it as the search of optimalsolutions in a serial AND/OR graph. Efficient methods of mapping a regular AND/OR graph into systolic arrays are developed. Cases are studied for transforming a problem in a nonserial formulation into aserial one.

**Key words:** AND/OR graph, dynamic programming, matrix multiplication, monadic, multistage graph, nonserial, parallel processing, polyadic, serial, systolic arrays

## 1 Introduction

Dynamic Programming (DP) is a powerful optimization methodology that is widely applicable to a large number of areas including optimal control, industrial engineering, and economics. Many practical problems involving asequence of interrelated decisions can be solved by DP efficiently. Bellman has characterized DP through the principle of optimality, which states that an optimal sequence of decision has the property that whatever the initial state and decision are, the remaining decisions must constitute an optimal decisiou sequence with regard to the state resulting from the first decision[1]. Subsequently, numerous efforts have been devoted to the rigorous mathematical framework and effective evaluation of DP problems[2-4].

In general, DP is an approach that yields a transformation of the problem into a more suitable form for optimization, but is not an algorithm for optimizing the objective function. Moreover, DP

Research supported by CIDMAC, a research unit of Purdue University, sponsored by Purdue, Cincinnati Millicron Corporation, Control Data Corporation, Cummins Engine Company, Ransburg Corporation and TRW.

can be interpreted differently depending on the computational approach. Bellman, Dreyfus, White, and many others viewed DP as a multistage optimization technique, that is, reducing a single $N$-dimensional problem to a sequence of $N$ one-dimensional problems[1,4]. The decisions that transform an initial state into a final state must be ordered in terms of stages, and functional equations relate state values in successive stages. The use of monotone sequential processes has been proved by Karp and Held to correspond naturally to DP[2]and has been further developed by Ibaraki[3] and Kumar[5]. On the other hand, Gensi and Montanari have shown that formulating a DP problem in terms of polyadic functional equations is equivalent to searching for a minimum-cost solution tree in an AND/OR graph with monotone cost function[6]. DP can also be formulated as a special case of the branch-and-bound algorithm, which is a general top-down OR-tree search procedure with dominance tests[7-9]. Lastly, nonserial DP has been shown to be optimal among all nonoverlapping comparison algorithms[10-11].

Although DP has long been recognized as a powerful approach to solving a wide spectrum of optimization problems, its applicability has been somewhat limited due to the large computational requirements. Recent advances in very-large-scale integration (VLSI) and multiprocessor technologies have provided feasible means of implementation. Casti, et al., have studied parallelism in DP[12]. Guibas, Kung and Thompson have proposed a VLSI algorithm for solving the optimal parenthesization problem[13]. Linear pipelines for DP have been described recently[14]. Clarke and Dyer have designed a systolic array for curve and line detection in terms of nonserial DP [15]. Wah, et al., have proposed parallel processing for branch-and-bound algorithms with dominance tests[16]. However, these studies were directed towards the implementation of a few special cases of DP formulations.

In this paper, we classify DP into monadic-serial, polyadic-serial, monadic-nonserial, and polyadic-nonserial. Potential parallelism and the corresponding systolic architectures are investigated for each class. Generally, a problem can be expressed in different DP formulations, and the efficiency and costs of implementation must be compared.

DP problems can be solved as the search of an optimal path in a multistage graph or as the search for an optimal solution in an AND/OR graph. We will adopt the graph search as a paradigm to illustrate the various approaches of DP. To take advantage of the regular and limited interconnections of systolic arrays, the graph should have a regular structure. For DP problems in serial formulations, the corresponding graph representations are serial; however, for nonserial problems, they must be converted into serial formulations before efficient implementations can be found. A type of regular graphs of special interest is the multistage graph, in which nodes are decomposed into stages, and nodes in one stage are connected to nodes in adjacent stages only. Figure 1 depicts two examples of multistage graphs.

## 2 Classification of dynamic programming formulations

DP formulation is characterized by a recursive functional equation whose left-hand side identifies a function name and whose right-hand

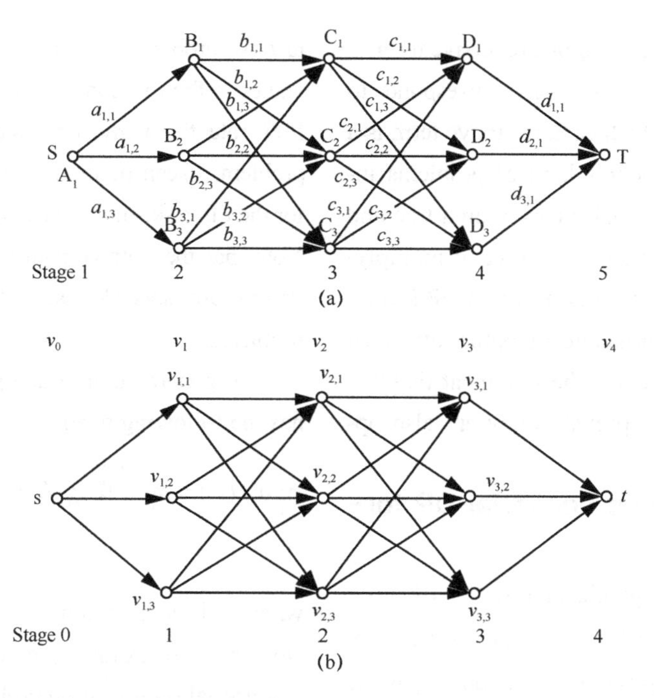


Figure1. (a) A multistage graph with five stages and three nodes in each intermediate stage. (The cost on each edge is constant.) (b) A multistage graph with four stages and three vertices in each stage. (The cost on each edge is a function of the nodes connected.)

side is an expression containing the maximization (or minimization) of values of some monotone functions. DP formulation can be classified according to the form of the functional equation.

## 2.1 Monadic versus polyadic DP formulations

For a general graph $G=(V, E)$, inwhich $V$ is the set of nodes and $E$ is the set of edges, let $c_{i,j}$ be the cost of edge $(i,j)\in E$. The cost of a path from source, $s$, to sink, $t$, is the sum of costs on the edges of the path. Define $f_1(i)$ as the minimum cost of a path from $i$ to $t$. Thus the cost of a path from $i$ to $t$ via a neighbor $j$ is $c_{i,j}+f_1(j)$. To find $f_1(i)$, paths through all possible neighbors must be compared. Hence,

$$f_1(i)=\min_j\left(c_{i,j}+f_1(j)\right) \tag{1}$$

This equation is termed a forward functional equation. Similarly, if $f_2(i)$ is defined as the minimum cost of a path from $s$ to $i$, then the functional equation becomes:

$$f_2(i)=\min_j\left(f_2(j)+c_{j,i}\right) \tag{2}$$

This equation is termed a backward functional equation. The formulations in Eq. (1) and Eq. (2) are monadic, that is, the cost of function involves one recursive term only.

Eq. (1) and Eq. (2) can be generalized to find the optimal path from any vertex $i$ to any other vertex $j$. The functional equation is:

$$f_3(i,j)=\min_k\left(f_3(i,k)+f_3(k,j)\right) \tag{3}$$

where $f_3(i,j)$ is the minimum cost of traversing from $i$ to $j$. This cost function is polyadic because it involves more than one recursive term.

Polyadic DP formulations correspond to three structures in which each recursive term is a subtree. Examples of this kind of problems include finding the optimal binary search tree and computing the minimum-cost order of multiplying a string of matrices. For polyadic DP formulations, Bellman's principle of optimality must be generalized to include the statement that "all subsequences of an optimal policy are also optimal."

## 2.2 Serial versus nonserial DP formulations

In a multistage graph, suppose $V = V_0 \cup \cdots \cup V_i \cup \cdots \cup V_N$, where $V_i, 0 \leqslant i \leqslant N$, is the set of nodes in stage $i$, $V_0=\{s\}$, $V_N=\{t\}$, and $v_{i,j_i}$ is the $j$'th node in $V_i$. An example is shown in Figure 1(b). Let $P$ be the set of all possible paths from $s$ to $t$. A path $p \in P$ from $s$ to $t$ is of the from $s\text{-}v_{2,j_2}\text{-}v_{3,j_3}\text{-}\cdots\text{-}t$. If the cost of edge $(v_{i,j_i}, v_{i+1,j_{i+1}})$ is $g_i(v_{i,j_i}, v_{i+1,j_{i+1}})$, then the minimum-cost path from $s$ to $t$ is:

$$\min_{p\in P} f(p) = \min_{p\in P} \sum_{i=0}^{N-1} g_i\left(v_{i,j_i}, v_{i+1,j_{i+1}}\right) \tag{4}$$

Optimization problems of the above form have terms that share one variable with its predecessor term and another one with its successor term. The problem has a serial structure and, consequently, is called a serial optimization problem. Many practical problems can be formulated in this way. For a traffic-control problem, $V_i$ can be the possible times for the traffic light to be in state $i$, and the cost on an edge is the difference in timing. For a circuit-design problem $V_i$ can be the possible voltages at point $i$, and the cost of an edge may be the corresponding power dissipation.For a fluid-flow problem, $V_i$ can be the possible pressure values in the $i$'th pump, and function $f$ may be the flow rate for a given pressure. For a scheduling problem, $V_i$ can be the possible task service times for the $i$'th task, and the edge cost reflects the delay. Note that the optimal-path problem in multistage graphs are special cases of serial optimization problems.

A general nonserial optimization problem has the following form:

$$\min_V f(V) = \min_V \left( \mathop{\Phi}_{i=0}^{N} g_i(E_j) \right) \qquad E_j = \bigcup_{\substack{f\in L \\ \varsubsetneq(0,\cdots,N)}} V_f \tag{5}$$

where $\Phi$ is a monotone function relating the $g_i$ stogether. For example, the following equation is a nonserial optimization problem:

$$\min_{v_j \in V_j} \{g_1(v_1,v_2,v_4) + g_2(v_3,v_4) + g_3(v_3,v_5)\} \tag{6}$$

$$V = \bigcup_{i=1}^{5} V_i$$

We have classified DP problems in terms of the recursive functional equations. Monadic and polyadic DP formulations are distinct approaches to representing various optimization problems, while serial and nonserial optimization problems are problems solvable by the DP formulation. We remark that the cost functions, $g_i$, are not restricted to be monadic, serial, and additive; however, they must be monotone in order for the principle of optimality to hold.

In the following sections, we investigate systolic architectures for the four classes of DP formulations. Although a given problem can be formulated in several ways, a systolic implementation favors the serial formulation, and a problem in a nonserial formulation may have to be transformed into a serial one before it can be mapped into systolic arrays.

# 3 Systolic array for monadic-serial DP formulations

Monadic-serial DP problems can be conveniently solved as the multiplication of astring of matrices. In this section, an efficient systolic design is presented. The proposed design does not exploit all potential parallelism of solving a given problem, especially when the number of stages is large. Other parallel designs using different formulations may allow a higher degree of parallelism and will be discussed later.

## 3.1 Solving monadic-serial DP problemsas strings of matrix multiplications

Recall that the search for a solution of a problem in a monadic-serialDP formulation can be viewed as finding a path in a multistage graph. For the multistage graph in Figure 1(a) and from Eq. (2), $f(C_1)$, the minimum cost from $C_1$to $t$, is:

$$f(C_1)=\min(c_{1,1}+d_{1,1},c_{1,2}+d_{2,1},c_{1,3}+d_{3,1}) \quad (7)$$

$f(C_2)$ and $f(C_3)$ are obtained similarly.

Eq. (7) is similar to an inner-product operation. If we define matrix multiplication in terms of a closed semi-ring ($R$, MIN, +, $+\infty$, 0), in which 'MIN' corresponds to addition and '+' corresponds to multiplication in conventional matrix multiplications [17], Eq. (7) becomes:

$$f(C)=C\cdot D=\begin{bmatrix} f(C_1) \\ f(C_2) \\ f(C_3) \end{bmatrix}=\begin{bmatrix} c_{1,1} & c_{1,2} & c_{1,3} \\ c_{2,1} & c_{2,2} & c_{2,3} \\ c_{3,1} & c_{3,2} & c_{3,3} \end{bmatrix}\begin{bmatrix} d_{1,1} \\ d_{2,1} \\ d_{3,1} \end{bmatrix} \quad (8)$$

Likewise, we have:

$$f(B)=B\cdot(C\cdot D)$$
$$f(A)=A\cdot(B\cdot(C\cdot D))$$

Thus solving the multistage-graph problem with a forward monadic DP formulation is equivalent to multiplying a string of matrices. The order of multiplications is reversed in backward monadic DP formulations.

For a multistage graph with $N$ stages and $m$ vertices in each stage, the computational complexity is $O(m^2N)$. For single-source and single-sink problems, the first and last matrices degenerate into row and column vectors, respectively.

## 3.2 Systolic array for string of matrix multiplications

A linear systolic array for evaluating problems in monadic-serial DP formulations with parallel inputs is described in this section. The following scheme is based on a combination of two methods of multiplying a matrix with a vector, one of which was discussed by Kung [18]. Figure 2(a) depicts a scheme for computing $(A\cdot(B\cdot(C\cdot D)))$ for the multistage graph in Figure 1(a). An iteration is defined as a shift-multiply-accumulate operation with respect to the time at which a row or columnof the input matrix enters a given processor. Note that the same iteration number are carried out at different times in different processors. The iteration numbers are indicated in Figure 2(a). In the first three iterations, $C\cdot D$ is evaluated. The control signal FIRST isone; $D$, the input vector, is serially shifted into the systolic array; and the result vector, $\{f(C_i),i=1,2,3\}$, remains stationary. At the end of the third iteration, FIRST is set to zero. In the following three iterations, $B\cdot(C\cdot D)$ is computed. Note that matrix $B$ is transposed, and the $i$'th column of matrix $B$ is fed into $P_i$. The input vector, $\{f(C_i), i=1,2,3\}$, remains stationary, while the result vector, $\{f(B_i), i=1,2,3\}$, is shifted. At the end of the sixth iteration, the output vector $\{f(B_i), i=1,2,3\}$ is formed. In the last three iterations, input vectors $A$ and $\{f(B_i), i=1,2,3\}$ are shifted into $P_1$ to form the final result.

For the systolic array in Figure 2(a), the data shifted alternate between the input vector and the result vector every three iterations. This alternation can be controlled by the processor structure of $P_i$ depicted in Figure 2(b). $R_i$ is a register that stores an element of the input vector, and $A_i$ is the accumulator that stores the temporary result of an element of the result vector. The data paths are controlled by control signals $\mathrm{ODD}_i$ and $\mathrm{MOVE}_i$. When the number of matrix multiplications is odd, $\mathrm{ODD}_i$ is one, hence $R_i$ is connected to the output, and the input vector is shifted along the pipeline. When the number of matrix multiplications is even, $\mathrm{ODD}_i$ is sero, $A_i$ is connected to the output, and the result vector is shifted. Atthe end of a matrix multiplication, the result vector generated becomes the input vector in the next iteration and is moved by the control signal $\mathrm{MOVE}_i$, from $A_i$ to $R_i$. Note that there is a one-cycle delay between switching the control signals for $P_{i+1}$ and $P_i$.

To search a multistage graph with $(N+1)$ stages and $m$ nodes in each intermediate stage (the first and last stages have one node each), it takes $Nm$ iterations with $m$ processors. There is

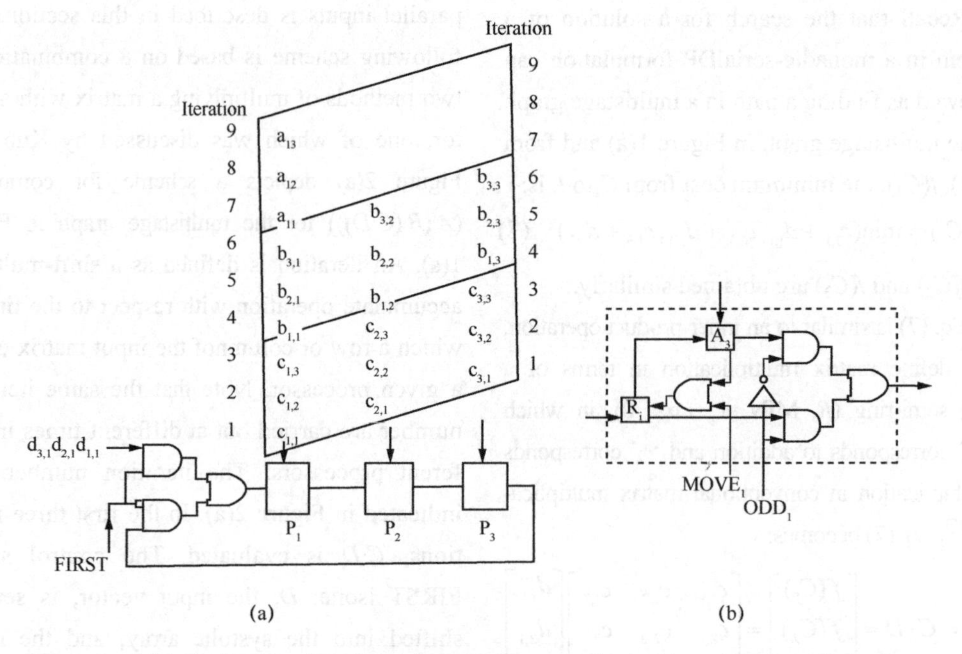


Figure 2. (a) A pipelined version of systolic array for computing a string of matrix multiplications; (b) Processor structure for $P_i$

no delay between feeding successive input matrices into the systolic array, and the processors are kept busy most of the time. In contrast, it takes $(N-2)m^2+m$ iterations to solve the problem with a single processor. Define PU, the processor utilization, as the ratio of the number of serial iterations to the product of the number of parallel iterations and the number of processors. PU for the above systolic array is:

$$\text{PU} = \frac{(N-2)m^2+m}{N\cdot m\cdot m} = \frac{N-2}{N} + \frac{1}{N\cdot m} \quad (9)$$

When $N$ and $m$ are large, PU is very close to 1.

Although the proposed systolic array is designed for matricesin which each element is a single constant, it can be extended to many practical sequentially-controlled systems, such as Kalman filtering, inventory systems, and multistage production processes, in which each matrix element is a vector with many quantized values. In this case, the potential parallelism could be very large.

The degree of parallelism of the proposed scheme is restricted by the limited number of input/output ports in a VLSI chip and the fact that the ratio of the computational overhead to the input/output overhead is relatively low in matrix-vector multiplications. The input/output bottleneck is due to the large number of edge costs that must be fed into the systolic array. For the serial optimalization problems formulated by Eq. (4) and illustrated in Figure l(b), the edge costs are expressed as functions of the nodes connected, and hence only the values of the nodes have to be input. This results in an order-of-magnitude reduction in the input overhead. An efficient and practical scheme with serial inputs has been developed [19], but is not shown here due to space limitation. In the improved scheme, the parallel input ports are replaced by a feedback bus, and only one serial input port is needed.

# 4 Parallel processing for polyadic-serial DP formulations

A polyadic-serial DP formulation can be solved by either a divide-and-conquer algorithm, the multiplication of a string of matrices, or the search of an AND/OR graph. The last two representations are related since the evaluation of a set of AND-nodes and their common parent is equivalent to computing an item in a matrix multiplication, that is, an inter-product of a row vector and a column vector. Divide-and-conquer algorithms will be discussed in Section 4.1, and the last two formulations will be discussed in Section 4.2.

The most efficient representation for a given problem is usually problem dependent. If the number of states in each stage is large and constant, then the matrix-multiplication method is preferable to an AND/OR-graph search, as more potential parallelism can be exploited. When the AND/OR graph is irregular, especially for nonserial DP problems, then the graph-search method is more beneficial.

## 4.1 Solving polyadic-serial DP problems by divide-and-conquer algorithms

Recall that a serial optimization problem can be solved as the multiplication of a string of matrices. However, a problem expressed in a monadic-serial formulation does not exploit all the potential parallelism because the order of matrix multiplications is fixed. On the other hand, there is more flexibility for parallelism when the problem is formulated in a polyadic equation because the matrices can be multiplied recursively by a divide-and-conquer algorithm.

One important issue in parallel divide-and-conquer algorithms is the granularity of parallelism[16]. This is the minimum size of a subproblem that is evaluated by a processor[1] in order to achieve the optimal performance, as measured byeither the PU or the $AT^2$ (or (area) × (computational time)$^2$) criteria. If the granularity is large, then the processors can be loosely coupled; otherwise, tight coupling is necessary. In this section, the lower-bound complexity for-solving polyadic-serial DP problems by parallel divide-and-conquer algorithms is derived with respect to the asymptotic PU and $AT^2$ criteria. The optimal numbers of processors to achieve these lower bounds are also studied.

Consider the polyadic-serial DP formulation in Eq. (3) for the multistage-path problem in Figure 1(b).

$$f_3(s,t)=\min_{k\in\{v_{2,1},v_{2,2},v_{2,3}\}}\{f_3(s,k)+f_3(k,t)\} \quad (10)$$

where $f_3(i,j)$ is the cost of the optimal path from $i$ to $j$,and $k$ is a node in stage 2 of the graph. In matrix notations, let $f_3(V_i,V_j)$ be a cost matrix, each element of which denotes the cost of the optimal path from a vertex in stage $i$ to a vertex in stage $j$. It is easy to see, for an intermediate stage $k$ between $i$ and $j$, that:

$$f_3(V_i,V_j)=f_3(V_i,V_k)\cdot f_3(V_k,V_j) \quad (11)$$

This formulation allows a string of matrix multiplications to be reduced to two smaller strings of matrix multiplications.

For simplicity in deriving the lower bound, all the matrices are assumed to have identical dimensions.

The fastest way to multiply $Nm$-by-$m$ matrices is to locate the matrices in the leaves of a complete binary tree of height $\lceil \log_2 N \rceil$. The $N$-stage graph problem can be solved in $O(m\lceil \log_2 N\rceil)$ time units with $\lceil N/2\rceil$ matrix-multiplication systolic arrays[20]. The PU and $AT^2$measures are relatively low for this approach.

It has been found that the processors are efficiently utilized if $\lceil N/(\log_2 N)\rceil$ processors are used[21-23]. However, the asymptotic effects of deviations from $\lceil N/(\log_2 N)\rceil$ processors have not been studied before. This is shown in the following proposition.

**Proposition 1:** Let PU($k$,$N$) be the processor utilization of multiplying a string of $N$ mairices using $k(N)$ synchronous systolic arrays, each of which computes a matrix multiplication. (PU does not measure the utilization of the processing elements in a systolic array, but measures the average fraction of time that a systolic array is used in matrix multiplications.) By defining $c_\infty=\lim\limits_{N\to\infty}\dfrac{k(N)}{N/\log_2 N}$, the normalized asymptotic processor utilization is:

$$\lim_{N\to\infty}\mathrm{PU}(k,N)\begin{cases}=0 & \text{if } c_\infty\to\infty\\ \geqslant\dfrac{1}{1+c_\infty} & \text{if } 0<c_\infty<\infty\\ =1 & \text{if } c_\infty=0\end{cases} \quad (12)$$

[1] Processors and systolic arrays are synonymous here.

The proof is omitted due to space limitation[19].

As an example in applying the results of Proposition 1, suppose that there are $\sqrt{N}$ processors, it is easy to show that $\lim_{N\to\infty}\sqrt{N}\log_2 N/N=0$. Hence $c_\infty=0$ in Eq. (12), and $\lim_{N\to\infty}\mathrm{PU}(k,N)=1$.

Since PU($k$,$N$) increases monotonically with decreasing $k$, PU($k$,$N$) alone is not adequate to measure the effects of parallel processing. A popular measure in VLSI complexity theory is the $\mathrm{AT}^2$ criterion. The following theorem proves the lower- bound $\mathrm{AT}^2$ complexity of divide-and-conquer algorithms for solving polyadic-serial DP problems. This lower bound is attained when $k(N)=N/(\log_2 N)$.

**Theorem 1:** Suppose that a string of $N$ $m$-by-$m$ matrices are multiplied by $S(N)$ processors in time $T(N)$ using adivide-and- conquer algorithm, and that each processor performs a multiplication of a pair of $m$-by-$m$ matrices in $T_1$ time units. Then $S(N)T^2(N)\geqslant\theta(N\log_2 N)T_1^2$, and equality holds when $S(N)=\theta(N/\log_2 N)$. ($\theta$ indicates the set of functions of the same order.)

**Proof:** The multiplication of a string of $M$ matrices by a divide-and-conquer algorithm can be represented as a complete binary tree with $M$ terminals. The number of matrix-multiplicalions, or the number of nonterminals, is $M-1$. A parallel divide-and-conquer algorithm for multiplying a string of $N$, $N>M$, matrices can roughly be divided into two phases: computation and wind-down. During the computation phase, all processors are kept busy until half of the number of inter-mediate matrices to be multiplied is less than the number or processors. There are $(N-1)-(S(N)-1)=N-S(N)$ nonterminals to be evaluated, and at least $(N/S(N)-1)T_1$ time units are required. In the wind-down phase, the results are combined together, and some processors would be idle. According to the data dependence, at least $\log_2 S(N)\cdot T_1$ time units are required in this phase. Therefore, the following lower bound of time complexity holds for any parallel divide-and-conquer algorithm.

$$T(N)\geqslant\left(\frac{N}{S(N)}-1+\log_2 S(N)\right)\cdot T_1 \qquad (13)$$

where $1\leqslant S(N)\leqslant N$. For simplicity, the constant term in Eq.(13) can be ignored without affecting the validity of the following proof. The $\mathrm{AT}^2$ lower bound is derived as:

$$S(N)T^2(N)\geqslant$$
$$((N)+2N\log_2 S(N)+S(N)\log_2^2 S(N))T_1^2 \qquad (14)$$

To find the order-of-magnitude minimum of Eq.(14), it is necessary to compare the following three cases. When $S(N)=\theta(N/\log_2 N)$, $S(N)T^2(N)=\theta((N\log_2 N)T_1^2)$. In contrast, when $S(N)<\theta(N/\log_2 N)$, the first term on the right-hand side of Eq. (14) is:

$$\frac{N^2}{S(N)}>\theta(N\log_2 N) \qquad (15)$$

When $S(N)>\theta(N/\log_2 N)$, the third term on the right-hand side of Eq. (14) is:

$$S(N)\log_2^2 S(N)>\theta(N\log_2 N) \qquad (16)$$

Since $\log_2^2 S(N)\geqslant\theta(\log_2^2 N)$. The above analysis shows that the $\mathrm{AT}^2$ complexity is $\Omega((N\log_2 N)T_1^2)$, and that $\theta(N/\log_2 N)$ is the optimal granularity to achieve this lower bound.

So far, the matrices are assumed to have identical dimensions. When this is not true, the order in which the matrices are multiplied together has a significant effect on the total number of operations. Finding the optimal order of multiplying a string of matrices with different dimensions is itself a polyadic-nonserial DP problem, the so-called secondary optimization problem[24-25]. Guibas, Kung, and Thompson have proposed a systolic array to solve the optimal parenthesization problem, which can be used to compute the minimum-cost order of multiplying a string of matrices[13]. Once the optimal order is found, the processors can be assigned to evaluate the matrix multiplicalions in the defined order and in an asynchronous fashion. In this sense, the tree of matrix multiplications can be treated as a dataflow graph.

## 4.2 Solving polyadic-serial DP problems by AND/OR graph searches

In this section, we discuss the evaluation of polyadic-serial DP problems as AND/OR graph searches. AND/OR graphs are naturally obtained by representing the DP problem using a problem-reduction method. The mapping of a regular AND/OR graph onto a systolic array is straightforward and will be illustrated in the next section.

Polyadic-serial problems are discussed with respect to the search of a multistage graph as formulated by Eq. (4). Suppose an $(N+1)$-stage graph, with stages from 0 to $N$ and $m$ nodes in each stage, is divided into $p$ subgraphs, each of which contains $N/p$ +1 consecutive stages. For simplity, assume that $N = p^{\ell}$, where $\ell$ is a non-negative integer. The minimum-cost path has to pass through one and only one vertex in stage 0, $N/p,\cdots,pN/p$ in the segmented graph. The cost of a path equals the sum of costs of the $p$ subpaths. If all the $m^2$subpaths from the $m$ vertices in stage $iN/p$ to the $m$ vertices in stage $(i+1)N/p$, $0 \leqslant i \leqslant p-1$, have been optimized, there are $m^{p+1}$ possible combinations of subpaths from stage 0 to stage$N$ that must be considered for the optimal path. Usinga divide-and-couquer algorithm, each subgraph with $N/p$+1 stages is further divided into $p$ smaller subgraphs. This partitioning process continues until each subgraph has one stage.

The partitioning process can be conveniently represented as an AND/OR graph, in which an AND-node corresponds to a subproblem sum, and an OR-node corresponds to alternative selections or comparisons[2]. In this case, we have a regular AND/OR graph of height $2\log_p N$, whose AND-nodes have $p$ branches ($p$-arc nodes) and whose OR-nodes have $m^{p-1}$ branches ($m^{p-1}$-arc nodes). Figure 3 shows an AND/OR graph that represents the reduction of the multistage-graph problem with $m$=2 and $p$=2 from three stages to one stage. The four nodes at the top of the AND/OR graph represents the four possible alternate paths in the reduced single-stage graph. The shortest path is obtained by a single comparison of these paths.

The relationship between DP and graph search was investigated by Martelli and Montanari[26] who showed that, in the case of polyadic cost functions, the solution of a DP problem can be obtained by finding a minimal-cost solution

[2] The definitions of AND and OR nodes are taken from [26]. The roles of the AND and OR nodes are reversed in Nilason's definition[27].

tree in anAND/OR graph. This equivalence allows various graph searching techniques to be translated into techniques for solvingDP problems. For those acyclic AND/OR graphs with positive arc costs, Martelli and Montanari have named them as additive[26], and have proposed top-down and bottom-up search algorithms. A similar algorithm, called AO*, for searching hypergraphs was discussed by Nilsson[27].

The above AND/OR graph representation of a polyadic DP problem can be considered as a folded AND/OR tree. It is easy to see that the efficiency of solving a DP problem by searching an AND/OR graph depends on the graph structure (parameter $p$). The following theorem analyzes the optimal structure.

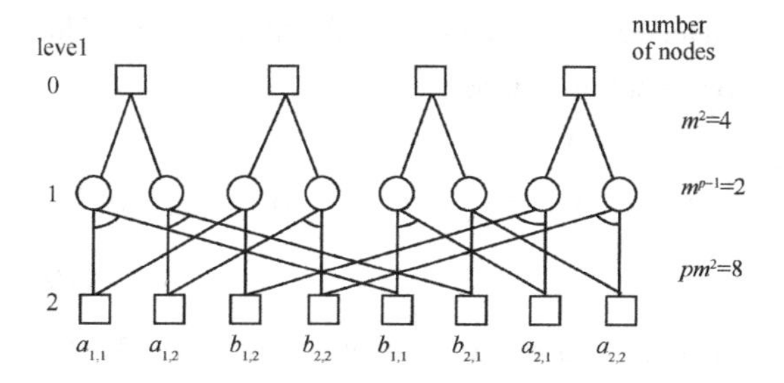


Figure 3. An AND/OR graph representation of the reduction in finding an optimal path in a 3-stage graph to a1-stage graph, ($m$=2; $p$=2; AND nodes are represented as circles and indicate summations; and OR nodes are represented as squares and indicate comparisons. The values in the terminal nodes are edge costs: costs between stages 1 and stages 2 are $a_{i,j}$, $i,j \in \{1,2\}$; costs between stages 2 and stages 3 are $b_{i,j}$.)

**Theorem 2:** If a serialDP problem is solved by searchinga regular AND/OR graph, the binary partition, namely, using 2-arc AND nodes, is optimal in the sense of minimizing the total-number of nodes in the AND/OR graph.

**Proof:** The total number of nodes in the AND/OR graph is derived here. First, consider the reduction from $p$+1 stages to one stage (see Figure 3 for $p$=2). The number of nodes in the bottom level is $pm^2$, as there are $pm^2$ cost values for all pairs of vertices between neighboring stages. Similarly, there are $m^2$ OR-nodes in Level 1 and $m^2m^{p-1}$AND-nodes in Level 2.

For the reduction from $N+1(=p^{\ell}+1)$ stages to one stage, there are $\log_p N$ levels of AND-nodes at odd levels, with $m^{p+1}, p\cdot m^{p+1}, \cdots, p^{(\log_p N)-1}m^{p+1}$ nodes, respectively. Similarly, thereare $(\log_p N)+1$ levels of OR-nodes at even levels with $m^2, p\cdot m^2, \cdots, p^{\log_p N}m^2$ nodes, respectively. Let $u(p)$ be the total number of nodes in the AND/OR graph:

$$u(p)=\sum_{i=0}^{(\log_p N)-1} p^i\cdot m^{p+1}+\sum_{j=0}^{\log_p N} p^j m^2 = \frac{N-1}{p-1}m^{p+1}+\frac{N\cdot p-1}{p-1}m^2 \tag{17}$$

To find the minimum $u(p)$, we relax the restriction that $p$ is an integer and evaluate the differential of $u(p)$.

$$\frac{\partial u(p)}{\partial p}=\frac{(N-1)\left(m^{p+1}\left((p-1)\ln_e m-1\right)-m^2\right)}{(p-1)^2} \tag{18}$$

From Eq. (18), it is seen that $\frac{\partial u(p)}{\partial p}\geqslant 0$ if $N\geqslant 1$, $p\geqslant 2$, and $m\geqslant 3$, or $N\geqslant 1$, $p\geqslant 3$, and $m\geqslant 2$. Considering $\frac{\partial^2 u(p)}{\partial p^2}$, we conclude that $u(p)$ increases monotonically when $N\geqslant 1$, $p\geqslant 2$, and $m\geqslant 3$. In other words, the binary partition with $p$=2 is optimal for solving regular multistage-graph problems in the sense of minimizing the total number of nodes in the corresponding AND/OR graph.

For an AND/OR graph, the larger the value of $p$ is, the less the principle of optimality is applied. In the extreme case, $p=N$, the corresponding AND/OR-graph search becomes a brute-force search, and the principle of optimality is never used. For irregular multistage-graph problems, the number of nodes in the AND/OR graph depends on the ordering of stage reduction. However, it is not difficult to demonstratethat binary partitioning is optimal. Assume that stages $i_1,\cdots,i_4$ with $m_1,\cdots,m_4$ nodes are to be reduced to two stages $i_1$ and $i_4$. If 3-arc AND-nodes are used, $m_1m_2m_3m_4$ comparisons are needed to eliminate stages 2 and stages 3. However, when 2-arc AND-nodes are used, $m_1m_3\,(m_2+m_4)$ comparisons are needed if stage 2 is eliminated first, and $m_2m_4(m_1+m_3)$ comparisons are needed if stage 3 is eliminated first. It is easy to see that using 3-arc AND-nodes requires more comparisonsas long as $m_i \geqslant 2, 1 \leqslant i \leqslant 4$. Furthermore, binary partitioning requires less additions since one addition is needed for each AND-node.

# 5 Parallel processing of non-serial DP problems

For nonserial optimization problems, it is more efficient to transform the problem into a serial formulation before implementing it in systolic arrays. In this section, we illustrate the transformation by some examples.

## 5.1 Monadic-nonserial DP formulations

It has been shown that unrestricted nonserial optimization problems are NP-hard, but problems with a favorable pattern of term interactions may be solved efficiently[10]. An approach to solve a monadic-nonserial problem with some structural properties is to first convert it into a monadic-serial problem, such as a multistage graph-search problem, and to map the serial formulation into systolic arrays.

For the nooserial formulation in Eq. (5), a multistage optimization procedure can be carried out separately for each variable $V_i$. Of course, this optimization must be performed on all values of the independent variables that "interact" with $V_i$. Let $V_{i_1},\cdots,V_{i_n}$ be the variables that are related to $V_i$ in one or more functional terms. The cost function can be written as:

$$f(v_1,\cdots,v_n)=\min_{v_i\in V_n}\{h_1(v_i,v_{i_1},\cdots,v_{i_n})+h_2(v_1,\cdots,v_{i-1},v_{i+1},\cdots,v_n)\} \tag{19}$$

where $h_2$ is a function independent of $V_i$. By denoting $h_{1,\text{opt}}$ as $\min\limits_{v_i\in V_i} h_1(v_i,v_{i_1},\cdots,v_{i_n})$, the cost function can be rewritten as:

$$f(v_1,\cdots,v_n)=\min_{V-V_1}\{h_{1,\text{opt}}(v_{i_1},\cdots,v_{i_n})+h_2(v_1,\cdots,v_{i-1},v_{i+1},\cdots,v_n)\} \tag{20}$$

A multistage optimization process is, therefore, a step-by-step elimination of all variables. The computational time and storage depend on the number of elements in the domain of function $h_1$. Eq. (20) can be treated as a monadic-serial form if the evaluation of $h_{1,\text{opt}}$ is done separately.

The method is illustrated by the following example. For instance, if $V = V_1 \cup \cdots \cup V_N$, and the objective function is:

$$f(V) = \min_{v_i \in V_j} \{ g_1(v_1, v_2, v_3) + g_2(v_2, v_3, v_4) + \cdots + g_{N-2}(v_{N-2}, v_{N-1}, v_N) \} \quad (21)$$

Let $h_1(v_2, v_3) = \min_{v_1 \in V_1} g_1(v_1, v_2, v_3)$, we have:

$$\min_V f(V) = \min_{V-V_1} \left\{ h_1(v_2, v_3) + \sum_{i=2}^{N-2} g_i(v_i, v_{i+1}, v_{i+2}) \right\} \quad (22)$$

If $h_k(v_{k+1}, v_{k+2})$ is defined as:

$$h_k(v_{k+1}, v_{k+2}) = \min_{v_k \in V_k} \{ h_{k-1}(v_k, v_{k+1}) + g_k(v_k, v_{k+1}, v_{k+2}) \} \quad (23)$$

Eq. (23) represents the minimum of the summation of the first $k$ terms of $f(V)$. After eliminating $k$ variables, $V_1, \cdots, V_k$, the remaining optimization problem becomes:

$$\min_V f(V) = \min_{V-(V_1 \cup \cdots \cup V_k)} \left\{ h_k(v_{k+1}, v_{k+2}) + \sum_{i=k+1}^{N-2} g_i(v_i, v_{i+1}, v_{i+2}) \right\} \quad (24)$$

The monadic DP procedure, thus, eliminates the variables in the order $V_1, \cdots, V_N$. If the variables $V_2$ and $V_3$ are treated as a single variable in a stage, and $m_k, 1 \leqslant k \leqslant n$, quantized values are allowed for $V_k$, there would be $m_2 \cdot m_3$ states in this stage, and $m_1 \cdot m_2 \cdot m_3$ steps are required to eliminate $V_1$, in which a step consists of a computation of function $f$, an addition, and a comparison operation. The process of eliminating the remaining variables is repeated until $V_{N-1}$ and $V_N$ remain. The optimal solution is obtained by comparing all values of $h_{N-2}(v_{N-1}, v_N)$. The total number of steps required to compute Eq. (21) is:

$$\sum_{k=1}^{N-2} (m_k \cdot m_{k+1} \cdot m_{k+2}) + m_{N-1} \cdot m_N \quad (25)$$

Although more operations are needed for monadic-nonserial DP problems than monadic-serial DP problems, the potential parallelism is higher. Further, there is no increase in delay in evaluating the transformed problem. With additional control, the linear systolic array presented earlier can be applied to evaluate monadic-nonserial DP problems.

## 5.2 Solving polyadic-nonserial DP problems by parallel AND/OR-graph search

AND/OR graphs can be sequentially searched in a breadth-first bottom-up fashion, which expands nodes by levels from the bottom up[27]. Since an acyclic AND/OR graph can be viewed as a folded tree, searching the AND/OR graph can be accomplished by searching the corresponding AND/OR tree. In a parallel AND/OR-tree search, the nodes in the tree are evaluated in parallel in a bottom-up fashion. The parallel architecture can be designed with a flexible interconnection, so a processor can be dynamically assigned when it is free, or can be designed with a limited interconnection, so a static evaluation order is maintained for a given problem. A dataflow processor is an example of the first alternative. We will investigate the second alternative here.

Parallel AND/OR search for polyadic-serial DP problems is a special case of that for polyadic-nonserial DP problems. For nonserial DP problems, the dependency between states is not restricted to successive stages, but may exist between states in arbitrary stages. In the corresponding AND/OR graphs, the arcs are not restricted to successive levels, but may run between any two arbitrary levels. It may be difficult to map an irregular AND/OR graph to a systolic array with a regular interconnection structure. The nonserial AND/OR graph may have to be transformed into a serial one before the mapping is done.

The strategy is illustrated by the problem of finding the optimal order of multiplying a string of matrices. For simplicity, consider the evaluation of the product of four matrices, $M = M_1 \times M_2 \times M_3 \times M_4$, where $M_i, 1 \leqslant i \leqslant 4$, is a matrix with $r_{i-1}$ rows and $r_i$ columns. Let $m_{i,j}$ be the minimum cost of computing $M_i \times \cdots \times M_j$. Clearly,

$$m_{i,j} = \min_{i \leqslant k < j} (m_{i,k} + m_{k+1,j} + r_{i-1} r_k r_j) \tag{26}$$

The salution to be found is $m_{1,4}$. This problem can be represented as a search of an AND/OR graph, where the AND-nodes denote additions and the OR-nodes denote comparisons (Figure 4).

The AND/OR graph in Figure 4 can be mapped directly into six processors connected by multiple broadcast buses. Each processor evaluates an OR-node and its immediate descendent AND-node(s). The broadcast structure is necessary because a processor has to communicate with multiple processors and not its neighbors alone. Let $T_d(k), 1 \leqslant k \leqslant N$, be the time to find the optimal order of multiplying $k$ matrices. Then,

$$T_d(k) = \begin{cases} T_d(\lceil k/2 \rceil) + \lfloor k/2 \rfloor & \text{if } k > 1 \\ 1 & \text{if } k = 1 \end{cases} \tag{27}$$

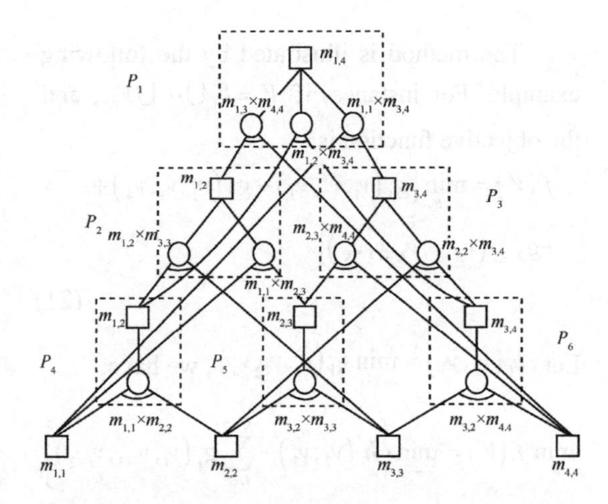


Figure 4. An AND/OR graph representation of finding the optimal order of multiplying a string of four matrices

This is true because, once the subproblems of size $\lceil k/2 \rceil$ are completed, the results can be used as inputs to subproblems of size larger than $\lceil k/2 \rceil$. In the following step, only subproblems of size $\lceil k/2 \rceil + 1$ can be completed, and the results will be available as inputs to subproblems of size larger than $\lceil k/2 \rceil + 1$. Thus it takes $\lfloor k/2 \rfloor$ steps to solve a subproblem of size $k$. In each step, two additions and two comparisons are performed.

**Proposition 2:** The solution to Eq. (27) is $T_d(N) = N$.

**Proof:** The proposition can be proved by induction.

Although the above scheme is fast, it requires alarge number of broadcast buses and may be difficult to implement when the problem size is large. To overcome this, we can transform the nonserial problem into a serial one, that is, convert the general AND/OR graph into a simpler graph in which all arcs connect nodes in successive levels. Suppose that an OR-node and its immediate parent are not located in adjacent levels, then the OR-node is connected to its parent via other intermediate nodes in adjacent levels. The additional connections are represented as dotted lines in Figure 5. This pipelined design is suitable for VLSI implementation because the interconnections can be mapped into a planar structure.

The computational time for the scheme in Figure 5 is analyzed here. Let $T_p(k), 1 \leqslant k \leqslant N$, be the time to find the optimal order of multiplying $k$ matrices. Once a subproblem of size $\lceil k/2 \rceil$ is solved, it takes $\lfloor k/2 \rfloor$ time units to transferthe result into the processor that evaluates the subproblem of size $k$. Analogous to the explanation for Eq. (27), it takes $\lfloor k/2 \rfloor$ steps to solve the subproblem of size $k$ after the results of subproblems of sise $\lceil k/2 \rceil$ are available. Consequently,

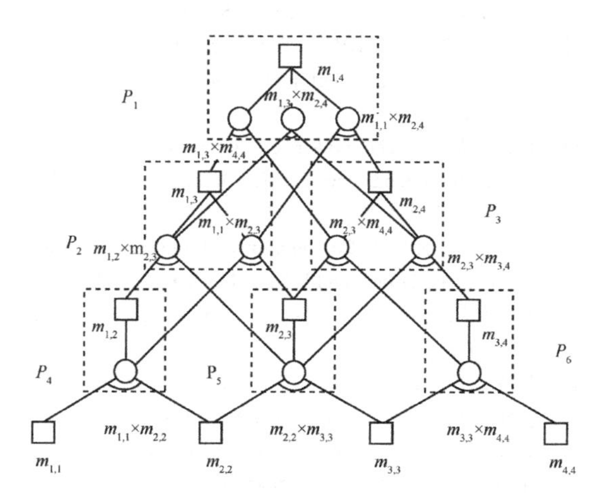


Figure 5. An AND/OR graph representation of finding the optimal order of multiplying a string of four matrices. (Communications are restricted to OR-nodes in adjacent levels of the graph and are indicated by dotted lines

$$T_p(k) = T_p(\lceil k/2 \rceil) + 2\lfloor k/2 \rfloor \tag{28}$$

**Proposition 3:** Suppose $T_p(1) = 2$, then the solution to Eq. (28) is $T_p(N) = 2N$.

**Proof:** The proposition can be proved by induction.

A systolic array usually demands that all operands for an operation arrive at a processor simultaneously and that the computations are carried out in a pipelined fashion. Recall from Theorem 2 that the optimal branching factors for AND andOR nodes are two and $m$ ($\geqslant 2$), respectively. Hence, it is necessary for two data items to arrive at an AND-node simultaneously, and that the OR-nodes are evaluated sequentially. Keeping the timing and Proposition 3 in mind, it is not difficult to design asystolic algorithm for this problem. In fact, the derived structure is the same as that proposed by Guibas, Kung and Thompson[13].

The above example demonstrates the relationship between an AND/OR-graph representation of polyadic DP problems and the corresponding systolic design. In general, starting from an AND/OR-graph, a systolic array with planar interconnections can be designed by first serializing links that connect nodes not in adjacent levels in the AND/OR graph, and by designing the appropriate control signals. As shown in the examples, the transformation may introduce additional delay and redundant hardware in the implementation.

## 6 Conclusions

Dynamic programming formulations have been classified according to the structure of the functional equations. A given problem can usually be formulated in multiple ways, hence it is important to compare the alternative implementations. The applicability of systolic processing is most suitable when the formulation is serial.

Many sequential decision problems have serial formulations that can be considered as searching a multistage graph. If there are a large number of states and/or quantized values in each stage, then a monadcic formulation is more appropriate, and the problem is efficiently solved as a serial string of matrix multiplications. On the other hand, if the number of stages is large, the problem should be put into a polyadic formulation. The matrices are grouped into a binary tree and multiplied by a divide-and-conquer algorithm. We have found the $AT^2$ lower bound for multiplying a string of *Nm*-by-*m* matrices, and have proved that dividing the string into $O(N/\log_2 N)$ groups and multiplying each by a systolic array is optimal in the sense of achieving this lower bound.

When the formulation is nonserial, it may be necessary to transform the problem into a serial formulation before an efficient implementation can be found. A monadic-nonserial formulation can be transformed into a monadic-serial one by grouping state variables. A problem in a polyadic-nonserial formulation can be represented as the search of an optimal solutionin an AND/OR graph, which can be transformed into an AND/OR graph for a serial problem by adding dummy nodes. The transformed AND/OR graph can be mapped directly into a planar systolic array by using appropriate control signals. The additional hardware and delay introduced is problem dependent.

We have not coosidered the use of branch-and-bound algorithms with dominance tests to solve dynamic programming problems. They are more suitable when a multiprocessor with a flexible interconnection or a dataflow system is used.

## References:


[1] BELLMAN R, DREYFUS S. Applied dynamic programming[M]. Princeton: Princeton University Press, 1962.

[2] KARP R, HELD M. Finite state processes and dynamic programming[J]. SIAM Journal on Applied Mathematics, 1967,15: 693-718.

[3] IBARAKI T. Solvable classes of discrete dynamic programming[J]. Journal of Mathematical Analysis and Applications, 1973, 43: 542-693.

[4] WHITE D. Dynamic programming[M]. Edinburgh: Oliver &Boyd, 1969.

[5] KUMAR V. A general bottom-up procedure for searching AND/OR graphs[J]. Proceedings of AAAI, 1984: 182-187.

[6] GENSI S, MONTANARI U, MARTELLI A. Dynamic programming as graph searching: an algebraic approach[J]. Journal of the ACM, 1981, 28(4): 737-751.

[7] MORIN T, MARSTEN R. Branch-and-bound strategies for dynamic programing[J]. Operations Research, 1976, 24(4): 611-637.

[8] IBARAKI T. The power of dominance relations in branch-and-bound algorithms[J]. Journal of the ACM, 1977, 24(2): 264-279.

[9] LI G J, WAH B W. Computational efficiency of parallel approximate branch-and-bound algorithms[C]// International Conference on Parallel Processing. [S.l.:s.n.], 1984: 473-480.

[10] BERTELÈ U, BRIOSCHI F. Nonserial dynamic programming[M]. New York: Academic Press, 1972.

[11] ROSENTHAL A. Dynamic programming is optimal for nonserial optimization problems[J]. SIAM Journal on Computing, 1982, 11(1).

[12] CASTI J, RICHARDSON M, LARSON R. Dynamic programming and parallel computers[J]. Journal of Optimization Theory and Applications, 1973, 12(4): 423-438.

[13] GUIBAS L, KUNG H, THOMPSON C. Direct VLSI implementation of combinatorial algorithms[C]// Conference on VLSI: Architecture, Design, and Fabrication.

[S.l.:s.n.], 1979: 509-525.

[14] VARMAN P, RAMAKRISHNAN V. Dynamic programming and transitive closure on linear pipelines[C]// International Conference on Parallel Processing. [S.l.:s.n.], 1984: 359-364.

[15] CLARKE M, DYER C. Systolic array for a dynamic programming application[C]// The 12th Workshop on Applied Imagery Pattern Recognition. [S.l.:s.n.], 1983.

[16] WAH B W, LI G J, YU C F. Multiprocessing of combinatorial search problems[J]. IEEE Computer, 1985.

[17] AHO A V, HOPCROFT J E,ULLMAN J D. The design and analysis of computer algorithms[M]. Boston: Addison-Wesley Publishing Co., 1974.

[18] KUNG H T. Lets design algorithms for VLSI systems[C]// Caltech Conference on VLSI. [S.l.:s.n.], 1979.

[19] LI G J, WAH B W. Parallel processingfor dynamic programming problems[R]. 1985.

[20] LI G J, WAH B W. The design of optimal systolic algorithms[J]. IEEE Transactions on Computers,1985, 34(1).

[21] KUCKD J. A survey of parallel machine organization and programming[J]. ACM Computing Surveys, 1977, 9(1): 29-59.

[22] BAUDET G, STEVENSON D. Optimal sorting algorithms for parallel computers[J]. IEEE Transactions on Computers, 1978, C-27(1): 84-87.

[23] SAVAGE G. Parallel algorithms for graph theoretic problems[D]. Champaign: University of Illinois Urbana-Champaign,1978.

[24] BRIOSCHI F, EVEN S. Minimizing the number of operations in certain discrete-variable optimization problems[J]. Operations Research, 1970, 18: 67-81.

[25] BERTELÈ U, BRIOSCHI F. A new algorithm for the solution of the secondary optimization in non-serial dynamic programming[J]. Journal of Mathematical Analysis and Applications, 1969, 27: 565-574.

[26] MARTELLI A, MONTANARI U. Additive AND/OR graphs[C]// International Joint Conference on Artificial Intelligence. [S.l.:s.n.], 1973:1-11.

[27] NILSSON N J. Principles of artificial intelligence[M]. Palo Alto: Tioga Publishing Company, 1980.

第2章

# 并行组合搜索

# 异常就是科研方向

到普渡大学以后，华云生教授要求我将我的硕士论文中提出的脉动阵列参数设计方法改进为寻找最优设计的方法，我将原来的脉动阵列算法设计抽象为一个整数规划问题，写了一篇论文 *The design of optimal systolic arrays*，后来发表在 *IEEE Transactions on Computers* 上，成为我执笔的引用次数最多的学术论文。当时 Systolic 算法还是较热门的研究方向，华云生教授也曾希望我继续做这个方向，研究卡尔曼滤波器的并行处理，但我已经被另一个完全不同的问题吸引住了。

华云生教授在加州大学伯克利分校读博士时，图灵奖得主 Richard M. Karp 教授是他的指导组成员（Committee Member），虽然他的导师 Ramamoorthy C.V.教授指导他的博士研究方向是分布式数据库，但 Karp 教授曾希望他做并行组合搜索方向的研究，因此他到普渡大学工作以后，就开始朝 Karp 教授建议的方向努力。美国计算机界有不少工程院院士，但只有 Donald E. Knuth 和 Karp 这两位教授是科学院院士，Karp 教授是 NP 完全理论的奠基人，基于对基础理论研究的偏爱，我对他有一种特殊的崇敬，因此我毫不犹豫地将“并行组合搜索”选定为我的博士论文研究方向。

组合搜索是典型的理论研究问题，而并行处理又是一个工程性很强的研究方向，做并行组合搜索研究有很大的风险。刚开始做这个方向研究时，我总是担心，我写的论文很可能搞理论的认为我做的是工程研究，而做工程的又认为我在讲理论问题，谁都不会喜欢。但我是一个夹在理论与工程之间的学者，天生就喜欢探索这种“理不理，工不工”的边缘科学，越做越来劲。最后的结果出乎我的意料，我投稿的论文，不论是期刊还是国际会议，做工程和做理论研究的学者都认可，几乎没有一篇被拒。从 1984 年到 1986 年的 3 年时间里，我发表的论文数量和论文质量远远超过周围的同学。

用并行处理来对付组合搜索的初衷是提高搜索效率，即缩短获得结果的计算时间。但在我的计算机仿真实验中，常常出现一些异常现象，要么得不到应有的加速，有时并行搜索还不如串行；要么“加速比”远远超过设定的处理机数。开始我一直怀疑我编写的仿真程序有问题，但怎么查也查不出来。在科研过程中，许多人期望实验结果与预想一致，从而证明自己的想法正确，这样容易出文章。但是如果出现“异常”，实验结果出乎意料，也不是坏事，这往往会提示我们新的研究方向。后来我花了 2 年时间潜心研究并行组合搜索中的各种异常现象，发现这种“异常”是非确定算法的本质特征。通过对 Karp 教授提出的分支估界（Branch and Bound）算法的性能分析，我找到了并行计算获得超线性加速的必要条件和保证不出现“坏的并行”的充分条件。所谓“坏的并行”就是“并行不如串行”或“增加处理机反而降速”。

我做博士研究的另一个体会是要把研究做到“底”。做研究要做到什么程度放手是个大问题，有些研究方向明明已经没有取得新成果的可能，而有些人还在不断啃别人啃过的馒头，浪费时间做些枝枝叶叶的修补工作。但更多的情况是浅尝辄止，刚做了初步研究就见异思迁，这等于给别人出题目。我的博士研究做了 4 年，坚持把并行组合搜索的性能分析做到底，对“与树（AND-Tree）”“或树（OR-Tree）”“与或图（AND/OR 图）”、具有或没有“优势关系（Dominance Relation）”的搜索、串行与并行的比较、不同处理数的性能比较等都做了深入的分析，证明了十几个定理，给出了各种情况下的性能上下界。1985 年 6 月，我在 *IEEE Computer* 杂志上发表的长文 *Multiprocessing of combinatorial search problems* 综述了我的博士论文的大部分内容，直到近几年还有人引用这篇论文。这篇论文还被 Vipin Kumar 教授编入合作的专著 *Parallel algorithms for machine intelligence and vision*。1990 年 1 月，我的博士论文的主要内容“Computational efficiency of parallel combinatorial OR-tree searches”才在 *IEEE Transactions on Software Engineering* 上发表。这是因为主要审稿人调动工作时，将我的投稿“压在箱子里”而忘了审稿，2 年后经催问才想起来审稿，由此可见 IEEE 出版机构的管理也有疏漏之处。自 2008 年起，在人工智能的顶级会议 AAAI 年会 和 IJCAI 上，每年都有 Symposium on Combinatorial Search（组合搜索专题讨论会），关于组合搜索并行处理的研究方向再次受到重视。

发表于Conference on Artificial Intelligence (AAAI)论文集，1984年

# How to cope with anomalies in parallel approximate branch-and-bound algorithms

Guo-jie LI, Benjamin W. WAH

School of Electrical Engineering, Purdue University, West Lafayette, Indiana 47907

**Abstract:** A general technique for solving a wide variety of search problems is the Branch-and-Bound (B&B) algorithm. We have adapted and extended B&B algorithms for parallel processing. Anomalies owing to parallelism may occur. In this paper sufficient conditions to guarantee that parallelism will not degrade the performance are presented. Necessary conditions for allowing parallelism to have a speedup greater than the number of processors are also shown. Anomalies are found to occur infrequently when optimal solutions are sought; however, they are frequent in approximate B&B algorithms. Theoretical analysis and simulations show that a best-first search is robust for parallel processing.

## 1 Introduction

The search for solutions in a combinatorially large problem space is very important in Artificial Intelligence (AI)[1]. Combinatorial-search problems can be classified into two types. The first type is decision problems that decide whether at least one solution exists and satisfies a given set of constraints. Theorem-proving, expert systems and some permutation problems belong to this class. The second type is optimization problems that are characterized by an objective function to be minimized or maximized and a set of constraints to be satisfied. Practical problems, such as traveling salesman, job-shop scheduling, knapsack, vertex cover, and game-tree search belong to this class.

A general technique for solving combinatorial searches is the B&B algorithm[2]. This is a partitioning algorithm that decomposes a problem into smaller subproblems and repeatedly decomposes until infeasibility is proved or a solution is found[2]. It can be characterized by four constituents: a branching rule, a selection rule, an elimination rule and a termination condition. The first two rules are used to decompose the problem into simpler subproblems and appropriately order the search. The last two rules are used to eliminate generated subproblems that are infeasible or that cannot lead to a better solution than an already-known feasible solution. Kumar et al. have shown that the B&B approach

Research was supported by National Science Foundation Grant ECS 81-05968.

provides a unified way of formulating and analyzing AND/OR tree searches such as SSS* and Alpha-Beta search[3]. The technique of branching and pruning in B&B algorithms to discover the optimal element of a set is the essence of many heuristic procedures in AI.

To enhance the efficiency of implementing B&B algorithms, approximations and parallel processing are two major approaches. It is impractical to use parallel processing to solve intractable problems with exponential complexity because an exponential number of processors must be used to solve the problems in polynomial time in the worse case. For these problems, approximate solutions are acceptable alternatives. Experimental results on vertex-cover, 0-1 knapsack and some integer-programming problems reveal that a linear reduction in accuracy may result in an exponential reduction in the average computational time[4]. On the other hand, parallel processing is applicable when the problem is solvable in polynomial time (such as finding the shortest path in a graph), or when the problem is NP-hard but is solvable in polynomial time on the average[5], or when the problem is approximately solvable in polynomial time (such as game-tree search).

Analytical properties of Parallel Approximate B&B (PABB) algorithms have been rarely studied. In general, a $k$-fold speedup (ratio of the number of iterations in the serial case to that of the parallel case) is sought when $k$ processors are used. However, simulations have shown that the speedup for PABB algorithms using $k$ processors can be (a) less than one-"detrimental anomaly"[6-7]; (b) greater than $k$-"acceleration anomaly"[6-7]; or (c) between one and $k$-"deceleration anomaly"[4,6-7]. Similar anomalous behavior has been reported by others. For instance, the achievable speedup for AND/OR-tree searches are limited by a constant (5 to 6) independent of the number of processors used (parallel-aspiration search) or $\sqrt{k}$ with $k$ processors (tree-splitting algorithm)[8]. So far, all known results of parallel tree searches showed that a near-linear speedup holds only for a small number of processors. It is desirable to discover conditions that preserve the acceleration anomalies, eliminate the detrimental anomalies and minimize the deceleration anomalies. The objectives of this paper are to provide conditions for achieving the maximum speedup and to find the appropriate parallel search strategy under which a near-linear speedup will hold for a considerable number of processors.

## 2 Parallel approximate branch-and-bound algorithms

Many theoretical properties of serial B&B algorithms have been developed[9], and a brief discussion is given here. In this paper minimization problems are considered. Let $P_i$ be a subproblem, i.e., a node in the state-space tree, and $f(P_i)$ be the value of the best solution obtained by evaluating all the subproblems decomposable from $P_i$. A lower bound, $g(P_i)$, is calculated for $P_i$ when it is created. If a subproblem is a feasible solution with the best objective-function value so far, the solution value becomes the incumbent $z$. The incumbent represents the best solution obtained so far in the process. During the computation, $P_i$ is terminated if:

$$g(P_i) \geqslant z \quad (1)$$

The approximate B&B algorithm is identical to the optimal algorithm except that the lower-bound test is modified to:

$$g(P_i) \geqslant \frac{z}{1+\epsilon} \quad \epsilon \geqslant 0, z \geqslant 0 \quad (2)$$

where $\epsilon$ is an allowance parameter. The final incumbent value $z_F$ obtained by the modified lower-bound test deviates from the optimal solution value, $z_O$, by:

$$\frac{z_F}{1+\epsilon} \leqslant z_O \leqslant z_F \quad (3)$$

Let $\boldsymbol{L}$ denotes the lower-bound cutoff test, that is, $P_j \boldsymbol{L} P_i$ means that $P_j$ is feasible solution and $f(P_j)/(1+\epsilon) \leqslant g(P_i)$, $\epsilon \geqslant 0$. For example, in Figure 1, $P_1' \boldsymbol{L} P_2$, since $91/1.1 < 85$, and similarly $P_4 \boldsymbol{L} P_1$. However, $P_4 \boldsymbol{L} P_2$ is false because $100/1.1 > 85$.

Ibaraki mapped breadth-first, depth-first and best-first searches into a general form called heuristic searches[9]. A heuristic function is used to define the order in which subproblems are selected and decomposed. The algorithm always decomposes the subproblem with the minimum heuristic value. In a best-first search, the lower-bound values define the order of expansion, hence the lower-bound function can be taken as the heuristic function. In a breadth-first search, subproblems with the minimum level numbers are expanded first. The level number can, thus, be taken as the heuristic function. Lastly, in a depth-first search, subproblems with the maximum level numbers are expanded first. The negation of the level number can be taken as the heuristic function.

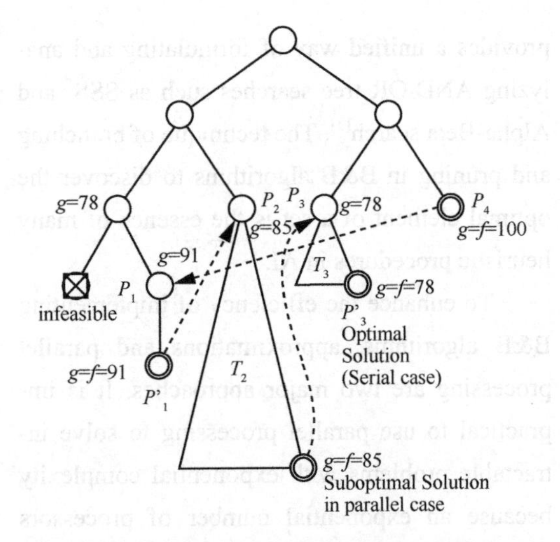


Figure 1. Example of an detrimental anomaly under a parallel depth-first search ( $\epsilon$ =0.1)

Branch-and-bound algorithms have inherent parallelism:

(a) Parallel selection of subproblems: A set of subproblems less than or equal to in size to the number of processors have to be selected for decomposition in each iteration. A selection function returns $k$ subproblems with the minimum heuristic values from $\boldsymbol{U}$, where $k$ is number of processors and $\boldsymbol{U}$ is the active list of subproblems. The selection problem is especially critical under a best-first search because a set of subproblems with the minimum lower bounds must be selected.

(b) Parallel branching: The subproblems assigned to the processors can be decomposed in parallel. In order for the processors to be well utilized, the number of active subproblems should be greater than or equal to $k$.

(c) Parallel termination test: Multiple infeasible nodes can be eliminated in each iteration. Further, multiple feasible solutions may be generated, and the incumbent may have to be updated in parallel.

(d) Parallel elimination test: If the incumbent is accessible to all the processors, the lower-bound test (Eq's (1) or (2) can be carried out in parallel.

The above sources of parallelism has been studied in MANIP, a multiprocessor implementing PABB algorithms with a best-first search and lower-bound tests[4].

## 3 Anomalies on parallelism

In this section anomalies are studied under lower-bound elimination and termination rules. The results on anomalies with dominance tests are shown elsewhere[10]. For simplicity, only the search for a single optimal solution is considered here.

A synchronous model of PABB algorithm is used. The incumbent is stored in a global register that can be updated concurrently. Active subproblems can be stored in a centralized list or multiple lists. The distinction lies in the memory configuration. When all the processors are connected to a centralized memory, the subproblem list is global to the processors. When each processor has a private memory, only the local subproblem list can be accessed. The sequence of operations performed in an iteration are selection, branching, feasibility and elimination tests, and inserting newly generated subproblems into the list(s). Let $T^c(k,\epsilon)$ and $T^d(k,\epsilon)$ denote the number of iterations required for expanding a B&B tree using centralized and $k$ subproblem lists respectively, where $k$ is the number of processors used, and $\epsilon$ is the allowance parameter.

An example of a detrimental anomaly is illustrated in Figure 1. In a serial depth-first search, subtree $T_2$ is terminated owing to the lower-bound test of $P'_1$: $f(P'_1)/(1+\epsilon)\leqslant g(P_2)$ where $\epsilon=0.1$. In a parallel depth-first search with two processors, a feasible solution, $P_4$, that terminates $P_1$ and $P'_1$ is found in the second iteration. As mentioned before, $P_2$ is not eliminated by $P_4$. Consequently, subtree $T_2$ has to be expanded that will eventually terminate subtree $T_3$. If the size of $T_2$ is much larger than the size of $T_3$, the time it takes to expand $T_2$ using two processors will be longer than the time it takes to expand $T_3$ using one processor. Note that the above anomaly does not happen in a best-first search because subtree $T_2$ is not expanded in both the serial and the parallel cases.

An example of an acceleration anomaly is shown in Figure 2. When a single processor with a death-first search is used, subtree $T$ win be expanded since $f(P_1)/(1+\epsilon)>g(P_2)$ where $\epsilon=0.1$. When two processors are used, $P_2$ and hence $T$ will be terminated by lower-bound tests with $P_3$: $f(P_3)/(1+\epsilon)<g(P_2)$. If $T$ is very large, an acceleration anomaly will occur.

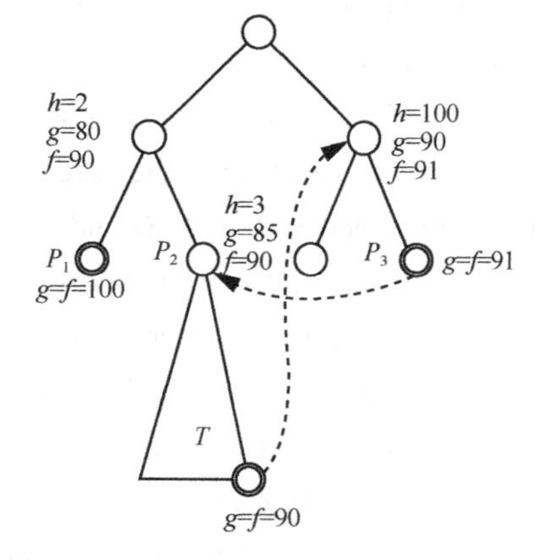


Figure 2. Example of an acceleration anomaly under a parallel depth-first search ( $\epsilon$ =0.1)

## 4 Generalized heuristic searches

Recall that the selection function uses the heuristic values to define the order of node expansions. In this section we show that detrimental anomalies are caused by the ambiguity in the selection rule. A generalized heuristic search is proposed to eliminate detrimental anomalies in a single subproblem list.

Consider the serial depth-first search. The subproblems are maintained in a last-in-first-out list, and the subproblem with the maximum level number is expanded first. When multiple subproblems have identical level numbers (heuristic values), the subproblem chosen by the selection function depends on the order of insertion into the stack. Suppose the rightmost son is always expanded and inserted first. Then the leftmost son will be the subproblem inserted last and expanded first in the next iteration.

In a parallel depth-first search with a single subproblem the mere extension of the serial algorithm may cause an anomalous behavior. For example, the order of expansion in a serial depth-first search for the tree in Figure 3 is *A*, *B*, *D*, *I*, *J*, *E*, etc. When two processors are used, nodes *B* and *C* are expanded in the second iteration that result in nodes *D*, *E*, *F*, *G* and *H*. Since these nodes have identical level numbers, any two of these nodes can be chosen for expansion in the next iteration by the conventional heuristic function discussed in Section 2. Suppose the nodes are inserted in the order *E*, *D*, *H*, *G* and *F*. Then nodes *F* and *G* will be selected and expanded in the third iteration. This may cause a detrimental anomaly if subtree $T_p$ is large. In fact, this is exactly the reason for the anomalies reported by Lai and Sahni[7].

To solve this problem, we must define distinct heuristic values for the nodes so that there is no ambiguity on the nodes to be chosen by the selection function. In this paper a path number is used to uniquely identify a node in a tree. The path number of a node is a sequence of *d*+1 integers that represent the path from the root to this node where d is the maximum number of levels of the tree. The path number $E = e_0e_1e_2\cdots e_d$ is defined recursively as follows. The root $P_0$ exists at level 0 and has a path number of $000\cdots0$. A node $P_{i_j}$ on level *l* which is the *j*-th son (counting from the left) of $P_i$ with path number $E_{P_i} = e_0e_1\cdots e_{l-1}000\cdots$ has path number $E_{P_{i_j}} = e_0e_1\cdots e_{l-1}j00\cdots$. As an example, the path numbers for the nodes in the tree of Figure 3 are shown.

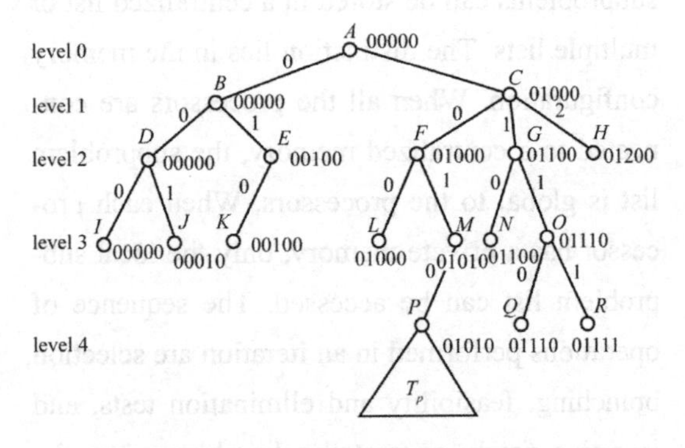


Figure 3. The path numbers of a tree

To compare path numbers, the relations "$>$" and "$=$" must be defined. A path number $E_1 = e_1^1e_2^1\cdots$ is less than another path number $E_2 = e_1^2e_2^2\cdots(E_1<E_2)$ if there exists $0\leqslant j\leqslant d$ such that $e_i^1=e_i^2$, $0\leqslant i<j$ and $e_j^1<e_j^2$. The

path numbers are equal if $e_i^1 = e_i^2$ for $0 \leqslant i \leqslant d$. For example, the path number 01000 is less than 01010. Note that nodes can have equal path numbers if they have the ancestor-descendant relationship. Since these nodes never coexist simultaneously in the list of active subproblems, the subproblems in the active list always have distinct path numbers.

The path number is now included in the heuristic function. The primary key is still the lower-bound value or the level number. The secondary or ternary key is the path number and is used to break ties in the primary key.

$$h(P_i) = \begin{cases} \text{(level number, path number)} , & \text{breadth-first search} \\ \text{(path number)} , & \text{depth-first search} \\ \text{(lower bound, level number, path number)} & \\ \text{or(lower bound, path number)} , & \text{best-first search} \end{cases} \tag{4}$$

For a best-first search, two alternatives are defined that search in a breadth-first or depth-first fashion for nodes with identical lower bounds. The heuristic functions defined above belong to a general class of heuristic functions that satisfy the following properties:

(a) $h(P_i) \neq h(P_j)$ if $P_i \neq P_j$, $P_i, P_j \in \boldsymbol{U}$

(all heuristic values in the active list are distinct) (5)

(b) $h(P_i) \leqslant h(P_j)$ if $P_j$ is a descendant of $P_i$ (heuristic values do not decrease) (6)

In general, any heuristic function with a tie-breaking rule that satisfy Eq's (5) and (6) will not lead to detrimental anomalies. Due to space limitation, the results are stated without proof in the following theorems. The proofs can be found in [10].

**Theorem 1:** Let $\epsilon = 0$, i.e., an exact optimal solution is sought. $T^c(k,0) \leqslant T^c(1,0)$ holds for parallel heuristic searches of a single optimal solution in a centralized list using any heuristic function that satisfies Eq's (5) and (6).

When approximations are allowed, detrimental anomalies cannot always be avoided for depth-first searches even though path numbers or other tie-breaking rules are used (see Figure 1). The reason for the anomaly is that lower-bound tests under approximation, $\boldsymbol{L}$, are not transitive. That is, $P_i \boldsymbol{L} P_j$ and $P_j \boldsymbol{L} P_k$ do not imply $P_i \boldsymbol{L} P_k$, since $f(P_i)/(1+\epsilon) \leqslant g(P_i)$ and $f(P_j)/(1+\epsilon) \leqslant g(P_k)$ implies $f(P_i)/(1+\epsilon)^2 \leqslant g(P_k)$ rather than $f(P_i)/(1+\epsilon) \leqslant g(P_k)$. In this case detrimental anomalies can be avoided for best-first or breadth-first searches only.

**Theorem 2:** $T^c(k,\epsilon) \leqslant T^c(1,\epsilon)$, $\epsilon > 0$, holds for parallel best-first or breadth-first searches for a single optimal solution when a heuristic function satisfying Eq's (5) and (6) is used.

Since the lower-bound function is used as the heuristic function in best-first searches, Eq's (5) and (6) are automatically satisfied if all the lower-bound values are distinct. Otherwise, path numbers must be used to break ties in the lower bounds. In Section 7, a more general condition will be given for best-first searches. For depth-first searches, the conditions of Theorem 2 are not sufficient, and the following condition is needed. For any feasible solution $P_i$, all nodes whose heuristic values are less than $h(P_i)$ cannot be eliminated by the lower-bound test due to $P_i$, that is, $f(P_i)/(1+\epsilon) \leqslant g(P_i)$ implies that $h(P_i) < h(P_j)$

for any $P_j$. Generally, this condition is too strong and cannot be satisfied in practice.

## 5 Necessary conditions to ensure acceleration anomalies in a single subproblem list

When an exact optimal solution is sought, acceleration anomalies may occur if a depth-first search is used or some nodes have identical heuristic values. This is characterized by the incomplete consistency between the heuristic and the lower-bound functions. A heuristic function, $h$, is said to be not completely consistent with $g$ if there exist two nodes $P_i$ and $P_j$ such that $h(P_i) > h(P_j)$ and $g(P_i) \leqslant g(P_j)$.

**Theorem 3:** Let $\epsilon = 0$. Assume that a single optimal solution is sought. The necessary condition for $T^c(k,0) < T^c(1,0)/k$ is that the heuristic function is not completely consistent with $g$.

For a breadth-first search, no acceleration anomaly will occur if the heuristic function defined in Eq. (4) is used. For a best-first search, acceleration anomalies may exist if the level number is not used in the heuristic function. It is important to note that the condition in Theorem 3 is not necessary when approximate solutions are sought. An example showing the existence of an acceleration anomaly when $h$ is completely consistent with $g$ is shown in Figure 2. A looser necessary condition is that $h$ is not completely consistent with the lower-bound test with approximation, that is, there exist $P_i$ and $P_j$ such that $h(P_i) > h(P_j)$ and $P_i L P_j$.

## 6 Multiple subproblem lists

When there are multiple subproblem lists, one for each processor, a node with the minimum heuristic value is selected for decomposition from each local list. This node may not belong to the global set of active nodes with the minimum heuristic values; however, the node with the minimum heuristic value will always be expanded by a processor as long as the nodes are selected in a consistent order when there are ties. Since it is easy to maintain the incumbent in a global data register, the behavior of multiple lists is analogous to that of a centralized list. However, the performance of using multiple lists is usually worse than that of a single subproblem list[4].

So far, we have shown conditions to avoid detrimental anomalies and to preserve acceleration anomalies under lower-bound tests only. The results are summarized in Table 1. The corresponding results when dominance tests are used will not be shown here due to space limitation[10].

## 7 Robustness of parallel best-first searches

The preceding sections have shown that best-first searches are more robust for parallel processing in the sense of avoiding detrimental anomalies and preserving acceleration anomalies. In this section we shown that best-first searches are more robust as far as deceleration anomalies are concerned.

Table 1. Summary of results for the elimination of detrimental anomalies and the preservation of acceleration anomalies in parallel B&B algorithms with lower-bound tests

| Allowance parameter | Subproblem lists | Search strategies | Suff. cond. to eliminate detrimental anomaly | Necessary cond. for acceleration anomaly |
|---|---|---|---|---|
| $\epsilon=0$ | single | all | I | II |
| | multiple | | no anomaly | |
| $\epsilon>0$ | single or multiple | breadth-first or best-first | I | exists |
| | | depth-first | anomaly | |

Conditions: I: heuristic function satisfies Eq's (5) and (6)

II: $h$ is not completely consistent with $g$

anomaly: the sufficient conditions are impractical

exists: the necessary conditions are too loose

Figure 4 shows the computational efficiency of a parallel optimal B&B algorithm using a best-first or a depth-first search for solving knapsack problems in which the weights, $w(i)$, are chosen randomly between 0 and 100 and the profits are set to be $p(i)=w(i)+10$. The assignment used is intended to increase the complexity of the problem. In the simulations each processor has a local memory. Load balancing is incorporated so that an idle processor with an empty subproblem list can get a subproblem from its neighbor. It is observed that the speedup is sensitive to $T(1, 0)$, and the speedup is better for best-first searches. For instance, when 64 processors are used, the average speedup is 48.8 for best-first searches and 27.9 for depth-first searches. Moreover, it should be noted that the generalized heuristic search presented in Section 4 cannot guarantee $T(k_1,0)<T(k_2,0)$, $k_1>k_2>1$, for depth-first and breadth-first searches. Similar results were observed for vertex-cover problems.

The following theorem gives the performance bound of parallel best-first searches. The maximum number of processors within which a near-linear speedup is guaranteed can be predicted.

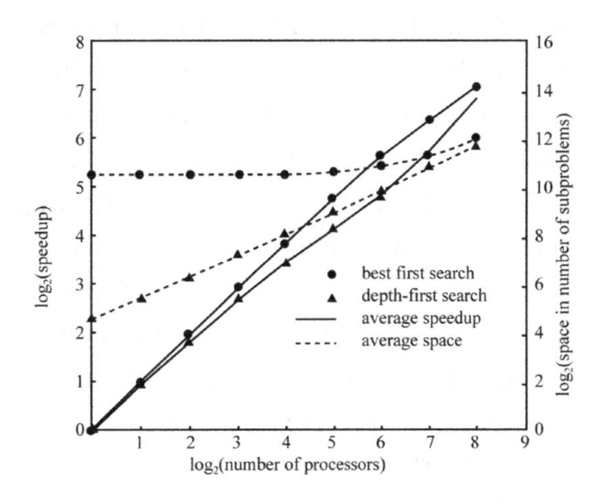


Figure 4. Average speedups and space requirements of parallel optimal B&B_ algorithms for 10 knapsack problems with 35 objects (average $T(1,0)$=15 180 for best-first searches; average $T(1,0)$= 15 197 for depth-first searches)

**Theorem 4:** For a parallel best-first search with $k$ processors, $\epsilon = 0$, and $g(P_i) \neq f^*$ if $P_i$ is not an optimal-solution node ($f^*$ is the optimal-solution value),

$$\frac{T(1,0)}{k} \leqslant T(k,0) \leqslant \frac{T(1,0)}{k} + \frac{k-1}{k} l \qquad (7)$$

where $l$ is the maximum number of levels of the B&B tree to be searched. Since the performance is not affected by using single or multiple subproblem lists, the superscript in $T$ is dropped.

Since $l$ is a polynomial function of (usually equal to) the problem size while $T(1, 0)$ is an exponential function of the problem size for NP-hard problems, the first term on the R.H.S. of Eq. (7) is much greater than the second term as long as the problem size is large enough. Eq. (7) implies that the near-linear speedup can be maintained within a considerable range of the number of processors for best-first searches. As an example, if $l = 50$, $T(1,0) = 10^6$ (for a typical traveling-salesman problem), and $k$=1 000, then $T(1\ 000, 0) < 1\ 049$. This means that almost linear speedup can be attained with 1 000 processors. Furthermore, it can be shown that there is always monotonic increase in performance for all $1 \leqslant k_1 < k_2 \leqslant \sqrt{T(1,0)/l}$ .For this example, there will not be any detrimental anomaly for any combinations of $1 \leqslant k_1 < k_2 \leqslant 141$ if the assumptions of Theorem 4 are satisfied.

Before ending this paper, it is worth saying a few words about the space required by parallel B&B algorithms. In the serial case, the space required by a best-first search is usually more than that required by a depth-first search. Somewhat surprisingly, the simulation results on 0-1 knapsack problems show that the space required by parallel best-first searches is not increased significantly (may also be decreased) until the number of processors is so large that a near-linear speedup is not possible. In contrast, the space required by parallel depth-first searches is almost proportional to the number of processors (Figure 4). Note that the space efficiency is problem-dependent. For vertex-cover problems, the space required by parallel best-first searches is not increased significantly regardless of the number of processors used.

## References:


[1] PEARL J. Heuristics[M]. Boston: Addison-Wesley Longman Publishing, 1984.

[2] LAWLER E L, WOOD D E. Branch-and-bound methods: a survey[J]. Operations Research, 1966, 14 (4): 699-719.

[3] KUMAR V, KANAL L. A general branch-and-bound formulation for understanding and synthesizing AND/OR tree-search procedures[J]. Artificial Intelligence, 1983, 21(1-2): 179-197.

[4] WAH B W, MA Y W. MANIP-a multicomputer architecture for solving combinatorial extremum-search problems[J]. IEEE Transactions on Computers, 1984, C-33(5): 377-390.

[5] SMITH D R. Random trees and the analysis of branch-and-bound procedures[J]. Journal of the ACM, 1984, 31(1): 163-188.

[6] IMAI M, FUKUMURA T, YOSHIDA Y. A parallelized branch-and-bound algorithm implementation efficiency[J]. Systems Computer Controls, 1979, 10 (3): 62-70.

[7] LAI T H, SAHNI S. Anomalies in parallel branch-and-bound algorithms[C]// 1983 International Conference on Parallel Processing. Washington: IEEE Computer Society Press, 1983: 183-190.

[8] FINKEL R, FISHBURN J. Parallelism in Alpha-Beta search[J]. Artificial Intelligence, 1982, 19(1): 89-106.

[9] IBARAKI T. Theoretical comparisons of search strategies in branch-and-bound algorithms[J]. International Journal of Computer and Information Sciences, 1976, 5(4): 315-344.

[10] LI G J, WAH B W. Computational efficiency of parallel approximate branch-and-bound algorithms[R]. 1984.

发表于*IEEE Computer*，1985年第6期

# Multiprocessing of combinatorial search problems

Benjamin W. WAH, Guo-jie LI, Chee-fen YU

Purdue University

Multiprocessing refers to the concurrent execution of processes (or programs) in a computer hardware complex with more than one independent processing unit[1]. Conventionally, multiprocessing is defined as a centralized computer system with central processors, input-output processors, data channels, and special-purpose processors.

With the advent of VLSI technology, it has become cost effective to include a large number of general- and special-purpose processors in a multiprocessing system. The definition of multiprocessing has been extended to systems such as multiprocessors, systolic arrays, and dataflow computers. In this sense, parallel processing and multiprocessing can be considered synonymous. Examples of such systems range from the IBM 360 computers with channels, Iliac IV, Staran, Cm*,Trac, MPP, and the Cray X-MP to the latest fifth-generation computer system.

## 1 Major issues of multiprocessing

In using a multiprocessing system to solve a given problem, tremendous effort can be spent in designing a good parallel algorithm. The objective is to obtain an algorithm with a speedup proportional to the number of processors over the best available serial algorithm. This must be done within the architectural constraints, and it involves trade-offs in computation time, memory space, and communications requirements.

The applicability of multiprocessing to the problem is also an important issue. The problem should be polynomially solvable by a serial computer. Intractable problems with complexity that is exponential in respect to problem size cannot be solved in polynomial time[1] unless an exponential number of processors are used[2]. This is, of course, technologically infeasible for large problems. For example, if the best serial algorithm requires $2^N$ micro seconds to solve a problem of size $N$, it would require $2^{60}$ microse-

[1] Saying that a problem is intractable with an exponential complexity implies that the best parallel algorithm cannot have a polynomial complexity with a polynomial number of processors. Supposing this is false, a serial simulation of this parallel algorithm can solve the problem in polynomial time, hence contradicting the fact that the problem has an exponential complexity.

conds, or 366 centuries, to solve a problem of size 60. Assuming that a linear speedup is possible, it would require $2^{50}$ processors to solve the problem in approximately one second, and $2^{40}$ processors to solve the problem in 20 minutes. For intractable problems, approximate solutions should be used in order to complete the algorithm in a reasonable time.

Multiprocessing is generally used to improve the computational efficiency of solving a given problem, not to extend the solvable problem space of the problem. Suppose that it takes $N^k$ units of time to solve a problem of size $N$, where $k$ is a constant ($\geqslant 1$) Assuming a linear speedup, a parallel algorithm with $N$ processors in $N^k$ units of time can solve problem of size $N^{(1+1/k)}$. For $k$=3, this size is $N^{1.33}$. Similarly, for a serial algorithm that takes $k^N$, $k>1$ units of time to solve a problem, the solvable problem size with $N$ processors in $k^N$ units oftime is $N+\log_k N$. For a large $N$, this size is approximately $N$.

Search problems. This article examines the use of multiprocessing in solving combinatorial search problems. Combinatorial search problems involve the search for one or more optimal or suboptimal solutions in a defined problem space. They can be classified as decision problems, which find solutions satisfying a given set of constraints, or as optimization problems, which seek solutions satisfying the constraints and also optimizing an objective function. Examples include proving theorems, playing games, evaluating a logic program, solving a database query, designing a computer system, assigning registers for a compiler, finding the shortest path in a graph, solving a mathematical programming problem, and searching for a permutation order to sort a set of numbers. These problems occur in a wide spectrum of engineering and science applications, including artificial intelligence and operations research.

A search problem can be represented as an acyclic graph[2] or as a search tree. These representations are characterized by a root node with no edge entering it and by one or more terminal nodes with no exiting edges. In a search graph, one or more edges can enter any node except the root; in a search tree, each node except the root has exactly one edge entering it.

An edge in a search graph represents an assignment of value to an unassigned parameter. This can be illustrated by the 0/1 knapsack problem, in which $N$ objects are to be packed into a knapsack. Object $i$ has a weight $w_i$, and the knapsack has a capacity $M$. If object $i$ is placed in the knapsack, then a profit $p_i$ is earned. The objective is to fill the knapsack so as to maximize profit. The unassigned parameters are the set of objects that have not been considered. In expanding a node, an object, say $i$, is selected, and two alternatives are created: (a) object $i$ is included in the knapsack, and (b) object $i$ is not included.

The nonterminal nodes in a search tree can be classified as AND nodes and OR nodes. An AND node represents a (sub) problem that is solved only when all its children have been solved. An example of an AND node is one that adds the solutions from all subtrees expanded from this node. In contrast, an OR node represents a (sub) problem that is solved if any

[2] An acyclic graph is one without cycles. In general, some graphs with cycles can be searched, but this topic is beyond the present scope.

one of its children are solved. (The definitions of AND and OR nodes are taken from Martelli and Montanari[3]; the roles of the AND and OR nodes in an AND/OR tree are reversed in Nilsson's definitions[4].) Expanding a 0/1 knapsack problem by choosing an object to be included or excluded corresponds to transforming the problem from one state to another until the goal state is achieved. In this sense, the resulting tree contains only OR nodes.

To facilitate the design of multiprocessing systems for solving a search problem, the problem is transformed into one of the following paradigms according to the functions of the nodes[4-5].

(1) AND tree: All nonterminal nodes in the search tree are AND nodes. An example is a divide-and-conquer algorithm that decomposes a problem into subproblems and solves the original problem by combining the solutions of the subproblems.

(2) OR tree: All nonterminal nodes in the search tree are OR nodes. Branch-and-bound algorithms that systematically prune unnecessary expansions belong to this class.

(3) AND/OR graph: The nonterminal nodes are either AND or OR nodes. Game trees and logic programs can be represented as AND/OR trees. Dynamic programming problems can be solved as acyclic AND/OR-graph searches.

A node is active if its solution value has not been found; otherwise, it is terminated. In a serial algorithm, the set of active nodes is maintained in a single list. A heuristic value defined by a heuristic function is computed for each node. The active node with the minimum heuristic value is always expanded first. A search is called a depth-first search if the negation of the level number is used as the heuristic function. In this case, the nodes in the active list are expanded in a last-in/first-out order. A search is called a breadth-first search if the level number is used as the heuristic function. In this case, the nodes in the active list are expanded in a first-in/first-out order. Lastly, a lower bound can be computed for each node in the active list. This represents the lower bound of the best solution that can be obtained from this node. By using the lower-bound function as the heuristic function, a best-first search expands the node with the minimum lower bound.

Dominance relations. To reduce the search space, unnecessary expansions can be pruned by dominance relations. When a node $P_i$ dominates another node $P_j$, it implies that the subtree rooted at $P_i$ contains a solution node with a value no more (or less) than the minimum (or maximum) solution value of the subtree rooted at $P_j$.

As an example, consider two assignments, $P_1$ and $P_2$, on the same subset of the objects to be packed into a knapsack in the 0/1 knapsack problem. If the total profit of the objects assigned to the knapsack for $P_1$ exceeds that of $P_2$ and the total weight of the objects assigned in $P_1$ is less than that of $P_2$, then the best solution expanded from $P_1$ dominates $P_2$.

A special case of the special class dominance tests is the class of lower-bound tests, which are used in branch-and-bound algorithms to solve minimization problems. If a solution with value $v$ has already been found, then all active nodes with lower bounds greater than $v$ can be terminated, since they would not lead to better solutions. The minimum of the solution values obtained at any time can be conveniently kept in a single location, called the incumbent.

Problem representation. A problem can be represented in multiple forms. For example, the knapsack problem can be represented in an OR tree and solved by a branch-and-bound algorithm[6] or formulated in dynamic programming and solved by an acyclic AND/OR-graph search. As another example, the search of the extrema from a set of numbers can be solved by either a divide-and-conquer (AND-tree) algorithm[7] or a decision-tree (OR-tree) search. In general, the search procedures for various representations are equivalent, in the sense that they generate the same solution(s). Kumar and Kanal have shown that various heuristic search procedures for state-space representations (e.g., A*, SSS*[4]), AND/OR-graph searches (e.g., AO*[4]), and game-tree searches (e.g., $\alpha$-$\beta$[8]) are equivalent to branch-and-bound searches with dominance tests[9].

Efficiency in solving a given problem depends on the representation. Although efficient search procedures have been established for some problems, the general question of deciding which representation leads to an efficient search is still open, especially in multiprocessing. Although combinatorial search algorithms have relatively large computational overheads, compared with the input-output overheads, efficient architectures for evaluating various search algorithms differ. It is difficult to map search algorithms to general-purpose architectures, since they have different architectural requirements.

Our objective is, therefore, to obtain the functional requirements of various search algorithms. Based on these requirements, a general-purpose architecture can be assessed as to whether it suits a given search algorithm and the most efficient way of mapping the algorithm can be developed. Special-purpose architectures can also be developed from the functional requirements.

In presenting the performance results in this article, we usually assume synchronous models, although the search algorithm can be evaluated asynchronously. The performance results for synchronous models form a lower bound to those of asynchronous models. The article also covers the effects of heuristic and approximation functions on time and space efficiency and the important problems in selecting the nodes for expansion and defining the level of granularity.

## 2 Divide-and-conquer algorithms

The divide-and-conquer method is a well-known AND-tree representation that is used to solve many combinatorial search problems. Since a subproblem cannot be solved until all its descendants have been solved, divide-and- conquer algorithms can be viewed as bottom-up search procedures of an AND tree. There is no problem in deciding which node to evaluate in any step, as every node in the tree must be evaluated.

Studies conducted on parallel processing of divide-and-conquer algorithms can be classified into three types.

In the first, multiprocessors are connected

in the form of a tree, especially a binary tree, to exploit the potential parallelism of divide-and-conquer algorithms[10]. A tree machine has a simple interconnection that is suitable for VLSI implementation. However, the root processor is often a bottleneck for such problems as merge sorting, and the fixed structure is not flexible enough.

The second approach uses a virtual tree machine[11] consisting of a number of processors with private memory. The processors are connected by an interconnection network, such as the binary *n*-cube, and a suitable algorithm to decide when and where each subproblem should be solved. The hierarchy of process communications in divide-and-conquer algorithms allows them to be mapped effectively onto this architecture.

The third type is a variation of the above approaches. Here, all processors are connected to a common memory through a common bus[12]. Since data must be shared during execution, the memory or bus can become a bottleneck. Multimodule memory, augmented by caches, has been proposed to reduce the memory and bus contentions.

The functional requirement for evaluating an AND tree is, therefore, an interconnected conglomerate of processors. It is important to determine the granularity of parallelism—the minimum size of a subproblem that a processor evaluates in order to achieve optimal performance. The generally used criterion is processor utilization, $KT^2$, or $AT^2$, where $K$ is the number of processors, $T$ is the computational time, and $A$ is the area of a VLSI implementation. If granularity is large, then the processors can be loosely coupled; otherwise, tight coupling, as in systolic arrays, might be necessary.

Studies of the complexity of divide-and-conquer algorithms in an SIMD model[12] and the conditions that assure optimal processor use[13] suggest, generally speaking, that a parallel search of an AND tree can roughly be divided into three phases: start-up, computation, and wind-down. In the start-up phase, the problem is split and the tasks diffuse through the network. During the computation phase, all processors are kept busy until the number of tasks in the system is less than the number of processors. In the wind-down phase, the results are combined, and some processors might be idle. Processor utilization depends on the ratio between the amount of time spent in the computation phase and that spent in the other phases. The time complexity of searching a binary AND tree of $N$ leaves can be formulated in the following recursive equation:

$$\begin{gathered} T(N) = S(N) + 2T(\frac{N}{2}) + C(N), N>1, \\ T(1) = O(1) \end{gathered} \quad (1)$$

where $S(N)$ and $C(N)$ are the complexity of the start-up and wind-down phases. The granularity that results in the optimal processor utilization is related to the complexity of $S(N)$ and $C(N)$. In finding the sum or the maximum of $N$ numbers, $S(N)+C(N)=O(1)$; in using $N/(\log_2 N)$, processors achieve the maximum utilization[14]. In sorting $N$ numbers, $S(N)+C(N)=O(N)$, and $\log_2 N$ processors should be used to maximize processor utilization. We have studied the asymptotic processor utilization and found that $N/\log_2 N$ is a threshold when $S(N)+C(N)=O(N)$[15]. For $k$ (a function of $N$) processors, the processor utilization is one, between zero and one, or zero when the limiting ratio of $k$ and $N/\log_2 N$ is zero, greater than zero, or approaching infinity, respectively.

Since processor utilization increases with a decreasing number of processors, it is not an adequate measure of the effects of parallel processing. A more appropriate measure is the $KT^2$ criterion, which considers both processor utilization and computational time. We have proven that the asymptotic optimal number of processors to minimize $KT^2$ in parallel divide-and-conquer algorithms is $\theta$ $(N/\log_2 N)$, when $S(N)+C(N)=O(1)$[15]. Simulations have verified that the optimal number of processors is either exactly $N/(\log_2 N-1)$ or very close to this value.

## 3 Branch-and-bound algorithms

A branch-and-bound algorithm is a systematic search of an OR tree[16]. It is characterized by four constituents: a branching rule, a selection rule, an elimination rule, and a termination condition.

The selection rule examines the list of active subproblems (nodes) and selects one for expansion based on the heuristic value. For a serial search, the minimum number of nodes is expanded under a best-first strategy, provided all lower bounds are distinct[16]. This is achieved at the expense of increased memory space, as there are a large number of concurrently active subproblems. The algorithm is terminated when all active subproblems have been either expanded or eliminated.

The elimination rule prunes unnecessary expansions by means of lower-bound and dominance tests. For lower-bound tests, the incumbent $z$ holds the value of the best solution found so far in the search. In minimization problems, a lower bound is calculated for each subproblem when it is created. A subproblem cannot lead to the optimal solution if its lower bound exceeds the incumbent; such subproblems can be eliminated from further consideration. This lower-bound test can be relaxed by defining an allowance function, $\epsilon(z)$. Subproblems with lower bounds greater than $z-\epsilon(z)$ are eliminated, resulting in a suboptimal solution that deviates from the optimal solution by at most $\epsilon(z_O)$, where $z_O$ is the value of the optimal solution[17]. An example of an allowance function is the relative error deviation; a subproblem is terminated if its lower bound is greater than $z/(1+\epsilon)$. An allowance function is very effective in reducing the computational complexity of branch-and-bound algorithms. We have found that for some NP-hard problems under best-first searches, a linear reduction in accuracy of the solution results in an exponential reduction in the computational overhead[18-19].

Each of the four constituents of a serial branch-and-bound algorithm can be implemented by means of parallel processing.

(1) Parallel selection of subproblems: Multiple subproblems with the smallest heuristic values can be selected for expansion.

(2) Parallel expansion of subproblems.

(3) Parallel termination tests and update of the incumbent.

(4) Parallel elimination tests: These include the lower-bound and dominance tests.

We have studied the performance bounds of parallel branch-and-bound search, assuming (1) that only lower-bound tests are active, (2) that there is a single shared memory, (3) that no approximations are allowed, (4) that the subproblems are expanded synchronously, and (5) that the heuristic function is unambiguous[20].

Let $T_b(k)$ (resp. $T_d(k)$) be the number of iterations required to obtain the optimal solution under a best-first (resp. depth-first) search with $k$ processors. The following bounds have been derived:

(1) For a parallel best-first search, if the value of optimal-solution nodes differs from the lower bounds of other nodes, then

$$\left\lceil \frac{T_b(1)-1}{k}+1 \right\rceil \leqslant T_b(k) \leqslant \left\lceil \frac{T_b(1)}{k}+\frac{k-1}{k}h \right\rceil \tag{2}$$

where $h$ is the maximum number of levels in the branch-and-bound tree.

(2) For a parallel depth-first search, if all solution nodes exist at level $h$, then

$$\left\lceil \frac{T_b(1)-1}{k}+1 \right\rceil \leqslant T_d(k) \leqslant \left\lceil \frac{T_d(1)}{k}+\frac{(k+1)\cdot(c+1)}{k}h \right\rceil \tag{3}$$

where $c$ is the number of distinct incumbents obtained during the search. A similar equation can be also derived for parallel breadth-first searches. Eq's (2) and (3) show almost a $k$-time reduction in the number of iterations when parallel processing is applied on the same search strategy and when $T_b(1)/k$ is large.

The best search strategy depends on the accuracy of the problem-dependent lower-bound function. Very inaccurate lower bounds are not useful in guiding the search; very accurate lower bounds prune most unnecessary expansions. In both cases, the number of subproblems expanded by depth-first and best-first searches does not differ greatly. A depth-first search is better because it requires less memory space, in proportion to the height of the search tree. When the accuracy of the lower-bound function is moderate, a best-first search performs better. In this case, a good memory management system is necessary to support the memory space required.

Several architectures based on implicit enumeration have been proposed for parallel processing of branch-and-bound algorithms. These architectures delegate a subproblem to each processor, which reports to its parent processor when the evaluation is complete[21]. The limited degree of communication causes some processors to work on tasks that a better interconnection network would eliminate. Moreover, implicit enumeration is wasteful. Imai et al.[22] and El-Dessouki and Huen[23] have investigated parallel branch-and-bound algorithms based on a general-purpose network architecture with limited memory space and slow interprocessor communication. They used depth-first search, due to memory limitations.

Problems more efficiently evaluated by a parallel best-first search require more complex architectures. The design problems are a selection of subproblems with the minimum lower bounds and management of the required large memory space.

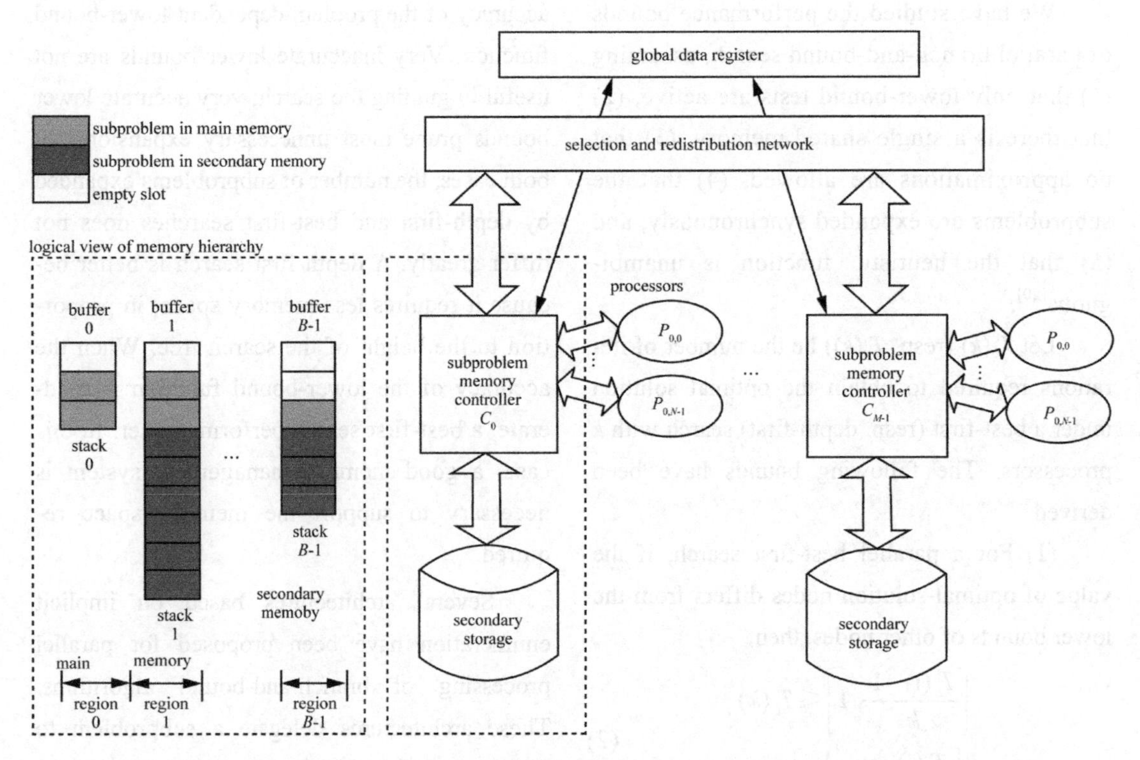


Figure 1. The architecture and logical structure of MANIP–a multiprocessor for parallel best-first search

MANIP-multiprocessor for parallel best-first search with lower bound tests only. Figure 1 shows the architecture of MANIP[18,24]. It consists of five major components: a selection and redistribution network, secondary storage, processors, global data register, and subproblem memory controllers.

The selection network selects subproblems with the minimum lower bounds for expansion in each iteration and connects the memory controllers for load balancing. Secondary storage holds excess subproblems that cannot be stored in the memory controllers. The memory controllers manage the local list of subproblems, maintain the secondary storage, and communicate with other controllers through the selection and redistribution network. The processors are general-purpose computers for partitioning subproblems and evaluating lower bounds. The global data register is accessible to all memory controllers and contains the value of the incumbent. To avoid contention during updates, this register can be implemented by a broadcast bus or a sequential associative memory. In the latter case, the minimum is found when the values of the feasible solutions are shifted out bit-serially and synchronously from all processors.

Two difficult issues must be solved in a parallel best-first search. First, the $k$ subproblems with the smallest lower bounds must be selected from the $N$ active subproblems in the system. Selection by software requires a time overhead of $O(N)$ in each iteration. A practical multistage selection network for selecting $k$ elements from $N$ elements requires $O(\log_2 N \cdot \log_2 k)$ time complexity and $O(N \cdot \log_2^2 k)$ hardware complexity[25].

A single-stage selection network can also be used. One or more subproblems with the minimum lower bounds in each processor are sent to the neighboring processors and inserted into their local lists. A maximum of $(k-1)$ shift-and-insert operations are needed to ensure that each processor has one of the $k$ subproblems with the smallest lower bounds[18]. Assuming that insertion is implemented in software, the time overhead in each iteration is $O(k \cdot \log_2 N)$. In all these cases, selection represents significant system overhead.

No-wait policy. Selection overhead is high; furthermore, the selection rule is based on a fallible lower-bound heuristic. Therefore, it might be more efficient not to follow the selection rule strictly; We propose a no-wait policy. Instead of waiting for one of the $k$ subproblems with the smallest lower bounds, each processor would expand the "most promising" subproblem in its local memory and initiate a fetch of the "most promising" subproblem from its neighbors. In this case, the most promising subproblem is the one with the minimum lower bound.

When the $k$ most promising subproblems are randomly distributed among the processors, the average fraction of processors containing one or more of the most promising subproblems is at least 0.63[18], resulting in a speedup proportional to $0.63k$. However, as expansion proceeds, the distribution might become nonrandom and require an interconnection network to randomize the distributions and balance the workload in the system. Experimental results on vertex-cover and knapsack problems have shown that the number of subproblems expanded increases by only about 10 percent when the above scheme replaces a complete selection. The performance is almost as good as that of a complete selection when the processors expand subproblems synchronously and perform one shift-and-insert operation for each subproblem expanded. The shift-and-insert operation can be overlapped with subproblem expansions and supported by a unidirectional ring network.

A second issue in implementing a best-first search lies in the management of the large memory space required. The multiprocessing model used to study this problem comprises a CPU, a main memory, a slower secondary memory, and a secondary-memory controller. The expected completion time of the branch-and-bound algorithm on this model is taken as the performance measure.

A direct implementation involving an ordered list of pointers to the subproblems results in poor locality of access, because the subproblems are not ordered by lower bounds in the virtual space. A better alternative is a special virtual memory that tailors its control strategies according to the locality of access[26]. However, this approach is inflexible, because the parameters of the control strategies are problem dependent. The inadequacies of these approaches are due, again, to strict adherence to the selection rule. We can also apply the no-wait policy here; it has resulted in the design of a modified branch-and-bound algorithm[27].

A modified algorithm. In this modified algorithm, the range of possible lower bounds is partitioned into b disjoint regions (Figure 1). The subproblems in each region are maintained in a separate list. The top portion of each list resides in the main memory, and the rest resides in the secondary memory. Due to the high overhead of secondary-storage accesses, subproblems in a list are expanded in a depth-first manner. To implement the no-wait policy, the modified selection rule chooses for expansion the subproblem in the main memory with the smallest lower bound. Since subproblems within a list are not sorted, the lower-bound elimination rule has to be modified.

Assuming that the new incumbent lies in the range of list $l$, all lists with indices greater than $l$ are eliminated. Subproblems in list $l$ with lower bounds greater than the incumbent are eliminated only when they are moved to the main memory during the expansion of list $l$. As a result, it is necessary to carry out the lower-bound test on each selected subproblem before it is expanded.

When one list is used, the modified algorithm is identical to a depth-first search; when infinity lists are used, it is identical to a best-first search. In general, as the number of lists increases, the number of subproblems expanded decreases and the overhead of the secondary-memory accesses increases. The number of lists should be chosen to maximize the overlap between computations and secondary-memory accesses. This overlap, in turn, depends upon the accuracy of the lower-bound function and the access times of the main and secondary memories. The accuracy of the lower-bound function is problem dependent and can be estimated from sample problems of the same type.

Experimental results on integer-programming and vertex-cover problems verify the usefulness of the modified algorithm. For vertex-cover problems, the lower-bound function is very accurate, so a depth-first search results in the best performance. For integer-programming problems, the lower-bound function is less accurate. As a result, more stacks (two to three) achieve best performance. The improvement in paging overhead over a direct implementation of the best-first search can exceed a factor of 100.

Experience with MANIP and prior studies show three functional requirements for efficient evaluation of branch-and-bound algorithms with only lower-bound tests: a loosely coupled interconnection of processors with load-balancing capability, a method of concurrent update, and broadcast of the incumbent.

Parallel dominance tests. When general dominance tests are used, it is necessary to keep the set of current dominating nodes (denoted by $N_d$) in memory. These are nodes that have been generated but not yet dominated. In general, $N_d$ can be larger than the set of active nodes. A newly generated node, $P_i$, has to be compared with all nodes in $N_d$ to see whether $P_i$ or any nodes in $N_d$ are dominated.

If $N_d$ is small, it can be stored in a bank of global data registers. However, centralized comparisons are inefficient when $N_d$ is large. A large

$N_d$ should then be partitioned into $k$ subsets, $N_d^{20},\cdots,N_d^{k-1}$, and distributed among the local memories of the $k$ processors. A subproblem, $P_{i,j}$, generated in processor $i$, is first compared with $N_d^i$; any subproblems in $N_d^i$ dominated by $P_{i,j}$ are removed. If $P_{i,j}$ is not dominated by a subproblem in $N_d^i$, it is sent to a neighboring processor and the process repeats. If it has not been dominated by any node in $N_d$, $P_{i,j}$ eventually returns to processor $i$ and is inserted into $N_d^i$.

The functional requirements for implementing parallel dominance tests depend on the size of $N_d$ and the structure of the dominance relation. When $|N_d|$ is small, broadcast buses or global registers carry important unstructured dominance tests, in which a dominance relation can exist between any pair of nodes. For structured dominance tests, it might be possible to partition the search tree and localize the dominance tests, but this poses additional complexity on the system architecture. On the other hand, when $|N_d|$ is large, it is necessary to partition $N_d$ into subsets and to perform the dominance tests in parallel. This results in tight coupling of the processors, because the transfer of newly generated nodes between processors must be synchronized and overlapped with computations.

Anomalies of parallelism in branch-and-bound algorithms. Since it is possible to overlap the communication overheads with computations for the various search strategies, the speedup of branch-and-bound algorithms can be measured by the ratio of the number of iterations of the best serial algorithm to that of the parallel algorithm under synchronous operations.

A $k$-fold speedup is expected when $k$ processors are used. However, simulations have shown that the number of iterations for a parallel branch-and-bound algorithm using $k$ processors can be more than the number of iterations of the best serial algorithm (this phenomenon is a detrimental anomaly); less than one-$k$th of the number of iterations of the best serial algorithm (an acceleration anomaly); or less than the number of iterations of the best serial algorithm, but more than one-$k$th of the number of iterations of the best serial algorithm (a deceleration anomaly)[20,28-29].

It is desirable to discover conditions that preserve acceleration anomalies, eliminate detrimental anomalies, and minimize deceleration anomalies.

Figure 2 gives an example of a detrimental anomaly. Let $g(P_i)$ be the lower bound of subproblem $P_i$ and $f(P_i)$ be the value of the best solution that can be obtained from $P_i$. Suppose that the best serial algorithm for the problem is a depth-first search. In a serial depth-first search, subtree $T_2$ is terminated by the lower-bound test of $P'_1$ as $f(P'_1)/(1+\epsilon)\leqslant g(P_2)$, when $\epsilon=0.1$. In a parallel depth-first search with two processors, a feasible solution, $P_4$, which terminates $P_1$ and $P'_1$, is found in the second iteration. Consequently, $P_2$ is not eliminated, since $P_1$ is not generated and $f(P_4)/(1+\epsilon)>g(P_2)$. Subtree $T_2$ has to be expanded; this eventually terminates subtree $T_3$. If $T_2$ is much larger than $T_3$, the time it takes to expand $T_2$ by using two processors exceeds the time to expand $T_3$ by using one processor.

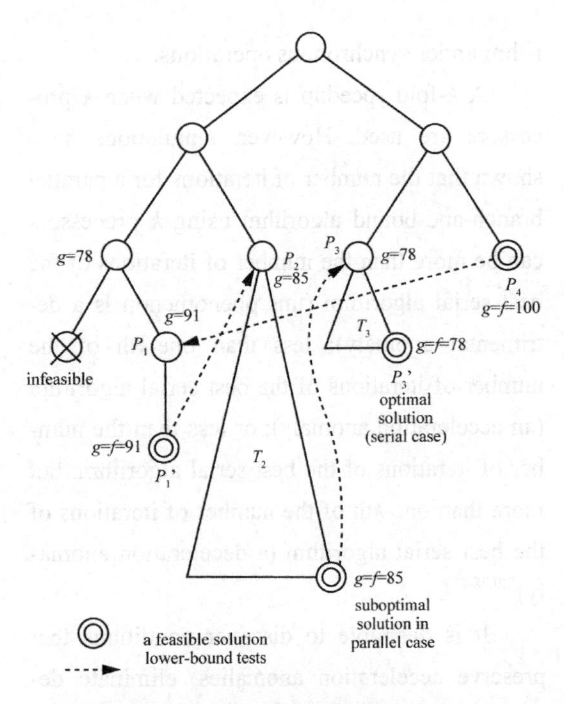


Figure 2. Example of a detrimental anomaly under a parallel depth-first search (allowance function $\epsilon$ =0.1)

Figure 3 shows an example of an acceleration anomaly. Subtree $T$ will be expanded in a serial depth-first search, as $f(P_3)/(1+\epsilon)>g(P_4)$ when $\epsilon=0.1$, but not in a parallel depth-first search with two processors, since $P_2$, and hence $T$, will be terminated by the lower-bound test with $P_6$: ($f(P_6)/(1+\epsilon)<g(P_4)$). If $T$ is very large, an acceleration anomaly will occur.

A heuristic function is unambiguous if all nodes in the search tree have distinct heuristic values. An elimination rule (lower-bound or dominance tests) is said to be consistent with the heuristic function if the elimination of $P_j$ by $P_i$ implies that $P_i$ is selected before $P_j$ in a serial search. Anomalies are caused by a combination of reasons: (1) there are multiple solution nodes; (2) the heuristic function is ambiguous; and (3) the elimination rule is not consistent with the heuristic function[2]. These conditions cause the tree to be searched in a different order in the serial and parallel cases.

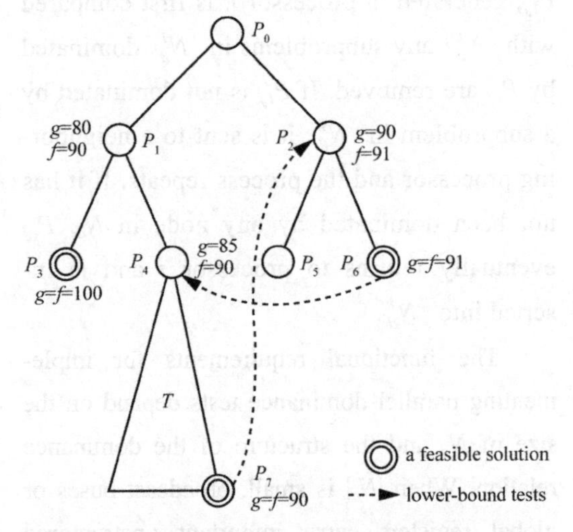


Figure 3. Example of an acceleration anomaly under a parallel depth-first search (allowance function $\epsilon$ =0.1)

We have discussed the conditions sufficient for eliminating detrimental anomalies and the conditions necessary to preserve acceleration anomalies in a previous article[20]; a brief summary is given here. Assume that the same search strategy is used in serial and parallel cases. For branch-and-bound algorithms with dominance tests, only a best-first search with the following conditions guarantees that detrimental anomalies will not occur:

(1) The heuristic function is unambiguous.

(2) Approximations are not allowed.

(3) The dominance relation is consistent with the heuristic function.

Ambiguity in the heuristic function can be resolved by augmenting the original heuristic function with a tie-breaking rule, say, by level and left-right orientation. For most problems, dominance relations that are consistent with the heuristic function can be designed. Acceleration anomalies can occur in one of following cases: when a breadth-first or depth-first search is used; when some nodes have identical lower bounds; when the dominance relation is inconsistent with the heuristic function; when multiple lists of subproblems are used; or when a suboptimal solution is sought.

## 4 AND/OR-tree search

Searching an AND/OR tree is more complex than searching an AND tree or an OR tree. An AND/OR tree is searched in two phases. The first is a top-down expansion, as in searching an OR tree; the second is a bottom-up evaluation, as in searching an AND tree. Due to the existence of both AND and OR nodes, a parallel search algorithm should combine the features of AND- and OR-tree searches. The presence of OR nodes demands that a good selection strategy be developed. The granularity of parallelism, like that of parallel divide-and-conquer algorithms, is an important consideration. Specific restrictions on a given problem, such as pruning rules, must be considered. These rules are usually more complicated, as more information is involved in the process.

When two AND/OR subtrees are searched concurrently, more work than necessary might be performed if pruning information obtained from one processor is unavailable to the other processor. The extra work is called information deficiency overhead. Pruning information can be exchanged by messages or through a common memory. Increased communication overhead needed for pruning is called information transfer overhead. In general, a tradeoff exists between the information-deficiency and information-transfer overheads. A good parallel AND/OR-tree search should weigh the tradeoffs—the merits of parallel processing against the communications overhead of obtaining the necessary pruning information.

Parallel $\alpha$–$\beta$ search. A two-person game between players MAX and MIN can be represented in a game tree in which the moves of MAX and MIN are put in alternate levels of the tree. In the corresponding AND/OR tree, OR modes represent board positions resulting from MAX's moves and AND nodes represent positions resulting from MIN's moves. All nonterminal MAX nodes take the maximum score of their children, while nonterminal MIN nodes take the minimum score. This minimax procedure is used to find the best move for the MIN player represented as the root[5].

A well-known technique to improve the efficiency of a minimax search is $\alpha$–$\beta$ pruning[4]. This technique uses two parameters, $\alpha$ and $\beta$, to define the search window. The $\alpha$ carries the lower bound of the MAX nodes; $\beta$ represents the upper bound of the MIN nodes. The game tree has solution values defined for the terminal nodes only and is searched in a depth-first fashion.

In expanding a MIN node, if the value returned by any of its children is less than $\alpha$, then this node can be pruned without further expansion. In this case, the value returned by this node to its parent—a MAX node—is less than $\alpha$ and another MAX node with value equal to $\alpha$ (according to the definition of $\alpha$) already exists. The $\beta$ is updated when a MIN node with a smaller value is found.

On the other hand, in expanding a MAX node, if the value returned by any of its children is greater than $\beta$, then this node can also be pruned. The $\alpha$ is updated when a MAX node with a larger value is found. The search is terminated when all nodes have been either pruned or expanded. The $\alpha$-$\beta$ search performs better when the initial search window is small.

The cost of searching a game tree depends on the distribution of values of the terminal nodes. The tree is said to have a best-case ordering if the first (or leftmost) branch from each node leads to the best value; it has a worst-case ordering if the rightmost branch from each node leads to the best value.

A number of parallel game-tree search techniques have been developed[30]. In the parallel aspiration search, the $\alpha$-$\beta$ window is divided into nonoverlapped subintervals, which are independently searched by multiple processors; Baudet reported that the maximum expected speedup is around five or six, regardless of the number of processors[31]. The speedup is limited because at least $W^{\lceil h/2 \rceil}+W^{\lfloor h/2 \rfloor}-1$ nodes must be evaluated for a uniform tree of depth $h$ and constant width $W$, even when $\alpha$ and $\beta$, are chosen to be the optimal minimax values[8]. Acceleration anomalies can also occur when the number of processors is small, say two or three.

Finkel and Fishburn have proposed a tree-splitting algorithm that maps a look-ahead tree onto a processor tree with the same interconnection structure[32]. The information-transfer overhead is small, due to the close match between the communications requirements and the interconnections. However, this is a brute-force search algorithm, and pruning is not considered in process assignments. The speedup drops to $\sqrt{k}$ under the best-case ordering, where $k$ is the number of processors.

In the mandatory-work-first scheme[33], the minimum tree searched in a serial algorithm is searched in parallel during the first phase. The resulting $\alpha$-$\beta$ window is used in the second phase, during which the rest of the tree is searched. This scheme performs better than the tree-splitting scheme under best-case ordering, but can be worse in worst-case ordering. In the latter case, many nodes pruned in the tree-splitting scheme might be visited in the second phase.

Another approach is to use a best-first search, such as the SSS* algorithm[34]. SSS* is effective in searching a randomly or poorly ordered tree, but requires more space and is not significantly better than an $\alpha$-$\beta$ search on strongly ordered trees. Kumar and Kanal have shown that the SSS* algorithm can be interpreted as a branch-and-bound procedure, and they have presented two parallel implementations of SSS*[35].

Previous approaches to parallel game-tree search have emphasized reduction of the information-transfer overhead, but paid little attention to information-deficiency overhead. We will consider the information-deficiency overhead in the illustrative context of the scheduling of parallel logic programs.

Parallel logic programs. Logic programming is a programming methodology based on Horn-clause resolution[36]. An example of a high-level language for logic programming is Prolog. Execution of a logic program can be considered as the search of an AND/OR tree[35,37].

The root represents the initial problem queried, the OR nodes represent (sub)goals, and the AND nodes represent clauses. All subgoals in the same body of a clause are children of an AND node. A (sub)goal (OR node) and its children display the choices of clauses with the same head. The terminal nodes denote clauses or subgoals that cannot be decomposed.

Searching an AND/OR tree for a logic program is quite different than searching other types of search trees. First, in contrast to extremum searches that find the best solution, solving a logic program corresponds to finding any or all solutions that satisfy the given conditions, the implicative Horn clauses, and the consistent binding of variables for the AND nodes. Second, the value of a node in the AND/OR tree for a logic program is either TRUE (success) or FALSE (failure). A node is usually selected for evaluation on the basis of a fixed order, such as the depth-first search. Third, a variable in a logic program can be bound to several values, and some subgoals might share a common variable.

An efficient search method must involve pruning. Two kinds of pruning exist here. In an AND pruning, if one of the children of an AND node is found to be FALSE, then all remaining children of this AND node can be pruned. Likewise, in an OR pruning, if one of the children of an OR node is found to be TRUE, then all remaining children of this OR node can be pruned. It should be noted that OR pruning applies only if the OR node shares no variables with its siblings.

Much research strives for parallel execution of logic programs. Conery and Kibler[37] have classified four kinds of parallelism of logic programs: AND parallelism, OR parallelism, stream parallelism, and search parallelism; they have also investigated AND parallelism. Furukawa et al.[38] and Ciepielewski et al.[39] have discussed OR parallelism, while Lindstrom et al.[40] have addressed stream parallelism and pipelined Prolog processors.

However, very few studies have addressed processor assignment as a means to reduce information-deficiency overhead. Below, we present an algorithm that schedules searches of nodes according to estimated probabilities of a terminal node being true and does not distinguish AND and OR parallelism.

A new scheduling algorithm. Consider the case in which all terminal nodes have the value TRUE. For a binary AND/OR tree of height $h$ ($h$ is even and the root is at level 0), the solution tree is found after $2^{h/2}$ terminal nodes have been visited, as shown in Figure 4(a). Once 1, 3, 9, and 11 have been visited, the root is determined to be true. In contrast, if all terminal nodes are FALSE, one can determine that the root is false by visiting $2^{h/2}$ terminal nodes (nodes 1, 2, 5, and 6 in Figure 4(a). These observations imply that when most of the terminal nodes in a subtree are TRUE, searching the subtree by assuming that its root is TRUE is more efficient; otherwise, the subtree should be searched by assuming that its root is FALSE.

For the AND/OR tree in Figure 4(a), we see that in a sequential search, if node 1 fails, then node 2 is examined; otherwise node 3 is examined next. That is, whether node 2 or node 3 is examined depends on the result of searching node 1. Similarly, the traversal of node 5 depends on the results of traversing nodes 1, 2, 3, and 4. A fail-token-flow graph, $G_f$, as depicted in

Figure 4(b) for the tree in Figure 4(a), can be drawn, according to this dependence information. A node (circle) in the graph is active only if it receives a fail-token from an incident edge. When a terminal node in the search tree is found to be false, a fail-token is sent along the direction of the corresponding edge. The coordinator (shaded box) in the graph represents a control mechanism that coordinates the activities of the connected blocks. When a fail-token is received from any of the incident edges of a coordinator, fail-tokens are sent to all directly connected nodes. At the same time, any node searched in the block directly connected to this coordinator can be terminated, because it does not belong to the solution tree. For example, when node 1 is found to be false, a fail-token is sent to node 2. If node 2 is found to be false, a fail-token is sent to coordinator $X_D$. Any node concurrently searched in block $D$ can then be terminated.

A simple parallel search strategy can be derived with the aid of $G_f$. To effectively search the tree, no more than $2^{h/2}$ processors are needed. A parallel depth-first search is applied in the first $h$ steps by generating all children of a selected AND node, but only the leftmost child of a selected OR node. As an example, nodes 1, 3, 9, and 11 in the search tree are assigned to four processors in the fourth step. This corresponds to generating fail-tokens to activate these nodes in $G_f$ (Figure 4(b)). If a node, say 3, is found to be FALSE, then a fail-token is generated and the idle processor is assigned to evaluate node 4. Close examination of Figure 4(b) shows that for each column of $G_f$ there must be at least one node with the value TRUE if a solution tree exists. When a node is found to belong to the solution tree, all nodes on the path from the initial start node to this node in $G_f$ must have failed. Processors for searching the AND/OR tree can be scheduled according to the state of execution in $G_f$ at any time.

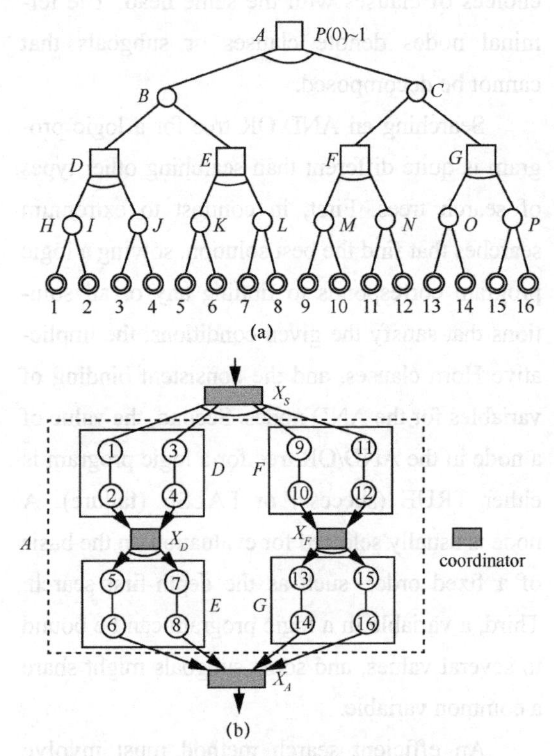


Figure 4. A binary AND/OR search tree. (a) AND nodes are represented as squared nodes; OR nodes are represented as circular nodes. (b) with high success probability and the corresponding fail-token-flow graph $G_f$

When the AND/OR tree is complete, and Pr($h$) (the probability that a terminal node is TRUE) is constant, Pr(0) (the probability for a solution tree is to be found from the root, which is assumed to be an OR node) can be shown to be close to one for Pr($h$) > 0.618. The threshold is 0.382 (=1 − 0.618) if the root is an AND node. In both cases, a node with the value TRUE can be found quickly in each column of $G_f$. As a result, the speedup is close to one.

On the other hand, if Pr($h$) is small, then the probability for a solution tree to exist at the root is close to zero and the above strategy is no longer suitable because a large number of nodes must be evaluated in each column of $G_f$. In this case, the scheduling should be done according to the success-token-flow graph, $G_s$. $G_s$, is the dual of $G_f$, in the sense that a success token replaces a fail token and the columns in $G_f$ are transposed to become the rows in $G_s$. Since searching for failure from an AND node is equivalent to searching for success from an OR node, the above scheduling algorithm can be extended with respect to $G_s$.

The token-flow graph obtained for the root of an AND/OR tree is modular and can be decomposed into modular token-flow subgraphs corresponding to all nonterminal nodes in the tree. If the probability of leading to a solution tree for a nonterminal node can be refined as the search progresses, the corresponding token-flow subgraph can be rederived. An idle processor can be scheduled according to the token-flow subgraph derived for the root of the given subtree. We have proposed a multiprocessor architecture, MALOP, which is based on an intelligent search strategy and effective scheduling[41].

In summary, the important issues in parallel AND/OR-tree search are the granularity of parallelism, the parallel selection of nodes for evaluation, and the intelligent pruning of unnecessary nodes. Processors should know the global state of search in order to select the nodes for expansion and should be able to tell other processors to prematurely terminate their tasks, when necessary. The architecture should support dissemination of this information.

# 5 Dynamic programming

Dynamic programming, a powerful optimization methodology, can apply to many areas, including optimal control, industrial engineering and economics[42]. In general, DP transforms the problem into a form suitable for optimization, but it is not an algorithm for optimizing the objective function. One can represent a problem solvable by DP as a multistage problem, a divide-and-conquer problem, or an acyclic AND/OR graph-search problem. Various computational approaches can be used, depending on the formulation and representation. We discuss DP problems separately here because they illustrate the effects of representation on the design of the supporting multiprocessing system.

A DP formulation is characterized by a recursive equation whose left-hand side identifies a function name and whose right-hand side is an expression containing the maximization (or minimization) of values of some monotone functions. Depending on the form of the functional equation, a DP formulation can be classified into four types: monadic-serial, polyadic-serial, monadic-nonserial, and polyadic-nonserial. Monadic and polyadic DP formulations are distinct approaches to representing various optimization problems; DP formulations can solve serial and nonserial optimization problems. Serial optimization problems can be decomposed into stages, and variables in one stage depend on variables in adjacent stages only. Problems such as sequential control, resource allocation, fluid flow, circuit design, and scheduling belong to this class. If variables in one stage are related to variables in other stages, the problem is a nonserial optimization problem. Examples include finding the optimal binary search tree and computing the minimum-cost order of multiplying a string of matrices.

To illustrate the concept of serial problems, consider the example of finding the shortest path in a multistage graph, as depicted in Figure 5(a). Let $c_{i,j}$ be the cost of edge $(i, j)$. The cost of a path from source $S$ to sink $T$ is the sum of costs on the edges of the path. Define $f_1(i)$ as the minimum cost of a path from $I$ to $T$. The cost of getting from $I$ to $T$ via neighbor $J$ is $c_{i,j}+f_1(j)$. To find $f_1(i)$, paths through all possible neighbors must be compared. Hence, the problem can be represented as

$$f_1(i) = \min_j \left[ c_{i,j} + f_1(j) \right] \tag{4}$$

This is a forward functional equation. The formulation is monadic; that is, the cost function involves one recursive term only.

From Eq. (4), $f(C_1)$, the minimum cost from $C_1$ to $T$ is

$$f(C_1) = \min \left\{ c_{1,1} + d_{1,1}, c_{1,2} + d_{2,1}, c_{1,3} + d_{3,1} \right\} \tag{5}$$

Eq. (5) can be interpreted as an inner-product operation in respect to addition and minimization. If we define matrix multiplication in terms of a closed semi-ring ($R$, MIN, +, $+\infty$ , 0) in which MIN corresponds to addition and + corresponds to multiplication in conventional matrix multiplications[7], then Eq. (5) becomes $f(C)=C{\cdot}D$, where $C$ is a cost matrix and $D$ is a cost vector. It is easy to see that searching the shortest path in a multistage graph with a forward monadic DP formulation is equivalent to multiplying a string of matrices, i.e., $A{\cdot}(B{\cdot}(C{\cdot}D))$.

The same problem can be generalized to find the optimal path from any vertex $i$ to any other vertex $j$. The functional equation is

$$f_2(i,j) = \min_k \left[ f_2(i,k) + f_2(k,j) \right] \tag{6}$$

where $f_2(i, j)$ is the minimum cost of getting from $I$ to $J$. This cost function is polyadic because it involves more than one recursive term. A divide-and-conquer formulation is a special case of polyadic-serial formulations.

AND/OR graphs can also be used to represent serial DP problems. Basic operations in comparisons of partial solutions over all alternatives are represented as OR nodes. AND nodes represent operations involving computations of a cost function, such as summations. Figure 5(b) shows an AND/OR graph for reducing the search of the shortest path in a three-stage graph with two nodes in each stage. Gensi and Montanari have shown that formulating a DP problem in terms of a polyadic functional equation is equivalent to searching for a minimal-cost solution tree in an AND/OR graph with monotone cost function[43].

AND nodes are represented as squared nodes and indicate summations; OR nodes are represented as circular nodes and indicate comparisons.

A nonserial DP problem can be represented in monadic or polyadic form[44]. A monadic-nonserial formulation is an extension of Eq. (4) in which the dependence of the functional term involves variables in more than one adjacent stage. A polyadic-nonserial formulation is usually represented in the form of an acyclic AND/OR graph in which edges can extend between any two arbitrary levels of the graph.

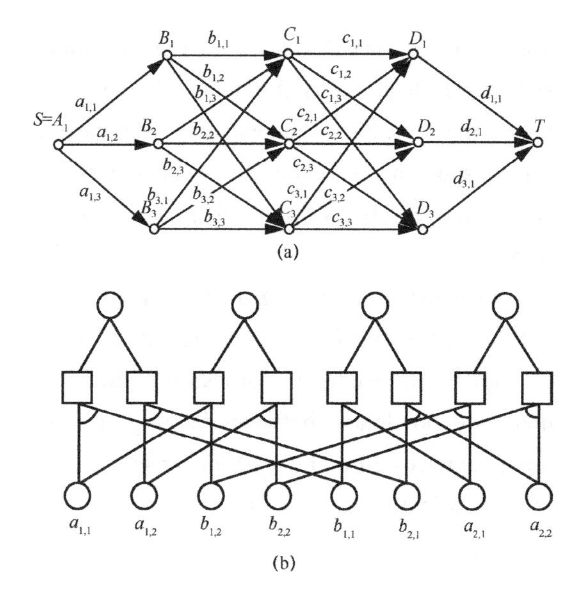


Figure 5. A graph with five stages and three nodes in each intermediate stage (a); an AND/OR-graph representation of the reduction in finding an optimal path in a three-stage graph (b). The problem in (b)is to find min $\{a_{i,j}+b_{j,k}\}$; $i,j,k\in\{1,2\}$

Parallel processing has been applied to DP problems. Guibas, Kung, and Thompson have proposed a VLSI algorithm for solving the optimal parenthesization problem[45], for which linear pipelines have also been proposed recently[46]. Clarke and Dyer have designed a systolic array for curve and line detection in terms of a nonserial formulation[47]. However, these designs were directed toward implementation of a few special cases of DP formulations.

The choice of an architecture to support a serial DP problem depends on the formulation. First, if the problem is represented in a polyadic form and considered a divide-and-conquer problem, the architecture discussed above under the divide-and-conquer heading can be applied. For example, the problem of finding the shortest path in a multistage graph can be considered as the multiplication of a string of matrices, which can be decomposed into the multiplication of two or more substrings of matrices.

Second, equivalence between polyadic representations and AND/OR graphs allows various graph-search techniques to be translated into techniques for solving DP problems. Sometimes, when the AND/OR graph is regular, it can be mapped directly into a systolic array[15].

Third, a problem can be represented in a monadic form and solved with a pipelining approach. This approach is suitable when many alternative partial solutions must be compared. Below, we illustrate this approach for evaluating the multiplication of a string of matrices.

Figure 6 depicts a scheme for computing $(A\cdot(B\cdot(C\cdot D)))$ for the multistage graph in Figure 5(a). An iteration is defined as a shift-multiply-accumulate operation in respect to the time at which a row or column of the input matrix enters a given processor in the systolic array. Note that the same iteration number is carried out at different times in different processors (iteration numbers are indicated in Figure 6(a).

In the first three iterations, $C\cdot D$ is evaluated. The control signal FIRST is one; $D$, the input vector is serially shifted into the systolic array; and the result vector, $\{f(C_i), i=1, 2, 3\}$, remains stationary. At the end of the third iteration, FIRST is set to zero. In the following three iterations, $B\cdot(C\cdot D)$ is computed. Note that matrix $B$ is transposed, and the $i$th column of matrix $B$ is fed into $K_i$. The input vector, $\{f(C_i), i=1, 2, 3\}$, remains stationary, while the result vector $\{f(B_i), i=1, 2, 3\}$, is shifted. At the end of the sixth iteration, the output vector $\{f(B_i), i=1, 2, 3\}$, is formed. In the last three iterations, input vectors A and $\{f(B_i), i=1, 2, 3\}$ are shifted into $P_1$ to

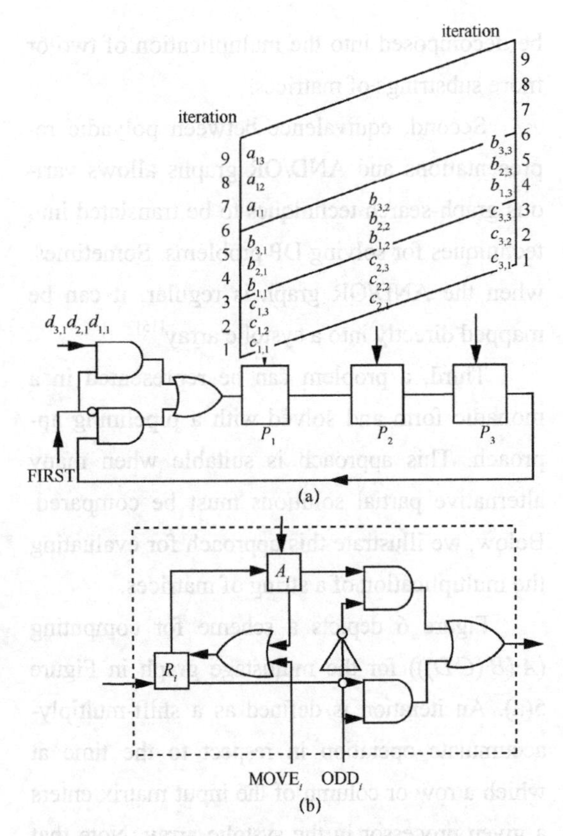


Figure 6. A pipelined version of a systolic array for multiplying a string of matrices (a); processor structure for $K_i$(b)

form the final result.

For the systolic array in Figure 6(a), shifted data alternates between the input vector and the result vector every three iterations. The processor structure of $K_i$, depicted in Figure 6(b), can control this alternation. $R_i$ is a register that stores an element of the input vector, and $A_i$ is the accumulator that stores the temporary result of an element of the result vector. Control signals ODD$_i$ and MOVE$_i$ control the data paths. When the number of matrix multiplications is odd, ODD$_i$ is one; hence, $R_i$ is connected to the output and the input vector is shifted along the pipeline. When the number of matrix multiplications is even, ODD$_i$ is zero, $A_i$ is connected to the output, and the result vector is shifted. At the end of a matrix multiplication, the generated result vector becomes the input vector in the next iteration and is moved, by the control signal MOVE$_i$, from $A_i$ to $R_i$.

In general, searching a multistage graph with $N$ stages and $k$ nodes in each stage takes $(N-1)k$ iterations with $k$ processors. Because there are no delays between feedings of the input matrices into the systolic array, a processor utilization is very close to one when $N$ and $k$ are large.

Few architectures solve nonserial DP problems directly. In an AND/OR graph representation of nonserial problems, edges may connect nodes at any two arbitrary levels. These graphs might have to be searched by an architecture with a flexible interconnection, such as a dataflow computer. Another approach is to transform the nonserial problem into a serial one and solve it with approaches developed for serial problems[15]. For problems in monadic-nonserial formulations, the dependence of variables can be removed by using one variable to represent the Cartesian product of several dependent variables. For problems in a polyadic-nonserial representation, such as an AND/OR graph, the dependence can be removed by replacing each edge that connects nodes not at adjacent levels with multiple edges that connect nodes at adjacent levels. This approach has been used in designing a systolic array for finding the optimal binary search tree[15].

Research in problem solving usually aims at developing better algorithms. Unnecessary combinatorial searches should be avoided, because they do not contribute to the quality of the solutions. Evidence of this is clear in the efforts to design optimal algorithms and to understand the reasoning process in artificial intelligence. However, searching becomes inevitable when a good algorithm has been developed and is an essential to many applications.

In this article, we have investigated the limitations of multiprocessing in solving combinatorial searches. The suitability of multiprocessing depends on problem complexity, problem representation, and the corresponding search algorithms. Problem complexity should be low enough that a serial computer can solve the problem. Problem representations are very important because they are related to the search algorithms.

However, the question of deciding which representation leads to an efficient search remains open. Moreover, efficient architectures to evaluate various search algorithms differ. That is why we have developed functional requirements for a given search algorithm, requirements that allow efficient mapping of a search algorithm on a general-purpose multiprocessor and development of special-purpose processors for searching.

In this article, we have not attempted to list all possible cases, but to illustrate the different approaches through examples (Table 1). We hope these guidelines and examples can help designers select appropriate multiprocessing systems for solving combinatorial search problems.

## References:


[1] SHAW A C. The logical design of operating systems[M]. Englewood Cliffs: Prentice-Hall, 1974.

[2] GAREY M R, JOHNSON D S. Computers and intractability: a guide to the theory of NP-completeness[M]. San Francisco: W. H. Freeman and Company, 1979.

[3] MARTELLI A, MONTANARI U. Additive AND/OR graphs[C]// The 3rd International Joint Conference on Artificial Intelligence. San Francisco: Morgan Kaufmann Publishers, 1973: 1-11.

[4] NILSSON N J. Principles of artificial intelligence[M]. [S.l.:s.n.], 1980.

[5] BARR A, FEIGENBAUM E A. The handbook of artificial intelligence[M]. Los Altos: William Kaufmann, 1981.

[6] HOROWITZ E, SAHNI S. Fundamentals of computer algorithms[M]. Potomac: Computer Science Press, 1978.

[7] AHO A V, HOPCROFT J E, ULLMAN J D. The design and analysis of computer algorithms[M]. Boston: Addison-Wesley Longman Publishing, 1974.

[8] KNUTH D E, MOORE R W. An analysis of alpha-beta pruning[J]. Artificial Intelligence, 1975, 6(4): 293-326.

[9] KUMAR V, KANAL L. A general branch-and-bound formulation for understanding and synthesizing AND/OR tree search procedures[J]. Artificial Intelligence, 1983, 21(1-2): 179-198.

[10] PETERS F. Tree machine and divide-and-conquer algorithms[C]// CONPAR'81: the Conference on Analysing Problem Classes and Programming for Parallel Computing. Heidelberg: Springer, 1981: 25-35.

[11] BURTON F W, HUNTBACH M M. Virtual tree machines[J]. IEEE Transactions on Computers, 1984, C-33(3): 278-280.

[12] HOROWITZ E, ZORAT A. Divide and conquer for parallel processing[J]. IEEE Transactions on Computers, 1983, C-32(6): 582-585.

[13] TANG C, LEE R C T. Optimal speedup of parallel algorithm based on the divide-and-conquer strategy[J]. Information Sciences, 1984, 32(3): 173-186.

[14] KUCK D J. A survey of parallel machine organization and programming[J]. ACM Computing Surveys, 1977, 9(1): 29-59.

[15] LI G J, WAH B W. Parallel processing for dynamic programming[C]// 1985 International Conference on Parallel Processing. Washington: IEEE Computer Society Press, 1985.

[16] LAWLER E L, WOOD D W. Branch-and-bound methods: a survey[J]. Operations Research, 1966, 14(4): 699-719.

Table 1. Functional requirements of different paradigms of search algorithms (The magnitudes of large and small granularities in various algorithms differ. Special interconnections include the tree architecture.)

| Algorithm | Functional requirements | | Tasks |
|---|---|---|---|
| Divide and conquer | Large granularity | Loosely coupled | Balance load |
| | Small granularity | Tightly coupled; special interconnections | Transfer control and data |
| Branch-and-bound | Lower-bound test only | Loosely coupled; broadcast capability | Balance load; share incumbent |
| | Dominance tests | Tightly coupled; shared memory | Balance load; share dominating nodes |
| Serial acyclic AND/OR-graph search | Large granularity | Loosely coupled; broadcast capability | Balance load; share state of evaluation |
| | Small granularity | Tightly coupled; special interconnections | Transfer control and data |
| Nonserial acyclic AND/OR-graph search | Large granularity | Dataflow processing | Share resources; coordinate tasks |
| | Small granularity | Map to serial AND/OR-graph search | Transfer control and data |

[17] IBARAKI T. Computational efficiency of approximate branch-and-bound algorithms[J]. Mathematical Operations Research, 1976, 1(3): 287-298.

[18] WAH B W, MA E Y W. MANIP-a multicomputer architecture for solving combinatorial extremum search problems[J]. IEEE Transactions on Computers, 1984, C-33(5): 377-390.

[19] WAH B W, YU C F. Probabilistic modeling of branch-and-bound algorithms[C]// 1985 IEEE Computer Software and Applications Conference. Washington: IEEE Computer Society Press, 1985.

[20] LI G J, WAH B W. Computational efficiency of parallel approximate branch-and-bound algorithms[C]// 1984 International Conference on Parallel Processing (ICPP'84). Washington: IEEE Computer Society Press, 1984: 473-480.

[21] DESAI B C. The BPU: a staged parallel processing system to solve the zero-one problem[C]// ICS 78. 1978: 802-817.

[22] IMAI M, FUKUMARA T, YOSHIDA Y. A parallelized branch-and-bound algorithm: implementation and efficiency[J]. System Computer Controls, 1979, 10(3): 62-70.

[23] EL-DESSOUKI O I, HUEN W H. Distributed enumeration on network computers[J]. IEEE Transactions on Computers, 1980, C-29(9): 818-825.

[24] WAH B W, LI G J, YU C F. The status of MANIP-a multicomputer architecture for solving combinatorial extremum-search problems[C]// The 11th Annual International Symposium on Computer Architecture. New York: ACM Press, 1984: 56-63.

[25] WAH B W, CHEN K L. A partitioning approach to the design of selection networks[J]. IEEE Transactions on Computers, 1984, C-33(3): 261-268.

[26] YU C F, WAH B W. Virtual-memory support for branch-and-bound algorithms[C]// 1983 IEEE Computer Software and Applications Conference. Washington:

IEEE Computer Society Press, 1983: 618-626.

[27] YU C F, WAH B W. Efficient branch-and-bound algorithms on a two-level memory hierarchy[C]// 1984 IEEE Computer Software and Applications Conference. Washington: IEEE Computer Society Press, 1984: 504-514.

[28] LI G J, WAH B W. How to cope with anomalies in parallel approximate branch-and-bound algorithms[C]// The 4th National Conference on Artificial Intelligence. Palo Alto: AAAI Press, 1984: 212-215.

[29] LAI T H, SAHNI S. Anomalies of parallel branch-and-bound algorithms[J]. Communications on the ACM, 1984, 27(6): 594-602.

[30] MARSIAND T A, CAMPBELL M. Parallel search of strongly ordered game trees[J]. ACM Computing Surveys, 1982, 14(4): 533-551.

[31] BAUDET G. The design and analysis of algorithms for asynchronous multiprocessors[R]. 1978.

[32] FINKEL R, FISHBURN J. Parallelism in alpha-beta search[J]. Artificial Intelligence, 1982, 19(1): 89-106.

[33] AKL S, BARNARD D, DORAN R. Design, analysis, and implementation of a parallel tree search algorithm[J]. IEEE Transactions on Pattern Analysis and Machine Intelligence, 1982, PAMI-4(2): 192-203.

[34] STOCKMAN G. A minimax algorithm better than alpha-beta?[J]. Artificial Intelligence, 1979, 12(2): 179-196.

[35] KUMAR V, KANAL L. Parallel branch-and-bound formulations for AND/OR tree search[J]. IEEE Transactions on Pattern Analysis and Machine Intelligence, 1984, PAMI-6(6): 768-778.

[36] KOWALSKI R. Logic for problem solving[M]. Amsterdam: North-Holland Publishing Company, 1979.

[37] CONERY J, KIBLER D. AND parallelism in logic programming[Z]. 1984.

[38] FURUKAWA K, NITTA K, MATSUMOTO Y. Prolog interpreter based on concurrent programming[C]// The 1st International Logic Programming Conference. [S.l.:s.n.], 1982: 38-44.

[39] CIEPIELEWSKI A, HARIDI S. Control of activities in the OR-parallel token machine[C]// IEEE International Symposium on Logic Programming. Piscataway: IEEE Press, 1984.

[40] LINDSTROM G, PANANGADEN P. Stream-based execution of logic programs[C]// The 1984 International Symposium on Logic Programming. Piscataway: IEEE Press, 1984: 168-176.

[41] LI G J, WAH B W. MALOP: a multicomputer architecture for solving logic programming problems[C]// 1985 International Conference on Parallel Processing. Washington: IEEE Computer Society Press, 1985.

[42] BELLMAN R, DREYFUS S. Applied dynamic programming[M]. Princeton: Princeton University Press, 1962.

[43] GENSI S, MONTANARI U, MARTELLI A. Dynamic programming as graph searching: an algebraic approach[J]. Journal of the ACM, 1981, 28(4): 737-751.

[44] BERTELE U, BRIOSCHI F. Nonserial dynamic programming[M]. New York: Academic Press, 1972.

[45] GUIBAS L, KUNG H, THOMPSON C. Direct VLSI implementation of combinatorial algorithms[C]// Caltech Conference on VLSI: Architecture, Design, Fabrication.[S.l.:s.n.], 1979: 509-525.

[46] VARMAN P, RAMAKRISHNAN V. Dynamic programming and transitive closure on linear pipelines[C]// 1984 International Conference on Parallel Processing. Washington: IEEE Computer Society Press, 1984: 359-364.

[47] CLARKE M, DYER C. Systolic array for a dynamic programming application[C]// The 12th Workshop Applied Imagery Pattern Recognition. [S.l.:s.n.], 1983.

发表于*IEEE Transactions on Computers*，1986年第6期

# Coping with anomalies in parallel branch–and–bound algorithms

Guo–jie LI, Benjamin W. WAH

**Abstract:** A general technique that can be used to solve a wide variety of discrete optimization problems is the branch-and-bound algorithm. We have adapted and extended branch-and-bound algorithms for parallel processing. The computational efficiency of these algorithms depends on the allowance function, the data structure, and the search strategies. Anomalies owing to parallelism may occur. In this correspondence, anomalies of parallel branch-and-bound algorithms using the same search strategy as the corresponding serial algorithms are studied. Sufficient conditions to guarantee no degradation in performance due to parallelism and necessary conditions for allowing parallelism to have a speedup greater than the number of processors are presented.

**Key words:** anomalies, approximation, branch-and-bound algorithms, heuristic search, lower-bound test, parallel processing

## 1 Introduction

The search for solutions in a combinatorially large problem space is a major problem in artificial intelligence and operations research. Generally speaking, these problems can be classified into two types. The first type is decision problems that decide whether at least one solution exists and satisfies a given set of constraints. Theorem-proving and evaluating queries for expert systems belong to this class. The second type is combinatorial extremum-search or optimization problems that are characterized by an objective function to be optimized and a set of constraints to be satisfied. Practical problems such as the traveling-salesman, warehouse-location, job-shop-scheduling, knapsack, vertex-cover, and integer-programming problems are examples in this class.

Exhaustive search is usually impractical and prohibitively expensive for solving large search problems, especially when the problem is NP-hard[1]. Studies on improving search efficiency is, thus, of considerable importance. Research has been conducted on designing a unified method for a wide variety of problems[2-4], the most general of which is the Branch-and-Bound (B&B) algorithm. This is a partitioning algorithm that decomposes a problem into smaller disjunctive subproblems and repeatedly decomposes until infeasibility is proved or a solution is found[5-6]. Backtracking, dynamic programming[7], and

This work was supported by the National Science Foundation under Grant ECS 81-05968.

AND-OR tree search[8] can be viewed as variations of B&B algorithms.

In implementing search algorithms for combinatorial searches, approximations and parallel processing are two major approaches to enhance their efficiency. Owing to the exponential nature of NP-hard problems, optimal solutions are usually infeasible to obtain. In practice, approximate solutions are acceptable alternatives. Our experimental results on the vertex-cover and 0-1 knapsack problems reveal that a linear reduction in accuracy results in an exponential reduction of the average computational time[9]. On the other hand, parallel processing is applicable when the problem is solvable in polynomial time, or when the problem is NP-hard but is solvable in polynomial time on the average[10], or the problem is heuristically solvable in polynomial time (such as game trees)[11]. It is impractical to use parallel processing to solve a problem with an exponential complexity because, in the worse case, an exponentialnumber of processors must be used to solve the problem in polynomial time.

Analytical properties of Parallel Approximate Branch-and-Bound (PABB) algorithms have been rarely studied. In general, a $k$-fold speedup (or ratio of the execution time in the serial case to that of the parallel case) is expected when $k$ processors are used. However, simulations have shown that the speedup for PABB algorithms using $k$ processors can be: a) less than one (called a detrimental anomaly)[12-14]; b) greater than $k$ (called an acceleration anomaly)[12-13]; or c) between one and $k$ (called a deceleration anomaly)[12-16]. It is desirable to discover conditions that preserve the acceleration anomalies, eliminate the detrimental anomalies, and minimize the deceleration anomalies.

One of the motivations of this correspondence is to improve Lai and Sahni's results on anomalies of parallel B&B algorithms[13]. First, Lai and Sahni have made an implicit assumption that all nonterminal nodes of the B&B tree have identical lower bounds, and hence the nonterminal nodes can be expanded in any order. Nearly all of their anomalies were discovered with an inconsistent selection strategy, namely, when $n_1$ processors were used, the expansion order of nonterminal nodes in a level was from left to right, but when $n_2$ processors were available, the expansion order was from right to left. We will show in Section 3 that all of their detrimental anomalies can be avoided by using a consistent selection order. Second, Lai and Sahni have only considered best-first searches and finding optimal solutions. However, our theoretical analysis and simulations have shown that anomalies are infrequent when optimal solutions are sought using best-first searches, while they are more frequent in approximate B&B algorithms with depth-first searches[17]. We will prove conditions that can avoid detrimental anomalies even when approximations are allowed. Since the anomalous behavior depends on the search strategies, we will investigate the anomalies with respect to the various search strategies separately. Lastly, Lai and Sahni have claimed that a near-linear speedup for parallel B&B algorithms with best-first search holds only for a "small" number of processors. On the contrary, we have shown that a near-linear speedup may hold for a large number of processors, and that the maximum number of processors to attain a near-linear speedup can be predicted[17].

The objective of this correspondence is to study conditions to cope with the anomalous behavior of B&B algorithms under approximations and parallel processing. Anomalies between the serial and parallel cases are studied with respect to the same search strategy. In general, anomalies should be studied with respect to the best serial algorithm and the best parallel algorithm (with possibly a different search strategy than that of the serial algorithm). However, conditions to resolve these anomalies would be problem dependent and may result in a large number of cases that cannot be enumerated. The conditions to resolve anomalies between $n_1$ and $n_2$ processors, $1 \leqslant n_1 < n_2$, are different from those presented here and are shown elsewhere[18]. These results on resolving anomalies are useful for designers to understand the existence of anomalies and to modify existing algorithms to prevent detrimental anomalies and enhance acceleration anomalies.

## 2 Parallel approximate branch-and-bound algorithms

Many theoretical properties of B&B algorithms have been developed by Kohler and Steiglitz[2] and Ibaraki[3,19]. A brief summary of these properties to solve a minimization problem were discussed elsewhere[20], and only the important properties that will be used in the following sections are stated here.

B&B algorithms can be characterized by four constituents: a branching rule, a selection rule, an elimination rule, and a termination condition. The first two rules are used to decompose problems into simpler subproblems and to appropriately order the search. The last two rules are used to eliminate generated subproblems that are not better than the ones already known. Appropriately ordering the search and restricting the region searched are the key ideas behind B&B algorithms.

The way in which $P_0$, the initial problem, is repeatedly decomposed into smaller subproblems can be represented as a finite rooted tree $\boldsymbol{B}=(\boldsymbol{P}, \boldsymbol{E})$ where $\boldsymbol{P}$ is a set of disjunctive subproblems, and $\boldsymbol{E}$ is a set of edges. The root of the tree is $P_0$. If a subproblem $P_{i_j}$ is obtained from $P_i$ by direct decomposition, then $(P_i, P_{i_j}) \in \boldsymbol{E}$. The level number of a node is the number of edges leading from the root to this node (the root is at Level 0). Let $f(P_i)$ be the value of the best solution obtained by evaluating all subproblems decomposable from $P_i$, $P_{i_j}$ be the $j$th subproblem directly decomposable from $P_i$, and $k_i$ be the total number of such subproblems ($k_i = |\{(P_i, x): (P_i, x) \in \boldsymbol{E}\}|$). Then $f$ satisfies

$$f(P_i) = \min_{j=1,\cdots,k_i} \left\{ f(P_{i_j}) \right\} \tag{1}$$

Each subproblem is characterized by a lower bound value that is computed from a lower bound function $g$. Let $\boldsymbol{T}$ be the set of all feasible solutions. The lower bound function satisfies the following properties:

1) $g(P_i) \leqslant f(P_i)$ for $P_i \in \boldsymbol{P}$

($g$ is a lower bound estimate of $f$) (2)

2) $g(P_i) = f(P_i)$ for $P_i \in \boldsymbol{T}$

($g$ is exact when $P_i$ is feasible) (3)

3) $g(P_i) \leqslant g(P_{i_j})$ for $(P_i, P_{i_j}) \in \boldsymbol{E}$

(lower bounds of descendents always increase) (4)

If a subproblem is a feasible solution with the best objective-function value so far, then the solution value becomes the incumbent $z$. The incumbent represents the best solution obtained so far in the process. Let $\boldsymbol{L}$ be the lower bound cutoff test. If a single solution is sought, the $P_j \boldsymbol{L} P_i$ means that $P_j$ is a feasible solution and that $f(P_j) < g(P_i)$. Using the best solution value obtained, $P_i$ is terminated during the computation if

$$g(P_i) \geqslant z \tag{5}$$

An approximate B&B algorithm is identical to the optimal B&B algorithm except that the lower bound testis modified. Hence, $P_j \boldsymbol{L} P_i$ if $f(P_j)/(1+\epsilon) \leqslant g(P_i)$ , $\epsilon \geqslant 0$ where $\epsilon$ is an allowance function specifying the allowable deviation of a suboptimal solution value from the optimal solution value. Using the incumbent, $P_i$ is terminated during the evaluation of the approximate B&B algorithm if

$$g(P_i) \geqslant \frac{z}{1+\epsilon}, \epsilon \geqslant 0, z \geqslant 0 \tag{6}$$

The final incumbent value $z_F$ obtained by the modified lower bound test deviates from the optimal solution value $z_O$ by

$$\frac{z_F}{1+\epsilon} \leqslant z_O \leqslant z_F \tag{7}$$

For example, suppose it were decided that a deviation of 10 percent from the optimum was tolerable. If a feasible solution of 150 is obtained, all subproblems with lower bounds of 136.4 (or 150/(1+0.1)) or more can be terminated, since they cannot lead to a solution that deviates by more than 10 percent from 150. This technique significantly reduces the amount of intermediate storage and time needed to arrive at a suboptimal solution.

Ibaraki has mapped breadth-first, depth-first, and best-first searches into a general form called the heuristic searches[3,21]. A heuristic function is defined to govern the order in which subproblems are selected and decomposed. The algorithm always decomposes the subproblem with the minimum heuristic value. In a best-first search, the lower bound values define the order of expansion, and the lower bound function can be taken as the heuristic function. In a breadth-first search, subproblems with the minimum level numbers are expanded first, and the level number can be treated as the heuristic function. Lastly, in a depth-first search, subproblems with the maximum level numbers are expanded first, and the negation of the level number can be taken as the heuristic function. If $\boldsymbol{U}$ is the current list of active subproblems in the process of expansion and $h$ is the heuristic function, then the selection function for a subproblem to be expanded in a serial B&B algorithm is

$$S_s(\boldsymbol{U}) = \left\{ P_{ji} \middle| h(P_i) = \min_{P_j \in \boldsymbol{U}} (h(P_j)) \right\} \tag{8}$$

B&B algorithms have inherent parallelism. Each of the four rules of a serial B&B algorithm can be enhanced by parallel processing.

1) Parallel Selection of Subproblems: In the parallel case, a set of subproblems less than or equal in size to the number of processors have to be selected for decomposition in each iteration. The selection problem is especially critical under a best-first search because a set of subproblems with the smallest lower bounds must be selected. The selection function in Eq. (8) becomes

$$s_P(\boldsymbol{U}) = \begin{cases} \underset{\substack{\text{where } h(P_i) < h(P_j), P_i, P_j \in \boldsymbol{U} \\ i \in \{i_1, \cdots, i_k\}, j \notin \{i_1, \cdots, i_k\}}}{\{P_{i_1}, \cdots, P_{i_k}\}} \quad , & \text{if } |\boldsymbol{U}| > k \\ \boldsymbol{U} \quad , & \text{if } |\boldsymbol{U}| \leqslant k \end{cases} \tag{9}$$

where $k$ is the number of processors. This returns the set of $k$ (or less) subproblems with the smallest heuristic values from $\boldsymbol{U}$.

2) Parallel Branching: The selected subproblems can be decomposed in parallel.

3) Parallel Termination Tests: Multiple infeasible nodes can be eliminated in each iteration. Further, multiple feasible solutions may be generated in an iteration, and the incumbent has to be updated in parallel to find a new incumbent.

4) Parallel Elimination Tests: The lower bound test [Eq's (5) or (6)] can be carried out in parallel by comparing lower bounds of multiple subproblems with the incumbent. However, the bounding functions are problem-dependent, and software implementation may be more flexible.

For simplicity, the case of searching a single optimal (or suboptimal) solution is discussed in this correspondence. The case in which all solutions are sought can be analyzed similarly. The parallel computational model used here consists of a set of processors connected to a shared memory. There is a single subproblem list shared by all processors. It is assumed that the processors operate synchronously in executing the steps of the PABB algorithm. In each iteration of the algorithm, multiple subproblems are selected and decomposed. The newly generated subproblems are tested for feasibility (and the incumbent updated if necessary), eliminated by lower bound tests, and inserted into the active list if not eliminated. In this model, eliminations are performed after branching instead of after selection as in Ibaraki's algorithm[19]. This reduces the memory space required for storing the active subproblems.

The shared memory in the proposed computational model may seem to be a bottleneck of the system. However, this model can be transformed into a second model in which all processors have a private memory and are connected by a ring network. The behavior of the PABB algorithm in the transformed model has been shown to be very close to that of the original model, except that very little interprocessor communication and interference are involved[16,20]. Since subproblems are decomposed synchronously and the bulk of the overhead is on branching operations, the number of iterations, which is the number of times that subproblems are decomposed in each processor, is an adequate measure in both the serial and parallel models. The speedup is thus measured by the ratio of number of iterations in the serial case to that of the parallel case.

## 3 Anomalies of PABB algorithms

In this section, some anomalies of the PABB algorithm are illustrated. Let $T(k,\epsilon)$ be the number of iterations required for expanding a B&B tree to find the first optimal (or suboptimal) solution, where $k$ is the number of processors and $\epsilon$ is the allowance function. Once the optimal solution is found, the time to drain the remaining subproblems from the active list is not accounted for here.

Figure 1 shows an example of a detrimental anomaly when approximations are allowed. In a serial depth-first search, subtree $T_2$ is terminated owing to the lower bound test of $P'_1$: $f(P'_1)/(1+\epsilon) \leqslant g(P_2)$ where $\epsilon = 0.1$. In a parallel depth-first search with two processors, a feasible solution $P_4$ is found first, and nodes $P_1$ and $P'_1$ are terminated owing to the lower bound test of $P_4$. Consequently, subtree $T_2$ has to be expanded, which will eventually prune subtree $T_3$. If the size of $T_2$ is much larger than that of $T_3$, the time taken to expand $T_2$ using two processors will be longer than that taken to expand $T_3$ using one processor. Strategies to handle these anomalies will be discussed in Section 5.2.

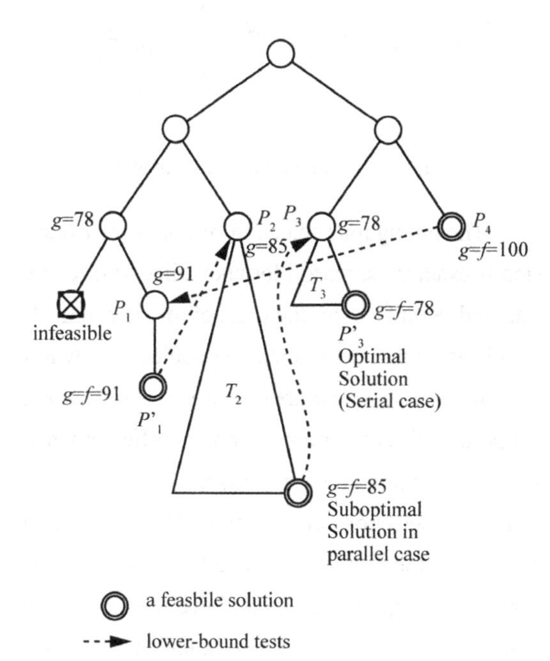


Figure 1. An example of a detrimental anomaly under a depth-first search with approximations ( $\epsilon$ =0.1)

Figure 2 shows an example of an acceleration anomaly when a depth-first search with approximations is used. When a single processor is used, subtree $T$ has to be expanded. However, when two processors are used, $P_2$ is expanded in the second iteration, and the feasible solution $P_4$ is found. Therefore, node $P_3$ and subtree $T$ will be eliminated by lower-bound tests with $P_4$. If subtree $T$ is large, then the speedup of using two processors over one processor will be much greater than two. Acceleration anomalies will be discussed in Section 6.

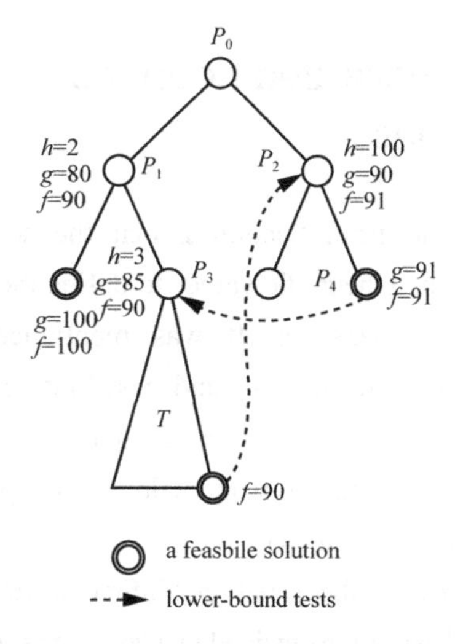


Figure 2. An example of an acceleration anomaly under a depth-first search or best-first with approximations ( $\epsilon$ =0.1)

Many examples to illustrate anomalies can be created for the various combinations of search strategies and allowance functions. However, the important consideration here is not in knowing that anomalies exist, but in understanding why these anomalies occur. Furthermore, it is desirable to preserve the acceleration anomalies and to avoid the detrimental anomalies. Our objective is to find the sufficient conditions to ensure that $T(k,\epsilon) \leqslant T(1,\epsilon)$, as well as the necessary conditions for $T(k,\epsilon) \leqslant T(1,\epsilon)/k$. The necessary conditions to eliminate detrimental anomalies and the sufficient conditions to preserve acceleration anomalies are not evaluated because they depend on the sequence of nodes expanded and the size of the resulting subtrees. Besides being impossible to enumerate due to the large number of possible combinations, these conditions are problem dependent and cannot be generalized to all problems.

## 4 Generalized heuristic searches

Recall from Section 2 that the selection function uses heuristic values to define the order of node expansions. It was mentioned that breadth-first, depth-first, and best-first searches are special cases of heuristic searches. These searches are potentially anomalous when parallel expansions are allowed.

Consider the serial depth-first search. The subproblems are maintained in a last-in-first-out list, and the subproblem with the maximum level number is expanded first. When multiple subproblems have identical level numbers (or heuristic values), the node chosen by the selection function depends on the order of insertion into the stack. If the rightmost child of a parent node is inserted first, then the leftmost child will be the node inserted last and expanded first in the next iteration.

In a parallel depth-first search, the mere extension of the serial algorithm may cause an anomalous behavior. For example, the order of expansion in a serial depth-first search for the tree in Figure 3 is *A*, *B*, *D*, *I*, *J*, *E*, etc. When two processors are used, nodes *B* and *C* are decomposed to nodes *D*, *E*, *F*, *G*, and *H* in the second iteration. Since these nodes have identical level numbers, the selection function can choose any two of these nodes in the next iteration. Suppose that they are inserted in the order *E*, *D*, *H*, *G*, and *F*. Then nodes *F* and *G* will be selected and expanded in the third iteration. This may cause an unexpected behavior as compared to the serial case. A similar example can also be developed for the best-first search when the lower bounds of nodes are identical.

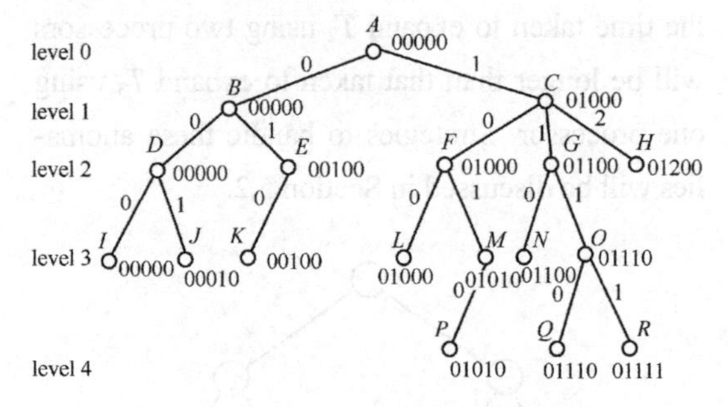

Figure 3. The path numbers of tree

The ambiguous selection of nodes for expansion is exactly the reason for anomalies reported by Lai and Sahni[13]. In their proof of Theorem 1, which states that detrimental anomalies can always exist when a larger number of processors are used, the nodes selected for expansion are different when a different number of processors are used. This change of selection order in their Theorem 1 (and almost all their other theorems) is based on the implicit assumption that nodes have identical lower bounds. In this case, the lower bounds are not useful in guiding the selection of subproblems. To have an accurate comparison when different number of processors are used, a consistent selection strategy must be used.

To resolve the ambiguity of the selection of subproblems, distinct heuristic values must be defined for the nodes. In this correspondence, a path number is proposed to uniquely identify a node. The path number of a node in a tree is a sequence of $d$+1 integers representing the path from the root to this node, where $d$ is the maximum number of levels of the tree. The path number $E = e_0e_1e_2\cdots e_d$ is defined, recursively, as follows. The root $P_0$ exists at Level 0 and has a path number of $E_0 = 000\cdots0$. A node $P_{i_j}$ on Level $l$, which is the $j$th child (counting from the left) of $P_i$ with path number $E_i = e_0e_1\cdots e_{l-1}000\cdots$, has path number $E_{i_j} = e_0e_1\cdots e_{l-1}j00\cdots$. As an example, the path numbers of all nodes in the tree of Figure 3 are shown outside the nodes.

To compare path numbers, the relations '>' and '=' must be defined. A path number $E_x = e_1^1e_2^1\cdots$ is less than another path number $E_y = e_1^2e_2^2\cdots (E_x < E_y)$ if there exists $0 \leqslant j \leqslant d$ such that $e_i^1 = e_i^2$, $0 \leqslant i < j$, and $e_j^1 < e_j^2$. The path numbers are equal if $e_i^1 = e_i^2$ for $0 \leqslant i \leqslant d$. For example, the path number 01000 is less than 01010. According to our definition of path numbers, nodes can have equal path numbers if they have the ancestor-descendant relationship. Since these nodes never coexist simultaneously in the active list of subproblems of a B&B algorithm, the subproblems in the active list always have unique path numbers.

The path number is now included in the heuristic function. The primary key is still the lower bound value or the level number. The secondary or ternary key is the path number and is used to break ties in the primary key.

$$h(P_i) = \begin{cases} \text{(level number, path number)}, & \text{breadth-first search} \\ \text{(path number)}, & \text{depth-first search} \\ \text{(lower bound, level number,} & \\ \text{path number) or} & \\ \text{(lower bound, path number)}, & \text{best-first search} \end{cases} \tag{10}$$

where the level number, path number, and lower bound are defined for $P_i$. For a best-first search, nodes with identical lower bounds can be searched in a breadth-first or depth-first fashion.

The heuristic functions defined above belong to a general class of heuristic functions satisfying the following properties:

$$h(P_i) \neq h(P_j) \text{ if } P_i \neq P_j,\ P_i, P_j \in \boldsymbol{P}$$

$$\text{(all heuristic values are distinct)} \tag{11}$$

$$h(P_i) \leqslant h(P_j) \text{ if } P_j \text{ is a descendant of } P_i$$

$$\text{(heuristic values do not decrease)} \tag{12}$$

In general, the results developed in the following sections are applicable to any unambiguous heuristic function satisfying Eq's (11) and (12), and are not restricted to the use of path numbers. For example, the lower bound can be used as the secondary key and the path number as the ternary key in a breadth-first search.

By using the heuristic function defined in Eq. (10), the study of anomalies for depth-first, best-first, and breadth-first searches can be unified. The results developed in the following sections apply to all these search strategies.

# 5 Sufficient conditions to eliminate detrimental anomalies

## 5.1 Parallel B& B algorithms with lower bound tests only

In this section, we show that any heuristic search with an unambiguous heuristic function can guarantee that $T(k,\epsilon) \leqslant T(1,\epsilon)$ when only lower bound tests with $\epsilon = 0$ are used. A basic node is the node with the smallest heuristic value in each iteration. Let $\Phi^1$ and $\Phi^k$ be the sets of nodes expanded in the B&B tree using one and $k$

processors, respectively. To show that $T(k, 0) \leqslant T(1, 0)$, it suffices to prove a) that at least one node belonging to $\Phi^1$ is expanded in each iteration of the parallel search; and b) that once all the nodes in $\Phi^1$ are expanded or terminated, the parallel heuristic search must terminate. The proof requires the following property on basic nodes.

**Lemma 1:** Let $P_i$ be a basic node, then for any node $P_j$ such that $h(P_j) < h(P_i)$, $P_j$ must be either expanded or terminated when $P_i$ is expanded.

**Proof:** Suppose that in the current active list, $\boldsymbol{U}$, $P_i \in \boldsymbol{U}$ is a basic node. Assume that there exists a node $P_j$ such that $h(P_j) < h(P_i)$ and that $P_j$ has not been expanded or terminated when $P_i$ is expanded. Since $P_i$ has the minimum heuristic value among the active nodes in $\boldsymbol{U}$, $P_j$ must not be active at that time. That is, $P_j$ is a descendant of some node $P_k$, $P_k \in \boldsymbol{U}$, and $h(P_i) < h(P_k)$. By Eq. (12), $h(P_i) < h(P_k) \leqslant h(P_j)$, which contradicts the assumption that $h(P_j) < h(P_i)$.

The following theorem proves that any unambiguous heuristic functions satisfying Eq's (11) and (12) are sufficient to eliminate detrimental anomalies.

**Theorem 1:** Let $\epsilon = 0$, i.e., an exact optimal solution is sought. $T(k,0) \leqslant T(1,0)$ holds for any parallel heuristic search with a heuristic function satisfying Eq's (11) and (12).

**Proof:** The proof is by contradiction. Suppose there exists a basic node $P_{i_1}$ in the parallel search such that $P_{i_1} \notin \Phi^1$ and that $P_{i_1} \in \Phi^k$ (see Figure 4). This means that either $P_{i_1}$ or its an ancestor is terminated by a lower bound test in the serial case. Hence, there must exist a feasible solution $P_{i_2} \in \Phi^1$ such that $f(P_{i_2}) \leqslant g(P_{i_1})$ and that $P_{i_2}$ has not been obtained when $P_{i_1}$ is expanded in the parallel case. It implies that a proper ancestor $P_{i_3} \in \Phi^1$ of $P_{i_2}$ exists in the serial case such that $h(P_{i_3}) \leqslant h(P_{i_1})$, and that $P_{i_2}$ is obtained before $P_{i_1}$ and terminates $P_{i_1}$. Since $P_{i_1}$ is a basic node in the parallel search, $h(P_{i_3}) < h(P_{i_1})$, and $P_{i_3}$ has not been expanded when $P_{i_1}$ is expanded in the parallel case, $P_{i_3}$ must be terminated according to Lemma 1. For the parallel search, there must exist a feasible solution $P_{i_4} \in \Phi^k$ such that $f(P_{i_4}) \leqslant g(P_{i_3})$, and that $P_{i_4}$ has not been obtained when $P_{i_3}$ is expanded in the serial case. Two cases are possible:

First, $P_{i_4}$ is not generated when $P_{i_3}$ is expanded in the serial case, i.e., a proper ancestor $P_{i_5} \in \Phi^k$ of $P_{i_4}$ exists when $P_{i_3}$ is active and $h(P_{i_5}) > h(P_{i_3})$. According to the properties of lower bound functions [Eq's (2), (3), and (4)], we have $f(P_{i_4}) \leqslant g(P_{i_3}) \leqslant f(P_{i_3}) \leqslant f(P_{i_2}) \leqslant g(P_{i_1})$. Moreover, in the parallel case, $P_{i_4}$ should be obtained before $P_{i_1}$ is expanded (otherwise, $P_{i_3}$ would not be terminated by $P_{i_4}$). Hence $P_{i_1}$ has to be terminated by $P_{i_4}$ in the parallel algorithm, which contradicts the assumption that $P_{i_1} \in \Phi^k$.

Second, $h(P_{i_5}) < h(P_{i_3})$, and $P_{i_5}$, as well as its descendant $P_{i_4}$, have been terminated in the serial case and not in the parallel case. We can then apply the above argument again to $P_{i_5}$ and eventually obtain a sequence of nodes $P_{i_1}, P_{i_2}, \cdots, P_{i_m}$ as depicted in Figure 4, in which $P_{i_m}$ is not terminated by any lower bound test. There are three possibilities:

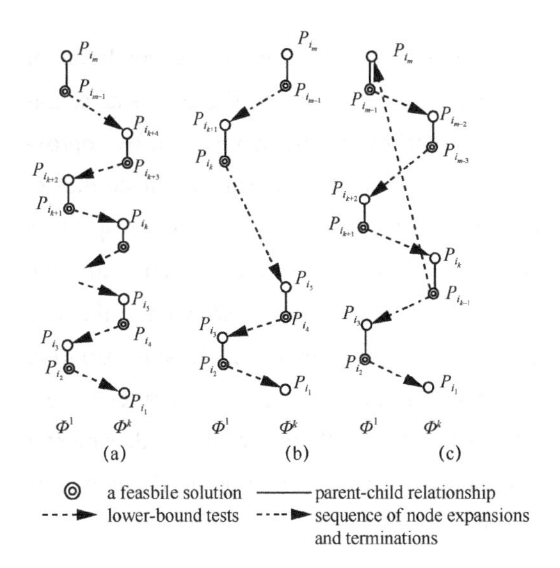


Figure 4. Proof of Theorem 1

1) The first node $P_{i_m}$ occurs in the serial case [Figure 4(a)]. Since $P_{i_{m-1}}$ is a feasible solution, we have: $g(P_{i_m}) \leqslant f(P_{i_{m-1}}) \leqslant g(P_{i_{m-2}}) \leqslant f(P_{i_{m-2}}) \leqslant g(P_{i_{m-3}}) \leqslant \cdots \leqslant f(P_{i_1}) \leqslant g(P_{i_1})$. Further, since $h(P_{i_{k+2}})<h(P_{i_k})$ (otherwise, $P_{i_k}$ could not have been terminated by $P_{i_{k+1}}$ in the serial case) and since $h(P_{i_{k+4}})<h(P_{i_{k+2}})$ (by the same argument as $h(P_{i_5})<h(P_{i_3})$ ), we have $h(P_{i_{k+4}})<h(P_{i_k})$. Repeating this, we get $h(P_{i_m})<h(P_{i_1})$. By Lemma 1, $P_{i_m}$ must be expanded in the parallel case and terminates $P_{i_1}$, which contradicts that $P_{i_1} \in \Phi^k$.

2) The first node $P_{i_m}$ occurs in the parallel case, Figure 4(b). Similar to the argument for $P_{i_4}$ discussed before, we can explain that $P_{i_{m-1}}$ has been obtained when $P_{i_1}$ is selected. Therefore, $P_{i_1}$ must be terminated in the parallel case, which contradicts that $P_{i_1} \in \Phi^k$.

3) There is a cycle of cutoffs such that $P_{i_{m-1}} \boldsymbol{L} P_{i_{m-2}}$, $P_{i_{m-3}} \boldsymbol{L} P_{i_{m-4}}, \cdots, P_{i_{k+1}} \boldsymbol{L} P_{i_k}$, and $P_{i_{k-1}} \boldsymbol{L} P_{i_m}$ where **L** denotes a lower bound cutoff test [Figure 4(c)]. By transitivity, we have $f(P_{i_m}) \leqslant f(P_{i_{m-1}}) \leqslant f(P_{i_{k-1}}) \leqslant f(P_{i_m})$, which implies that $f(P_{i_m}) = f(P_{i_{m-1}}) = f(P_{i_{k-1}})$. The heuristic value of all nodes of the cycle are less than $h(P_{i_1})$, so a feasible solution has been obtained before $P_{i_1}$ is selected. Thus, $P_{i_1}$ must be terminated in the parallel case, which contradicts that $P_{i_1} \in \Phi^k$.

So far, we have proved that at least one node in $\Phi^1$ is expanded in each iteration of a parallel heuristic search. Since approximation is not allowed, the optimal-solution node cannot be eliminated by lower bound tests. Hence during a parallel heuristic search, once all nodes in $\Phi^1$ are either expanded or terminated, the optimal-solution node must be found. The remaining unexpanded nodes do not belong to $\Phi^1$ because their lower bounds are greater than the optimal solution. The parallel heuristic search is thus completed at this time.

The above theorem shows that detrimental anomalies can be avoided for depth-first, breadth-first, and best-first searches with $\epsilon = 0$ by augmenting the heuristic function with an unambiguous function. As a special case, for a best-first search in which all nodes have distinct lower bounds, the node with the smallest lower bound can always be selected from the priority queue. In this case, the path numbers do not have to be used, and no detrimental anomaly will occur. Due to space limitation, the performance bounds of the parallel best-first search will not be shown here[17-18].

## 5.2 Parallel B&B algorithms with approximations

When parallel approximate B&B algorithms are considered, Theorem 1 is no longer valid all the time (see the example in Figure 1). The reason for the detrimental anomaly is that **L**, the lower bound tests under approximation, are not transitive. That is, $P_i\boldsymbol{L}P_j$ and $P_j\boldsymbol{L}P_k$ do not imply $P_i\boldsymbol{L}P_k$, since $f(P_i)/(1+\epsilon)\leqslant g(P_j)$ and $f(P_j)/(1+\epsilon)\leqslant g(P_k)$ implies $f(P_i)/(1+\epsilon)^2\leqslant g(P_k)$ rather than $f(P_i)/(1+\epsilon)\leqslant g(P_k)$. Somewhat surprisingly, it is possible that $\Phi^1$ and $\Phi^k$ are almost disjoint, and most of the nodes in $\Phi^1$ are not expanded in the parallel case. The following theorem shows that detrimental anomalies can be avoided for a best-first search.

**Theorem 2:** $T(k,\epsilon)\leqslant T(1,\epsilon)$, $\epsilon>0$, holds for best-first searches if the heuristic function satisfies Eq's (11) and (12).

**Proof:** The key idea of this proof is to show that a detrimental anomaly cannot occur although transitivity of lower bound tests is not valid here. Suppose that there exists a basic node $P_{i_1}$ in $\Phi^k$ and not in $\Phi^1$. There are two cases as described in the proof of Theorem 1 (see Figure 4).

First, assume that $h(P_{i_5})>h(P_{i_3})$. Since the relations $P_{i_4}\boldsymbol{L}P_{i_3}$ and $P_{i_2}\boldsymbol{L}P_{i_1}$ exist, and $P_{i_3}$ is selected before $P_{i_1}$ in the serial case (for a best-first search), it is true that $g(P_{i_3})\leqslant g(P_{i_1})$. This implies that $f(P_{i_4})/(1+\epsilon)\leqslant g(P_{i_1})$. $P_{i_4}$ can be shown to be available before $P_{i_1}$ is expanded by the same argument as in the proof of Theorem 1. Hence, $P_{i_1}$ must be eliminated by $P_{i_4}$ in the parallel case, a contradiction!

Second, assume that $h(P_{i_5})<h(P_{i_3})$. The argument here is similar to the proof of Theorem 1 except that the lower bound test to be used is $f(P_i)/(1+\epsilon)\leqslant g(P_j)$ and not $f(P_i)\leqslant g(P_j)$.

For depth-first searches, the condition of Theorem 2 is not sufficient. Figure 1 is an example of a detrimental anomaly caused by approximations. In breadth-first searches, since the sequences of feasible solutions with respect to serial and parallel cases are the same, and the minimal feasible solution is selected as the new incumbent if more than one feasible solution are obtained in an iteration of a parallel search, approximations will not result in detrimental anomalies when the condition in Theorem 2 is satisfied.

## 6 Necessary conditions for acceleration anomalies

In this section, the necessary conditions for $T(k, 0)<T(1, 0)/k$ are developed. One condition is based on the complete consistency of heuristic functions. A heuristic function $h$ is said to be consistent (respectively, completely consistent) with the lower bound function $g$ if $h(P_i)<h(P_j)$ implies that $g(P_i)\leqslant g(P_j)$, (respectively, $g(P_i)<g(P_j)$) for all $P_i, P_j\in \boldsymbol{P}$. A heuristic function $h$ is said to be not completely consistent with $g$ if there exist two nodes $P_i$ and $P_j$ such that $h(P_i)>h(P_j)$ and $g(P_i)\leqslant g(P_j)$. Note that if there are nodes with equal lower bounds, then the heuristic function for a serial best-first search is consistent, but not completely consistent, with the lower bound function.

**Theorem 3:** The necessary condition for $T(k, 0)<T(1, 0)/k$ is that the heuristic function $h$ is not completely consistent with $g$.

**Proof:** An acceleration anomaly does not exist if $\Phi^1\subset\Phi^k$ because at least $\lceil|\Phi^1|/k\rceil$

iterations are needed to expand the nodes in $\Phi^1$. Hence, the proof is based on the assumption that a node $P_{i_1} \in \Phi^1$ exists and that $P_{i_1} \notin \Phi^k$. This means that $P_{i_1}$ is terminated by a lower bound test in the parallel case. That is, there is a feasible solution node $P_{i_2} \in \Phi^k$ such that $f(P_{i_2}) \leqslant g(P_{i_1})$, and $P_{i_2}$ does not exist in the serial case when $P_{i_1}$ is expanded. Let $P_{i_3}$ be the immediate parent of $P_{i_2}$. Referring to Figure 5, two cases are possible.

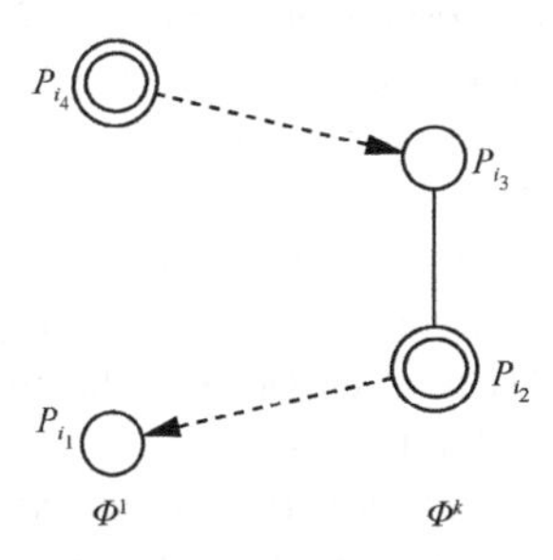


Figure 5. Proof of Theorem 3 (the notation used is similar to those of Figure 4)

First, $h(P_{i_3}) > h(P_{i_1})$. This means that $P_{i_3}$ has not been generated when $P_{i_1}$ is selected in the serial case. Since $P_{i_2} \boldsymbol{L} P_{i_1}$, thus $g(P_{i_3}) \leqslant g(P_{i_2}) = f(P_{i_2}) \leqslant g(P_{i_1})$, which implies that $h$ is not completely consistent with $g$. Second, $h(P_{i_3}) < h(P_{i_1})$. In order for $P_{i_1}$ to exist in the serial case, $P_{i_4} \in \Phi^1$ must exist such that $P_{i_4} \boldsymbol{L} P_{i_3}$ and that $h(P_{i_4}) < h(P_{i_3})$. By the transitivity of lower bound tests, $P_{i_4} \boldsymbol{L} P_{i_1}$. This contradicts the assumption that $P_{i_1} \in \Phi^1$.

The significance of Theorems 1 and 3 is in showing that acceleration anomalies may exist and that detrimental anomalies can be prevented for depth-first searches when no approximation is allowed and an unambiguous heuristic selection function is used. For a best-first search without approximation, detrimental anomalies can be prevented by using an unambiguous heuristic selection function; however, acceleration anomalies may exist when there are nonsolution nodes of the B&B tree with lower bounds equal to the optimal solution value, since these nodes have heuristic values that are not completely consistent with their lower bound values. For a breadth-first search without approximation, no acceleration anomaly occurs because at least one node belonging to $\Phi^1$ must be expanded in each iteration of a parallel breadth-first search.

It is important to note that the conditions in Theorem 3 are not necessary when approximate solutions are sought. An example in Figure 2 shows the existence of an acceleration anomaly when $\epsilon = 0.1$ and $h$ is completely consistent with $g$ (since a best-first search is used and all lower bounds are distinct). In this case, the additional necessary condition is that the lower bound test with approximation is inconsistent with $h$; that is, there exist $P_i$ and $P_j$ such that $h(P_i) > h(P_j)$ and that $P_i \boldsymbol{L} P_j$. Clearly, this necessary condition is weak and can be satisfied easily.

# 7 Conclusion

Branch-and-bound algorithms belong to a general class of algorithms that can solve a wide variety of combinatorial-search problems. Two major ways to improve the search efficiency are guiding the search by a heuristic function (or selection rule) and narrowing the search space (or elimination rules). Anomalies due to parallelism may occur because of the identical heuristic values of different subproblems and the inconsistencies between the selection and elimination rules.

In this correspondence anomalies between the serial and parallel cases for the same search strategy are studied. A method is proposed to assign a distinct heuristic value to each node and to uniquely identify a node in the B&B tree. It has been proved that detrimental anomalies can be prevented for a best-first or breadth-first search by using an unambiguous heuristic function, even when approximations are allowed. For a depth-first search, detrimental anomalies can only be prevented when an exact optimal solution is sought. On the other hand, acceleration anomalies may occur when one of the following conditions is true: 1) a depth-first search is used; 2) a best-first search is used, and some nonsolution nodes have lower bounds equal to the optimal solution value; or 3) a suboptimal solution is sought.

Although our results have been proved with respect to a system in which all subproblems are maintained in a single list, they apply to a system in which multiple subproblem lists are used. When there are multiple lists, one for each processor, a subproblem with the minimum heuristic value is selected from each local list for decomposition. This subproblem may not belong to the global set of active subproblems with the minimum heuristic values, but the subproblem with the minimum heuristic value will always be expanded by one of the processors. Further, when multiple lists are used, it is not difficult to maintain a global incumbent in a global data register[16,20]. Hence the behavior of using multiple lists is analogous to that of a centralized list.

Due to the limitation of space, deceleration anomalies, anomalies due to dominance relations, and anomalies when $k_1$ and $k_2$, $1 \leqslant k_1 < k_2$, processors are used are not discussed here. These results can be found elsewhere[17-18].

## References:


[1] GAREY M R, JOHNSON D S. Computers and intractability: a guide to the theory of NP-completeness[M]. San Francisco: W. H. Freeman and Company, 1979.

[2] KOHLER W, STEIGLITZ K. Characterization and theoretical comparison of branch-and-bound algorithms for permutation problems[J]. Journal of the ACM, 1974, 21(1): 140-156.

[3] IBARAKI T. Theoretical comparisons of search strategies in branch-and-bound algorithms[J]. International Journal of Computer and Information Sciences, 1976, 5(4): 315-344.

[4] SEKIGUCHI Y. A unifying framework of combinatorial optimization algorithms; tree programming and its validity[J]. Journal of the Operation Research Society of Japan, 1981, 24(1): 67-94.

[5] LAWLER E L, WOOD D W. Branch-and-bound methods: a survey[J]. Operations Research, 1966, 14(4): 699-719.

[6] MITTEN L. Branch-and-bound methods: general formulation and properties[J]. Operations Research, 1970, 18(1): 24-34.

[7] MORIN T L. Branch-and-bound strategic for dynamic programming[J]. Operations Research, 1976, 24(4): 611-627.

[8] KUMAR V, KANAL L. A general branch-and-bound formulation for understanding and synthesizing AND/OR tree-search procedures[J]. Artificial Intelligence, 1983, 21(1-2): 179-197.

[9] WAH B W, YU C F. Probabilistic modeling of branch-and-bound algorithms under a best-first search[J]. IEEE Transactions on Software Engineering, 1985, SE-11(9): 922-934.

[10] SMITH D R. Random trees and the analysis of branch-and-bound procedures[J]. Journal of the ACM, 1984, 31(1): 163-188.

[11] WAH B W, LI G J, YU C F. Multiprocessing of combinatorial search problems[J]. IEEE Computer, 1985, 18(6): 93-108.

[12] IMAI M, FUKUMURA T, YOSHIDA Y. A parallelized branch-and-bound algorithm implementation and efficiency[J]. Systems Computer Controls, 1979, 10(3):

62-70.

[13] LAI T H, SAHNI S. Anomalies in parallel branch-and-bound algorithms[J]. Communications of the ACM, 1984, 27(6): 594-602.

[14] MOHAN J. Experience with two parallel programs solving the traveling-salesman problem[C]// 1983 International Conference on Parallel Processing. Washington: IEEE Computer Society Press, 1983: 191-193.

[15] WAH B W, MA Y W. MANIP-a parallel computer system for implementing branch-and-bound algorithms[C]// The 8th Annual Symposium on Computer Architecture. Washington: IEEE Computer Society Press, 1981: 239-262.

[16] WAH B W, LI G J, YU C F. The status of MANIP-a multicomputer architecture for solving combinatorial extremum-search problems[C]// The 11th Annual International Symposium on Computer Architecture. New York: ACM Press, 1984: 56-63

[17] LI G J, WAH B W. Computational efficiency of parallel approximate branch-and-bound algorithms[C]// International Conference on Parallel Processing. Washington: IEEE Computer Society Press, 1984: 473-480.

[18] LI G J. Parallel processing of combinatorial search problems[D]. West Lafayette: Purdue University, 1985.

[19] IBARAKI T. Computational efficiency of approximate branch-and-bound algorithms[J]. Mathematics of Operations Research, 1976, 1(3): 287-298.

[20] WAH B W, MA Y W. The architecture of MANIP-a multicomputer architecture for solving combinatorial extremum-search problems[J]. IEEE Transactions on Computer, 1984, C-33(5): 377-390.

[21] PEARL J. Heuristics: intelligent search strategies for computer problem solving[M]. Boston: Addison-Wesley Longman Publishing, 1984.

发表于*Journal of Parallel and Distributed Computing*, 1990年第10期

# Optimal parallel evaluation of AND trees

Benjamin W. WAH[1], Guo-jie LI[2]

1. Department of Electrical and Computer Engineering and the Coordinated Science Laboratory, University of Illinois at Urbana-Champaign, 1101 West Springfield Avenue, Urbana, Illinois 61801

2. Institute of Computing Technology, Chinese Academy of Sciences, Beijing, People's Republic of China

**Abstract:** AND-tree evaluation is an important technique in artificial intelligence and operations research. An example is the divide-and-conquer algorithm, which can be considered the evaluation of a precedence graph consisting of two opposing AND trees. In this paper, the optimal degree of parallelism for evaluating any given AND tree is quantitatively analyzed. The efficiency analysis is based on both preemptive and nonpreemptive critical-path scheduling algorithms. It is found that the optimal degree of parallelism depends on the complexity of the problem to be solved, the shape of the precedence graph, and the task-time distribution along each path. The major results consist of showing the optimality of the preemptive critical-path scheduling algorithm for evaluating any given AND tree on a fixed number of processors, and tight bounds on the number of processors within which the number of processors leading to the optimal processor-time efficiency can be sought efficiently.

## 1 Introduction

A wide class of problems arising in artificial intelligence, operations research, decision making, and various scientific and engineering fields involves finding a solution of a problem which is made up of a large number of subproblems. Solving such a problem can be represented as an AND-tree computation. Examples include evaluating arithmetic expressions, searching possible solution trees of logic programs, evaluating functional programs, scheduling operations in assembly lines, finding the extremum, merge-sorting, and quick-sorting.

There are two kinds of AND trees, intrees and outtrees, which specify the precedence relationships among nodes. In an intree (resp. outtree), each node has at most one immediate successor (resp. predecessor), and the root is an exit node (resp. entry node). Every node is reachable from the entry node for an outtree or can reach the exit node for an intree. In recursive computations, such as divide-and-conquer algorithms[1-4], a problem is partitioned into smaller and distinct subproblems, and the solutions found for the subproblems are combined into a solution for the original problem. The procedure is applied recursively until the subproblems are so small that they can be solved directly. In this way, the evaluation can be viewed as a process with two

Research supported by National Aeronautics and Space Administration Contract NCC 2-481 and National Science Foundation Grant MIP 89-10584.

phases, the decomposition of subproblems based on an outtree and the composition of results based on an intree. Hence, the precedence graph is composed of an in tree and an outtree. We call this particular precedence graph an outin tree. Deterministic programs can be represented by outin trees. Functional programming deals exclusively with AND graphs[5], and data-flow graphs are AND graphs[6]. In this paper, AND trees and outin trees are used synonymously.

Outin trees have the following characteristics. First, they are single-entry single-exit precedence graphs without cycles. Second, in contrast to general forests, an outin tree has one outtree and one intree, and a one-to-one correspondence of all the leaves in the two trees. We call these leaves the leaves of the out in tree. Third, the subtasks generated by a task in an outtree are usually less complex; likewise, the parent subtasks of a task in an intree are generally less complex. In general, outin trees can be considered special cases of single-entry single-exit acyclic precedence or data-flow graphs. We focus our studies on outin trees in this paper, although some of our results can be generalized to general acyclic single-entry, single-exit precedence graphs.

Figure 1 illustrates an outin tree, which reflects the precedence relationships among tasks in a merge-sort algorithm to sort six numbers. The nonterminal nodes in the outtree represent decompositions, each of which splits a (sub) list into two smaller sublists, whereas the nonterminal nodes in the intree part represent composition, each of which generates a sorted list based on two smaller sorted sublists.

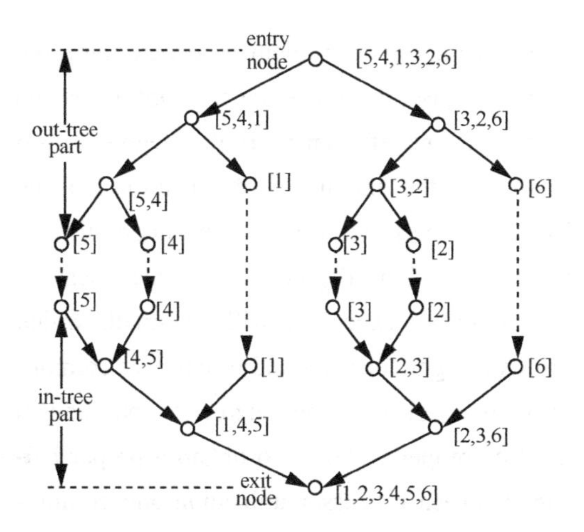


Figure 1. Outin-tree representation of the sort-merge algorithm

Evaluation of outin trees naturally suggests implementation on parallel computers due to the independence of subproblems. AND-tree evaluations are important in the parallel evaluation of logic programs[7]. Studies conducted on parallel computers for executing divide-and-conquer algorithms can be classified into three types. First, multiprocessors that are connected in the form of a tree, especially a binary tree, can be used to exploit the potential parallelism of divide-and- conquer algorithms[8-10]. A second approach is the virtual-tree machine[11], which consists of a number of processors that are connected by an interconnection network, such as a binary $n$-cube network, and a suitable algorithm to decide when and where each subproblem should be solved. The third approach is a variation of the above approaches using a common memory. All processors are connected to the memory by a common bus[12].

To evaluate an AND tree in parallel, it is necessary to schedule the subproblems to achieve high throughput and processor utilization. An important problem is to determine the proper degree of parallelism that optimizes the processor-time efficiency. If the degree is too small, then the granularity of execution may be large, and the processors may be underutilized. In contrast, if the degree is too large, then the granularity of execution will be small, which results in tight coupling and prohibitive communication overhead. The degree of parallelism must be properly chosen to obtain a proper balance between processor utilization and communication overhead.

In previous studies, one can find different arguments on the issue of granularity. Some researchers advocate a fine grain, while others suggest a coarse grain. For example, in the FFP machine[13], a small grain is chosen for supporting ease and generality of parallel computations. In contrast, in Rediflow[14], large-grain parallelism is used to minimize communication overheads. Tang and Lee have derived the speedup ratios (ratio of order-of-magnitude time complexity of a sequential divide-and-conquer algorithm to the order-of-magnitude time complexity of the same algorithm in a parallel SIMD computer model) for various node complexities of any given outin tree[15]. They have also derived the order-of-magnitude number of processors that can achieve the optimal speedup for various node complexities.

In this paper, we quantitatively analyze the optimal degree of parallelism for evaluating any given outin tree using both preemptive and nonpreemptive scheduling policies. We identify the factors that influence the optimal degree of parallelism, in particular, the relationship between degree of parallelism and problem complexity. In Section 2, we prove that the preemptive critical-path scheduling algorithm is optimal in minimizing the completion time for scheduling any given outin tree. We also develop tight bounds on the number of processors within which the processor-time efficiency can be optimized. The case using preemptive critical-path scheduling is discussed in Section 3, while the case using nonpreemptive critical-path scheduling is discussed in Section 4.

In the analysis shown in this paper, it is assumed (a) that the precedence graph is known and is in the form of an outin tree; (b) that the computation time of each task is deterministic and known; (c) that the computational overheads of tasks are monotonically decreasing in the intree part and are monotonically increasing in the outtree part when nonpreemptive scheduling algorithms are used; (d) that the parallel processors are identical; (e) that an asynchronous model of parallel computation is assumed: a new task in the ready queue is scheduled whenever a task completes; (f) that all idle processors, if any, must be assigned to compute an available executable task; and (g) that the overhead involved in scheduling a task is negligible.

# 2 Scheduling parallel outin-tree evaluations

## 2.1 Terminologies and background

The precedence graph of an outin tree is oriented such that the entry node is at the top of the figure and the exit node is at the bottom. An arc is assumed to be always directed toward the bottom of the graph. The length of a node is the sum of the task execution times for nodes in the longest path from this node to the exit node. Figure 2 is an example of an outin tree. In this figure, the number inside a node is its task execution time, and the number next to a node is its length.

The execution time of a task can be interpreted as either its maximum processing time or its expected processing time. In the former case, the worst-case time to complete the schedule is considered, while in the latter case the length of the schedule represents a rough estimate on the average time of computation. In some outin-tree problems, the execution time of each task can be predicted quite accurately. For example, in evaluating arithmetic expressions, the time to execute a primitive operation, such as a multiplication, is known. In other cases, the average execution times may have to be estimated from statistics or from previous experience. In all cases, the communication overhead is nontrivial when preemptions are allowed, and the task time should also include the overhead of preemptions.

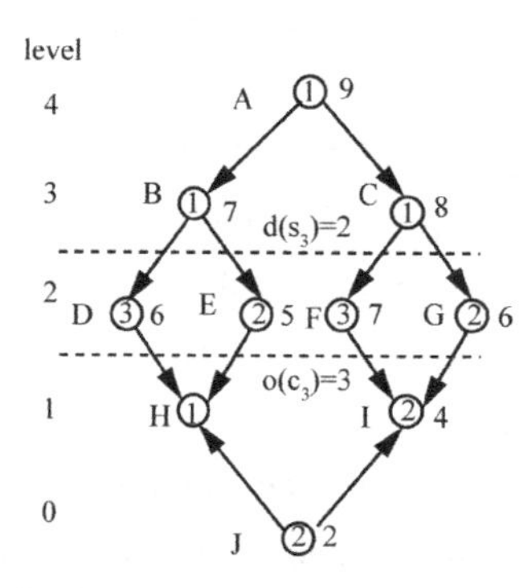


Figure 2. Task precedence graph as outin tree

Our goal is to choose an algorithm that minimizes $C_{max}$, the maximum completion time, for scheduling an outin tree on a set of $P$ identical processors, and to find the number of processors that optimizes the processor-time efficiency based on this scheduling algorithm. Our scheduling problem is similar to the $P$/tree/$C_{max}$ scheduling problem in which tree precedence graphs are considered[16-18]. Note that one must start with the optimal scheduling algorithm in order to find the optimal degree of parallelism.

If preemption is allowed, $P$/preemption, in tree/$C_{max}$ can be solved optimally either by Muntz and Coffman's Critical-Path Scheduling (CPS) algorithm in $O(N)$ time[19], or by other polynomial-time algorithms[20]. Besides being efficient and optimal, the CPS algorithm is easy to implement and, consequently, is one of the most common scheduling algorithms[21-22]. In the CPS algorithm, the next job chosen is the one with the longest length of unexecuted jobs. This longest path is called the critical path. If preemption is not allowed, then optimal scheduling algorithms have been obtained only for two cases: (a) all tasks have equal execution times and the precedence relationships are in the form of an in tree (Hu's algorithm[23]) and (b) when two processors are used[24]. Hu's optimal scheduling algorithm is indeed a CPS algorithm. Many other cases have been proved to be NP-hard[25-26].

In the case that the precedence graph is a tree, that all processors are identical, and that each task requires $t_i$, $0 < t_i \leqslant t_{max}$, units of time to complete, the nonpreemptive CPS algorithm turns out to be almost-optimal in the sense that

$$T_p(k) \leqslant T_{np}(k) \leqslant T_p(k) + t_{max} \qquad (1)$$

where $T_{np}(k)$ and $T_p(k)$ are, respectively, the completion times required by the nonpreemptive and preemptive CPS algorithms using $k$ processors[27]. Some researchers have strived for nonpreemptive scheduling algorithms in order to solve scheduling problems with tree precedence[18, 22, 28]. Garey, Dolev, et al. have studied the scheduling of forests consisting of intrees and outtrees[28-29]. Given a fixed number of processors, polynomial-time algorithms with high complexities for finding an optimal schedule of these forests have been developed.

In this research, we are interested in a special, but widely used, form of precedence trees, namely, the outin trees. With respect to a fixed number of processors, evaluating an outin tree can be divided into the splitting, all-busy, and combining phases. In the splitting phase, the problem is decomposed, and the number of busy processors is increased from one up to at most $k-1$, where $k$ is the maximum number of processors available. In the combining phase, the subproblems are composed, and the number of busy processors is decreased from at most $k-1$ to one. During these two phases, some processors are idle. In contrast, in the all-busy phase, all the $k$ processors are busy. Schindler has proved that the schedule of a precedence graph is optimal if either the computations can be completed in only the all-busy or combining phase or the precedence graph can be partitioned into the all-busy and combining phases by a "height-line"[30]. We show later that this result can be extented to scheduling outin trees, and that the CPS algorithm guarantees the optimal preemptive scheduling and near-optimal nonpreemptive scheduling of outin-tree evaluations.

To analyze the properties of preemptive CPS, in short, PCPS, algorithms, it is more convenient to represent a task (a node of the outin tree) of execution time $t_i$ by a chain task, which is $t_i$ element tasks (or element nodes, or in short, e-tasks or e-nodes), each of which has one unit of execution time (see Figure 3). We use a subscript $i$ in the task identifier to indicate the $i$th e-task in the chain task. Hence, $F_2$ is the second e-task of task F. The new outin tree is called the element-outin tree (or e-outin tree). For each chain task, the e-task farthest from the exit e-node of the e-outin tree is called a task-head e-node. It is easy to verify that the length of the task-head e-node is the same as the length of the original multiunit task. Two e-nodes are said to be in the same e-level of the e-outin tree if their lengths are identical; that is, the e-level number of an e-node is equal to its length assuming that the exit e-node is in Level 0. To distinguish between nodes and levels in the original outin tree (as exemplified by Figure 2) and in the e-outin tree (as exemplified by Figure 3), we use tasks and levels with respect to the original outin tree and e-task and e-level with respect to the e-outin tree in subsequent discussions.

There is another variation of preemptive scheduling algorithms called General Scheduling (GS) discipline[19], which is strongly related to preemptive critical-path scheduling. In the GS algorithm, each processor in the system is considered to have a certain amount of computing capacity rather than as a discrete unit, and this computing capacity can be assigned to tasks in any amount between zero and the equivalence of one processor. For example, if we assign half of a processor to task $P_i$ with execution time $t_i$, then it will take $2 \cdot t_i$ units of time to complete $P_i$.

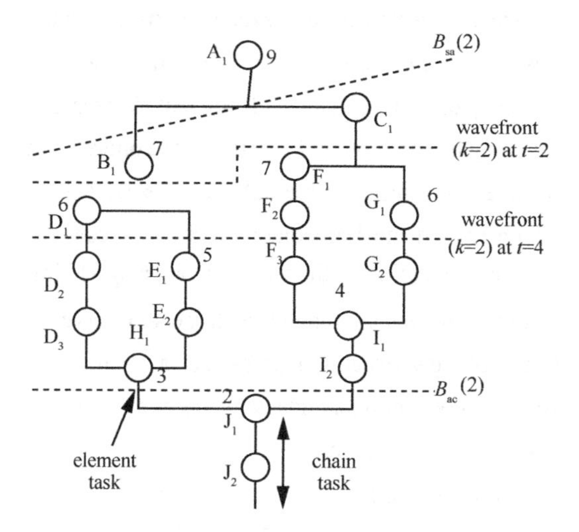


Figure 3. Task precedence graph as e-outin tree using chain tasks (all tasks require unit times)

In scheduling a given outin tree using the GS discipline, one processor is assigned to each of the $k$ e-nodes farthest from the exit e-node. If there is a tie in the lengths among $u$ e-nodes for the last $v$, $u > v$, processors, then $v/u$ of a processor is assigned to each of these $u$ e-nodes. Each time when either (a) a chain task of the e-outin tree is completed or (b) a point is reached where, if we continue with the present assignment, some e-nodes will be computed at a faster rate than other e-nodes that are farther from the exit e-node then the processors are reassigned to the remaining tree according to the CPS principle. Situation (b) occurs when an e-node that is being computed has the same length as that of some unexecuted task-head e-node(s). In this case, one (or part of a) processor must be assigned to the unexecuted task-head e-node. The GS discipline is illustrated in Figure 4.

Muntz and Coffman have proved the equivalence between the GS and PCPS algorithms[19]. That is, if preemptions were permitted, then the "processor-sharing" capability is not needed for optimal scheduling. To illustrate this equivalence, Figure 5 shows the preemptive schedule for the corresponding e-outin tree in Figure 3.

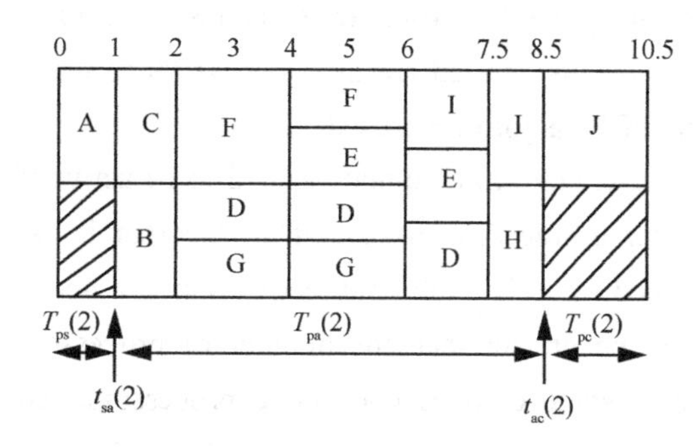


Figure 4. Timing diagram for general scheduling (processor sharing) using two processors for the e-outin tree in Figure 3

Figure 5. Timing diagram for preemptive CPS scheduling using two processors for the e-outin tree in Figure 3

In practice, preemptions are usually restricted to the beginning of a time unit, so the overhead of a practical PCPS algorithm is equal to that of Hu's algorithm, which assumes that tasks have unit execution times. From Eq. (1), we have[19]

$$T_{\mathrm{p}}(k)=T_{\mathrm{gs}}(k) \leqslant T_{\mathrm{h}}(k) \leqslant \left\lceil T_{\mathrm{gs}}(k)+1 \right\rceil = \left\lceil T_{\mathrm{p}}(k)+1 \right\rceil \quad (2)$$

where $T_{\mathrm{h}}(k)$ and $T_{\mathrm{gs}}(k)$ are the completion times required by Hu's and GS algorithms, respectively. Eq. (2) shows that the behavior of the GS algorithm is very close to that of any PCPS algorithm which only allows preemptions at the beginning of a time unit. In subsequent discussions, the results are derived without any restriction on the allowable times for preemptions. Moreover, we use the GS discipline as a means for analyzing the properties of PCPS algorithms. The results derived are the same as those when the PCPS algorithm is used.

At time $t$, an e-node is said to be active if either a processor or part of a processor is assigned to it. The total number of active e-nodes may be greater than the number of processors since some e-nodes may share processors. All active e-nodes form a wavefront in the e-outin-tree evaluation. Two particular times of the wavefront are of special interest: $t_{sa}(k)$ and $t_{ac}(k)$. The computation enters the all-busy phase at $t_{sa}(k)$ and enters the combining phase at $t_{ac}(k)$. In both times, the wavefronts serve as phase boundaries. We call the former phase boundary $B_{sa}(k)$ and the latter $B_{ac}(k)$.

For the task graph in Figure 2, if the PCPS algorithm is employed, then $t_{sa}(2)=1$ and $t_{ac}(2)=8.5$ (see Figures 4 and 5). The corresponding phase boundaries $B_{sa}(2)$ and $B_{ac}(2)$ are indicated in Figure 3.

If a preemptive (resp. nonpreemptive) CPS algorithm is applied, then the computational times required by $k$ processors to complete the splitting, all-busy, and combining phases are denoted by $T_{ps}(k)$, $T_{pa}(k)$, and $T_{pc}(k)$ (resp. $T_{nps}(k)$, $T_{npa}(k)$, and $T_{npc}(k)$).

In the intree part of an e-outin tree, each e-node is associated with a path to the exit e-node, while each e-node in the outtree part may be associated with more than one path to the exit e-node. The longest path from any given e-node to the exit e-node is selected as the execution path through this e-node. For an e-node, if more than one such longest path exists, then a left-to-right tie-breaking rule may be used to break the tie. In the outtree part, if an e-node has $q$ immediate successors, one of which (called the immediate execution successor) is in its execution path, then the other $q-1$ immediate successors (called path heads) serve as heads of new execution paths. As a result, each active e-node corresponds to a unique execution path to the exit e-node.

For example, in Figure 3, the execution path from $A_1$ is $(A_1,C_1,F_1,F_2,F_3,I_1,I_2,J_1,J_2)$ and $B_1$ is the path head of the execution path $(B_1,D_1,D_2,D_3,H_1,J_1,J_2)$. Note that when $k$ processors are used, at most $k-1$ path heads are active in the splitting phase.

Let $A(t,k)$ be the set of active e-nodes at time $t$ when $k$ processors are used. This set can be divided into two classes in terms of the lengths of the corresponding execution paths. At time $t$, the active e-nodes whose execution paths are the shortest among all active e-nodes belong to a subset $A_s(t,k)$ and lie in a single e-level, called the minimal active e-level. The other active e-nodes belong to another subset $A_h(t,k)$. (If $t$ and $k$ are obvious in the context, they will be omitted for brevity.) For example, in Figure 3, when $t$ is 2, e-nodes $D_1$ and $G_1$ belong to $A_s(2,2)$, and e-node $F_1$ belongs to $A_h(2,2)$.

## 2.2 Optimality of the PCPS algorithm

In the following four propositions, we show

the properties of the PCPS algorithm. We then prove a new result in Theorem 1 that the PCPS algorithm is optimal in minimizing the completion time for evaluating any given outin tree using a fixed number of processors.

During the evaluation of an outin tree, the active e-nodes are executed at different rates depending on whether the assigned processor is shared or not. Let $r_i(t)$ be the processing rate in e-node per unit time for active e-node $i$ at time $t$. The following proposition distinguishes the processing rates under various conditions.

**Proposition 1:** During a parallel evaluation of an outtin tree with $k$ processors, $r_i(t)$, the processing rate of active e-node $i$ at time $t$, satisfies the conditions

$$r_i(t)=\begin{cases}=1,\text{if e-node } i\in A_h(t,k) or \left|A(t,k)\right|\leqslant k\\ <1,\text{otherwise}\end{cases}$$

where $\left|A(t,k)\right|$ is the cardinality of $A(t,k)$, the set of active e-nodes.

**Proof:** This follows from the GS strategy immediately.

Proposition 1 reflects the following facts. First, in the splitting and combining phases, the processing rate for any e-node is one; that is, an e-node is processed in each time unit. Second, if an active e-node is not in the minimal active e-level, then one processor (rather than part of a processor) has to be assigned to it; that is, its corresponding processing rate is one. Third, in the all-busy phase, if the number of active e-nodes is equal to the number of processors, then the processing rates for e-nodes in the minimal active e-level is also one. It is only when the number of active e-nodes is larger than the number of processors used that the processing rates of e-nodes in the minimal active e-level are less than one.

Let $h_{\max}$ be the length of the critical path in the outin tree to be evaluated. Since at least one time unit is needed to complete each e-node, it is evident for any scheduling algorithm that

$$T(k)\geqslant h_{\max} \tag{3}$$

where $T(k)$ is the completion time using $k$ processors under a preemptive or nonpreemptive scheduling discipline. The following proposition shows the relationship between $T_p(k)$ and the shape of the phase boundary.

**Proposition 2:** (a) $T_p(k)>h_{\max}$ implies that all active e-nodes on the phase boundary $B_{ac}$ are located in the same e-level. (b) If an active e-node on the phase-boundary $B_{ac}$ belongs to $A_h$, then $T_p(k)=h_{\max}$.

**Proof:** (a) By Proposition 1, $T_p(k)>h_{\max}$ implies that at least one node in the critical path must share a processor with another node(s) in the all-busy phase. This means that at time $t$ in the all-busy phase, some active node in the critical path must belong to $A_s(t,k)$. According to situation (b) of the GS strategy, when an active node in the critical path has the same length as that of some unexecuted task-head node(s), one (or part of) processor must be assigned to the unexecuted task-head node. That is, during the evaluation of an AND tree, the active node in the original critical path still has the longest execution path among all nodes in the remaining unexecuted AND tree. As a result, at time $t$ in the all-busy phase, all active nodes must be in the same level and hereafter all their successors in the same level are evaluated simultaneously and at the same rate. It implies that all active nodes on the phase boundary $B_{ac}$ are located in the same level. (b) In this case, at any time during the computation, the active nodes in the critical path belong to $A_h$. From Proposition 1, all nodes in the critical path are executed at the rate of one element task per unit time; hence, $T_p(k)=h_{max}$.

The example in Figures 3 and 4 illustrates Property (a) above. At time $t=4$, e-node $G_1$ in the critical path completes and "enters" the set $A_s$, and hereafter all active e-nodes are in the same e-level. Proposition 2 reflects the fact that preemptive scheduling algorithms distribute work uniformly among the available processors, thereby reducing the computational time required in the combining phase. This is the reason for a preemptive schedule to be shorter than or equal to the no preemptive counterpart.

To investigate the optimal scheduling of outin trees, we need to examine the phase boundaries when a different number of processors are used. The following proposition compares two boundaries with respect to $k$ and $k+1$ processors. In subsequent discussions, the phase boundary in a single e-level means that all active e-nodes on this boundary have execution paths with the same length.

**Proposition 3:** If phase boundary $B_{ac}(k+1)$ is in a single e-level, then phase boundary $B_{ac}(k)$ must be in a single e-level.

**Proof:** This proposition can be proved by contradiction. Let $P_e$ be the entry node in the outin tree to be considered. By Proposition 2, the number of e-nodes from $P_e$ to the phase boundary $B_{ac}(k+1)$ along the critical path is less than $(h_{max}-T_{pc}(k+1))$ (see Figure 6).

$$T_p(k+1)$$
$$=T_{ps}(k+1)+T_{pa}(k+1)+T_{pc}(k+1)\geqslant h_{max}$$

or

$$(h_{max}-T_{pc}(k+1))\leqslant(T_{ps}(k+1)+T_{pa}(k+1)) \quad (4)$$

Suppose when $k$ processors are used and after $(T_{ps}(k+1)+T_{pa}(k+1))$ time units, there is an active e-node $P_k$ belonging to $A_h(k)$ in the critical path. According to Proposition 1, during this time interval, the processing rate in the critical path is always one, so the number of e-nodes from $P_e$ to $P_k$ is equal to $(T_{ps}(k+1)+T_{pa}(k+1))$, which is larger than $(h_{max}-T_{pc}(k+1))$. By our assumption, when $k+1$ processors are used, the phase boundary $B_{ac}(k+1)$ is in a single level. This means that when $k$ processors are used, the wavefront after $(T_{ps}(k+1)+T_{pa}(k+1))$ time units must be below the phase boundary $B_{ac}(k+1)$. The above argument implies that the number of e-nodes evaluated by $k$ processors after $(T_{ps}(k+1)+T_{pa}(k+1))$ time units would be larger than that evaluated by $k+1$ processors in the same time interval. The latter is indicated by the shaded area in Figure 6. This is impossible since, in the first $T_{ps}(k+1)$ time units, the number of e-nodes evaluated by $k$ processors is equal to that evaluated by $k+1$ processors and, in the subsequent $T_{pa}(k+1)$ time units, the number of e-nodes evaluated by $k$ processors is less than $(k+1)T_{pa}(k+1)$. Hence, if the phase boundary $B_{ac}(k+1)$ is in a single level, then when $k$ processors are used and after $(T_{ps}(k+1)+T_{pa}(k+1))$ time units, the wavefront must be in a single level and above $B_{ac}(k+1)$; that is, there are at least $k+1$ active nodes in the wavefront at that time. Hereafter, the wavefront must move down level by level according to the PCPS algorithm, so the phase boundary $B_{ac}(k)$ must be in a single level.

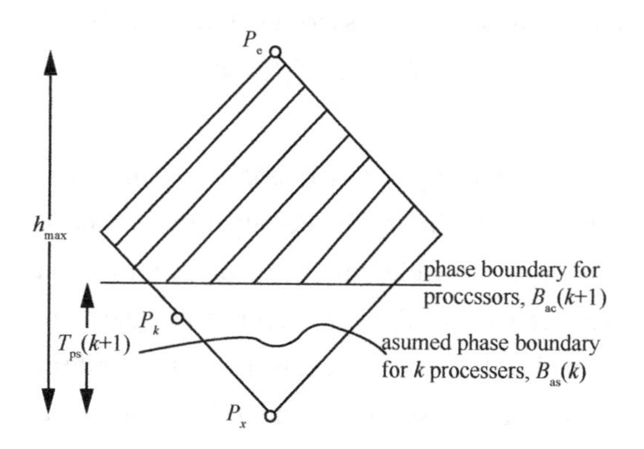


Figure 6. Proof of Proposition 3

If the phase boundary $B_{ac}(k+1)$ is not in a single e-level, then there are $k'$, $0<k'<k+1$, active e-nodes belonging to $A_h(k+1)$ on this phase boundary. During the splitting and all-busy phases, we can partition the $(k+1)$ processors into two groups. The first group consists of $k'$ processors which only evaluate e-nodes in the $k'$ execution paths from the topmost $k'$ path heads to the $k'$ e-nodes. The other $(k+1-k')$ processors evaluate all e-nodes other than those in the aforementioned $k'$ execution paths. Accordingly, we can prove the following proposition.

**Proposition 4:** If an outin tree is evaluated by the PCPS algorithm, then $T_{ps}(k+1) \geqslant T_{ps}(k)$, and $T_{pc}(k+1) \geqslant T_{pc}(k)$.

**Proof:** First, it is easy to see that $T_{ps}(k+1) \geqslant T_{ps}(k)$, since when $k+1$ processors are used, the computation cannot enter the all-busy phase during the first $T_{ps}(k)$ time units. Second, the proof of Proposition 3 shows that if the phase boundary $B_{ac}(k+1)$ is in a single level, then $T_{pc}(k+1) \geqslant T_{pc}(k)$. Therefore, we only need to show that when $B_{ac}(k+1)$ is not in a single level, this inequality still holds.

Suppose $P_{k+1}$ is a node on the phase boundary $B_{ac} \times (k+1)$ and also on the critical path; i.e., $T_{pc}(k+1)$ is equal to the length from $P_{k+1}$ to the exit node $P_x$ (see Figure 7). When $k$ processors are used, if $P_{k+1}$ becomes active at time $t$, we claim that at this time either nodes on $B_{ac}(k+1)$ or their predecessors (rather than successors) should be active. That is, the wavefront should be totally above or in $B_{ac}(k+1)$. The claim is certainly true if the wavefront is in a single level. Otherwise, $P_{k+1}$ belongs to $A_h(t,k)$. According to Proposition 1, along the critical path from $P_e$ to $P_{k+1}$, the processing rate is always one, so $t = T_{ps}(k+1) + T_{pa}(k+1)$.

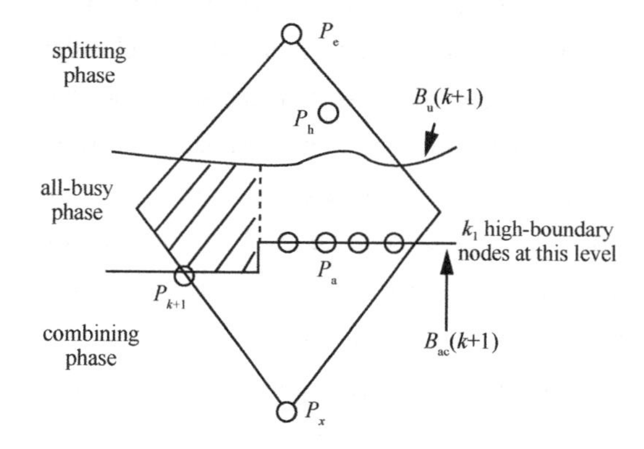


Figure 7. Proof of Proposition 4

Suppose that $P_a$ is one of the $k_1$ nodes on the phase boundary $B_{ac}(k+1)$, which belongs to $A_h(t,k+1)$ (see Figure 7). We call these $k_1$ nodes high-boundary nodes because all their predecessors are evaluated at unit rate when $k+1$ processors are used.

Assume that $P_h$, one of the topmost $k_1$ path heads which is active in the splitting phase, is a predecessor of $P_a$. When $k$ processors are used, $P_h$ cannot be evaluated earlier than the time it was evaluated when $k+1$ processors were used because the processing rate cannot be greater than one. Likewise, when $k$ processors are used, $P_a$ and the other $k_1-1$ high-boundary nodes cannot be evaluated earlier than time $t$.

If at time $t$ and when $k$ processors are used some predecessors of the high-boundary nodes share processors with other tasks, then they must be in the minimal active level, $A_s(t,k)$: that is, all wavefronts when $k$ processors are used are above $B_{ac}(k+1)$ at this time. When $k$ processors are used, if the $k_1$ execution paths from the topmost $k_1$ path heads to the $k_1$ high-boundary nodes are evaluated without processor sharing, then all nodes except those in the $k_1$ execution paths, as indicated by the shaded area in Figure 7, are evaluated by $k-k_1$ processors, and all nodes in the same level within the shaded area should share processors. Since less processors are available, the processing rate when $k$ processors are used should be smaller than that when $k+1$ processors are used. At time $t$, the minimal active level when $k$ processors are used must also be above $B_{ac}(k+1)$.

Having proved our claim, we found that at time $t$ there are more than $k$ active nodes in the wavefront when $k$ processors are used. This implies that $T_{pc}(k)<T_{pc}(k+1)$.

The following theorem shows that the PCPS algorithm can be used to find the optimal preemptive schedule for outin trees.

**Theorem 1:** PCPS is a minimum- completion- time scheduling algorithm for any given outin tree and a fixed number of identical parallel processors.

**Proof:** Let $\Phi_{ps}(k)$ and $\Phi_{pc}(k)$ be the total amount of idle times in the splitting and combining phases when the PCPS algorithm is applied and $k$ processors are used. Clearly,

$$T_p(k)=\frac{\Phi_{ps}(k)+T(1)+\Phi_{pc}(k)}{k} \tag{5}$$

Minimizing $T_p(k)$ implies minimizing $(\Phi_{ps}-\Phi_{pc})$. In the PCPS algorithm, once an e-node is available, that is, its predecessor node has been finished, a processor is assigned to it immediately. The time spent in the splitting phase for any schedule cannot be shorter than that in the PCPS algorithm. This means that $\Phi_{ps}$ for the PCPS algorithm is the minimum.

We now consider $\Phi_{pc}$. If the phase boundary $B_{ac}$ is not in a single e-level, then $T_p(k)=h_{max}$ according to Proposition 2. That is, the PCPS algorithm achieves the minimum computational time $T(k)$ according to Eq. (3). Hence, we only need to consider the case when the phase boundary $B_{ac}$ is in a single e-level. This boundary is indicated by line $B$ in Figure 8.

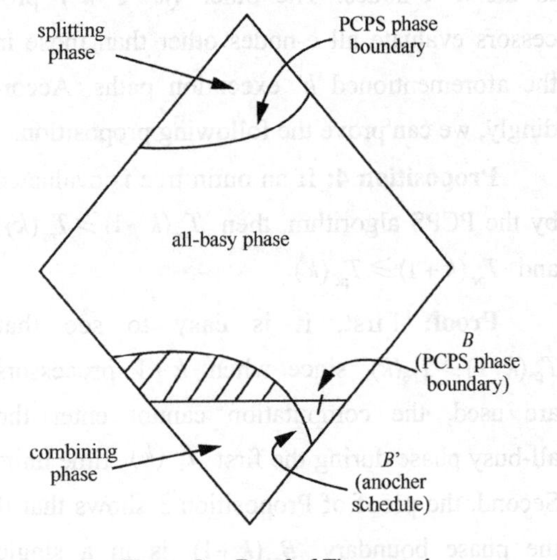


Figure 8. Proof of Theorem 1

Suppose that an arbitrary scheduling algorithm is used; the corresponding phase boundary is denoted by $B'$. Note that it is impossible for all e-nodes on boundary $B'$ to be beneath boundary $B$; otherwise, at least one processor is idle before the wavefront achieves $B'$, which implies that $B'$ is not a phase boundary. In other words, at least one e-node on boundary $B'$ is above or on boundary $B_{ac}$. Similarly, it is impossible for all e-nodes of boundary $B'$ to be above boundary $B$.

Let $N_{pc}$ and $N'_{c}$ be the amount of task times in the combining phase of the PCPS and any other scheduling algorithm. Let $T'_{c}(k)$ and $\Phi'_{c}(k)$ be the computational time and the total idle time in the combining phase when an arbitrary scheduling algorithm using $k$ processors is adopted. If $T_{pc}(k)$ equals $T_c'(k)$, then $N_{pc}(k) \geqslant N'_{c}(k)$; hence, $(\Phi'_{c} - \Phi_{pc}) = (N_{pc} - N'_{c}) \geqslant 0$. If $T_{pc}(k)$ is less than $T'_{c}(k)$, since at least one e-node on boundary $B'$ is beneath or on boundary $B_{ac}$, $(N'_{c} - N_{pc})$ cannot be larger than the amount of task times for e-tasks beneath $B'$ and above $B$ (the shaded area in Figure 8). Since, from Proposition 1, the processing rate for any path in the combining phase is one, then after $(T'_{c} - T_{pc})$ time units, all e-nodes in the shaded area in Figure 8 must be completed. As less than $k$ e-nodes can be completed during a time unit in the combining phase, the amount of e-nodes in the shaded area must be less than $k(T'_{c} - T_{pc})$. Therefore,

$$(\Phi'_{c} - \Phi_{pc}) = \left[k(T'_{c} - T_{pc}) - (N'_{c} - N_{pc})\right] > 0$$

This means that $\Phi_{pc}$, the total idle time in the combining phase, is also minimum for the PCPS algorithm. In summary, $\Phi_{ps}$ and $\Phi_{pc}$ are minimum for the PCPS algorithm, which implies that the PCPS algorithm is the minimum-completion-time algorithm.

The above proof does not imply that the PCPS algorithm is the unique optimal algorithm. However, it proves that the amount of idle times introduced by the PCPS algorithm is the minimum. In the following section, we find, with respect to this algorithm, the range on the number of processors that can optimize the processor-time efficiency.

## 3 Optimal degree of parallelism in preemptive scheduling

The criteria that can be used to define the optimal degree of parallelism are $U$, speedup, $k \cdot T^2$, and $A \cdot T^2$, where $U$ is the processor utilization, $k$ is the number of processors, $T$ is the computational time, and $A$ is the area of a VLSI implementation.

The complexity of divide-and-conquer algorithms in a SIMD model and the conditions to assure the optimal processor utilization have been studied[12]. It is found that processor utilization increases monotonically with decreasing number of processors, which means that processor utilization is the maximum when one processor is used. Hence, processor utilization is not an adequate measure for the effects of parallel processing.

Tang and Lee use the speedup ratio as a measure of effectiveness of scheduling a divide-and-conquer algorithm and develop the optimal order-of-magnitude number of processors to achieve the optimal speedup[15]. However, their results do not lead to the optimal number of processors for evaluating any given outin tree because they are only true asymptotically.

The measure used in this paper for deriving the optimal degree of parallelism for evaluating a given outin tree is $k \cdot T^2$. This measure considers both processor utilization and computational time, since

$$k \cdot T^2(k) = \frac{T(1)T(k)}{U} = \frac{k \cdot T^2(1)}{\text{speedup}^2},$$

where $U = \frac{\text{speedup}}{k}$ and $\text{speedup} = \frac{T(1)}{T(k)}$.

To minimize $k \cdot T^2$ means to reduce the computational time and to maximize the processor utilization. $k \cdot T^2$ is linearly related to $A \cdot T^2$ if the area of connection wires is proportional to the area of processing elements. As both computational time and processor utilization are important in many applications, $k \cdot T^2$ is a good criterion to use. In other applications, such as real-time processing, the completion time may be more critical and processor utilization is a secondary consideration. In this case, a different optimization criterion may have to be used.

Note that finding the number of processors to minimize $k \cdot T^2$ is equivalent to finding the number of processors to maximum the speedup. Hence, the order-of-magnitude results derived by Tang and Lee[15] on optimizing with respect to the speedup ratio also apply to our case on optimizing with respect to the $k \cdot T^2$ measure. An advantage on using the $k \cdot T^2$ measure rather than the speedup is that it is a product rather than a ratio, which simplifies the derivation of results in this paper.

Our goal in this paper is to find the optimal degree of parallelism for evaluating any given outin tree with respect to the $k \cdot T^2$ measure; that is, given an outin tree, we need to choose $k$ to minimize $k \cdot T^2$, or given a fixed $k$, we need to determine the type of outin trees (their shapes, complexities, etc.) and its proper size that can be solved most efficiently by this system. We derive our result with respect to scheduling any given outin tree and not with respect to the order-of-magnitude behavior of all outin trees.

It is difficult to find directly the optimal degree of parallelism with respect to $k \cdot T^2$ because the optimal degree depends on the execution time of each task and the shape of the outin tree. We attempt to find the optimal degree of parallelism via an intermediate variable, the total idle time. Let $\Phi_p(k)$ (resp. $\Phi_{np}(k)$) be the total amount of idle times when a preemptive (resp. nonpreemptive) scheduling algorithm with $k$ processors is used. $\Phi_p(k)$ takes into account the idle times in both the splitting and the combining phases. Clearly, $\Phi_p(k) = [\Phi_{ps}(k) + \Phi_{pc}(k)]$ and

$$k \cdot T(k) = T(1) + \Phi(k) \tag{6}$$

Eq. (6) holds for both preemptive and nonpreemptive scheduling algorithms.

The total idle time $\Phi_p(k)$ is related to both $k$ and $k \cdot T^2$. The following two lemmas show the difference between the total idle times when different numbers of processors are used.

**Lemma 1:** When an outin tree is evaluated by the PCPS algorithm,

$$\left[\Phi_p(k+1) - \Phi_p(k)\right] \leqslant h_{\max} \tag{7}$$

where $h_{\max}$ is the length of the critical path.

**Proof:** Two cases are possible.

(a) $T_p(k+1)=h_{max}$. The computational time for evaluating any AND tree cannot be less than $h_{max}$ regardless of $k$, i.e., $T_p(k) \leqslant h_{max}$. Hence, we can derive Eq. (7) using Eq. (6).

(b) $T_p(k+1) > h_{max}$. In this case, all active nodes at time $t_{ac}(k+1)$ are in a single level from Proposition 2. According to Proposition 3 and its proof, when $k$ processors are used, $B_{ac}(k)$, the phase boundary, is in a single level that is beneath $B_{ac}(k+1)$. Hence, each processor has identical amounts of idle times in both cases of using $k$ and $k+1$ processors except that the additional $(k+1)$th processor becomes idle first in the case of using $(k+1)$ processors. That is,

$$\Phi_{pc}(k+1)-\Phi_{pc}(k)=T_{pc}(k+1) \tag{8}$$

In the splitting phase, the idle time of the $(k+1)$th processor is equal to $T_{ps}(k+1)$, which is not included in the case of $k$ processors. It is certain that

$$\Phi_{ps}(k+1)-\Phi_{ps}(k)=T_{ps}(k+1) \tag{9}$$

Since at least one e-node in the critical path has been evaluated in the all-busy phase in this case, and the processing rate is one in the splitting and all-busy phases (according to Proposition 1), we have

$$\begin{aligned}&\Phi_p(k+1)-\Phi_p(k)\\&=\left[T_{ps}(k+1)+T_{pc}(k+1)\right]<h_{max}\end{aligned}$$

**Lemma 2:** Suppose that an outin tree is evaluated by the PCPS algorithm, then

$$\left[\Phi_p(k+1)-\Phi_p(k)\right] \geqslant \left[T_{ps}(k)+T_{pc}(k)\right] > 0 \tag{10}$$

**Proof:** Adding Eq's (8) and (9) results in the condition

$$\begin{aligned}T_{ps}(k+1)+T_{pc}(k+1)&=\left[\Phi_{ps}(k+1)+\Phi_{pc}(k+1)\right]\\&\quad-\left[\Phi_{ps}(k)+\Phi_{pc}(k)\right]\\&=\Phi_p(k+1)-\Phi_p(k)\end{aligned}$$

From Proposition 4, we know that

$$T_{ps}(k+1)+T_{pc}(k+1) \geqslant T_{ps}(k)+T_{pc}(k) > 0$$

The lemma is, therefore, proved.

**Lemma 3:** Suppose that an outin tree is evaluated by the PCPS algorithm, then

$$T_p(k+1) \leqslant T_p(k)$$

**Proof:** A sketch of the proof is given here. Suppose that the given outin tree is optimally scheduled for $k$ processors. Two cases are possible. First, if $T_p(k)=h_{max}$, then $T_p(k+1)=h_{max}$, and $T_p(k+1)=T_p(k)$. Second, if $T_p(k)>h_{max}$, then at least one node in the critical path must share a processor with another node(s) in the all-busy phase according to Proposition 1. This means that some of these nodes can be scheduled in the $(k+1)$th processor, resulting in a shorter or equivalent all-busy phase when $k+1$ processors are used. This implies that $T_p(k+1) \leqslant T_p(k)$.

The above lemmas reveal that when the number of processors used are increased, the total idle times must increase, and the difference of the total idle times with respect to $k_1$ and $k_2$ processors is bounded by $(k_2-k_1)[T_{ps}(k)+T_{pc}(k)]$ and $(k_2-k_1)\cdot h_{max}$. From these facts, we can determine the conditions under which $k \cdot T^2$ is either monotonically increasing or decreasing with respect to $k$. The following theorem shows the relation between $\Phi_p(k)$ and $k \cdot T^2$.

**Theorem 2:** Suppose that an outin tree is

evaluated by the PCPS algorithm. $k\cdot T_p^2(k)$ is monotonically increasing with $k$ if $\left[\Phi_p(k+1)-\Phi_p(k)\right]>T_p(k)/2$. $k\cdot T_p^2(k)$ is monotonically decreasing with $k$ if $\left[\Phi_p(k+1)-\Phi_p(k)\right]<T_p(k)/(2+1/k)$.

**Proof:** By Eq. (6), we get

$$
\begin{aligned}
&(k+1)T_p^2(k+1)-k\cdot T_p^2(k)\\
&=\left[T_p(1)+\Phi_p(k+1)\right]^2/(k+1)-\left[T_p(1)+\Phi_p(k)\right]^2/k\\
&=\left(k\left[T_p(1)+\Phi_p(k+1)\right]^2-k\left[T_p(1)+\Phi_p(k)\right]^2-\left[T_p(1)+\Phi_p(k)\right]^2\right)\Big/k(k+1) \qquad (11)\\
&=\left(\left[\Phi_p(k+1)-\Phi_p(k)\right]\left[2T_p(1)+\Phi_p(k+1)+\Phi_p(k)\right]-k\cdot T_p^2(k)\right)\Big/(k+1)\\
&=\left(\left[\Phi_p(k+1)-\Phi_p(k)\right]\left[(k+1)T_p(k+1)+k\cdot T_p(k)\right]-k\cdot T_p^2(k)\right)\Big/(k+1)
\end{aligned}
$$

Since from Lemma 2, $\Phi_p(k+1)>\Phi_p(k)$, hence

$$
\begin{aligned}
(k+1)T_p(k+1)&=\left[T_p(1)+\Phi_p(k+1)\right]\\
&>\left[T_p(1)+\Phi_p(k)\right]=kT_p(k)
\end{aligned}
\qquad (12)
$$

From Eq's (11) and (12), we conclude that if $\left[\Phi_p(k+1)-\Phi_p(k)\right]>T_p(k)/2$, then $(k+1)T_p^2(k+1)>kT_p^2(k)$. From Lemma 1 and Eq's (11) and (3), we obtain the following condition.

If

$$\left[\Phi_p(k+1)-\Phi_p(k)\right]<\frac{T_p(k)}{2+1/k}$$

then

$$
\begin{aligned}
&\left[(k+1)T_p^2(k+1)-k\cdot T_p^2(k)\right]\\
&<\left[\frac{k\cdot T_p(k)\left[(k+1)T_p(k+1)+k\cdot T_p(k)\right]}{(2k+1)\qquad(k+1)}-\frac{k\cdot T_p^2(k)}{(k+1)}\right]\\
&<\left[\frac{k\cdot T_p(k)}{2k+1}\frac{(2k+1)T_p(k)}{(k+1)}-\frac{k\cdot T_p^2(k)}{(k+1)}\right]=0
\end{aligned}
$$

Theorem 2 restricts the region within which we need to find a value $k$ that minimizes $k\cdot T_p^2(k)$. In other words, the approximate condition that adding a processor will not degrade the processor-time efficiency is that all processors should be busy at least half of the time.

| 0 | 1 | 2 | 3 | 4 | 5 | 6 | 7 | 8 | 9 | 10 | 11 |
|---|---|---|---|---|---|---|---|---|---|---|---|

| A | C | F | G | I | J |
|---|---|---|---|---|---|
| (idle) | B | D | E | H | (idle) |

Figure 9. Timing diagram for nonpreemptive CPS scheduling using two processors for the e-outin tree in Figure 2

In the example shown in Figures 2~5 and 9, $T_p(1)=18$, $\Phi_p(1)=0$, $T_p(2)=10.5$, $\Phi_p(2)=3$, $T_p(3)=9$, and $\Phi_p(3)=9$. (Readers are suggested to schedule this outin tree with three processors.) Since $(\Phi_p(2)-\Phi_p(1))=\ 3<T_p(1)/(2+1/1)=6$, and $\left[\Phi_p(3)-\Phi_p(2)\right]=6>\ T_p(2)/2=5.25$, according to Theorem 2, we conclude that the use of two processors minimizes $k\cdot T^2$ for this outin tree.

A question about the monotonicity of $\left[k\cdot T_p^2(k+1)-k\cdot T_p^2(k)\right]$ now arises naturally. If $\left[k\cdot T_p^2(k+1)-k\cdot T_p^2(k)\right]$ is increasing monotonically with $k$, then $k\cdot T_p^2(k)$ is a unimodal function of $k$, and the optimal value of $k$ can be found easily. This monotonicity is proved in the following theorem.

**Theorem 3:** Suppose that an outin tree is evaluated by the PCPS algorithm, then $k \cdot T_{\mathrm{p}}^2(k)$ is a concave function of $k$, that is, $k \cdot T^2(k)$ achieves the minimum when $k = k'$, and $k \cdot T^2(k)$ is monotonically decreasing (resp. increasing) with $k$ when $k<k'$ ( resp. $k>k'$ ).

**Proof:** To show $k \cdot T_{\mathrm{p}}^2(k)$ is a concave function of $k$, we need to prove that its second-order difference is positive, namely, $\left[(k+2)T_{\mathrm{p}}^2(k+2)-(k+1)T_{\mathrm{p}}^2(k+1)\right]>[(k+1)\cdot T_{\mathrm{p}}^2(k+1)-k\cdot T_{\mathrm{p}}^2(k)]$ . Let $\Delta(k\cdot T_{\mathrm{p}}^2(k))$ denote $\left[(k+1)T_{\mathrm{p}}^2(k+1)-k\cdot T_{\mathrm{p}}^2(k)\right]$. Then

$$\Delta\left(k\cdot T_{\mathrm{p}}^2(k)\right)=k\left[T_{\mathrm{p}}^2(k+1)-T_{\mathrm{p}}^2(k)\right]+T_{\mathrm{p}}^2(k+1) \tag{13}$$

$$\Delta\left((k+1)T_{\mathrm{p}}^2(k+1)\right)=(k+1)\left[T_{\mathrm{p}}^2(k+2)-T_{\mathrm{p}}^2(k+1)\right]+T_{\mathrm{p}}^2(k+2) \tag{14}$$

Subtracting Eq. (13) from Eq. (14) and applying Eq. (6) yield

$$\begin{aligned}&\Delta\left((k+1)T_{\mathrm{p}}^2(k+1)\right)-\Delta\left(k\cdot T_{\mathrm{p}}^2(k)\right)\\&=(k+2)\left[T_{\mathrm{p}}^2(k+2)-T_{\mathrm{p}}^2(k+1)\right]-\\&k\left[T_{\mathrm{p}}^2(k+1)-T_{\mathrm{p}}^2(k)\right]=\left[T_{\mathrm{p}}(k+2)+T_{\mathrm{p}}(k+1)\right]\times\\&\left[\Phi_{\mathrm{p}}(k+2)-\Phi_{\mathrm{p}}(k+1)-T_{\mathrm{p}}(k+1)\right]\\&-\left[T_{\mathrm{p}}(k+1)+T_{\mathrm{p}}(k)\right]\times\\&\left[\Phi_{\mathrm{p}}(k+1)-\Phi_{\mathrm{p}}(k)-T_{\mathrm{p}}(k+1)\right]\end{aligned} \tag{15}$$

From Eq's (3) and (15) and Lemmas 1, 2, and 3, we conclude that $\{\Delta((k+1)T_{\mathrm{p}}^2(k+1))-\Delta(k\cdot T_{\mathrm{p}}^2(k))\}>0$ .

We have found the condition under which $k\cdot T_{\mathrm{p}}^2$ is increased or decreased based on the intermediate variable, $\Phi_{\mathrm{p}}(k)$ , and that $k\cdot T_{\mathrm{p}}^2$ is a concave function. Next, we determine the number of processors such that $k\cdot T_{\mathrm{p}}^2$ is minimum for any given outin tree.

Note that in the original outin tree, each node is a multiunit task, and tasks in a level may have different lengths. If there are $m(i)$ tasks in level $i$, then there are $m(i)$ paths from level $i$ to the exit node. Among these paths, the minimum length is denoted by $l(i)$ . Similarly, we can define the depth of a node as the sum of task times along a path from the entry node to and including this node, and denote the shortest depth from the entry node to level $i$ by $d(i)$ . For example, in Figure 2, node A is in level 4, and node J is in level 0. Nodes B and C are in level 3. The depths of nodes B and C are both 2; hence, $d(3)=2$ .

Given $k$ processors, we can find $c_k$ , a particular level in the intree of the original outin tree, such that $m(c_k)$ , the number of tasks in this level, is less than $k$ , but $m(c_k+1)\geqslant k$ . This particular level is called the minimum-all-busy level. Note that the entry node has the maximum level number and the exit node has the minimum level number. Likewise, in the outree, there is a level called the maximum-all-busy level and denoted by $s_k$ , such that $m(s_k)<k$ and $m(s_k-1)\geqslant k$ . For the outin tree in Figure 2, $s_3=3$ , $c_3=1$ , $d(s_3)=2$ , and $l(c_3)=3$ .

Note that the minimum-all-busy level (resp. maximum-all-busy level) does not correspond to the phase boundary $B_{\mathrm{ac}}$ (resp. $B_{\mathrm{sa}}$ ). However. by recognizing the minimum-all-busy and maximum-all-busy levels, we can roughly estimate the locations of the phase boundaries. Recall from Proposition 2 that $T_{\rm p}(k)=h_{\max}$ if the phase boundary $B_{\rm ac}$ is not in a single e-level. In this case, $(k+1)T_{\rm p}^2(k+1)>k\cdot T_{\rm p}^2(k)$. To achieve the minimum $k\cdot T_{\rm p}^2(k)$, the number of processors should be reduced until the phase boundary appears in a single e-level. (When $B_{\rm ac}(k)$ is not in a single e-level but $B_{\rm ac}(k-1)$ is, $k\cdot T_{\rm p}^2(k)$ may be minimum.) This observation shows that the use of the minimum-all-busy level to estimate $T_{\rm pc}$ is accurate in most cases. The same argument applies to $T_{\rm pc}$. The following lemma shows that the shortest length from the minimum-all-busy ( resp. maximum-all-busy) level to the exit (resp. entry) node gives the lower-bound computational time in the combining (resp. splitting) phase.

**Lemma 4:** Suppose that an outin tree is evaluated by $k$ processors. If $s_k$ and $c_k$ are the maximum-all-busy and minimum-all-busy levels, then (a) $T_{\rm ps}(k)\geqslant d(s_k)$, and (b) $T_{\rm pc}(k)\geqslant l(c_k)$. Further, if the phase boundary $B_{\rm ac}(k)$ lies in a single e-level, then $T_{\rm pc}=l(c_k)$.

**Proof:** In part (a), assume that there are $m$ tasks in the maximum-all-busy level, $s_k$, where $m<k$. Hence, the computational process can enter the all-busy phase only if some task in the minimum-all-busy level has finished. That is, the computational time required in the splitting phase cannot be less than $d(s_k)$, the depth of the shortest path from the entry node to the minimum-all-busy level $s_k$. Likewise, we can prove part (b). Lastly, if the phase boundary $B_{\rm ac}(k)$ lies in a single level, then all nodes in $B_{\rm ac}(k)$ complete simultaneously. From then on, each e-node in the combining phase is allocated an entire processor, and no tasks got delayed due to processor sharing. As a result, the all-busy phase ends when it enters the minimum-all-busy level, and $T_{\rm pc}(k)=l(c_k)$.

This lemma is illustrated by the example in Figure 2. Suppose that three processors are used, the maximum-all-busy level contains tasks B and C, and the minimum-all-busy level contains tasks H and I. $T_{\rm ps}(3)$ (resp. $T_{\rm ps}(3)$) cannot be less than 2 (resp. 3) because if either task B or C, which are associated with the shortest depth from the entry node to this level, is not finished, then the computation cannot enter the all-busy phase. Likewise, if task H has been assigned to a processor, then the computation must have entered the combining phase. As another example, the minimum-all-busy level when two processors are used contains J. Since $B_{\rm ac}(2)$ lies in a single e-level, $T_{\rm pc}(2)=l(c_2)=2$.

For an arbitrary outin tree and $k>1$,

$$T_{\rm ps}\geqslant t_{\rm en},\ T_{\rm pc}\geqslant t_{\rm ex},\ \text{and}\ \Phi_{\rm p}\geqslant(t_{\rm en}+t_{\rm ex})\quad(16)$$

where $t_{\rm en}$ and $t_{\rm ex}$ are the task times of the entry and exit nodes, respectively. When more than one processor is used, some processors must be idle when the entry and exit nodes are evaluated. If the times spent in evaluating the entry and exit nodes dominate all other computations, then parallel processing is definitely inefficient. The following corollary identifies the condition under which sequential computation is better than parallel processing.

**Corollary 1:** Suppose that an outin tree is scheduled by the PCPS algorithm and $(t_{\rm en}+t_{\rm ex})>T(1)/2$, then sequential processing, i.e., $k=1$, achieves the minimum $k\cdot T^2$.

**Proof:** This follows from Theorem 2 immediately.

Having proved a series of propositions, lemmas, and theorems, the main theorem for deriving the optimal degree of parallelism under the PCPS scheduling algorithm can be obtained now. In the following theorem, the region on $k$ in which we can find the optimal degree of parallelism for evaluating any given outin tree is given.

**Theorem 4:** Suppose that an AND tree is evaluated by the PCPS algorithm and $k>1$, then

$$(k+1)T_p^2(k+1)>k\cdot T_p^2(k) \quad \text{if} \quad k>\frac{2T_p(1)}{h_{max}} \tag{17}$$

and

$$(k+1)T_p^2(k+1)>k\cdot T_p^2(k) \text{ if}$$

$$k<\left\{\frac{T_p(1)+t_{en}+t_{ex}}{2h_{max}}-\frac{1}{2}\right\} \tag{18}$$

**Proof:** Let $\Delta\Phi_p(k)=\left[\Phi_p(k+1)-\Phi_p(k)\right]$.

From Lemma 2.

$$\Delta\Phi_p(k)\geqslant T_{ps}(k)+T_{pc}(k) \tag{19}$$

Since the idle time of each processor cannot be larger than $(T_{ps}+T_{pc})$. we have

$$\Phi_p(k)\leqslant(k-1)\left[T_{ps}(k)+T_{pc}(k)\right] \tag{20}$$

By Theorem 2 and Eq's. (6), (19), and (20), the condition that guarantees the monotonic increase of $k\cdot T_p^2(k)$ with $k$ is

$$(k+1)T_p^2(k+1)>k\cdot T_p^2(k)$$

$$\text{if}\left[T_{ps}(k)+T_{pc}(k)\right]>\frac{T_p(1)+(k-1)\left[T_{ps}(k)+T_{pc}(k)\right]}{2k},$$

or equivalently,

$$(k+1)T_p^2(k+1)>k\cdot T_p^2(k) \text{ if}$$

$$k>\left(\frac{T_p(1)}{T_{ps}(k)+T_{pc}(k)}-1\right) \tag{21}$$

On the other hand, when all tasks but those in the critical path can be completed by $(k-1)$ processors during $h_{max}-\left[T_{ps}(k)+T_{pc}(k)\right]$ time units, i.e.,

$$(k-1)>\frac{T_p(1)-h_{max}}{h_{max}-\left[T_{ps}(k)+T_{pc}(k)\right]} \tag{22}$$

the phase boundary $B_{ac}$ must not be in a single e-level. As discussed before, the number of processors to achieve the optimal processor-time efficiency cannot be larger than the RHS of Eq. (22) plus one. By Lemma 4 and Eq's (21) and (22),

$$(k+1)T_p^2(k+1)>k\cdot T_p^2(k) \quad \text{if}$$

$$k>\min\left\{\frac{T_p(1)}{d(s_k)+l(c_k)}-1,\frac{T_p(1)}{h_{max}-\left[d(s_k)+l(c_k)\right]}\right\} \tag{23}$$

The condition described in Eq. (17) is obtained from Eq. (23).

Note that $\Delta\Phi_p(k)\leqslant h_{max}$, and that $k\cdot T_p(k)>\left[T_p(1)+t_{en}+t_{ex}\right]$ according to Lemma 1 and Eq's (6) and (16). By Theorem 2, the following result can be derived.

$$(k+1)T_p^2(k+1)<k\cdot T_p^2(k) \quad \text{if}$$

$$h_{max}<\frac{T_p(1)+t_{en}+t_{ex}}{2k+1}$$

which is equivalent to Eq. (18).

To find the number of processors that achieves the optimal processor-time efficiency, we need to search the small region of $k$ defined by Theorem 4. The lower and upper bounds of this region are, respectively,

$\{T_p(1)/(2h_{max})-1\}$ and $\{2T_p(1)/(h_{max})+1\}$. Note that we have not made any assumption about the distribution of task times in deriving these bounds. Since $k\cdot T_p^2$ is a concave function of $k$ (Theorem 3), the desirable number of processors can be found efficiently by a binary search. The binary search can be completed within about $\log_2(T_p(1)/h_{max})$ steps. Each step in the binary search tests whether $\Delta(k\cdot T_p^2(k))$ is positive. If it is, then a smaller value of $k$ will be checked in the next step, otherwise, a larger $k$ will be tested.

For any $k$ inside the search region, the phase boundary $B_{ac}(k)$ is in a single e-level; hence the location of $B_{ac}(k)$ can be uniquely determined without knowing the detailed schedule. Accordingly,

$$\Phi_p(k)=k\left[T_{ps}(k)+T_{pc}(k)\right]-\left[N_{ps}(k)+N_{pc}(k)\right]$$

where $N_{ps}(k)$ and $N_{pc}(k)$ are the amount of task times in the splitting and combining phases. From Lemma 4, $T_{pc}(k)=l(c_k)$, and $T_{ps}(k)$ can be found directly from the e-outin tree. As a result,

$$k\cdot T_p^2(k)=\frac{\left[T_p(1)+\Phi_p(k)\right]^2}{k}$$

For instance, suppose that $N$ items need to be sorted. It is well known that $T(1)=N\cdot\log_2 N$ if a merge-sort algorithm is used. In this case, the overhead in the intree part dominates that of the outtree part. For the intree part, $h_{max}=N+N/2+\cdots+1=2N-1$, so the lower and upper bounds of the search region can be determined from Theorem 4, which are close to $(\log_2 N)/4$ and $\log_2 N$, respectively. Since there are only $(3\cdot\log_2 N)/4$ candidate values in this search region, $\log_2\log_2 N$ steps of a binary search can guarantee to find the optimal number of processors for parallel mergesorting. For problems such as evaluating numerical or logic expressions and finding the maximum (or minimum) value, all task times are identical. Theorem 4 predicts that the optimal degree of parallelism is between $N/(2\cdot\log_2 N)$ and $2N/\log_2 N$. Figure 10 shows the simulation results of applying a nonpreemptive CPS algorithm to a binary intree of 4 096 terminal nodes and $t_i=1$ for all $i$. Since all tasks have unit execution times, the performance of the nonpreemptive CPS algorithm is very close to that of GS algorithm (see Eq. (2)). In this example, $k\cdot T^2$ is minimum when 431 processors are used, which is between $N/(2\cdot\log_2 N)$ (=170) and $2N/\log_2 N$ (=683).

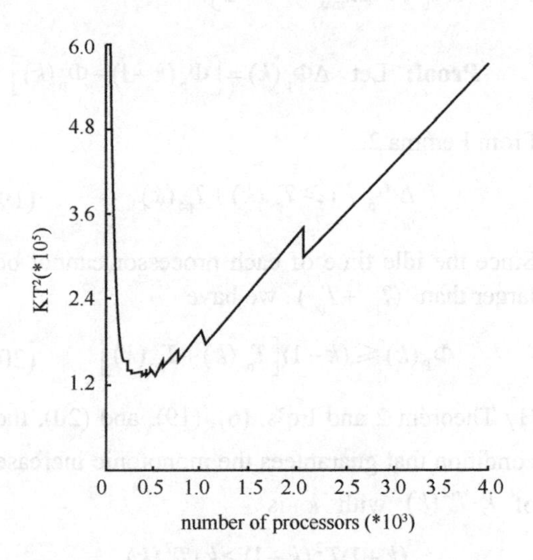


Figure10. Simulation results to find the number of processors to achieve the optimal processor-time efficiency in evaluating an in tree which has 4 096 leaves and unit execution times for all nodes using a non-preemptive CPS algorithm

The above analysis reveals that the number of processors for optimizing processor-time efficiency in evaluating any given outin tree is related to the following parameters, all of which can be obtained easily from the original outin tree:

(a) $T(1)$, the time required by a sequential evaluation, which is the sum of all task times in the outin tree;

(b) $h_{max}$, the length of the critical path;

(c) $d(s_k)$, the depth from the entry node to $s_k$, the maximum-all-busy level; and

(d) $l(c_k)$, the length from $c_k$, the minimum-all-busy level, to the exit node (recall that $c_k$ and $s_k$ depend on $k$).

$T(1)/h_{max}$ reflects the shape of the outin tree, while $T(1)/d(s_k)$ and $T(1)/l(s_k)$ reflect the distribution of the task times. If the outin tree is "wide" and nearly balanced, i.e., $T(1)/h_{max}$ is large, then a large number of processors are more appropriate. Further, if tasks in levels closer to the entry and exit nodes have long execution times, i.e., $T_p(1)/[d(s_k)+l(c_k)]$ is small, then the degree of parallelism should be small in order to optimize the processor-time efficiency. Both $T(1)/h_{max}$ and $T(1)/[d(s_k)+l(c_k)]$ are related to the problem complexity.

The influence of problem complexity on the optimal degree of parallelism is again shown in the next section with respect to nonpreemptive scheduling algorithms.

## 4 Optimal degree of parallelism in nonpreemptive scheduling

Nonpreemptive CPS algorithms are similar to the PCPS algorithm except that preemption is not allowed. In the nonpreemptive CPS algorithm, one processor is assigned to each of the $k$ nodes farthest from the exit node. If there is a tie in lengths among more than one node, then a left-to-right tiebreaking rule is used to assign a processor to one of these nodes. When a task of the outin tree is completed, the free processor is assigned to the node farthest from the root in the remaining outin tree to be evaluated. Figure 9 illustrates a schedule obtained by a nonpreemptive CPS algorithm.

In general, nonpreemptive scheduling is more practical due to the smaller task-switching overheads; however, it is more difficult to predict its performance and determine the optimal degree of parallelism. The difficulty lies in the anomalous behavior of nonpreemptive CPS algorithms. Graham has proved that if an AND tree is evaluated twice by using $k_1$ and $k_2$ processors, respectively[31], then

$$\frac{T_{np}(k_1)}{T_{np}(k_2)} \leqslant \left\{1+\frac{k_2-1}{k_1}\right\}$$

The above inequality implies that the anomaly $T_{np}(k+1)/T_{np}(k) < k/(k+1)$ is possible. In other words, $k \cdot T_{np}^2(k)$ is generally not a concave function of $k$ and cannot be searched by a binary search or other efficient search methods. This phenomenon is illustrated in Figure 10.

In a special case, if the execution times of tasks of an outtree are monotonically decreasing as the tree is decomposed, then it is shown in Lemma 5 below that $\Phi_{np}(k_2) > \Phi_{np}(k_1)$ holds for $k_2 > 2k_1$. Likewise, the same relation holds for the case when the execution times of tasks of an intree are monotonically increasing as the tree is composed. Using the conditions developed for $\Phi_{np}(k_2)$ and $\Phi_{np}(k_1)$, we show in Theorem 5 the conditions under which $k \cdot T^2$ is monotonically increasing or decreasing with $k$. These results demonstrate that the optimal degree of parallelism of outin tree computations based on a nonpreemptive CPS algorithm can be bounded in a relatively small region. The assumption on monotonic distribution of task times is valid in many divide-and-conquer algorithms.

**Lemma 5:** Suppose that an outin tree is scheduled by a nonpreemptive CPS algorithm and that $t_i > t_j$ if task $i$ is a predecessor (resp. successor) of task $j$ in the outtree (resp. intree) part, then

$$\left[\Phi_{np}(k_2)-\Phi_{np}(k_1)\right] \geqslant \left\{(k_2-k_1)\left[T_{ps}(k_1)+T_{pc}(k_1)\right]-k_1 t_{npa}(k_1)\right\}>0$$

if $k_2 > 2k_1$ (24)

$$\left[\Phi_{np}(k_2)-\Phi_{np}(k_1)\right] \leqslant \left\{(k_2-k_1)\left[T_{ps}(k_2)+T_{pc}(k_2)\right]+k_2 t_{npa}(k_2)\right\}>0$$

if $k_2 > k_1$ (25)

where $t_{npa}(k)$ is the longest task time among all tasks in the all-busy phase when $k$ processors are used.

**Proof:** Since the same behavior appears in the splitting phase under both preemptive and nonpreemptive scheduling, and only tasks in the all-busy phase are involved in the proof of Eq. (1)[27], it is easy to refine Kaufman's results as follows.

$$T_p(k) \leqslant T_{np}(k) \leqslant T_p(k)+t_{npa}(k) \tag{26}$$

By Eq. (26), we have

$$k_2 T_{np}(k_2) \geqslant k_2 T_p(k_2) \tag{27}$$

and

$$k_1 T_{np}(k_1) \leqslant k_1 T_p(k_1)+k_1 t_{npa}(k_1) \tag{28}$$

In Lemma 2, we have shown that $\Phi_p(k+1)-\Phi_p(k) \leqslant \left[T_{pc}(k)+T_{ps}(k)\right]$. From Proposition 4 and Eq. (6), we get, for $k_2 > k_1$,

$$\left[k_2 T_p(k_2)-k_1 T_p(k_1)\right]=\left[\Phi_p(k_2)-\Phi_p(k_1)\right] \geqslant \left[()k_2-k_1\left(T_{ps}(k_1)+T_{pc}(k_1)\right)\right] \tag{29}$$

By Eq's (27)~(29) and the hypothesis of monotonicity of the task times, which guarantees that $(T_{ps}(k_1)+T_{pc}(k_1)) > t_{npa}(k_1)$, Eq. (24) is proved. The proof for Eq. (25) is analogous.

We should point out that the above lemma holds for the case in which a part of the phase boundary is in the intree and another part is in the outtree. The above lemma is true because the task time of a node in the all-busy phase is less than either $T_{ps}$ or $T_{pc}$ from the monotonicity assumption on task times; that is, $[T_{ps}(k_1)+T_{pc}(k_1)] > t_{npa}(k_1)$ is always true regardless of the location of the phase boundary.

Similar to Theorem 2, we first study the relationship between $k \cdot T^2$ and the idle times. The following theorem gives the conditions under which $k \cdot T^2$ is monotonically increasing or decreasing based on the intermediate variable $\Phi_{np}(k)$.

**Theorem 5:** Suppose that an outin tree is scheduled by a nonpreemptive CPS algorithm and that $t_i > t_j$ if task $i$ is a predecessor (resp. successor) of task $j$ in the outtree (resp. intree) part, then

$k_2 T_{np}^2(k_2) > k_1 T_{np}^2(k_1)$ if

$$\left[\Phi_{np}(k_2)-\Phi_{np}(k_1)\right] > \left[\frac{k_2-k_1}{2} T_{np}(k_1)\right] \tag{30}$$

and $k_2 > 2k_1$

$k_2 T_{np}^2(k_2) < k_1 T_{np}^2(k_1)$ if

$$\left[\Phi_{np}(k_2)-\Phi_{np}(k_1)\right]<\left[\frac{2(k_2-k_1)k_1}{2k_1+3k_2}T_{np}(k_1)\right] \quad (31)$$

and $k_2>k_1$

**Proof:** From Eq. (6), we have

$$\left[k_2T_{np}^2(k_2)-k_1T_{np}^2(k_1)\right]$$
$$=\left\{\frac{\left(T_{np}(1)+\Phi_{np}(k_2)\right)^2}{k_2}-\frac{\left(T_{np}(1)+\Phi_{np}(k_1)\right)^2}{k_1}\right\}$$
$$=(k_1\{\left[\Phi_{np}(k_2)-\Phi_{np}(k_1)\right]\cdot[2T_{np}(1)+\Phi_{np}(k_2)+$$
$$\Phi_{np}(k_1)]\}-(k_2-k_1)\left[k_1T_{np}(k_1)^2\right])/k_1\cdot k_2 \quad (32)$$

From Lemma 5, if $k_2>2k_1$, then $\Phi_{np}(k_2)>\Phi_{np}(k_1)$, which implies that $k_2T_{np}(k_2)>k_1T_{np}(k_1)$ and that $(2T_{np}(1)+\Phi_{np}(k_2)+\Phi_{np}(k_1))>2k_1T_{np}(k_1)$. Hence, we obtain Eq. (30) from Eq. (32).

On the other hand, from Lemma 5,

$$\Phi_{np}(k_2)\leqslant[\Phi_{np}(k_1)+(k_2-k_1)\times$$
$$\left(T_{ps}(k_2)+T_{pc}(k_2)\right)+k_2t_{npa}(k_2)]$$

Note that in intrees and outin trees, $t_{ex}$, the execution time of the exit node, must be included in $h_{max}$, and $t_{ex}>t_{npa}$, hence,[1]

$$t_{npa}(k_2)<h_{max}/2 \quad (33)$$

Moreover, $\left[T_{ps}(k_2)+T_{pc}(k_2)\right]\leqslant h_{max}\leqslant T_{np}(k_1)$.

Thus,

$$\left(2T_{np}(1)+\Phi_{np}(k_2)+\Phi_{np}(k_1)\right)$$
$$\leqslant\left(2k_1T_{np}(k_1)+(k_2-k_1)T_{np}(k_1)+\frac{k_2T_{np}(k_1)}{2}\right)$$
$$=\left(k_1+\frac{3k_2}{2}\right)T_{np}(k_1)$$

and Eq. (30) follows from Eq. (31).

The main theorem to find the optimal degree of parallelism can be derived from Theorem 5. Before this theorem is proved, the following lemma is needed.

**Lemma 6:** For any given out in tree, suppose that both the PCPS and the nonpreemptive CPS algorithms are applied, then $[T_{nps}(k)+T_{npc}(k)]\leqslant[T_{ps}(k)+T_{pc}(k)+t_{npa}(k)]$.

**Proof:** By observing the PCPS and nonpreemptive CPS algorithms, it is easy to see that $T_{nps}(k)=T_{ps}(k)$. From Kaufman's proof of Eq. (1)[27], we have

$$T_{npc}(k)<\left[T_{pc}(k)+t_{npa}(k)\right] \quad (34)$$

The example in Figures 2~5 and 10 illustrates this lemma. Here, $\left[(T_{nps}+T_{npc})-(T_{ps}-T_{pc})\right]=1$, which is less than $t_{npa}(=3)$.

**Theorem 6:** Suppose that an outin tree is scheduled by a nonpreemptive CPS algorithm and that $t_i>t_j$ if task *i* is a predecessor (resp. successor) of task *j* in the outtree (resp. intree) part, then *k*, the number of processors that minimizes $k\cdot T_{np}^2(k)$, is bounded between $[T_{np}(1)+t_{en}+t_{ex}]/(8h_{max})$ and $3T_{np}(1)/[d(s_k)+l(c_k)-2t_{npa}(k)]$.

**Proof:** From Lemma 5 and Theorem 5, we obtain, for $k_2>2k_1$, that

$$k_2T_{np}^2(k_2)>k_1T_{np}^2(k_1) \text{ if}$$
$$\left\{(k_2-k_1)\left[T_{ps}(k_1)+T_{pc}(k_1)\right]-k_1t_{npa}(k_1)\right\}>\left\{\frac{k_2-k_1}{2}T_{np}(k_1)\right\} \quad (35)$$

From Eq. (6), Lemma 6, and the fact that $\Phi_{np}(k_1)<\left\{k_1\left[T_{nps}(k_1)+T_{npc}(k_1)\right]\right\}$, the condition

[1] For outin trees, $t_{npa}(k)<h_{max}/3$ because $t_{en}$ and $t_{ex}$, the execution times of entry and exit nodes, must be included in $h_{max}$, and $[t_{en}-t_{en}]>2\ t_{npa}$.

in Eq. (35) can be rewritten as

$$\begin{aligned}&\{(k_2-k_1)[T_{ps}(k_1)+T_{pc}(k_1)]-k_1t_{npa}(k_1)\}\\&>\left\{\frac{k_2-k_1}{2k_1}\left[T_{np}(1)+k_1\left(T_{ps}(k_1)+T_{pc}(k_1)+t_{npa}(k_1)\right)\right]\right\}\\&\text{for } k_2>2k_1\end{aligned} \tag{36}$$

A simple algebraical manipulation of Eq. (36) yields the condition

$$k_1>\frac{T_{np}(1)}{T_{ps}(k_1)+T_{pc}(k_1)-\dfrac{k_2+k_1}{k_2-k_1}t_{npa}(k_1)} \quad \text{for } k_2>2k_1 \tag{37}$$

When $k_2=3k_1$, Eq. (37) becomes

$$k_1>\frac{T_{np}(1)}{T_{ps}(k_1)+T_{pc}(k_1)-2t_{npa}(k_1)} \tag{38}$$

Since $[(k_2+k_1)/(k_2-k_1)]$ is monotonically decreasing with increasing $k_2$, Eq. (38) and the condition $k_2>3k_1$ imply Eq. (37). Note that Eq. (38) assures that for any $k$ larger than max $\left[3, 3T_{np}(1)/\left(T_{ps}(k)+T_{pc}(k)-2t_{npa}(k)\right)\right]$, we can find $k'<k/3$ such that $k'T_{np}^2(k')<kT_{np}^2(k)$. As a result, all numbers larger than $\left[3T_{np}(1)/\left(T_{ps}(k)+T_{pc}(k)-2t_{npa}\right)\right]$ can be excluded from consideration. The upper bound of the region containing the optimal number of processors is obtained by applying Lemma 4.

From Lemma 5, Theorem 5, and the fact that $\left[T_{ps}(k)+T_{pc}(k)\right]<h_{max}$, we get

$$k_2T_{np}^2(k_2)<k_1T_{np}^2(k_1) \text{ if } \left\{h_{max}+\frac{k_2t_{npa}(k_2)}{k_2-k_1}\right\}<\left\{\frac{2k_1T_{np}(k_1)}{2k_1+3k_2}\right\} \text{ and } k_2>k_1 \tag{39}$$

When $k_2=2k_1$, by Eq's (6), (16), and (33), the condition in Eq. (39) can be rewritten as

$$k_1<\frac{T_{np}(1)+t_{en}+t_{ex}}{8h_{max}} \tag{40}$$

This gives the lower bound of the optimal number of processors; that is, for any $k$ less than $(T_{np}(1)+t_{en}+t_{ex})/(8h_{max})$, we can find $k'=2k$ such that $k'T_{np}^2(k')<kT_{np}^2(k)$.

As an example, we can determine the area within which the number of processors that optimizes processor-time efficiency can be found for the parallel merge-sort of $N$ elements. In this problem, the computational overhead in the intree is dominant, so only the part of the intree has to be considered in the scheduling. From Theorem 6, the lower bound of the search region is $(\log_2 N)/16$, since $T_{np}(1)=N\cdot\log_2 N$ and $h_{max}<2N$. If $N$ is large enough, then $[d(s_k)+l(c_k)-2t_{npa}(k)]$ will be larger than 1.5$N$; hence, the upper bound of the search region is $2\cdot\log_2 N$. In contrast, recall that the search region to this problem is bounded by $(\log_2 N)/4$ and $\log_2 N$ when preemptive scheduling is used.

Comparing these bounds with those in Theorem 4, we see that the range within which the optimal number of processors using nonpreemptive scheduling can be found is larger than that of preemptive scheduling. Moreover, $k\cdot T^2$ is not monotonically decreasing or increasing with $k$ for nonpreemptive scheduling; i.e., $k\cdot T^2$ is not a unimodal function of $k$. Hence, an exhaustive search is required to find the optimal degree of parallelism in the region bounded by Theorem 6.

## 5 Conclusions

In this paper, we have (a) proved that the critical-path scheduling algorithm is optimal in minimizing the completion time of evaluating any given outin tree on a fixed number of identical parallel processors when preemptions are allowed, and (b) derived tight bounds on the number of processors within which the processor-time efficiency of the parallel evaluation of any given outin tree under preemptive and non-preemptive scheduling can be optimized. According to our efficiency analysis, we found that the optimal degree of parallelism depends on the problem complexity, the shape of the precedence graph (balanced or skewed), and the task-time distribution along each path (random or monotonic).

The complexity of each node in the outin tree is an important factor that influences the optimal degree of parallelism. If the overhead of each node is high with respect to the size of its input (such as $\theta(2^n)$ for inputs of size $n$) and a large number of processors are used, then the processor-time efficiency must be poor regardless of the capacity of the interconnection network. In this case, the time needed to evaluate a subproblem is increased rapidly during the decomposition phase in the outtree and the composition phase in the in tree. Hence, the root and exit nodes of the tree are obvious bottlenecks. In contrast, if the complexity of each node is a constant, then the root and exit nodes are not bottlenecks. Examples of this kind of problems include finding the maximum and evaluating an arithmetic expression. Here, a computing system with a large number of processors is appropriate, and a large speedup is expected. Tree-structured computer architectures[9,13] and virtual-tree computers[11] are good candidates in these applications. In cases in which the overhead of each node is small with respect to the size of its inputs (such as $\theta(n)$ or $\theta(\log n)$ for inputs of size $n$), then the time needed to evaluate a subproblem is increased slowly during the decomposition phase in the outtree and the composition phase in the intree. A computing system with a moderate number of processors is more cost-effective. For example, to sort 4 000 elements by a parallel merge-sort algorithm, using 10 to 12 processors is a good choice.

The shape of the outin tree is another important factor to be considered. Let $T_{\mathrm{p}}(1)/h_{\max}$ be "average width" of an outin tree. The optimal degree of parallelism is found to depend strongly on the average width. If the outin tree is "wide," then the degree of parallelism is high and the granularity can be small. On the other hand, if the outin tree is "narrow," then the degree of parallelism is low and the granularity is necessarily large. In order to achieve a high processor-time efficiency, a narrow tree may have to be restructured to arrive at a different representation.

Lastly, the distribution of task times is an important factor that influences the processor-time efficiency. For many practical problems, especially when divide-and-conquer algorithms are used, the precedence graph is nearly balanced and the task times of all nodes in each level are approximately equal. In this case, well-balanced workloads with overlapped process communications can be assigned to processors connected in a SIMD model. The optimal degree of parallelism is, therefore, close to the theoretical one predicted in this paper. On the other hand, for problems represented as irregular outin trees and for problems in which the overhead of each task may be data dependent, it is important that the underlying computer architecture support the dynamic distribution of workload. For a computer system with a high degree of parallelism, an efficient interconnection network is needed. In a computing system with a low degree of parallelism, an effective load balancing mechanism is necessary. The correct analysis for the latter system should, therefore, include the communication overhead, as process communications may not overlap with computations. Consequently, the optimal degree of parallelism may be less than the theoretical value predicted in this paper.

## References:


[1] BENTLEY J L, SHAMOS M I. Divide and conquer for linear expected time[J]. Elsevier, 1978, 7(2): 87-91.

[2] BENTLEY J L. Multidimensional divide-and-conquer[J]. Communications of the ACM, 1980: 51-59.

[3] BENTLEY J L, HAKEN D, SAXE J B. A general method for solving divide-and-conquer recurrences[J]. ACM SIGACT News, 1980, 12(3): 36-44.

[4] MONIER L. Combinatorial solutions of multidimensional divide-and-conquer recurrences[J]. Journal of Algorithms, 1980, 1(1): 60-74.

[5] BACKUS J W. Can programming be liberated from the von Neumann style? A functional style and algebra of programs[J]. Communications of the ACM, 1978, 21: 613-641.

[6] DENNIS J B. Data flow supercomputers[J]. Computer, 1980, 13(11): 48-56.

[7] CONERY J S, KIBLER D F. AND parallelism and nondeterminism in logic programs[J]. New Generation Computing, 1985, 3(1): 43-70.

[8] BENTLEY J L, KUNG H T. A tree machine for searching problems[C]// Proceedings of the International Conference on Parallel Processing. Piscataway: IEEE Press, 1979: 257-266.

[9] HARRIS J A, SMITH D R. Simulation experiments of a tree organized multicomputer[C]// Proceedings of the 6th Annual Symposium on Computer Architecture. New York: ACM Press, 1979: 83-89.

[10] PETERS F J. Tree machine and divide-and-conquer algorithms[C]// Proceedings of the International Conference on Parallel Processing. Heidelberg: Springer Press, 1981: 25-35.

[11] BURTON F M, HUNTBACH M M. Virtual tree machines[J]. IEEE Transactions on Computer, 1984, C-33(3): 278-280.

[12] HOROWITA E, ZORAT A. Divide-and-conquer for parallel processing[J]. IEEE Transactions on Computers, 1983, C-32(6): 582-585.

[13] MAGO G A. Making parallel computations simple: the FFP machine[C]// Proceedings of the 30th IEEE Computer Society International Conference. Piscataway: IEEE Press, 1985: 424-428.

[14] KELLER R M, LIN F C H, TANAKA J. Rediflow multiprocessing[C]// Proceedings of the 28th IEEE Computer Society International Conference. Piscataway: IEEE Press, 1984: 410-417.

[15] TNAG C Y, LEE R C T. Optimal speeding up of parallel algorithm based on the divide-and-conquer strategy[J]. Information Sciences, 1984, 32(3): 173-186.

[16] COFFMAN E G. Computer and job-shop scheduling theory[J]. Oral Surgery Oral Medicine Oral Pathology, 1976, 5(2): 143-149.

[17] CONWAY R W, MAXWELL W L, MILLER L W. Theory of scheduling[M]. Massachusetts: Addison-Wesley, 1967.

[18] GRAHAM R L, LAWLER E L, LENSTRA J K, et al. Optimization and approximation in deterministic sequencing and scheduling: a survey[J]. Annals of Discrete Mathematics, 1979, 5(1): 287-326.

[19] MUNTZ R R, COFFMAN E G. Preemptive scheduling of real-time tasks on multiprocessor systems[J]. Journal of the ACM, 1970, 17(2): 324-338.

[20] GAREY M R, JOHNSON D S. Scheduling tasks with nonuniform deadlines on two processors[J]. Journal of the ACM, 1976, 23(3): 461-467.

[21] LLOYD E L. Critical path scheduling with resource and processor constrains[J]. Journal of the ACM, 1982, 29(3): 781-811.

[22] KUNDE M. Nonpreemptive LP-scheduling on homogeneous multiprocessor systems[J]. Siam Journal on Computing, 1981, 10(1): 151-173.

[23] HU T C. Parallel sequencing and assembly line problems[J]. Operations Research, 1961, 9(6): 841-848.

[24] COFFMAN E G, GRAHAM R H. Optimal scheduling for two processors systems[J]. Acta Informatica, 1972, 1(3): 200-213.

[25] LENSTRA J K, RINNOOY KAN A, BRUCKER P. Complexity of machine scheduling problems[J]. Annals of Discrete Mathematics, 1977(1): 343-362.

[26] ULLMAN J D. NP-complete scheduling problems[J]. Journal of Computer and System Sciences, 1975, 10(3): 384-393.

[27] KAUFMAN M. An almost-optimal algorithm for the assembly line scheduling problem[J]. IEEE Transactions on Computers, 1974, C-23(11): 1169-1174.

[28] DOLEV D, WARMUTH M K. Profile scheduling of opposing forests and level orders[J]. SIAM Journal on Algebraic Discrete Methods, 1985, 6(4): 665-687.

[29] GAREY M R, JOHNSON D S, TARJAN R E, et al. Scheduling opposing forests[J]. SIAM Journal on Algebraic Discrete Methods, 1983, 4(1): 72-93.

[30] SCHINDLER S. On optimal scheduling for multiprocessor systems[C]// Proceedings of the Princeton Conference on Information Science and Systems. [S.l.:s.n.], 1972: 219- 223.

[31] GRAHAM R L. Bounds for certain multiprocessing anomalies[J]. Bell System Technical Journal, 1966, 45(9): 1563-1581.

发表于*IEEE Transactions on Software Engineering*,1990年第1期

# Computational efficiency of parallel combinatorial OR-tree searches

Guo-jie LI, Benjamin W. WAH

**Senior Member, IEEE**

**Abstract:** The performance of parallel combinatorial OR-tree searches (or OR-tree searches in short) is analytically evaluated in this paper. This performance depends on the complexity of the problem to be solved, the error allowance function, the dominance relation, and the search strategies. The exact performance may be difficult to predict due to the nondeterminism and anomalies of parallelism. We derive the performance bounds of parallel OR-tree searches with respect to the best-first, depth-first, and breadth-first strategies, and verify these bound by simulations. We show that a near-linear speedup can be achieved with respect to a large number of processors for parallel OR-tree searches. Using the performance bound developed, we derive sufficient conditions for assuring that parallelism will not degrade the performance, and necessary conditions for allowing parallelism to have a speedup greater than the ratio of number of processors. These bounds and conditions provide the theoretical foundation for determining the number of processors to assure a near-linear speedup.

**Key words:** anomalies, approximation, combinatorial search, dominance test, heuristic search, lower-hound test, OR-tree, parallel processing, performance bounds

## 1 Introduction

In a wide class of combinatorial search problems in artificial intelligence, operations research, decision making, and various scientific and engineering fields, it is necessary to find one or more, optimal, suboptimal, or feasible solutions in a large problem space. A combinatorial search, in short a search, enumerates some or all elements of the problem space until the solutions are found. Combinatorial search problems can be classified into two types. The first type is decision problems that decide whether at least one solution exists and satisfies a given set of constraints[1]. Theorem-proving and expert systems belong to this class. The second type is combinatorial extremum-search or optimization problems which are characterized by an objective function to be minimized or maximized and a set of constraints to be satisfied. Many practical problems, such as finding the shortest path, planning, finding the shortest tour of a traveling salesman, job-shop scheduling, packing a knapsack, vertex cover, and integer programming,

This work was supported in part by the National Science Foundation under Grants MIP 85-19649 and MIP 88-10584 and by the National Aeronautics and Space Administration under Grant NCC 2-481.

belong to this class.

Approximations and parallel processing are two major approaches to enhance the efficiency of combinatorial OR-tree searches. Owing to the exponential nature of many search problems, optimal solutions are usually infeasible to obtain. In practice, approximate solutions are acceptable alternatives. Experimental results on vertex-cover, 0-1 knapsack, and some integer-programming problems reveal that a linear reduction in accuracy may result in an exponential reduction of the average computational time[2]. On the other hand, parallel processing is applicable when the problem is solvable in polynomial time, or when the problem is NP-hard, but is solvable in polynomial time on the average[3], or when the problem is heuristically solvable in polynomial time. Parallel processing is generally useful for improving the computational efficiency of solving a given problem, and not in substantially extending the solvable problem size of the problem[4].

A search problem can be represented as either a tree or a graph. The nonterminal nodes in a search tree (or graph) can be classified as AND-nodes and OR-nodes. An AND-node represents a problem (or subproblem) that is solved only if all its descendent nodes have been solved. An OR-node represents a problem (or subproblem) that is solved only if any of its immediate descendents is solved. Based on these two kinds of nodes, a combinatorial search can be classified into an OR-tree search, an AND-tree search, and an AND/OR-tree search[4].

An OR-tree is a state-space tree in which all nonterminal nodes are OR-nodes. It is important because it represents nondeterminism, which is a natural property of many decision problems. Moreover, methods for solving OR-tree-search problems provides a basis to solving many general AND/OR-tree (or AND/OR-graph) search problems. A lot of the heuristic search procedures, such as A*, B*, AO*, SSS*, and dynamic programming, can be formulated as a general branch-and-bound (B&B) procedure[5-7]. Likewise, evaluating a logic program can be represented as an OR-tree search[8].

An OR-tree can be represented as an OR-graph in which all common nodes in the tree are grouped into a single node. Although more compact in representation, searching OR-graphs is inefficient in a parallel processing system in which the communication overhead is dominant. Frequent broadcasts of information from a parent node to multiple descendent nodes that it connects have to be made when the parent and its descendants are searched in different processors. In the corresponding OR-tree search, these broadcasts are eliminated because state information is only passed between the parent and one of the descendents that it activates. This savings in communication overhead is achieved at the expense of performing redundant search in the tree. A second reason for preferring OR-trees over OR-graphs is that, for some problems, the redundant nodes in the search tree are not apparent at program design time, and have to be found, usually at great overhead, at run time. For these reasons, we focus only on the efficiency of OR-tree searches in this paper.

OR-parallelism and its efficiency has been studied extensively in the literature[9-11]. One of the important issues addressed is linear scaling, that is, when a large number of processors are used, how can an OR-tree search be scheduled without any reprogramming such that the speed of computation increases in direct proportion to the number of computing elements. Rao and Kumar analyzed linear scaling in terms of an iso-efficiency function[10]. It has been reported that linear scaling does not usually hold for parallel game-tree searches[12-13] and parallel processing of forward-chaining rule-based expert systems[14]. Simulation results have also revealed that using more processors in parallel OR-tree searches might degrade the performance or might result in superlinear speedup when the communication overhead is ignored[15-17]. A number of conditions and methods to cope with anomalous behavior of branch-and-bound algorithms have been developed[11,15,18-20].

This paper presents new results on the efficiency of evaluating parallel OR-tree searches in the presence of anomalies. Results are derived with respect to finding one of the optimal (or suboptimal) solutions in the problem space, although they can be generalized to finding all optimal solutions. Our results are useful for designers to understand the existence of anomalies and modify existing algorithms to cope with these anomalies. They can also be used to determinate a good search strategy and predict the maximum number of processors under which a near-linear speedup can be achieved. Our results are derived with respect to the conventional best-first, depth-first, and breadth-first search strategies. Performance is measured in terms of the number of nodes expanded for a given number of processors. Modified search strategies, such as the IDA*[21], RTA*[22], and modified A* search using a virtual memory system[23], consider tradeoffs between computational resources and evaluation efficiency. The derivation of performance bounds for these modified strategies is much harder as performance has to be evaluated with respect to the actual execution time, rather than the number of nodes expanded.

## 2 Parallel combinatorial OR-tree searches

A combinatorial OR-tree search procedure can be characterized by four constituents: a branching rule, a selection rule, an elimination rule, and a termination condition. The first two rules are used to decompose problems into simpler subproblems and to appropriately order the search. The last two rules are used to eliminate subproblems generated that are not better than the ones already known. Appropriately ordering the search and restricting the region searched are key ideas behind any OR-tree search algorithm.

In contrast to conventional B&B algorithms, combinatorial OR-tree search algorithms allow approximate lower-bound and dominance tests as general elimination rules, and a generalized heuristic function as the selection rule. These rules are briefly explained in this section[24].

Let $P$, $E$, and $T$ be, respectively, sets of subproblems, edges, and feasible solutions in the OR-tree. Let $P_i$ be the $i$th subproblem, $P_{i_j}$ be the $j$th subproblem decomposable from $P_i$, and $f(P_i)$ be the value of the best solution obtained by evaluating all subproblems decomposable from $P_i$. For minimization problem, the lower-bound function satisfies the following properties.

$$g(P_i) \leqslant f(P_i) \quad \text{for} \quad P_i \in P \tag{1}$$

$$g(P_i) = f(P_i) \quad \text{for} \quad P_i \in T \tag{2}$$

$$g(P_i) \leqslant g(P_{i_j}) \quad \text{for} \quad (P_i, P_{i_j}) \in E \tag{3}$$

If a subproblem is a feasible solution with the best objective-function value so far, then its solution value becomes the incumbent $z$. The incumbent represents the best solution obtained so far in the process. During the search of an OR-tree, an active node $P_i$ is terminated if

$$g(P_i) \geqslant \frac{z}{1+\epsilon}, \quad \epsilon \geqslant 0, \quad z \geqslant 0 \tag{4}$$

where $\epsilon$ is an allowance function specifying the allowable deviation of a suboptimal value from the exact optimal value, and an active node is a subproblem that has been generated but not expanded. (In searching for an optimal solution, the corresponding elimination condition is $g(P_i) \geqslant z$.) The final incumbent value $z_F$ obtained by the approximate lower-bound test deviates from $z_O$, the optimal solution value, by

$$\frac{z_F}{1+\epsilon} \leqslant z_O \leqslant z_F, \quad \epsilon \geqslant 0, \quad z_O \geqslant 0 \tag{5}$$

Approximations significantly reduce the amount of intermediate storage and the time needed to arrive at a suboptimal solution. In the following sections, $L$ denotes the lower-bound cutoff test, that is, $P_jLP_i$ means that $P_j$ is a feasible solution and that $f(P_j)/(1+\epsilon) \leqslant g(P_i)$, $\epsilon \geqslant 0$.

Dominance tests are powerful elimination rules that are systematically applied in dynamic programming algorithms to reduce their complexity of enumeration[25]. In contrast to lower-bound tests which compare the incumbent with the lower bounds of all active nodes, dominance tests use a set of parameters to compare pairs of nodes in the search tree. These nodes can be either nonterminal or terminal nodes, and either active or expanded nodes. In this sense, lower-bound tests can be regarded as a special case of dominance tests. A lot of the well-known elimination rules, such as $\alpha$-$\beta$ pruning in game-tree searches, are actually dominance tests. Dominance tests, in the best case, can reduce the problem complexity from exponential to polynomial. For example, the shortest-path and the two-stage flowshop scheduling problems require exponential time using lower-bound tests only, but can be solved in linear time on the average with dominance tests[26].

A dominance relation is a binary relation such that $P_iDP_j$ implies that $P_i$ dominates $P_j$[27-28]. Hence, for minimization problems, the subtree rooted at $P_i$ contains a solution node whose value is no more than the minimum solution value of the subtree rooted at $P_j$. Thus if $P_i$ and $P_j$ are generated and $P_iDP_j$, then $P_j$ can be terminated. As an example, in the knapsack problem, it is necessary to pack objects into a knapsack of fixed capacity such that the total profit is maximized. Subproblem $P_i$ dominates Subproblem $P_j$ if the two subproblems involve the same set of objects, and the total profit of $P_i$ is larger than or equal to that of $P_j$, while the total weight of $P_i$ is smaller than or equal to that of $P_j$. Some results on the parallel implementation of dominance tests have been reported[4,16].

Another method to reduce the search space is to order the subproblems expanded. If the list of active subproblems is maintained in a first-in/first-out order, then the algorithm is called a breadth-first search. If the list is maintained in a last-in/first-out order, then the algorithm is called a depth-first search. Lastly, if the list is maintained in increasing order of lower bounds, then the search algorithm is called a best-first search. Ibaraki mapped these searches into a general form called heuristic searches[1][24]. A heuristic function defines the order in which subproblems are selected and decomposed. The algorithm always decomposes the subproblem with the minimum heuristic value. In a best-first search, the lower-bound values define the order of expansion. Therefore, the lower-bound function can be taken as the heuristic function. In a breadth-first search, subproblems with the minimum level numbers are expanded first. The level number can be taken as the heuristic function. Lastly, in a depth-first search, subproblems with the maximum level numbers are expanded first. The negation of the level number can be taken as the heuristic function.

A generalized heuristic function can be used to unify depth-first, breadth-first, and best-first searches. A heuristic function is said to be monotone when:

$$h(P_i) < h(P_j),\ \text{if}\ P_j\ \text{is a descendant of}\ P_i \tag{6}$$

Ibaraki also proved that, for any heuristic function, there exists an equivalent monotone heuristic function[24]. A heuristic function is said to be unambiguous when:

$$h(P_i) \neq h(P_j),\ \text{if}\ P_i \neq P_j P_i,\ P_j \in P \tag{7}$$

[1] The definition of general heuristic searches used in this paper is taken from Ibaraki's definition[24]. Depth-first, breadth-first, and best-first searches are considered as special cases of heuristic searches. This is different from the conventional definition of heuristic searches in artificial intelligence.

Notice that the heuristic functions defined conventionally for best-first, depth-first, and breadth-first searches are ambiguous because more than one node in the active list can have the same heuristic value. In subsequent sections, the properties on monotonicity and unambiguity are discussed with respect to predicating the performance and coping with anomalies of parallel OR-tree searches.

To resolve the ambiguity on the selection of subproblems, distinct heuristic values must be defined for the nodes to allow ties to be broken. A path number can be used to define an unambiguous heuristic function[11,16]. The path number of a node in a tree is a sequence of $(h+1)$ integers representing the path from the root to this node, where $h$ is the maximum number of levels of the tree. The path number $E=e_0,e_1,\cdots,e_h$ is defined recursively as follows. The root $P_0$ exists at Level 0 with a path number $000\cdots0$. A node $P_{i_j}$ on Level $L$, which is the $j$th child (counting from the left and starting from zero) of $P_i$ with path number $E_i=e_0e_1\cdots e_{L-1}000\cdots$, has path number $E_{i_j}=e_0e_1\cdots e_{L-1}j00\cdots$. According to this definition, nodes can have equal path numbers only if they have the ancestor-descendant relationship. Since these nodes never coexist simultaneously in the active list, the subproblems in the active list always have distinct path numbers. For example, the path numbers of all nodes in the tree in Figure 1 are shown next to the nodes. Relations "＞" and "＜" on path numbers are defined in the usual sense as those of integers.

In best-first and breadth-first searches, the path number can be included in the general heuristic function as a secondary key to resolve ties in the primary key, which is still the lower-bound value or the level number. In depth-first searches, the path number serves as a primary key. That is,

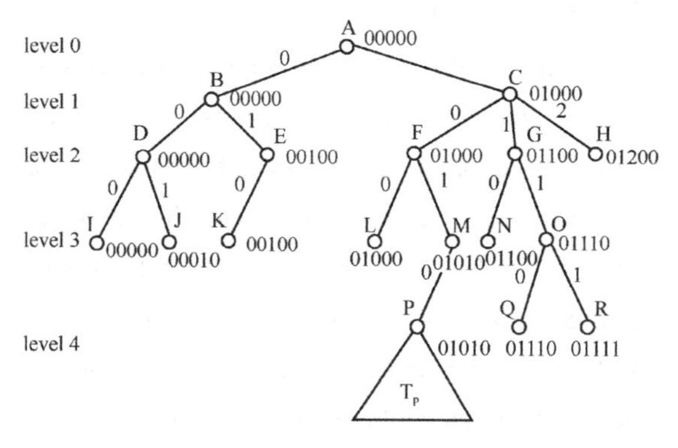


Figure 1. The path numbers of a tree

$$h(P_i)=\begin{cases}\text{(level number, path number)} & \text{breadth-first search}\\ \text{(path number, level number)} & \text{depth-first search}\\ \text{(lower bound, level number, path number)} & \\ \text{or (lower bound, path number, level number)} & \text{best-first search}\end{cases} \quad (8)$$

where the level number, path number, and lower bound are defined for $P_i$. For a best-first search, nodes with identical lower bounds can be searched in a breadth-first or depth-first fashion. In general, unambiguous heuristic functions are not restricted to the use of path numbers. Any tie-breaking rule can be adopted as long as Eq. (7) is satisfied.

The definition of the path number dictates that a partial order exists among successors of any given node in the search tree, and such an order is used consistently by both the sequential and parallel algorithms. Such a partial order implies that the same algorithm for decomposing a given node into successors is consistently used in both the sequential and parallel search. An attribute for defining this partial order may be needed if successors of a given node are generated in a random order. The lower-bound values can serve to define this partial order.

OR-tree searches have inherent parallelism. Each of the four rules of serial OR-tree search algorithms can be implemented by parallel processing. When $k$ processors are used, if the number of active subproblems is greater than or equal to $k$ in an iteration, then $k$ subproblems with the $k$ smallest heuristic values can be selected for decomposition; otherwise, all active subproblems can be selected. Parallel computer architecture for OR-tree searches have been studied elsewhere[16,29].

## 3 Bounds on computational efficiency of parallel OR-tree searches

To predict the number of processors needed to assure a near-linear speedup, we need to derive the bounds on computational efficiency of a parallel OR-tree search. The results in this section indicate the relationship among the number of nodes expanded in a parallel search, the number of processors used, and the complexity of the problem to be solved.

### 3.1 Model of efficiency analysis

In analyzing the performance bounds, a synchronous model is assumed, that is, all processors must finish the current iteration before proceeding to the next iteration. The performance is difficult to evaluate if the parallel search algorithm were evaluated asynchronously. The performance results for synchronous models form a lower bound to that of asynchronous models.

The parallel computational model used here consists of a set of processors connected to a shared memory. In each iteration, multiple subproblems are selected and decomposed. The newly generated subproblems are tested for feasibility (and the incumbent updated if necessary), eliminated by (exact or approximate) lower-bound tests and dominance tests, and inserted into the active list(s) if not eliminated. In this model, eliminations are performed after branching instead of after selection as in Ibaraki's algorithm[27] in order to reduce the memory space required.

The shared memory in the above computational model may seem to be a bottleneck of the system. However, this model can be transformed into a second model in which all processors have a private memory and are connected by a ring network. We have found that the efficiency of parallel OR-tree search algorithms in the transformed model is very close to that of the original model, and the performance is not affected by whether the active subproblems are kept in a single list or multiple lists[11,29]. Since subproblems are decomposed synchronously and the bulk of the overhead is on branching operations, the number of iterations, which is the number of times that subproblems are decomposed in each processor, is an adequate measure in both the serial and parallel models. The speedup is thus measured by the ratio of the number of iterations with respect to the different number of processors used. Once the optimal solution is found, the time to drain the remaining subproblems from the list(s) is not accounted for since this overhead is negligible as compared to that of branching operations.

The results proved in subsequent sections show the performance bounds of parallel best-first, depth-first, and breadth-first searches, respectively. The proofs of these theorems require the following definitions on essential nodes and basic nodes. A node expanded in a serial OR-tree search under a given heuristic function is called an essential node for that heuristic function; otherwise, it is called a nonessential node. The speedup of a parallel OR-tree search under a given heuristic function depends on the number of essential nodes selected in each iteration. An iteration is said to be perfect if the number of essential nodes selected is equal to the number of processors; otherwise, it is said to be imperfect. We denote $T_b(k,\epsilon)$, $T_d(k,\epsilon)$, and $T_r(k,\epsilon)$ as the number of iterations required to find a single optimal (or suboptimal) solution using $k$, $k \geqslant 1$, processors and an allowance function $\epsilon$ in a best-first, depth-first, and breadth-first search, respectively. The subscripts $b$, $d$, or $r$ are omitted when the context refers to more than one search strategy.

A basic node is the node with the smallest heuristic value in each iteration. It is easy to show that if $P_i$ is a basic node and $P_j$ is any node with a smaller heuristic value, then $P_j$ must be either expanded or terminated when $P_i$ is expanded[16].

Let $\Phi^k$, $k \geqslant 1$, be the set of nodes expanded in the OR-tree using $k$ processors. In fact,

$\Phi^1$ is a set of essential nodes. The following theorem states that any heuristic search with an unambiguous heuristic function can be guaranteed to expand at least one node in $\Phi^1$ in each iteration of the parallel search, when an exact optimal solution is sought and dominance tests are inactive.

**Theorem 1:** Let $\epsilon = 0$ and $D= I$ , i.e., an exact optimal solution is sought and dominance tests are inactive. For any parallel heuristic search with a heuristic satisfying Eq. (6) and Eq. (7), all basic nodes are essential nodes, i.e., all basic nodes belong to $\Phi^1$.

**Proof:** The proof is omitted due to space limitation[11,20].

## 3.2 Parallel update of incumbent

When the search is done in parallel, the way in which the incumbent is updated plays an important role in determining the lower bound of $T(k,\epsilon)$ and in coping with anomalies in approximate search. When more than one feasible solution is obtained in an iteration, updating the incumbent with the minimum feasible solution does not always result in a smaller search overhead and better solution. This phenomenon is demonstrated in an OR-tree search in Figure 2(a) with $\epsilon = 0.1$. The suboptimal solution found in a serial depth-first search is $F_7$ (with value 70), and subtree $T$ is eliminated. When the same tree is searched by four processors using a parallel depth-first search, five feasible solutions, $F_1,\cdots,F_5$, are generated in the fourth iteration. If the minimum among the five feasible solutions is chosen as the incumbent, $F_5$(with value 75) will become the incumbent, which eliminates $P_4$ and $P_5$ by approximate lower-bound tests Eq. (4). As a result, Subtree $T$, which may be very large, has to be expanded. To avoid this problem, the following algorithm updates the incumbent in a consistent fashion when multiple feasible solutions are obtained in an iteration.

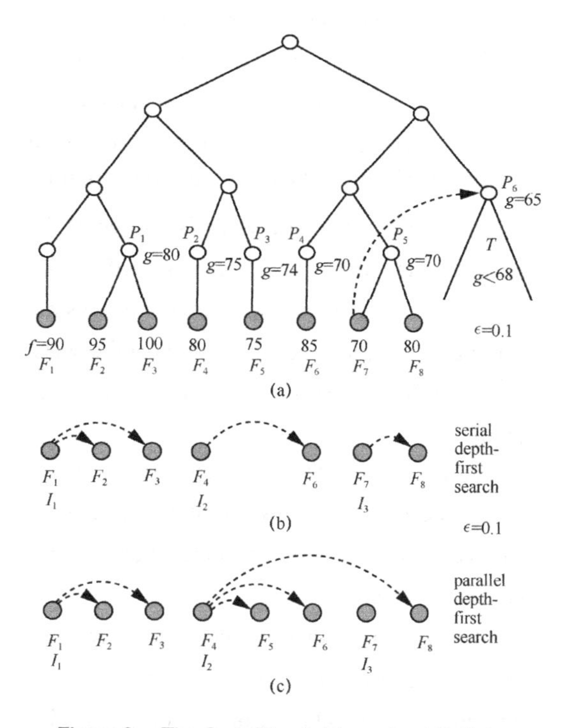


Figure 2. The Seria/Parallel Incumbent-Update algorithm. (Nodes pointed to by dashed arrows are eliminated by lower-bound tests. $I_1$, $I_2$, and $I_3$ are three distinct incumbents obtained during the serial and parallel searches.) (a) Complete search tree with feasible solutions indicated as shaded nodes. (b) Sequence of incumbents generated and pruned in a sequential depth-first search. (c) Sequence of incumbents generated and pruned using the Serial/Parallel Incumbent-Update algorithm

Serial/Parallel Incumbent-Update Algorithm:

a) When multiple feasible solutions are obtained in an iteration, they are compared to the incumbent. A feasible solution is eliminated if either its value is larger than or equal to the current incumbent, or the lower bound of its parent is not less than $[\text{incumbent}/(1+\epsilon)]$. If all feasible solutions are eliminated, then exit.

b) If there is one or more feasible solution whose value is less than the current incumbent and the lower bound of its parent is less than $[\text{incumbent}/(1+\epsilon)]$, then the feasible solution with the smallest heuristic value is chosen as the new incumbent and eliminated from further consideration. If there are no remaining feasible solutions, then the algorithm exists; otherwise, the entire Incumbent-Update algorithm is repeated.

Since the number of incumbents in an OR-tree search is usually small, it rarely happens that two or more incumbent-updates are needed in the same iteration.

To uniquely define the sequence of feasible solutions generated under a given heuristic function and a fixed number of processors, a feasible solution $F_i$ is said to appear before another feasible solution $F_j$ in the sequence of feasible solutions if either $F_i$ and $F_j$ are obtained in the $t_i$th and $t_j$th, $t_i < t_j$, iterations, or they are obtained in the same iteration and $h(F_i) < h(F_j)$. The following theorem proves the correctness of the Incumbent-Update algorithm.

**Theorem 2:** In an OR-tree search with $D=I$, the sequence of distinct incumbents obtained in the serial and parallel searches are identical regardless of $E$ and the number of processors used if the following conditions are satisfied: a) a feasible solution $F_i$ appearing before another feasible solution $F_j$ implies that $h(F_i) < h(F_j)$ in both the serial and parallel cases, and b) the Serial/Parallel Incumbent-Update algorithm is used.

**Proof:** The theorem is proved by induction. Let the sequence of all possible feasible solutions as ordered by the corresponding heuristic values be $F_1, \cdots, F_n$. The initial feasible solutions (or incumbents) in the serial and parallel cases before the search begins are identical. (An infinite value is taken as the first incumbent if there is no heuristic method to find the initial feasible solution.) Suppose that the sequence of the first $k$ distinct incumbents in the serial and parallel searches are identical and that $F_i$ is the $k$th incumbent. Let $F_j$, $j > i$, be the ($k$+1)th distinct incumbent in the serial case. We now show in the parallel search that any feasible solution $F_x$, $i < x < j$, cannot become an incumbent and that $F_j$ must be the ($k$+1)th distinct incumbent. Note that for all possible feasible solutions $F_x$, $i < x < j$, either a) $f(F_x) > f(F_i)$, or b) $g(P_x) \geqslant f(F_m)/(1+\epsilon)$, $m \leqslant i$, where $P_x$ is the parent of $F_x$. In Case a), $F_x$ will be eliminated by $F_i$ in the parallel case. In Case b), $F_x$ must also be eliminated by $F_i$ in the parallel search according to Eq. (3) and Step a) of the Seria/Parallel Incumbent-Update algorithm. Since $f(F_j) < f(F_i)$ and $g(P_j) < f(F_i)/(1+\epsilon)$, $F_j$ cannot be eliminated in the parallel search, where $P_j$ is the parent of $F_j$. Further, $h(F_j)$ has the smallest heuristic values among those of all possible feasible solutions $F_y$, $j \leqslant y \leqslant n$, hence $F_j$ has to be selected as ($k$+1)th incumbent in the parallel search.

The example in Figure 2 illustrates the Serial/Parallel Incumbent-Update algorithm and demonstrates the result in Theorem 2. In this OR-tree, using a depth-first search satisfies the condition that $F_i$ appearing before $F_j$ implies that $h(F_i) < h(F_j)$. The sequence of feasible solutions obtained in the parallel depth-first search with four processors is $F_1,\cdots,F_8$. During the parallel search, five feasible solutions, $F_1,\cdots,F_5$, are obtained in the fourth iteration. According to the Serial/Parallel Incumbent-Update algorithm, $F_2$, $F_3$, and $F_5$ are eliminated in Step a), and $F_4$ becomes the second incumbent. In the next iteration, $F_6$ and $F_8$ are eliminated, and $F_7$ is the final incumbent. Hence the sequence of incumbents is $F_1$, $F_4$, and $F_7$, which is identical to that of the serial case.

To derive the performance bounds of parallel depth-first and breadth-first searches, the following corollary is needed.

**Corollary 1:** In a parallel OR-tree search with $D=I$, all essential nodes under a given heuristic function must be expanded if the following conditions are satisfied: a) the sequences of distinct incumbents are the same in the serial and parallel cases, and b) $P_iLP_j$ and $P_k$ is the parent of $P_i$ imply that $h(P_k) < h(P_j)$.

**Proof:** Suppose that an essential node $P_1$ is eliminated by Node $P_2$, and that $P_3$ is the parent of $P_2$, then $h(P_3)$ must be less than $h(P_1)$ from Condition a). Since in a serial search, the nodes are expanded according to the heuristic values, $P_3$ should be selected before $P_1$. From Condition b), the feasible solution $P_2$ must be obtained in the serial case as well, hence $P_1$ must also be eliminated and cannot be an essential node. A contradiction!

Note that Condition a) of Theorem 2 and Condition b) of Corollary 1 are not satisfied for all OR-trees and search strategies even if the Serial/Parallel Incumbent-Update Algorithm is used. In particular, we have found that for $\epsilon \geqslant 0$, the conditions are satisfied for all OR-trees searched by a breadth-first search, and an OR-tree with feasible solutions at the bottom-most level and searched by a depth-first search. The proof for the case using a depth-first search is shown in Corollary 2. The proof for the case with a breadth-first search is straightforward and is not shown here. In both cases, the same sequence of distinct incumbents to be generated in the serial and parallel searches can be maintained by the Incumbent-Update Algorithm. Counterexamples can be designed to show that the conditions are invalid when an OR-tree with an arbitrary structure is searched by a best-first or a depth-first search.

## 3.3 Parallel best-first searches

The performance bounds of a parallel best-first search are shown in the following theorem.

**Theorem 3:** Suppose that $\epsilon = 0$, $D=I$. Let the root be in Level 0 and $h$ be the maximum number of levels in an OR-tree. If the heuristic function satisfies Eq. (6) and Eq. (7), then the following bounds hold for the parallel best-first search with $k$ processors.

$$\left(\frac{T'_b(1,0)-1}{k}+1\right) \leqslant T_b(k,0) \leqslant \left(\frac{T_b(1,0)}{k}+\frac{k-1}{k}h\right) \tag{9}$$

where $T'_b(1,0)$ is the number of essential nodes in a serial best-first search with lower bounds less than the optimal-solution value.

**Proof:** All the required iterations to find an optimal solution can be classified into either perfect or imperfect, and imperfect iterations cause degradation in performance in parallel processing. The proof of the upper bound centers on finding the maximum number of imperfect iterations. Let $h_{min}(x)$ be the level with the minimum level number in which some active essential nodes reside in the $x$th iteration, which is an imperfect iteration. For levels less than $h_{min}(x)$, all active nodes are nonessential. We show that Iteration $x$ may be imperfect only if all essential nodes in $h_{min}(x)$ are selected for expansion. Suppose that Iteration $x$ is imperfect and that an essential node, say $P_i$, in $h_{min}(x)$ is not selected for expansion, then this contradicts the selection rule of a heuristic search, since in this case at least one node with a heuristic value greater than $h(P_i)$ is selected. (Note that all nonessential nodes have heuristic values greater than those of essential nodes.) Thus after Iteration $x$ is carried out, $h_{min}(x)$ must be increased by at least one, that is, $h_{min}(x+1) \geqslant (h_{min}(x)+1)$.

As $h_{min}(x+1)$ must be less than $h$, the maximum number of levels, and the root is defined at Level 0, there can be at most $h$ imperfect iterations. When $\epsilon=0$ and $D=I$, it is true that once all essential nodes are expanded, the optimal solution must be found. Recall that in each iteration at least one essential node must be selected according to Theorem 1. The upper bound of $T_b(k,0)$ satisfies

$$T_b(k,0) \leqslant \left( \left\lceil \frac{T_b(1,0)-h}{k} \right\rceil + h \right) \leqslant \left( \frac{T_b(1,0)}{k} + \frac{k-1}{k} h \right) \tag{10}$$

To find the lower bound of $T_b(k,0)$, we know that when $\epsilon=0$ and $D=I$, all essential nodes with lower bounds less than the optimal-solution value must be expanded in a parallel best-first search. On the other hand, the first iteration during which the root node is expanded is imperfect for $k>1$. Hence

$$T_b(k,0) \geqslant \left( \frac{T'_b(1,0)-1}{k} + 1 \right)$$

The bounds in Eq. (9) are tight in the sense that examples can be generated to achieve one of these bounds. In the case that a feasible solution must be located in levels greater than or equal to $h'$, then $h'$ iterations are necessary, and $T_b(k,0) \geqslant \max(h', (T_b'(1,0)-1)/k+1)$. In this paper, we assume that $[(T_b'(1,0)-1)/k+1)] \geqslant h'$. Notice that if we know $L_{max}$, the maximum level in which at least one essential node exists, i.e., there is no essential node in levels larger than $L_{max}$, then the upper bound can be tightened by substituting $L_{max}$ for $h$ in Eq. (9). In practice, the information on $L_{max}$ is not available in advance[16,18].

In Eq. (9), $h$ is a function of the problem size. However, $T_b(1,0)$ and $T_b'(1,0)$ reflect the complexity of the problem to be solved and are unknown before the solution is found. $T_b(1,0)$ can be estimated, as in the analysis of $\alpha$-$\beta$ algorithms[12,30-31], by defining a branching factor $\alpha$.

$$\alpha = \lim_{h\to\infty} [T_b(1,0)]^{1/h} \text{ i.e., } \alpha^h \approx T_b(1,0)$$

The branching factor measures the average number of branches of an essential node and can be estimated statistically. For example, $\alpha$ is close to 1.1 for knapsack problems when all profits and weights are independent and are generated from uniform distributions. $T_b'(1,0)$ can be estimated by sampling methods from the problem characteristics and the distribution on the number of subproblems generated.

Let $w$ be $T_b(1,0)/h$. $w$ can be viewed as the "average width" of an OR-tree, which only consists of essential nodes. Eq. (10) can be rewritten as

$$\frac{T_b(1,0)}{T_b(k,0)} \geqslant \frac{k \cdot w}{w+(k-1)} \tag{11}$$

Eq. (11) shows that if $w \gg k$, then the speedup is close to $k$; whereas if $w \ll k$, then the lower-bound speedup is close to $w$.

Eq. (9) shows that a near-linear speedup can be achieved when a large number of processors are used. As an example, if $h$=50, $T_b(1,0)=10^6$ (for a typical traveling-salesman problem), and $k$=1 000, then $T_b(1\ 000,0) \leqslant 1\ 049$. This means that a near-linear speedup can be attained with one thousand processors. In Table 1, the theoretical bounds derived above are compared with the simulation results of parallel best-first and depth-first searches for solving two 35-object knapsack problems. In generating the knapsack problems, the weights, $w(i)$, were chosen randomly between 0 and 100 with a uniform distribution, and the profits were set to be $p(i) =(w(i)+10)$. This assignment is intended to increase the complexity of the randomly generated problems. The results demonstrate that the bounds on parallel best-first searches are very tight, hence its performance can be predicted accurately. Table 1 also shows that the speedup depends strongly on $w$. In Case 1, $w \approx 2\ 023$ and a near-linear speedup of $0.93k$ is achieved with 256 processors. In Case 2, $w \approx 188$ and a speedup of $0.48k$ is obtained with 256 processors. Other simulation results also demonstrate a similar behavior.

From Theorem 3, it is easy to determinate the maximum number of processors to guarantee a near-linear speedup. Assume that the speedup required is $T_b(1,0)/T_b(k,0) \geqslant (\eta k)$ , $0<\eta<1$ . From Eq. (9),

$$T_b(k,0) \leqslant \left(\frac{T_b(1,0)}{k} + \frac{k-1}{k}h\right) \leqslant \frac{T_b(1,0)}{\eta k}$$

or

$$k \leqslant \left(\frac{1-\eta}{\eta h} T_b(1,0) + 1\right)$$

For instance, if $\eta$=0.9, $h$= 50, and $T_b(1,0)=10^6$, then $k \leqslant 2\ 223$. That is, a minimum of $0.9k$ speedup is obtained if 2 223 or less processors are used.

Note that in parallel best-first searches with $\epsilon = 0$, essential nodes can be eliminated only if their lower-bounds are equal to the optimal-solution value. When $\epsilon > 0$, other essential nodes can also be eliminated because it is possible for a feasible solution whose value is slightly larger than the optimal-solution value to be found early in the parallel case, while this solution is found quite late or even not found in the serial case. A looser lower bound with respect to approximate best-first searches is derived in Section 4.3(3).

### 3.4 Parallel depth-first searches

The performance of a parallel depth-first search can be analyzed in terms of the generalized heuristic function. The following theorem shows the performance bounds of parallel depth-first OR-tree searches. The range of bounds on parallel depth-first searches are larger than that of parallel best-first searches.

Table 1. Comparison between theoretical bounds and simulation results on parallel best-first and depth-first searches for Knapsack problems with 35 objects. ($T_b'(1,0)= T_b(1,0)$. During depth-first searches, $c$=22 in case 1 and $c$=12 in case 2)

| Num of proc. | Case 1 | | | | Case 2 | | | |
|---|---|---|---|---|---|---|---|---|
| | Lower bound | No. of iterat. | Upper bound | Speedup | Lower bound | No. of iterat. | Upper bound | Speedup |
| Parallel Best-First Search | | | | | | | | |
| 1 | 70 790 | 70 790 | 70 790 | 1.000 | 6 566 | 6 566 | 6 566 | 1.000 |
| 2 | 35 395 | 35 405 | 35 413 | 1.999 | 3 283 | 3 292 | 3 301 | 1.995 |
| 4 | 17 698 | 17 705 | 17 724 | 3.998 | 1 642 | 1 659 | 1 668 | 3.958 |
| 8 | 8 849 | 8 857 | 8 880 | 7.993 | 821 | 836 | 852 | 7.854 |
| 16 | 4 425 | 4 436 | 4 458 | 15.959 | 411 | 431 | 444 | 15.234 |
| 32 | 2 213 | 2 230 | 2 247 | 31.745 | 206 | 226 | 240 | 29.053 |
| 64 | 1 107 | 1 124 | 1 141 | 62.982 | 103 | 129 | 138 | 50.899 |
| 128 | 554 | 575 | 588 | 123.117 | 52 | 78 | 87 | 84.179 |
| 256 | 277 | 298 | 312 | 237.557 | 26 | 53 | 61 | 123.887 |
| Parallel Depth-First Search | | | | | | | | |
| 1 | 70 790 | 70 790 | 70 790 | 1.000 | 6 566 | 6 582 | 6 582 | 1.000 |
| 2 | 35 395 | 35 630 | 35 787 | 1 987 | 3 283 | 3 488 | 3 513 | 1.887 |
| 4 | 17 698 | 18 044 | 18 285 | 3.923 | 1 642 | 1 940 | 1 978 | 3.393 |
| 8 | 8 849 | 8 884 | 9 534 | 7.968 | 821 | 1 161 | 1 211 | 5.669 |
| 16 | 4 425 | 4 460 | 5 159 | 15.872 | 411 | 777 | 827 | 8.471 |
| 32 | 2 213 | 2 247 | 2 971 | 31.504 | 206 | 584 | 635 | 11.271 |
| 64 | 1 107 | 1 143 | 1 877 | 61.934 | 103 | 485 | 539 | 13.571 |
| 128 | 554 | 592 | 1 330 | 119.578 | 52 | 219 | 491 | 30.055 |
| 256 | 277 | 316 | 1 057 | 224.019 | 26 | 90 | 467 | 73.133 |

**Theorem 4:** For a parallel depth-first search with $k$ processors, $\epsilon=0$, $D$=$I$, and a generalized heuristic function $h(P_i)$ = (path number, level number), then

$$\left(\frac{T'_b(1,0)-1}{k}+1\right)\leqslant T_d(k,0) \leqslant\left(\frac{T_d(1,0)}{k}+\frac{k-1}{k}[(c+1)h-c]\right) \tag{12}$$

where $c$ is the number of the distinct incumbents obtained during the serial depth-first search, and $T_b'(1,0)$ is the number of essential nodes in a serial best-first search with lower bounds less than the optimal-solution value.

**Proof:** The sequence of iterations obtained during a serial depth-first search can be divided into ($c$+1) subsequences according to the $c$ distinct monotonically decreasing incumbents obtained. Let the $c$ feasible solutions and their corresponding parents be denoted by $F_1,\cdots,F_c$, and $P_1,\cdots,P_c$. Further, assume that $F_1,\cdots,F_c$ are obtained in the $i_1$th,$\cdots$,$i_c$th iterations, respectively. Hence iterations from 1 to $i_1$ belong to the first subsequence, and iterations from $i_j$+1 to $i_{j+1}$, belong to the $j$th subsequence.

Consider the first subsequence of iterations. Suppose that the $j$th, $1\leqslant j\leqslant i_1$, iteration is imperfect. This imperfect iteration occurs because either less than $k$ nodes are selected due to insufficient active subproblems, or some nodes are expanded in the parallel depth-first search but are eliminated in the serial case. In the latter case the heuristic values of these nodes must be greater than $h(P_1)$ because the incumbents used initially in the serial and parallel cases were identical, and if the heuristic values of all expanded nodes in an iteration were less than $h(P_1)$, then this iteration is necessarily a perfect iteration. Let $h_{\min}(x)$ be the minimum level in which some active nodes with heuristic values less than $h(P_1)$ reside in the $x$th imperfect iteration. Similar to the proof of Theorem 3, we have

$$h_{\min}(x+1) \geqslant (h_{\min}(x)+1)$$

Consequently, after at most $h$ imperfect iterations, $F_1$, the first feasible solution better than the initial incumbent, must be found.

Analogous to the above argument, we can prove that in the $i$th, $1 < i \leqslant c$, subsequence of iterations, at most $h$ imperfect iterations will be encountered before a better feasible solution is obtained. During the last subsequence of iterations, since the optimal solution has been generated, all iterations are imperfect only if less than $k$ nodes are selected in each iteration. In other words, an imperfect iteration implies that all currently active nodes are selected and expanded, and only descendents of these nodes can be active in the next iteration. Hence no active node remains after at most $h$ imperfect iterations in the last subsequence. The previous analysis shows that at most $(c+1)h$ imperfect iterations can appear in a parallel depth-first search. Since at least one node in each iteration in the parallel case belongs to $\Phi^1$, the upper bound of $T_d(k,0)$ can be derived as

$$T_d(k,0) \leqslant \left(\frac{T_d(1,0)-(c+1)h}{k}+(c+1)h\right)$$

In the above discussion, the expansion of the root is counted in each of the $(c+1)$ subsequences. Since the root is only expanded once, the above upper bound should be compensated by the additional number of times that the root is expanded Eq. (12).

The lower bound on $T_d(k,0)$ can be proved easily because all essential nodes in a serial best-first search with lower bounds less than the optimal solution must be expanded in the parallel depth-first search.

For problems such as integer programming and 0-1 knapsack, all feasible solutions are located in the bottom-most level of the OR-tree. In this case, the following corollary shows that all essential nodes of a serial depth-first search must be expanded in a parallel depth-first search, and a tighter lower bound is obtained.

**Corollary 2:** In searching an OR-tree using a parallel depth-first search and a heuristic function of (path number, level number), if all feasible solutions are in Level $h$, and the Serial/Parallel Incumbent-Update strategy is used, then

$$\left(\frac{T_d(1,\epsilon)-1}{k}+1\right) \leqslant T_d(k,\epsilon) \qquad (13)$$

where $h$ is the maximum number of levels of the OR-tree.

**Proof:** First we show that under the given assumptions, $P_iLP_j$ implies that $h(P_k) < h(P_j)$, where $P_i$ is in Level $h$, $P_j$ is in Level $j$, $j < h$, and $P_k$ is the parent node of $P_i$. Suppose that $P_iLP_j$ and $h(P_j) < h(P_k)$, then when $P_k$ is expanded, an ancestor of $P_j$, say $P_a$, has to be expanded simultaneously from the defined selection rule and Eq. (6). Note that when $P_a$ is active, $h(P_a)<h(P_k)$, that $P_k$ is in Level $(h-1)$, and that $P_a$ is in a level less than $(h-1)$. Let $P_m$ be the first common ancestor of $P_i$ and $P_j$. By carrying out the above argument repeatedly, $P_m$ has a child $P_a'$ which is expanded simultaneously with an indirect descendant $P_k'$, where $P_a'$ is an ancestor of $P_a$, and $P_k'$ is an ancestor of $P_k$. It is also seen that $P_k'$ is at a level number larger than that of $P_a'$. Thus when $P_m$ is expanded, all descendant nodes between $P_m$ and $P_k'$ must be generated simultaneously, which is an impossible situation.

The above argument also shows that a feasible solution $P_x$ obtained before $P_i$ implies that $h(P_x)<h(P_i)$. By using the Serial/Parallel Incumbent-Update strategy discussed in Section 3.2, the sequences of distinct incumbents are the same in the serial and parallel cases according to Theorem 2. By Corollary 1, all essential nodes must be expanded in a parallel depth-first search. The lower bound of $T_d(k,\epsilon)$ is derived immediately.

The bounds in Theorem 4 are tight in the sense that we can construct examples to achieve the lower-bound and upper-bound computational times. These degenerate cases occur rarely. Simulations have revealed that for a number of OR-tree search problems, $T_d(k,0)$ may be very close $T_b(k,0)$. The second part of Table 1 shows the simulation results of solving two cases of the 0-1 knapsack problems using a depth-first search. Note that when the number of processors is large, the number of essential nodes in each imperfect iteration of the parallel depth-first search is usually larger than one. In contrast to the upper bound in Eq.(12), which was derived with the assumption of one essential node in each imperfect iteration, $T_d(k,0)$ may be much smaller than the upper bound.

Although $c$, the number of distinct incumbents, is unknown until the solution is found, $c$ is usually small and can be estimated when integral solutions are sought. It has been observed that $c$ is less than 10 for vertex-cover problems with less than 100 vertices. For most integer programming problems, $c\approx 1$. In these cases, the bounds given in Theorem 4 are tight.

## 3.5 Parallel breadth-first searches

The following theorem presents the performance bounds on parallel breadth-first searches.

**Theorem 5:** For a parallel breadth-first search with $k$ processors, $D=I$, and a generalized heuristic function satisfying Eq. (6) and Eq. (7), and using the Serial/Parallel Incumbent-Update strategy, then

$$\left(\frac{T_r(1,\epsilon)-1}{k}+1\right) \leqslant T_r(k,\epsilon) \leqslant \left(\frac{T_r(1,\epsilon)}{k}+\frac{k-1}{k}(c+h)\right) \tag{14}$$

where $c$ is the number of distinct incumbents obtained during the serial breadth-first search.

**Proof:** For a parallel breadth-first search, there are two possible cases in which imperfect iterations may exist. First, when the number of active subproblems is less than $k$, an imperfect iteration occurs and may happen at most $h$ times from our discussion in Theorems 3 and 4. Second, imperfect iterations may occur if there exist some nodes that are eliminated by a feasible solution in the serial case but are expanded simultaneously with the parent of this feasible solution in the parallel case. From Theorem 1, it is easy to show that at least one essential node must be expanded in a parallel iteration. $T_r(k,\epsilon)$ is, thereby, bounded by

$$T_r(k,\epsilon) \leqslant \left(\frac{T_r(1,\epsilon)}{k} + \frac{k-1}{k}(c+h)\right)$$

On the other hand, in breadth-first searches, all essential nodes have to be expanded in the parallel case (Theorem 2 and Corollary 1). Thus $T_r(k,\epsilon)$ has the following lower bound.

$$\left(\frac{T_r(1,\epsilon)-1}{k}+1\right) \leqslant T_r(k,\epsilon)$$

Since the performance bounds of $T_r(k,0)$ are tighter than those of $T_d(k,0)$, the performance of parallel breadth-first searches can be predicted more accurately.

## 4 Coping with anomalies in parallel OR-tree searches

Up to now, the efficiency of parallel OR-tree searches have been little studied. When comparing the efficiency between using $k_1$ processors and $k_2$, $1 \leqslant k_1 < k_2$, processors, a $k_2/k_1$-fold speedup (ratio of the number of iterations in the two cases in our model) is expected. However, simulations have shown that the speedup can be a) less than one (called a detrimental anomaly)[15,32-33]; or b) greater than $k_2/k_1$ (called an acceleration anomaly) [15,32]; or c) between one and $k_2/k_1$ (called a deceleration anomaly) [15,32-33]. So far, all known results on parallel OR-tree searches showed a near-linear speedup for only a small number of processors.

The objective of this section is to develop, using the results in Section 3, conditions to cope with the anomalous behavior of parallel OR-tree search algorithms in the possible presence of approximations and dominance tests. Anomalies are studied with respect to the same search strategy. In general, anomalies should be studied with respect to the best serial algorithm and the best parallel algorithm (with possibly a different search strategy than that of the serial algorithm). However, conditions to resolve these anomalies would be problem dependent and may result in a large number of cases that cannot be enumerated. Anomalies are also studied with respect to the assumption that all idle processors are used to expand active subproblems. In fact, detrimental anomalies cannot happen if some processors can be kept idle in the presence of active subproblems. The number of processors to be kept idle is problem dependent and is very difficult to find without first solving the problem.

The conditions to resolve anomalies are described with respect to serial-to-parallel processing, which is processing between using one and $k$ processors, and parallel-to-parallel processing, which is processing between using $k_1$ and $k_2$, $1 < k_1 < k_2$ processors. The conditions to cope with serial-to-parallel anomalies and parallel-to-parallel anomalies are different. These conditions we developed are useful for designers to understand the existence of anomalies and to modify existing algorithms to prevent detrimental anomalies and enhance acceleration anomalies.

## 4.1 Anomalies of parallel OR-tree searches

In this section, some anomalies on parallel OR-tree searches are illustrated. A single list of subproblems is assumed, and generalizations to multiple lists are discussed later.

For the OR-tree in Figure 3(a), the order of nodes expanded in a serial depth-first search is

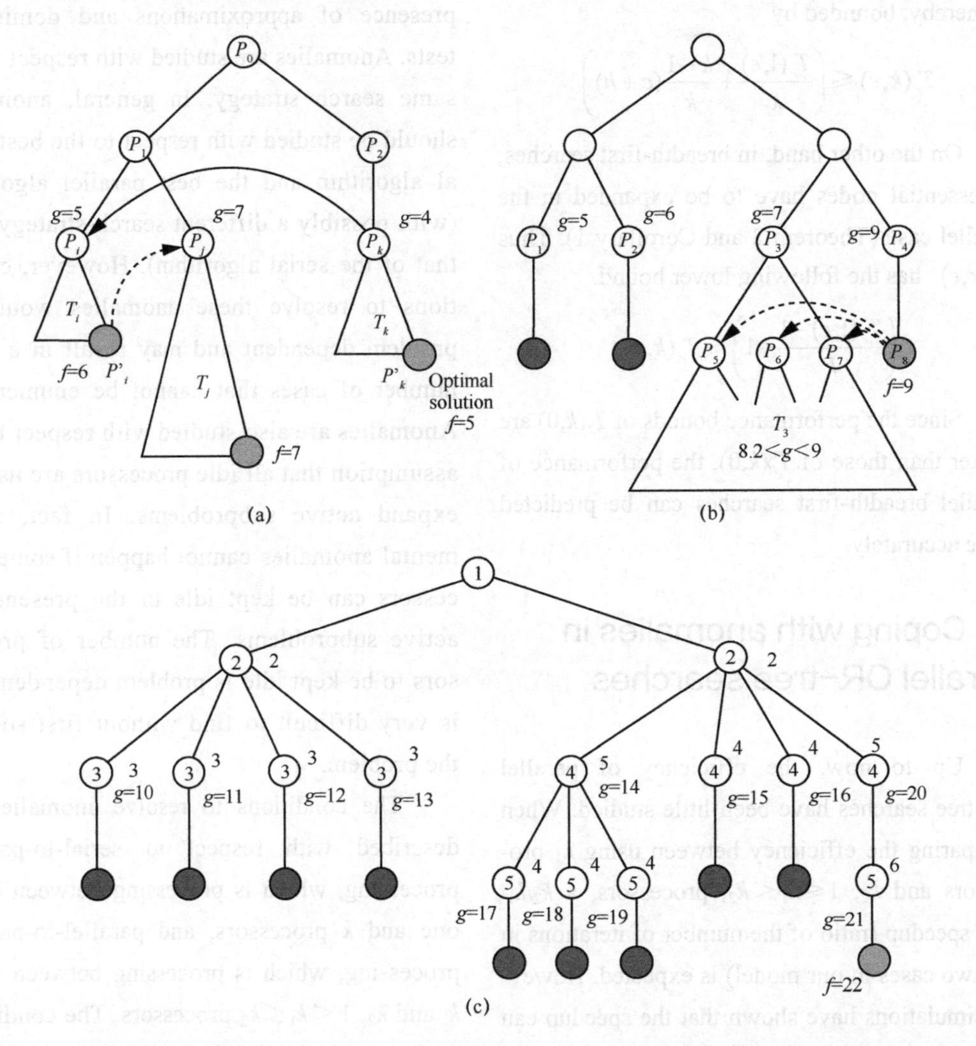


Figure 3. Examples of detrimental anomalies. (Nodes pointed to by solid arrows are eliminated by dominance tests; nodes pointed to by dashed arrows are eliminated by lower-bound tests. Shaded nodes indicate feasible solutions. Dark nodes indicate infeasible solutions.) (a) With lower boundand dominance tests, $T_d(2,0) > T_d(1,0)$. (b) With approximate lower-bound tests, $T_b(3,0.1) > T_b(2,0.1)$; $T_d(3,0.1) > T_d(2,0.1)$. (c) Without lower-bound and dominance tests in a depth-first or best-first search $T(4,0) = 5$, $T(5,0) = 6$. (The number inside the node is the evaluation order using four processors; the number outside the node is the evaluation order using five processors.)

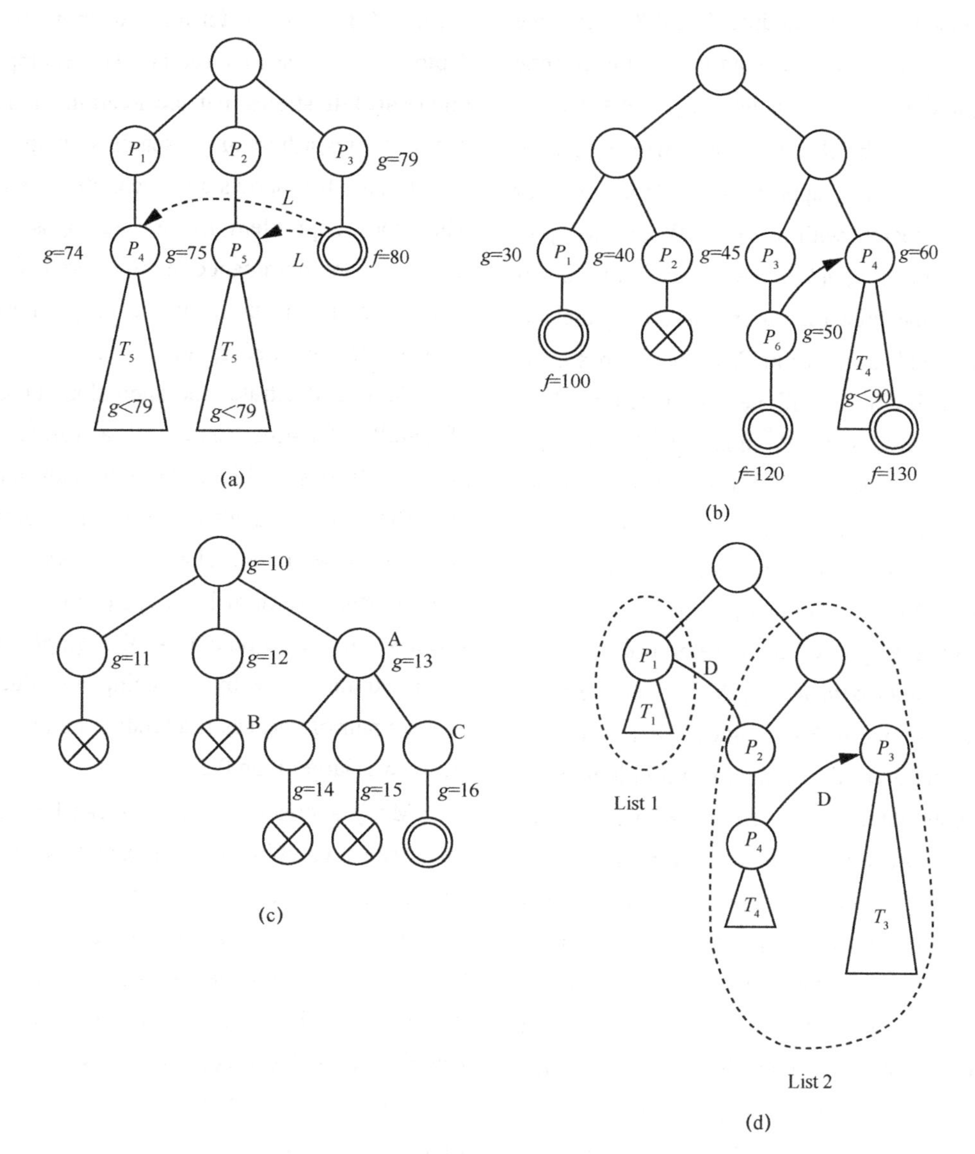


Figure 4. Examples of acceleration anomalies. (Double nodes indicate feasible solutions. Crossed out nodes indicate infeasible solutions.) (a) With approximate lower-bound tests; $T_b(2,0.1)/T_b(3,0.1) > 3/2$; $T_d(2,0.1)/T_d(3,0.1) > 3/2$. (b) With dominance tests only; dominance tests are consistent with $g$ and $h$ for all search strategies; $T(2,0)/T(3,0) > 3/2$. (c) Without lower-bound and dominance tests for all search strategies; $T(2,0)/T(3,0)=5/3 > 3/2$. (d) With two subproblem lists and local dominance tests only; $T(1,0)/T(2,0)=2$

$P_0$, $P_1$, $P_i$, all nodes in $T_i$ resulting in a feasible solution $P_i$' with value 6, $P_k$, and all nodes in $T_k$ resulting in the optimal solution of $f(P_k')=5$. $P_j$ is terminated by the lower-bound test with $P_i$'. In contrast, when two processors are used, $P_1$ and $P_2$ are expanded concurrently. $P_i$ is terminated as a result of dominance by $P_k$. Since $T_i$ is terminated, $T_j$ is expanded next. If $T_j$ is large, then the combined time of expanding $T_j$ and $T_k$ using one processor can be smaller than the combined time of expanding $T_j$ and $T_k$ using two processors.

Figure 3(b) illustrates a detrimental anomaly under an approximate depth-first or best-first search with $\epsilon = 0.1$. When two processors are used, $f(P_8)$, the optimal solution, is found in the fourth iteration. Assuming that the lower bounds of nodes in $T_3$ are between 8.2 and 9, all nodes in $T_3$ will be eliminated by lower-bound tests with $P_8$ since $[9/(1+\epsilon)] < 8.2$. When three processors are used, $P_3$ is expanded in the third iteration. $P_5$, $P_6$, and $P_7$ are generated and will be selected in the next iteration. If $T_3$ is large, $T(2,\epsilon) < T(3,\epsilon)$ will occur.

An example of an acceleration anomaly with an approximate depth-first or best-first search is shown in Figure 4(a). When three processors are used, the optimal solution is found in the second iteration, and $P_4$ and $P_5$ are eliminated. If two processors are used, subtrees $T_4$ and $T_5$ have to be expanded. $T(2,0.1)/T(3, 0.1)$ will be much larger than 3/2 if $T_4$ and $T_5$ are very large.

Figure 4(b) illustrates another example of acceleration anomalies with dominance tests under a best-first, depth-first, or breadth-first search strategy. When three processors are used, $P_4$ will be dominated by $P_6$, and four iterations are required to complete the search. In contrast, when two processors are used, $P_4$ is expanded before $P_6$ is generated. If no dominance relation exists between $P_6$ and the descendants of $P_4$ (which is possible), then $T_4$ has to be searched, and $T(2,0) / T(3,0) > 3/2$.

Anomalies may occur even when both the lower-bound and dominance tests are inactive. Figure 3(c) shows a detrimental anomaly in a depth-first or best-first search. The example in Figure 4(c) illustrates that acceleration anomalies may occur regardless of the search strategy.

Figure 4(d) shows an acceleration anomaly when the subproblem list is decomposed into local lists, and dominance relations are restricted to subproblems in the local lists only. This phenomenon is described in Section 4.3(4).

Table 2 illustrates the anomalous behavior of parallel OR-tree search of a vertex-cover problem. In this problem, the minimum number of vertices to cover all edges in an undirected graph is to be found. That is, all edges in the graph emanate from at least one of the included vertices. In the simulations, the graphs were generated randomly by assigning an edge between a pair of vertices if a random number generated was larger than 0.2.

Many anomalous examples can be created for various combinations of search strategies, allowance functions, and elimination rules. However, the important point here is not in knowing that anomalies exist, but in understanding why these anomalies occur and in developing strategies to cope with these anomalies. It is desirable to find the sufficient conditions to ensure that $T(k_2,\epsilon) \leqslant T(k_1,\epsilon)$ as well as the necessary conditions for $T(k_1,\epsilon)/T(k_1,\epsilon) \geqslant k_2/k_1$. The necessary conditions to eliminate detrimental anomalies are not evaluated because they are problem dependent. A condition necessary to avoid detrimental anomalies depends on the sequence of nodes expanded and the size of the resulting subtrees. There are many possible combinations, and it is difficult to enumerate

Table 2. Examples of anomalous behavior of parallel OR-tree searches for a vertex-cover problem with 80 vertices, $\epsilon$ =0, and using a depth-first search(*c*=3). (S-P indicates serial-to-parallel anomalies; P-P indicates parallel-to-parallel anomalies)

| $k$ | $T_d(k,0)$ | Speedup $T_d(1,0)/T_d(k,0)$ | $k/(k-1)$ | Speedup $T_d(k-1,0)/T_d(k,0)$ | Anomaly |
|---|---|---|---|---|---|
| 1 | 4 134 | 1.000 | | | |
| 2 | 2 065 | 2.002 | 2.000 | 2.002 | S-P、P-P acceleration |
| 3 | 1 474 | 2.805 | 1.500 | 1.401 | |
| 4 | 1 042 | 3.967 | 1.333 | 1.415 | P-P acceleration |
| 5 | 854 | 4.841 | 1.250 | 1.220 | |
| 6 | 717 | 5.766 | 1.200 | 1.191 | |
| 7 | 626 | 6.604 | 1.167 | 1.145 | |
| 8 | 516 | 8.012 | 1.143 | 1.213 | S-P, P-P acceleration |
| 9 | 481 | 8.595 | 1.125 | 1.073 | |
| 10 | 438 | 9.438 | 1.111 | 1.098 | |
| 11 | 398 | 10.387 | 1.100 | 1.101 | P-P acceleration |
| 12 | 365 | 11.326 | 1.091 | 1.090 | |
| 13 | 380 | 10.879 | 1.083 | 0.961 | P-P detrimental |
| 14 | 298 | 13.872 | 1.077 | 1.275 | P-P acceleration |
| 15 | 271 | 15.255 | 1.071 | 1.100 | S-P, P-P acceleration |
| 16 | 257 | 16.086 | 1.067 | 1.054 | S-P acceleration |

them for a given problem. Further, the necessary conditions developed for one problem cannot be generalized to other problems. For a similar reason, sufficient conditions to preserve acceleration anomalies are not evaluated.

## 4.2 Coping with serial-to-parallel anomalies

In this section, we discuss results when dominance tests are active. Some results on coping with serial-to-parallel anomalies without dominance tests have been derived elsewhere and are only summarized here[11,16].

(1) Sufficient condition to eliminate serial-to-parallel detrimental anomalies

(a) Finding an exact optimal solution: In this section, we show that $T(k,0)\leqslant T(1,0)$ holds if the heuristic function is monotone and unambiguous, and the dominance relation satisfies some consistency requirements on the lower-bound and heuristic functions.

The following concept on transitivity of lower-bound and dominance tests is needed. Recall that $P_iDP_k$ implies that $f(P_i)\leqslant f(P_k)$, but the converse is false because some nodes are incomparable (otherwise, the number of active nodes can always be reduced to one). When lower-bound and dominance tests are used together, it is important to note that dominance tests are not transitive with lower-bound tests. That is, $P_iLP_j$ and $P_jDP_k$ do not imply $P_iLP_k$. Similarly, $P_iDP_j$, and $P_jLP_k$, do not imply $P_iDP_k$. In both cases, only $f(P_i)\leqslant f(P_k)$ can be deduced.

To combine the dominance and lower-bound tests, conditions are defined for a special class of dominance relations. A dominance relation $D$ is said to be consistent with a heuristic function $h$ if $P_iDP_j$, implies that $h(P_i)<h(P_j)$ for all $P_i$, $P_j\in P$. A dominance relation $D$ is said to

be consistent with the lower-bound function if $P_iDP_j$, implies that $g(P_i) \leqslant g(P_j)$ for all $P_i, P_j \in P$. To show that $T(k,0) \leqslant T(1,0)$, it suffices to prove: i) that at least one node belonging to $\Phi^1$ is expanded in each iteration of the parallel search, and ii) that once all the nodes in $\Phi^1$ are expanded or terminated, the parallel heuristic search must terminate.

**Theorem 6:** $T(k,0) \leqslant T(1,0)$ holds for heuristic searches that satisfy Eq. (6) and Eq. (7) and that has dominance relation $D$ which is consistent with the lower-bound function $g$ and the heuristic function $h$.

**Proof:** The proof is not shown due to space limitation[19-20].

The requirement on the consistency of $D$ with $g$ is satisfied in many practical problems, such as the shortest-path problem[5], the traveling-salesman problem[25], the $n$-job two-machine mean-completion-time flowshop problem[34], and the $n$-job one-machine scheduling problem with deadlines[35]. However, the requirement on the consistency of $D$ with $h$ may not be satisfied in general. The detrimental anomaly illustrated in Figure 3(a) for a depth-first search is caused by the inconsistency of $D$ with $h$. The dominance relation for the 0-1 knapsack problem is consistent with the upper-bound function (instead of the lower-bound function for maximization problems). However, the conventional definition of dominance relations for the 0-1 knapsack problem may result in detrimental anomalies, since when the profit and weight of $P_i$ are the same as the corresponding profit and weight of $P_j$, it is possible that either $P_iDP_j$ or $P_jDP_i$. If $h(P_i) < h(P_j)$ happens but $P_jDP_i$ is true, then $D$ is inconsistent with $h$. For the knapsack problem, detrimental anomalies can be avoided if the dominance relation is redefined as follows. For $P_i$ and $P_j$ defined on a given subset of objects, $P_iDP_j$ if i) the profit of $P_i$ is larger than or equal to that of $P_j$ and the total weight of $P_i$ is less than that of $P_j$; or ii) the profit of $P_i$ is larger than that of $P_j$ and the total weight of $P_i$ is less than or equal to that of $P_j$; or iii) the profit and weight of $P_i$ are equal to the corresponding profit and weight of $P_j$ and the heuristic value of $P_i$ is less than that of $P_j$.

In general, Theorem 6 pinpoints the problem when detrimental anomalies happen and serves as a useful guideline to eliminate them.

(b) Finding an approximate solution: When approximations are allowed, Theorem 6 is not always true. The reason for the detrimental anomaly is that $L$, the lower-bound elimination rules under approximation, are not transitive. That is, $P_iLP_j$ and $P_jLP_k$ do not imply $P_iLP_k$, since $f(P_i)/(1+\epsilon) \leqslant g(P_j)$ and $f(P_j)/(1+\epsilon) \leqslant g(P_k)$ imply $f(P_i)/(1+\epsilon)^2 \leqslant g(P_k)$ rather than $f(P_i)/(1+\epsilon) \leqslant g(P_k)$. Although the lower-bound tests are not transitive, the conditions in Theorem 6 are sufficient to avoid detrimental anomalies for best-first searches.

**Theorem 7:** $T_b(k,\epsilon) \leqslant T_b(1,\epsilon)$, $\epsilon > 0$, holds for best-first searches when $h$, the heuristic function, satisfies Eq. (6) and Eq. (7), and $D$, the dominance relation, is consistent with $h$ and $g$.

**Proof:** The proof is omitted due to space limitation[11,20].

To avoid detrimental anomalies in approximate breadth-first and depth-first searches, the sufficient conditions are more restricted than those of Theorem 7. A possible set of sufficient conditions are shown below.

**Theorem 8:** Suppose i) that the heuristic function satisfies Eq. (6) and Eq. (7), ii) that the dominance relation is consistent with $h$ and $g$, and iii) that the sequences of feasible solutions obtained in the serial and parallel cases are identical, then $T(k,\epsilon) \leqslant T(1,\epsilon)$, $\epsilon \geqslant 0$.

**Proof:** Suppose that $P_1$, a basic node in the parallel search, is eliminated in the serial case by a feasible solution $P_2$ and that $P_3$ is a parent of $P_2$, then $h(P_3) < h(P_1)$. From the definition of basic nodes and the assumption that the sequences of feasible solutions are the same in the serial and parallel cases, $P_3$ must be expanded in the parallel case, and $P_2$ is obtained before $P_1$ is expanded. Hence $P_1$ has to be eliminated by $P_2$ in the parallel case. A contradiction!

In general, the sequences of distinct incumbents in the serial and parallel cases are usually not identical. However, for breadth-first searches and some special cases of depth-first searches, such as the case when all feasible solutions are in the bottommost level of the OR-tree, the conditions of Theorem 8 are satisfied if the Serial/Parallel Incumbent-Update algorithm is used in both the serial and parallel cases (Theorem 2).

(2) Necessary conditions to allow serial-to-parallel acceleration anomalies

In this section, the necessary conditions for $T(k,0) < T(1,0)/k$ are developed. One of these conditions is based on the complement of the special class of dominance relations defined in Section 4.2(1). A dominance relation, $D$, is said to be inconsistent with $h$ if there exist two nodes $P_i$ and $P_j$ such that $P_i D P_j$ and $h(P_i) > h(P_j)$. Another condition is based on the complete consistency of heuristic functions. A heuristic function $h$ is said to be consistent (resp. completely consistent) with the lower-bound function $g$ if $h(P_i) < h(P_j)$ implies that $g(P_i) \leqslant g(P_j)$ (resp. $g(P_i) < g(P_j)$) for all $P_i, P_j \in P$. A heuristic function, $h$, is said to be not completely consistent with $g$ if there exist two nodes $P_i$ and $P_j$ such that $h(P_i) > h(P_j)$ and $g(P_i) \leqslant g(P_j)$. Note that if $g(P_i)=g(P_j)$ is allowed, then the heuristic function for a best-first search is consistent, but not completely consistent, with the lower-bound function.

**Theorem 9:** The necessary condition for $T(k,0) < T(1,0)/k$ is either i) that the heuristic function is not completely consistent with $g$, or ii) that the dominance relation is inconsistent with $h$.

**Proof:** The proof is omitted due to space limitation[11,20].

According to Theorem 5, it is easy to show that if $\epsilon = 0$ and dominance tests are inactive, then no serial-to-parallel acceleration anomalies exist when a breadth-firsts earch is used. However, if a best-first search is used, then the heuristic function is not completely consistent with the lower-bound function when multiple subproblems can have identical lower bounds. Hence acceleration anomalies can occur, but detrimental anomalies can be prevented by the use of an unambiguous heuristic function. The depth-first search has a similar behavior.

Note that the conditions in Theorem 9 are not necessary when approximate solutions are sought; that is, acceleration anomalies may occur in this case even though $h$ is completely consistent with $g$ and the dominance tests are consistent with $h$. The corresponding necessary conditions are studied in Section 4.3(3).

### 4.3 Coping with general parallel-to-parallel anomalies

In Sections 3.3 through 3.5, we have derived three theorems for the performance bounds with respect to different search strategies. From these results, we can investigate the relative efficiency between using $k_1$ and $k_2$, $1<k_1<k_2$, processors. First, we discuss the simple cases with $\epsilon=0$ and $D=I$ in Sections 4.3(1) and 4.3(2). Then strategies for cases with approximate lower-bound and dominance tests are studied in Sections 4.3(3) and 4.3(4).

(1) Sufficient conditions to eliminate parallel-to-parallel detrimental anomalies when $\epsilon=0$ and $D=I$: First, we derive a sufficient condition to assure the monotonic increase in computational efficiency with respect to the number of processors.

**Corollary 3:** Suppose that a parallel best-first search satisfies the assumptions of Theorem 3, then $T_b(k_2,0)\leqslant T_b(k_1,0)$ when

$$\frac{T_b(1,0)}{h}\geqslant\frac{k_1k_2}{r'_b k_2-k_1}\quad\text{and}\quad r'_b>\frac{k_1}{k_2},\quad 1<k_1<k_2 \tag{15}$$

where $r'_b=T'_b(1,0)/T_b(1,0)$.

**Proof:** From Theorem 3, at least $((T'_b(1,0)-1)/k_1+1)$ iterations are needed for $k_1$ processors, and at most $(T_b(1,0)/k_2+(k_2-1)/k_2h)$ iterations are needed for $k_2$ processors to find an optimal solution in a given OR-tree using a best-first search. The sufficient condition for $T_b(k_2,0)\leqslant T_b(k_1,0)$ is

$$\left(\frac{T_b(1,0)}{k_2}+\frac{k_2-1}{k_2}h\right)\leqslant\left(\frac{T'_b(1,0)-1}{k_1}+1\right)$$

To derive Eq. (15), we use a stronger condition such that

$$\left(\frac{T_b(1,0)}{k_2}+h\right)\leqslant\frac{T'_b(1,0)}{k_1}$$

For the example in Figure 3(c), $T_b(1,0)=15$, $k_1=4$, $k_2=5$, $h=4$, and $r'_b=1$. Eq. (15) is not satisfied, hence an increase in the number of processors from four to five may not ensure an improvement in performance. As another example, for $h=50$, $T_b(1,0)=10^6$, and $r'_b=1$, there will not be any detrimental anomalies for any combinations of $k_1$ and $k_2$ such that $1\leqslant k_1<k_2\leqslant 141$.

As mentioned in Section 3.3, the term $T_b(1,0)/h$ can be viewed as the average width of the essential OR-tree consisting of essential nodes only. Intuitively, detrimental anomalies can be prevented if the essential OR-tree searched is wide enough. On the other hand, for a given problem, the average width is fixed. If $k_2$ is not too large, $k_2/k_1$ is sufficiently large, and $r'_b\approx 1$, then $T_b(k_2,0)$ will be less than $T_b(k_1,0)$. If $r'_b\ll 1$, that is, there are a large number of active nodes with lower bounds equal to the optimal-solution value, then the condition in Eq. (15) may not be satisfied, and detrimental anomalies may happen.

Similarly, for a depth-first search, the corresponding sufficient conditions can also be determined. The conditions derived are more restricted than Eq. (15) because the range on $T_d(k,0)$ are larger (Theorem 4). To simplify the sufficient conditions, the following bounds on

$T_d(k,0)$ are used.

$$\frac{T'_b(1,0)}{k} \leqslant T_d(k,0) \leqslant \left(\frac{T_d(1,0)}{k} + (c+1)h\right)$$

**Corollary 4:** Let $r'_b = T'_b(1,0)/T_d(1,0) \leqslant 1$. In a parallel depth-first search that satisfies the assumptions of Theorem 4, $T_d(k_2,0) \leqslant T_d(k_1,0)$ when

$$\frac{T_d(1,0)}{h} \geqslant \frac{(c+1)k_1k_2}{r'_bk_2 - k_1} \text{ and } r'_b > \frac{k_1}{k_2}, \quad 1 < k_1 < k_2 \tag{16}$$

where $c$ is the number of the distinct incumbents obtained during the serial depth-first search.

**Proof:** The proof is similar to that of Corollary 3.

From Corollary 4, we can conclude that the existence of parallel-to-parallel detrimental anomalies in depth-first searches depends on $T'_b(1,0)$, $r'_d$, and $c$. If $r'_d \approx 1$, $c$ is small, and $T'_b(1,0)$ is very large, then Eq. (16) will be satisfied. Our simulation results reveal that for some problems, such as the 0-1 knapsack and vertex-cover problems, $T_d(1,0)$ is close to $T'_b(1,0)$, hence $r'_d \approx 1$. Moreover, if the feasible-solution values must be integers, then $c$ is often small. For this kind of problems, detrimental anomalies can be prevented for parallel depth-first searches when $T'_b(1,0)$ is large and $k_2$ is relatively small. However, the range of parallel processing within which no detrimental anomalies occur for depth-first searches is smaller than that for best-first searches.

Corollary 4 can be verified by using the example in Table 2 in which $r'_b \approx 1$, $c=3$, $h=80$, and $T_d(1,0)=4\,134$. From Eq. (16), we found that no detrimental anomalies occur when $k_2 \leqslant 4$. The detrimental anomaly observed when $k$ is 13 is beyond the predicted range.

The following theorem proves that detrimental anomalies do not occur when a breadth-first search is used. Note that this theorem is not a corollary of Theorem 5.

**Theorem 10:** In a parallel breadth-first search, If $D = I$ and the generalized heuristic function satisfies Eq's (6) and (7), then $T_r(k_2,0) \leqslant T_r(k_1,0)$, $1 < k_1 < k_2$.

**Proof:** Suppose that there are $m$ imperfect iterations when $k_2$ processors are used. The sequence of iterations can be divided into $(m+1)$ subsequences, each of which ends in an imperfect iteration except (possibly) the last one. Consider an arbitrary subsequence, say subsequence $j$, which contains $x$ perfect iterations and ends in an imperfect iteration. The theorem can be proved by showing that, when $k_1$ processors are used, the total number of iterations required to expand some of the nodes while eliminating other nodes in subsequence $j$ must be greater than or equal to $(x+1)$.

To account for the iterations expanded in subsequence $j$ when $k_1$ processors are used, we adopt the convention that any node expanded in conjunction with nodes in subsequence $(j+1)$ is not counted. When $k_2$ processors are used, an imperfect iteration occurs in a breadth-first search only when either a) the number of active nodes is less than $k_2$, or b) at least one new incumbent is found in this iteration, and one or more nodes expanded in this iteration have lower-bound values larger than the new incumbent. Let $P_i$ be the node having the smallest heuristic value in the imperfect iteration of subsequence $j$ considered, and $P_j$ be the node with the largest heuristic value in subsequence $j$. If the imperfect iteration in subsequence $j$ belongs to Case a) above, then when $k_1$ processors are used, all nodes with heuristic values larger than $h(P_j)$ cannot be expanded simultaneously with $P_i$, because all of them are descendents of nodes belonging to the imperfect iteration. If the imperfect iteration in subsequence $j$ belongs to Case b) above, then when $k_1$ processors are used, node $P_i$ also cannot be expanded simultaneously with nodes having heuristic values larger than $h(P_j)$; otherwise, the number of nodes expanded in this iteration would be larger than $k_1$. Therefore, Node $P_i$ has to be included in subsequence $j$ when counting the number of iterations under $k_1$ processors.

Since there are $x$ perfect iterations in subsequence $j$, from Corollary 1, all $k_2 \cdot x$ essential nodes must be expanded when $k_1$ processors are used. Among nodes in subsequence $j$, the minimum number of nodes that must be expanded when $k_1$ processors are used is $(k_2 \cdot x+1)$. Hence the minimum number of iterations with respect to $k_1$ processors is

$$\left\lceil \frac{k_2 \cdot x+1}{k_1} \right\rceil \geqslant (x+1),\quad 1<k_1<k_2$$

Since the above inequality holds for any subsequence, we obtain $T_r(k_2,0) \leqslant T_r(k_1,0)$.

(2) Necessary conditions to allow parallel-to-parallel acceleration anomalies when $\epsilon=0$ and $D$=$I$: From Theorem 3, we can derive a necessary condition for acceleration anomalies with respect to $k_1$ and $k_2$ processors in parallel best-first searches.

**Corollary 5:** In a parallel best-first search that satisfies the assumptions of Theorem 3,

$$T_b(k_1,0)/T_b(k_2,0) > k_2/k_1$$

only if

$$\left(T_b(1,0)-T'_b(1,0)\right) > \left(k_2-1-(k_1-1)h\right)$$

$$\text{for } 1<k_1<k_2 \qquad (17)$$

**Proof:** From Theorem 3, when $k_1$ processors are used, at most $T_b(1,0)/k_1+(k_1-1)h/k_1$ iterations are needed. When $k_2$ processors are used, at least $(T'_b(1,0)-1)/(k_2+1)$ iterations are needed. The necessary condition follows directly.

Note that this necessary condition cannot be obtained from the looser lower bound $T_b(k,0) \geqslant T_b(1,0)/k$ derived by Lai and Sahni[15].

Figure 4(b) illustrates the acceleration anomaly that $T_b(2,0)/T_b(3,0)>3/2$, which obviously satisfies Eq. (17). Usually, if $k_1$ and $k_2$ are close to each other and $h$ is large, then acceleration anomalies may occur in practice, even if a best-first search is adopted and all lower bounds are distinct. However, for best-first searches, $T_b(k_1,0)/T_b(k_2,0)$ cannot be much larger than $k_2/k_1$ due to the tight bounds on $T_b(k,0)$.

For depth-first searches, the necessary condition for acceleration anomalies is as follows.

**Corollary 6:** In a parallel depth-first search that satisfies the assumptions of Theorem 4,

$$T_d(k_1,0)/T_d(k_2,0) > k_2/k_1$$

only if

$$\left(T_d(1,0)-T'_b(1,0)\right) >$$

$$\left(k_2-1-(k_1-1)\left[(c+1)h-c\right]\right)$$

$$\text{for } 1<k_1<k_2 \qquad (18)$$

**Proof:** The proof is similar to that of Corollary 5.

Obviously, the necessary condition in Eq. (18) is readily satisfied, and $T_d(k_1,0)/T_d(k_2,0)$ may be much greater than $k_2/k_1$. Table 2 shows that acceleration anomalies occur frequently. If all solutions are located at the bottommost level of the OR-tree, then the corresponding necessary condition is simplified as (from Corollary 2):

$$\left((c+1)h-c\right)>\frac{k_2-1}{k_1-1},\quad 1<k_1<k_2 \tag{19}$$

Analogous to Corollaries 5 and 6, we can derive from Theorem 5 the necessary condition for acceleration anomalies to exist in a parallel breadth-first search. That is,

$$(c+h)>\frac{k_2-1}{k_1-1},\quad 1<k_1<k_2 \tag{20}$$

(3) Coping with parallel-to-parallel anomalies in approximate searches: It should be noted that Theorem 7 is no longer valid when comparing $T_b(k_1,\epsilon)$ and $T_b(k_2,\epsilon)$. Figure 3(b) gives a counterexample showing that $T_b(3,0.1) > T_b(2,0.1)$ even if all lower bounds are distinct. This happens because a node with a larger lower bound (not a basic node) may be expanded before nodes with smaller lower bounds in the parallel case, and nodes with smaller lower bounds may be eliminated by approximate lower-bound tests.

Analogous to the proof of Theorem 3, the upper bound on $T_b(k,\epsilon)$ can be derived. To find the lower bound on $T_b(k,\epsilon)$, let $f_0$ be the optimal-solution value and $\text{MINT}_b(\epsilon)$ be the minimum number of nodes that are expanded in the approximate best-first search. $\text{MINT}_b(\epsilon)$ represents the number of nodes whose lower bounds are less than $f_0/(1+\epsilon)$, since these nodes must be expanded in the best case. $T_b(1,\epsilon)$ may not achieve $\text{MINT}_b(\epsilon)$ because essential nodes may be eliminated by approximate lower-bound tests in the parallel search.

$\text{MINT}_b(\epsilon)$ may be estimated from the distribution of the number of subproblems with respect to lower bounds. From simulations on the 0-1 knapsack and vertex-cover problems, it was observed that the distributions are exponential. In this case, let $a^g\partial g$, $a>1$, be the number of subproblems whose lower bounds are between $g$ and $(g+\partial g)$. It is easy to show that

$$T_b(1,0)=\int_{-\infty}^{f_0}a^g\partial g=\frac{a^{f_0}}{\log_e a}$$

$$\text{MINT}_b(\epsilon)=\int_{-\infty}^{f_0/(1+\epsilon)}a^g\partial g=\left(T_b(1,0)\right)^{1/(1+\epsilon)}$$

From the above analysis, we get

$$\left(\frac{\text{MINT}_b(\epsilon)-1}{k}+1\right)\leqslant T_d(k,\epsilon)\leqslant\left(\frac{T_d(1,\epsilon)}{k}+\frac{k-1}{k}h\right) \tag{21}$$

Similarly, for depth-first searches,

$$\left(\frac{\text{MINT}_b(\epsilon)-1}{k}+1\right)\leqslant T_d(k,\epsilon)\leqslant \left(\frac{T_d(1,\epsilon)}{k}+\frac{k-1}{k}\left[(c+1)h-c\right]\right) \tag{22}$$

**Corollary 7[2]:** In parallel best-first searches that satisfy the assumptions of Theorem 3 with the exception that $\epsilon>0$, $T_b(k_2,\epsilon)\leqslant T_b(k_1,\epsilon)$, when

$$\frac{T_b(1,\epsilon)}{h}\geqslant\frac{k_1k_2}{r_bk_2-k_1}\ \text{and}\ r_b>\frac{k_1}{k_2},\quad 1<k_1<k_2 \tag{23}$$

[2] The proof is similar to that of Corollaries 3 and 5.

where $r_b = \text{MINT}_b(\epsilon)/T_b(1,\epsilon)$. Similarly, $T_b(k_1,\epsilon)/T_b(k_2,\epsilon) > k_2/k_1$ when

$$(T_b(1,\epsilon) - \text{MINT}_b(\epsilon)) > (k_2 - 1 - (k_1 - 1)h),\quad 1 < k_1 < k_2 \tag{24}$$

If $(k_1+1)/k_2 < r_b$, then we can predict that no detrimental anomalies occur if $k_2 \leqslant \sqrt{T_b(1,\epsilon)/h}$. As Eq. (24) is quite loose, it is often satisfied. When $k_1$=1 and $k_2$=$k$, the corresponding necessary condition becomes

$$(T_b(1,\epsilon) - \text{MINT}_b(\epsilon)) > (k-1) \tag{25}$$

For depth-first searches with approximate lower-bound tests, the following corollary shows the required conditions.

**Corollary 8:** In parallel depth-first searches that satisfy the assumptions of Theorem 4 with the exception that $\epsilon > 0$, $T_d(k_2,\epsilon) \leqslant T_d(k_1,\epsilon)$ when

$$\frac{T_d(1,\epsilon)}{h} \geqslant \frac{(c+1)k_1k_2}{r_dk_2 - k_1} \text{ and } r_d > \frac{k_1}{k_2},\quad 1 < k_1 < k_2 \tag{26}$$

where $r_d = \text{MINT}_b(\epsilon)/T_d(1,\epsilon)$. Similarly, $T_b(k_1,\epsilon)/T_b(k_2,\epsilon) > k_2/k_1$ when

$$(T_d(1,\epsilon) - \text{MINT}_b(\epsilon)) > (k_2 - 1 - (c+1)(k_1 - 1)h),\quad 1 < k_1 < k_2 \tag{27}$$

When $k_1$=1 and $k_2$=$k$, the corresponding necessary condition becomes

$$(T_d(1,\epsilon) - \text{MINT}_b(\epsilon)) > (k-1) \tag{28}$$

If all feasible solutions are located at the bottommost level of the OR-tree and the Serial/Parallel Incumbent-Update strategy is used, then a weaker sufficient condition to eliminate detrimental anomalies can be derived from Corollary 2. That is,

$$\frac{T_d(1,\epsilon)}{h} > \frac{(c+1)k_1k_2}{k_2 - k_1} \tag{29}$$

In this case, the necessary condition to allow acceleration anomalies is the same as that stated in Eq. (19).

**Corollary 9:** Suppose that a parallel breadth-first search satisfies the assumptions of Theorem 5 and that the Serial/Parallel Incumbent-Update Strategy is used, then $T_r(k_2,\epsilon) \leqslant T_r(k_1,\epsilon)$, and the necessary condition to allow acceleration anomalies is the same as (20).

**Proof:** If the Serial/Parallel Incumbent-Update strategy is used, then Theorem 10 holds regardless of the value of $\epsilon$.

(4) Coping with parallel-to-parallel anomalies under dominance tests: When dominance tests are considered, the above corollaries are no longer valid. Figure 4(b) shows an example in which none of the necessary conditions discussed previously is met, but the acceleration anomaly that $T_b(2,0)/T_b(3,0) > 3/2$ still occurs. The reason for the anomaly is that $P_iDP_j$ does not imply the dominance of $P_i$ over the descendants of $P_j$. For instance, in a 0/1 knapsack problem, $P_iDP_j$ only if the weight of $P_i$ is less than or equal to that of $P_j$, and $P_i$ and $P_j$ are defined on the same subset of objects. However, for descendants of $P_j$, their weights may be greater than that of $P_i$, and the dominance relation may not exist. One way to alleviate the detrimental anomalies in this case is to update the dominance test as follows. If a newly generated node is not dominated, then it is necessary to test

whether any of its ancestors is dominated by an active node (it is not necessary to check all expanded nodes). If so, the newly generated node can be eliminated. Although additional overhead may be incurred in this strategy, the dominance relation may be applied more often, and more nodes may be pruned. When the number of active nodes is not very large, the new dominance-test procedure is acceptable, and detrimental anomalies can be reduced.

When dominance tests are applied with approximate lower-bound tests, the sufficient conditions would be the conjunction of the corresponding ones in both cases. In general, there is no realistic sufficient condition to avoid detrimental anomalies. In contrast, the necessary conditions are the disjunction of the corresponding necessary conditions in both cases. These conditions are loose and are satisfied in most cases.

### 4.4 Multiple subproblem lists

When there are multiple subproblem lists, one for each processor, a node with the minimum heuristic value is selected from each local list for decomposition. This node may not belong to the global set of active nodes with the minimum heuristic values. It is not difficult to maintain a global incumbent in a global data register and broadcast it to the processors. Hence when dominance tests are inactive ($D=I$), all the theorems and corollaries derived in Sections 3 and 4 are applicable, and the behavior of using multiple lists is analogous to that of a centralized list.

When dominance tests are active, these tests can be restricted to the local subproblem lists or can be performed globally. If global dominance tests are applied, then the behavior is similar to that of a centralized list. On the other hand, if dominance tests are performed for subproblems within each local list, then it is possible that $P_iDP_j$ exists in the serial case and $P_j$ is not terminated in the parallel case because $P_i$ and $P_j$ are stored in different processors. As a result, both detrimental and acceleration anomalies may occur. For instance, in Figure 4(d), an acceleration anomaly may happen because $P_2$ is eliminated in the serial case and not terminated in the parallel case, hence $T_3$ is expanded in the serial case and pruned in the parallel case. Note that this acceleration anomaly will not appear in case of a single subproblem list.

Before leaving this section, we must point out that the reasons for the anomalies discussed in Sections 4.2 and 4.3 are not exactly the same. Therefore, some conditions obtained in Section 4.2 cannot be derived directly from the corresponding conditions in Section 4.3 by setting $k_1=1$. The sufficient conditions for the latter case are usually stronger, while the necessary condition are weaker.

## 5 Comparison of best-first, depth-first, and breadth-first searches

In this section, we answer the question on whether a parallel best-first search is the best search strategy as compared to a depth-first or breadth-first search when a constant number of processors are used. This fact has been established for serial searches when all subproblems can fit in the main memory[4]. However, anomalies have been found in parallel searches. This is illustrated in Figure 5 with two processors. Six iterations are needed to complete the depth-first and breadth-first searches, whereas seven iterations are required for a best-first search. Anomalies usually occur when the total number of iterations is small. Analogous to the proof of corollaries in Section 4, we can derive for the same problem to be solved the sufficient conditions that assure $T_b(k,0)<T_r(k,0)$ and $T_b(k,0)<T_d(k,0)$.

$$\left(T_r(1,0)-T_b(1,0)\right)>(k-1)(h-1) \quad (30)$$

and

$$\left(T_d(1,0)-T_b(1,0)\right)>(k-1)(h-1) \quad (31)$$

Eq. (31) is valid if all feasible solutions are located in the bottommost level of the OR-tree. In general, if $I_d$, the number of imperfect iterations in a parallel depth-first search, is greater than $(k-1)h$, then a parallel best-first search will run faster than a parallel depth-first search. Since the average number of essential nodes in an imperfect iteration is almost the same in both strategies, hence $T_b(k, 0) < T_d(k, 0)$ when $I_d > h$[20].

Simulation results have demonstrated that the number of imperfect iterations in a depth-first search is usually larger than that in a best-first search, especially when the number of processors used is relatively large. Therefore, the speedup in the depth-first search drops quickly. In Case 2 of Table 1, when more than two processors are used in the depth-first search, the speedup is less than 0.9$k$. In contrast, in a best-first search solving the same problem, a 0.9$k$ speedup is attained when 32 processors are used.

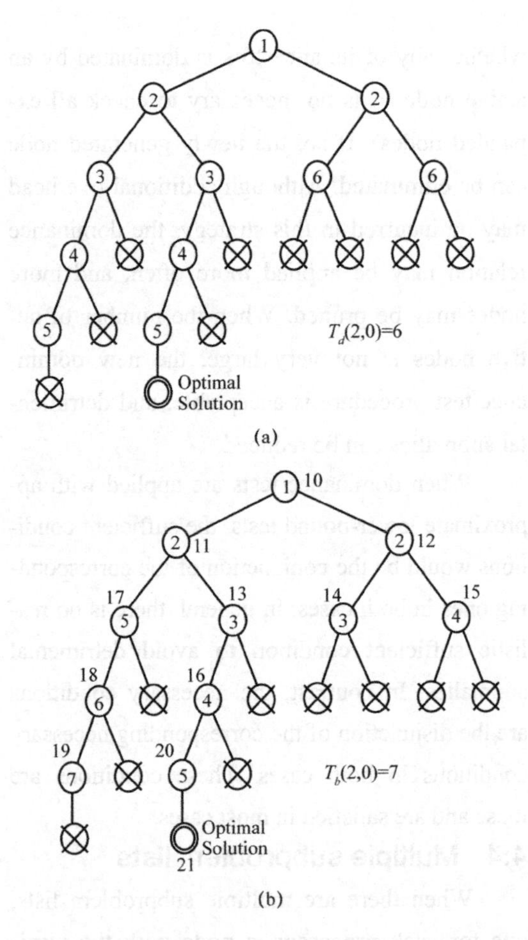


Figure 5. Anomaly in performance between depth-first and best-first searches. (a) Parallel depth-first search with two processors (number in each node is the selection order). (b) Parallel best-first search with two processors (number in each node is the selection order, number outside each node is the lower bound)

From the viewpoint of coping with anomalies, a breadth-first search is a conservative strategy. There are neither detrimental anomalies nor serial-to-parallel acceleration anomalies when dominance tests are inactive. In contrast, a depth-first search is an adventurous search strategy. It may gain a superlinear speedup but suffers from the risk of detrimental anomalies when approximate solutions are sought. For best-first searches, serial-to-parallel detrimental anomalies can be avoided, while acceleration anomalies may occur even if an approximate solution is sought. Generally, linear speedups can be achieved in a larger range of the number of processors for parallel best-first searches than for depth-first and breadth-first searches. In this sense, the best-first search is more robust for parallel processing.

On the other hand, since the best-first search requires the secondary memory to maintain the large number of active nodes, the total time, including time spent on data transfers between the main and secondary memories, to solve a problem should be taken as a measure of efficiency. From this point of view, a best-first search may not always be desirable. Simulations have shown that the best OR-tree search strategy depends on the accuracy of the problem-dependent lower-bound function[2]. Very inaccurate lower bounds are not useful in guiding the search, while very accurate lower bounds will prune most unnecessary expansions. In both cases, the number of subproblems expanded by depth-first and best-first searches will not differ greatly, and a depth-first search is better as it requires less memory space. When the accuracy of the lower-bound function is moderate, a best-first search gives a better performance. In this case, either a good memory management system[23] is necessary to support the large memory space required, or a modified best-first strategy, such as the IDA*[21], is applied.

In Figure 6, the average speedups of parallel best-first and depth-first searches for ten knapsack problems with thirty-five objects are compared. Note that for knapsack problems, the lower-bound function is accurate. As a result, the performance of using the depth-first search is close to that of using the best-first search. (On the average, $T_d(1,0) = 15\,197$, and $T_b(1,0) = 15\,180$.) The speedups in best-first searches are a little larger than those of depth-first searches. When the number of processors are very large such that nodes in each level of the OR-tree can be expanded simultaneously, then $T_d(k, 0)=T_b(k, 0) = h$, where $h$ is the height of the tree. Therefore, the two curves on speedups will coincide eventually.

Finally, we compare the space requirements between depth-first and best-first searches. In a serial search, the space required by a best-first strategy is usually more than that required by a depth-first strategy. Somewhat surprisingly, simulation results on 0-1 knapsack problems show that the space required by a parallel best-first strategy is not increased significantly (but may be reduced) until the number of processors is so large that a near-linear speedup cannot be kept. In contrast, the space required by a parallel depth-first search is almost proportional to the number of processors (Figure 6). Note that the space efficiency is problem-dependent. For vertex-cover problems, the space required by a parallel best-first search is almost constant for the entire range of the number of processors.

# 6 Conclusions

In this paper, we have derived the performance bounds of parallel best-first, depth-first, and breadth-first OR-tree searches, respectively. These bounds provide the theoretical foundation

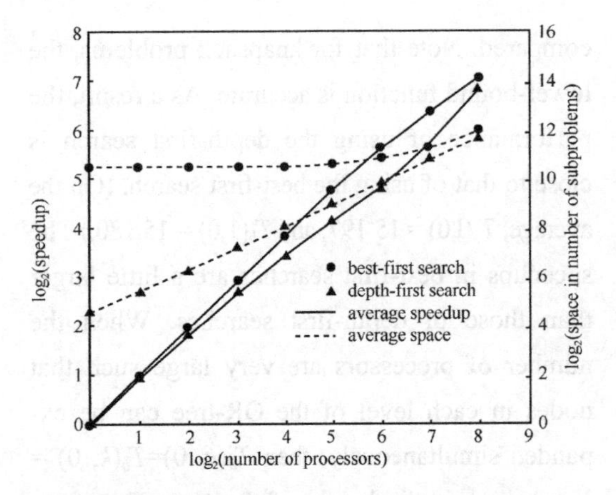


Figure 6. Average speedup and space requirements of parallel branch-and-bound algorithms for ten knapsack problems with 35 objects

to determine the number of processors in order to assure a near-linear speedup. It is found that for best-first searches, the speedup is related to the problem complexity, which is reflected by $T_b(1,0)/h$, where $T_b(1,0)$ is the number of iterations in a sequential optimal best-first search, and $h$ is the maximum height of the tree. To guarantee a near-linear speedup, the number of processors must be much less than $T_b(1,0)/h$. For depth-first and breadth-first searches, the speedups are related to the number of incumbents obtained during the search in addition to the problem complexity. Since the performance bounds on best-first searches are tighter than those on depth-first and breadth-first searches, the range within which a near-linear speedup is maintained is usually larger for best-first searches.

The anomalous behavior of parallel OR-tree searches has been studied thoroughly. Anomalies are caused by a combination of the following reasons: a) there are multiple solution nodes; b) the heuristic function is ambiguous; c) the elimination rule is not consistent with the heuristic function; d) the tree structure causes imperfect iterations when multiple processors are used; and e) the feasible solutions are not generated in the same order when different number of processors are used. The existence of a combination of these conditions causes the tree to be searched in a different order when a different number of processors is used. We have analytically investigated the conditions to eliminate detrimental anomalies and to preserve acceleration anomalies with respect to different search strategies. A summary of the results proved in this paper are shown in Table 3.

A best-first search is found to be a robust search strategy in the sense of the large range of the number of processors within which a linear speedup is achieved. However, the best OR-tree search strategy depends on the accuracy of the problem-dependent lower-bound function. A best-first search is more suitable for parallel processing when the accuracy of the lower-bound function is moderate and the memory-management scheme is efficient. On the other hand, a depth-first search is more efficient when either the lower-bound function is very accurate in pruning unnecessary searches, or the heuristic function is inaccurate in guiding the search.

## References:

[1] SIMON H A, KADANE J B. Optimal problem-solving search: all-or-none solutions[J]. Artificial Intelligence, 1975, 6(3): 235-247.

[2] WAH B W, YU C F. Stochastic modeling of branch-and-bound algorithms with best-first search[J]. IEEE Transactions on Software Engineering, 1985, 11(9): 922-934.

[3] SMITH D R. Random trees and the analysis of

Table 3. Conditions to cope with anomalies in parallel or-trees searches

| Allow. function | Dom. rel. | Search strategy | Suff. cond. to elim. detrimental anom. | | Nec. cond. to allow acceleration anom. | |
|---|---|---|---|---|---|---|
| | | | (1vs.$k$) | ($k_1$ vs.$k2$) | (1vs.$k$) | ($k_1$ vs.$k2$) |
| $\varepsilon$=0 | $D$=$I$ | bfs | Eq's (6), (7) | Cor.3 | Th.9 | Cor.5 |
| | | dfs | | Cor.4 | | Cor.6 |
| | | brfs | | Th.10 | | Eq. (20) |
| | $D$≠$I$ | all | Th.6 | anom. | | exist |
| $\varepsilon$>0 | $D$=$I$ | bfs | Eq's (6), (7) | Cor.7 | Eq. (25) | Cor.7 |
| | | dfs | Th.8 | Cor.8 | Eq. (26) | Cor.8 |
| | | brfs | Th.5 | Cor.9 | Cor.9 | Cor.9 |
| | $D$≠$I$ | bfs | Th.7 | anom. | exist | exist |
| | | dfs | Th.8 | | | |
| | | brfs | Th.8 | | | |

Keys: bfs: best-first search;
dfs: depth-first search;
brfs: breadth-first search;
anom.: when multiple subproblem lists and local dominance tests are used, sufficient conditions are impractical;
exist: necessary conditions are too loose


branch-and-bound procedures[J]. Journal of the ACM, 1984, 31(1): 163-188.

[4] WAH B W, LI G J, YU C F. Multiprocessing of combinatorial search problems[J]. Computer, 1985, 18(6): 93-108.

[5] IBARAKI T. The power of dominance relations in branch-and-bound algorithms[J]. Journal of the ACM, 1977, 24(2): 264-279.

[6] KUMAR V, KANAL L N. A general branch and bound formulation for understanding and synthesizing and/or tree search procedures[J]. Artificial Intelligence, 1983, 21(1-2): 179-198.

[7] NAU D S, KUMAR V, KANAL L. General branch and bound, and its relation to A* and AO*[J]. Artificial Intelligence, 1984, 23(1): 29-58.

[8] KOWALSKI R. Logic for problem solving[M]. The Netherlands: North-Holland, 1979.

[9] CIEPIELEWSKI A, HARIDI S. Execution of bag of on the OR-parallel token machine[C]// International Conference on 5th Generation Computer Systems. [S.l.:s.n.], 1984: 551-560.

[10] KUMAR V, RAO V N. Parallel depth-first search on multiprocessors Part II: analysis[J]. International Journal of Parallel Programming, 1987, 16(6).

[11] LI G J, WAH B W. Coping with anomalies in parallel branch-and-bound algorithms[J]. IEEE Transactions on Computers, 1986, 35(6): 568-573.

[12] BAUDET G M. On the branching factor of the alpha-beta pruning algorithm[J]. Artificial Intelligence, 1978, 10(2): 173-199.

[13] FINKEL R A, FISHBURN J P. Parallelism in alpha-beta search[J]. Artificial Intelligence, 1982, 19(1): 89-106.

[14] FORGY C L, GUPTA A, NEWELL A, et al. Initial assessment of architectures for production systems[C]// Conference on Artificial Intelligence. Piscataway: IEEE Press, 1984: 116-120.

[15] LAI T H, SAHNI S. Anomalies in parallel branch-and-bound algorithms[J]. Communications of the ACM, 1984, 27(6): 594-602.

[16] LI G J, WAH B W. Computational efficiency of parallel approximate branch-and-bound algorithms[C]// International Conference on Parallel Processing. Piscataway: IEEE Press, 1984: 473-480.

[17] RAO V N, KUMAR V. Superlinear speedup in parallel state-space search[R]. 1988.

[18] LAI T H, SPRAGUE A. Performance of parallel branch-and-bound algorithms[C]// International Conference on Parallel Processing. Piscataway: IEEE Press,

1985: 194-201.

[19] LI G J, WAH B W. Computational efficiency of parallel approximate branch-and-bound algorithms[R]. West Lafayette: Purdue University, 1984.

[20] LI G J. Parallel processing of combinatorial search problems[D]. West Lafayette: Purdue University, 1985.

[21] KORF R E. Depth-first iterative deepening: an optimal admissible tree search[J]. Artificial Intelligence, 1985, 27(1): 97-109.

[22] KORF R E. Real-time heuristic search: first results[C]// The 6th National Conference on Artificial Intelligence. New York: ACM Press, 1987: 133-138.

[23] YU C F, WAH B W. Efficient branch-and-bound algorithms on a two-level memory system[J]. IEEE Transactions on Software Engineering, 1988, 14(9): 1342-1356.

[24] IBARAKI T. Theoretical comparisons of search strategies in branch-and-bound algorithms[J]. International Journal of Computer and Information Sciences, 1976, 5(4): 315-343.

[25] MORIN T, MARSTEN R. Branch-and-bound strategies for dynamic programming[J]. Operations Research, 1976, 24(4): 611-627.

[26] IBARAKI T. On the computational efficiency of branch-and-bound algorithms[J]. Journal of the Operations Research Society of Japan, 1977, 20(1): 16-35.

[27] IBARAKI T. Computational efficiency of approximate branch-and-bound algorithms[J]. Mathematics of Operations Research, 1976, 1(3): 287-298.

[28] KOHLER W, STEIGLITZ K. Characterization and theoretical comparison of branch-and-bound algorithms for permutation problems[J]. Journal of the ACM, 1974, 21(1): 140-156.

[29] WAH B W, MA Y W. MANIP-a multicomputer architecture for solving combinatorial extremum problems[J]. IEEE Transactions on Computers, 1984, C-33(5): 377-390.

[30] KNUTH D E, MOORE R W. An analysis of alpha-beta pruning[J]. Artificial Intelligence, 1975, 6(4): 293-326.

[31] PEARL J. The solution for the branching factor of the alpha-beta pruning algorithm and its optimality[J]. Communications of the ACM, 1982, 25(8): 559-564.

[32] IMAI M, FUKUMURA T. A parallelized branch-and-bound algorithm implementation and efficiency[J]. Systems Computer Controls, 1979, 10(3): 62-70.

[33] MOHAN J. Experience with two parallel programs solving the traveling-salesman problem[C]// International Conference on Parallel Processing. Piscataway: IEEE Press, 1983: 191-193.

[34] CONWAY R W, MAXWELL W L, MILLER L W. Theory of scheduling[M]. Massachusetts: Addison-Wesley, 1967.

[35] SAHNI S. Algorithms for scheduling independent tasks[J]. Journal of the ACM, 1976, 23(1): 116-127.

# Parallel iterative refining A* search

Guo-jie LI[1], Benjamin W. WAH[2]

1. National Research Center for Intelligent Computing Systems, Beijing, P. R. China

2. Coordinated Science Laboratory University of Illinois, 1101W. Springfield, Avenue Urbana, IL 61801

**Abstract:** In this paper, we study a new search scheme called Iterative Refining A* (IRA*). The scheme solves a class of combinatorial optimization problems whose lower-bound values and heuristic solutions of partial problems are easy to compute. An example of problems in this class is the symmetric traveling-salesman problem for a set of fully-connected cities. The scheme consists of a sequence of guided depth-first searches, each searching for a more accurate solution than the previous one. It is an extension from Kerf's IDA* search, except that it prunes by approximation rather than by thresholding. In IRA*, global upper bound (or incumbents), rather than lower-bound information, is passed between the successive depth-first searches. The time required by IRA* for finding optimal solutions to problems in this class is very close to that required by a best-first search, while the memory space required is the same as that of a depth-first search. We derive the performance bounds of parallel IRA* and found that it is more robust and has much better worst-case behavior than a parallel depth-first search.

**Key words:** approximation, combinatorial optimization, iterative-deepening A*, iterative-refining A* (IRA*), parallel processing, performance bound

## 1 Introduction

NP-hard combinatorial optimization problems (COPs) require an exponential amount of computation time and memory space in the worst case in solving them. It has been proved that the A* algorithm with an admissible lower-bound function expands the minimum number of nodes in finding the optimal solution or a solution with a guaranteed deviation from the optimal solution. Our objective in this paper is to find an efficient and robust parallel search scheme for solving COPs so that the memory space required is bounded, and that the ratio of the total number of nodes expanded in the best sequential (or A*) algorithm to that of the proposed (parallel) search scheme is maximum.

There are a number of recent results on parallel combinatorial searches[1-2]. Most of them are focused on load balancing, variable sharing, and reducing communication overheads. A number of studies were devoted to finding the best

This research was supported in part by National Aeronautics and Space Administration under Grant NCC 2-481 and in part by National Science Foundation under Grant MIP 88-10584.

speedup of a parallel algorithm as compared to a sequential algorithm using the same search strategy. For instance, near-linear or superlinear speedups have been observed for a parallel depth-first search as compared to a sequential depth-first search[1]. Similar behavior has been observed for parallel IDA* search as compared to a sequential one[3]. These issues are important in parallel processing; however, they do not indicate the actual speedup achieved in solving the given problem. The speedup achieved by parallel processing must be compared against the best sequential algorithm for solving the given problem and not the sequential version of the parallel algorithm. Using the number of nodes expanded as a measure of performance, the best sequential performance is achieved by the best-first strategy. In this paper we characterize the aggregate speedup achieved as a function of the speedup achieved by parallel processing (using the same search strategy) and the slowdown resulted due to the use of a more space-efficient algorithm (such as depth-first search) than the best-first search.

In general, combinatorial optimization problems can be roughly divided into two classes according to whether feasible solutions can be found easily. In the first class, finding feasible solutions may be as hard as finding optimal ones. Examples of problems in this class include integer programming, solving 15-puzzle problems, and finding vertex covers. These problems can be solved efficiently by IDA*[4]. In the second class, it is relatively easy to find feasible solutions and hence upper-bound values to partially evaluated problem instances. Examples of problems include the symmetric traveling-salesman problem on fully-connected cities and many unconstrained scheduling problems. In this paper, we address problems in the second class. Without loss of generality, we discuss our results with respect to minimization problems.

In developing an efficient parallel search strategy, it is necessary to decompose the search into tasks so that the processing resources are scheduled to search independent subtrees. It is important that memory and time limitations be observed, that anomalies in parallelism be addressed, and that the scheme be applicable to a wide variety of problems. The iterative refining A* (in short, IRA*) search proposed in this paper are developed with these objectives in mind.

In general, parallelism of COPs can be exploited in two levels: intra-node and inter-node levels. Current techniques in parallel processing, such as pipelining and array processing, are suitable for intra-node parallelism. In contrast, inter-node parallelism can be broadly classified into four types.

(a) In tree-splitting or tree-decomposition, each processor has a local list for storing active nodes, each constituting a disjoint part of the search tree. When a processor finishes evaluating all nodes in the local list, it gets an unsearched node from another processor. The major issues here are in finding efficient load balancing and variable-sharing (or information communication) strategies and in coping with anomalies in parallelism.

(b) In parallel window searches, each processor expands the nodes whose lower bounds fall into a given window, namely, a lower-bound interval. This approach is suitable for bottom-up searches, such as game-tree search. but suffers from the high overhead of node distribution found in top-down searches.

(c) In parallel iterative searches, the tree searched is expanded in successive iterations and may overlap with each other. In the original IDA* algorithm[4], thresholds are defined so all nodes with values less than a given threshold are searched in an iteration. Trees in successive iterations are searched sequentially. These trees can be searched in parallel. Parallel iterative search has the benefit of obtaining a desirable approximate solution early in the process and in communicating these solutions to all trees searched in parallel.

(d) The last type of inter-node parallelism is based on exploiting multiple heuristics and has not been addressed seriously in the literature. In this form, multiple heuristics are run in parallel. This takes advantage of the large fluctuations in computation times of various heuristics in generating feasible solutions.

The parallel IRA* search scheme studied in this paper utilizes the four types of inter-node parallelism discussed above and addresses three important issues: memory-space limitation, detrimental anomalies in parallel search[5], and achieving a performance as close as possible to the theoretical minimum number of nodes expanded in a best-first search. It performs better than a pure depth-first search or a guided depth-first search, which only address the issue on memory limitation. We derive bounds on performance of parallel IRA* searches. These bounds show that the IRA* algorithms are more robust than pure depth-first searches with respect to parallel processing and approximate computations.

In this paper, we evaluate our results using the traveling-salesman problem. These problems are generated by calling the random number generator rand(), which randomly determines the location of cities on a 100-by-100 Cartesian plane. All cities are assumed fully connected, and distances between cities are symmetric; that is, the triangular inequality is satisfied. The lower-bound function used here is computed by the spanning-tree method, and the upper bounds by a hill-climbing heuristic. Since the method for computing the upper bound generates a feasible solution, the global minimum upper bound is the same as the incumbent.

## 2 A basic principle of parallel processing

In general, two criteria can be used to measure the efficiency of parallel solution of a COP.

Type I: Minimize $T(P)$

Subject to $\alpha \geqslant \alpha$ , and $M(P) \leqslant M_{max}$

Type II: Maximize $\alpha$

Subject to $T(P) \leqslant T_{max}$ and $M(P) \leqslant M_{max}$

where $T(P)$ and $M(P)$ are the computation time and memory space spent by the algorithm using P processors; $\alpha$ and $\alpha_r$ are, respectively, the achieved and desired accuracies of the solution; and $T_{max}$ and $M_{max}$ are the time and space constraints. In this paper, the criterion used is of Type I.

Let $T_{best}(1)$ be the computation time required by the best sequential algorithm for solving a given problem. Since $T_{best}$ is fixed, Type I criterion is equivalent to

$$\text{maximize } \frac{T_{\text{best}}(1)}{T(P)} \text{ subject } \alpha \geqslant \alpha \text{ , and } M(P) \leqslant M_{\max}$$

Let $T(1)$ be the underlying sequential algorithm corresponding to the parallel algorithm to be measured. Then,

$$\frac{T_{\text{best}}(1)}{T(P)} = \frac{T_{\text{best}}(1)}{T(1)} \cdot \frac{T(1)}{T(P)} = R \cdot S \tag{1}$$

where $R = \frac{T_{\text{best}}(1)}{T(1)}$ and $S = \frac{T(1)}{T(P)}$. It is clear that $R$ characterizes the efficiency of the underlying sequential algorithm, and $S$ characterizes the efficiency of the parallel implementation of the underlying sequential algorithm. A desirable objective of a parallel search is that it achieves the maximum of $R{\cdot}S$ rather than only $R$ or $S$ individually, and that it satisfies the constrains listed in Type I or II criterion.

A parallel best-first search may achieve the maximum $R{\cdot}S$. However, it violates the memory-space constraints unless the COP solved is very small. For the original IDA* algorithm, the average $S$ achieved for different problem instances may be close to $P$, although anomalous behavior occurs, as reported by Rao, Kumar, and Ramesh[3]. $R$ achieved, however, is small for solving the second class of COPs, such as TSPs. For the conventional depth-first search, $R$ achieved is quite small, and $S$ is unpredictable due to its anomalous behavior: acceleration anomalies may result in superlinear speedups, and detrimental anomalies are frequent. The basic idea in this paper is, therefore, on obtaining both $R$ and $S$ as large as possible and on making $S$ predictable under the restriction of memory space.

# 3 Effects of initial upper-bounds

Most search algorithms including A* can be formulated as a version of a branch-and-bound algorithm (B&B). In the B&B computation for solving a minimization problem, the global minimum upper bound, or incumbent, is initially set to a given value $u_0$ (or infinity) and monotonically decreases until it achieves $f_0$ , the optimal solution, which remains unchanged until the end of computation. In the following sections, the incumbent initially obtained before starting the computation of an iteration is called the initial upper bound of the iteration.

In general, a combinatorial search process can be split into two phases:

(a) computation to obtain an optimal solution;

(b) computation to verify its optimality.

To analyze the performance independent of the details of the underlying computer architecture, we use the number of nodes expanded as the performance measure throughout this paper. Let $T_f$ and $T_v$ be the numbers of nodes expanded before and after the first optimal solution is found. Let $T_b$ be the number of nodes expanded in an A* search. Obviously, the total number of nodes expanded in any search algorithm, $T$, satisfies

$$T = T_f + T_v \geqslant T_b$$

In an A* algorithm, the best-first search strategy is adopted, so $T_f = T_b$, and $T_v = 0$. This implies that the initial upper bound has no impact on the A* algorithm. However, in depth-first searches, $T_f$ is usually much larger than $T_v$ and $T_b$. Table 1 shows $T_f$, $T_v$, and $T_b$ for 5 symmetric TSPs of 20-cities. These examples show that in most cases, $T_f$ is the dominant pant of search time for depth-first searches. To reduce the computation time of depth-first searches, the most important concern is to make $T_f$ as small as possible.

Table 1. Comparison of $T_f$ and $T_v$ of guided depth-first searches for 5 symmetric 20-city TSPs

| | Initial Seeds | | | | |
|---|---|---|---|---|---|
| | 1 | 2 | 3 | 4 | 5 |
| $T_f$ | 54 806 | 44 143 | 181 299 | 146 335 | 106 305 |
| $T_v$ | 12 981 | 13 483 | 42 658 | 10 602 | 29 547 |
| $T_b$ | 16 235 | 23 491 | 77 292 | 13 867 | 48 062 |

For general heuristic searches, $T_f$ depends mainly on the accuracy of the guidance function. while $T_v$ depends uniquely on the accuracy of the lower-bound function. The only way to reduce $T_v$ is to design more accurate lower-bound functions, which is difficult since it is problem dependent. In a depth-first search, since no heuristic function is used to guide the selection of nodes, the upper-bound function has a strong influence on $T_f$, especially when approximations are applied.

So far, a lot of work has been carried out on improving the guidance and lower-bound functions, but the impact of the upper-bound function has not been studied seriously. In this paper, we do not attempt to compare different upper-bound functions, but rather study the influence of the initial upper bound on the performance of a search.

We have extensively studied the computational behavior of guided depth-first searches[1] for solving TSPs and knapsack problems. It is found that a decrease of the initial upper bound results in a significant decrease in the computation time in most cases. When an optimal solution is sought, a smaller initial upper bound will always have a positive effect, i.e., reducing $T_f$ for the depth-first search. Due to space limitation, this claim is stated without proof in Theorem 1.

**Theorem 1:** Suppose that $A(u_1)$ and $A(u_2)$ are B&B algorithms using the different initial upper bounds $u_1$ and $u_2$ and that $u_1 > u_2$. Assume that dominance relation is consistent with the lower-bound function[2]. Let $T(u_1)$ and $T(u_2)$ be the computation time to find the optimal solution by $A(u_1)$ and $A(u_2)$, respectively. Then $T(u_1) \geqslant T(u_2)$.

Theorem 1 holds for any search strategies. However, a smaller initial upper bound does not always guarantee a reduction in computation time when approximate B&B algorithms are employed. Anomalies may happen, although they do not happen frequently.

## 4 Iterative refining A* search

Having realized that the initial upper bound has a vital influence on $T_f$ in depth-first searches, we need to answer the next question: how to obtain a small initial upper bound with the minimum overhead. This can be found by an efficient heuristic algorithm, which finds a feasible solution to the problem. This approach is not

[1] In a guided depth-first search, the children of a node are searched in monotonic order of lower bounds.

[2] A dominance relation $D$ is said to be consistent with the lower-bound function if $P_iDP_j$ implies that $g(P_i) < g(P_j)$, where $g(P_i)$ and $g(P_j)$ are nodes in the B&B search tree. More detailed discussion on dominance relations can be found in the reference [5].

taken in this paper as it is problem dependent. Our goal is to develop a general problem-independent scheme for finding initial upper bounds efficiently.

Note that the initial upper bound cannot be assigned arbitrarily. If the assigned initial upper bound is less than $f_0$, then it is not a true upper bound but a threshold. Thresholds have been used in IDA* search[4] and can be selected optimally to minimize the total search overhead[6]. However, the criterion for their selection is not based on minimizing the overhead for finding an initial upper bound.

An approximate B&B algorithm can be used to find suboptimal solutions of COPs. During an approximate B&B procedure, a node $P_i$ generated will be terminated if

$$g(P_i) \geqslant \frac{u(P_i)}{1+\varepsilon}, \varepsilon \geqslant 0 \quad (2)$$

where $u(P_i)$ is a global upper bound for $P_i$ and $g(P_i)$ is its lower bound. $u^f$, the final suboptimal solution obtained by the approximate B&B algorithm, deviates from $f_0$, the optimal solution value, by

$$\frac{u^f}{1+\varepsilon} \leqslant f_0 \leqslant u^f, \varepsilon \geqslant 0 \quad (3)$$

Usually, $\varepsilon_f (= (u^f - f_0)/f_0)$, the final error achieved, is much smaller than the allowed error $\varepsilon$. Since $f_0$ is unknown in advance, $g^f$, the global minimum lower bound, instead of $f_0$, is often used to get $\varepsilon_g$, the upper bound of the error of the suboptimal solution obtained. Note that $\varepsilon_g = (u^f - g^f)/g^f$. If $\alpha$, the worst-case accuracy of the solution, is defined as $f_0/u^f$, then $\alpha = 1/(1+\varepsilon)$.

Approximations significantly reduce the amount of time needed to arrive at suboptimal solutions. Let $A(\varepsilon_1)$ and $A(\varepsilon_2)$ be B&B algorithms with $\varepsilon_1$ and $\varepsilon_2$, respectively, and $\varepsilon_1 < \varepsilon_2$. If a suboptimal solution $u^f$ is found by both algorithms, then in most cases. $A(\varepsilon_2)$ finds $u^f$ earlier than $A(\varepsilon_1)$ does, though anomalous behavior may appear when a depth-first search is employed.

The above claim has been verified experimentally and is demonstrated in Figure 1. A symmetric TSP is solved by a guided depth-first search with four different error allowances, 0.2, 0.1, 0.05 and 0. The computational behaviors are measured in terms of the rates of decrease of the global upper bounds. Figure 1 shows that the larger the error allowance is, the faster the global upper bound decreases. For instance, when $\varepsilon = 0.2$, 240 nodes are expanded to reach the global upper-bound value 369.2, while 138 000 nodes are expanded in order to reach the same global upper-bound value when $\varepsilon = 0$.

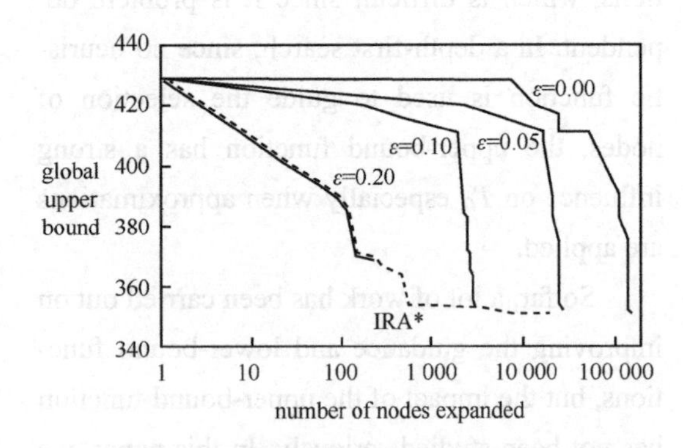


**Figure 1. The effect of error allowance on global upper-bound for guided depth-first searches. (TSP of 20-cities, initial seed= 4)**

Comparing the four curves and considering the effect of the initial upper bounds, we can arrange the depth-first searches in such a way that the approximation search with a large $\varepsilon$ is implemented first; the suboptimal solution obtained is then transferred to the next iteration as the initial upper bound, in which a smaller $\varepsilon$ is used. As a result, the time for solving a COP can be reduced significantly!

For the above example, the same problem can be solved in four iterations with the error allowances decreasing from 0.2 to 0. It takes 241 node expansions to finish the first iteration; the second iteration with $\varepsilon=0.1$ and initial incumbent 369.2 takes 491 node expansions; the third iteration with $\varepsilon=0.05$ and initial incumbent 353.3 takes 2 419 node expansions; finally, the fourth iteration with initial incumbent 351.8 takes 14 229 node expansions to find the optimal solution 351.0 and complete the search. The computational behavior of this implementation is reflected by the dotted curve in Figure 1, which shows an order-of-magnitude improvement in performance (see Case 4 in Tables 2 and 3). Since the suboptimal solution is refined iteratively, we call this scheme an iterative refining A* (in short, IRA*) search.

Table 2. Performance comparison of various search algorithms for 5 20-cities TSPs. ($T_g$: time for guided depth-first search; $T_r$: time for IRA* search; and $T_b$: time for best-first search)

| | Initial Seeds | | | | |
|---|---|---|---|---|---|
| | 1 | 2 | 3 | 4 | 5 |
| $T_f$ | 67 787 | 57 626 | 223 957 | 156 937 | 135 852 |
| $T_r$ | 20 828 | 51 043 | 88 864 | 17 380 | 54 889 |
| $T_b$ | 16 235 | 23 491 | 77 292 | 13 867 | 48 062 |
| $T_r/T_b$ | 1 028 | 2.17 | 1.15 | 1.25 | 1.14 |

Table 3. Time of each iteration in IRA* for 5 20-cities TSPs. $T(\varepsilon)$ is the time needed in an iteration of IRA* to achieved an approximation degree $\varepsilon$

| Approx. Searches | Initial Seeds | | | | |
|---|---|---|---|---|---|
| | 1 | 2 | 3 | 4 | 5 |
| $T(0.2)$ | 769 | 210 | 407 | 241 | 303 |
| $T(0.1)$ | 1 071 | 2 291 | 2 329 | 491 | 1 279 |
| $T(0.05)$ | 1 886 | 9 055 | 8 836 | 2 419 | 5 245 |
| $T(0.0)$ | 17 102 | 39 487 | 77 292 | 14 229 | 48 062 |
| $T(0.00)/T(0.05)$ | 9.17 | 4.36 | 8.75 | 5.89 | 9 016 |

The IRA* search scheme is outlined in Figure 2. The stopping criterion can be the number of iterations, the final error allowance, or other conditions. An upper-bound function can be used to compute $u_1^0$, the initial incumbent of the root. The key of the IRA* search scheme is in refining, i.e., choosing the proper strategy to reduce the error allowance. The goal of refining the computation except the last iteration is to find an optimal or desired suboptimal solution as early as possible. Once the desired solution is found, the remaining time is only used to verify optimality.

The sequence of refinements must be chosen properly. If the difference in error allowances between successive iterations is too large, then the final upper bound found in each iteration may have little influence on the next iteration. In contrast, if the difference is too small, then the global upper bound may remain unchanged in some iterations; hence, these iterations do not contribute to improving the solution. From our experience, four iterations are sufficient for the problems we have tested, and either the optimal or a good suboptimal solution is found in the iteration with $\varepsilon=0.05$.

```
INPUT:      initial error allowance ε_1 ;
            initial incumbent u_1^0 if available;
            problem parameters;
OUTPUT:     Optimal or suboptimal solution;
BEGIN
  WHILE (stopping-criterion is not satisfied) {
    DO approximate depth-first search from the root
       of the tree with ε_i and initial incumbent
       u_i^0
    UNTIL no active node remains in the stack or
       the time constraint is exceeded;
    IF (time constrain is exceeded) THEN
       report the achieved solution and the error
       bound ε_g ;
    ELSE {
       (a) record the suboptimal solution and ε_g ;
       (b) set new error allowance ε_{i+1} < ε_i ;
       (c) set u_{i+1}^0 , the initial incumbent of the
           next iteration, equal to u_i^f , the final
           incumbent in this iteration;
    }
  }
END
```

Figure 2. The IRA* search scheme

Table 2 compares the performance among a guided depth-first search, IRA*, and a best-first search[3] for 5 20-city symmetric TSPs[4]. Due to space limitations, other results which exhibit similar behavior are not shown. The results show that the performance of IRA* is very nice. In all instances except when the initial seed is 2, the search times required by IRA* are only 14 to 28 percent more than that for the best-first search, and the space required is linear. To our knowledge, no other general search scheme has achieved this high performance.

We now explain the reason why IRA* is efficient and analyze its performance. Note that TSP is a strict NP-hard problem, whose computation time grows exponentially when the accuracy of the solution is increased. Table 3 demonstrates this claim. The time required in an approximation search with $\varepsilon = 0.05$ is only about 20 percent of $T_b$. On the other hand, since the lower-bound function for TSP based on the spanning-tree method is not very tight, it is quite possible that a node provides a new incumbent but its lower bound is relatively loose. Thus, an optimal or a good suboptimal solution is usually found before the last iteration, and the overhand of IRA* is quite small.

The computational behavior of the approximate depth-first searches are related to the sequences of incumbents, which are problem-instance dependent. Therefore, a theoretical analysis of the performance of IRA* is very difficult. The following theorem shows the minimum time required by IRA*.

Throughout this paper, all analysis are based on the assumption that the number of nodes whose lower bound is less than $g$ increases exponentially with $g$. Empirically, we have verified this assumption for strict NP-hard COPs, such as symmetric TSPs for fully connected cities[7]. This assumption holds true in a large range of lower-bound values, from the minimum lower-bound value to some value larger than the optimal solution. Let $n_0$ and $B_0$ be, respectively, the numbers of nodes whose lower bounds are minimum and are less than the optimal-solution

[3] $T_b$ is obtained in the last iteration of IRA*, since $u_k^0 = f_0$ . It is very difficult to solve a large symmetric TSP by a best-first search due to the space constraint.

[4] For a symmetric TSPs and TSPs on incompletely connected graphs, problems with several hundred cities can be solved. The solvable size of TSPs on symmetric and fully connected graphs is quite small. Even for 20 cities, the number of nodes expanded in the search tree may exceeds one million.

value. $n_0$ is one for most COPs, i.e., only the root of the B&B search tree has the minimum lower bound.

**Lemma 1:** Assume that $n(g)$, the number of nodes whose lower bounds are less than $g$ in the B&B tree, increases exponentially with $g$ within the range $g_r < g < L$, and that $n(g_r) = n_0$, then

$$n(g) = n_0\left(\frac{B_0}{n_0}\right)^{\left(\frac{g-g_r}{f_0-g_r}\right)} \quad (4)$$

where $f_0$ and $g_r$ are the optimal-solution and minimum lower-bound values, respectively, and $L > f_0$.

**Proof:** According to the assumption. we can denote $n(g)$ as an exponential function.

$$n(g) = M \cdot A^g \quad (5)$$

where $M$ and $A$ are constants. From the following conditions,

$$M \cdot A^{gr} = n_0 \quad (6)$$

$$M \cdot A^{f_0} = B_0 \quad (7)$$

we get $A = \left(\frac{B_0}{n_0}\right)^{\left(\frac{1}{f_0-g_r}\right)}$ and $M = n_0\left(\frac{B_0}{n_0}\right)^{\left(\frac{-g_r}{f_0-g_r}\right)}$.

In terms of the above result, we can analyze the best-case behavior of IRA*.

**Theorem 2:** Let $\varepsilon_i$ be a series of error allowances, $\varepsilon_i > \varepsilon_{i+1}$, $1 \leqslant i < k$, and $\varepsilon_k \geqslant 0$. Let $T_r$ be the computation time required by IRA* for solving a COP and $n_0$=1. Then $T_r \geqslant \sum_{i=1}^{k} B_0^{r_i}$, where $r_i$ satisfies $\frac{f_0}{1+\varepsilon_i} - g_r = r_i(f_0 - g_r)$.

**Proof:** Consider iteration $i$ in which the error allowance is $\varepsilon_i$. From Eq. (2), all nodes whose lower bound is less than $\frac{f_0}{1+\varepsilon_i}$ have to be expended; thus, at least $n\left(\frac{f_0}{1+\varepsilon_i}\right)$ nodes are expanded in iteration $i$. From Lemma 1,

$$n\left(\frac{f_0}{1+\varepsilon_i}\right) = B_0^{\left(\frac{\frac{f_0}{1+\varepsilon_i}-g_r}{f_0-g_r}\right)} \quad (8)$$

Note that, in the above equation, $n(\cdot)$ is a function notation. Let $\frac{f_0}{1+\varepsilon_i} - g_r = r_i(f_0 - g_r)$. Eq. (8) shows that at least $B_0^{r_i}$ nodes have to be expanded in iteration $i$. The theorem is proved by considering $k$ iterations.

Only a loose upper bound of $T_r$ can be derived due to its anomalous behavior. However, our results in Table 3 demonstrate that most of the overhead of IRA* is incurred in the last iteration. If the optimal solution is found before this, then $T_r$ is usually close to $T_b$, since all nodes expanded in the last iteration have to be expanded in a best-first search as well. In general, this is not the worst case. We now discuss the worst-case behavior of IRA*.

**Theorem 3:** Let $k$ be the number of iterations of an IRA*, $\varepsilon_k$ be 0, and the optimal solution be found in the last iteration. Let $t_j$ be the computation time required in the $j$th iteration, and $n_0$ be 1. Then $t_k < B_0^{r_{k+1}}$, and $t_{k-1} < B_0^{r_{k-1}+1}$, where $r_k = \frac{\varepsilon_{k-1} f_0}{f_0 - g_r}$ and $r_{k-1} = \frac{(\varepsilon_{k-2} - \varepsilon_{k-1}) \cdot f_0}{(1+\varepsilon_{k-1})(f_0 - g_r)}$.

**Proof:** Since the optimal solution is found in the last iteration and using the approximate lower-bound test defined in Eq. (2), we can claim that $u_{k-1}^0 = u_{k-2}^f < (1+\varepsilon_{k-2}) f_0$; otherwise, the optimal solution must be found before the $(k-1)$th iteration. Similarly, $u_{k-1}^f < (1+\varepsilon_{k-1}) \cdot f_0$, and $g_{k-1}^f < u_{k-1}^0 / (1+\varepsilon)$. The worst case happens when the incumbent is unchanged until the last node is expanded in the $(k-1)$th iteration. This implies that

$$t_{k-1} < n\left(\left(\frac{1+\varepsilon_{k-2}}{1+\varepsilon_{k-1}}\right)\cdot f_0\right)$$

From Eq. (5), we have

$$t_{k-1} < M \cdot A^{\left(\frac{(1+\varepsilon_{k-2})\cdot f_0}{1+\varepsilon_{k-1}}\right)}$$

Hence,

$$t_{k-1} < B_0^{\left(1+\frac{(\varepsilon_{k-2}-\varepsilon_{k-1})\cdot f_0}{(1+\varepsilon_{k-1})(f_0-g_r)}\right)} \tag{9}$$

Applying the above analysis to the $k$-th iteration, we get

$$t_k < B_0^{\left(1+\frac{\varepsilon_{k-1} f_0}{f_0-g_r}\right)} \tag{10}$$

The theorem follows from Eq's (9) and (10) and the definition of $r_i$, $i = k-1, k$.

Theorems 2 and 3 are of importance only in verifying the performance of IRA* and not in predicting it, as $B_0$ is problem-instance dependent and is difficult to estimate. Some parameters in the above theorems can be estimated. For example, $g_r$ is easy to obtain by abstracting the information of the root node. From our experience, $f_0$ is close to the mean of the root's lower-bound and upper-bound values. More accurate estimate of $f_0$ can be obtained by using statistical extreme-value distribution theorem. According to this theorem, TSP can be viewed as $(n-1)!/2$ samples and each upper bound as a local minimum, and the optimal solution satisfies the Weibull distribution.

To verify this upper bound, we examine a 20-city TSP with $u_k^0 = 398.792$, $g_r = 303.165$, $f_0 = 395.792$, $T_b = 552\,101$, and $t_k = 756\,385$. Since $B_0$ is not available, we use $T_b$ instead of $B_0$ in order to derive a loose upper bound of $t_k$. In this example, $u_{k-1}^f = u_k^0 = 398.792$; hence, the achieved $\varepsilon_{k-1}$ is $(u_k^0 - f_0)/f_0 = 3/f_0$, and $r_k$ is 0.033. From Theorem 3, we get $t_k = 756\,385 < T_b^{1+r_k} = 854\,093$.

The sequence of $\varepsilon$ used in IRA* is called a refinement schedule. A good refinement schedule should be determined in terms of the distribution of nodes by lower bounds and the accuracy of the initial incumbent. The design of an optimal refinement schedule is open at this time. In this paper, a static refinement schedule is adopted. We have examined a number of dynamic refinement schedules and found that the search performance of IRA* for solving TSPs is not sensitive to changes in the refinement schedule, since the computation time of the last iteration is dominant and is close to $T_b$ in most cases.

## 5 Parallel iterative refining A*

In studying parallel IRA*, we focus on improving its speedup with respect to the best sequential algorithm. To simplify the analysis, we assume that the overheads of load balancing and communication are negligible. We assume that the computational model is a loosely coupled multicomputer, and that the B&B tree is split into a number of subtrees that are searched in parallel. We found earlier that when multiple lists are used, a parallel search may risk searching a large number of extra nodes[5]. Our objective here is to reduce this risk. We further assume that the parallel computation is synchronized by iterations. Note that the search trees in different iterations have different sizes.

Before presenting the theoretical analysis, we need to explain the representation of the search tree. The search trees used in the literature are usually depicted by depth. For the sake of understanding IRA*, the search trees used here are depicted by cost (see Figure 3). Note that nodes in the same cost level do not mean that they are located in the same depth of the original tree.

In iteration *i*, the global upper bound decreases from $u_i^0$ to $u_i^f$ (it is possible that $u_i^f = u_i^0$). When a parallel IRA* is implemented, the speedup is related to the interval $\frac{u_i^0 - u_i^f}{1+\varepsilon_i}$. Similar to the discussion in Section 3, we define $r_i$ as $\frac{u_i^0 - u_i^f}{(1+\varepsilon_i)(f_0 - g_r)}$. Since $(u_i^0 - u_i^f) < (f_0 - g_r)$, we have $0 \leqslant r_i \leqslant r_{\max} \leqslant 1$. $r_{max}$ reflects the maximum threshold interval among all iterations.

Define $C_i = \frac{n\left(\frac{u_i^0}{1+\varepsilon_i}\right)}{n\left(\frac{u_i^f}{1+\varepsilon_i}\right)}$. From Lemma 1, we have $C_i = A^{r_i(f_0-g_r)}$. Let $C_{\max} = \max_i(C_i)$, namely, $C_{\max} = A^{r_{\max}(f_0-g_r)}$. $C_{max}$ reflects the ratio between the numbers of nodes expanded with different thresholds and is an important parameter for characterizing the behavior of IRA* searches. In a conventional depth-first search, $r_{max}=1$; hence, $C_{max}=B_0$ if $n_0=1$.

In the following discussion, notations (1) and (*P*) are used to denote the parameters used in serial and in parallel computing.

**Theorem 4:** Assume that the search tree is decomposed into multiple subtrees, that the processors are synchronized in node expansions, that $u_i^0(P) = u_i^0(1)$, and that $u_i^f(P) = u_i^r(1)$. Let $T_r(1)$and $T_r(P)$ be the computation times required for the IRA* algorithms using one and P processors, respectively. The speedup, $\frac{T_r(1)}{T_r(P)}$, are limited within the following bounds.

$$\max\left(1, \frac{P}{C_{\max}}\right) \leqslant \frac{T_r(1)}{T_r(P)} \leqslant C_{\max} \cdot P \qquad (11)$$

**Proof:** The proof is in two parts. Due to space limitation, we show only the part for finding the lower-bound speedup; the part for finding the upper-bound speedup is similar.

In finding the lower-bound speedup, since the parallel IRA* algorithm is synchronized by iterations, $T_r(P) = \sum_{i-1}^{k} t_i(P)$. For the *i*th iteration in both serial and parallel cases, the nodes whose lower bounds are less than $\frac{u_i^f}{1+\varepsilon_i}$ must be expanded. These nodes are called essential nodes, while other nodes expanded are called nonessential nodes. In the worst case, no nonessential node is expanded in the serial case but all nonessential nodes are expanded in the parallel case. This situation is possible when $u_i^f$, the final incumbent of iteration *i*, is not found until $\frac{1}{P} \cdot n\left(\frac{u_i^0}{1+\varepsilon_i}\right)$ essential nodes have been expanded in the serial case, and all these nodes are assigned to processor 1 in the parallel case if $P > C_i$. In addition to essential nodes, all nonessential nodes are expanded in the parallel case, and these nonessential nodes are assigned to processor 2 to processor *P*. Figure 3 shows the worst case for parallel IRA* in which $P > C_i$ and

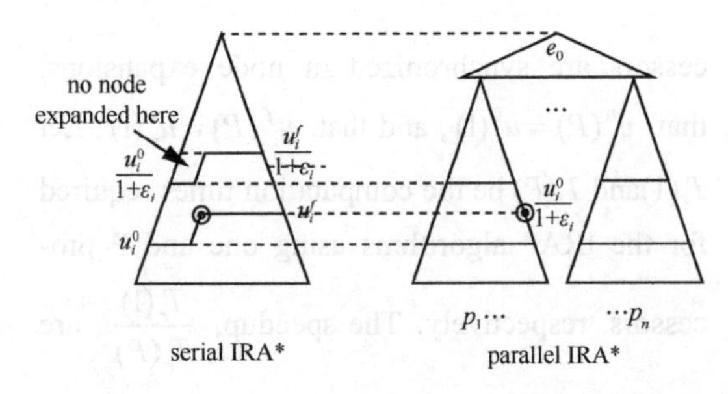


Figure 3. The worst case behavior of parallel IRA*

$$t_i(1)=n\left(\frac{u_i^f}{1+\varepsilon_i}\right),\quad t_i(P)=\frac{1}{P}\cdot n\left(\frac{u_i^0}{1+\varepsilon_i}\right)$$

Without loss of generality, the initial steps for generating $P$ nodes in the parallel IRA* search and the ceiling function is omitted in this proof. This means that the overhead of synchronization is ignored. To derive the upper bound of minimum speedup, i.e., the maximum speedup in the worst case, we neglect the overhead of load balancing. Accordingly,

$$\frac{t_i(1)}{t_i(P)}=\frac{n\left(\frac{u_i^f}{1+\varepsilon_i}\right)}{\left(\frac{1}{P}\right)\cdot\left(n\frac{u_i^0}{1+\varepsilon_i}\right)}\quad \text{if } P>C_i \tag{12}$$

From Eq. (12) and the definition of $C_1$, we have

$$\frac{P}{C_i}\leqslant\frac{t_i(1)}{t_i(P)}\quad \text{if } P>C_i \tag{13}$$

When $P \leqslant C_i$, it is possible that in the parallel case processor 1 expands the same nodes as expanded in the serial case, and all nodes expanded by other processors are unnecessary. This implies that

$$1\leqslant\frac{t_i(1)}{t_i(P)}\quad \text{if } P\leqslant C_i \tag{14}$$

The lower bound is proved by combining Eq's (13) and (14), and using the definition of $C_{max}$.

Note that the speedup is dominated by the last iteration in which the threshold interval is restricted within $\varepsilon_{k-1}f_0$, and oftentimes $u_k^0=u_k^f$. This implies that $C_k$ is very small and is very likely equal to 1. Consequently, in practice, the bounds of speedup are tighter than those given in Theorem 4, especially when load balancing is applied.

The above theorem shows the robustness of parallel IRA*. Since $C_{max}$ is small in IRA*, the condition that $P > C_{max}$ is generally satisfied. Parallel IRA* also avoids the detrimental anomalies and poor slowdown often found in parallel depth-first searches; that is, at least $\frac{P}{C_{max}}$ times speedup can be gained. The cost of this gain is in the loss of large super-linear speedups, although super-linear speedup less than $C_{max}P$ is still achievable. In contrast, the possible speedups in conventional parallel depth-first search may range from l to $B_0 \cdot P$.

# 6 Simulation results of parallel IRA*

To verify the theoretical results derived above, we have simulated the performance of parallel IRA* (PIRA *) and parallel guided depth-first searches (PGDFS) using 16 processors. Results for 5 symmetric TSPs of 20 fully-connected cities are shown here. Other results are not shown due to space limitation.

The parallel search is implemented as follows. Initially, a serial best-first search is carried out to produce 16 active nodes, which are distributed into 16 separated lists. A parallel search is then carried out with the 16 nodes as the roots of subtrees. In each step, a processor expands one node, provided there are 16 or more active nodes among the lists. If a list is empty, it gets a node from another list. Since we are focusing on the speedup (*SP*) of the parallel search, we use the number of steps as our performance measure. The refinement schedule adopted in the parallel IRA* search is the same as that in the serial one.

The simulation results for PIRA* and PGDFS are shown in Table 4. The sequential search times, $T_b$, $T_r$ and $T_g$, are taken from Table 2. $T_g(P)$ and $T_r(P)$ are the times required by PGDFS and PIRA*, respectively. Since the sets of the nodes expanded in a serial search and the corresponding parallel search are not identical, we denote the total number of nodes expanded in all processors in PGDFS (resp. PIRA*) by $T_g^1(\text{resp.} T_r^1)$.

Let $H$ be the ratio of the number of the nodes expanded in a serial search and that in the corresponding parallel search. $H$ reflects the amount of extra work or savings in a parallel search as compared to the serial one. During a parallel search, especially in the first and last few steps, some processors may be idle as there are insufficient active nodes. The parameter $U$ characterizes the speedup of $T_r^1(\text{resp.} T_g^1)$ over $T_r(P)(\text{resp.} T_g(P))$. The simulation results show that in almost every step all processors are busy. That is, $U$ is very close to 16. Using the above definitions and recalling $R$ and $S$ defined in Section 2, we have the following relations among these parameters.

$$SP = R \times S = R \times (H \times U) \tag{15}$$

Table 4. Simulation results on parallel IRA* and PGDFS for solving 5 20-city TSPs using 16 processors

| Init. Seed | 1 | 2 | 3 | 4 | 5 |
|---|---|---|---|---|---|
| Sequential Best-First Search | | | | | |
| $T_b$ | 16 235 | 23 491 | 77 292 | 13 867 | 48 062 |
| IRA* and Parallel IRA* | | | | | |
| $T_r$ | 20 828 | 51 043 | 88 864 | 17 380 | 54 889 |
| $T^1_r$ | 26 879 | 48 404 | 87 739 | 177 783 | 55 025 |
| $T_r(16)$ | 1 697 | 3 043 | 5 501 | 1 127 | 3 457 |
| $R_r$ | 0.78 | 0.46 | 0.87 | 0.80 | 0.88 |
| $H_r$ | 0.77 | 1.05 | 1.01 | 0.98 | 1.00 |
| $U_r$ | 15.84 | 15.91 | 15.95 | 15.78 | 15.92 |
| $S_r$ | 12.27 | 16.77 | 16.15 | 15.42 | 15.88 |
| $SP_r$ | 9.57 | 7.72 | 14.05 | 12.30 | 13.90 |
| GDFS and Parallel GDFS | | | | | |
| $T_g$ | 67 787 | 57 626 | 223 957 | 156 937 | 135 852 |
| $T^1_g$ | 122 035 | 49 340 | 219 463 | 42 032 | 85 028 |
| $T_g(16)$ | 7 643 | 3 099 | 13 732 | 2 643 | 5 340 |
| $R_g$ | 0.24 | 0.41 | 0.35 | 0.09 | 0.35 |
| $H_g$ | 0.56 | 1.17 | 1.02 | 3.73 | 1.60 |
| $U_g$ | 15.97 | 15.92 | 15.98 | 15.90 | 15.92 |
| $S_g$ | 8.87 | 18.60 | 16.31 | 59.38 | 25.44 |
| $SP_g$ | 2.12 | 7.58 | 5.63 | 5.25 | 9.00 |

Tables 4 demonstrates that PIRA * is much more efficient and robust than PGDFS. For the 5 TSP instances, the speedups (*SP*) range from 7.72 to 14.05 in PIRA*, with an average speedup[5] of 12.03. For PGDFS, the speedups range from 2.12 to 9.00, with an average speedup of 5.51. In PIRA*, the number of nodes expanded is close to that of the serial IRA*. As a result, poor slowdown in performance is avoided, and *S*, the speedups with respect to the same search

[5] The average speedup is computed by dividing the summation of $T_b$'s by the summation of $T_r(16)$'s (or $T_g(16)$'s) for the 5 TSPs.

Table 5. Summary of various speedups for 10 TSP instances using 16 processors and random seeds ranging from 1 to 10. ($T_b$: time for best-first search; $T_g$: time for guided depth-first search; $T_r$: time for IRA* search)

| Average Speedup | $\frac{T_b(1)}{T_r(16)}$ | $\frac{T_b(1)}{T_g(16)}$ | $\frac{T_g(16)}{T_r(16)}$ | $\frac{T_g(1)}{T_g(16)}$ | $\frac{T_r(1)}{T_r(16)}$ | $\frac{T_g(1)}{T_r(16)}$ |
|---|---|---|---|---|---|---|
| Method 1 | 11.42 | 5.06 | 2.26 | 14.34 | 14.69 | 32.38 |
| Method 2 | 11.37 | 5.67 | 3.37 | 20.26 | 14.96 | 64.45 |

scheme, are close to the number of processors (ranging from 12.27 to 16.77). Obviously, all these speedups obtained by simulations are within the bounds derived in Section 5. In PGDFS, the values of parameter $H$ is more diverse than those in PIRA *; hence, anomalies occur more frequently ($S$ ranges from 8.87 to 59.38). Even in the case when $S$ equals 59.38, the real speedup ($SP$) is only 5.25, since $R$ is too small.

Table 5 compares different methods for computing speedups. We focus in this paper on $T_b(1)/T_r(P)$, namely, the speedup of the parallel algorithm as compared to the best sequential algorithm. This is shown in column 2 in Table 5. We see that PIRA* has a speedup of around 11 as compared to a sequential best-first search.

Method 1 computes the speedup as the ratio of the total computation times of the 10 problem instances using different search strategies. Method 2 computes the speedup as the average speedup of individual problem instances.

For research on parallel algorithms, what is important is the speedup based on the same search strategy. This is illustrated by the results in columns 5 and 6 in Table 5, which show that the speedups are closer to linear and can even be superlinear. This linear or superlinear speedup has been observed frequently by researchers in this area. However, these speedups do not represent the merits of parallel processing in actually solving the original problem.

The speedups achieved by PIRA* as compared to traditional sequential methods based on guided depth-first search are more dramatic. These are shown in the last column in Table 5, which show that a speedup of over 32 can be achieved.

# 7 Caveats

The effectiveness of the IRA* algorithm depends on the complexity of the COP to be solved. For problems with fully polynomial-time approximation algorithms, such as the knapsack problem, the IRA* algorithm may not be effective. A necessary condition for IRA* to be effective is that the computation time of the problem instance to be solved must be reduced substantially by approximate depth-first searches. Second, since IRA* is based on depth-first searches, it acquires the drawbacks and anomalous behavior of approximate depth-first searches. Overall, anomalies happen much less frequently in IRA* than in depth-first searches because we are performing a sequence of depth-first searches, each of which may compensate the anomalous behavior of another.

## References:

[1] KUMAR V, GOPALAKRISHNAN P S, KANAL L N. Parallel algorithm for machine intelligence and vision[M]. New York: Springer-Verlag, 1990.

[2] KUMAR V, RAMESH K, RAO V N. Parallel best-first search of state-space graphs: a summary of result[C]// The 7th AAAI National Conference on Artificial Intelligence. Palo Alto: AAAI Press, 1988: 122-127.

[3] RAO V N, KUMAR V, RAMESH K. A parallel implementation of iterative-deepening-A*[C]// The 6th National Conference on Artificial Intelligence. Palo Alto: AAAI Press, 1987: 878-882.

[4] KORF R E. Depth-first iterative deepening: an optimal admissible tree search[J]. Artificial Intelligence, 1985, 27(1): 97-109.

[5] LI G J, WAH B W. Computational efficiency of combinatorial OR-tree searches[J]. IEEE Transactions on Software Engineering, 1990, 16(1): 13-31.

[6] WAH B W. MIDA*: an IDA* search with dynamic control[R]. Urbana: University of Illinois, 1991.

[7] WAH B W, CHU L C. TCA*: a time-constrained approximate A* search algoritlun[C]// The 2nd International IEEE Conference on Tools for Artificial Intelligence. Piscataway: IEEE Press, 1990: 314-320.

发表于《计算机研究与发展》1992年第12期

# 采用有效切分的负载平衡

孙凝晖，李国杰

国家智能计算机研究开发中心，北京 100080

**摘　要：**负载平衡是影响分布式处理效率的关键。对于静态的负载平衡，特别是数值计算中的负载平衡，已有不少研究成果，但是对于并行组合搜索等人工智能问题，由于负载的动态特性，现有的负载平衡算法的实际效果并不好。本文讨论动态的负载平衡，着重分析负载平衡对并行组合搜索效率的影响。本文的结果表明，任务切分的有效性是影响负载平衡时间开销的重要因素。我们在共享存储型多处理机ENCORE Multimax 520上模拟分布式计算，把有效切分策略应用于求解N皇后问题的并行深度优先搜索算法和求解旅行推销员问题的并行分枝限界算法中，获得了比已有结果更好的并行效率。

**关键词：**负载平衡；有效切分；分布式处理；并行组合搜索

## 1　引言

对于并行计算机系统和并行算法，负载平衡是充分利用系统处理能力和获得高的并行效率不可缺少的条件。共享存储的计算机系统通过处理机调度算法使得进程和 thread 在各处理机均匀分布，分布式计算机系统通过作业调度和进程迁移最大限度地利用系统的处理能力。并行算法的负载以子任务的形式表示，子任务可以是用户进程或 thread，也可以是其他形式，如并行搜索中待扩展的搜索树结点。并行算法的负载平衡需要由用户对各处理机的任务量通过数据转移进行平衡，使得算法总的时间开销最小。本文研究并行算法的负载平衡的效率。

组合搜索是人工智能和运筹学的重要研究领域，由于并行组合搜索的动态和非确定性[1]，以低的时间开销做到负载的平衡有许多困难，所以研究并行搜索中的负载平衡有重要的理论意义与实际价值。

并行搜索算法一般有两种控制策略：集中式控制和分布式控制。两种策略在共享存储型和消息传送型多处理机上都可以实现，但并行效率有很大差别。集中式控制策略需要一个大的复杂的数据结构作为共享任务数据结构，如队列、堆、栈等，每个处理机从中获取子任务执行，这种方式适用于共享存储型多处理机求解。对于共享任务队列机制，只有任务粒度较大时，才有负载平衡问题[2]，即尽可能保证各处理机同时结束，但任务粒度太小时，并行存取冲突十分严重[3]。

分布式控制策略任务队列采用分散的数据结构，将求解的问题划分成若干个子问题，分配给各个处理机求解。各处理机存取局部任务队列，并由通信和同步协调运行，空闲处理机通过负载平衡算法从“忙”处理机获得任务继续执行[4]。在共享存储型多处理机上，由于共享任务队列有严重的并行存取冲突，搜索算法经常采用分散任务队列的形式，即物理上共享而逻辑上分布；同时由于通过共享存储器的通信，时间开销较小，负载平衡只受算法本身影响，便于研究算法本身的效率。而消息传送型多处理机中，存储器物理上的分布决定了任务队列逻辑上的分布，负载平衡除受算法本身影响外，数据传送还要受处理机拓扑结构、路径选择、网络带宽的影响。

在共享存储型多处理机上实现分布式算法是实现人工智能问题并行处理的一种有效途径，因此我们在分布式树划分并行机制和共享存储器体系结构下，研究并行组合搜索的负载平衡算法。从这一研究得出的关于负载平衡算法本身对性能影响的结论也适用于松散耦合并行计算机。

我们在 Multimax 520 并行机上进行研究。Multimax 520 是 ENCORE 公司生产的共享存储器型多处理机，采用基于总线的对称式结构，可以配置 2~20 个 NS 32032 系列的微处理机。我们在 Multimax 520 上实现了 $N$ 皇后问题的并行深度优先搜索，找到了有效切分的负载平衡算法，并且应用于属于组合优化的求解旅行推销员问题的并行分枝限界搜索[5]。

## 2 并行组合搜索的负载平衡模式

由于并行组合搜索的动态性和非确定性，绝对的负载平衡是不可能也不必要的，因为搜索中 1 个子任务可能派生出若干新的子任务，派生过程是动态和不可预测的，因此每一时刻的平衡往往被下一时刻子任务的派生和消亡所破坏，导致无效的负载平衡过程。与负载有关的因素很多，而且有些因素无法预测，不可能精确地估计处理机的负载量。

因此我们只能追求负载在一定限度内的动态相对平衡。负载平衡算法要能够把大的负载量分散开来，并且保证每次操作都是有效的，尽量减少调度本身的开销。要做到这一点，对处理机负载大小的估计至关重要，也十分困难。

分布式并行组合搜索算法，初始时将搜索树分解成若干子树，分配到多个并行进程的局部任务队列（局部表）上，各个进程执行局部搜索，当某个进程的局部表为空时，才进行负载平衡。

负载平衡算法的基本框架主要由任务请求和任务切分组成[1]，任务请求算法用于确定向哪个进程请求任务，任务切分算法用于决定转移多少负载量。在共享存储型多处理机上实现的负载平衡算法包括如下共享数据结构。

- 共享任务表：用于进程之间的任务转移。
- 进程状态标志：表明进程的忙、闲、结束或没有获得负载。
- 进程负载平衡标志：表明进程是否被请求任务，以及请求任务的进程号。
- 同步锁。

任务请求算法主要有静态、随机和动态 3 种策略，它们分别构成固定请求、随机请求和均匀请求 3 种任务请求算法。

- 固定请求：向固定的进程请求任务，例如在环状结构中向左邻居进程请求任务。这种方法的优点是算法简单，实现开销小；缺点是大的负载集中于若干进程时，任务量分布不均匀，导致负载平衡的任务转移次数增多，从而开销增大。
- 随机请求：随机向某个进程请求任务，在信息量不足的情况下，随机常常是比较好的方法。
- 均匀请求：Rao 和 Kumar[1]提出用动态的任务请求策略来减少负载平衡的通信开销。算法设立了任务请求指针，指向负载被转移的当前进程，每当有任务请求时，任务请求算法从任务请求指针所指的进程开始，在环状结构上向右寻找“忙”进程，随后将指针指向这个进程。这种方法保证每个进程被请求任务的概率相等，大的负载可以很快平衡到空闲进程，以利于任务量的均匀分布，从而使负载平衡频率降低，通信量减少。

当一个进程的局部表为空时，首先执行任务请求算法。任务请求算法假定多进程以虚拟的环状拓扑结构连接。每个进程周期性地察看各自的负载平衡标志，若存在任务请求，则执行任务切分算法。并行组合搜索的子任务以结点形式存放在局部表中，任务切分将局部表中部分结点转移至共享任务表，同时启动空闲进程。空闲进程将子任务从共享任务表移至局部表中，继续执行搜索过程。从进程 $j$ 中转移任务到进程 $i$ 的负载平衡过程如下所示。

| 进程 $i$： | 进程 $j$： |
|---|---|
| 搜索 | 搜索 |
| 空闲，置进程状态标志 | 循环查看负载平衡标志 |
| 任务请求算法寻找“忙”进程 | 执行任务切分算法 |
| 没有则结束，置进程负载平衡标志 | 转移任务至共享任务表 |
| 等待唤醒 | 启动进程 $i$ |
| 从共享任务表转移任务至局部表 | 继续搜索 |
| 继续搜索 | |

## 3 负载平衡的时间开销

我们以 $N$ 皇后问题为例研究负载平衡算法。$N$ 皇后问题是在 $N{\cdot}N$ 的棋盘上放置 $N$ 个皇后，使它们彼此不在同一行、同一列和同一对角线，这是一种不用启发式信息的限制满足问题。求解 $N$ 皇后问题常采用深度优先的分枝限界算法，搜索树第 $h$ 层的结点代表放置第 $h$ 个“皇后”时可以放置的棋盘位置。$N$ 皇后问题属于组合判定问题，随着问题规模（$N$）的扩大，搜索树扩展的结点数呈指数增长，寻找问题的所有可行解对应于“与”树的搜索。

利用 $N$ 皇后问题本身的“与并行”性可以容易地实现并行深度优先搜索。初始时，将搜索树划分成若干相互独立的子树，并且尽可能使子树代表的任务量均匀分布，各个子树由相应的处理机执行深度优先搜索。当某个处理机空闲时，进行负载平衡。由于这种方法和串行搜索扩展的结点数及淘汰的结点数相同，即总的搜索任务量大小相同，不存在异常搜索问题，所以负载平衡算法的时间开销是影响并行效率的决定性因素。

我们首先以 $N$ 皇后问题为例，分析负载平衡的时间开销及其原因。任务请求算法采用均匀请求，任务切分算法简单地将深度优先搜索中所保留的搜索路径的结点栈一分为二，形成两棵搜索子树。表 1 显示了有负载平衡和没有负载平衡情况下的运行情况，处理机个数为 4，问题大小 $N$ 表示 $N{\cdot}N$ 棋盘。

可见，有负载平衡和无负载平衡的并行组合搜索加速比相差无几，说明采用均匀请求任

**表 1 有无负载平衡的性能比较**

| $N$ | 10 | 11 | 12 | 13 | 14 |
|---|---|---|---|---|---|
| 可行解个数/个 | 724 | 2 680 | 14 200 | 73 712 | 365 596 |
| 串行时间/s | 3.76 | 21.14 | 125.3 | 815.22 | 5 513.6 |
| 无负平衡并行时间/s | 1.18 | 6.83 | 39.96 | 257.28 | 1 702.64 |
| 有负载平衡并行时间/s | 1.18 | 6.6 | 38.02 | 244.28 | 1 633.56 |
| 无负平衡加速比 | 3.19 | 8.1 | 3.14 | 3.17 | 3.21 |
| 有负平衡加速比 | 3.19 | 3.2 | 3.3 | 3.34 | 3.38 |

务请求算法的负载平衡时间开销仍然过大，通信开销抵消了任务并行处理带来的好处。另外，$N$=14 时，可行解的个数为 365 569，搜索树扩展的结点数为 26 992 956，淘汰的结点数为 348 166 472，所以总任务量相对于通信量足够大，有效的负载平衡算法应该使得求解 $N$ 皇后问题的并行效率非常接近 100%。

负载平衡的时间开销等于任务转移次数和一次平衡过程的时间开销之积。在共享存储器体系结构下，任务转移只是结点数据的复制，一次平衡过程的时间开销较小，因此我们研究的主要目的是从算法的角度降低负载平衡的频率，即任务转移次数。

上述负载平衡效率不高的主要原因是存在无效的负载平衡。在迄今已知的方法中，任务切分是简单地以搜索树结点个数衡量任务量。但有的结点扩展后会形成很大的搜索树，有的结点扩展后会立刻被淘汰，即每个子任务的计算时间差别很大。因此，简单的结点数不能反映任务量的真正大小，机械的切分方法不能保证任务均匀分布和转移足够大的任务量，由此产生许多无效的负载平衡，增加了负载平衡的通信开销。

## 4 有效切分策略

并行组合搜索中负载平衡的关键在于对任务量的正确估计，保证任务切分时任务量尽量均匀分布，即保证切分的有效性。组合搜索是非确定性算法，任务量以搜索树结点的形式表示。结点的任务量依赖于结点扩展次序、启发式函数和当时的最佳可行解，因此对每个结点代表的任务量很难估计。对于不同的问题，切分策略各不相同，但都必须遵循一个基本原则：充分利用结点中包含的信息近似估计任务量，使得任务量相似的结点均匀分布，也就是做到切分的有效性。

另外，如果切分的处理机的任务量较小，负载平衡的时间开销常常大于任务平衡带来的时间节省。向题本身的并行度随任务量大小的变化而变化，当问题任务量变小时，并行度随之降低，多余的处理机既浪费资源，又干扰正常工作，增加通信开销。因此，切分时适时撤销多余的处理机，可以使得负载平衡的实际效果更好。我们总结了 3 条一般规则：一是切分时任务量已经小于“最小平衡任务量”，撤销请求任务进程；二是切分时有两个空闲进程对切分进程进行任务请求，撤销其中一个；三是任务请求算法寻找“忙”进程时，连续 $k$ 个进程处于空闲或结束状态。例如，$k=P-P/4$（$P$ 为进程个数，4 个 CPU），即最多 $P/4$ 个进程“忙碌”，不进行切分，撤销请求任务进程。

对于 $N$ 皇后问题，深度优先搜索待扩结点栈的元素按搜索树深度从大到小排列。为了提高负载平衡的效率，首先改进深度优先搜索。在每个待扩展结点的结构中加入子结点信息。压栈前先计算结点代表的棋盘位置的下一行中可放的“皇后”位置，利用棋盘中已放“皇后”的信息，去掉同行、同列、同对角线的位置，即在搜索树中预先向前看一层，先进行限制满足检验，将代表必须产生的子结点的棋盘位置作为附加信息放在待扩结点的结构中，以备扩展结点之用。这部分工作在原来的算法中也是必须进行的，没有增加额外的开销。

负载平衡进行任务切分时，将搜索子树中深度最小的那层待扩结点，按结点数一分为二，其中一部分结点代表的子树被转移给空闲进程。改进前由于这些结点未进行限制满足检验，被淘汰的概率很大，由此导致无效的负载平衡。改进后，小任务量结点（即经过对已知条件的限制满足检验可以淘汰的结点）和没有子结点的结点被预先淘汰，排除了大多数无效的负载平衡。由于每个待扩结点都保留其必须扩展的子结点，而且这些待扩结点搜索路径的历史相同，只是结点所在层上棋盘位置不同，所以它们保留的子结点信息是近似的，即必须扩展的子结点数目相似，因此每个结点代表的任务量也是相似的，保证结点数目的均分可以接近任务量的均分。

表 2 显示了采用有效切分策略的负载平衡算法求解 *N* 皇后问题的效率（4 个处理机）。

算法改进后，由于避免了许多不必要的限制满足检验，串行算法的运行时间也有很大降低。采用有效切分的负载平衡算法的并行搜索，平均加速比从 3.282 增加到 3.964，平均效率达到 99.1%，效率增长了 20%。平均负载平衡频率由 6.75 次降到 2.65 次，即每个处理机少进行 4.3 次负载平衡，有效切分策略的负载平衡时间开销很小。由于操作系统多用户环境进程切换的影响，算法统计的运行时间和真正的 CPU 时间略有差距，所以出现了有的问题的加速比略大于处理机数目的现象。

一般来说，问题大小不变时，处理机数目增多，任务转移次数要增多，负载平衡的时间开销增大，导致并行效率降低。但是表 3 的结果显示，采用我们提出的有效切分策略，用不同数目的处理机求解 *N* 皇后问题的并行效率非常高（4 个以上 CPU 采用分时模拟）。

从表 3 可以看出，非有效切分策略负载平

表 2　有效切分负载平衡的效率

| *N* | 10 | 11 | 12 | 13 | 14 |
|---|---|---|---|---|---|
| 改进前串行时间/s | 3.76 | 21.14 | 125.3 | 815.22 | 5 513.6 |
| 改进后串行时间/s | 2.96 | 15.06 | 83.4 | 489.76 | 3 071.54 |
| 改进前并行时间/s | 1.18 | 6.6 | 36.2 | 244.28 | 1 633.56 |
| 有效切分策略并行时间/s | 0.78 | 3.8 | 21.1 | 121.42 | 752.2 |
| 改进前加速比 | 3.19 | 3.2 | 3.3 | 3.34 | 3.38 |
| 有效切分策略加速比 | 3.8 | 3.96 | 3.95 | 4.03 | 4.08 |
| 改进前任务转移次数/次 | 15 | 29 | 16 | 33 | 42 |
| 有效切分策略任务转移次数/次 | 6 | 7 | 9 | 20 | 11 |

表 3　不同处理机数的有效切分负载平衡的效率

| 处理机数目 | 1 | 2 | 3 | 4 | 8 | 16 | 20 |
|---|---|---|---|---|---|---|---|
| 方法 | 非有效切分策略 | | | | | | |
| 并行时间/s | 5 513.6 | 3 241.52 | 2 230.74 | 1 633.56 | 859.96 | 426.38 | 343.42 |
| 加速比 | 1.0 | 1.701 | 2.472 | 3.375 | 6.412 | 12.931 | 16.055 |
| 效率 | 100% | 85.05% | 82.4% | 84.375% | 80.15% | 80.819% | 80.275% |
| 任务转移次数/次 | 0 | 5 | 37 | 42 | 83 | 273 | 284 |
| 方法 | 有效切分策略 | | | | | | |
| 并行时间/s | 3 071.54 | 1 538.14 | 1 026.14 | 751.98 | 380.44 | 192.22 | 155 |
| 加速比 | 1.0 | 1.997 | 2.993 | 4.085 | 3.074 | 15.979 | 19.816 |
| 效率 | 100% | 99.85% | 99.78% | 102.12% | 100.92% | 99.87% | 99.08% |
| 任务转移次数/次 | 0 | 4 | 17 | 24 | 54 | 99 | 114 |

衡开销较大，20 个处理机时任务转移频率达到 14.2 次，使得并行搜索只能获得 80%左右的并行效率。20 个处理机时有效切分策略的并行加速比几乎保持线性增加，采用 20 个处理机处理组合搜索问题得到的实际加速为 19.816 倍，据我们所知，还没有在文献中报道过（请注意，求 $N$ 皇后问题所有解没有异常行为）。对于不同数目的处理机（2~20 个），并行搜索负载平衡的任务转移频率分别为 2.5、5.67、6.125、5.75、5.281、5.025 次，近似稳定。这说明采用有效切分策略时，均匀请求算法的负载平衡算法可以高效地解决 $N$ 皇后问题。

由于不同问题有不同的特点，找到通用的有效切分方法十分困难。有效切分的基本原则是尽量利用问题中可得到的信息，使任务量相似的子任务均匀分布，切分出的任务量足够大，保证问题的并行度和处理机数目相适应。

## 5 有效切分策略应用于组合优化问题

求解组合优化问题（如 TSP）多采用启发式搜索，利用搜索树的“或”并行性，使用并行分枝限界算法实现。启发式搜索就会有结点淘汰，并行搜索会带来异常搜索开销（并行搜索扩展了串行搜索不扩展的结点），使得并行算法的总任务量不等于串行算法的总任务量。为了研究负载平衡的开销，我们将相对加速比作为可比因素。由于单位时间扩展的结点数可以反映并行算法的效率，我们将 $R(k)$是 $k$ 个处理机的实际运行时间，$V(k)$是 $k$ 个处理机扩展的结点总数，则相对加速比定义为：

$$\mathrm{rsp}(k)=\frac{V(k)/R(k)}{V(1)/R(1)}=\frac{R(1)}{R(k)}\cdot\frac{V(k)}{V(1)}=\mathrm{sp}(k)\cdot c(k)$$

其中，$\mathrm{sp}(k)$是实际加速比，$c(k)$是多余因子，反映异常搜索的大小。

使用并行分枝限界算法求解 TSP 时，从 OPEN 表空间开销和维护的时间开销考虑，搜索树结点一般分为两类：简单结点代表求解的一个子问题；复合结点代表求解的多个子问题，不同的复合结点包含的子问题个数不同。结点在搜索树中的深度不同，它们的任务量也不相同。

通常采用的切分方法将 OPEN 表中的待扩结点按结点数一分为二，不区分简单结点和复合结点，也不考虑结点的深度。有效切分策略充分考虑结点中包含的信息，尽量做到任务量的均分。一般而言，搜索树深度大的结点的任务量小于深度小的结点的任务量，因为它们的启发式下界值较大，淘汰的可能性也较大。另外，复合结点的任务量一般大于简单结点的任务量，由于包含的子问题个数不同，复合结点之间任务量差别很大，所以切分时要考虑复合结点包含的子问题个数的不同，同时将结点按深度排序后交叉切分，即深度从小到大，每一深度的结点，轮流分配相同数目的子问题给两个处理机，保证同样深度的子问题均匀分布。当局部表长度过小时，还要将复合结点预先展开成多个简单结点，但这样也会增加任务转移的开销。

表 4 显示了用两种任务切分算法在 4 个处理机的 Multimax 520 上求解 20 个城市 TSP 问题的效率，时间单位是 s，上标 1 和 2 分别代表一般切分方法和有效切分方法。

其中串行算法是启发式导向的深度优先分枝限界算法（GDFS），即结点深度相同时执行最佳优先策略。$R(k)$是并行算法的实际运行时间，rsp 是相对加速比，$m$ 是任务转移次数。实验表明，平均任务转移频率从 21 次降为 5.8 次，平均相对加速比从 3.657 升为 4.119，即有效切分策略在 4 个处理机时效率提高了 12.62%。

表 4　一般切分与有效切分性能的比较

| 实例 | 1 | 2 | 3 | 4 | 5 | 6 | 7 | 8 |
|---|---|---|---|---|---|---|---|---|
| $R(1)$ | 10.74 | 35.1 | 22.1 | 178.76 | 9.76 | 52.68 | 200.32 | 44.02 |
| $R^1(4)$ | 2.48 | 10.9 | 5.38 | 65.34 | 2.84 | 14.76 | 55.36 | 8.22 |
| $R^2(4)$ | 2.34 | 10.64 | 5.1 | 40.76 | 2.7 | 18.68 | 52.34 | 6.82 |
| $\mathrm{rsp}^1$ | 3.46 | 4.34 | 3.9 | 2.65 | 3.86 | 3.64 | 4.07 | 3.36 |
| $\mathrm{rsp}^2$ | 3.73 | 4.44 | 4.12 | 4.21 | 4.09 | 3.93 | 4.37 | 4.06 |
| $m^1$ | 45 | 54 | 60 | 255 | 35 | 110 | 164 | 79 |
| $m^2$ | 31 | 32 | 12 | 36 | 25 | 40 | 14 | 23 |

组合优化问题扩展结点部分的时间开销较大，负载平衡只需转移若干结点数据即可，时间开销较小，因此负载平衡的时间开销对实际时间的影响较小，在处理机数目较少的情况下，有效切分策略虽使负载平衡的频率明显降低，但是效率提高的余地并不大。因为并行搜索是动态过程，无法精确估计局部结点的任务量，增加任务量的估计因素，会增加负载平衡本身的时间开销。有效切分策略在组合优化中的效果不如组合判定那么明显，但好处仍是显然的。表 5 显示了使用不同数目的处理机（2~20 个）求解 30 个城市 TSP 问题的结果，采用一般切分方法在 20 个处理机时，任务转移次数达到 1 303，出现许多无效负载平衡，相对加速比只有 12.56。采用有效切分策略的并行效率非常高，使用 20 个处理机求解 TSP 问题得到的实际加速为 18.338 倍。

## 6　结束语

负载平衡是影响分布式并行组合搜索效率的关键。负载平衡算法由任务请求和任务切分组成。均匀请求策略可以大大降低负载平衡的频率，但存在无效负载平衡问题，不能保证并行组合搜索的线性加速。采用有效切分策略可以明显地降低任务转移的次数，减少负载平衡的开销，适用于求解 $N$ 皇后问题和旅行推销员问题，提高了并行搜索的效率。但是由于任务量难以精确估计及并行搜索中异常搜索开

表 5　有效切分负载平衡用于求解 TSP 问题

| 处理机目 | 1 | 2 | 4 | 8 | 16 | 20 |
|---|---|---|---|---|---|---|
| 方法 | 一般切分方法 | | | | | |
| 并行时间/s | 2 144.36 | 1 138.28 | 643.88 | 437.02 | 288.14 | 196.3 |
| 相对加速比 | 1.0 | 1.929 | 6.727 | 6.727 | 10.969 | 12.56 |
| 任务转移次数/次 | 0 | 122 | 869 | 869 | 1 078 | 1 005 |
| 方法 | 有效切分策略 | | | | | |
| 并行时间/s | 2 144.36 | 122.74 | 626.1 | 391.56 | 204.34 | 177.1 |
| 相对加速比 | 1.0 | 1.954 | 3.997 | 7.642 | 14.872 | 18.338 |
| 任务转移次数/次 | 0 | 8 | 37 | 101 | 206 | 280 |

销的影响，有效切分策略应用于组合优化的效果不如组合判定那么明显。而且寻找通用的有效切分方法十分困难，依赖于具体的问题特性。有效切分的基本原则是使得任务量相似的子任务均匀分布，切分出的任务量足够大，保证问题的并行度和处理机数目相适应。

## 参考文献：


[1] RAO V N, KUMAR V. Parallel depth-first search, Part I: implementation, Part II: analysis[J]. International Journal of Parallel Programming, 1988.

[2] NI L M, TZEN T H. Dynamic loop scheduling for shard memory multiprocessors[C]// International Conference on Parallel Processing. Piscataway: IEEE Press, 1991.

[3] 孙凝晖, 李国杰. 优先队列的并行插入和删除[J]. 计算机研究与发展, 1993(3): 52-61.

[4] KUMAR V, RAO V N, RAMESH K. Parallel depth-first search on the ring architecture[C]// International Conference on Parallel Processing. Piscataway: IEEE Press, 1988.

[5] KUMAR V, KANAL L. Parallel branch-and bound formulations for and/or tree search[J]. Transactions on Pattern Analysis and Machine Intelligence, 1984, 6(6).

# 第3章

# 智能计算机

# 智能计算机的初级阶段

30年前，人工智能处在第二波高潮，我也算是那一波浪潮的“弄潮儿”之一。我在智能计算机（Intelligent Computer）领域做的第一件事是与华云生教授共同编著了一本IEEE Computer Society出版社出版的“Tutorial” *Computers for artificial intelligence applications*。“Tutorial”就是导师辅导材料，也许称为“自学教材”更合适。每当一个新的技术兴起，国内就会冒出一大堆“二流”作者赶时髦的新作，很少见到严肃的学者编选便于自学的论文汇集。美国各个研究方向都有Tutorial，其中IEEE Computer Society出版社出版的最多。当你想进入一个新领域或新研究方向时，认真阅读一本好的Tutorial是条捷径，但编著一本Tutorial是件很难的事，需要从基本概念开始，由浅入深地收集最合适的经典论文，既要易懂，又要有权威性。在20世纪80年代，互联网还不太流行，没有像今天这么方便的搜索引擎。当时我只能在图书馆一篇一篇地查阅各种期刊和会议文集，一页一页地复印，我记得我复印的论文堆起来超过一米高。我要从几百篇论文中精心挑选，争取所选的论文能将读者引入智能计算机这个新的领域。此书于1986年年初出版后，连续3年都是IEEE Computer Society出版社的畅销书。我回国以后还不断收到IEEE Computer Society出版社用支票寄来的稿费。

编写了关于智能计算机的Tutorial之后，许多出版社约我写智能计算机方面的论文。我和华教授先后在*SIGART Newsletter*、*Proceedings of the IEEE*、*IEEE Transactions on Systems, Man, and Cybernetics*等期刊上发表了几篇关于智能计算机的综述论文。我们编著的Tutorial没有采用“智能计算机”的标题，而是称之为“适于智能应用的计算机”。在后来的综述论文中，我们也是称之为*Special purpose computer architecture for AI*。日本学者开始研制“第五代计算机”后的一段时间内，智能计算机受到吹捧，人们开口闭口都是“Prolog机”“LISP机”。但我们一直认为研制通用的智能计算机还不到时候，当时做的所谓的智能计算机实际上都是适合某些智能应用的专用计算机。现在看来，我们的判断是明智的，即使到了30年后的今天，仍然不具备研制通用智能计算机的条件。

在这些综述论文中，我和华教授明确指出：“设计智能系统的关键在于对要求解的问题的理解，而不仅仅是高效的软件和硬件。利用基于常识、高层的元知识、更好的知识表示获得的启发式信息比改善计算机结构可以获得更大的性能提高。是否用硬件实现一个给定的算法取决于问题的复杂性和该问题出现的频率。计算机结构师的角色是选择好的知识表示、识别开销密集型任务、学习元知识、确定基本操作，用软硬件支持这些任务”。这些观点对今天的智能计算机研究可能还有借鉴意义。

上一波人工智能浪潮的热点是符号处理（Symbolic Processing）和并行逻辑推理，主要研究面向智能应用的语言、编程和并行加速。经过30年的努力，集成电路的性能飞速提高，算力大大增强，加上互联网带来的数据资源的极大丰富和人工神经网络算法的突破，计算机处理视觉、听觉的感知能力达到了实用水平，因此近年来人工智能的发展方向转向低层的感知应用。但是，高层的逻辑推理仍然是智能的主要特征，回顾总结30年前的智能计算机研究成果很有必要，今天的学者应当读一些30年前关于智能计算机的论文。

1986年，我国启动863计划，在计算机领域设立了智能计算机主题，显然国家的初衷是要研制智能计算机。在我担任国家智能计算机研究开发中心主任之前，1989年10月我给国家科学技术委员会的领导写了一份报告，阐述了我对智能计算机和第四代计算机的看法，提出863计划应重点发展并行处理技术。这份报告指出：“智能机的发展必须以VLSI计算机（第四代计算机）的技术为基础。有些同志可能认为跳过第四代计算机直接发展所谓的第五代计算机是一条捷径。但这只是一种空想。如果把人工智能看成是一朵花，则它的根是计算机技术。而计算机技术有它自己的特殊发展规律，充分利用第四代计算机已成熟的技术是发展智能机必须要考虑的一条重要原则”。

1991年9月17日，全国第一次人工智能与智能计算机学术会议在北京召开。我在这次大会上做了特邀报告，题目是《我们的近期目标——计算机智能化》。这次报告在国内第一次以“顶天立地——发展智能计算机的战略”的标题提出了“顶天立地”发展战略。报告指出：“开展智能计算机研究必须同时在两条战线上进行工作。一方面要努力突破传统计算机甚至图灵机的限制，探索关于智能机的新概念、新理论和新方法；另一方面要充分挖掘传统计算机的潜力，在目前计算机主流技术的基础上实现计算机的智能化”。如同实现共产主义要经过社会主义初级阶段一样，现在还处在实现“智能计算机”的初级阶段。国家智能计算机研究开发中心后来的主要工作是发展高性能计算机，逐步实现计算机的智能化。

20多年后，中国科学院计算技术研究所的陈云霁、陈天石兄弟研制出“寒武纪”智能加速芯片，我国智能计算机的研制上了一个新台阶，但仍然还处在智能计算机的初级阶段。

发表于International Symposium on Computer Architecture (ISCA)论文集，1984年

# The status of MANIP-a multicomputer architecture for solving combinatorial extremum-search problems

Benjamin W. WAH, Guo-jie LI, Chee-fen YU

School of Electrical Engineering, Purdue University, West Lafayette, IN 47907

**Abstract:** In this paper, we report the status of study on MANIP, a parallel computer for solving combinatorial extremum-search problems. The most general technique that can be used to solve a wide variety of these problems on a uniprocessor system, optimally or suboptimally, is the branch-and-bound algorithm. We have adapted and extended branch-and-bound algorithms with lower-bound tests for parallel processing. Three major problems are identified in the design: interconnection networks for supporting the selection of subproblems, anomalies in parallelism and virtual-memory support. A unidirectional ring network has been shown to be the most cost-effective selection network. We provide sufficient conditions so that anomalies can be eliminated under certain conditions and show that anomalies are unavoidable otherwise. Lastly, we develop a modified branch-and-bound algorithm so that the amount of memory space required under a best-first search is significantly reduced.

**Key words:** anomalies, approximations, dominance criteria, heuristic search, NP-hard problems, parallel branch-and-bound algorithms, ring network, virtual memory

## 1 Introduction

MANIP is a parallel computer architecture for processing combinatorial extremum-search problems by branch-and-bound algorithms with lower-bound tests[1]. In this paper, we report the current status of research on MANIP. The optimization problems considered are characterized by an objective function to be minimized or maximized and a set of constraints to be satisfied. A special class is the class of NP-hard problems[2]. There are no known optimal algorithms to solve these problems in a time that increases polynomially with the problem size. Many practical problems belong to this class. Examples include traveling salesman, warehouse location, job-shop scheduling, graph partitioning and mathematical programming.

Owing to the apparently exponential running times of algorithms for NP-hard problems,

This research was supported by National Science Foundation Grant ECS 81-05968.

optimal solutions are usually infeasible to obtain. In practice, approximate solutions are often acceptable alternatives. The simplest approach is to solve small problems optimally, and to apply heuristics to solve large problems sub-optimally. However, it is difficult to assess the goodness of results. Polynomial-time approximation algorithms with guaranteed error bounds have been developed for some problems[3], However, efficient algorithms for slightly different problems are usually different and cannot be generalized. A general algorithm that can solve, optimally or sub-optimally (with guaranteed error bounds), a wide variety of these problems is, thus, preferred. A good candidate is the branch-and-bound algorithm.

A branch-and-bound algorithm is a partitioning algorithm. It decomposes a problem into smaller subproblems and repeatedly decomposes until infeasibility is proved or a solution is obtained[4]. Branch-and-bound algorithms will be discussed in Section 2. A parallel version of the algorithm is presented in Section 3, In Section 4, anomalies of parallel branch-and-bound algorithms with lower-bound tests are studied. Sufficient conditions to avoid some of these anomalies are discussed. The problem on the exponential space requirement will be addressed in Section 5.

## 2 Serial branch-and-bound algorithms

Many theoretical properties of branch-and-bound algorithms have been developed[5-9]. The way in which a problem is repeatedly decomposed into smaller subproblems can be represented as a finite rooted tree $\boldsymbol{B}=(\boldsymbol{P}, \boldsymbol{E})$, where $\boldsymbol{P}$ is a set of subproblems, and $\boldsymbol{E}$ is a set of edges. The root of the tree is $P_0$. If a subproblem $P_{i_j}$ is obtained from $P_i$ by decomposition, then $(P_i, P_{i_j}) \in \boldsymbol{E}$. The level number of a node is the number of edges leading from the root to this node (the root is at level 0). Let $f(P_i)$ be the value of the best solution obtained by evaluating all the subproblems decomposable from $P_i$, let $P_{i_j}$ be the $j$-th subproblem decomposable from $P_i$, and let $k_i$ be the number of such subproblems (i.e., $k_i = |\{(P_i, x): (P_i, x) \in \boldsymbol{E}\}|$). Then $f$ satisfies:

$$f(P_i) = \min_{j=1,\cdots,k_i} \{f(P_{i_j})\} \qquad (1)$$

Each subproblem is characterized by a value that is computed from a lower-bound function $g$. The lower-bound function satisfies the following properties:

(a) $g$ is a lower-bound estimate of $f$ (2)

(b) g is exact when $P_i$ is feasible (3)

(c) lower bounds of descendant nodes always increase (4)

Branch-and-bound algorithms can be characterized by four constituents: a branching rule, a selection rule, an elimination rule and a termination condition. The first two rules are used to decompose problems into simpler subproblems and appropriately order the search. The last two rules are used to eliminate generated subproblems that are not better than the ones already known.

Only lower-bound elimination and termination rules are used in MANIP. The elimination of subproblems due to dominance tests are studied elsewhere[10]. For simplicity, only the search for a single optimal solution is considered here. The computational efficiency for searching all optimal solutions can be investigated similarly.

## 2.1 Selection rules

This examines the list of active subproblems and selects one for expansion. If the list is maintained in a first-in/first-out order, the algorithm is called a breadth-first search. If the list is maintained in a last-in/first-out order, the algorithm is called a depth-first search. Lastly, if the list is maintained in increasing order of lower bounds, the algorithm is called a best-first search. Ibaraki mapped these searches into a general form called heuristic searches[8]. A heuristic function is defined which governs the order in which subproblems are selected and decomposed. The algorithm always decomposes the subproblem with the minimum heuristic value. In a best-first search, the lower-bound values define the order of expansion. Therefore, the lower-bound function can be taken as the heuristic function. In a breadth-first search, subproblems with the minimum level numbers are expanded first. The level number can, thus, be taken as the heuristic function. Lastly, in a depth-first search, subproblems with the maximum level numbers are expanded first. The negation of the level number can be taken as the heuristic function. If $\boldsymbol{U}$ is the current list of active subproblems in the process of expansion and $h$ is the heuristic function, the search function for a subproblem to expand in a serial branch-and-bound algorithm is:

$$s_S(U)=\{P_i \mid h(P_i)=\min_{P_j\in U} h(P_j)\} \tag{5}$$

## 2.2 Lower-bound elimination and termination rules

A lower bound is calculated for a subproblem when it is created. If a subproblem is a feasible solution with the best objective-function value so far, the solution value becomes the incumbent $z$. The incumbent represents the best solution obtained so far in the process. In minimization problems, if the lower bound of a subproblem exceeds the value of the incumbent, this subproblem can be pruned because it will not lead to an optimal solution. The decomposition process continues until all subproblems are either expanded or pruned. During the computation, $P_i$ is terminated if:

$$g(P_i) \geqslant z \tag{6}$$

Let $\boldsymbol{L}$ denotes the lower-bound cutoff test, that is, $P_j L P_i$ means that $P_j$ is a feasible solution and $f(P_j) \leqslant g(P_i)$.

The above lower-bound test for obtaining an exact optimal solution can be relaxed in order to obtain a suboptimal solution with guaranteed accuracy[4]. Suppose it were decided that a deviation of 10% from the optimum was tolerable. If a feasible solution of 150 is obtained, all subproblems with lower bounds of 136.4 (or 150/(1 + 0.1)) or more can be terminated since they cannot lead to a solution that deviates by more than 10% from 150. This technique significantly reduces the amount of intermediate storage and the time needed in order to arrive at a suboptimal solution. Define an allowance function $\varepsilon(z)$: $\boldsymbol{R}\to\boldsymbol{R}$ (set of reals) such that $P_i$ is terminated if:

$$g(P_i) \geqslant z-\varepsilon(z) \tag{7}$$

The final incumbent value $z_F$ that is obtained by using the modified lower-bound test deviates from the optimal solution value $z_O$ by[7]:

$$z_F-\varepsilon(z_F) \leqslant z_O \leqslant z_F \tag{8}$$

Examples of often used allowance functions are:

$\varepsilon(z)=\varepsilon \geqslant 0$ (absolute error deviation) and (9)

$\varepsilon(z)=\frac{\varepsilon z}{1+\varepsilon}$, $\varepsilon \geqslant 0$, $z \geqslant 0$ (relative error deviation) (10)

Properties of these functions are similar. In this paper, the function for relative error deviation is assumed.

## 2.3 Serial branch-and-bound algorithm (single solution and lower-bound tests)

$g$ = lower-bound function;

$\varepsilon$ = allowance function;

$s_S$ = serial search function;

$z$ = incumbent value;

$U$ =set of active subproblems.

(1) (Initialize): $z \leftarrow \infty$; $U \leftarrow \{P_0\}$;

(2) (Select): If $U=\emptyset$ then go to Step 7 else let $P_i \leftarrow s_S(U)$, $U \leftarrow U-\{P_i\}$; if $g(P_i) \geqslant z-\varepsilon(z)$ then go to Step 2;

(3) (Decompose): Generate sons $\{P_{i_1}, P_{i_2}, \cdots, P_{i_{k_i}}\}$ of $P_i$; $U \leftarrow U \cup \{P_{i_1}, \cdots, P_{i_{k_i}}\}$;

(4) (Feasibility test): For all $j \in \{1, \cdots, k_i\}$, if $P_{i_j}$ is a feasible solution then $U \leftarrow U-\{P_{i_j}\}$ and let $z \leftarrow \min[z, f(P_{i_j})]$;

(5) (Lower-bound test): For all $j \in \{1, \cdots, k_i\}$, if $g(P_{i_j}) \geqslant z-\varepsilon(z)$ then $U \leftarrow U-\{P_{i_j}\}$;

(6) (Terminate): Go to Step 2;

(7) (Halt): $f(P_0)$ satisfies $z-\varepsilon(z) \leqslant f(P_0) \leqslant z$.

In the above algorithm, eliminations are done after branching instead of after selection as in Ibaraki's algorithm[7]. This reduces the memory space required for storing the active subproblems.

# 3 Parallel branch-and-bound algorithms

Branch-and-bound algorithms have inherent parallelism. Each of the four rules of branch-and-bound algorithms can be implemented by parallel processing.

(a) Parallel selection of subproblems: In the parallel case, a set of subproblems less than or equal in size to the number of processors have to be selected for decomposition in each iteration. The selection problem is especially critical under best-first search because a set of subproblems with the minimum lower bounds must be selected. The selection function (Eq. (5)) becomes:

$$s_p(U)=\begin{cases}\underset{\substack{\text{where } h(P_i)<h(P_j), P_i, P_j \in U \\ i \in (i_1, \cdots, i_k), j \notin (i_1, \cdots, i_k)}}{\{P_{i_1}, \cdots, P_{i_k}\}} & , \text{if } |U|>k \\ U, \text{if } |U| \leqslant k \end{cases} \tag{11}$$

where $k$ is the number of processors. This returns $k$ subproblems with the minimum heuristic values from $U$.

(b) Parallel branch: The subproblems assigned to the processors can be decomposed in parallel. In order for the processors to be well utilized, the number of active subproblems should be greater than or equal to $k$.

(c) Parallel termination test: Multiple infeasible nodes can be eliminated in each iteration. Further, multiple feasible solutions may be generated, and the incumbent may have to be updated in parallel.

(d) Parallel elimination test: The lower-bound test (Eq's (6) or (7)) can be sped up by comparing lower bounds of multiple subproblems with the incumbent. However, the execution of the bounding function is problem-dependent, and software implementation is more flexible.

Desai has proposed to use implicit enumeration to find the optimal solution of NP-hard problems[11]. The system dedicates one processor to each subproblem, and this processor reports to its parent processor when the evaluation is completed. Implicit enumeration is time-consuming and wasteful for large problems.

Imai et al. studied parallel branch-and-bound algorithms with a depth-first search in a multiprocessor with shared memory[12]. Depth- first search is used owing to memory limitations. Similarly, Dessouki and Huen studied the same algorithm on a general purpose architecture with a slow communication network[13]. Depth-first search is not effective in minimizing the execution time because a serial best-first search expands the minimum number of nodes when all the lower bounds are distinct[4]. However, a best-first search is space-consuming because the number of active subproblems is large. Currently, with the availability of VLSI technology, larger and inexpensive memories and faster communication media, reducing the execution time becomes an important problem.

MANIP was proposed for implementing parallel branch-and-bound algorithms with a best-first search and lower-bound tests[1,14]. The system was designed with the following objectives:

(1) it should be modularly expandable to include a very large number of processors;

(2) the design must have high performance, and the cost should be kept low by replicating simple cells;

(3) it should use distributed control so that a controller would not become the bottleneck.

There are multiple subproblem memory controllers connected by a selection network and a secondary-storage redistribution network (Figure 1). Each of the $m$ controllers is connected to $n$ processors and expands $n$ subproblems in parallel.

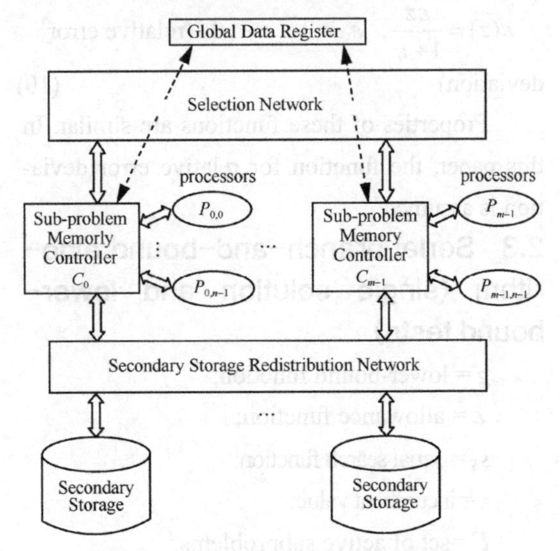

Figure 1. The architecture of MANIP

The controller is also responsible for managing the local list of subproblems in ascending order of lower bounds, communicating with other processors through the networks, updating the incumbent and maintaining the virtual memory operating system. Parallel selection is achieved through the selection network. In order to select $m{\cdot}n$ subproblems with the minimum lower bounds for processing in each iteration, $n$ subproblems with the minimum lower bounds are selected from each controller and sent to the neighboring controller through a ring network. The subproblems received from the neighboring controller is inserted into the local list. The process is repeated $m-1$ times, and it is proved that each of the controller has $n$ of the $m{\cdot}n$ subproblems with the minimum lower bounds. However, a complete selection using $m-1$ shifts is time-consuming. Assuming random distribution of the $m{\cdot}n$ required subproblems, over 70% of the required subproblems can be selected with only one shift[14]. The unidirectional ring network is found to be the most cost-effective selection network[15]. The incumbent is maintained in a global data register. The concurrent update of the incumbent is facilitated by a sequential associative memory that finds the minimum of the feasible solutions when they are shifted out bit-serially and synchronously from the controllers. Since the register containing the incumbent is accessible to all the processors, lower-bound elimination and termination tests can be done in parallel.

## 4 Computational efficiency of parallel branch-and-bound algorithms

### 4.1 Anomalies on parallelism

From experimental results on vertex-cover, 0-1 knapsack and integer-programming problems, the reduction in computational time of branch-and-bound algorithms with lower-bound tests and best-first search is approximately linear on the average with respect to the number of processors[1,14]. However, simulations have revealed that the number of iterations for parallel branch-and-bound algorithms using $k$ processors can be (a) more time than that of a single processor—“detrimental anomaly”[12,16-17]; or (b) less than one-$k$'th of the number of iterations of a single processor — “acceleration anomaly”[12,16,18-19]. It is desirable to discover conditions that preserve the acceleration anomalies and eliminate the detrimental anomalies.

Let $T^c(k,\varepsilon)$ and $T^d(k,\varepsilon)$ denote the number of iterations required for expanding a branch-and-bound tree using centralized and $k$ subproblem lists respectively, where $k$ is the number of processors used, and $\varepsilon$ is the allowance function. Assuming that subproblems are decomposed synchronously, the number of iterations is measured by the number of times that subproblems are decomposed in each processor. The distinction between centralized and multiple lists lies in the memory configuration. When all the processors are connected to a centralized memory, the subproblem list is global to the processors. When each processor has a private memory, only the local subproblem list can be accessed. The configuration of MANIP (Figure 1) with $m$–1 shifts in each iteration is equivalent to a system with a centralized subproblem list.

An example of a detrimental anomaly is illustrated in Figure 2. In a serial depth-first search, subtree $T_2$ is terminated owing to the lower-bound test of $.P'_1 : f(P'_1)/(1+\varepsilon) \leqslant g(P_2)$. where $\varepsilon$=0.1. In a parallel depth-first search with two processors, a feasible solution, $P_4$, that terminates $P_1$ and $P'_1$ is found first. Consequently, subtree $T_2$ has to be expanded that will eventually terminate subtree $T_3$. If the size of $T_2$ is much larger than the size of $T_3$, the time it takes to expand $T_2$ using two processors will be longer than the time it takes to expand $T_3$ using one processor.

An example of an acceleration anomaly is shown in Figure 3. When a single processor with a depth-first search is used, $T_4$ will be expanded. When two processors are used, $P_4$ and hence $T_4$ will be terminated by lower-bound tests. If $T_4$ is very large, an acceleration anomaly will occur.

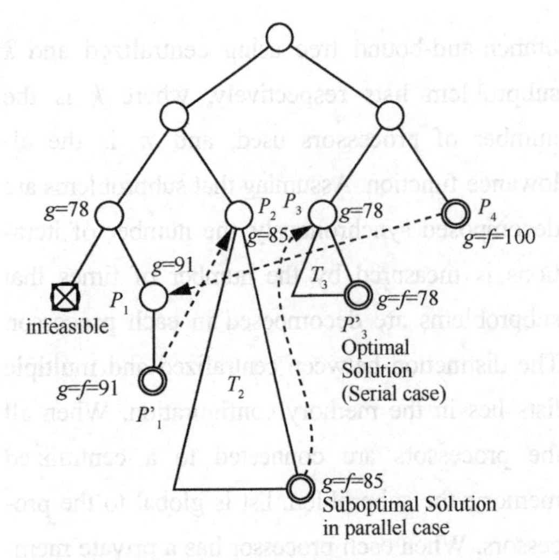


Figure 2. Counterexample for $T^c(2,0.1) \leqslant T^c(1,0.1)$ using a depth-first search

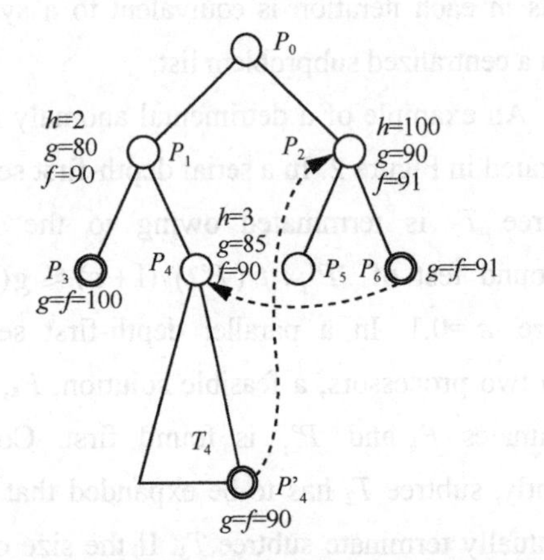


Figure 3. Example of an acceleration anomaly $T^c(1,0.1) > T^c(1,0.1)/2$ using a depth-first search

The important consideration here is not in knowing that anomalies exist, but in understanding why these anomalies occur. Furthermore, it is desirable to preserve the acceleration anomalies and to avoid the detrimental anomalies. Our objective is to find the sufficient conditions to ensure that $T^c(k,\varepsilon) \leqslant T^c(1,\varepsilon)$ and $T^d(k,\varepsilon) \leqslant T^d(1,\varepsilon)$, as well as the necessary conditions for $T^c(k,\varepsilon) \leqslant T^c(1,\varepsilon)/k$ and $T^d(k,\varepsilon) \leqslant T^d(1,\varepsilon)/k$. Note that when there is one processor, only one memory system is used, and $T^c(1,\varepsilon) = T^d(1,\varepsilon)$.

## 4.2 Generalized heuristic searches to eliminate detrimental anomalies in a single subproblem list

Recall in Sections 2 and 3 that the selection function uses heuristic values to define the order of node expansions. It was mentioned that breadth-first, depth-first and best-first searches are special cases of heuristic searches. These searches are potentially anomalous when parallel expansions are allowed.

Consider the serial depth-first search. The subproblems are maintained in a last-in-first-out list, and the subproblem with the maximum level number is expanded first. When multiple subproblems have identical level numbers (heuristic values), the node chosen by the selection function depends on the order of insertion into the stack. Suppose the rightmost son is always expanded and inserted first. Then the leftmost son will be the node inserted last and expanded first in the next iteration.

In a parallel depth-first search, the mere extension of the serial algorithm may cause an anomalous behavior. For example, the order of expansion in a serial depth-first search for the tree in Figure 4 is *A*, *B*, *D*, *I*, *J*, *E*, etc. When two processors are used, nodes *B* and *C* are expanded in the second iteration that result in nodes *D*, *E*, *F*, *G* and H. Since these nodes have identical level numbers, the search function can choose any two of these nodes for expansion in the next iteration. Suppose the nodes are inserted in the order *E*, *D*, *H*, *G* and *F*. Then nodes *F* and *G* will be selected and expanded in the third iteration. This may cause unexpected behavior as compared to the serial case and is exactly the reason for the anomalies reported in [16].

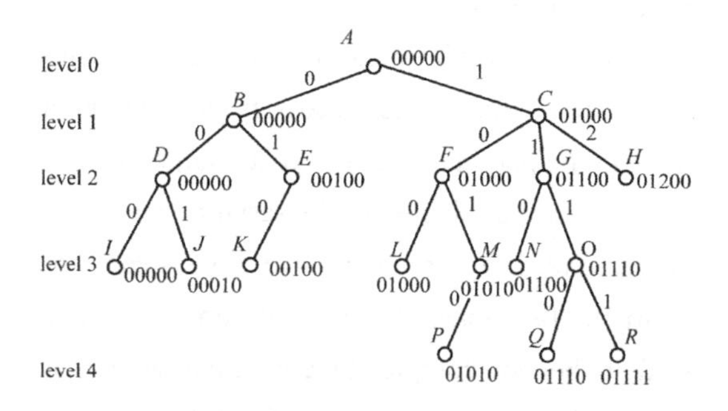


Figure 4. The path numbers of a tree

To solve this problem, we must define distinct heuristic values for the nodes so that there is no ambiguity on the nodes to be chosen by the selection function. In this paper, a path number is used to uniquely identify a node in a tree. The path number of a node is a sequence of $d+1$ integers that represent the path from the root to this node where $d$ is the maximum number of levels of the tree. The path number $E = e_0e_1e_2 \cdots e_d$ is defined recursively as follows. The root $P_0$ exists at level 0 and has a path number of $000\cdots0$. A node $P_{i_j}$ on level $l$ which is the $j$-th son (counting from the left) of $P_i$ with path number $E_{P_i} = e_0e_1 \cdots e_{l-1}000\cdots$has path number $E_{P_{i_j}} = e_0e_1 \cdots e_{l-1}j00\cdots$. As an example, the path numbers for the nodes in the tree of Figure 4 are shown.

To compare path numbers, the relations "$>$" and "$=$" must be defined. A path number $E_1 = e_1^1e_2^1\cdots$ is less than another path number $E_2 = e_1^2e_2^2\cdots$ ($E_1 < E_2$) if there exists $0 \leqslant j \leqslant d$ such that $e_i^1 = e_i^2$, $0 \leqslant i \leqslant j$, and $e_j^1 < e_j^2$. The path numbers are equal if $e_i^1 = e_i^2$ for $0 \leqslant i \leqslant d$. For example, the path number 01000 is less than 01010. Note that nodes can have equal path numbers if they have the ancestor-descendant relationship. Since these nodes never coexist simultaneously in the list of active subproblems, the subproblems in the active list always have distinct path numbers.

The path number is now included in the heuristic function. The primary key is still the lower-bound value or the level number. The secondary or ternary key is the path number and is used break ties in the primary key.

$$h(P_i)=\begin{cases}(\text{level number, path number}) \\ \quad \text{for breadth-first search} \\ (-\text{level number, path number}) \\ \quad \text{for depth-first search} \\ (\text{lower bound, level number, path number}) \text{ or} \\ (\text{lower bound,} -\text{level number, path number}) \\ \quad \text{for best-first search}\end{cases} \tag{12}$$

For a best-first search, two alternatives are defined that search in a breadth-first or depth-first fashion for nodes with identical lower bounds. The heuristic functions defined above belong to a general class of heuristic functions that satisfy the following properties:

$$\text{(a) } h(P_i) \neq h(P_j) \text{ if } P_i \neq P_j,\ P_i, P_j \in \boldsymbol{P} \tag{13}$$

All heuristic values in the active list are distinct.

$$\text{(b) } h(P_i) < h(P_j) \text{ if } P_j \text{ is a descendant of } P_i \tag{14}$$

(heuristic values always increase)

In general, any heuristic function that satisfy Eq's (13) and (14) will not lead to detrimental anomalies. Due to space limitation, the results are stated without proof in the following theorems. The proofs can be found in [10].

**Theorem 1:** Let $\varepsilon=0$, i.e., an exact optimal solution is sought. $T^c(k,0)\leqslant T^c(1,0)$ holds for parallel heuristic searches of a single optimal solution in a centralized list using any heuristic function that satisfies Eq's (13) and (14).

When approximations are allowed, detrimental anomalies cannot always be avoided (see Figure 2). The reason for the anomaly is that lower-bound tests under approximation, $L$, are not transitive. That is, $P_iLP_j$: and $P_jLP_k$ do not imply $P_iLP_k$, since $f(P_j)/(1+\varepsilon)\leqslant g(P_j)$ and $f(P_j)/(1+\varepsilon)^2\leqslant g(P_j)$ implies $f(P_j)/(1+\varepsilon)^2\leqslant g(P_k)$ rather than $f(P_i)/(1+\varepsilon)\leqslant g(P_k)$. Detrimental anomalies can be avoided for best-first or breadth-first searches only.

**Theorem 2:** $T^c(k,\varepsilon)\leqslant T^c(1,\varepsilon)$, $\varepsilon>0$, holds for parallel best-first or breadth-first searches for a single optimal solution when a heuristic function satisfying Eq's (13) and (14) is used.

Since the lower-bound function is used as the heuristic function in best-first searches, Eq's (13) and (14) are automatically satisfied if all the lower-bound values are distinct. Otherwise, path numbers must be used to break ties in the lower bounds. For breadth-first and depth-first searches, the conditions of Theorem 2 are not sufficient, and the following condition is needed. For any feasible solution $P_i$, all nodes whose heuristic values are less than $h(P_i)$ cannot be eliminated by the lower-bound test due to $P_i$ that is, $f(P_i)/(1+\varepsilon)\leqslant g(P_j)$ implies that $h(P_i)<h(P_j)$ for any $P_j$. Generally, this condition is too strong and cannot be satisfied in practice.

## 4.3 Necessary conditions to ensure acceleration anomalies in a single subproblem list

In order to allow the occurrence of acceleration anomalies, it is necessary to expand a node in the parallel case such that a large number of nodes are terminated in the serial case. This is characterized by the complete consistency of heuristic functions. A heuristic function, $h$, is said to be not completely consistent with $g$ if there exist two nodes $P_i$ and $P_j$ such that $h(P_i)>h(P_j)$ and $g(P_i)\leqslant g(P_j)$.

**Theorem 3:** Let $\varepsilon=0$ and assume that a single optimal solution is sought. The necessary condition for $T^c(k,0)<T^c(1,0)/k$ is that the heuristic function is not completely consistent with $g$.

The theorem implies that acceleration anomalies may exist and detrimental anomalies can be prevented. It is important to note that the condition in Theorem 3 is not necessary when approximate solutions are sought. An example showing the existence of an acceleration anomaly when $h$ is completely consistent with $g$ is shown in Figure 3. The additional necessary condition is that, $h$ is not completely consistent with the lower-bound test with approximation, that is, there exist $P_i$ and $P_j$ such that $h(P_i)>h(P_j)$ and $P_iLP_j$.

## 4.4 Multiple subproblem lists

When there are multiple subproblem lists, one for each processor, a node with the minimum heuristic value is selected for decomposition from each local list. This node may not belong to the global set of active nodes with the minimum heuristic values, however, the node with the minimum heuristic value will always be expanded by a processor as long as the nodes are selected in a consistent order when there are ties. Since it is easy to maintain the incumbent in a global data register, the behavior of multiple lists is analogous to that of a centralized list. However, the performance of using multiple lists is usually worse than that of a single subproblem list[14].

### 4.5 Robustness of parallel best-first searches

Although a parallel best-first search does not always give the best performance, the following performance bounds can be proved [10]:

**Theorem 4:** For a parallel best-first search with $k$ processors, $\varepsilon=0$ and $g(P_i)\neq f^*$ if $P_i$ is not an optimal-solution node,

$$\frac{T(1,0)}{k}\leqslant T(k,0)\leqslant\frac{T(1,0)}{k}+\frac{k-1}{k}l \qquad (15)$$

where $l$ is the maximum number of levels of the branch-and-bound tree to be searched. Since the performance is not affected by using single or multiple subproblem lists, the superscript in $T$ is dropped.

As an example, if $l=50$, $T(1,0) = 10^6$ (for a typical traveling-salesman problem), and $k=$ 1 000, then $T(1\ 000,0)\leqslant 1\ 049$. This means that almost linear speedup can be attained with 1 000 processors.

In this section, we have shown conditions to avoid detrimental anomalies and to preserve acceleration anomalies. The results are summarized in Table 1. These results have to be extended when dominance tests are allowed. Due to space limitations, these results will not be presented here[10].

## 5 Virtual-memory support for best-first searches

Virtual memory is essential in supporting the large space requirements of best-first searches[20]. In this section, the alternatives for supporting best-first searches on systems with limited main-memory space are discussed. It is found that a direct implementation exhibits weak locality (Section 5.1). Modification of the virtual memory to tailor to the characteristics of the algorithm is effective but inflexible (Section 5.2). Lastly, the modification of the search algorithm in order to enhance the amount of locality exhibited is found to be the most viable approach (Sections 5.3).

### 5.1 Direct implementation on a conventional virtual memory

A direct implementation uses a list of subproblems and a priority queue[1] of pointers to these subproblems. In each iteration, the subproblem with the minimum lower bound is deleted from the list. This generates new subproblems that are inserted back. The priority queue of pointers is used to maintain the ordering required for a best-first search as restoring a sorted list of pointers is less costly than maintaining a sorted list of subproblems.

As subproblems are not ordered by lower bounds in the list, the subproblem chosen for expansion is equally likely to be in any slot in the list. This implies weak locality, and a direct implementation is inefficient.

---

[1] The priority queue can be implemented by a heap which is a complete binary tree such that the value of each node is at least as small as that of its descendants. Although hardware implementation such as systolic arrays can be used, the problem of mapping subproblems onto the secondary memory is not solved.

Table 1. Summary of results for the elimination of detrimental anomalies and the preservation of acceleration anomalies in parallel branch-and-bound algorithms

| Allowance function | Subproblem lists | Search strategies | Suff. cond. to eliminate detrimental anomaly | Necessary cond. for acceleration anomaly |
|---|---|---|---|---|
| $\varepsilon = 0$ | single | all | I | II |
| | multiple | | no anomaly | |
| $\varepsilon > 0$ | single or multiple | breadth-first or best-first | I | exists |
| | | depth-first | anomaly | |

Conditions:
I: heuristic function satisfies Eq's (13) and (14).
II: $h$ is not completely consistent with $g$.
anomaly: the sufficient conditions are impractical.
exists: the necessary conditions are too loose

## 5.2 Modified virtual memory

The proposed virtual-memory system is depicted in Figure 5[21]. Subproblems in the secondary storage are organized as a $B^+$-tree[2]. Each leaf of the tree is a page and contains one or more subproblems. Since the non-terminal nodes of the tree are pointer nodes that are much smaller than leaf nodes, a substantial portion of the pointer tree may be kept in main memory. This reduces the number of disk accesses required to access a subproblem in the secondary storage.

The main memory contains a partial list of subproblems and a heap of pointers to subproblems in the partial list. Newly generated subproblems are inserted into the memory list. When the main-memory space is exhausted, subproblems are moved to the $B^+$-tree in the secondary storage. In a best-first search, the subproblem with the smallest lower bound is always expanded. It is prohibitively expensive to compare the smallest subproblem in the memory list with the smallest subproblem in the $B^+$-tree each time a subproblem is expanded. This can be solved by keeping a portion of the first $B^+$-tree page in the main memory.

In inserting a subproblem into a normal $B^+$-tree, the page into which the subproblem is inserted is read into the main memory. The subproblem is inserted into the page image in main memory, and the page image is written back onto disk. A more efficient scheme may be obtained by setting the block size to an integral number of disk sectors greater than or equal to the size of a subproblem and using a bit map in main memory that shows the status of blocks of all the pages. Inserting a subproblem into a page, therefore, consists of searching for an empty block in the bit map and writing the block when the disk head is properly positioned.

The design of the virtual-memory operating system depends on the replacement algorithm and the page size. An efficient replacement algorithm should maximize the number of sub-

[2] A B-tree of order $m$ is a search tree which is either empty or satisfies the following properties[22]: (i) the root node has at least two children; (ii) each node contains at least $m$ keys and $m+1$ pointers; and (iii) each node contains at most $2m$ keys and $2m+1$ pointers. A $B^+$-tree is a variant of the B-tree in which all records reside in the leaves. The upper levels are organized as a B-tree and serve as an index to locate a record. The leaf nodes are linked from left to right for easy sequential processing.

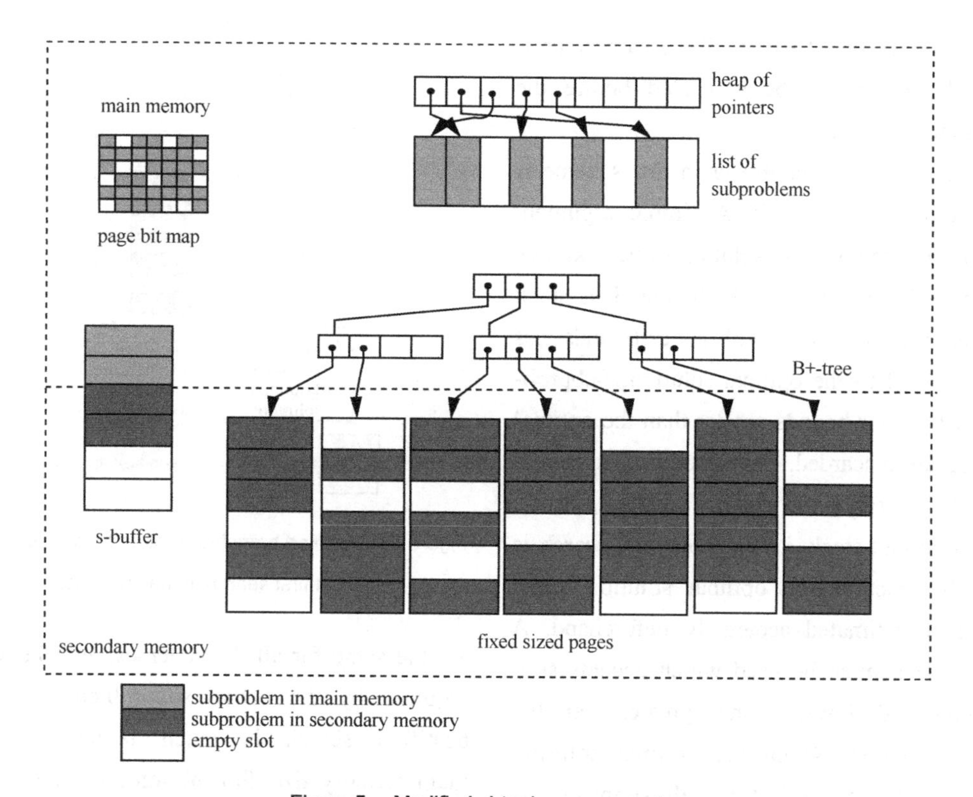


Figure 5. Modified virtual-memory system

problems inserted into each page and avoid replacing subproblems that will be expanded in the immediate future. Analysis and simulation results have shown that for integer-programming problems, between 70% to 90% of the subproblems in main memory with the largest lower bounds should be replaced each time. For knapsack problems, between 30% to 50% of the subproblems should be replaced. The page size should be chosen to minimize the disk traffic. A suitable page size for both integer-programming and knapsack problems is found to be between 60% to 90% of the maximum size of the subproblem list in main memory.

## 5.3 Modified branch-and-bound algorithms

One flaw of the modified virtual-memory system is that whenever a page is full, it has to be split into two. This copying of subproblems between pages introduces a substantial overhead that may be avoided by allowing the logical pages to vary in size.

The range of possible lower bounds is partitioned into a set of disjoint regions, each of which is implemented as a variable-sized stack. An active subproblem generated is pushed into the appropriate stack. Subproblems within a stack are not ordered by lower bounds. Thus, it is prohibitively expensive to access the subproblem with the least lower bound in a stack. Instead, the top of the first non-empty stack is loaded into the main memory and expanded in a depth-first fashion. The effect of this relaxation on the order of expansion is small. Suppose the optimal solution is in stack $m$. At worst, all the subproblems in stack $m$ have to be expanded before the process terminates.

The important parameter in this scheme is the width of each stack. A simple argument shows that if the optimal-solution value is known, the optimal number of stacks is one. This stack accepts all subproblems with lower bounds less than or equal to the optimal solution. Subproblems with lower bounds greater than the optimal solution are discarded. This stack can be implemented entirely in main memory as a last-in-first-out stack, i.e., a depth-first search is used. In practice, the optimal solution value cannot be estimated accurately beforehand. A second stack must be used which accepts subproblems with lower bounds greater than the estimated value. When the optimal solution resides in the second stack, all subproblems there must be examined. To limit the number of subproblem examined when this happens, the second stack should further be partitioned into smaller stacks. The scheme called modified depth-first search implements an approximation of a best-first search and is illustrated in Figure 6.

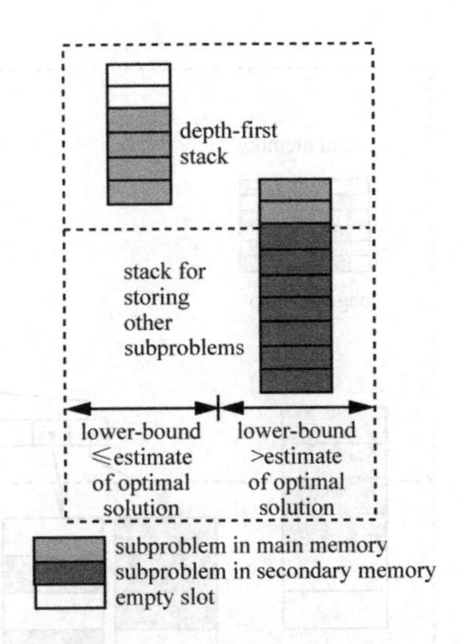


Figure 6. Modified branch-and-bound algorithm with a depth-first search in the first stack

## 5.4 Comparison of various schemes

In this section, we compare the different virtual-memory-support schemes for branch-and-bound algorithms by deriving the lower-bound estimates on paging overhead and presenting the simulation results on integer-programming problems.

Let $x$ be the number of subproblems expanded and $y$ be the number of subproblems generated but never expanded (i.e., the number of terminated subproblems). Assume that $x$ and $y$ are the same for all the schemes, and the space required by the active subproblems under a best-first search is much larger than the main-memory size. For the three schemes, let $a_i$ and $h_i$, $i\in$ {Direct Implementation (DI), Modified Virtual Memory (MVM), Modified Depth-First Search (MDF)}, be the page size and the ratio of page size to subproblem size respectively. Further, let the sum of the average seek and rotational delays of a disk be $s$, and the disk transfer overhead be $t$ seconds per byte.

For direct implementation, assume that the heap of pointers resides in main memory. Let $g_{DI}$ be the probability that a subproblem chosen for expansion resides in the secondary storage, $g_{DI}$ is not one because a subproblem to be expanded may reside in main memory in a space vacated by an expanded subproblem. Therefore, an integral number of pages containing $g_{DI}x$ subproblems are read from disk. On the other hand, $g_{DI}x+y$ subproblems are written to disk. Out of these subproblems, $g_{DI}x$ subproblems are written into slots vacated by subproblems being expanded, thus giving rise $g_{DI}x\lceil 1/h_{DI}\rceil$ page writes. The remaining $y$ subproblems that occupy $y/h_{DI}$ pages must be first read from disk and later written out. The disk traffic is:

$$p_{DI} = 2\left(g_{DI}x\left\lceil \frac{1}{h_{DI}} \right\rceil + \frac{y}{h_{DI}}\right)(s + a_{DI}t) \qquad (16)$$

For the modified virtual-memory scheme, assume that the page containing the set of subproblems with the smallest lower bounds resides completely in main memory, and let $g_{MVM}$ be the probability that a subproblem being expanded is in this page. Then $(g_{MVM}x+y)$ subproblems are written to disk, and $g_{MVM}x$ subproblems are read from disk. On the average, each $B^+$-tree page is three-quarters full. The average number of pages read in is $\frac{g_{MVM}x}{((3/4)h_{MVM})}$. Let $f_{MVM}$ be the average number of subproblems inserted into a page during a replacement, and assume that no pages are split in a replacement. The disk traffic is:

$$p_{MVM} = \left(\frac{4g_{MVM}x}{3h_{MVM}} + \frac{g_{MVM}x + y}{f_{MVM}}\right)(s + a_{MVM}t) \qquad (17)$$

For the modified branch-and-bound algorithm, let the boundary between the first and the second stacks be the value of the optimal solution. Then $y$ subproblems are written to disk. The disk traffic is:

$$p_{MDF} = \frac{y}{h_{MDF}}(s + a_{MDF}t) \qquad (18)$$

When all the $h_i$'s and $a_i$'s, $i \in$ {DI, MVM, MDF}, are equal, a comparison of the different equations shows that the modified depth-first search has the best lower-bound performance. $p_{MDF}$ is always smaller than $p_{MVM}$ as $f_{MVM}$ must be smaller than $h_{MVM}/2$ in order to avoid a page split. Likewise, $p_{MDF}$ is always smaller than $p_{DI}$ because $g_{DI}$ is non-zero.

Simulation results on 17-variable 17-constraint integer-programming problems are shown in Figure 7. Each subproblem requires a memory size of 1 476 bytes. A main-memory size of 128 Kbytes (10% of the virtual space) is assumed. Cases with small and large pages are shown, and the improvement is smaller for small pages. Only the results with the optimal page size is shown for the modified virtual-memory scheme. With large pages, there is a 100-times reduction of paging overhead for the modified depth-first search as compared to a direct implementation, and a 60-times reduction as compared to a modified virtual-memory scheme. The improvement is within 20 times for small pages. From these results, we can also infer that when the memory space required by a subproblem is small (such as knapsack problems), the improvement will be even better.

## 6 Conclusion

In this paper, we have shown some research results on the parallel execution and virtual-memory support of (approximate) branch-and-bound algorithms. Sufficient conditions are proposed so that parallel execution of branch-and-bound algorithms under lower-bound tests always results in reduction in execution time. It is shown that the improvement for parallel execution can be greater than the number of processors under some very loose necessary conditions. A best-first search is found to be a good

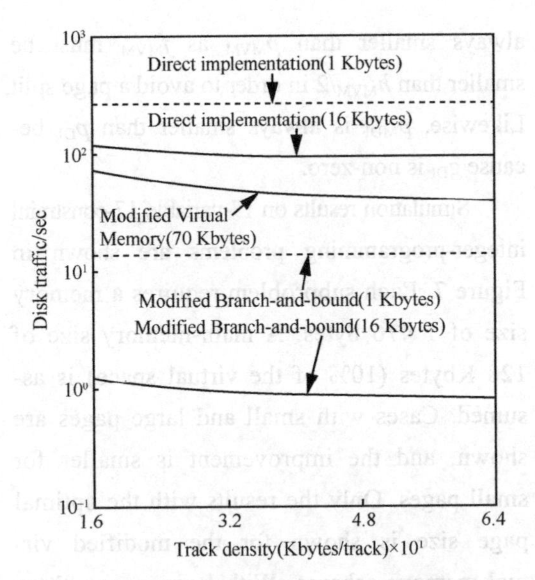


Figure 7. Simulation results for various virtual-memory schemes to support branch-and-bound algorithms on 17 by 17 integer-programming problems (main memory size is 128 Kbytes, page size is shown on each curve)

search strategy that can guarantee little degradation in performance with increasing degrees of parallelism. We have also presented alternatives for the virtual-memory support. It is found that by relaxing the requirements of branch-and-bound execution, over 100-times reduction in paging overhead can be attained for integer-programming problems as compared to a direct implementation on a conventional virtual-memory system.

## References:


[1] WAH B W, MA Y W. MANIP-a parallel computer system for implementing branch-and-bound algorithms[C]// The 8th Annual Symposium on Computer Architecture. Washington: IEEE Computer Society Press, 1982: 239-262.

[2] GAREY M R, JOHNSON D S. Computers and intractability, a guide to the theory of NP-completeness[M]. San Francisco: W. H. Freeman and Company, 1979.

[3] SAHNI S. General techniques for combinational approximation[J]. Operations Research, 1977, 25(6): 920-936.

[4] LAWLER E L, WOOD D. Branch-and-bound methods: a survey[J]. Operations Research, 1966(14): 699-719.

[5] NILSSON N J. Principles of artificial intelligence[M]. Palo Alto: Tioga Publishing Company, 1980.

[6] KOHLER W, STEIGLITZ K. Characterization and theoretical comparison of branch-and-bound algorithms for permutation problems[J]. Journal of ACM, 1974, 21(1): 140-156.

[7] IBARAKI T. Computational efficiency of approximate branch-and-bound algorithms[J]. Mathematics of Operations Research, 1976, 1(3): 287-289.

[8] IBARAKI T. Theoretical comparisons of search strategies in branch-and-bound algorithms[J]. International Journal of Computer & Information Sciences, 1976, 5(4): 315-344.

[9] IBARAKI T. The power of dominance relations in branch-and-bound algorithms[J]. Journal of ACM, 1977, 24(2): 264-279.

[10] LI G J, WAH B W. Computational efficiency of parallel approximate branch-and-bound algorithms[C]// 1984 International Conference on Parallel Processing. Washington: IEEE Computer Society Press, 1984: 473-480.

[11] DESAI B C. The BPU, a staged parallel processing system to solve the zero-one problem[C]// Proceeding of ICS'78. [S.l.:s.n.], 1978: 802-817.

[12] IMAI M, FUKUMURA T, YOSHIDA Y. A parallelized branch-and-bound algorithm implementation and efficiency[J]. Systems, Computers, Controls, 1979, 10(3): 6270.

[13] EL-DESSOUKI O, HUEN W H. Distributed enumeration on network computers[J]. IEEE Transactions on Computers, 1980, C-29(9): 818-825.

[14] WAH B W, MA Y W. MANIP-a multicomputer architecture for solving combinatorial extremum-search problems[J]. IEEE Transactions on Computers, 1984, C-33(5): 377-390.

[15] WAH B W, CHEN K L. A partitioning approach to the design of selection networks[J]. IEEE Transactions on Computers, 1984, C-33(3): 261-268.

[16] LAI T H, SAHNI S. Anomalies in parallel branch-and-bound algorithms[J]. Communications of the ACM, 1983, 27(6): 183-190.

[17] MOHAN J. Experience with two parallel programs solving the traveling-salesman problem[C]// 1983 International Conference on Parallel Processing. Washington: IEEE Computer Society Press, 1983: 191-193.

[18] FULLER S H, OUSTERHOUT J K, RASKIN L, et al. Multi-microprocessors: an overview and working exam-

ple[J]. Proceedings of the IEEE, 1978, 66(2): 216-228.

[19] WEIDE B W. Modeling unusual behavior of parallel algorithms[J]. IEEE Transactions on Computers, 1982, C-31(11): 1126-1130.

[20] WAH B W, YU C F. Probabilistic modeling of branch-and-bound algorithms[C]// The 6th International Computer Software and Applications Conference. [S.l.:s.n.], 1982: 647-653.

[21] YU C F, WAH B W. Virtual-memory support for branch-and-bound algorithms[C]// The 7th International Computer Software and Applications Conference. [S.l.:s.n.], 1983: 618-626.

[22] COMAR D. The ubiquitous B-tree[J]. Computing Surveys, 1979, 11(2): 121-137.

发表于International Conference on Parallel Processing论文集，1985年

# MANIP-2: a multicomputer architecture for evaluating logic programs

Guo-jie LI, Benjamin W. WAH

School of Electrical Engineering, Purdue University, West Lafayette, IN 47907

**Abstract:** Logic programs are conventionally evaluated by brute-force depth-first search. To avoid unnecessary searching, an intelligent search strategy that guides the search by heuristic information is desirable. In this paper, the evaluation of a logic program is modeled as the search of an AND/OR tree. A heuristic function using the ratio of the success probability of a subgoal to the estimated overhead of evaluating the subgoal is found to be useful in guiding the search of logic programs. An optimal search strategy that minimizes the expected overhead is proposed and analyzed. The optimal strategy requires a large run-time computational or storage overhead. An efficient heuristic search strategy that can be implemented on a multiprocessor architecture is presented.

**Key words:** AND/OR tree, AND parallelism, heuristic search, logic programming, OR parallelism, pruning, success probability

## 1 Introduction

Logic programming is a programming methodology based on Horn-clause resolution[1]. The efficiency of solving a logic programming problem depends strongly on the many ways of representing the logic program. Evaluating a logic program can be considered as the search of an AND/OR tree[1-2]. The root is labeled by the initial problem to be queried; the OR nodes represent (sub)goals; and the AND nodes represent clauses[1]. All subgoals in the same body of a clause are children of an AND node. A (sub)goal (OR node) and its children display the nondeterministic choices of clauses with the same head. The terminal nodes denote clauses or subgoals that cannot be decomposed. Figure 1(a) shows an example of the AND/OR-tree representation of a logic program. In general, a logic program without any inference loop can be viewed as an acyclic AND/OR graph. A finite AND/OR tree is obtained from the AND/OR graph by duplicating common descendent nodes.

The AND/OR tree in Figure 1(a) can be represented more clearly in Figure 1(b) as a high-level OR tree involving the selection of all combinations of clauses, and multiple low-level AND trees representing the solution tree resulted

Research supported by CIDMAC, a research unit of Purdue University, sponsored by Purdue, Cincinnati Milicron Corporation, Control Data Corporation, Cummins Engine Company, Ransburg Corporation, and TRW.

[1] The definitions of AND and OR nodes are taken from [3]. The roles of the AND and OR nodes are reversed in Nilsson's definition[4].

from clauses selected in the OR tree. Parallel processing can be applied to evaluate the multiple solution trees in parallel (OR-parallelism), or can be applied to search a solution tree in parallel (AND-parallelism). Note that the number of edges in this representation could be much large than that of an AND/OR-tree representation.

AND-parallelism involves the simultaneous execution of subgoals in a clause. Since subgoals within a clause can share variables, the binding of variables of concurrently executing subgoals must be coordinated to avoid conflicts of a shared variable being bound to more than one value. AND-parallelism is limited by the measures to avoid conflicts. One approach to avoid conflicts is to annotate variables to indicate which subgoals can bind values to specific variables and which cannot[5]. In particular only one subgoal, called producer, is allowed to bind a value to a variable. Conery proposed a complex, non-annotated, process-structured system that dynamically monitors variables and continually develops data dependency networks to control the order of execution of subgoals, never allowing two potential producers for the same

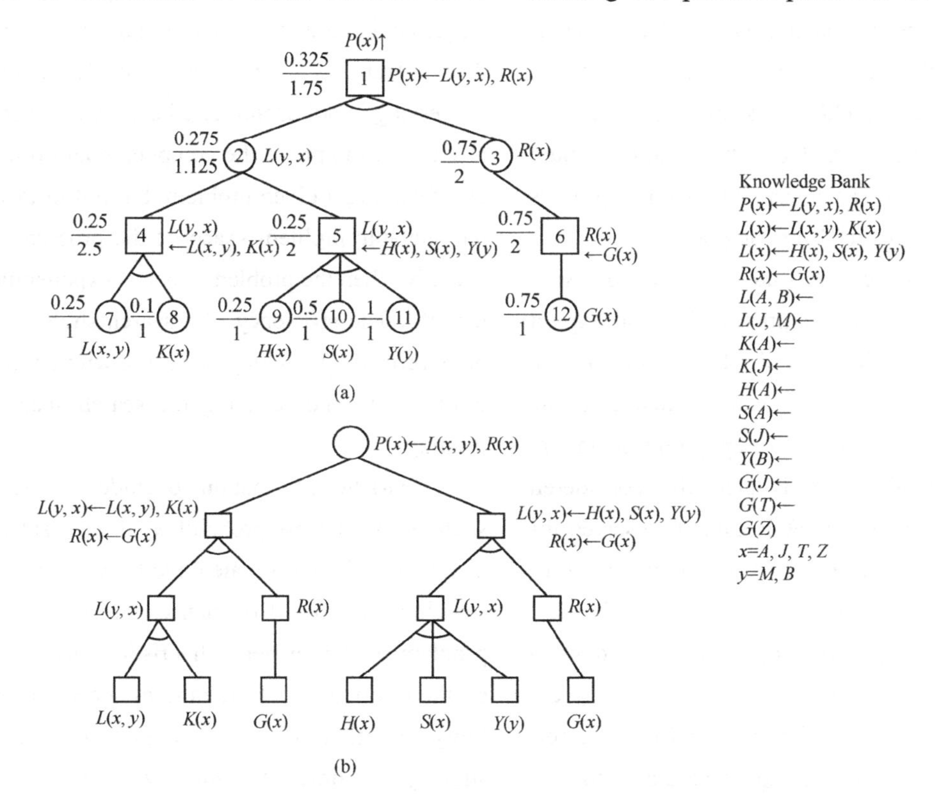


Figure 1. (a) An example of a logic program represented as an AND/OR tree (OR nodes are represented as circular nodes; AND nodes are represented as squared nodes. In the logic program, *P*: happy; *L*: like; *R*: rich; *K*: kind; *H*: handsome; *S*: strong; *Y*: young; *G*: goodjob; *A*: Ares; *B*: Betty; *J*: John; *M*: Mary; *T*: Tom; *Z*: Zeus. The numbers outside each node are the ratio of success probability to expected search cost.) (b) The same logic program represented as a hierarchy of OR tree followed by AND trees

variable to execute in parallel[2]. DeGroot described a method to obtain restricted. AND-parallelism by compile-time creation of a parallel execution-graph expression for each program clause[6].

In OR-parallelism all subgoals are independent of each other, and consistency checks for shared variables needed in AND-parallelism are avoided[1,7]. However, an OR-tree representation is inefficient due to the large number of branches needed as compared to that of an AND/OR tree. Given an initial query, ←A, B, with *n* ways of solving A and *m* ways of solving B, the OR tree contains $n \times m$ branches, whereas an AND/OR tree contains $n+m$. To improve the efficiency of an OR-tree search of logic programs, several models that modify pure OR-parallelism, such as introducing process bundles[8] and bagof[9], are proposed.

In this paper, we study parallel processing for an AND/OR-tree representation and exploit both AND-parallelism and OR-parallelism. The search strategy developed can be extended to the corresponding AND/OR-graph representation. The search algorithm is generally considered different from an AND/OR game-tree search for the following reasons. First, in contrast to combinatorial-extremum searches that find the best solution, solving a logic program corresponds to finding any (or all) solution(s) satisfying the given conditions, the implicative Horn clauses, and the consistent binding of variables for the subgoals. Second, the value of a node in the AND/OR tree for a logic program is either TRUE (success) or FALSE (failure). The selection of a node for evaluation is usually based on a fixed order, and heuristic information to guide the search is not available. Third, a variable in a logic program can be bound to several values, and some subgoals may share a common variable. For example, in Figure 1, some subgoals share variable *x*, and *x* can be bound to any of Ares, John, Tom, and Zeus. For a particular variable in a subgoal, a subset of its possible values may be allowed. In contrast, the nodes in a game tree are independent. Lastly, pruning rules for evaluating the AND/OR tree of a logic program are different from $\alpha$-$\beta$ pruning due to the binary values returned by the terminal nodes.

Since the search space of a logic program is large, an intelligent search strategy that guides the search is very important[10]. Parallel processing is not useful here because it is generally used to improve the computational efficiency of solving a given problem, but not to extend the solvable problem space of the problem, especially when the problem space is exponentially large[11]. The efficiency of a search strategy can be improved by guiding the search with heuristic information, and reducing the search space by pruning.

Heuristic information to guide the search, such as the success probability of each subgoal or clause and the estimated overhead (or cost) of searching a subgoal or clause, remain an open problem. In this paper, a heuristic search based on the information of the ratio of success probability to estimated cost is studied. This search strategy is aimed to minimize the expected search cost and the dynamic run-time overhead of evaluating logic programs.

Pruning is used to eliminate unnecessary expansions when it is determined that a solution tree cannot be found from a subgoal or clause.

Two kinds of pruning exist. In AND-pruning, if one successor of an AND node for a given binding of values to variables is known to be FALSE, then all its remaining successors for the same binding can be pruned. Likewise, if one solution is sought, then OR-pruning can be applied to prune other successors of an OR node once one of its successors is known to be TRUE. In this paper, it is assumed that one solution tree is sought from an AND/OR-tree representation of a logic program, and hence both AND-pruning and OR-pruning can be applied.

Pruning and parallelism are conceptually illustrated in Figure 2(a). In a sequential depth-first search, if Node 1 fails, then Node 2 will be examined next, otherwise Node 3 will be examined. Similarly, the traversal of Node 5 depends on the results of traversing Nodes 1, 2, 3, and 4. This dependence information can be represented in a fail-token-flow graph, $G_f$, as depicted in Figure 2(b) for the tree in Figure 2(a). A node (circle) in the graph will be active only if it receives a fail-token from an incident edge. When a terminal node b found to be FALSE, a fail-token is sent along the direction of the corresponding edge. The coordinator (shaded box) in the graph coordinates the activities of the connected blocks. When a fail-token is received from any incident edge of a coordinator, fail-tokens arc sent to all directly connected nodes. At the same time, any node searched in the block directly connected to this coordinator can be terminated because it does not belong to the solution tree. For example, when Node 1 is found to be FALSE, then a fail-token is sent to Node 2. If Node 2 is found to be FALSE, then a fail-token is sent to Coordinator $x_D$. At this time, any node concurrently searched in Block D can be terminated. When a solution tree is found, there is one node in each column of $G_f$ that return TRUE. $G_f$ can be used to represent pruning in AND-parallelism when the success probability is high and most of the terminal nodes are TRUE.

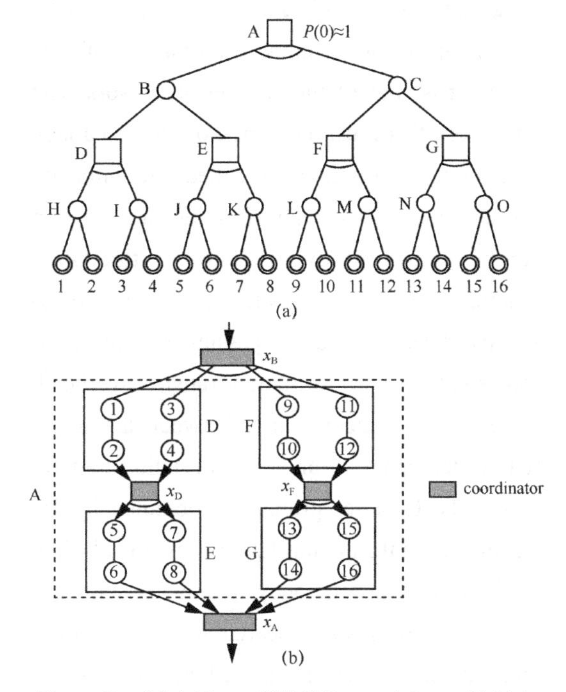


Figure 2. (a) A binary AND/OR search tree with high success probability (circular nodes represent OR nodes; squared nodes represent AND nodes.) (b) The corresponding fail-token-flow graph, $G_f$

On the other hand, when the success probability of the goal is low, most of terminal nodes are FALSE. The search for the inexistence of a solution tree in OR-parallelism can now be represented by the success-token-flow graph, $G_s$. $G_s$ is the dual of $G_f$ in the sense that a fail-token is replaced by a success-token, and the columns in $G_f$ are transposed to become the rows in $G_s$. Referring to Figure 2(a), Nodes 1, 2, 5, and 6 are assigned to four processors. If any of Nodes 1, 2,

5, and 6 succeeds, then a success-token is generated, and the next connected node is assigned to as idle processor. Since most of the terminal nodes are FALSE, the search will be completed when a small number of nodes have been searched in each column of $G_s$.

When two AND/OR subtrees are searched in parallel, more work than necessary might be performed if the pruning information of one processor is unavailable to other processors. The extra work that must be carried out due to a lack of pruning information is called the information-deficiency overhead. Pruning information can be exchanged by messages or through a common memory. This increased communication overhead needed for passing this information is called the information-transfer overhead. In general, a tradeoff exists between the information-deficiency and information-transfer overheads. If better pruning is obtained by increasing the information-transfer overhead, then the information-deficiency overhead will decrease. A good parallel search algorithm should consider these tradeoffs and reduce the run-time overheads by proper assignments of tasks to processors.

Several parallel models of logic programs and the corresponding multiprocessing architectures have been studied[6-7,9,12-15]. Nearly all these architectures were based on ad hoc search strategies and scheduling methods. In this paper, we propose MANIP-2[2], a multicomputer architecture to evaluate logic programs. However, the goal here is not in describing the details of an architecture, but in giving the theoretical foundation of the necessary search algorithm upon which the architecture is based. The emphasis of this paper is in showing the reasonableness of heuristic searching and the feasibility of an effective scheduling method.

## 2 Heuristic information for searching AND/OR trees

The useful heuristic information to guide the search include the predicted success probability of a solution tree being found from a subgoal or clause, and the associated average cost of finding the solution tree.

The success probability of a node (or alternatively a branch) in an AND/OR-tree representation of a logic program is an a priori probability that reflects the possibility of finding a solution tree over all unifications from this node. These probabilities are used to guide the search initially, and will be improved after more dynamic pruning information is obtained.

For a terminal node with variable $x$, its success probability is $m(x)/n(x)$, where $n(x)$ is the total number of values that variable $x$ can acquire, and $m(x)$ is the number of values acquired by $x$ in this terminal node. It is assumed that all values in the domain of a variable are equally likely to be assigned to a subgoal. When a subgoal shares more than one variable with other subgoals, its success probability can be computed as the product of the success probabilities of the variables if all variables are independent. In general, the success probability of a node cannot be directly determi-

[2] MANIP is multicomputer architecture proposed earlier to evaluate parallel branch-and-bound algorithms with a best-first search, which is an OR-tree heuristic search guided by lower bounds of subproblems[16].

nated by the success probabilities of its immediate descendents, which may be correlated (the descendent nodes may contain shared clauses or subgoals that renders them dependent). It may have to be evaluated from actual information in the knowledge base. For example, in Figure 1(a), four values can be bound to $x$, and two can be bound to $y$. For the eight combinations of values of $x$ and $y$, only two of them exist in the knowledge base. Hence, the success probability of $L(x, y)$ is 0.25.

Assuming that a nonterminal node $K$ has two immediate descendents, $K_1$ and $K_2$, let $P(\cdot)$ be the a priori success probability of a node. Then

$$P(K)=\begin{cases}P(K_1)\cdot P(K_2\mid K_1), & K \text{ is AND}\\ P(K_1)+P(K_2)-P(K_1)\cdot P(K_2\mid K_1), & K \text{ is OR}\end{cases} \quad (1)$$

Eq. (1) can be generalized to nodes with more than two descendents. The computation of the a priori conditional probabilities can be complex due to the shared variables.

The success probabilities of a logic program can also be either assigned initially by the designer, or determinated by statistic collected during execution. In the latter case, no a prior probability is available before the program is executed, and all branches are assumed to have equal possibility of success. A deterministic search strategy, such as a depth-first search, has to be used initially. For example, in Figure 1, the success probability of the clause $L(y, x)\leftarrow L(x, y)\cdot K(x)$ is the probability that both $L(x, y)$ and $K(x)$ succeed. However, $L(x, y)$ and $K(x)$ are dependent, and the success probability of $L(y, x)$ would be difficult to compute. Statistic collected show that the success probability is 0.25. Other success probabilities in Figure 1 were computed by enumerations of all combinations of binding of variables.

Since unification has a linear complexity[17-18], the search cost can be defined by the number of nodes traversed before a solution tree is found to exist or not exist. The cost for searching a subtree depends on the structure and dependence of the subtrees, the query used, and the partial search results already obtained. One way is to define an average search cost based on the estimated probabilities of success. For node $K$ with descendents $K_1$ and $K_2$ and assuming that $K_1$ is searched first, the average search cost is

$$C(K)=C(K_1)+\begin{cases}C(K_2\mid K_1)\cdot P(K_1), & K \text{ is AND}\\ C(K_2\mid K_1)\cdot(1-P(K_1)), & K \text{ is OR}\end{cases} \quad (2)$$

Simulations have shown that the average search cost depends on the structure of the tree, but is quite insensitive to changes in the success probability. A complete binary AND/OR tree with unitary search cost at the terminal nodes was assumed. The simulation results depicted in Figure 3 indicate the relationship between the success probabilities of the root of trees with height 12 and 16, respectively, and the associated average search cost. The average cost is the smallest when $p$ is either zero or one, and is maximum when $p$ is around 0.5. Moreover, the difference between the maximum and the minimum coats is relatively small, reflecting the insensitivity of the expected search cost with respect to the success probability.

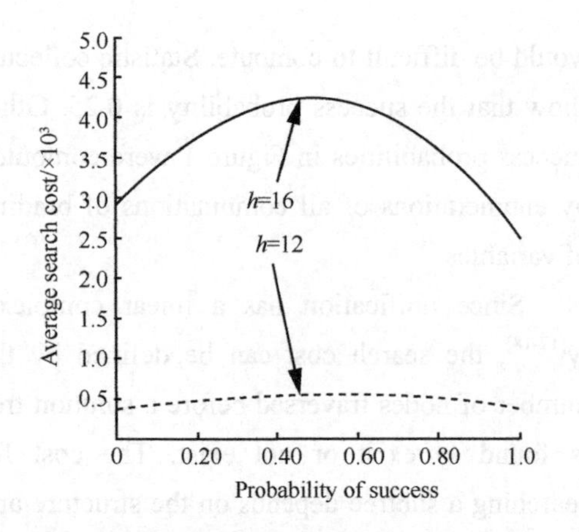


Figure 3. A plot of the search cost versus the success probability of the root

The average search cost of a subgoal is difficult to be formulated mathematically because it is related to the dependence of descendent subgoals (due to the shared clauses and subgoals). Moreover, the average search cost depends on the search strategy and the order that values are bound to variables, which in turn are driven by the average search costs. Hence the search costs would have to be initially estimated from statistic collected by a given search strategy. As better search costs are obtained, the search will become more efficient, and better estimates on the search costs can be obtained. For example, in Figure 1, the average search cost for $L(y, x)$ was computed by averaging the search cost to verify the result of $L(y, x)$ for all combinations of values of $x$ and $y$. The search strategy used in computing the costs in Figure 1 assumed a left-to-right traversal of the descendents.

The expected search cost of a subgoal represents an average over all possible queries, all possible paths leading to this subgoal, and all possible combinations of values of variables for a given search strategy. However, when the path leading to a given subgoal is known, the minimum cost and the associated success probability of obtaining a solution tree are better measures to guide the search. Of course, this will result in an enormously large amount of stored information for each subgoal that renders the scheme impractical.

## 3 Heuristic search for logic programs

Let $P(x)$ and $C(x)$ be the probability of success and the associated average search cost for node $x$. Define the criteria $\Phi_a$ and $\Phi_o$ for any node $x$ as

$$\Phi_a(x)=\frac{P(x)}{C(x)} \text{ (}x\text{ is descendent of an OR node)} \tag{3}$$

$$\Phi_o(x)=\frac{1-P(x)}{C(x)} \text{ (}x\text{ is descendent of an AND node)} \tag{4}$$

Simon and Kadane have studied the optimal OR-tree search and have proved that the search sequence $b=b_1,\cdots,b_n$ is optimal if $\Phi_a(b_i)\geqslant\Phi_a(b_{i+1})$, where the $b_i$ are descendents of an OR node $b$ with precedence relationships[19]. Barnett has extended their results to optimal search from AND nodes with the assumption that all immediate descendents are independent[20], Garey has proved that the optimal sequence of performing a set of tasks until one of them fails or all tasks are fulfilled is in descending order of $\Phi_a$[21].

For a pure OR-tree or AND-tree search, the search order is well defined by a single heuristic function. However, when an AND/OR tree is searched, there are two criteria, $\Phi_a$ and $\Phi_o$ to order the AND and OR nodes, respectively. Hence a complete order cannot be defined for all active nodes. To resolve this problem, we can decompose an AND/OR tree into a hierarchy of a single OR tree, each terminal of which is an AND tree (Figure 1(b)). The following theorem relates the criteria $\Phi_a$ and $\Phi_o$ and defines an optimal search order for an AND/OR tree with dependent nodes.

**Theorem 1:** Suppose that an OR node $K$ has $n$ immediate descendent AND nodes, $K_1,\cdots, K_n$, and that the AND node $K_i$, $1\leqslant i\leqslant n$, has $i_m$ immediate descendent OR nodes, $K_{i_1},\cdots, K_{i_m}$. If $\Phi_a(K_i)\geqslant\Phi_a(K_{i+1})$, then the expected search cost $C(K)$ is minimum when all descendents of $K_i$ are searched before $K_{i+1}$.

**Proof:** Suppose that $K_{i_1}$ is found to be TRUE, then the conjunction of the remaining subgoals of $K_i$, namely, $K_{i_1},\cdots, K_{i_m}$, forms a new AND node, $K'_{i_2}$. The conditional probability of $K_{i_1},\cdots, K_{i_m}$ being TRUE, given that $K_{i_1}$ is TRUE is

$$P(K'_{i_2})=P(K_{i_2},\cdots,K_{i_m}|K_{i_1})=\frac{P(K_i)}{P(K_{i_1})} \tag{5}$$

To get the optimal strategy, we need to compare the probability-to-cost ratios in respect to node $K'_{i_2}$ and $K_{i+1}$. Using Eq's (2) and (5),

$$\Phi_a(K'_{i_2})=\frac{P(K'_{i_2})}{C(K'_{i_2})}=\frac{P(K_i)/\mathrm{P}(K_{i_1})}{C(K_{i_2},\cdots,K_{i_m}|K_{i_1})}\geqslant$$

$$\frac{P(K_i)}{C(K_i)}\geqslant\frac{P(K_{i+1})}{C(K_{i+1})}=\Phi_a(K_{i+1}) \tag{6}$$

Eq. (6) implies that the remaining subgoals $K_{i_1},\cdots,K_{i_m}$ should be unified first before $K_{i+1}$. If subgoal $K_{i_2}$ is found to be TRUE, then the above proof can be applied again to show that the remaining subgoals of $K_i$ should be unified before $K_{i+1}$.

Theorem 1 shows that the optimal strategy of searching an AND/OR tree is to select the most promising solution tree with the largest $\Phi_a$ value among all possible solutions trees and to examine AND nodes in this solution tree in descending order of $\Phi_o$ values. The search is switched to the next best solution tree (with the next largest $\Phi_a$ value) if the first solution tree fails.

The key issue in performing the above optimal strategy is to find the most promising solution tree with the largest $\Phi_a$ value, and for the solution tree selected, the node with the largest $\Phi_a$ value. If $h$ the height of a complete AND/OR tree, is taken as a measure of the problem size, then it is unlikely that a polynomial-time algorithm exists for finding the most promising solution tree. Let $B_o$ and $B_a$ be the numbers of branches of each OR and AND node, respectively. If $B_a = 1$, then there are $B_o^{h/2}$ possible solution trees, each of which consists of one node. To get the maximum $\Phi_a$ over all solution trees, at least $B_o^{h/2}-1$ comparisons are needed. This is the lower-bound complexity for computing the largest $\Phi_a$ When $B_a$=1. In general, if $B_a>1$, then before Theorem 1 can be applied, the AND/OR tree has to be transformed into a hierarchy of an OR tree, each terminal of which is an AND tree. The number of terminal nodes in the transformed tree has a lower bound of $O(B_oB_a)^{h/2}$. Hence to select the largest $\Phi_a$ and $\Phi_o$ values would require an exceedingly large amount of computational time. Another approach is to store these values associated with each node in the AND/OR tree and to retrieve the decisions in real time. Unfortunately, this would require a large amount of storage space with a tower bound at the same complexity as stated above. Therefore, it is unrealistic to apply the optimal search strategy in respect to an AND/OR-tree search.

Owing to the intractable complexity of propagating the $\Phi_a$ and $\Phi_o$ values of all terminal nodes to the root in implementing the optimal AND/OR-tree search, an efficient top-down heuristic search is proposed here. As stated in Section 2, each node in the AND/OR tree can be assigned an estimated success probability and the associated expected search cost. These values, together with the information of the path leading from the root to this node, can be used to compute a heuristic value for the node. The search will be guided by the heuristic values.

A solution tree is a tree such that all nonterminals are AND nodes and all terminals are OR nodes. To minimize the search effort for a solution tree, it is necessary to first find one of the FALSE terminals in this tree, since the search can be terminated once this node is found. This method will be used to compute the heuristic values for AND nodes. From the duality between AND and OR nodes, a failure tree is a tree such that all nonterminals are OR nodes and all terminals are AND nodes. A failure tree is the dual of a solution tree, such that the entire AND/OR tree is FALSE if all nodes in the failure tree are FALSE. To stop the search of a failure tree as early as possible, it is necessary to verify one of the terminals is TRUE. This method will be used to compute the heuristic values for OR nodes.

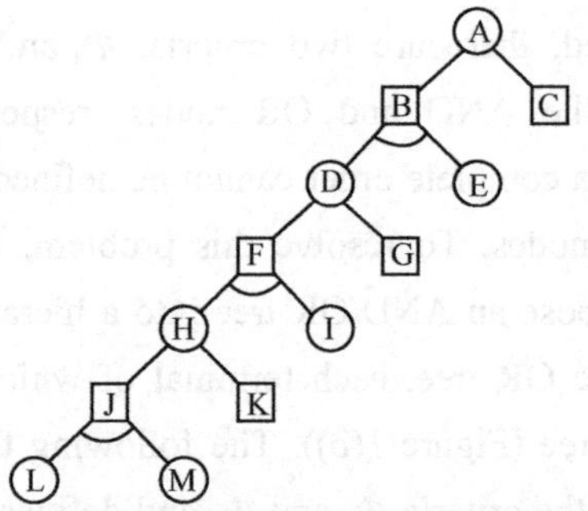

Figure 4. Computation of the heuristic values (OR nodes are represented as circular nodes; AND nodes are represented as squared nodes)

Referring to the partial binary AND/OR tree in Figure 4, suppose that AND-node F is to be searched. For the goal to be TRUE, Nodes E and F must be TRUE. This is the information that can be extracted directly from the path leading from the root to F. The heuristic value of Node F will be the ratio of the probability of verifying Node A to be TRUE to the associated search cost. This can be computed as the ratio of the probability of success of Nodes E and F to the total average search cost of verifying that Nodes E and F are TRUE or any of them is FALSE. Note that Nodes E and F may be dependent. Other information, such as nodes searched from E, would be also crucial to computing the heuristic value for F. However, this information cannot be extracted directly from the path, and would require a high run-time overhead to maintain, hence will not be considered here. Similarly, the heuristic value of AND-node J is based on information about Nodes B, I, and J. In general, the heuristic value $\Phi_a(x)$ of an AND node $x$ is based on information extractable from the path leading from the root to $x$ such that the goal be verified to be TRUE. The information on the complete path from the root to any AND node must be maintained with each AND node.

To compute $\Phi_a(x)$ for an OR node $x$, the information extractable from the path leading from the tool to $x$ such that the goal can be proved to be FALSE will be used. Complete information on the path from the root must also be maintained for each OR node. Referring to Figure 1, in computing the heuristic value for OR-node H, the goal will be FALSE if Nodes C, G, and H are FALSE. Note that these nodes may be dependent.

The following heuristic-search algorithm, BAO, is a top-down best-first search algorithm that uses heuristic information about a particular set of solution trees for a given AND node and a particular set of failure trees for a given OR node. It is assumed that the values of $\Phi_a$ are available for each set of clauses with the same head (an OR node) in the logic program. Similarly, the values of $\Phi_a$ for each subgoal in a clause (an AND node) are also available. It is further assumed that the values bound to a variable are independent and in a fixed order, and that the values of $\Phi_a$ and $\Phi_o$ are applicable to all instantiation. The following procedure is applied iteratively by binding each variable to a constant value (or to a set of constant values that could be the domain of the variable) until the goal is proved to be TRUE or FALSE. There is an Active List containing tasks in decreasing order of heuristic values. Without loss of generality, assume that the goal node, S, is an OR node.

BAO—Heuristic Search Algorithm for one solution tree:

(1) (Initialization): Initiate search from S. If S is known to be TRUE or FALSE, then stop. Otherwise, create a task for S with the information that S is an OR node, and compute its heuristic value. The task is inserted into the Active List.

(2) (Decomposition): Select a task T from the Active List with the maximum heuristic value. If T is a ground node, then go to Step 3. If T contains a variable that is not bound, a constant value (or a set of constant values that could be the domain of the variable) is bound to the variable. Decompose the task (an AND or OR node) into its immediate descendent tasks. The new tasks, with their heuristic values the information about path from S, and the values bound to variables are inserted into the Active List in the proper order. Go to Step 2.

(3) (Pruning): For Task T under consideration, the following steps will be carried out based on values returned by the ground node.

(a) The subgoal or clause ($T_a$) on the path from the root to T that is nearest to the root and becomes TRUE or FALSE is found.

(b) The information on $T_a$ is incorporated into all tasks in the Active List. Subtasks in tome tasks can be removed based on $T_a$. If all subtasks within a task W are removed, then pruning will be carried out recursively on W.

(c) The values successfully or unsuccessfully bound to variables are broadcast to all active tasks.

(d) The heuristic values for all active tasks are updated. The active tasks are reordered if necessary.

(4) (Termination): If S is terminated, then return success with the bound value. If S cannot be bound to any new value, then return failure. Otherwise, go to Step 2.

There are several considerations when Algorithm BAO is implemented. First, a single list of active tasks is kept, and the AND and OR nodes are not distinguished. Depending on the heuristic values and whether the node concerned an AND or OR node, the objective is to either prove that the goal is TRUE or prove that it is FALSE. Second, pruning performed in Step (3) requires a large overhead because the entire Active List has to be updated and reordered. However, the merits on the number of tasks eliminated and the better heuristic values generated and doubtful. To avoid this run-time overhead, pruning can be performed after a task is selected from the Active List (Step 2). In this case, all pruning information will be recorded in a common list. Of course, this may result in some unnecessary tasks in the Active List. Third, the computation of the heuristic values could be complex when all dependency of subgoals and variables are considered. In practice, some simplifying assumptions may be made in combining the heuristic values together. For example, a heuristic function to compute the success probability and cost of a conjunction of subgoals could be the product of the success probabilities and the sum of the associated costs of the subgoals. Lastly, the algorithm investigates many possible solution trees in parallel by switching from one to another based on the heuristic information obtained. This results in a large number of tasks in the Active List, which is a fundamental problem in heuristic searches. To reduce the storage space required, static analysis can be performed to arrange the clauses with the same head and the subgoals in each clause in a proper order, and to use a depth-first strategy to search the AND/OR tree. Of course, the order defined will be based on the average over all possible queries and all possible ways of reaching a particular subgoal.

The BAO algorithm is illustrated in Figure 5 by explaining snapshots of solving the logic program in Figure 1. The cost of each unification

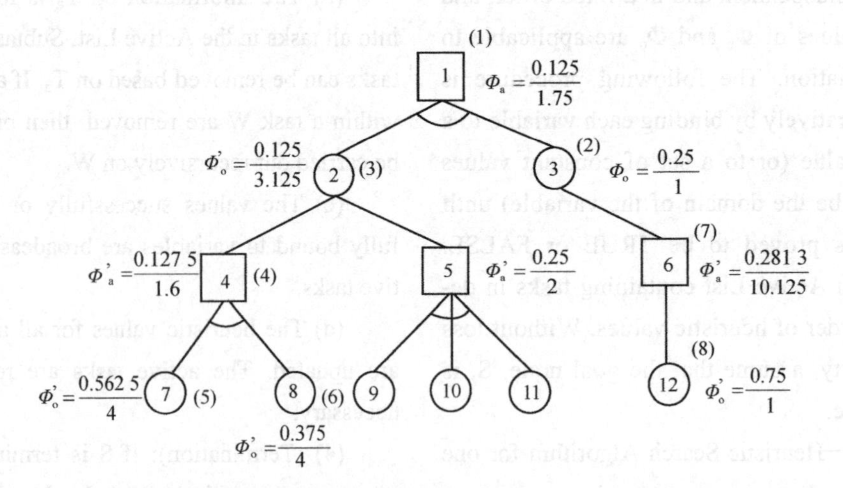


Figure 5. Illustration of the BAO procedure using the same example shown in Figure 1 (OR nodes are represented as circular nodes; AND nodes are represented as squared nodes. The number in parenthesis outside each node shows the order that this node is examined in the BAO procedure.)

is assumed to be unity. The query is "Who is happy?" Node 1 is unified first. For simplicity, suppose that all successful bindings are equally likely, and that $x$ is first bound to John. Let the updated heuristic values be $\Phi'_a$ and $\Phi'_o$, respectively. After decomposition, subgoals $L(y, x)$ (Node 2) and $R(x)$ (Node 3) with the corresponding heuristic values are inserted into the Active List. Since $\Phi'_o(3) > \Phi'_o(2)$, Node 3 is selected, and Node 5 is created. In computing, $\Phi'_a(6)$, the goal is expected to be TRUE, implying that Node 2 and 6 are TRUE. Hence the heuristic information of Node 2 must be included. The correlation between Nodes 2 and 6 may be complex. For simplicity, Nodes 2 and 6 are assumed to be independent. $P'(6)$ (resp. $C'(6)$), the new success probability (resp. new average cost) of Node 6, is the product (resp. summation) of the original success probabilities (resp. original average costs) of Nodes 2 and 6. $\Phi'_a(6)$ becomes 0.281 3/10.125. As $\Phi'_o(2) > \Phi'_a(6)$, Node 2 is selected in the next iteration, and Nodes 4 and 5 are created. Intuitively, Node 6 is likely to be TRUE, but to terminate the search for the goal as soon as possible, the active node that may fail first should be investigated. In computing $\Phi'_a(4)$ and $\Phi'_a(6)$, the heuristic information on Node 3 must be included. For example, $\Phi'_a(4)$ is computed as (0.75×0.25)/(2.5+3). Suppose that variable $y$ is first bound to Mary. Node 4 is next selected, $\Phi'_a(4)$ is the largest among all active nodes. This means that the possible solution trees involving Node 4 is the most promising. Nodes 7 and 8 are inserted into the Active List in iteration 4. Note that in computing $\Phi'_o(7)$ and $\Phi'_o(8)$, $\Phi'_a(5)$ rather than $\Phi'_a(6)$ is used. The reason for this is that the goal is expected to be FALSE if Node 7 or 8 is selected, and the failure probability and cost of Node 5 are needed in this case, but the information on Node 3 (which is included in $\Phi'_a(5)$) does not affect the decision on Node 7 or 8. $\Phi'_a(7)$ is computed as (0.75×0.75)/(1+3). Once Nodes 7 and 8 are instantiated in Iterations 5 and 6, respectively, Node 2 is known to be TRUE by values returned from the pound terms. Node 5 can thus be pruned by OR-pruning if a single solution tree is sought. In the last two iterations Nodes 6 and 12 are unified, and the solution is "John is happy".

## 4 Parallel heuristic search of logic programs

In this section, we study the problem associated with the parallel processing of the AND/OR-tree representations of logic programs. By minor modifications of MANIP[22], parallel heuristic search of logic programs can be carried out efficiently.

The first problem in parallel processing is the unification of shared variables. When AND-parallelism is involved, variable binding conflicts must be prevented. A lot of efforts have been devoted to solving this problem[2,5-6,23]. We do not attempt to propose a new method to overcome this problem. Instead, we assume one or more of the following conditions: (a) that the subgoal do not share variables; (b) that the shared variables are uninstantiated; (c) that a producer subgoal has bound one or more values to a shared variable and forwards them to many consumer subgoals. In any of these cases, processors can carry out tasks independently.

The second problem is the granularity of

parallelism. If the granularity is small, then the information-transfer overhead will be high. In contrast, it the granularity is large, then the information-deficiency overhead will be high, and the degree of parallel processing may be small. The proper granularity depends on the capacity of the communication network, the number of processors, and the relative overhead between the top-down unification of variables and the bottom-up return of solution values. In searching a logic program, if the cost of a subgoal selected is smaller than the defined granularity, then the subgoal is considered as an indivisible operation and processed by a single processor.

Once the granularity is determined, the next problem is to determine the scheduling of the *k* parallel processors. Given a list of active ordered by decreasing probability-to-cost ratios, the problem is to determine the number of processors to evaluate each task in parallel. To minimize the expected completion time, the problem can be formulated into a complex integer-programming optimization problem. Moreover, processors have to be rescheduled again once a given solution tree is found to be TRUE or FALSE. Considering the facts that the scheduling algorithm is complex and that static scheduling is not feasible, we decide to assign a free processor to each task in the Active List. When a task has reached its minimum granularity, it will be processed by a single processor until completion.

Another problem is on resource sharing. For a set of processors evaluating a solution tree, it is necessary to search those subtrees with the largest failure-probability-to-cost ratios. Likewise, for sets of processors examining different solution trees, the possible solution trees with the largest success-probability-to-cost ratios must be evaluated. Further, load balancing must be carried out to keep all processors busy. The ring network in MANIP is adequate for load balancing the processors.

The parallel selection problem has been studied thoroughly in MANIP. In the parallel evaluation of branch-and-bound algorithms, a set of subproblems with the minimum lower bounds must be expanded. The selection overhead has been found to be high; furthermore, the selection rule is based on a fallible lower-bound heuristic. Therefore, it might be more efficient not to follow the selection rule strictly. A no-wait policy has been proposed[11,16]. Instead of waiting for one of the *k* subproblems with the smallest lower bounds, each processor would expand the "most promising" subproblem in its local memory and initiate a fetch of the "most promising" subproblem from its neighbors. In this case, "most promising" subproblem is the one with the minimum lower bound. Evaluations for MANIP have shown when the *k* most promising subproblems are randomly distributed among the processors, the average fraction of processors containing one or more of the most promising subproblems is at least 0.63, resulting in a speedup proportional to 0.63k. Further, with occasional redistribution of subproblems with small lower bounds using a ring network, the performance is almost as good as that of a complete selection.

The no-wait policy can be applied here to schedule processors for evaluating subgoals in its local memory. Subgoals with large failure-probability-to-cost ratios (resp. large success-probability to-cost ratios) and a suitable granularity can be sent to neighboring processors connected on the ring network when AND-parallelism (resp. OR-parallelism) is considered. Load balancing is, therefore, carried out automatically with the shuffle and selection of subgoals.

Yet another problem is on the communication of pruning information from one processor to another processor. When the result on a common subgoal is found, it must be communicated to other processors to stop the processing of a subset of the eliminated solution trees and allow the heuristic values of related tasks to be updated. Similarly, when either a solution tree is found or the goal is proved to be FALSE, all processors should stop further processing. These pruning information are more complicated than the incumbent in the parallel branch-and-bound algorithm implemented in MANIP. In this case, a bus is necessary to broadcast to all processors a subgoal or clause found to be TRUE or FALSE. To minimize the information broadcast, the subgoal or clause should correspond to the nonterminal node closest to the root in the AND/OR tree. Other processors receiving this information must update all tasks in its Active List by reordering the tasks according to the new heuristic values computed and by eliminating tasks that cannot lead to a solution tree. The complete path from the root to each active task must be maintained to allow the pruning information to be incorporated. The overhead for the propagation of pruning information is extensive and may not be beneficial because the probability-to-cost ratios may be fallible. The no-wait policy is again applied here to continue the evaluation of tasks according to previously computed heuristic values. Information received on subgoals will be used to eliminate unnecessary work when the task is selected.

The last problem on the implementation of a heuristic search lies in the management of the large memory space required. In our study of MANIP, it was found that a direct implementation involving ordered list of pointers to the subproblems results in poor locality of access, because the subproblems are not ordered by lower bounds in the secondary memory. A specially designed virtual memory that tailors its control strategies to the access behavior of the algorithm was found to be inflexible. The inadequacies of these approaches are due, again, to the strict adherence to the selection rule. A better solution is to use the no-wait policy to implement a modified heuristic search in each processor. In the modified heuristic search, the range of possible lover bounds (in this case, the range of probability-to-cost ratios) is partitioned into *b* disjoint regions. The subproblems in each region are maintained as a separate list. The top portion of each list resides in the main memory, and the rest resides in the secondary memory. Due to the high overhead of secondary-storage accesses, subproblems in a list are expanded in a depth-first manner. Only subproblems in the main memory are candidates for selection. The modified algorithm is identical to a depth-first search when one list is used, and is identical to a pure heuristic search when infinity lists are used. In general, as the number of lists increases, the number of subproblems expanded decreases and the overhead of the secondary-memory accesses increases. The number of lists should be chosen to maximize the overlap between computations and secondary-memory accesses. Experience on branch-and-bound algorithms showed that two to three lists are adequate.

## 5 Conclusions

In this paper, we have extended the architecture of MANIP for the parallel processing of logic programs. A logic program is assumed to be represented in the form of an AND/OR tree. The results that we have obtained can be summarized as follows.

(1) OR-parallelism and AND-parallelism have been unified into OR-parallelism. The objective of an OR-tree is to select a solution tree and to prove that the goal is TRUE; whereas in an AND-tree search, the objective is to find a set of subgoals to prove that the goal is FALSE. Both types of searches require only one of the correct descendents to be selected and can be considered as OR-tree searches. When the goal is likely to be TRUE, OR-parallelism should be used. In contract, when the goal is likely to be FALSE, AND-parallelism should be used.

(2) Heuristic information using success probabilities and average overheads of evaluation have been defined to guide the search of logic programs. These information can be generated statically; however, they represent a prior information that do not take into account the query used, the dynamic pruning information obtained, and the values of variables being bound. They are useful to roughly differentiate between tasks that are likely to lead to solution trees and those that might not. Moreover, they define whether the goal is likely to be TRUE or FALSE.

(3) An optimal heuristic search strategy that minimizes the expected overhead of obtaining one solution tree is derived. The search is guided by the probability-to-cost ratios of subgoals.

(4) The architecture of MANIP, proposed earlier for the parallel evaluation of branch-and-bound algorithms, has been extended to implement a heuristic search of logic programs. Problems on selection and virtual-memory support have been considered. The strict adherence to the heuristic search is found to be unrewarding because the probability-to-cost ratios may be fallible, and the overheads of selecting tasks according to these ratios are high. These overheads include the update of the ratios when new pruning information is received, and the selection of tasks from other processors or the secondary memory according to these ratios.

## References:


[1] KOWALSKI R. Logic for problem solving[M]. Amsterdam: North-Holland Publishing Company, 1979.

[2] CONERY J, KIBLER D. AND parallelism in logic programming[C]// The 8th International Joint Conference on Artificial Intelligence. Los Altos: William Kaufman, 1983: 539-543.

[3] MARTELLI A, MONTANARI U. Additive AND/OR graphs[C]// The 3rd International Joint Conference on Artificial Intelligence. Los Altos: William Kaufman, 1937: 1-11.

[4] NILSSON N J. Principles of artificial intelligence[M]. Palo Alto: Tioga Publishing Company, 1980.

[5] CLARK K, GREGORY S. PARLOG: a parallel logic programming language[R]. London: Imperial College, 1993.

[6] DEGROOT D. Restricted AND-parallelism[C]// 1984 International Conference on Fifth Generation Computer Systems. Amsterdam: North-Holland Publishing Company, 1984: 471-478.

[7] MOTOOKA T, TANAKA H, AIDA H, et al. The archi-

tecture of a parallel inference engine-PIE[C]// 1984 International Conference on Fifth Generation Computer Systems. Amsterdam: North-Holland Publishing Company, 1984: 479-488.

[8] YASUHARA H, NITADORI K. ORBIT: a parallel computing model of Prolog[J]. New Generation Computing, 1984(2): 277-288.

[9] CIEPIETEWSKI A, HARIDI S. Execution of bagof on the OR-parallel token machine[C]// 1984 International Conference on Fifth Generation Computer Systems. Amsterdam: North-Holland Publishing Company, 1984: 551-560.

[10] PEARL J. Heuristics[M]. Boston: Addison-Wesley, 1984.

[11] WAH B W, LI G J, YU C F. Multiprocessing of combinatorial search problems[J]. IEEE Computer, 1985: 93-108.

[12] FURUKAWA K, NITTA K, MATSUMOTO Y. Prolog interpreter based on concurrent programming[C]// The 1st International Logic Programming Conference. [S.l.:s.n.], 1982.

[13] KASIF S, KOBLI M, MINKER J. PRISM: a parallel inference system for problem solving[C]// The 8th International Joint Conference on Artificial Intelligence. Los Altos: William Kaufman, 1983: 544-546.

[14] HASEGAWA R, AMAMIYA M. Parallel execution of logic programs based on dataflow concept[C]// 1984 International Conference on Fifth Generation Computer Systems. Amsterdam: North-Holland Publishing Company, 1984: 507-516.

[15] STOLFO S J, MIRANKER D P. DADO: a parallel processor for expert systems[M]. Washington: IEEE Computer Society Press, 1984: 74-82.

[16] WAH B W, MA E Y W. MANIP-a multicomputer architecture for solving combinatorial extremum search problems[J]. IEEE Transactions on Computers, 1984, C-33(5): 377-390.

[17] PATERSON M, WEGMAN M. Linear unification[J]. JCSS, 1978(16): 158-167.

[18] MARTELLI A, MONTANARI U. An efficient unification algorithm[J]. ACM Transactions on Programming Languages and Systems, 1982(4): 258-282.

[19] SIMON H A, KADANE J. Optimal problem-solving search: all-or-none solutions[J]. Artificial Intelligence, 1975(6): 235-247.

[20] BARNETT J. Optimal searching from AND nodes[C]// The 8th International Joint Conference on Artificial Intelligence. Los Altos: William Kaufman, 1983: 780-788.

[21] GAREY M. Optimal task sequencing with precedence constrains[J]. Discrete Mathematics, 1973(4): 37-56.

[22] WAH B W, LI G J, YU C F. The status of MANIP-a multicomputer architecture for solving combinatorial extremum-search problems[C]// The 11th Annual International Symposium on Computer Architecture. New York: Association for Computing Machinery, 1984: 56-63.

[23] SBAPIRO E. A subset of concurrent Prolog and its Interpreter[R]. 1983.

发表于*Proceedings of the IEEE*，1989年第4期

# Computers for symbolic processing

Benjamin W. WAH, Matthew B. LOWRIE, Guo-jie LI

**Abstract:** In this paper, we provide a detailed survey on the motivations, design, applications, current status, and limitations of computers designed for symbolic processing. Symbolic processing applications are computations that are performed at the word, relation, or meaning levels. A major difference between symbolic and conventional numeric applications is that the knowledge used in symbolic applications may be fuzzy, uncertain, indeterminate, and ill represented. As a result, the collection, representation, and management of knowledge is more difficult in symbolic applications than in conventional numeric applications. We survey various techniques for knowledge representation and processing, from both the designers' and users' points of view. The design and choice of a suitable language for symbolic processing and the mapping of applications into a software architecture are then presented. We examine the design process of refining the application requirements into hardware and software architectures and discuss state-of-the-art sequential and parallel computers designed for symbolic processing.

## 1 Introduction

The development of the programming language IPL in the 1950s by Newell, Shaw, and Simon was a pioneering effort on symbolic processing by computers[1]. Data structures of unpredictable shape and size could be manipulated conveniently by programs written in IPL. Many of the early symbolic programs, including the Logic Theorist and the General Problem Solver, were written in IPL. The invention of Lisp in 1958 by John McCarthy further enhanced some of the programming tasks for symbolic processing. The language featured the use of conditional expressions recursively, representation of symbolic information externally by lists and internally by linked lists, and representation of program and data using the same data structures[2].

Recent advancements in applications of computers suggest that the processing of symbols rather than numbers will be the basis for the next generation of computers. This is highlighted by the numerous research efforts in Japan, Europe, and the United States[3-5]. Symbolic processing has been applied in a wide spectrum of areas; among them are pattern recognition, natural language processing, speech understanding, theorem proving, robotics, computer vision, and expert systems. Researchers in artificial in-

This work was supported in part by National Aeronautics and Space Administration Grant NCC 2-481 and National Science Foundation Grant MIP 85-19649. The research of M.B. Lowrie was also supported by a Ph.D. scholarship from AT&T Bell Laboratories.

telligence, database, programming languages, cognitive science, psychology, and many others have addressed overlapping issues within the area of symbolic processing.

Conventional computers have been designed with tremendous numeric processing power as their focus, rather than symbolic processing power. The disparity between symbolic and numeric operations, therefore, calls for different architectures for symbolic processing and innovative research in computers for symbolic processing. A review of the state of the art in computers for symbolic processing is presented in this paper. The discussion proceeds in a top-down fashion. The relevant features and characteristics of symbolic processing are first presented. A perspective on the role of techniques and methodologies involved in the design process are discussed. Hardware and software architectures in different levels of design are classified. The general view of computers designed and/or used for a symbolic processing application is depicted in Figure 1. The section discussing each portion is indicated in the figure.

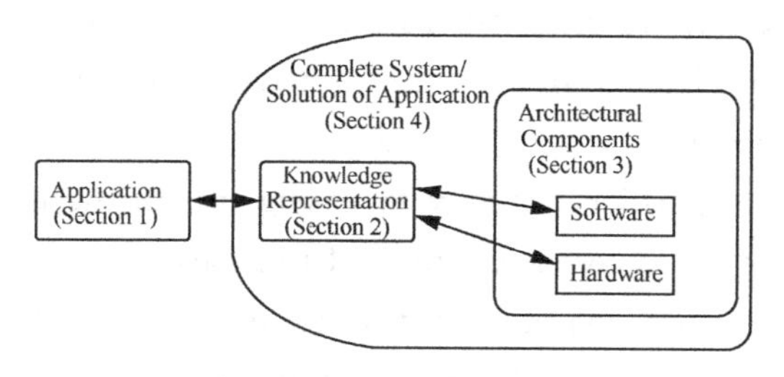


Figure 1. Overview of this paper

In Section 1, a classification of general computations is developed. From this classification, a definition of symbolic processing is derived in Section 1.1. Typical symbolic processing applications and their characteristics are discussed in Section 1.2.

Knowledge representation and knowledge processing are two important characteristics of solutions to a symbolic processing problem. Knowledge representation refers to the technique for representing data and information in a computer and is discussed in Section 2.1. Knowledge processing refers to the technique for controlling the manipulation of knowledge in the system and is the topic of Section 2.2.

The design of a computer relies on various concepts and strategies for implementing knowledge processing techniques. Section 3 emphasizes the architectural concepts behind the design of symbolic processing systems. Software architectures are covered in Section 3.1, and hardware architectures are studied in Section 3.2.

A complete system for symbolic computation is the result of the application of design philosophy, architectural components, and available technology. Complete systems are the topic of Section 4. The status of many existing and experimental systems are discussed and compared.

Future symbolic processing systems will evolve as new concepts and technologies develop. Section 5 outlines some recent research that is likely to impact the design of symbolic processing systems in the future.

## 1.1 Classification of computations

One of the fundamental debates on intelligent behavior has been related to the explanation of what symbols are. A number of scientists view human beings and computers as physical symbolic systems that produce through time an evolving collection of symbolic structures. In their 1975 Turing award lecture, Newell A and Simon H stated a general scientific hypothesis—the physical symbol system hypothesis[6]:

"A physical symbol system has the necessary and sufficient means for general intelligent action."

By "necessary", they mean that any system that exhibits general intelligence will prove upon analysis to be a physical symbol system. By "sufficient", they mean that any physical symbol system of sufficient size can be organized further to exhibit general intelligence. Research on Artificial Intelligence (AI) addresses the sufficiency of physical symbol system for producing intelligent action, while investigators in cognitive psychology attempt to demonstrate the necessity of having a physical symbol system wherever intelligence is exhibited. Although empirical in nature, the continuous accumulation of empirical evidence on the above hypothesis in the last 30 years has formed the basis of much research on AI and cognitive science.

Since our focus is on computers for symbolic processing, we will first classify computations performed on computers. The definition of symbolic processing used in this paper is derived from this classification. There are five classes of computations: analog, numeric, word, relational, and meaning. These classes are based on the primary unit of storage in the computation.

Analog: The analog class of computation encompasses those computations that have continuous variables as the parameters of the functions it performs. This is not the primary area of computation in a digital computer, as digital computers use digital memory. In the computations discussed in the context of digital computers, this layer of computation primarily entails the measurements of parameters from the environment.

Numeric: In this class of computation, the primary unit upon which functions are performed represents magnitude. Many applications of computers fall into this category; functions on memory elements containing integers, floating point numbers, ⋯ are numeric.

Word: In this class of computation, the parameters of functions are words that do not necessarily have quantitative value. Text processing is such an example.

Relational: In relational computations, functions operate on relations among words; that is, the primary unit of storage to be operated on are groups of words that have some relational interpretation.

Meaning: Very little research has been done on techniques for automated computation at the meaning level. The primary unit of evaluation is an interrelated series of relations that represent semantics and meaning.

A few examples to illustrate the various classes of computations are shown in Figure 2. In Figure 2(a), a standard supercomputer application, weather forecasting, is presented. The computation begins with analog measurements of the atmosphere (arc 1). These measurements are then converted to numeric entities (arc 2). The majority of computation occurs in the numeric stage (arc 3, representing the con version of numbers into different sets of numbers), with conversion to meaning done at the very end (arc 4, which may be done by humans instead of the computer). An example that uses the full spectrum of the classes of computations would be story comprehension from speech in put. The flow of data would appear as in Figure 2(b). Computation may also flow in the opposite direction. Robot control is an example of this, as depicted in Figure 2(c).

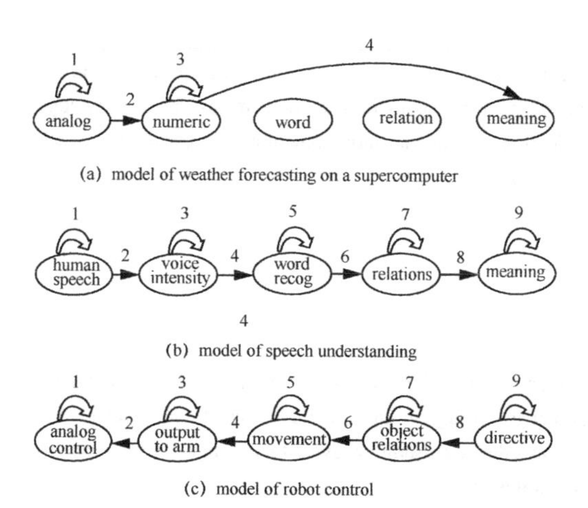


Figure 2. Examples to illustrate model of symbolic processing applications

The design of a computer system can be viewed as a problem in which performance with respect to the problem to be solved is to be maximized subject to cost constraints. Computations in more abstract classes are usually carried out by transformations into computations in more definite classes. For example, database queries function primarily at the word and relation level of computation. However, if we wish to know the average salary in a database of employees, numeric computations would be required. The design of this system may not, however, benefit from the inclusion of fast arithmetic units, because the performance gain may not be large enough to counteract the loss of other features that are necessitated by the cost constraint. The difference between this database system and a system that would include fast arithmetic units is in the emphasis of the computations.

A computer that is aimed at functioning at a more abstract level of computations should be able to perform computations that are found at a more definite level of computations. For instance, the database computer above should be able to perform fluid dynamics calculations. When this is done, however, the efficiency may be much poorer than a computer of comparable cost that is aimed at numeric calculations.

Using this classification, a concise description of what is meant by "symbolic processing" is possible. Symbolic processing is defined as those computations that are performed at the same or more abstract level than the word level. We call computers that are designed with specific orientation to carry out operations at or more abstract than the word level symbolic processing computers.

## 1.2 Characteristics of symbolic processing applications

In this section, symbolic processing applications and their overall features are presented. A few applications and their characteristics are presented in Table 1. The features that characterize general symbolic processing applications are listed below[7]. These features do not always apply to every instantiation of a symbolic processing problem and are intended as design guidelines that target general symbolic processing techniques.

Table 1. Some symbolic processing application

| Application | Characteristics |
| --- | --- |
| Problem solving<br>general<br>specific<br>• programming/compilation<br>• text processing<br>• human interface | User in puts problem, system attempts to solve; user encodes solution of problem; meta-knowledge for specific problems is well understood; small-gain parallelism is predominant |
| Database management<br>• variety of applications<br>• often an integral part of larger systems | Organization of information for retrieval; efficient algorithms to consider all run-time possibilities are too complex; meta-knowledge is application dependent; large potential for parallelism, both small-grain and large-grain |
| Expert systems<br>diagnosis<br>• medical<br>• plant disease<br>• computer system errors<br>design assistance<br>• architecture<br>• computer architecture<br>• computer chips<br>personal systems<br>• business<br>• finance<br>• wine tasting<br>others | III-structured collection of facts, inferences … as knowledge-intensive program in specific domain[8]; knowledge and meta-knowledge are usually provided by designers; large potential for parallelism |
| Natural language processing<br>understanding<br>generation<br>translation | Translate natural language to machine representation; translate machine representation to natural language; translate between two forms of natural language[9-11] |
| Computer vision<br>signal processing<br>pattern recognition<br>image understanding | Primarily numeric at the signal-processing level; patterns viewed as sentences—symbolic in nature; higher-level reasoning at the image-understanding level[12-13] |
| Learning<br>• experimentation<br>• deduction<br>• knowledge acquisition | Ability to adapt to environment to improve system efficiency; fundamental to symbolic processing[14-15] |

• Incomplete knowledge: Many applications are nondeterministic; it is not possible to predict the flow of computation in advance. This is due to in complete knowledge and understanding of the application. This lack of complete knowledge may also lead to dynamic execution, which refers to the possibility of new data structures and functions being created during the actual solution of the problem. In addition, data structures used in solution of the problem may be arbitrarily large, thereby necessitating dynamic allocation of memory, tasks, and other resources. System design should meet the need to cope with dynamic and non deterministic execution. An architecture that can adapt to more efficiently perform computations not anticipated at design time is referred to as an open system.

- Knowledge processing: A computation can be viewed as manipulations on a set of data. In Section 1.1, computations are classified on the basis of the nature of the data operated on. The nature of the operations performed on the data depends on the application and the storage technique for the data. Figure 3 shows a possible view of a computation and illustrates that it is processing knowledge about the computation, whether that be algorithms, techniques for evaluation, or some other reasoning technique, that controls and dictates the manipulation of data. At a level above this, meta-knowledge includes the extent and origin of the domain knowledge of a particular object, the reliability of certain information, and the possibility that an event will occur. In other words, meta-knowledge is knowledge about knowledge[16-18]. Meta-level knowledge can be considered to exist in a single level or in a hierarchy[19]. In fact, there can be an arbitrary number of levels, each serving to direct the use of knowledge at the lower levels.

Meta-knowledge can be classified as deterministic or statistical according to correctness and performance considerations[20]. Deterministic meta-knowledge refers to the knowledge about precede nee relationships, which results from a better understanding of the problem and helps to reduce the resource and time complexity. Statistical meta-knowledge can be used to order object-level actions in advance for efficient operations.

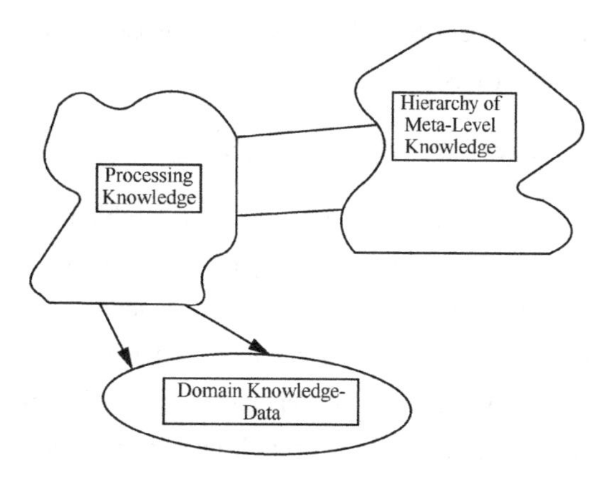


Figure 3. Knowledge processing

Rather than adding more heuristics to improve performance, more meta-knowledge about effective use of existing heuristics can be collected and developed. Meta-knowledge can also account for the formalization of belief, default reasoning, inference in changing situations, and others[16].

Symbolic primitives: A general symbolic application may contain primitive symbolic operations. Typical operations are comparison, sorting, selection, matching, and logical operations such as union, negation and intersection, transitive closure, pattern retrieval and recognition. These operations may be performed at more than one level of computation (such as word or relation). Higher levels of computation may also contain complicated "primitive" operations such as unification.

Parallel and distributed processing: Many symbolic applications exhibit a large potential for parallelism. Parallelism may be categorized into AND-parallelism and OR-parallelism. In AND-parallelism, a set of necessary and independent tasks are executed concurrently. OR-parallelism is a technique used to shorten the processing time in non-deterministic computations by evaluating alternatives at a decision point simultaneously.

## 2 Symbolic processing

A symbolic application is a problem in which the inputs and outputs are symbolic. Symbolic processing refers to the techniques employed by the system for finding the solutions of the application. The characteristics of symbolic applications have been discussed in the last section. The emphasis of this section is on the features of symbolic processing as they relate to the design of computers. In Section 2.1, techniques for representing knowledge are discussed. The issues involved in the control of knowledge processing are presented in Section 2.2.

### 2.1 Knowledge representation

In order to design an efficient computer for a given application, it is necessary to characterize the programs that will run on the computer. A primary decision in the solution of symbolic processing problems is the knowledge representation to be used[21-22]. The issues involved include the selection of the appropriate symbolic structures to represent knowledge, and the appropriate reasoning mechanisms to both answer questions and assimilate new information. There are four criteria to evaluate a knowledge representation scheme: flexibility, user-friendliness, expressiveness, and efficiency of processing. Flexibility, user-friendliness, and expressiveness are required to simplify the tasks of programming and comprehension. The efficiency or tractability of a knowledge-representation scheme dictates the efficiency of the solution to the application. Much of the research in this area represents a tradeoff between expressiveness and tractability.

Despite a great deal of effort devoted to research in knowledge representation, very little scientific theory is available to either guide the selection of an appropriate representation scheme for a given application or transform one representation into a more efficient one. Although a number of knowledge-representation schemes have been proposed, none is clearly superior to the others for all applications.

The following sections present two attributes for comparing knowledge representations: local versus distributed aspects, and declarative versus procedural. The classical knowledge-representation schemes are also evaluated on the basis of these features.

(1) Features of knowledge representations

• Local versus distributed representations: In a local representation, each conceptual datum is stored in a separate hardware unit. A word or item of data stored in a register is an example of local storage for that data item. As a result, the data are simple to read, update, and understand. Unfortunately, if any hardware unit fails, all knowledge contained in that unit is lost to the system. Most current systems, symbolic and numeric, utilize local representations for individual pieces of data.

In a distributed representation, a piece of knowledge is represented over many units, and each unit may be shared among multiple pieces of knowledge which correspond to features of multiple concepts or data items. The advantage of such a representation is that it is fault tolerant. If a small proportion of units fail, the integrity of the distributed data undergoes little change. This property is very attractive for practical implementations. Distributed representations also allow for a great deal of parallelism in computation[15]. However, they are usually harder for a user to understand and modify.

Table 2 summarizes the salient characteristics of local and distributed representations. It should be noted that there is not a concrete boundary between local and distributed features. Some features in a knowledge-representation scheme are local, while others are distributed. At one extreme is a standard implementation of a simple Lisp program, for instance, which can be thought of as a hierarchy of local representations. The program is stored as one unit of information. The data structures used by the program are also stored as a single entity. Finally, each piece of data within a data structure is stored in one memory location. On the other extreme is a standard implementation of a neural network. A predicate logic program, however, is neither fully local nor fully distributed. The complete program is not a single entity but a set of logic statements. Each statement, however, is an example of a local representation. This is considered further in Section 2.1-(2).

Table 2. Attributes of local and distributed representations

| Attribute | Local | Distributed |
| --- | --- | --- |
| Storage technique | Each data stored in dedicated hardware | Data represented over multiple units |
| Ease of understanding | Easy for humans to comprehend | Difficult for humans to interpret |
| Modification of stored data | Simple | More difficult |
| Fault tolerance | Loss of hardware results in loss of all stored data in this unit | Loss of small proportion of units does not seriously damage integrity of data |

- Declarative versus procedural representations: The issue of distributed versus local representations concerns the methodology for representing information in the computer. In contrast, the issue of procedural versus declarative representations distinguishes between techniques for representing the processing knowledge and processing methodologies employed by the computer programs.

A program written in a declarative representation consists of a set of domain-specific facts or statements and a technique for inferring knowledge from these statements. It is, therefore, characterized as a set of statements of knowledge about the problem domain. Examples of declarative representations include pure predicate logic and production systems (to be discussed in Section 2.1-(2)).

In a procedural representation, program statements consist of steps to be taken in the solution of the problem, which are statements of knowledge about how to solve the task. Examples of procedural program representations include the C language and Lisp.

Declarative representations are user-oriented and emphasize correctness and user-friendliness of programs. They are referentially transparent: the meaning of the whole can be derived solely from the meaning of the parts, independent of its historical behavior. This may increase programmer productivity[23] and result in tremendous potential for parallelism[24-25].

Unfortunately, programs written in declarative representations are often inefficient to evaluate due to nondeterminism, implicit control aspects, and inconsistent knowledge. It is hard to add domain-specific knowledge and meta-level knowledge to declarative programs. The difficulty with using declarative representations to solve symbolic problems lies in determining how to use the facts stored in the program's data structures, not in deciding how to store them.

Procedural programs are not as user friendly as declarative ones because the programmer must specify all control knowledge. In addition, the validity of a procedural statement often relies heavily on other procedural statements in the program, which complicates both the creation and modification of software. The loss in flexibility in a procedural programming environment is counteracted by the inherent gain in ease of representing control knowledge. Procedural schemes allow the specification and direct interaction of facts and heuristic information, thereby eliminating wasteful search. Meta-knowledge can also be easily included in procedures. Overall, procedural representations are as much concerned with the technique and efficiency of the computation as with the ease of representing; the domain knowledge.

The salient features of declarative and procedural representations are summarized in Table 3. As with distributed versus local features of a representation, practical knowledge- representation schemes may have both procedural and declarative features.

(2) Classical knowledge representation schemes: In this section, the classical knowledge-representation schemes are described and evaluated with respect to the features of local versus distributed, and declarative versus procedural qualities. Those that have received the greatest attention in elude predicate logic, production systems, semantic networks, frames, procedural languages, and fully distributed representations.

• Predicate logic: Predicate logic studies the relationship of implication between assumptions and conclusions. Logic often seems a natural way to express certain notions, and there are standard methods of determining the meaning of expression in logic formalism[26]. Logic is useful for exploring the epistemological problems that determine how the observed facts can be represented in the memory of a computer without being concerned with the use of the knowledge. The major disadvantage of logic stems from the separation of representation and processing.

• Production systems: Production systems use collections of rules to solve problems. These

Table 3. Attributes of declarative and procedural representations

| Attribute | Declarative | Procedural |
|---|---|---|
| Emphasis | Knowledge of domain | Knowledge of solution |
| Technique | Domain-specific statements | Solution techniques |
| Orientation | User friendliness; ease of understanding | Efficiency of solutions;<br>Ease of representing control knowledge |
| Parallelism | Natural, but countered by unnecessary search | Constrained and often user specified |
| Control | Transparent to the user | Specified by the user |

rules consist of condition and action parts, or antecedent and con sequent parts[27]. It has been found that production systems provide a useful mechanism for controlling the interaction between statements of declarative and procedural knowledge. For this reason, production systems have been used extensively in expert systems and knowledge engineering. Unfortunately, the expressive power of production systems is limited. Some researchers have argued that rule-based expert systems cannot achieve expert-level behavior[28]. Another problem with production systems is their inefficiency due to high control overhead.

• Semantic networks: A semantic network is a directed graph whose nodes represent objects, concepts, or situations, and whose arcs represent relationships between nodes[29]. The basic inference mechanism in semantic networks is "spreading activation". The idea here has a clear neural inspiration: certain concepts in memory become a source of activation, and activation spreads in parallel to related concepts. The significance of this graphical representation is in allowing certain kinds of inference to be performed by simple graph-search techniques. Yet simple semantic networks can only express a collection of variable-free assertions. Several authors have shown that semantic networks can be extended so that they have the same expressive power as predicate logic[30]. Frequently, semantic networks are used as data structures for manipulation using other knowledge- representation schemes (such as Lisp, which is a procedural representation).

• Frame representation: Frame representations employ a data structure for representing stereotypical situations[31]. The frame-description form is mainly an elaboration of the semantic-network one. Its emphasis is on the structure of types themselves (called frames) in terms of their attributes (called slots). A frame in eludes declarative and procedural information in predefined internal relations. Attached to each frame is various heuristic information, such as a procedure on how to use the frame. Although many issues about the possible implementations of frame-based systems are unresolved, the basic idea of frame-like structuring of knowledge appears promising and has appeared in various forms in many conventional languages.

• Procedural representations: In a procedural representation, a knowledge base is viewed as a collection of modules expressed in a procedural language, such as Lisp or C. The procedural scheme is capable of representing heuristic knowledge and performing extended logical inferences, such as plausible reasoning. Due to the elimination of wasteful search, this representation scheme can be carried out efficiently. However, it is often limited by the available constructs. Conventional Fortran or Pascal programs, for example, have been found to be inadequate in supporting efficient symbolic processing.

• Connectionist representations: A connectionist representation is a form of distributed representation: concepts are represented over a number of modules or units. When presented with input, units that have a positive correlation with an input feature activate, and those with negative correlation exhibit inhibitory signals. In this fashion, input can be recognized as a function of connection strengths among units (see also Section 3.2-(2))[15]. Distributed representations allow automated procedures for learning concepts and representations and have great potential for parallelism in computation. Their major drawback lies in the difficulty of interpreting the system state and the internal representations. Additionally, the programming of these computers often requires a lengthy training period.

A given representation may exhibit local or distributed and declarative or procedural aspects at different levels of the representation. Table 4 summarizes the characteristics of these representations and categorizes the representations by the hierarchy of knowledge representation inherent in the technique.

Table 4. Examples of knowledge-representation schemes

| Representation | Level of representation | Characterization |
|---|---|---|
| Logic | Variable | Local/declarative |
| | Statement/relation | Local/declarative |
| | Program | Distributed/declarative |
| Production system | Variable | Local/declarative |
| | Statement/relation | Local/either |
| | Program | Distributed/declarative |
| Semantic networks | Node | Local/declarative |
| | arc/relation | Local/declarative |
| | Network | Local/procedure |
| | Program | Distributed/declarative |
| Frames | Variable | Local/declarative |
| | Statement | Local/either |
| | Slots | Local/either |
| | Frame | Local/declarative |
| | Program | Distributed/declarative |
| Procedural | Variable | Local/either |
| | Statement | Local/procedural |
| | Program | Local/procedural |
| Connectionist | Connection strength | Local/distributed |
| | Propagation technique | Local/procedural |
| | Data and knowledge | Distributed/declarative |

## 2.2 Knowledge processing

Different reasoning methods are associated with different knowledge-representation schemes and require different architectural supports. Table 5 shows the classical knowledge- representation paradigms and their respective reasoning techniques.

Table 5. Reasoning techniques

| Representation | Typical reasoning technique |
|---|---|
| Logic | Resolution (unification) |
| Production rules | Forward/backward chaining |
| Semantic networks | Spreading activation |
| Frames | Procedural attachments |
| Procedural | Control flow |
| Connections | Propagation of excitation |

It is argued that humans use logic-like reasoning in the domain of rational knowledge and apply memory-based reasoning for perceptual actions. For over 30 years, logic-like deduction has been the dominant paradigm in AI research. This paradigm has been applied to a wide range of problems, especially expert systems. Although intelligent behavior often resembles logic-like reasoning with limited search, the intensive use of memory to recall specific episodes from the past (rather than rules) could be another foundation of machine reasoning[32]. Memory-based reasoning (or case-based reasoning) does not use rules, but attempts to solve the problem by direct reference to memory. The Connection Machine is an example of a machine designed for memory-based reasoning although it can also be programmed to perform logic-like reasoning[33].

In the use of a knowledge representation, the knowledge processing technique must be tailored to cope with the application requirements. The greatest need, in symbolic processing applications, is the ability to deal with uncertain, incomplete, or conflicting information. Techniques for dealing with this problem are discussed in Section 2.2-(1). In Section 2.2-(2), methods for exploiting parallelism are discussed.

(1) Uncertain, incomplete, and inconsistent knowledge processing: The techniques for dealing with these problems in knowledge processing are detailed in Figure 4. The rest of the section is devoted to a brief discussion of the entries in the figure.

• Uncertain knowledge: Conventional knowledge-representation techniques based on predicate calculus and related methods are not well suited for representing common-sense knowledge. Explicit and implicit quantifiers are fuzzy, and the standard inference methods mentioned earlier make no provision for dealing with uncertainty. Two types of uncertainty have been studied. One comes from noisy data and the fuzzy meaning of symbols; the other is associated with uncertain inference rules.

Methods and theories of capturing uncertainty have been examined in recent years. Probability and Bayesian statistics is the fundamental basis of most approaches to this problem. This takes many forms reflecting different issues. Approaches and techniques include fuzzy logic[34-35], confidence factors[36], Dempster and Shafer's theory of plausible inference[37], odds[38], and endorsements[39].

Dealing with uncertain knowledge is most frequently handled by two principal components. The first is a translation system for representing the meaning of propositions and other semantic

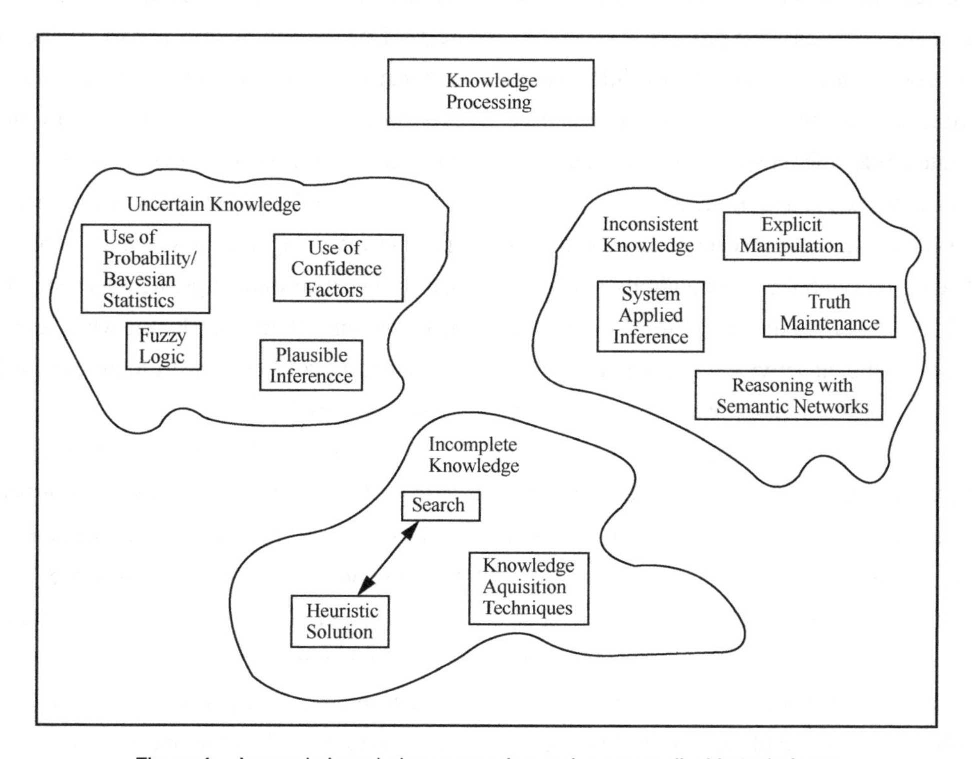


Figure 4. Issues in knowledge processing and some applicable techniques

entities. The second is an inferential system for arriving at an answer to a question that relates to the information resident in a knowledge base. Application of Bayesian statistics to expert systems follows this approach[38]. A Confidence Factor (CF), or certainty factor, is used to decide among alternatives during a consultation session. A CF of a rule is a measurement of the association between premises and actions. A positive CF indicates that the evidence confirms hypothesis, while a negative CF negates the hypothesis.

Dempster and Shafer's theory of plausible inference provides a natural and powerful methodology for representing and combining evidence. Ignorance and uncertainty are directly represented in belief functions and remain throughout the combination process.

Endorsements are records of information that affect a hypothesis' certainty. They can be propagated over inferences, but in a manner that is sensitive to the context of the inference.

• Incomplete and inaccurate knowledge: A key feature of symbolic computations is nondeterminism, which results from the fact that almost any intelligent activity is likely to be poorly understood. This implies that no systematic, direct algorithms for solving the problems are available. When a problem becomes well understood and can be solved by a deterministic algorithm, the solution of the problem is no longer considered as "intelligent"[40].

The starting point of conventional computations is a deterministic algorithm. Since most symbolic processing applications are knowledge intensive, such a deterministic algorithm may not exist. Efficient solution of the problem, therefore, requires continual refinement of the computation technique and may employ various knowledge-acquisition techniques. When knowledge of the application domain is incomplete or uncertain, heuristic solutions are utilized[41-45]. A heuristic is knowledge capable of suggesting plausible actions to follow, or implausible ones to avoid. It is desirable to use concise and accurate domain-specific knowledge and meta-knowledge. Unfortunately, this information is difficult to acquire in practice and, if available, may be fallible or tremendously large in size.

The nondeterministic nature of computations and fallibility of heuristic guiding may lead to anomalies of parallelism. As a result, when multiple processors are used, one or more of the processors may be guided by the heuristic-guidance function into a part of the search tree that is not explored in the same order as that in sequential processing. This out-of-order exploration of the search tree, coupled with the pruning of undesirable nodes, may result in a speedup (as compared to sequential processing) that is less than one or greater than the number of processors. Some results on when anomalies occur and how to cope with them can be found in the literature[46-49].

In addition to heuristics, several new forms of logic for belief and knowledge have been introduced. Traditional reasoning methods suffer from the problem of logic omniscience[50]. Logic omniscience refers to the assumption that agents are sufficiently intelligent that they know all valid formulas. Thus, if an agent knows $p$, and that $p$ implies $q$, then the agent must know $q$. In real life, people are certainly not omniscient. The newly introduced forms of logic are more suitable than traditional logic for modeling beliefs of humans (or computers) with limited reasoning capabilities[51-52]

• Inconsistent knowledge processing: Traditional logic is monotonic. Monotonicity implies that new axioms may be added to the list of provable theorems only when they are consistent. Nonmonotonic reasoning provides a more flexible and complete logic system, as well as a closer model to human-thought processes. The motivations for nonmonotonic reasoning can be classified into two general areas: default reasoning, and reasoning in a changing environment[53].

Default reasoning can be broken into two distinct areas: exceptions to the rule, and autoepistemic logic. As the name implies, exception to the rule allows relations that contradict more general relations. A statement with most as a relation will not add any information to a monotonic logic system. Nonmonotonic systems allow a representation that in eludes exceptions to the general rule, without eliminating the validity of the computing environment.

Autoepistemic logic (a.k.a. circumscription and closed-world assumption) allows conclusions to be reached about relations for which no facts exist in the database. This involves the assumption that all relevant knowledge is in the database (closed-world assumption).

In a monotonic system, no modification of existing knowledge and data can be made without restarting all inference processes and results. In a world where new discoveries and revisions of previous beliefs is the norm, this is a poor model for a large knowledge/data-based system. Accommodating a changing environment is particularly important when default reasoning is used. A statement inferred by default may be corrected in light of additional evidence.

The distinguishing feature of different techniques for dealing with inconsistent knowledge is the method for handling correction of the knowledge base. These methods include explicit encoding, system applied inference, semantic networks, and truth maintenance. In explicit encoding, the programmer is responsible for writing code that will update the database when a new statement is added that may conflict with other statements[54]. In a system employing system applied inference, the system has user-encoded functions that automatically search for inconsistencies in the knowledge base[55]. McCarthy and Hayes have indicated how actions might be described using modal operators like "normally" and "consistent"[56]. Sandewall used a deductive representation of nonmonotonic rules based on a primitive called UNLESS[57]. Reasoning with semantic networks is another technique for ordering inferences and default reasoning, although they have been criticized for lacking a clear inference technique and for not being a sufficiently formal logic system. In Doyle's Truth Maintenance System (TMS), the reasons for program beliefs are recorded and maintained. These beliefs can be revised when discoveries contradict assumptions[58]. Improvements have been explored by de Kleer in his Assumption-based Truth Maintenance Systems (ATMS)[59-61]. In the IBM YES/MVS expert system, inconsistent deductions are automatically removed and new consequences are then computed in accordance with the changed facts[62].

(2) Parallel Knowledge Processing: Humans are often thought of as the most efficient symbolic processing engines. Some researchers claim that symbolic problems can, therefore, be most effectively solved using techniques similar to those employed in the human brain. Observations of human intelligence suggest that human knowledge can be divided into perceptual and rational knowledge, each of which may involve different degrees of parallelism. In the perceptual stage of cognition, such as vision and speech understanding, massive parallel processing is possible due to the large number of independent data and simple control. Only limited parallelism can be exploited for rational knowledge. In other words, the degree of parallelism that could benefit the high-level reasoning is relatively small. Therefore, this kind of task should be solved by trying to accumulate heuristics, rather than trying to exploit parallelism.

Unfortunately, early experiences with symbolic multiprocessor architectures, such as Hearsay-II[63], Eurisko[64], and multiprocessor implementation of forward changing rule-based expert systems[65], have shown that parallel symbolic programs exhibit small speedups[66]. This has led to the possibly incorrect conclusion that symbolic programs written for sequential execution have low potential for parallelism.

The considerations of parallel knowledge processing are distinguished by four features: deterministic and nondeterministic parallelism, granularity of parallelism, data-and control-level parallelism, and user-and system-defined parallelism. These features are summarized in Table 6. Designs of parallel symbolic processors are presented in detail in Sections 3 and 4.

Table 6. Issues on parallel processing

| Issue | Definition | Comments |
|---|---|---|
| Deterministic and Nondeterministic | Concurrent execution of multiple units of computation, all of which are necessary for job completion.<br>Multiple potential solutions evaluated in parallel; parallelism used to replace or augment backtracking | Low overhead guarantees speedup; tasks must be independent; pure functional programming is deterministic.<br>Easy to implement—always independent; nondeterministic nature may lead to anomalies in parallelism |
| Granularity | Size of units of computation to be executed by a single functional unit | Difficult to determine; a function of knowledge representation, problem complexity, the shape of the search graph, distribution of processing times, and the dynamic nature of the problem[67] |
| Data level and Control level | Data stored one element per processor, program executed in SIMD fashion.<br>Independent control for parallel tasks | Can be used for large database operations, sort, set operations, statistical analysis, ... [68]. Can be implemented in memory—referred to as active memory[69].<br>Major type of parallelism used; MIMD systems; detection of parallelism can be more difficult than in numeric programs; nondeterministic nature requires dynamic mapping |
| User defined and System defined | Portions of program specified by users which can execute in parallel.<br>Parallelism detected and exploited automatically by the compiler or run-time software and hardware | In numeric processing, DOACROSS is a typical example. The FUTURE construct in Multilisp is a symbolic construct[70].<br>Fully distributed representations allow massive system-level parallelism; some systems may employ both—the user only aids in indicating available parallelism |

# 3 Architecture concepts for symbolic processing

With the symbolic processing application characterized and the representation technique for the solution of problems in that application selected, it is possible to choose the appropriate features and attributes for a computer system to solve problems in that application. An architectural component of a processing system is defined as a hardware or software structure that supports the solution of the application. In this section, current software and hardware architectures which are useful in symbolic processing systems are discussed. This section focuses on the specific architectures for symbolic processing and the way that they relate to fundamental design concepts.

Software architectures are comprised primarily of software languages and environments for encoding solutions to the application. Selection of a software environment imposes certain features that the software and hardware must support. The design process considers software and hardware implementations of the required features and selection of the best alternative, and is driven by a tradeoff between cost and expected performance improvement.

## 3.1 Software architectures

The area of software architectures for symbolic processing encompasses two important facets. The first is the design of appropriate software facilities, tools, and languages for the symbolic application, while the second facet concerns the tools used for mapping a symbolic application into software.

The discussion of software architectures is organized in the following manner. First, the process of designing software facilities and languages is analyzed. Following this is an overview of the most frequently researched and utilized programming paradigms for symbolic processing. Finally, the problem of mapping applications into software is overviewed.

• The design of software languages: The objective behind a software language is to provide software support and implementation of the knowledge representation(s) employed. As with the choice of the knowledge-representation scheme, the major goals are ease and ability to represent the solution of the application and the promotion of efficient execution of the algorithm. Once a technique, or techniques, for representing knowledge is selected, the major features of the language become apparent. For example, the choice of a logic representation dominates the characteristics of the Prolog language.

The software technique for implementing these features may not be as clear, however. For instance, Lisp was a procedural language developed for symbolic processing. In its design, functional programming with recursion and list-structured data were selected. These are not obvious choices, but they arise out of the use of a procedural representation. The selection of these features is also made on the basis of ease of representation and efficiency of processing. As another example, additional "impure" features may be added to the implementation of a knowledge-representation scheme for efficiently supporting computation and/or providing flexibility to the user. CUTs and side effects in Prolog are well-known examples. This aspect can be seen as a procedural addition to a logic representation, which is done to support the efficient implementation of logic programming.

The extension of conventional von Neumann computer languages for symbolic processing is an issue that has been explored extensively. By their nature, conventional computer languages are based on procedural representations. By examining the characteristics of symbolic processing, to be discussed in Section 3.1-(1), the desirable features to incorporate into a conventional programming language become apparent. Such features in elude data structures, symbolic primitives, recursion, and others. Conventional languages which were designed for numeric processing, such as Fortran, have not proven to provide adequate support for symbolic processing. In particular, the languages are not sufficiently flexible to enable simple encoding of very complex symbolic operations. For this reason, the new and less conventional languages for symbolic processing are emphasized in this section.

As was observed in the preceding section, the emphasis in the design of new representations for symbolic processing problems has been focused in the area of adding declarative and distributed features to existing representation schemes. Part of the motivation for this emphasis is referential transparency, or freedom from side effects. This relieves some of the programming burden of the users, allowing easier programming of complex applications. In the following sections, three paradigms for the design of software languages which promote referential transparency are discussed: functional languages[71-73], rule-based languages[30,74-75], and object-oriented languages[76-79].

(1) Functional Programming Languages: The functional programming approach does not employ states, program counters, or other sequence-related computational constructs. A program is a function in the mathematical sense. The program, or function, is applied to the input, and the function is evaluated to the desired output. A functional approach can be thought of as a language based on Lambda Calculus; operators are applied to data or results of further function evaluations. John McCarthy's conception of list processing is viewed as a pioneering effort in this area[80]. Examples of functional language in elude pure Lisp[2], Backus' FP[71], Hope[81], Val[82], and Id[83].

In a functional language, the meaning of an expression is independent of the history of any computation performed prior to the evaluation of the expression (referential transparency). Precedence restrictions occur only as a result of function application. Notions such as side effects and shared memory do not exist in functional programs. The lack of side effects results in the determinacy property that is so valuable in parallel processing. Regardless of the order of computations of the arguments of a function, the same result (assuming termination) is guaranteed. Hence, all arguments and distinct elements in dynamically created structures in a functional program can be evaluated concurrently. For example, consider a simple program for computing the average of numbers in a list $s$:

average($s$) = div (sum ($s$), count($s$))

If we attempt to evaluate average (1.(2.(3.nil))), the computation of sum (1.(2.(3.nil))) can clearly proceed independently of the computation of count (1.(2.(3.nil))). The key point is that parallelism in fundamental languages is implicit and supported by their underlying semantics. There is no need for special message-passing constructs, synchronization primitives, or constructs for specifying parallelism. It has been reported that implementation of functional languages on a parallel computer seems easier than on a sequential computer[84].

Programming in functional languages facilitates specification or prototyping, prior to development of efficient programs. With a satisfactory specification, it is possible to develop an efficient program through program transformation. The idea is that the program specification should be systematically refined to produce the program. Because functional languages are referentially transparent, they can be refined as familiar mathematical forms. Another advantage of functional programming is that it can represent high-order functions; a function can be passed as an argument. A comparison of functional programming with von Neumann programming is presented in Table 7.

Pure Lisp is a functional language. Many dialects of Lisp, however, are not purely functional. Operations on global variables, property lists, input/output, and other features incorporated in these dialects create side effects. This is seen as necessary to support efficient computation by avoiding recomputation of functions whose results are required in more than one place and to support convenient input/output. Unfortunately, the property of referential transparency is lost in most practical Lisp languages. Moreover, precedence restrictions are represented not only by functional calls, but also in procedures.

With the presence of side effects, it is not straightforward to identify the parallel tasks as in a pure functional language. Users are required to identify independent tasks with special primitives. Several parallel Lisp languages have been proposed and implemented. Multilisp, developed by Halstead R at MIT and implemented on a 128-processor Butterfly parallel processor, includes the usual Lisp side-effect primitives for altering data structures and changing the values of variables[85]. Concurrency is introduced by means of the pcall and future constructs[86]. Both utilize an implicit fork-join. For example, (pcall ABC) will result in the concurrent evaluation of expressions A, B, and C; while (future *X*) immediately returns a pseudo location for the value of *X* and creates a task to concurrently evaluate *X*. The use of future allows concurrency between the computation of a value and the use of that

Table 7. Functional versus von Neumann programming

| Functional Programming | von Neumann Programming |
|---|---|
| Programs are composed only of other programs | Programs contain programs, expressions, and variables |
| Programs can be freely built from others | Programs are composed only with common data storage |
| Same program can treat objects of different structure and size | Changing size or structure of data means changing the program |
| There is a strong theoretical background about programs. Programs may be proven as in mathematics | Few general practical theorems exist about programs. Proving correctness of a program is extremely difficult |

value. The primitive future was introduced because the use of pcall alone did not provide a great deal of parallelism[87].

Proponents of functional languages believe that their simplicity and elegance will conduce to more orderly, more rigorous, more verifiable, and ultimately more efficient programming. Opponents worry about losing expressiveness as a result of the expression-evaluation-only model. The crucial disadvantage of functional programming is that it is difficult to represent the inherent nondeterminism in AI problems. The recursive formulation and leftmost outermost reduction of functional programs enable depth-first search naturally, but it is difficult to write a heuristic search program by a pure functional language, since heuristic search is inherently history-sensitive. In fact, heuristic search programs written in Lisp include many "setq" and "do" statements that are not pure functional primitives[88]. Due to the inability of representing nondeterminism and the inefficiency of dealing with large data structures, pure functional languages are often less suitable for general symbolic applications. Their usefulness for deterministic symbolic applications is, however, significant.

(2) Rule-Based Languages: There are two major forms of rule-based languages available: logic and production systems. The languages associated with these representations are referred to as rule-based since both emphasize the relation between a condition and an inference or rule.

• Logic: In its modest form, a logic program is the procedural interpretation of Horn clauses or predicate logic[26,30]. Some ideas of logic programming, like automatic backtracking, have been used in the early AI languages QA3, PLANNER, and MICRO-PLANNER[89-90]. The more con temporary language, Prolog, is based on logic programmer[91-92]. Logic programming is a reasoning-oriented or deductive programming environment. Logic programming has received considerable attention because of its choice as the core computer language for the Fifth Generation Computer System Project in Japan[93].

The motivation of logic programming is to separate knowledge from control. However, logic programming implementations often include extralogical primitives to improve their run-time efficiency and flexibility in specification. For example, in Prolog, the CUT predicate is an extralogical control mechanism to define a similar construct as the if-then construct in conventional languages. In addition, variables in a logic program are often nondirectional, meaning that a variable does not have to be defined as an input or output variable at compile time, and its mode can be changed at run time depending on the context. As a result, dependencies among subgoals are not defined at compile time, and static detection of parallelism is very difficult. The solution is to require the users to specify the parallel processible tasks. In Parlog[94], every argument has a mode declaration that states whether the argument is input (?) or output (^). In Concurrent Prolog[95], a "read-only" annotation (?) is used. Users can also distinguish between "parallel AND" and "sequential AND" by using "," and "&", respectively.

Constructs can also be introduced to restrict parallelism until certain preconditions are satisfied. An example is a guard clause that has been adopted in Parlog. A guarded clause has the format: $h$: $-$ $g$ | $b$, where $g$ is the guard of the clause and $b$ is its body. Subgoals in the body can only be evaluated when all subgoals in the guard have succeeded and values bound have been committed to the body.

User specification of parallelism certainly detracts from the objective of declarative programming. This is a problem even in the Restricted AND-Parallelism (RAP) model[96]. Although the user does not have to explicitly specify parallelism, the user must be aware of the underlying computational model. Both mode declarations in Parlog and read-only annotations in Concurrent Prolog impose a fixed execution order on subgoals, which may be inefficient. Choosing the proper subgoals in the guard is sometimes difficult and is not guided by any general principle. The distinction between "sequential AND" and "parallel AND", which is a linear order, is not sufficient to specify all precedence relationships, which form a partial order. Owing to the non deterministic nature of AI applications, users can not identify all parallel processible tasks perfectly. A better symbolic processing language should utilize both compile and run-time detection of parallelism.

• Production system: The other major form of rule-based language which promotes the separation of knowledge from control is based on production-system representation[97-98]. A production-system program consists of a set of data and a set of rules that can act on the data. A rule is composed of a Left-Hand Side (LHS) and a Right-Hand Side (RHS). The LHS is the antecedent or situation and represents the conditions necessary for applying the rule. The conditions are in the form of a Boolean combination of clauses[97]. The RHS is called the consequent and indicates a set of changes to the data memory to be performed when the conditions of its LHS are met. Thus, a production system can be viewed as a combination of matching the logic condition and modifying the data in a procedural fashion. Strategies are required for matching data conditions with LHS of rules and for resolving conflicts when more than one rule has a match. The conflict set is the set of antecedents and their bindings that match elements in the working memory. Production systems operate in a recognize-act cycle. The recognize cycle computes the conflict set, while the act cycle selects one matching production and acts on it.

One of the popular programming environments for implementation of a production-system representation is the OPS5 system[97]. The OPS5 system highlights the issues involved in designing production systems. OPS5 employs data typing. The working memory (data) is viewed as a separate entity from the production memory, where the rules are stored. OPS5 employs a Rete match algorithm which computes the conflict set but does not select one production to act on. The algorithm works by storing the matching condition in the form of a tree. After the recognize-act cycle, rather than recomputing the entire conflict set, the tree is updated via tokens, which reflect the addition or deletion of elements from the working memory[97,99]. OPS5 has two conflict-resolution strategies: LEX and MEA, LEX orders the conflict set on the basis of the recency of the time tags corresponding to the working memory elements that match the condition elements of the production rule. In contrast, MEA orders the conflict set using the recency of the working memory element that matches the first condition of the matching conditions (even if it is not a maximum).

Production systems provide a natural programming paradigm for "if-then" programming environments, such as those employed in expert systems. Unfortunately, algorithms with iterations and recursions are difficult to encode. In addition, rules are independent, and structural organization of programs requires special attention by users. Thus, it is difficult to develop large programs using production systems[100].

• On functional versus rule-based languages: The advantages and disadvantages of functional languages stem from the procedural and formal mathematical nature of Lambda Calculus. In contrast, rule-based languages have complementary advantages and disadvantages which stem from their declarative nature.

The properties of non directionality of inputs and outputs, dynamic bin ding of variables, and nondeterminism make logic languages more expressive. On the other hand, with the aid of high-order functions, which permits quantification over individual data items as well as predicates and functions, functional programming enables more concise programs. Programs with high-order functions are easier to understand and reason about.

Although logic languages are more expensive, their implementation in a parallel environment is more difficult due to the nondirectionality of variables. This flexibility complicates the detection of parallelism at compile time and results in the dynamic behavior of execution of logic programs. Current Prolog systems also lack a means to describe the termination of computations on conceptually in finite data structures and the concept of "evaluation" of function invocations, which makes the logic base non transparent. In contrast, the run-time behavior of pure functional programs is much simpler to control than that of first-order logic programs, particularly in a parallel context. Techniques, such as graph reduction and data flow, have been studied for parallel evaluation of pure functional languages.

Obviously, it would be advantageous if the simple control mechanism of functional languages could be applied to support languages with the great expressive power of logic languages. Considerable efforts have been devoted to combine functional and logic programming[75], Some researchers are trying to simplify logic languages by introducing directionality of information in logic programs[101]. This approach will degrade the expressive power of logic programs to that of first-order functional programs.

The alternative approach is to extend functional languages so that they have the expressive power of logic languages while retaining the underlying functional simplicity. The addition of unification to the Hope language is one example[102]. Subrahmanyam et al. have proposed FUNLOG, a language that integrates functional and logic programming. FUNLOG provides the programmer with the flexibility of choosing between a backtrack-free computational framework and a logic computational framework. Semantic unification has been introduced to serve as a basis for achieving the integration of function and logic, and can be used to replace the conventional unification procedure in logic programming[103]. TABLOG, a new approach to logic programming designed by Malachi et al., is based on first-order predicate logic with equality and combines rule-based and functional programming. The use of this richer and more flexible syntax overcomes some of the short-comings of the Prolog syntax[104]. Other languages that combine features of Prolog and Lisp include LOGLISP, QLOG, POPLOG, Qute, and Lambda Prolog[101].

(3) Object-Oriented Languages: New languages and programming systems are being developed to radically simplify AI programming. Object-oriented programming holds promise as a programming framework that can be extended to concurrent systems, databases, and knowledge bases.

In conventional software, data and procedures are the main focus of the representation and are treated as separate entities. The choice of procedures and data is made by the programmer. In an object-oriented system, there is only one entity: the object. Objects may be manipulated like data, or describe manipulation such as a procedure, or both Processing is performed by sending and receiving messages to and from the object that possesses the appropriate information. A selector in the object specifies the kind of operation. Message sending is uniform, and a message represents only what the sender wants (or the result returned) but does not include information on how to accomplish that. Objects respond to messages using their own procedures (called methods) for performing operations. Since all communication is done via messages, one method may not "call" another method. The technique for representing the procedural knowledge can be any knowledge-representation scheme, although most implementations of object-oriented languages have employed procedural representations of control knowledge within the object.

In addition to objects and messages, object-oriented languages may also employ the concepts of class and instance. A class is a description of similar types of objects. Using classes, attributes of objects may be shared. In addition, classes provide a mechanism for inheritance or implicit sharing. Inheritance is used to define objects that are almost like other objects. Thus, classes provide an interface for the programmer to interact with the definition of objects.

Data abstraction is an important principle that is entailed through message sending. Object-oriented languages support both the management and collection of data through abstract data types, and the composition of abstract data types through an inheritance mechanism.

The requirement of typed data abstraction with inheritance is explicit and definitive, suggesting that object-oriented programming should be characterized by the nature of its type mechanisms rather than by the nature of its communication mechanisms. In a sense, object-oriented programming can be defined as

object-oriented = data abstraction + data types +type inheritance

The object-oriented programming paradigm is a methodology mainly for organizing knowledge domains but is permissive in its methodology for communication. The message/object model provides no new leverage for expressing concurrent problems. Concurrent models, operating systems, and coordination tools can be built from such lower-level objects as process, queues, and semaphores.

Early exploration of object-oriented programming was found in Simula[105]. A more contemporary object-oriented language which has received a great deal of attention is Smalltalk[76]. A variety of object-oriented languages includes LOOPS[78], Actor[106], CommonObjects[77], OIL[107], and others[79]. Recently, CommonLoops has been suggested as a standard for object-oriented extensions to Common Lisp[108].

The Actor model, developed by Hewitt at MIT, is a formalization of the ideas of object-oriented language that also considers the added effect of parallelism[109]. An actor is the analogue of a class or type instance. Computations in the Actor model are partial orders of events, inherently parallel and having no assignment commands. The language Act3, based on the Actor model, combines the advantages of object-oriented programming with those of functional programming[110]. The Apiary network architecture has been proposed to support the Actor model[111-112].

(4) Mapping Applications into Software: Software development, an active area of research in software engineering, is the process of mapping the application into a language chosen for the given symbolic application. The process begins with the selection of a solution technique. This decision includes the choice of a knowledge representation and a method within that representation for solving the problem. Naturally, the available languages would greatly impact this choice. These choices may be referred to as requirement analysis in life-cycle models of software development[113-114].

Software development environments can be classified into four generations[100]: discrete tools, toolboxes, lifecycle support and knowledge-based tools, and intelligent life-cycle support. Discrete tools were typical in the 1960s and 1970s and refer to the development of individual tool-like debuggers. Toolboxes refer to integrated packages of tools, the most prevalent example being Interlisp[115]. Lifecycle support and knowledge-based tools are being developed in the 1980s. Life-cycle support refers to software-development environments suitable for each stage in the design cycle, while knowledge-based tools try to incorporate domain knowledge to provide interactive assistance to the programmer. Finally, intelligent life-cycle support, a topic for future research[100], provides knowledge-based support for all stages of the software-development cycle in an integrated manner. Software engineering environment for distributed software development are also an increasingly important area of research[114].

With increased software complexity, verification, validation, and the enforcement of a structured discipline leading to reliable software are very important problems. Although a number of symbolic languages, such as Prolog, have been criticized for lack of structure, a programming style can be followed so that the resulting program is hierarchically developed, as in conventional structured programming languages, such as Pascal. There are mixed feelings on verifying and validating software written in symbolic processing languages .If these programs are treated as an algorithm and the requirements are well specified, then verification and validation are similar to those of programs written in traditional languages. Techniques such as test-case generation and path testing can be applied. However, it is difficult to test the validity of the knowledge used, since the knowledge may be heuristic and fuzzy in nature. The same criticism can be made about programs written in traditional languages. In this case, the validity of the program will largely depend on the experience of the experts and the procedures used in deriving the knowledge. Systematic knowledge-capture tools will help but will not guarantee proper collection and maintenance of consistent knowledge from multiple experts.

## 3.2 Hardware architectural support for symbolic processing

The choice of the knowledge-representation schemes and the software languages largely dictates the desirable hardware architectures. In this section, the desirable intermediate hardware designs of a symbolic processing architecture are discussed. These hardware designs can support language-specific features, or primitive symbolic operations such as sorting and pattern matching. In Table 8, some of the features that require hardware or software support in Lisp and Prolog are shown.

As with languages, hardware features can incorporate many well-established design philosophies, such as pipelining, parallel processing, microprogramming, and redundancy. Consider, for example, the hardware support of type checking in Lisp. In Figure 5, the role that design concepts and design requirements play in a hardware architecture is illustrated. The figure is intended to identify the role of concepts for approaching this design problem. It is not meant to say that hardware designers utilize such an approach in a real design.

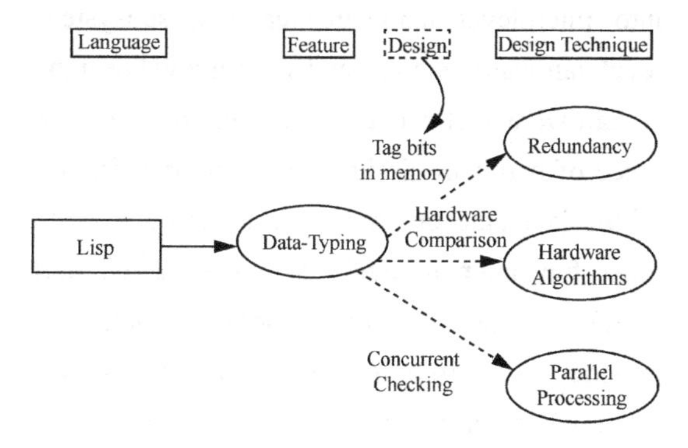


Figure 5. Hardware support for data typing

Table 8. Features of two example languages that can be supported by hardware

| Lisp | Prolog |
|---|---|
| Data typing | Condition matching |
| Function calls | Database functions |
| Recursion | Search |
| List structures—garbage collection | • search strategy |
| Individual commands—car, cdr, etc | • backtrack mechanism |
| Parallelism—future, etc | Unification |
| Application support | Parallelism |
| • application dependent | modes, guards, etc |
| • database support | Application support |
| • transitive closure | |
| • and others | |

Once a set of hardware and software alternatives has been enumerated, a subset of them must be selected for incorporation in to the complete system. This will depend largely on the design philosophy of the system (see Section 4). The selection of competing structures (alternative designs that perform the same function) is made on the basis of anticipated gain in performance versus anticipated cost, both of which may be difficult to estimate, especially when the structures are not commercially available components. An approximate model is often used to guide the selection.

The current hardware architectures used to support symbolic computation can be classified into microlevel hardware features, subsystem level hardware features, and system level designs. A hardware architecture at the microlevel is a piece of hardware designed to support a feature of the language or processing techniques at its most fundamental level. It is specialized and does not, in general, provide useful computations outside of its role in the system. A subsystem-level architecture is an architecture that performs a complete and useful function by itself, but is often included as a portion of a larger system. A system-level design provides a complete hardware/software solution to a symbolic processing application. Microlevel and subsystem level designs are discussed below, while system-level designs are presented in Section 4.

(1) Microlevel Hardware Features: A microlevel hardware architecture is a unit dedicated to the support of a specific symbolic processing technique used in the system. As the number of features in symbolic applications and their languages is quite large, the number of corresponding possibilities for microlevel architectures is very large. Some microlevel architectures and the features that they support are presented in Table 9.

Table 9. Functions that can be supported by microlevel architectures

| Function | Example Architectures |
|---|---|
| Function calls, Recursion | Hardware stacks, Register windows, Fast memory techniques |
| Data typing | Memory tags, Concurrent tag-checking hardware |
| Sorting | VLSI sorter |
| Set intersection | Marker-passing systems such as NETL |
| Pattern matching | Systolic arrays, Content addressable memories, Finite state automata |
| Best-matching | Value-passing system such as THISTLE and neural networks |
| Garbage collection | Hardware pointers, reference counters |

Five microlevel architectures are discussed in detail in this section: stacks, data tags, garbage collection, pattern matching, and unification. The remainder of the section is devoted to a discussion of emerging technologies employed in the construction of microlevel hardware.

• Hardware stacks and fast stack-access techniques: Stack architectures support function calls. This is especially useful for Lisp and other functional programming languages. In the Symbolics 3600 computer[116], there are three stacks: the control stack, the binding stack, and the data stack, that are used to support tail recursion and shallow binding. To speed access to the control stack, the top several (up to four) virtual-memory pages of the stack are held in a dedicated fast-access 1K-word memory, referred to as the stack buffer. The stack buffer contains all of the current function environment (frame) plus as many of the older frames as fit. A second stack buffer contains an auxiliary stack for servicing page faults and interrupts without disturbing the primary buffer.

In ALPHA, a commercially available Lisp computer produced in Japan[117], a hardware stack is divided into four 2K-word physical blocks. Variables or arguments in a function of the Lisp program are stored in different locations of cell space or stack. A great deal of time is required to search for a free variable, especially in the case of deep binding. To speed the evaluation of the function, the hardware stack is designed to support value caching. The value for a variable is fetched from the environment stack in the first access in a given function evaluation and stored in the value cache. Subsequent accesses will refer to values in the value cache. When the function exits, the variables it used in the value cache are marked invalid. Virtual stacks are also used to avoid the overhead of using a single stack and having to swap the entire stack in process switching.

Fast stack operations are also useful for implementing Prolog. Data are often pushed on and pulled from stacks in backtracking operations. However, the frames of a caller clause may be deeply buried in the stacks; hence, a stack architecture may not be adequate. In the Personal Sequential Interference (PSI, a product of the Japanese Fifth Generation Computer Systems Project) Machine, a cache memory is employed. Several operations suited to stack access are carried out in the cache memory.

• Tagged memory: A conventional von Neumann computer does not distinguish between data and program. Both are stored as fixed-size binary words. Meaning is not inherent in the contents of storage but defined by the program manipulating the storage. A tagged architecture, however, relies on self-identifying representation at all levels of storage. Although tagging has been employed since the 1960s, early design considered tagging as a relatively unimportant and expensive peripheral concept. Tagged memory can be a key feature in symbolic processing computers today. Symbolic architectures often require identification of different types of physical and abstract data types, including integer, character, event, garbage, and others. During processing, it is necessary to identify the different operands employed in the computation. For this reason, the tagging of data to improve real-time type checking is appropriate.

The most common hardware support for data tagging is the allocation of extra bits in each word to represent its type. Data-type checking at run time may be supported by additional hardware and overlapped with regular processing. The speed of symbolic computers is often linked to how effectively they emulate a tagged- memory architecture[118]. Special hardware for data-type tagging has been estimated to increase system performance by as much as an order of magnitude in Lisp computers. Data tagging also supports garbage collection, facilitates better register utilization, reduces memory traffic, and simplifies the design of cooperating parallel processors and specialized functional units[119]. Data tagging is also essential in untyped languages since the programmer does not specify the type of instruction used. For instance, the programmer need not specify the type of an add instruction as integer, long, real, or double. The type of adder used at runtime depends on the types of operands detected at run time, which are specified by the corresponding tags.

Data tags can be used to represent information other than just data-type. In the Classifier Machine[120], no addresses are used at all. Only tags are used to connect classifiers to each other. The no-addressing technique employed makes the Classifier Machine startingly different from classical von Neumann architectures.

• Garbage collection: Garbage collection refers to the process of identifying memory cells whose contents are no longer useful for the computation in progress. In this case, memory cells are contiguous groups of at least one memory word. The process involves marking these cells as available for future use and compacting free memory into contiguous blocks. It has been estimated that 10 to 30 percent of the execution time in large LISP programs is spent on garbage collection. The implication on interactive or real-time systems is great, as garbage collection often requires large continuous segments of time from the CPU.

The initial techniques for automatic garbage collection centered around the use of reference counts[80]. Each cell had an extra field that indicated the number of times the cell had been referenced. The reference count was updated each time a pointer to the cell was created or destroyed. When the reference count reached zero, the cell could be reclaimed as garbage. These techniques have the advantage of intuitive simplicity and distributing the processing overhead along the processing of the task. However, the extra space and time required for use of the reference counts can be high, although some of the overhead can be shifted to compile time[121]. In addition, there is no way to reclaim cyclic structures. Generation scavenging is an important technique which reduces the overhead rate of garbage collection by using different rates for memory areas of different age (or generation)[122].

More recent research in the area of garbage collection has focused on parallel garbage-collection methods. Parallel garbage collection is garbage collection that is performed concurrently with program execution. Two processing entities are involved: the Mutator and the Collector. The Mutator is responsible for program execution, while the Collector is responsible for garbage collection. The techniques often center around the use of coloring cells[123-124], or the division of memory space into (two) distinct regions[125-126]. Tagging can be a useful microlevel hardware feature.

Parallel garbage-collection processors can be designed with very simple and fast components, without becoming a bottleneck of the system. In a design proposed by Hibino[127], the collector processor cycle time was 200 ns—six times faster than typical processor cycle times.

• Pattern matching hardware support: Addition and multiplication are the mainstay of scientific computations. Similarly, pattern matching is the basic operation of symbolic processing. A pattern matcher may be employed for two major tasks (among other things): finding entries in a database and choosing an operation to execute next. For example, determining the applicable rule in a production system is a pattern matching problem. Empirical results show that 90 percent of execution time in a production system employed for expert systems can be spent in the matching phase[65]. Therefore, hardware pattern-matching support can simplify the programming task and improve run-time efficiency.

In most symbolic representations, symbols are represented in the form of strings. Conventional string-matching hardware can be classified into four categories. The first approach is associative memory. Although straightforward, it is difficult to operate on strings of variable lengths. The second technique is cascaded logic-memory arrays, also called a cellular array[128]. Each character in a pattern-string is stored in a cell and is compared with a character in the input string. A third technique is the Finite-State-Automata (FSA) method that uses a transition table to perform complex string matching. Finally, there is the dynamic programming technique that uses statistical characteristics of the general pattern in order to determine the parameter table for promixity matching. A survey of techniques for hardware support of pattern matching can be found in [129]. More recent techniques for hardware support have also been proposed[130-131].

Pattern matching in some symbolic processors differs from conventional database retrievals because many symbolic applications contain widely varying field lengths and uncertainties in data may forbid exact matching of patterns. The pattern matching hardware for a symbolic processor must be tailored to the representation(s) for which it will be used.

In matching under uncertainties, best-matching is required. Best-matching structures search for the pattern which best matches the defined objective. Best-matching using associative memories has been explored[132]. Neural networks also provide potential for performing best-matching (see Section 3.2-(2)). Kanerva's Sparse Distributed Memory (SDM) is a system designed for best-matching[133]. The proposed prototype consists of a virtual memory that is addressed by a 1 024-bit address, and a small physical memory. Each word in the physical memory has the 1 024-bit address for this word and the data fields. When a memory address is given, all locations in the physical memory with addresses that differ by less than 450 bits of the given address are accessed, and the corresponding data fields are combined together into a single response. The 450 bits are chosen so that approximately 0.1 percent of the memory words will respond on the average for a physical memory of 4 M words.

In a semantic-network representation, pattern matching and other functions can be performed in parallel using a marker-passing operation[134]. A high-bandwidth communication channel is important for this type of pattern matching.

In a forward chaining rule-based production system, the objects to be matched are constants, and multipattern multiobject pattern matching is required. The Rete Match Algorithm is an efficient solution to this problem[99]. A number of hardware implementations of this technique have been proposed, including tree architectures[135-136], SIMD Cellular Array Processor (CAP)[137], and tagged token data-driven multiprocessor[138]. It has been pointed out that the key architectural requirements to support the Rete

Match Algorithm in parallel production systems are the use of a memory to maintain information across multiple recognize-act cycles, and the proper choice of granularity of parallelism[139].

• Unification hardware: Unification, a form of pattern matching，is the fundamental technique in logic programming. It determines whether two terms can be made textually identical by finding a set of substitutions for variables in the terms, and replaces all occurrences of each variable by that variable's substitution. In general, both terms to be resolved in unification are allowed to contain variables; hence, unification can be thought of as a bidirectional pattern matching operation[30]. Since unification is applied extensively and is known to consume over 60 percent of the execution time in sequential logic execution, it is desirable that additional hardware or firmware support be available.

The primitive operations in unification are: 1) search for the called clause, 2) fetch of arguments of the caller and called predicates, and 3) examination of equivalence of arguments. In order to carry out unification in hardware, fast memory access is required. To support dynamic memory allocation, an efficient garbage collection technique is also required. Finally, hardware support for data-type checking can also speed performance.

Research in this area has concentrated on string-matching hardware[140], uniprocessor machines[141], special unification chips[142], and pipelined unification[143]. To reduce the required memory space and improve performance, the use of structure sharing[144] and techniques for structure copying[145] have also been explored.

The Parallel Inference Engine (PIE) developed at the University of Tokyo employs special hardware, referred to as UNIRED, for unification and reduction. UNIRED may be characterized by the following features: tagged memory, high-speed local memory that can be accessed in parallel, parallel hardware stacks, and dedicated internal buses. The unify processor fetches a goal from a memory module and candidate clauses from definition memory. The unify processor then attempts to unify them, generates new goals, and returns these goals to the memory module.

Parallel unification is also an area of great interest. Parallel unification can be performed either by unifying each term pair in two atoms simultaneously, or by finding many possible unifications concurrenty[93]. Unfortunately, the unification problem is proven to be log-space-complete in the number of processors. This means that it is not possible to perform parallel unification in $O(\log^k n)$ time using a polynomial number of processors for any constant $k$, where $n$ is the total number of nodes and edges of the directed acyclic-graph representation of the clauses. It has also been shown, however, that near linear speedup can be achieved in parallel unification. Thus, unification algorithms are parallelizable from a practical perspective[146]. Array architectures for parallel unification have been proposed, such as the Cellular Array Processor (CAP)[137], and others[140], A mesh connected array of unifiers has been proposed to exploit AND-parallelism in unification and may achieve superlinear speedup[147].

• VLSI and emerging technologies: Very Large Scale Integrated circuit or VLSI technology has been a major factor in the cost reduction and increased functionality of symbolic processing systems. The high degree of space-time complexity in AI and symbolic computations has necessitated the use of both parallel processing and VLSI technology. The development of specialized microelectronic functional units is among the major objectives of the Japanese Fifth Generation Computing project[148], MCC[149],and DARPA's Strategic Computing projects[150].

VLSI technologies allow a single-chip computer to be realized. Although many functions can be implemented on a single chip, the size of the chip and the number of input/output pins are usually limited, and chip area has to be carefully allocated to achieve the highest performance. Reduced Instruction Set Computers (RISC) is a highly popular design approach that carries out only the most frequently used instructions in hardware and the less popular instructions in software[151]. The smaller chip area required by the control unit of RISC computers as compared to that of complex-instruction-set computers enables incorporation of these registers onto a single chip. It has been found that a large set of registers on the chip is a good design tradeoff to reduce the overhead of swapping registers in context changes.

SOAR, or Smalltalk on a RISC, is a project to develop a RISC chip for Smalltalk-80[152]. The SOAR design details a 32-bit NMOS microprocessor containing 35 700 transistors and runs roughly 400 ns per instruction. Cycle time may be decreased to 290 ns if 3-μ lines were used. FAIM-1 is another project that designs RISC chips to carry out specialized functions in the system[153].

Besides implementing RISC on a single chip, specialized symbolic processing functions can be carried out in hardware as well. An example is the Texas Instruments Lisp chip with over 500K transistors on a 1-$cm^2$ chip implementing approximately 60 percent of the functions in a Texas Instruments Explorer[154].

Many existing computers for symbolic computation employ VLSI technology. The major building block of the Connection Machine CM-1 and CM-2 is a custom VLSI chip containing 16 processor cells[33]. The chip is implemented on a CMOS die about 1 $cm^2$ in area. There are approximately 50 000 active devices. Although each add takes approximately 21 μs, an aggregate maximum rate of 2 500 MIPS or 5 000 MFLOPS can be achieved with 64K processors implemented in 4K processor chips and 4K floating-point chips.

Cellular array structures are a widely studied technique that can take advantage of the available VLSI technology and exploit data-level parallelism in many symbolic processing problems. The Cellular Array Processor (CAP) is an example in this class[137]. A systolic cellular hardware design has been explored for performing unification[140].

There are also emerging technologies that will likely become cost effective in the near future for implementing computers for symbolic processing. These include GaAs circuits, Wafer-Scale Integration (WSI), and optical computing techniques.

GaAs circuits have similar design requirements as conventional semiconductor circuits but are much faster. Switching speeds on the order of 10 ps have been reported in High Electron

Mobility Transistor (HEMT) GaAs circuits[155]. Gate propagation delays are typically on the order of 200 ps. Unfortunately, fabrication of GaAs circuits is subject to limited size and greater numbers of defects, which adds to the already high cost[156].

WSI refers to the integration of multiple circuits on the same wafer in order to avoid the high performance and cost penalty of off-chip connections. As chip yield is low, the yield of a complete wafer would be negligible. Techniques such as Focused Ion Beam (FIB) repair are utilized to increase yield of wafers. Low yields on WSI may make it more suitable for implementing a distributed knowledge-representation scheme, such as a neural network, in which the loss of a small fraction of the distributed knowledge may not be critical. When GaAs and WSI are combined, it is possible to implement a 32-bit GaAs processor on a single wafer[156]. Design of such a processor for production systems has been explored[157].

Optical processing can be used in a fashion similar to silicon gating. Switching speeds on the order of 5 to 10 ps are possible in optical gates. Optical circuits do not have the penalty of capacitance[158] and can communicate with low propagation delay and no interference. Optical processing has been developed in the form of arrays of light rays[159-160] and optical crossbars[161]. Use of optics in storage media may greatly improve performance in symbolic systems with erratic memory behavior. It has been proposed that optical techniques have the potential for improving data-and knowledge-base processing speeds by two orders of magnitude[159].

In Table 10, the microlevel architectures presented in this section are summarized. Associated with each type of microlevel architecture is one of the main applications for that type of hardware, and an example of the significance of architectures for support of that feature.

Table 10. Microlevel architectures and their significance

| Architecture | Significance |
| --- | --- |
| Stacks | For function calls and value binding; more than one stack may be used |
| Tagging | Data type checking; up to an order of magnitude speed improvement for Lisp |
| Garbage Collection | Reclamation of usable storage; accounts for 10 to 30 percent run-time in Lisp |
| Pattern Matching | Fundamental operation; up to 90% of execution time in production systems |
| Unification | Type of pattern matching for logic; over 60% of execution time in sequential logic programs |

(2) Subsystem Level Architectures: The subsystem level architecture represents an intermediate level between microlevel and complete-system designs. In this section, we identify three techniques for classifying the subsystem-level architectures that address different types of knowledge processing. Three different types of subsystems will emerge from this analysis: data- and knowledge-base machines, inference engines, and neural networks.

Control-flow, data-flow, and demand-flow are three important approaches that are used in the design of subsystem level architectures as well as the complete system[162]. Their definitions and relative advantages and disadvantages

Table 11. Control-driven versus data-driven versus demand-driven computations

| | Control Flow (control-driven) | Data Flow (data-driven) | Reduction (demand-driven) |
|---|---|---|---|
| Definition | Conventional computation; statements are executed when a token of control indicates that they should be evaluated | Eager evaluations; statements are executed when all of their operands are available | Lazy evaluation; statements are only executed when their result is required for another computation |
| Advantages | Full control<br>Complex data and control structures are easily implemented | Very high potential parallelism<br>High throughput<br>Free from side effects | Only required instructions are executed<br>High degree of parallelism<br>Easy manipulation of data structures |
| Disadvantages | Less efficient<br>Difficulty in programming<br>Difficulty in preventing run-time error | Time lost waiting for unneeded arguments<br>High control overhead<br>Difficulty in manipulating data structures | Does not support sharing of objects whose local state changes<br>Time to propagate demand tokens |

are summarized in Table 11.

Knowledge representation plays an important role in the way that a complete system integrates its components. The Japanese FGCS project emphasizes logic representations and stresses the development of separate knowledge-base and inference engines, which are integrated after their development[163]. This viewpoint of the system architecture is illustrated in Figure 6(a).

An alternative perspective is seen from the design of the Connection Machine[33], which closely reflects a semantic-network knowledge representation and memory-based reasoning. All the knowledge in the system is embodied in a large collection of facts; no intentional knowledge, or rules, are employed. For this type of system, the database is implemented directly on the architecture, and inferences are carried out in software and message exchanges. This perspective of system design is illustrated in Figure 6(b).

The last design perspective is that of a hierarchical nature as shown in Figure 6(c). An inference machine handles all meta-level inferences, while a knowledge-base computer deals with rules manipulating domain knowledge. A separate database computer carries out search and selection operations on the domain knowledge.

• Database architectures: Early studies on specialized database architectures emphasized the use of parallelism. Such designs include CASSM[164], RAP[165], and DIRECT[166]. Some later systems, such as the Connection Machine, are designed with massive parallelism for symbolic applications and can be applied for a number of specialized database functions[68]. A comparison of several early parallel database computers and a survey of commercially available database computers can be found in [167] and [168]. Commercial manufacturers include Britton-Lee, Hitachi[169], International Computers Limited[170], and Teradata[171].

These systems may not function well for some applications because the bottleneck in database retrieval has found to be disk input/output and not processor cycles. As a result, very intelligent database processing using application--dependent knowledge and indexing may be preferable to massive parallelism[172].

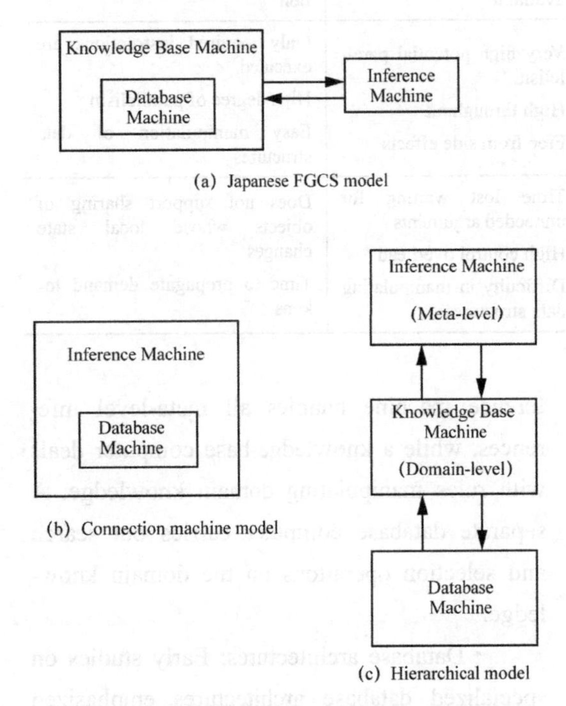


Figure 6. Relationship among knowledge-base machine, inference machine, and database machine

• Knowledge-base architectures: The objectives and requirements of a knowledge-base computer are different from those of a database architecture. The most prominent differences are noted in Table 12. An evolving knowledge-base subsystem should have a mechanism for either rejection of or truth maintenance for the insertion of inconsistent data or rules. Support for the inference mechanism is also desirable and may take the form of an automatic rule-selection mechanism, logic support, or special hardware for operations such as joins and projections of relations. Finally, the interface to the host computer should preferably be intelligent and may draw on the resources of the knowledge-base system.

Table 12. Differences between databases and knowledge bases

| Issue | Database | Knowledge Base |
|---|---|---|
| Contents | Collection of data and facts | Higher level of abstraction<br>Classes of objects |
| Complexity | Stored items are simple | Stored items are complex relations |
| Time dependence | Data changes over time | Knowledge changes less frequently with exception in situation knowledge |
| Size | Large number of facts | Fewer relations on classes of objects |
| Use | Operational purposes | Analysis, planning, ... |

The issues in the design of a knowledge-base computer include:

a) Storage and manipulation of intentional and extensional data: Extensional data refers to data representing facts, that is, statements with no quantified variables. Intentional data refers to general facts or rules. In Figure 6(c), the extensional and intentional data are stored and operated on separately. In Figures 6(a) and 6(b), they are processed by the same physical entity.

b) Relational operations: As above, hardware and software support must be provided.

c) Hierarchical storage: Knowledge can be classified into categories by degree of generality: fact through the most general meta-knowledge. The access characteristics are highly dependent on the type of knowledge accessed. A hierarchical storage for meta-knowledge may be used for efficiently exploiting the knowledge structure.

d) Access-control algorithms: A knowledge-base subsystem may be required to control access to its contents. The main issues here are security, integrity, and concurrency control.

e) Parallel and Distributed processing: Database updates are history-sensitive. It remains an open issue as to the best technique(s) for exploiting parallel and distributed processing in knowledge-base systems. Data-flow may be a good concept to apply.

In many applications, a database computer may be integrated with an existing host to form a knowledge-base computer. An example is shown in the Intelligent Information Resource Assistant developed at System Development Corporation[173]. The system consists of a Britton-Lee IDM 600 backend database computer, a Xerox 1100 workstation as a logic-based deductive engine, and a VAX 11/780 computer as a file and print server. This prototype has demonstrated that a knowledge-base computer can be easily constructed from existing hardware components.

The Japanese FGCS project has developed Delta, a combined knowledge and database computer. The motivation behind a specialized design is that the integrated system calls for a performance that cannot be met by commercial components. The system consists of a control processor, relational database engine, and a hierarchical memory. The control processor translates commands (received from an interface processor that communicates with a parallel inference machine) into subcommands that the relational database engine can perform. The relational database engine operates through a data-path with the hierarchical memory, which is composed of semiconductor and magnetic disk storage[174].

• Hardware support for inference engines: Inference engines are a key component of knowledge processing architectures. Their structure is highly dependent on the knowledge representation and programming language employed.

An important problem in designing hardware support for inference engines is the architectural supports for searching the knowledge base. Deduction and search have been the dominant paradigms for machine inference over the last 30 years. As discussed in Section 2, development of superior heuristics combined with efficient hardware is the best approach. The following are some of the key issues in research on search architectures (More general issues on parallel processing have been discussed in Table 6).

a) Prediction of performance: A major difficulty in developing search-based inference engines is the inability to estimate their performance without the actual execution of the search. This is due to both the nondeterministic nature of searches and anomalies in parallel search algorithms[49].

b) Space-time tradeoff: There is a space-time tradeoff in using heuristic knowledge. Very accurate heuristic functions may require greater amount of space and computation time than less accurate ones. This relationship, as well as the given architectural constraints, must be understood for the design of effective search subsystems.

c) Architectural support for machine learning: Heuristic functions used in search algorithms should be improved over time by automatic learning methods. Architectural support for nonmonotonic processing may be also helpful.

d) Management of large memory space: Heuristic search strategies may require large amounts of memory space to store the intermediate results and heuristic information. Tradeoff with respect to the effectiveness of a search strategy and the corresponding overhead incurred must be considered. Techniques for efficient memory management tuned to the search behavior are vital in such a system.

e) Granularity of parallelism: The proper choice of granularity is difficult to determine at design time due to the dynamic nature of the problem. Granularity may have to be varied at run time when more is known about the application.

f) Scheduling and load balancing: Due to nondeterminism in many search problems, direct mapping of a sequential search strategy into a parallel system may not result in the best performance. The key to effective scheduling is the proper order of execution, not just keeping the available processors busy. Counter to intuition, depth-first search is sometimes preferable to best-first search in real systems when memory constraints are considered[175]. Conditions for the sequential search strategy may also have to be relaxed to accommodate the architectural constraints[49]. For instance, the selection of the subproblem with the minimum heuristic value in a heuristic search may not be desirable when the overhead of selecting subproblems distributed in local memories of multiple processors is high. Selecting the local subproblem with the minimum heuristic value will suffice in most cases.

g) Communication of pruning information: When the search space is explored in parallel, excess computation may be performed if pruning information cannot be shared among the processors. In general, a tradeoff exists between search efficiency and communication overhead.

Microlevel hardware features discussed in Section 3.2-(2) are often components of the inference subsystem. Data-flow, control-flow, and demand-flow techniques (see Table 11) have been employed in designing hardware supports for inference engines of Lisp and logic[176-178]. An example of a combined data-driven and demand-driven approach is demonstrated in ALICE[179] and Rediflow[180]. Although the theory behind data-driven and demand-driven computations appears very promising, no clear solution has been developed for many of the design considerations such as the proper granularity. It remains an open problem as to the proper tradeoff among demand-flow, data-flow, and control-flow in an inference computer.

• Artificial neural networks: Automated computations based on a neural-network design philosophy originated many years ago. A pioneering contributor was Rosenblatt who developed the concept of perceptrons[181]. Recently, a great deal of interest has been revived in this area. Besides capturing the imagination by modeling computers after the human brain, artificial neural networks, or in short neural networks, offer a high potential for automated learning. The focus of storing information in the connections between neurons is why they may be referred to as connectionist systems.

A neural network can be viewed as containing eight components: processing units, state of activation, output function, pattern of connectivity, propagation rule, activation rule, learning rule, and an environment[182]. These are summarized in Table 13. The first design consideration is on the representation in the set of processing units. There are two possible approaches: local and distributed. A local representation allocates one concept to its own unit. Hopfield's network solving the traveling-salesman problem is an example of a local representation on a neural network[183]. While this may be an efficient solution to the problem, it does not perform learning. In contrast, in a distributed representation, each unit participates in the storage of many concepts[184]. In this way, a processing unit in a distributed representation stores an abstract feature at the microlevel[182].

In a neural network, each unit has an associated activation level. This activation level can be analog, analog and bounded, or discrete. Its output is a function of its activation level. Output functions are usually threshold or sigmoidal functions. The output of each unit is distributed to a set of processing units. The emphasis that the outputs from other units have on a unit is determined by the weight of the connection between units. Finally, a node's activation is modified by the activation rule which is a function of inputs, their relative connection strengths, and the current activation level.

Classes of neural networks may be discerned by their learning paradigms[182]. One class is associative learning in which the network learns associations between the inputs and their desired outputs. The second class is regularity detectors, which learn to recognize interesting pattern on the input. An overview of designs for neural networks of both classes can be found in the references[185-186]. Learning techniques in alternative neural network strategies are also extensively reviewed in the references[187].

Neural networks have been proposed for a variety of applications, although they are not yet

Table 13. Artificial neural network components

| Component | Description |
|---|---|
| Processing units | Three types; input, output, and hidden |
| State of activation | Vector of the activation levels of the units in the system |
| Output function | Function on the activation level of a unit which produces the units' output; may vary between units, but most systems are homogeneous |
| Pattern of connectivity | The connections are what determines the performance and function of the system |
| Propagation rule | A way of combining outputs of units and patter ns of connectivity into an input for each unit; usually is a weighted sum of the inputs and the excitatory ( + ) and inhibitory (–) connection strengths |
| Activation rule | A function for determining the new activation level of a unit on the basis of current activation and inputs to the unit |
| Learning rule | Three types: develop new connections, abandon old connections, modify weights; only the last has been pursued; almost all learning rules are based on the Hebbian learning rule |
| Environment | In which the computing engine functions |

widely used in practice. A neural network can be easily adapted as associative memory with capability for in exact matching. Its strengths include both speed and accuracy in the presence of noise[188]. At a more complex level, neural networks can be used for speech recognition[189], and vision tasks such as letter recognition[190]. Neural networks have been applied for solving combinatorial search problems, such as the traveling-salesman problem[183,191], although solutions for large search problems on neural networks are not yet of as high quality as those of good digital algorithms[192]. Neural networks are also useful for some strategy-learning tasks. Experiments have been performed for a balance-control system[193].

The design of neural networks is still plagued by a number of difficult problems. First, a neural network must be trained for a given application and must be retrained when the system parameters change. There is no systematic method to generalize a neural network trained for one application and apply it for another application. Second, all the learning algorithms known today require extensive amount of training for good performance. For example, over 9 000 learning sweeps, each of which has 40 phases, are required to train a shifter network using the Boltzmann Machine learning algorithm[194]. Moreover, the learning speed depends on the configuration of the neural network, which cannot be selected systematically. Dedicated hardware to emulate various configurations of neural networks and map the inner-loop operations into analog instead of digital circuits can improve the learning speed significantly. A million times improvement in speed has been demonstrated using hardware emulation[195]. Third, extremely large neural networks cannot be built with the current technology. There are numerous industrial efforts in building neural networks. For example, AT&T's R3 chip has a word size of 1 048 bits and performs learning by back propagation. It was estimated that 4.5 million neurons can be built on one wafer by 2010 using 0.25 micron lithography and 250 million transistors[196]. Consequently, it is unlikely that a complete symbolic processor as intelligent as the human brain (with over 10 billion biological neurons) can be built with neural networks alone in the near future. Lastly, with the limited size of neural networks, it is necessary to partition the problem so that part of it can be learned by traditional methods and the other part by neural processing. However, knowledge representations in neural networks are drastically different from procedural and declarative representations in conventional symbolic processors. Systematic methods to integrate them are still missing.

## 4 Complete systems

Up to this point, we have focused on individual techniques for effective symbolic computations. In this section, complete systems for solution of symbolic applications are classified into single-processor systems, parallel computers, and connectionist systems. As in the preceding sections, we are interested in analyzing the role of individual components in the design of complete systems.

There are two prevailing trends in designs. With the increasing complexity of many symbolic processing applications, there is a shifting emphasis on knowledge representations, as well as software languages and support, away from strongly procedural techniques and toward distributed and declarative computing environments. Parallel processing is another obvious trend in computer-system design for symbolic processing.

Depending on the starting point of the design, the system can be classified as following a top-down or a bottom-up approach. Top-down design begins with the specification and analysis of the application. A knowledge representation is then designed, tailored to the needs of the application. From this, a language is designed, and the system is mapped into software, microlevel and subsystem-level structures. This process may have to undergo many iterations if it is discovered that the functional requirements of the design cannot be implemented using the current technology. The FGCS project can be viewed as a system designed using a top-down design approach.

In contrast, a bottom-up design first selects the technology and design options such as data- or demand-driven calculation. A language and application suitable for implementation with these concepts are then sought. As above, the design process may have to undergo many iterations if it is found that the applications cannot be suitably supported by the computer designed. ZMOB[197] and the Butterfly Multiprocessor[198] are instances designed using this technique.

A shortcut to the top-down approach is to compromise between the top-down and bottom-up approaches, resulting in a middle-out design approach[7]. The middle-out approach begins with the selection of an appropriate and well established knowledge-representation scheme that is the most suitable for the application. The representation scheme should already have a well-developed programming environment that can be modified later for the needs of this specific system. Primitives for concurrent execution may be added to allow users to annotate concurrent tasks in the application. The hardware and software architectures are then designed, and the selection of features to incorporate includes consideration of previously used and designed structures. The middle-out approach can further be classified into top-first and bottom-first. In a top-first middle-out approach, the designers start with a well-defined knowledge-representation scheme and tailor it to the given application. ALICE and FAIM-1 are examples of architectures designed using this philosophy. In a bottom-first middle-out approach, the designers first develop the architecture to support a well-established representation scheme before mapping the application to the scheme chosen. DADO is an example developed using this approach.

The overall design process of a computer for symbolic processing is iterative (see Figure 7). In mapping from application specifications to hardware and software technologies, the designers iterate in proposing and selecting knowledge-representation schemes, algorithms, and physical designs until a feasible and cost-effective mapping is found. Applicable design tools include methods for synthesis, analysis, optimization, heuristic designs, validation, simulation, and testing. During the design process, the designers have to address issues related to theory, programmability, design methodology, design tradeoffs, effective control, and emerging technologies.

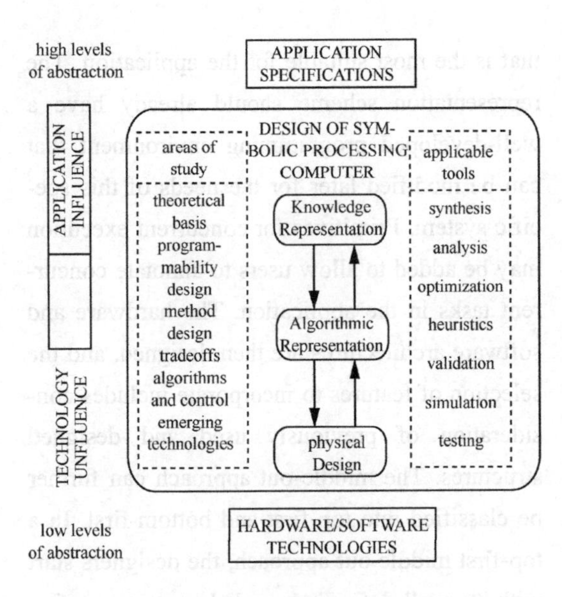


Figure 7. Perspective of design of a computer for symbolic processing applications

## 4.1 Single-processor symbolic computers

Lisp has enjoyed the longest tenure in the main stream of languages for symbolic processing and has led to the greatest number of computers devoted to its execution. The earliest Lisp computers were PDP-6s, followed by the PDP-10s and PDP-20s of Digital Equipment Corporation[2]. The half-word instructions and stack operations in these computers were particularly well suited to Lisp. A great deal of work has also been done specifically for garbage collection on the PDP-10s and 20s.

The MIT AI Laboratory introduced the design of a Lisp computer called CONS in 1976[199]. This was soon followed by CADR in 1978, their second generation Lisp computer. This computer was the basis for many commercial Lisp computers, all introduced in 1981. These computers were the Symbolics LM2, the Xerox 1100 Interlisp computer, and the Lisp Machines Incorporated Series IU CADR. Some notable recent commercial and research-oriented Lisp computers are shown in Table 14.

In addition to providing special hardware to improve efficiency, commercial Lisp computers also provide an integrated software development environment, such as KEE and ART, that allows programmers to develop, debug, and maintain large Lisp Programs. A recent trend is that these development environments can be implemented efficiently in software (rather than microcode) on high-speed general-purpose workstations. It is likely that in the future special-purpose Lisp computers will be used to execute rather than develop Lisp programs.

The design of special-purpose Lisp computers continues to be a popular research area. Many experimental computer designs have been reported[200-209]. A single-chip design to support Scheme, a dialect of Lisp, has been demonstrated in the Scheme-79 and Scheme-81 chips[210]. Scheme-79 was limited by its implementation of a register file and slow programmable array logic. A redesign, Scheme-81, has proven to be much faster.

The popularity of rule-based systems has come at a time when parallel processing also became highly popular. As a result, most architectures designed for supporting these languages have been parallel computers. However, the study of single-processor rule-based systems is still important because a parallel system can be limited by the speed of its inferences, and the building block of a parallel system is likely to be a single-processor symbolic computer.

The three most notable designs for executing logic languages are SRI's pipelined Prolog

Table 14. Notable single-processor symbolic computers

| Machine | Year/Status* | Primary Language | Features |
|---|---|---|---|
| Scheme-79, Scheme-81 | 1981/PO | Scheme | Single chip; tail recursion; lexical scoping |
| Lisp Machines Lambda | 1983/CD | Zetalisp, LMLisp | NuBus—multiprocessor capability; stack orientation |
| Symbolics 3600 | 1983/C A | Zetalisp, Flavors | Tagged memory; stack-buffer, hardware garbage collection; single-address instruction |
| Tektronix 4400 | 1984/C A | Smalltalk, Franz Lisp | Lower-end AI workstation; Motorola 68010/20 |
| TI Explorer | 1984/CA | Common Lisp | NuBus; tagged memory; microprogrammed; mega-chip version has 60% of processor implemented in one chip with 550 000 transistors[150] |
| Fujitsu ALPHA | 1983/CA | Utilisp | Value cache; hardware stack; virtual stack |
| FGCS PSI-I, PSI-II | 1985/PO | KL0 (Logic) | PSI-II: CMOS-GaAs; TTL; 200 ns cycle time; cache; stacks; copying for structure data; hardware unification; tagged data; interpretive execution; 150 KLIPS average speed |
| FCCS CHI-I, CHI-II | 1986/PO | Current Mode Logic (CML) | CHI-II: CMOS-GaAs; TTL; 170 ns cycle time; cache; about 400 KLIPS average speed for append operation |
| SRIs Pipelined Prolog Processor | 1984/SI | Prolog | Pipelined execution; microprogrammed controller; interleaved memory; FCFS module queues |
| SOAR | 1984/PO | Smalltalk-80 | RISC; hardware support for expensive procedure calls in Smalltalk-80; tagged and untagged instructions; large number of registers; automatic storage reclamation; direct object addressing; fast type checking |

*Status Codes: (Tables 14-19) CA: commercially available; CD: discontinued commercial system; PO: prototype operational; HS: hardware simulated; UC: under construction; SI: simulations completed; PD: paper design

processor[211], the Personal Sequential Inference (PSI) Machine of the Japanese FCCS Project[212], and the Cooperative High-Speed Inference (CHI) Machine, also of the Japanese FGCS Project[213]. PSI, an integrated workstation with an execution speed of 30 KLIPS, is intended to be a software development tool for the project. PSI has been redesigned into PSI-II and has a performance of 150 KLIPS on the average[213]. CHI was designed with speed in mind, and took on a less constrained design technique. The original CHI was also redesigned into CHI-II, with an estimated improvement in performance from 280 to 400 KLIPS for the APPEND operation[213].

Single-processor support for production systems has focused on additional data memories[214] and RISC architectures[65].

Object-oriented languages have been implemented on sequential processors. The Xerox 1 100 family of computers were one of the first workstations on which Smalltalk-80 was built. Smalltalk-80 has been implemented on a single chip using the RISC approach in the Smalltalk-On-A-RISC (SOAR) project[215]. There is no microcode or fine-grained addressing hardware, and few multicycle instructions.

## 4.2 Parallel symbolic processors

In this section, we classify parallel symbolic processors in terms of their representations or programming techniques employed. Section 4.2-(1) discusses the methods of communication and synchronization in parallel symbolic processors. The rest of this section is devoted to discussions on parallel symbolic processors for functional, rule-based, and object-oriented representations.

(1) Communication and synchronization

• Communication: Message passing, marker passing, and value passing are three predominant communication methods in parallel symbolic processors.

Message passing is the conventional way of communication in which the information to be communicated is formulated into a message and sent over the interconnection network. The computing elements are generally complex, and the communication costs are high. Message passing is popular in many parallel symbolic processors.

Marker passing refers to the transfers of single-bit markers from one processor to another. A marker indicates the presence of a given property, and a set of markers indicate the conjuction of a set of properties. Each processor is simple and can store a few distinct marker bits. There is never any contention: if two markers arrive at the same destination, they are simply ORed together. The basic inference operation performed is, therefore, set intersection. Marker passing is especially suitable for implementing semantic networks and recognition problems in hardware. One such system is NETL[216-217]. The Connection Machine was originally designed as a marker-passing system, but was modified later to carry out more powerful processing and communication features as well as support of virtual processors[134].

The last method of communicating information is value passing. In this form, information is passed as continuous quantities. Only simple operations are performed on these numbers. The salient feature of this approach is that if several values arrive simultaneously at one point, they are combined into a single output by a mathematical function. Consequently, there is no contention in information transfer. Examples of value passing systems in elude the Boltzmann machine[217] and other neural computation systems[218]. Iterative relaxation techniques for problems such as low-level vision, speech understanding, and optimization all seem to be suited to value-passing architectures.

• Synchronization: Synchronization refers to the control of concurrent accesses to shared items in a parallel processing system. It is important in message-passing systems because messages may result in contention for shared resources. In contrast, synchronization is not critical in marker-passing or value-passing systems because there is a predefined method of passing markers and combining values so that contention will not occur.

Synchronization is important when there are shared data items. In a program written in a procedural language, the order of statements dictates the order of execution. If two statements share a variable, the first is to be executed prior to the second. Therefore, synchronization control is implicitly defined by the order of statements when data sharing is necessary. In contrast, the order of execution in a program written in a pure declarative language is not defined. When two tasks share a common variable, the order of execution is indeterminate. As a result, explicit specification of synchronization control is needed when data sharing is present. However, most declarative languages lack facilities for explicit specification of synchronization control.

Synchronization can be carried out by shared memory or message passing. Shared memory is popular and has been used in systems such as Aquarius[219], Concurrent Lisp Machine[220], Concert MultiLisp Multiprocessor[221], and the Parallel Inference Engine[222]. Blackboard architectures and shared variables are two applicable techniques for shared-memory synchronization.

The blackboard model was originally developed for abstracting features of the HEARSAY-II speech understanding system[223-224]. There are three components: a set of knowledge sources, a blackboard, and control. The knowledge processing technique to solve the problem is partitioned into separate knowledge sources. The data, including input/output and partial solutions, are stored in the blackboard. The blackboard may be partitioned into smaller blackboards, forming a hierarchy of solution spaces. Knowledge sources manipulate the data in the blackboard in order to reach a solution. The only communication mechanism between these knowledge sources is the blackboard. A monitor is present to ensure that only one knowledge source is changing the blackboard at any time.

A more powerful blackboard architecture has been proposed in which control information (or meta-knowledge) is allocated a separate blackboard[225]. This approach is more flexible and suits the nondeterministic nature of symbolic processing.

Synchronization may also be achieved through shared-memory variables. Lisp languages that have been modified for parallel processing often contain shared variables for synchronization. Multilisp provides a mechanism for waiting for values to be generated in the future. As in other languages, procedure activations may not be well nested, and a process can terminate prior to an activation that it began. This is a problem that must be addressed by the programming system[226].

Single assignment languages, such as pure Prolog and pure dataflow languages, do not require careful synchronization of shared memory variables since a variable may be written only once[227]. In Prolog, the technique is to try to delay process reduction until information is available in order to make a better decision. In Guarded Horn Clauses (GHC)[228], the current kernel language of the Japanese FGCS project, OR-parallelism is not exploited. A strict synchronization rule suspends a subgoal if it tries to modify its parent environment. This simplifies the implementation of the language but results in a less expressive language[229].

Synchronization in a message passing system is accomplished through a protocol implemented in hardware or software. In a standard message passing environment, the messages may be of arbitrary complexity. This is more appropriate, for example, in systems using an object-oriented programming technique. Actor is a paradigm for systems with message passing of this nature[109,112]. When an actor receives a message, it performs predefined primitive actions. In this sense, actor systems are inherently parallel. The Apiary architecture is based on actors[111]. Other message passing systems for symbolic applications in elude the Contract Net system[230] and the Rand Distributed Air Traffic Control System[231].

(2) Parallel functional programming computers: The majority of special-purpose parallel processors designed to support functional languages are oriented toward Lisp. Examples include Concert[221], EM-3[232], and a multi-microprocessor Concurrent Lisp system developed at Kyoto University[220] .In all these systems, users are required to specify to some extent the tasks to be decomposed. Compilers for automatic detection of parallelism in sequential Lisp programs is an area of active research. Table 15 presents some of the more publicized parallel systems for functional programming techniques.

The majority of computers that were designed for general-purpose applications have only a few features specifically appropriate to symbolic processing. As a result, the inference engine and knowledge base are not separated and are almost exclusively implemented by sophisticated software structures. Lisp is added as one of the several languages to accompany their parallel

Table 15. Notable parallel functional programming computers

| Machine | Year/Status* | Primary Language | Interconnection | Communication | Features |
|---|---|---|---|---|---|
| Butterfly | 1985/CA | Multilisp | Butterfly Switch | Shared memory | 256 MC68000-series PEs; homogeneous, tightly coupled; general-purpose multiprocessor |
| iPSC | 1986/CA | Common Lisp | Hypercube | Message passing | 256 Intel 80286 and 80386 processors; no shared environment; user decomposes program into concurrent processes that communicate by messages; general-purpose multiprocessor |
| Connection Machine | 1986/C A | *Lisp, CM-Lisp | Hypercube | Message passing | Model CM-2 has 4 096 bit-serial processors; users annotate Lisp programs for SIMD or multiple SIMD parallel processing; C*, Fortran, and Paris (CM-2 assembly language) are also supported |
| Concurrent Lisp Machine | 1983/PO | C-Lisp | Multiple buses | Shared memory | 17 MC68000-series PEs; special cell interface; control stack; garbage collector |
| EM-3 | 1984/PO | EMLISP | Modified Delta network | Message passing | List-like data-driven language; 16 MC68000-series PEs; special router chip; control for function evaluation |
| Concert | 1986/PO | Multilisp | Ringbus | Shared memory | 32-64 MC68000-series PEs; network is segmented bus in shape of ring |
| Rediflow | 1984/SI | Functional Equation Language (FEL) | Mesh or richer connections | Message passing | Demand/data-driven; loosely coupled; hardware support for load balancing; distributed garbage collection |
| Alice | -/UC | Hope, Lisp, Prolog | Cluster of processors, ring buffer | Message passing | Transputer as basic processor; reference counter for garbage collection |

* Refer to Table 14 for explanation of status codes

computers. A Lisp compiler is used for decomposing tasks for parallel processing, and users are required to annotate tasks in various degrees. Examples of commercial multiprocessors supporting Lisp include the Butterfly[85], Connection Machine[20], and the Intel iPSC concurrent computer[233].

For the Connection Machine, special chips containing 16 bit-serial processors and router circuits were developed. Lisp in CM-2 allows users to specify a parallel variable (pvar), which is a first-class object with value for each processor in the computer[234]. The primitive pvar can be accessed con currently (with possible masks) by all local or remote processors in a SIMD or multiple SIMD mode. CM-Lisp is a dialect of Common Lisp extended to allow fine-grained, data-oriented parallel processing. It provides higher-level data abstractions called "zappings", which are similar in structure to arrays or hash tables. Broadcasts, reductions, and combinations can be specified.

Other general-purpose computers with limited support for parallel symbolic processing, such as ZMOB, are being developed at universities[197]. These computers can be viewed not just as symbolic processing computers, but also as general-purpose computers that are appropriate for both numeric and symbolic computations.

(3) Parallel logic architectures: In this section, parallel systems suitable for evaluating logic programs are presented. A summary of notable projects is shown in Table 16.

Unification and search are two key features in evaluating logic programs. Architectures that emphasize efficient search of logic programs

Table 16. Notable parallel computers for logic representations

| Machine | Year/Status* | Interconnection | Communication | Features |
|---|---|---|---|---|
| BAGOF | 1984/PD | Bus | Shared memory | OR-parallelism; separate static and dynamic memory; token pool |
| MANIP-2 | 1985/PD | Global broadcast bus | Message passing | Cluster of PEs with local memory; distributed selection; heuristic guiding and pruning |
| Aquarius | -/UC | Bus and crossbar | Shared memory | Heterogeneous MIMD; 16 PEs; synchronization through Goodman Cache; crossbar to shared memory modules; special Prolog, floating-point, and I/O processors |
| Parallel Inference Engine (PIE) | 1984/SI | Switching network | Shared memory | 100 s to 1 000 s of inference units; goal rewriting model; OR-parallelism; sequential AND processing; Activity Controllers to control inference tree; Unify Processors connected to Definition Memory containing program |
| Parallel Inference Machine-Reduction (PIM-R) | 1986/HS | Multistage network | Shared memory | Many Inference Modules connected to Structure Memory units through network; structure copying |
| Parallel Inference Machine-Dataflow (PIM-D) | 1986/HS | Multistage network | Shared memory | Multiple PEs connected to Structure Memory; unfolding interpreter; asynchronous communication; streams for nondeterministic control |
| PIM-I | -/UC | Hierarchy | Shared memory message passing | 100 PEs; 8-PE clusters interconnected with shared memory and parallel cache |

*Refer to Table 14 for explanation of status codes

include the BAGOF architecture[235] and MANIP-2[236]. The MANIP-2 architecture is particularly interesting due to its emphasis on heuristic parallel search strategies.

There are two significant parallel logic systems developed at universities. The Aquarius multiprocessor developed at University of California, Berkeley, emphasizes a coupling of intensive numeric calculations and symbolic manipulations[219]. It intends to utilize parallelism at all levels of computation and considers cost as secondary to performance considerations. Another university project is the Parallel Inference Engine (PIE), being developed at the University of Tokyo[237]. The target is 1 000 processors, and a speedup of 170 has been estimated for 256 processors. PIE utilizes OR-parallelism only.

Probably the most massive effort at the development of parallel logic systems is contained in the Japanese Fifth Generation Computing System project (FGCS). The project discerns three major areas of development: problem solving and inference machines (hardware), knowledge-base management systems (software and algorithms), and an intelligent man-machine interface[238]. The project is divided into three stages. The initial stage explores basic computer technology and processing techniques. This stage has been completed. The middle stage is for the development of subsystems, and the construction of experimental subsystems. The final stage is devoted to the development of the complete system.

The initial-stage designs of the Parallel Inference Machines (PIM) were based on two concepts: reduction and dataflow (PIM-R, and PIM-D, respectively)[174,239]. The architectures for these computers were similar, but the technique for evaluation reflected these two philosophies. The hardware of PIM-R and PIM-D have both been simulated.

PIM-I is a hardware design for the intermediate stage of the FCCS project. The target speed for the 100-processor PIM-I is 10 to 20 MLIPS, with target speed of 200-500 KLIPS for the individual processors[213]. The machine language for this computer will be KL1-B, which is based on Guarded Horn Clauses. The software development will be done on a network of PSI systems (multi-PSI)[240].

(4) Parallel systems for production systems computations: The exploration of computers for production systems has been carried out primarily at universities. Table 17 presents a summary of these projects.

The DADO1 and DADO2 projects[241-242] at Columbia University develop a class of computers based on tree architectures. The upper-level nodes synchronize and select rules, intermediate nodes match and store rules, and the leaves are for the working memory.

Another project at Columbia University is the Non-Von computer (Non-Von-1 is an earlier version)[243-244]. Unlike DADO2, Non-Von connects smaller processing elements in a binary tree, which are subject to the control of large processing elements. Most of the pattern matching tasks that are done in the working memory have small granularity and are more suitable to be executed on a large number of small processing elements.

Table 17. Notable parallel machines for production systems

| Machine | Year/Status* | Interconnection | Communication | Features |
|---|---|---|---|---|
| DADO1, DADO2, | 1986/PO | Binary tree | Message passing | DADO2; 1 023 8-bit processors; 16K user memory; two modes; MIMD, and multiple SIMD; special I/O circuits |
| Non-Von | 1985/PO | Binary tree with leaf connections; connections to Large PEs (LPEs) | Message passing | Binary SIMD tree of Small PEs; leaves are connected in mesh; LPEs connected by network, with connections to high level nodes in tree; intelligent disk drives connected to LPEs |
| PSM | 1986/SI | Shared buses | Shared memory | 32-64 processors; parallel Rete Match algorithm; PEs connected to memory modules through cache; local memory; hardware task scheduler |

* Refer to Table 14 for explanation of status codes

Finally, the PSM computer is a large-grain machine that is specifically designed to support the OPS5 system and a parallel Rete-Match algorithm. Simulations have shown promising speedups, and that 32 processors are sufficient to exploit most parallelism for this system[139].

Numerous studies on strategies of mapping productions systems to multiprocessors can be found in the references[214,245-247].

(5) Parallel object-oriented architectures: Most development work on object-oriented programming has been on computers not specifically designed for object-oriented computation, such as the Intel iPSC. Two notable multiprocessors are designed specifically for object-oriented computations. FAIM-1[153] is a multiprocessor with special RISC processors connected by a hexagonal mesh. OIL, an intermediate language, was first developed for interfacing modules. However, the scope of the project has recently been changed to exclude the development of OIL and base programming of the computer on MultiScheme. The Dragon project is another object-oriented design project, but supports only 10 processors[248]. Table 18 shows a summary of the notable projects.

## 4.3 Connectionist processing

The connectionist implementations focus on the correlation between nodes in a graph containing the knowledge and have been designed

Table 18. Notable parallel computers for object-oriented computation

| Machine | Year/Status* | Interconnection | Communication | Features |
|---|---|---|---|---|
| FAIM-1 | -/UC | Hexagonal mesh | Message passing | PEs = Hectagons; heterogeneous shared-memory multiprocessor: Instruction Stream memory, Post Office communication processor, Evaluation Processor, and others; three-port switch at edge of array for I/O and wrapping of connections |
| Dragon | -/UC | Bus | Shared memory | Up to 10 32-bit workstation/processors; tightly coupled; associative cache at each processor |
| Apiary | 1980/PD | Single-stage network | Message passing | Implements Actor model; computations in the Actor model are partial orders of events with no assignment commands |

* Refer to Table 14 for explanation of status codes

primarily for semantic networks. Other connectionist designs of interest are the artificial neural networks. Unfortunately, current technology for design of artificial neural networks have precluded their development in a role greater than that of subsystem.

The four designs of connectionist system for the implementation of semantic networks correspond to the three types of message passing environments. NETL utilizes the most elementary processing elements in a marker passing system[217,249]. THISTLE is a similar design, but employs value passing instead[217]. The Connection Machine can be programmed to simulate marker passing in semantic networks and value passing in artificial neural networks using massive data-level parallelism[20]. Finally, SNAP relies on message passing[250]. These machine are summarized in Table 19.

## 4.4 Summary

The purpose of this section has been to give a top-level perspective of existent special-purpose computers designed for symbolic processing. Sequential, parallel, and connectionist processing are three fundamental approaches to processing techniques that are appropriate for various knowledge representations. Thus, the first attribute of distinction for complete systems will be made on this basis.

The other primary discerning feature is the overall processing technique. Current systems define the processing technique in one of two ways. The first way is by knowledge representation, for example, computers for parallel logic computation. The second processing technique is based on the programming paradigm, such as functional or object-oriented paradigm. In either case, the processing technique can be used as the second attribute for distinguishing existent systems.

Systems can be further distinguished by various design decisions, and micro/macrolevel architectures employed. For an overall perspective, however, it is not necessary to further classify these systems in order to understand the state of the art in symbolic processing systems.

Figure 8 classifies the computers presented in this section, on the basis of the above criteria. A number of systems have been designed for diverse symbolic applications and cannot be un-

Table 19. Parallel connectionist systems

| Machine | Year/Status* | Interconnection | Communication | Features |
|---|---|---|---|---|
| NETL | 1979/PD | Multi-level switching network | Marker passing | For semantic networks; million processors, each can store 16 markers; simple PEs; only Boolean functions |
| THISTLE | 1983/PD | Multi-level switching network | Value passing | For semantic networks; similar to NETL, only with 8-bit value passing |
| Connection Machine | 1986/CA | Hypercube | Message passing | General-purpose SIMD and multiple SIMD processing; can be programmed for marker passing operations in semantic networks and simulating artificial neural networks |
| SNAP | 1985/PD | Mesh with global bus | Message passing | Square array of identical processors; CAM in each PE for relationships between nodes; Communication Unit; Processing Unit |

* Refer to Table 14 for explanation of status codes

iquely classified into one category. The Connection Machine is one of the notable examples in this class.

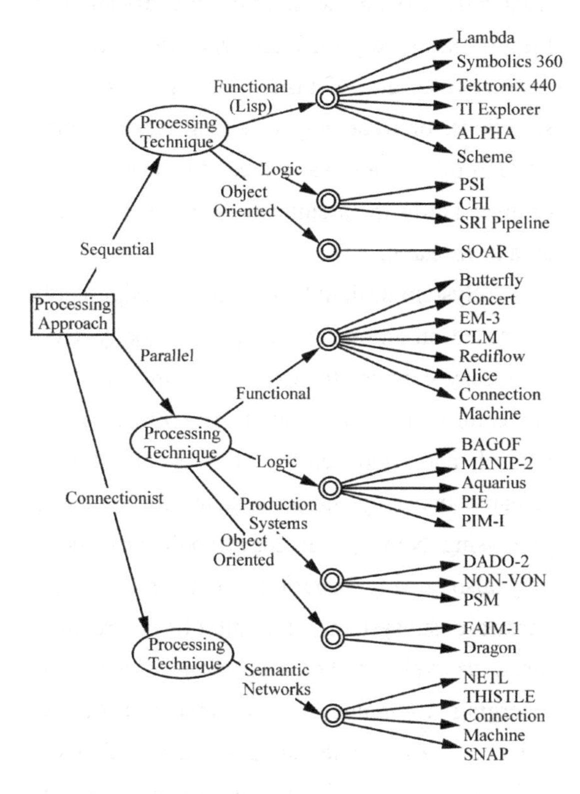


Figure 8. Complete symbolic processing systems

## 5 Research direction

We have presented in this paper an extensive discussion and analysis of the state of the art in computer solution of symbolic processing problems. We conclude in this section by indicating some of the research areas where advancement will most likely benefit fast and efficient computer solution of these applications.

• Technologies: The basis for any computer system is the technology in which it is implemented. The design of a system is often driven by its cost; hence, the fastest technologies, subject to cost constraints, are used. New and emerging technologies may give higher performance but are often prohibitively expensive. The candidates that will likely become cost effective in the near future include GaAs circuits[155,157], Wafer-Scale Integrated (WSI) circuits[156], analog-digital VLSI circuits, and optical computing techniques[158-160].

These emerging technologies offer tremendous potential for increasing the processing speeds of current computers. This extension in the limit of processing power is valuable, especially for real-time systems. However, the most that can be expected from these technologies is about one to two orders of magnitude speed improvement in the next ten years. They will not greatly impact the size or type of symbolic applications that are addressed today. Many of these applications involve huge search spaces of an exponential size; one to two orders of magnitude increase in computational speed will do little to extend the size of a solvable instance of such a problem[49].

• Algorithms: Research in the area of application-specific algorithms will have the greatest potential for speeding the solution of the given application. The development of new and improved algorithms for an application can be seen as finding alternative ways to incorporate knowledge about the application domain into the computer solution. In this way, advancement of symbolic processing capabilities in the area of application-specific algorithms is tightly linked to advancement in the area of knowledge representations.

• Knowledge representations: Most new knowledge representations for symbolic processing have emphasized declarative and distributed features in order to reduce programming complexity. These representation schemes may have to be modified or extended to tailor to the applications and computational environment. The addition of temporal features and nonmonotonicity would be helpful.

A major problem in the area of knowledge representations is the lack of an overall technique to guide the evaluation and selection of a knowledge-representation scheme. Research in this area could prove extremely valuable. Learning techniques for incorporating new knowledge about application domains into current solutions in a knowledge-intensive application may also have a great impact on symbolic processing. Artificial neural networks and connectionist representations are examples of incorporating automated learning techniques into the design philosophy from the knowledge-representation level.

• Software architecture: Software architectures are highly dependent on research in the area of knowledge representations. Generation of new software environments, tools, and languages will probably rely on amalgamation of known knowledge-representation techniques. Software development systems and automated intelligent programming assistants represent prime areas for advancement of symbolic programming. The problems of program verification and validation and continuous maintenance of symbolic programs are important related topics.

• Hardware architectures: As with software, hardware architectures are often based on known design techniques such as parallel processing and pipelining. Innovation for new architectural concepts may be caused by the availability of new and emerging technologies.

New hardware architectures are best utilized for operations that the computer performs frequently. Counter to intuition, identification of these tasks is very difficult. Operations may be instructions, parts of instructions, groups of instructions, or frequently recurring tasks. Identification of new and valid areas for development of new hardware architectures is an important area of research.

• System design: System-level design is often based on an overall design philosophy. Systems may contain, for example, a mix of data- and control-flow computation. The proper mix of control, data, and demand flow is one area of research that may impact systems for symbolic processing. New systems for symbolic processing may also greatly benefit from integrating new hardware subsystems and microlevel architectures, as well as the integration of new and emerging technologies. The major difficulty, however, lies in integrating designs with radically different knowledge representations. Combination of distributed representation offered by artificial neural networks and procedural or declarative representation offered by standard computers is an interesting area for development.

## References:

[1] NEWELL A, SHAW J C, SIMON H A. Programming the logic theory machine[C]// 1957 Western Joint Computer Conference: Techniques for Reliability. New York: ACM Press, 1957: 230-240.

[2] MCCARTHY J. History of Lisp[J]. ACM SIGPLAN Notices, 1978, 13(8): 217-223.

[3] ESPRIT: Europe challenges U.S. and Japanese competitors[J]. Future Generation Computer Systems, 1984, 1(1): 61-69.

[4] SCHAEFER D, FISCHER J. Beyond the supercom-

puter[J]. IEEE Spectrum, 1982, 19(3): 32-37.

[5] TRELEAVEN P C, LIMA I G. Japan's fifth-generation computer systems[J]. Computer, 1982, 15(8): 79-88.

[6] NEWELL A, SIMON H. ACM Turing award lecture: computer science as an empirical inquiry: symbols and search[J]. Communications of the ACM, 1976, 19(3): 113-126.

[7] WAH B W. Guest editor's introduction: new computers for artificial intelligence processing[J]. Computer, 1987, 20(1): 10-15.

[8] HAYES-ROTH F, WATERMAN D A, LENAT D B. Building Expert Systems[R]. 1983.

[9] Special issue on natural language processing[J]. Proceedings of the IEEE, 1986(6).

[10] ANDREWS H L. Speech processing[J]. Computer, 1984, 17(10): 315-324.

[11] TUCKER A B. A perspective on machine translation: theory and practice[J]. Communications of the ACM, 1984, 27(4): 322-329.

[12] Special issue on computer vision[J]. Proceedings of the IEEE, 1987.

[13] FU K S. Syntactic methods in pattern recognition[M]. New York: Academic Press, 1974.

[14] MICHALSKI R S, CARBONELL J G, MITCHELL T M. Machine learning: an artificial intelligence approach[M]. Palo Alto: Tioga Publishing Company, 1983.

[15] RUMMELHART D E, MCCLELLAND J L. Parallel distributed processing: explorations in the microstructure of cognition[M]. Cambridge: MIT Press, 1986.

[16] AIELLO L, CECCHI C, SARTINI D. Representation and use of metaknowledge[J]. Proceedings of the IEEE, 1986, 74(10): 1304-1321.

[17] DAVIS R, BUCHANAN B. Metal-level knowledge: overview and applications[C]// The 5th International Joint Conference on Artificial Intelligence. Los Altos: William Kaufman, 1977: 920-928.

[18] GENESERETH M R. An overview of meta-level architecture[C]// 1983 National Conference on Artificial Intelligence. Palo Alto: AAAI Press, 1983:119-124.

[19] BOWEN K. Meta-level programming and knowledge representation[J]. New Generation Computing, 1985, 3(4): 359-383.

[20] Thinking Machines Corporation. Connection machine model CM-2 technical summary[R]. 1987.

[21] KING M, ROSNER M. Scanning the issue: the special issue on knowledge representation[J]. Proceedings of IEEE, 1986, 74(10): 1299-1303.

[22] MCCALLA G, CERCONE N. Special Issue on knowledge representation[J]. Computer, 1983, 16(10).

[23] WINOGRAD T. Frame representations and the declarative procedural controversary[M]// BOBROW D G, COLLINS A. Representation and understanding: studies in cognitive science. New York: Academic Press, 1975: 185-210.

[24] HALSTEAD R H. Implementation of MULTILISP: LISP on multiprocessor[C]// The 1984 ACM Conference on LISP and Functional Programming. New York: ACM Press, 1984: 9-17.

[25] KURZWEIL R. What is artificial intelligence anyway?[J]. American Scientist, 1985, 73(3): 258-264.

[26] KOWALSKI R. Predicate logic as a programming language[M]// MASON R E A. IFIP information processing. Amsterdam: North-Holland Publishing Company, 1974: 569-574.

[27] NEWELL A, SIMON H A. Human problem solving[M]. Englewood Cliffs: Prentice-Hall, 1972.

[28] DREYFUS H, DREYFUS S. Why expert systems do not exhibit expertise[J]. IEEE Expert, 1986, 1(2): 86-90.

[29] QUILLIAN M R. Word concepts: a theory and simulation of some basic semantic capabilities[J]. Behavioral Science, 1967, 12(5): 410-430.

[30] KOWALSKI R. Logic for problem solving[M]. Amsterdam: North-Holland Publishing Company, 1979.

[31] MINSKY M. A framework for representing knowledge[M]// WINSTONP H. The psychology of computer vision. New York: McGraw-Hill Companies, 1975.

[32] STANFILL C, WALTZ D. Toward memory-based reasoning[J]. Communications of the ACM, 1986, 29(12): 1213-1228.

[33] HILLIS W D. The connection machine[M]. Cambridge: MIT Press, 1985.

[34] ZADEH L A. Fuzzy sets[J]. Information and Control, 1965(8): 338-353.

[35] ZADEH L A. Approximate reasoning based on fuzzy logic[C]// The 6th International Joint Conference on Artificial Intelligence. Los Altos: William Kaufman, 1979: 1004-1010.

[36] BUCHANAN B G, SHORTLIFFE E H. Rule-based experts programs: the MYCIN experiments of the Stanford heuristic programming project. Reading[Z]. Boston: Addison-Wesley Longman Publishing, 1984.

[37] SHAFER G. A mathematical theory of evidence[M]. Princeton: Princeton University Press, 1976.

[38] DUDA R O, HART P E, NILSSON N J. Subjective Bayesian methods for rule-based inference systems[C]// American Federation of Information Processing Societies: 1976 National Computer Conference. Arlington: AFIPS Press, 1976: 1075-1082.

[39] COHEN P R, GRINBERG M R. A theory of heuristic reasoning about uncertainty[J]. The AI Magazine, 1983: 17-24.

[40] SIMON H A. Whether software engineering needs to be artificially intelligent[J]. IEEE Transactions on Software Engineering, 1986, SE-12(7): 726-732.

[41] LENAT D B. The ubiquity of discovery[C]// The 5th International Joint Conference on Artificial Intelligence. Los Altos: William Kaufman, 1977: 1093-1105.

[42] LENAT D B. The nature of heuristics[J]. Artificial Intelligence, 1982, 19(2): 189-249.

[43] PEARL J. Heuristics: intelligent search strategies for computer problem solving[M]. Boston: Addison-Wesley Longman Publishing, 1984.

[44] PEARL J. Some recent results in heuristic search theory[J]. IEEE Transactions on Pattern Analysis and Machine Intelligence, 1984, PAMI-6(1): 1-13.

[45] SIMON H A. Search and reasoning in problem solving[J]. Artificial Intelligence, 1983, 21(1-2): 7-29.

[46] LAI T H, SAHNI S. Anomalies in parallel branch-and-bound algorithms[J]. Communications of the ACM, 1984, 27(6): 594-602.

[47] LI G J, WAH B W. Computational efficiency of parallel approximate branch-and-bound algorithms[C]// 1984 International Conference on Parallel Processing (ICPP'84). Washington: IEEE Computer Society Press, 1984: 473-480.

[48] LI G J, WAH B W. How good are parallel and ordered depth-first searches?[C]// 1986 International Conference on Parallel Processing (ICPP'86). Washington: IEEE Computer Society Press, 1986: 992-999.

[49] WAH B W, LI G J, YU C F. Multiprocessing of combinatorial search problems[J]. Computer, 1985, 18(6): 93-108.

[50] HINTIKKA J. Impossible possible world vindicated[Z]. 1975.

[51] PAGIN R, HALPERN J. Belief, awareness, and limited reasoning: preliminary report[C]// The 9th International Joint Conference on Artificial Intelligence. Los Altos: William Kaufman, 1985: 491-501.

[52] HALPERN J, MOSES Y. A guide to the modal logics of knowledge and belief: preliminary[C]// The 9th International Joint conference on Artificial Intelligence. Los Altos: William Kaufman, 1985: 480-490.

[53] BOBROW D G, HAYESP J. Special issue on non-monotonic logic[J]. Artificial Intelligence, 1980, 13(1-2).

[54] FIKES R E, NILSSON N J. STRIPS: a new approach to the application of theorem proving to problem solving[J]. Artificial Intelligence, 1971, 2(3-4): 189-208.

[55] HEWITT C E. Description and theoretical analysis of planner: a language for proving theorems and manipulating models in robot[R]. 1972.

[56] MCCARTHY J, HAYES P. Some philosophical problems from the standpoint of artificial intelligence[J]. Machine Intelligence, 1969(4): 463-502.

[57] SANDEWALL E. An approach to the frame problem, and its implementation[J]. Machine Intelligence, 1972(7): 195-204.

[58] DOYLE J. A truth maintenance system[J]. Artificial Intelligence, 1979, 12(3): 231-272.

[59] KLEER J. An assumption-based TMS[J]. Artificial Intelligence, 1986, 28(2): 127-161.

[60] KLEER J. Extending the ATMS[J]. Artificial Intelligence, 1986, 28(2): 163-196.

[61] KLEER J. Problem solving with the MMS[J]. Artificial Intelligence, 1986, 28(2): 197-224.

[62] SCHOR M. Declarative knowledge programming: better than procedural[J]. IEEE Expert, 1986, 1(1): 36-43.

[63] FENNELL R D, LESSER V R. Parallelism in artificial intelligence problem solving: a case study of Hearsay[J]. IEEE Transactions on Computers, 1977, C-26(2): 98-111.

[64] LENAT D B. Computer software for intelligent systems[J]. Scientific American, 1984, 251(3): 204-213.

[65] FORGY C L, GUPTA A, NEWELL A, et al. Initial assessment of architectures for production systems[C]// 1984 National Conference on Artificial Intelligence. Palo Alto: AAAI Press, 1984: 116-120.

[66] KIBLER D F. Parallelism in AI programs[C]// The 9th International Joint Conference on Artificial Intelligence. Los Altos: William Kaufman, 1985: 53-56.

[67] LI G J, WAH B W. Optimal granularity of parallel evaluation of AND-trees[C]// 1986 Fall Joint Computer Conference. Washington: IEEE Computer Society Press, 1986: 297-306.

[68] WALTZ D L. Applications of the connection machine[J]. Computer, 1987, 20(1): 85-97.

[69] FAHLMAN S E. Parallel processing in artificial intelligence[J]. Parallel Computing, 1985, 2(3): 283-286.

[70] YAMAGUCHI Y, TODA K, YUBA T. A performance evaluation of a Lisp-based data-driven machine (EM-3)[C]// The 10th Annual Symposium on Computer Architecture. Piscataway: IEEE Press, 1983: 363-369.

[71] BACKUS J. Can programming be liberated from the von Neumann style? A functional style and algebra of programs[J]. Communications of the ACM, 1978, 21(8): 613-641.

[72] DARLINGTON J, P. HENDERSON P, TURNER D. Functional programming and its applications[M]. Cambridge: Cambridge University Press, 1982.

[73] HENDERSON P. Function Programming, Application and Implementation[M]. Englewood Cliffs: Prentice-Hall, 1980.

[74] CLARK K L, TARNLUNDS A. Logic programming[M]. New York: Academic Press, 1982.

[75] DEGROOTD, LINDSTROMG. Logic Programming[Z]. 1985.

[76] GOLDBERG A J, ROBSON D. Smalltalk-80: the language and its implementation[Z]. 1983.

[77] SNYDER A. Object-oriented programming for common Lisp[R]. 1985.

[78] STEFIK M, BOBROW D G. Object-oriented programming: themes and variations[J]. AI Magazine, 1986, 6(4): 40-62.

[79] WEGNER P, SHRIVER B. Special issue on object-oriented programming workshop[J]. ACM SIGPLAN Notices, 1986, 21(10).

[80] MCCARTHY J. Recursive functions of symbolic expressions and their computation by machine[J]. Communications of the ACM, 1960, 3(4): 184-195.

[81] BURSTALL R M, MACQUEEN D B, SANNELLA D T. HOPE: an experimental applicative language[C]// The 1980 ACM conference on LISP and Functional Programming. New York: ACM Press, 1980: 136-143.

[82] MCGRAW J R. Data flow computing: Software development[J]. IEEE Transactions on Computers, 1980, C-29(12): 1095-1103.

[83] ARVIN D, GOSTELOW K, PLOUFFE W. An asynchronous programming language and computing machine[R]. Irvine: University of California, 1978.

[84] DARLINGTON J Functional programming (Chapter 5)[M]// CHAMBERS F B, DUCE D A, JONES G P. Distributed computing. London: Academic Press, 1984.

[85] HALSTEAD R. Parallel symbolic computing[J]. Computer, 1986, 19(8): 35-43.

[86] CROMARTY A S. What are current expert system tools missing?[C]// The 30th IEEE Computer Society International Conference (COMPCON'85). Washington: IEEE Computer Society Press, 1985: 411-418.

[87] HALSTEAD R H. An assessment of Multilisp: Lessons from experience[J]. Parallel Programming, 1986, 15(6): 459-501.

[88] WINSTON P H, HORN B. Lisp (second edition)[M]. Boston: Addison-Wesley Longman Publishing, 1984.

[89] BARR A, FEIGENBAUM E A. The handbook of artificial intelligence[M]. Los Altos: William Kaufmann, 1981.

[90] SUSSMAN G J, MCDERMOTT D V. From PLANNER to CONNIVER: a genetic approach[C]// 1972 Fall Joint Computer Conference. New York: ACM Press, 1972: 129-137.

[91] CLOCKSIN W F, MELLISH C S. Programming in Prolog[M]. Berlin: Springer-Verlag, 1981.

[92] WARREN D H, PEREIRA L M, PEREIRA F. Prolog-the language and its implementation compared with Lisp[C]// 1977 Symposium Artificial Intelligence and Programming Languages. New York: ACM Press, 1977: 109-115.

[93] MOTO-OKA T, STONE H S. Fifth-generation computer systems: a Japanese project[J]. Computer, 1984, 17(3): 6-13.

[94] CLARK K, GREGORY S. Note on system programming in PARLOG[C]// 1984 International Conference on Fifth Generation Computer Systems. Amsterdam: North-Holland Publishing Company, 1984: 299-306.

[95] SHAPIRO E, TAKEUCHI A. Object oriented programming in concurrent Prolog[J]. New Generation Computing, 1983, 1(1): 25-48.

[96] DEGROOT D. Restricted AND-parallelism[C]// 1984 International Conference on Fifth Generation Computer Systems. Amsterdam: North-Holland Publishing Company, 1984: 471-478.

[97] BROWNSTON L, FARRELL R, KANT E, et al. Programming Expert Systems in OPS5[Z]. 1985.

[98] NEWELL A. Production systems: models of control structures[M]// CHASE W G. Visual information processing. New York: Academic Press, 1975.

[99] FORGY C L. Rete: a fast algorithm for the many pattern/many object pattern match problem[J]. Artificial Intelligence, 1982, 19(1): 17-37.

[100] RAMAMOORTHY C V, SHEKHAR S, GARG V. Software development support for AI programs[J]. Computer, 1987, 20(1): 30-40.

[101] REDDY U S. On the relationship between logic and functional languages[M]// DEGROOT D, LINDSTROME G. Logic programming. Englewood Cliffs: Prentice-Hall, 1985.

[102] DARLINGTON J, FIELD A J, PULL H. The unification of functional and logic languages[R]. 1985.

[103] SUBRAHMANYAM P A, YOU J H. FUNLOG: a computational model integrating logic programming and functional programming[M]// DEGROOT D, LINDSTROM G. Logic programming. Englewood Cliffs: Prentice-Hall, 1985.

[104] MALACHI Y, MANNA Z, WALDINGER R. TABLOG: a new approach to logic programming[M]// DEGROOT D, LINDSTROM G. Logic programming. Englewood Cliffs: Prentice-Hall, 1985.

[105] DAHL O J, NYGAARD K. SIMULA: an ALGOL-based simulation language[J]. Communications of the ACM, 1966, 9(9): 671-678.

[106] AGHA G. Actor: a model of concurrent computation in distributed systems[M]. Cambridge: MIT Press, 1986.

[107] DAVIS A L, ROBISON S V. The FAIM-1 symbolic multiprocessing system[C]// The 30th IEEE Computer Society International Conference (COMPCON'85). Washington: IEEE Computer Society Press, 1985: 370-375.

[108] BOBROWD G, KAHN K, KICZALES G, et al. CommonLoops: Merging common Lisp and object-oriented programming[J]. ACM SIGPLAN Notices, 1986, 21(11): 12-29.

[109] HEWITT C. Viewing control structure as patterns of passing messages[J]. Artificial Intelligence, 1977, 8(3): 323-364.

[110] AGHA G, HEWITT C. Concurrent programming using actors: exploiting large-scale parallelism[J]. Lecture Notes in Computer Science, 1985(206): 19-41.

[111] HEWITT C. The apiary network architecture for knowledgeable systems[C]// The 1980 ACM Conference on LISP and Functional Programming. New York: ACM Press, 1980: 107-117.

[112] HEWITT C, LIEBERMAN H. Design issues in parallel architectures for artificial intelligence[C]// The 28th IEEE Computer Society International Conference (COMPCON'84). Washington: IEEE Computer Society Press, 1984: 418-423.

[113] RAMAMOORTHY C V, PRAKASH A, TSAI W T, et al. Software engineering: problems and perspectives[J]. Computer, 1984, 17(10): 191-210.

[114] SHATZ S M, WANG J P. Introduction to distributed-software engineering[J]. Computer, 1987, 20(10): 23-32.

[115] TEITELMAN W, MASINTER L. The Interlisp programming environment[J]. Computer, 1981, 14(4): 25-33.

[116] MOON D A. Symbolics architecture[J]. Computer, 1987, 20(1): 43-52.

[117] HAYASHI H, HATTORI A, AKIMOTO H. ALPHA: a high-performance Lisp machine equipped with a new stack structure and garbage collection system[C]// The 10th Annual Symposium on Computer Architecture. Piscataway: IEEE Press, 1983: 342-348.

[118] DEERI M F. Architectures for AI[J]. Byte, 1985: 193-206.

[119] FEUSTELE A. On the advantages of tagged architecture[J]. IEEE Transactions on Computers, 1973, C-22(7): 644-656.

[120] BURKS A W. Keynote of CONPAR86[J]. Lecture Notes in Computer Science, 1986(237): 1-17.

[121] BARTH J M. Shifting garbage collection overhead to compile time[J]. Communications of the ACM, 1977, 20(1): 513-518.

[122] UNGAR D. Generation scavenging: a non-disruptive high performance storage reclamation algorithm[J]. ACM SIGPLAN Notices, 1984, 9(3): 157-167.

[123] DIJKSTRA E W, LAMPORT L, MARTIN A J, et al. On-the-fly garbage collection: an exercise in cooperation[J]. Communications of the ACM, 1978, 21(11): 966-975.

[124] KUNG H, SONG S. An efficient parallel garbage collection systems and its correctness proof[R]. 1977.

[125] BAKER H G. Optimizing allocation and garbage collection of spaces[M]// WINSTON P H,BROWN R H. Artificial intelligence: an MIT perspective. Cambridge: MIT Press, 1979.

[126] LIEBERMAN H, HEWITT C. A real-time garbage collector based on the lifetimes of objects[J]. Communications of the ACM, 1983, 26(6): 419-429.

[127] HIBINO Y. A practical parallel garbage collection algorithm and its implementations[C]// The 7th Annual

Symposium on Computer Architecture. New York: ACM Press, 1980: 113-120.
[128] LI G J, WAH B W. The design of optimal systolic arrays[J]. IEEE Transactions on Computers, 1985, C-34(1): 66-77.
[129] MUKHOPADHYAY A. Hardware algorithms for non-numeric computation[J]. IEEE Transactions on Computers, 1979, C-28(6): 384-394.
[130] FALOUTOS C. Access method for text[J]. ACM Computing Surveys, 1985, 17(1): 49-74.
[131] TAKAHASI K, YAMADA H, NAGAI H, et al. A new string search hardware architecture for VLSI[C]// The 13th Annual Symposium on Computer Architecture. Piscataway: IEEE Press, 1986: 20-27.
[132] RAMAMOORTHY C V, TURNER J L, WAH B W. A design of a fast cellular associative memory for ordered retrieval[J]. IEEE Transactions on Computers, 1978, C-27(9): 800-815.
[133] KANERVA P. Parallel structures in human and computer memory[R]. 1986.
[134] FAHLMANS E, HINTON G E. Connectionist architecture for artificial intelligence[J]. Computer, 1987, 20(1): 100-109.
[135] SHAW D E. On the range of applicability of an artificial intelligence machine[R]. 1985.
[136] STOLFO S J, MIRANKER D P. The DADO production system machine[J]. Journal of Parallel and Distributed Computing, 1986 (2): 269-296.
[137] BROOKS R, LUM R. Yes, an SIMD machine can be used for AI[C]// The 9th International Joint Conference on Artificial Intelligence. Los Altos: William Kaufman, 1985: 73-79.
[138] GAUDIOT J L, LEE S, SOHN B. Data-driven multiprocessor implementation of the Rete match algorithm[R]. 1987.
[139] GUPTA A, FORGY C L, NEWELL A, et al. Parallel algorithms and architectures for rule-based systems[C]// The 13th Annual Symposium on Computer Architecture. Piscataway: IEEE Press, 1986: 28-37.
[140] SHOBATAKE Y, ASIO H. A unification processor based on uniformly structured cellular hardware[C]// The 13th Annual Symposium on Computer Architecture. Piscataway: IEEE Press, 1986: 140-148.
[141] NAKAZAKI R, KONAGAYA A, HABATA S, et al. Design of a high speed Prolog machine (HPM)[C]// The 12th Annual Symposium on Computer Architecture. Piscataway: IEEE Press, 1985: 191-197.
[142] ROBINSON P. The SUM: an AI co-processor[J]. Byte, 1985, 10(6): 169-180.
[143] NAKAGAWA H. AND Parallel Prolog with divided assertion set[C]// The 1984 International Symposium on Logic Programming. Piscataway: IEEE Press, 1984: 22-28.
[144] BRUYNOOGHE M. The memory management of PROLOG implementations[M]// CLARK K, TARNLUND S A. Logic programming. New York: Academic Press, 1982: 83-89.
[145] MELLISHC S. An alternative to structure sharing in the implementation of a Prolog interpreter[M]// CLARKK, TARNLUNDS A. Logic programming. New York: Academic Press, 1982: 99-106.
[146] VITTER J S, SIMONS R A. New classes for parallel complexity: a study of unification and other complete problems[J]. IEEE Transactions on Computers, 1986, C-35(5): 403-418.
[147] SHIH Y, RANI K B. Large scale unification using a mesh-connected array of hardware unifiers[C]// 1987 International Conference on Parallel Processing. State College: Pennsylvania State University Press, 1987: 787-794.
[148] FUCHI K. The direction the FGCS project will take[J]. New Generation Computing, 1983, 1(1): 3-9.
[149] FISCHETTI M A. A review of progress at MCC[J]. IEEE Spectrum, 1986, 23(3): 76-82.
[150] DARLINGTON J, REEVE M. ALICE and the parallel evaluation of logic programs[Z]. 1983.
[151] PATTERSOND A. Reduced instruction set computers[J]. Communications of the ACM, 1985, 28(1): 8-21.
[152] UNGAR D, PATTERSON D. What price Smalltalk?[J]. Computer, 1987, 20(1): 67-74.
[153] ANDERSON J M, COATES W S, DAVIS A L, et al. The architecture of FAIM-1[J]. Computer, 1987, 20(1): 55-65.
[154] MATTHEWS G, HEWES R, KRUEGER S. Single-chip processor runs Lisp environments[J]. Computer Design, 1987, 26(9): 69-72, 74.
[155] LARSON L E, JENSEN J F, GREILING P T. GaAs high-speed digital IC technology: an overview[J]. Computer, 1986, 19(10): 21-28.
[156] MCDONALD J F, GREUB H J, STEINVORTH R H, et al. Wafer scale interconnections for GaAs packaging-applications to RISC architecture[J]. Computer, 1987, 20(4): 21-35.

[157] LEHR T F, WEDIG R G. Toward a GaAs realization of a production-system machine[J]. Computer, 1987, 20(4): 36-49.

[158] WEST L C. Picosecond integrated optical logic[J]. Computer, 1987, 20(12): 34-47.

[159] BERRA P B, TROULLINOS N B. Optical techniques and data/knowledge base machines[J]. Computer, 1987, 20(10): 59-70.

[160] GAYLORD T K, VERRIEST E I. Matrix triangularization using arrays of integrated optical Givens rotation devices[J]. Computer, 1987, 20(12): 59-67.

[161] MCAULAY A D. Spatial-light-modulator interconnected computers[J]. Computer, 1987, 20(10): 45-58.

[162] VEGDAHL S R. A survey of proposed architectures for the execution of functional languages[J]. IEEE Transactions on Computers, 1984, C-33(12): 1050-1071.

[163] AMAMIYA M, HAKOZAKI K, YOKOI T, et al. Mew architecture for knowledge based mechanisms[C]// 1981 International Conference on Fifth Generation Computer Systems. Amsterdam: North-Holland Publishing Company, 1982: 179-188.

[164] SU S Y W. Associative programming in CASSM and its applications[C]// The 3rd International Conference on Very Large Databases. Washington: IEEE Computer Society Press, 1977: 213-228.

[165] OZKARAHAN E A, SCHUSTER S A, SMITH K C. RAP-an associative processor for database management[C]// 1975 National Computer Conference. Arlington: AFIPS Press, 1975: 379-388.

[166] DEWITT D J. DIRECT-a multiprocessor organization for supporting relational database management systems[J]. IEEE Transactions on Computers, 1979, C-28(6): 395-406.

[167] HAWTHORN P B, DEWITT D J. Performance analysis of alter native database machine architectures[J]. IEEE Transactions on Software Engineering, 1982, SE-8(1): 61-75.

[168] MALABARBAF J. Review of available database machine technology[C]// Proceedings of IEEE Trends and Applications. [S.l.:s.n.], 1984: 14-17

[169] TORII S, KOJIMA K, YOSHIZUMI S, et al. A database system architecture based on a vector processing method[C]// The 3rd International Conference on Data Engineering. Washington: IEEE Computer Society Press, 1987: 182-189.

[170] BABB E. Implementing a relational database by means of specialized hardware[J]. ACM Transactions Database Systems, 1979, 4(1): 1-29.

[171] NECHES P M. Hardware support for advanced data management systems[J]. Computer, 1984, 17(11): 29-40.

[172] STONE H S. Parallel querying of large databases: a case study[J]. Computer, 1987, 20(1): 11-21.

[173] KELLOGG C. Intelligent assistants for knowledge and information resources management[C]// The 8th International Joint Conference on Artificial Intelligence. Los Altos: William Kaufman, 1983: 170-172.

[174] MURAKAMI K, KAKUTA T, ONAI R, et al. Research on parallel machine architecture for fifth-generation computer systems[J]. Computer, 1985, 18(6): 76-92.

[175] YU C F. Efficient Combinatorial search algorithms[D]. West Lafayette: Purdue University, 1986.

[176] AMAMIYA M, TAKESUE M, HASEGAWA R, et al. Implementation and evaluation of a list--processing-oriented data flow machine[C]// The 13th Annual Symposium on Computer Architecture. Piscataway: IEEE Press, 1986: 10-19.

[177] ITO N, et al. The architecture and preliminary evaluation results of the experimental parallel inference machine PIM-D[C]// The 13th International Symposium Computer Architecture. Piscataway: IEEE Press, 1986: 149-156.

[178] MURAKAMI K. Research on parallel machine architecture for fifth generation computing systems[J]. Computer, 1985, 18(6): 76-92.

[179] DARLINGTON J, REEVE J. ALICE: a multi-processor reduction machine for the parallel evaluation of applicative languages[C]// The 1981 Conference on Functional Programming Languages and Computer Architecture. New York: ACM Press, 1981: 65-74.

[180] KELLER R M, LIN F C H, TANAKA J. Rediflow multiprocessing[C]// The 28th IEEE Computer Society International Conference (COMPCON'84). Washington: IEEE Computer Society Press, 1984: 410-417.

[181] ROSENBLATT F. Principles of neurodynamics[M]. [S.l.:s.n.], 1962.

[182] RUMELHART D D, HINTON G, MCCLELLAND J L. A general framework for parallel distributed processing[M]// RUMELHARTD E, MCCLELLANDJ L. Parallel distributed processing: explorations in the microstructure of cognition. Cambridge: MIT Press, 1986.

[183] HOPFIELD J J, TANK D W. Disordered systems and biological organization[M]. Heidelberg: Springer, 1986.

[184] HINTON G, MCCLELLAND J L, RUMELHART D D. Distributed representations[M]// RUMEL D E, MCCLELLAND J L. Parallel distributed processing: explorations in the microstructure of cognition. Cambridge: MIT Press, 1986.

[185] Defense Advanced Research Project Agency. DARPA neural network study[Z]. 1988.

[186] LIPPMANNR P. An introduction to computing with neural nets[J]. IEEE ASSP Magazine, 1987: 4-22.

[187] HINTON G E. Connectionist learning procedures[R]. 1987.

[188] DECKER J S. Neural network models of learning and adaptation[J]. Physica D Nonlinear Phenomena, 1986, 22(1-3): 216-232.

[189] SEJNOWKSI T J, ROSENBERG C R. NETtalk: a parallel network that learns to read aloud[R]. 1986.

[190] MCCLELLAND J L, RUMELHART D D, HINTON G. The appeal of parallel distributed processing[M]// RUMELHARTD E, MCCLELLAND J L. Parallel distributed processing: explorations in the microstructure of cognition. Cambridge: MIT Press, 1986.

[191] HOPFIELD J J, TANK D W. Computing with neural circuits: a model[J]. Science, 1986, 233(4764): 625-633.

[192] BAUME B. Towards practical ‘neural’ computation for combinatorial optimization problems[C]// AIP Conference Proceedings 151 on Neural Networks for Computing. New York: American Institute of Physics Inc., 1986: 53-58.

[193] ANDERSON C W. Strategy learning with multilayer connectionist representations[C]// The 4th International Workshop on Machine Learning. [S.l.:s.n.], 1987: 103-114.

[194] HINTON G, SEJNOWSKI T J. Learning and relearning in Boltzmann machines[M]// RUMEL D E, MCCLELLAND J L. Parallel distributed processing: explorations in the microstructure of cognition. Cambridge: MIT Press, 1986.

[195] ALSPECTOR J, ALLEN R B. A neuromorphic VLSI learning system[M]// LOSELEBEN P. Advanced research in VLSI: proceeding of 1987 Stanford conference. Cambridge: MIT Press, 1987.

[196] FAGIN F. How far can we go with proven technology?[Z]. 1987.

[197] WEISER M, KOGGE S, MCELVANY M, et al. Status and performance of the ZMOB parallel processing system[C]// The 29th IEEE Computer Society International Conference (COMPCON'85). Washington: IEEE Computer Society Press, 1985: 71-73.

[198] BROWN C M, ELLIS C S, FELDMAN J A, et al. Research with the Butterfly multicomputer[J]. Rochester Research Review, 1984: 3-23.

[199] KNIGHT T. The CONS microprocessor[Z]. 1974.

[200] DEUTSCH P. Experience with a microprogrammed Interlisp systems[C]// MICRO-11: Proceedings of the 11th Annual Workshop on Microprogramming. Piscataway: IEEE Press, 1978: 128-129.

[201] GOTO E, IDA T, HIRAKI K, et al. FLATS, a machine for numerical, symbolic and associative computing[C]// The 6th International Joint Conference on Artificial Intelligence. Los Altos: William Kaufman, 1979: 1058-1066.

[202] GREENFIELD N, JERICHO A. A Professional's personal computer system[C]// The 8th Annual Symposium on Computer Architecture. Piscataway: IEEE Press, 1981: 217-226.

[203] GRISS M, SWANSON M. MBALM/1700: a microprogrammed Lisp machine for the Burroughs B1726[C]// MICRO-10: Proceedings of the 11th Annual Workshop on Microprogramming. Piscataway: IEEE Press, 1977: 15-25.

[204] NAGAO M, TSUJII J I, NAKAJIMA K, et al. Lisp machine NK3 and measurement of its performance[C]// The 6th International Joint Conference on Artificial Intelligence. Los Altos: William Kaufman, 1979: 625-627.

[205] PUTTKAMER E. A microprogrammed Lisp machine[J]. Microprocessing and Microprogramming, 1983, 11(1): 9-14.

[206] SANSONNET J, BOTELLA D, PEREZ J. Function distribution in a list-directed architecture[J]. Microprocessing and Microprogramming, 1982, 9(3): 143-153.

[207] SANSONNET J P, CASTAN M, PERCEBOIS C. M3L: a list-directed architecture[C]// The 7th Annual Symposium on Computer Architecture. New York: ACM Press, 1980: 105-112.

[208] SANSONNET J P, CASTAN M, PERCEBOIS C, et al. Direct execution of Lisp on a list-directed architecture[C]// The 1982 Symposium Architectural Support for Programming Languages and Operating Systems. New York: ACM Press, 1982: 132-139.

[209] TAKI K, KANEDA Y, MAEKAWA S. The experimental Lisp machine[C]// The 6th International Joint Conference on Artificial Intelligence. Los Altos: William Kaufman, 1979: 865-867.

[210] SUSSMAN G J, HOLLOWAY J, STEEL G L, et al. Scheme-79-Lisp on a chip[J]. Computer, 1981, 14(7): 10-21.

[211] TICK E, WARREN D H D. Towards a pipelined Prolog processor[J]. New Generation Computing, 1984, 2(4): 323-345.

[212] TAKI K, YOKOTA M, YAMAMOTO A, et al. Hardware design and implementation of the personal sequential inference machine (PSI)[C]// 1984 International Conference on Fifth Generation Computer Systems. Amsterdam: North-Holland Publishing Company, 1984: 398-409.

[213] UCHIDA S. Inference machines in FGCS project[C]// 1987 International Conference on Very Large Scale Integration (VLSI'87). [S.l.:s.n.], 1985.

[214] LENAT D B, MCDERMOT T J. Less than general production system architectures[C]// The 5th International Joint Conference on Artificial Intelligence. Los Altos: William Kaufman, 1977: 923-932.

[215] UNGAR D, BLAU R, FOLEY P, et al. Architecture of SOAR: Smalltalk on a RISC[C]// The 11th Annual International Symposium Computer Architecture. New York: ACM Press, 1984: 188-197.

[216] FAHLMAN S E. NETL: a system for representing and using real-world knowledge[Z]. 1979.

[217] FAHLMAN S E, HINTON G E. Massively parallel architectures for AI: NETL, THISTLE, and BOLTZMANN machines[C]// The 3rd National conference on Artificial intelligence. Palo Alto: AAAI Press, 1983: 109-113.

[218] HOPFIELD J J, TANK D W. Neural computation of decisions in optimization problems[J]. Biological Cybernetics, 1985, 52(3): 1-25.

[219] DESPAIN A M, PATT Y N. Aquarius: a high performance computing system for symbolic/numeric applications[C]// The30th IEEE Computer Society International Conference (COMPCON'85). Washington: IEEE Computer Society Press, 1985: 376-382.

[220] SUGIMOTO S, AGUSA K, TABATA K, et al. A multi--microprocessor system for con current Lisp[C]// 1983 International Conference on Parallel Processing (ICPP'83). Washington: IEEE Computer Society Press, 1983: 135-143.

[221] HALSTEAD R, ANDERSON T, OSBORNE R, et al. Concert: design of a multiprocessor development system[C]// The 13th Annual Symposium on Computer Architecture. Piscataway: IEEE Press, 1986: 40-48.

[222] GOTO A, TANAKA H, MOTO-OKA T. Highly parallel inference engine PIE: goal rewriting model and machine architecture[J]. New Generation Computing, 1984, 2(1): 37-58.

[223] ERMAN L D, HAYES-ROTH F, LESSER V R, et al. The Hearsay-II speech-understanding system: Integrating knowledge to resolve uncertainty[J]. ACM Computing Surveys, 1980, 12(2): 213-253.

[224] NII H P. Blackboard systems, blackboard systems from a knowledge engineering perspective[J]. AI Magazine, 1986(6): 82-106.

[225] HAYES-ROTH B. A blackboard architecture for control[J]. Artificial Intelligence, 1985, 26(3): 251-321.

[226] HALSTEAD R, LOAIZA J. Expection handling in Multilisp[C]// 1985 International Conference on Parallel Processing (ICPP'85). Washington: IEEE Computer Society Press, 1985: 822-830.

[227] LINDSTROM G, PANANGADEN P. Stream-based execution of logic programs[C]// 1984 International Symposium on Logic Programming. Piscataway: IEEE Press, 1984: 168-176.

[228] UEDA K. Guarded horn clauses[R]. 1985.

[229] TAKEUCHI A, FUKUKAWA K. Parallel logic programming languages[C]// The 3rd International Conference on Logic Programming. Heidelberg: Springer, 1986: 242-254.

[230] SMITHR G. The contract net protocol: high-level communication and control in a distributed problem solver[J]. IEEE Transactions on Computers, 1980, C-29(12): 1104-1113.

[231] CAMMARATA S,MCARTHUR D, STEEB R, et al. Strategies of cooperation in distributed problem solving[C]// The 8th International Joint Conference on Artificial Intelligence. Los Altos: William Kaufman, 1983: 767-770.

[232] YAMAGUCHI Y, TODA K, HERATH J, et al. EM-3: a Lisp-based data-driven machine[C]// 1984 International Conference on Fifth Generation Computer Systems. Amsterdam: North-Holland Publishing Company, 1984: 524-532.

[233] BILLSTROM D, BRANDENBURG J, TEETER J. CCLISP on the iPSC concurrent computer[C]// The 6th National conference on Artificial intelligence. Palo Alto: AAAI Press, 1987: 7-12.

[234] Thinking Machines Corporation. Lisp Reference Manual(version 4.0)[R]. 1987.
[235] CIEPIELEWSKI A, HARIDI S. Execution of bagof on the OR-parallel token machine[C]// 1984 International Conference on Fifth Generation Computer Systems. Amsterdam: North-Holland Publishing Company, 1984: 551-560.
[236] LI G J, WAH B W. MANIP-2: A multicomputer architecture for evaluating logic programs[C]// 1985 International Conference on Parallel Processing (ICPP'85). Washington: IEEE Computer Society Press, 1985: 123-130.
[237] MOTO-OKA T, TANAKA H, AIDA H, et al. The architecture of a parallel inference engine (PIE)[C]// 1984 International Conference on Fifth Generation Computer Systems. Amsterdam: North-Holland Publishing Company, 1984: 479-488.
[238] MOTO-OKA T. Overview to the fifth generation computer system project[C]// The 10th Annual Symposium on Computer Architecture. Piscataway: IEEE Press, 1983: 417-422.
[239] ITO N, SHIMIZU H, KISHI M, et al. Dataflow based execution mechanisms of parallel and concurrent Prolog[J]. New Generation Computing, 1985, 3(1): 15-41.
[240] ICHIYOSHI N, MIYAZAKI T, TAKI K. A distributed implementation of flat GHC on the multi-PSI[C]// The 4th International Conference on Logic Programming. Cambridge: MIT Press, 1987: 257-275.
[241] STOLFO S J, SHAW D E. DADO: a tree-structured machine architecture for production systems[R]. 1982.
[242] STOLFO S J. Initial performance of the DADO-2 prototype[J]. Computer, 1987, 20(1): 75-84.
[243] HILLYERAND B K, SHAW D E. Execution of OPS5 production systems on a massively parallel machine[J]. Parallel Distributed Computing, 1986, 3(2): 236-268.
[244] IBRAHIM H A H, KENDER J R, SHAW D E. Low-level image analysis tasks on fine-grained tree-structured SIMD machines[J]. Journal of Parallel and Distributed Computing, 1987, 4(6): 546-574.
[245] OFLAZER K. Partitioning in parallel processing of production systems[C]// 1984 International Conference on Parallel Processing (ICPP'84). Washington: IEEE Computer Society Press, 1984: 92-100.
[246] TENORIO M F M, MOLDOVAN D I. Mapping production systems into multiprocessors[C]// 1985 International Conference on Parallel Processing (ICPP'85). Washington: IEEE Computer Society Press, 1985: 56-62.
[247] UHR L M. Parallel-serial production systems[C]// The 6th International Joint Conference on Artificial Intelligence. Los Altos: William Kaufman, 1979: 911-916.
[248] MONIER L, SIDHU P. The architecture of the dragon[C]// The 29th IEEE Computer Society International Conference (COMPCON'85). Washington: IEEE Computer Society Press, 1985: 118-121.
[249] FAHLMAN S E. Design sketch for a million-element NETL machine[C]// The 1st Annual National Conference on Artificial Intelligence. Palo Alto: AAAI Press, 1980: 249-252.
[250] MOLDOVAN D I, TUNG Y W. SNAP: a VLSI architecture for artificial intelligence processing[J]. Journal of Parallel and Distributed Computing, 1985, 2(2): 109-131.

发表于*IEEE Transactions on Systems, Man, and Cybernetics*，1989年第4期

# A survey on the design of multiprocessing systems for artificial intelligence applications

Benjamin W. WAH[1], Guo-jie LI[2]

1. Senior Member, IEEE

2. Member, IEEE

**Abstract:** Some important issues in designing computers for Artificial Intelligence (AI) processing are discussed. The issues discussed are divided into three levels: the representation level, the control level, and the processor level. The representation level deals with the knowledge and methods used to solve the problem and the means to represent it. The control level is concerned with the detection of dependencies and parallelism in the algorithmic and program representations of the problem, and with the synchronization and scheduling of concurrent tasks. The processor level addresses the hardware and architectural components needed to evaluate the algorithmic and program representations. Solutions in each level are illustrated by a number of representative systems. Design decisions in existing projects on AI computers are classical into the top-down, bottom-up, and middle-out approaches.

## 1 Introduction

In recent years, Artificial Intelligence (AI) techniques have been widely used in various applications, such as natural-language understanding, computer vision, and robotics. As AI applications move from the laboratories to the real world and as AI software grows in complexity, the computational throughput and cost are increasingly important concerns. The conventional von Neumann computers are not suitable for AI applications because they were designed mainly for sequential and deterministic numeric computations. Extensive efforts have been devoted to investigate and develop efficient AI architectures[1]. This paper provides a state-of-the-art assessment of AI-oriented systems and discusses the major issues involved in such designs.

### 1.1 Characteristics of AI computations

To develop a special-purpose computer to support AI applications, the requirements of these applications must be fully understood. Many conventional numeric algorithms are well analyzed, and bounds on their computational

This work was supported by the National Aeronautics and Space Administration under Contract NCC 2-481.

performance have been established. In contrast, many AI applications are characterized by symbolic processing, nondeterministic computations, dynamic execution, large potential for parallel and distributed processing, management of extensive knowledge, and an open system.

Symbolic processing: Data are generally processed in symbolic form in AI applications. Primitive symbolic operations, such as comparison, selection, sorting, matching, logic set operations (union, intersection, and negation), contexts and partitions, transitive closure, and pattern retrieval and recognition, are frequently used. At a higher level, symbolic operations on patterns such as sentences, speech, graphics, and images may be needed.

Nondeterministic computations: Many AI algorithms are nondeterministic, that is, planning in advance the procedures to execute and to terminate with the available information is impossible. This is attributed to a lack of knowledge and a complete understanding of the problem; it may result in exhaustively enumerating all possibilities when the problem is solved or in a controlled search through a solution space.

Dynamic execution: With a lack of complete knowledge and anticipation of the solution process, the capabilities and features of existing data structures and functions may be defined and new data structures and functions created while the problem is actually being solved. Further, the maximum size for a given structure may be so large that it is impossible to allocate the necessary memory space ahead of time. As a result, when the problem is solved, memory space and other resources may have to be dynamically allocated and deallocated, tasks may be dynamically created, and the communication topology may be dynamically changing.

Large potential for parallel and distributed processing: In parallel processing of deterministic algorithms, a set of necessary and independent tasks must be identified and processed concurrently. This class of parallelism is called AND parallelism. In AI processing, the large degree of nondeterminism offers an additional source of parallel processing. Tasks at a nondeterministic decision point can be processed in parallel. This latter class is called OR parallelism.

Knowledge management: Knowledge is an important component in reducing the complexity of solving a given problem: more useful knowledge means less exhaustive searching. However, many AI problems may have very high inherent complexity, hence the amount of useful knowledge may also be exceedingly large. Further, the knowledge acquired may be fuzzy, heuristic, and uncertain in nature. The representation, management, manipulation, and learning of knowledge are, therefore, important problems to be addressed.

Open system: In many AI applications, the knowledge needed to solve the problem may be incomplete because the source of the knowledge is unknown at the time the solution is devised, or the environment may be changing and cannot be anticipated at design time. AI systems should be designed with an open concept and allow continuous refinement and acquisition of new knowledge.

In general, two basic approaches are available for improving the computational efficiency of processing AI tasks: having heuristic knowledge to guide searches and using faster computers. In the following sections, these approaches will be discussed.

## 1.2 Heuristic searches

The key performance-related feature of AI computations is their nondeterminism, which results from a lack of complete understanding of the solution process. In other words, when a problem becomes well understood and can be solved by a deterministic algorithm, we usually cease to consider it "intelligent" although the problem may still be symbolic[2].

The starting point of conventional computations is deterministic algorithms, whereas efficient deterministic algorithms to solve a given AI problem is result from the knowledge accumulated and the gradual refinement of the computations. This involves the succinct choice of an appropriate knowledge-representation scheme, learning mechanisms to acquire the related knowledge, and a suitable architecture to support the computations. Good heuristics designed from previous experience may allow a complex problem to be solved efficiently, even on a serial processor.

Since the mid-1960's, the AI community has realized that inference alone was often inadequate to solve real-life problems. To enhance the performance of AI algorithms, they must be augmented with knowledge and metaknowledge of the problem domain in addition to formal reasoning methods. Metaknowledge refers to the control information to guide the search. This realization gave birth to knowledge engineering and knowledge-based systems, the field of applied AI[3]. Since the knowledge stored in any knowledge-based system may be incomplete and inaccurate, combinatorial searches are still needed.

## 1.3 Faster technologies and parallel processing

An AI computer system must support both knowledgebase management and heuristic searches. Faster technologies and parallel processing are means to improve the computational efficiency. For many applications, such as natural-language understanding and computer vision, the current achievable performance is much lower than that needed. For example, according to DARPA's Strategic Computing proposal, it was estimated that an equivalent of one trillion yon Neumann computer operations per second were required to perform the vehicle-vision task at a level that would satisfy the long-range objective of the Autonomous Vehicle Project[4]. At best, current sequential computers of reasonable cost achieve processing rates below 100 million operations per second, which implies at least $10^4$ times improvement in performance are required.

Newer technologies can help in designing faster computers. For example, using GaAs High-Electron-Mobility Transistors (HEMT's), it was estimated for a computer with over 500 000 gates operating at 77 K and 15 levels per pipeline stage, the cycles times were predicted to be 2.7 ns with 5 W and 3 200 gates per chip, and 2.0 ns with 20 W and 5 200 gates per chip, respectively[5]. In contrast, a liquid-cooled Cray 2 supercomputer built using ECL technologies has eight levels per pipeline stage, more than 500 000 gates, and operate at 300 K and 4.1 ns/cycle. The delay of one ECL gate level is approximately translated into 1.5 GaAs HEMT gate levels; hence correcting the cycle time of the Cray 2 supercomputer into HEMT technologies and 15 levels results in 5.1-ns cycle time for the Cray 2 computer. In short, there is a factor of two in using the newer technologies available today.

Another way to reduce the cycle time is to reduce the interconnect delay. It was estimated that with GaAs HEMT operating at 2-ns cycle time, the switching, fan-out, and interconnect delays were approximately 2, 10.5, and 87.5 percent of the cycle time, respectively[5]. Although superconductivity can be used to reduce the interconnect delays, it is less desirable with GaAs technologies due to the high impedance in the gates, and more desirable with ECL technologies. When combined, these newer technologies available today may allow improvement in the cycle times of one to two orders of magnitude.

The trend in design AI computers has been toward applying faster technologies and parallelism to process computation-intensive AI tasks. Examples of parallel AI systems currently available or under research/development include Alice, Aquarius, Butterfly, Concurrent Lisp machine, Connection Machine, Dado, FAIM-1, FFP, iPSC, Japanese Fifth Generation Computer System (FGCS), NETL, Non-Von, Rediflow, Soar, Spur, and ZMOB[1]. Some of these computers, such as the Aquarius, Butterfly, iPSC, and ZMOB, were designed for both numeric and symbolic processing.

Recently, there is another trend to design small-grain massively parallel architectures for AI applications. These architectures are sometimes called connectionist systems; they are composed of a very large number of simple processing elements. Knowledge of a given entity in such systems are distributed on a number of processing elements and links, and each processor or link may be shared by multiple entities. The use of connections rather than memory cells as the principal means to store information leads to the name "connectionism"[6]. The resemblance to neurons in a brain also results in the name "neural networks" Many computers can simulate connectionist systems. An example is the Connection Machine developed by Thinking Machines Inc., which can perform neural-network simulations two to three orders of magnitude faster than serial machines of comparable cost[7-8].

The high performance in many parallel AI computers is achieved through associative processing and "data-level parallelism". This approach is suitable for operations on large databases, such as sorting, set operations, statistical analysis, and associative pattern matching. Yet data-level parallelism alone is not enough. For general AI applications involving heuristic searches, control-level parallelism should be involved. Unfortunately, early experience with multiprocessor architectures for Hearsay-II[9], Eurisko[10], OPS-5[11], and others have led to a belief that parallel AI programs will not have a speedup of more than one order of magnitude. A possibly revolutionary approach to designing parallel languages and systems for AI processing may be needed.

One misconception in parallel processing is to use the total computing power of a parallel system to characterize the rate at which a given AI application is processed. Due to the nondeterminism in AI computations, a high computing power does not always imply a shorter completion time. Since most AI applications involve heuristic searches, resources may be devoted to fruitless searches, which use more computing power but do not help to decrease the time to find a solution. In fact anomalies may happen such that increasing the degree of parallelism may even increase the completion time in nondeterministic searches[12-13]. What is important is how to allocate resources so only useful tasks are performed. The question of solving an AI problem in a parallel processing environment is still largely unanswered.

Another misconception about parallel processing is that it can be used to extend the solvable problem size of AI problems. Due to the high complexity of AI problems, parallel processing is useful in improving the computational efficiency, but not in extending the solvable problem size[13]. For example, a problem of size $N$ and complexity $N^k$ can be solved in $N^k$ time units by a sequential processor. Assuming that $N$ processors are used, the new problem size $X$ that can be solved in the same amount of time satisfies

$$N \times N^k = X^k$$

The left side of the equation represents the total computing power in $N^k$ units of time with $N$ processors, and the right size represents the number of operations to be performed in solving a problem of size $X$. Solving the previous equation yields

$$X = N^{1+1/k}$$

Table 1 summarizes the results for other cases. It is assumed that the size of the problem solved by a sequential processor is $N$, that the number of parallel processors ranges from 1 to $2^N$, that linear speedup is achievable, and that the same amount of time is allocated to both sequential and parallel processing. The first column shows the complexities of solving the problem optimally, and the other columns show the corresponding sizes of the same problem that can be evaluated in the same amount of time for various number of processors. The extension in problem size is minimal when the problem involved is complex. This is evident in the last row in which the problem solved has exponential complexity. In this case only a logarithmic increase in problem size is achieved when a polynomial number of processors are used, and a linear increase is resulted with exponential number of processors.

In essence, parallel processing alone cannot circumvent the difficulty of combinatorial explosion. The power of multiprocessing should not

Table 1. Relative problem sizes solvable in a fixed amount of time assuming linear speedup*

| Complexity to Find Optimal Solution | Number of Processors | | | | | |
|---|---|---|---|---|---|---|
| | 1 | $N$ | $N^2$ | $N^3$ | $N^k$ | $2^N$ |
| $N$ | $N$ | $N^2$ | $N^3$ | $N^4$ | $N^{k+1}$ | $N2^N$ |
| $N^2$ | $N$ | $N^{1.5}$ | $N^2$ | $N^{2.5}$ | $N^{k/2+1}$ | $N2^{N/2}$ |
| $N^3$ | $N$ | $N^{1.33}$ | $N^{1.67}$ | $N^2$ | $N^{k/3+1}$ | $N2^{N/3}$ |
| $N^k$ | $N$ | $N^{1+1/k}$ | $N^{1+2/k}$ | $N^{1+3/k}$ | $N^2$ | $N2^{N/k}$ |
| $2^N$ | $N$ | $N+\log N$ | $N+2\cdot\log N$ | $N+3\cdot\log N$ | $N+k\cdot\log N$ | $2N$ |

* Problem size, when sequential processing is used, is $N$

be overemphasized and must be combined with heuristic information to solve complex AI problems. Currently methods for combining heuristic information and massive parallelism are still largely unknown. The publication in 1985 of the Sixth Generation Computing System development proposal shows a serious intention in Japan to go beyond the current FGCS activities and address the AI aspects of computations[14].

### 1.4 Design issues of parallel AI architectures

The essential issues in designing a computer system to support a given AI application can be classified into the representation level, the control level, and the processor level. The representation level deals with the knowledge and methods used to solve a given AI problem and the means to represent it. Design issues related to the representation level are discussed in Section 2. The control level is concerned with the detection of dependencies and parallelism in the algorithmic and program representations of the problem. Design issues related to the control level are presented in Section 3. The processor level addresses the hardware and architectural components needed to evaluate the algorithmic and program representations. Issues related to the processor level arc discussed in Section 4. Examples of issues in each level are shown in Table 2.

Developing an AI architecture requires solutions to many issues in each level. Yet some of these issues are still open at this time. In this paper, we do not provide an exhaustive survey of all reported projects and their relevant issues. Instead, we discuss some important issues in the three levels and illustrate the solutions by a number of representative systems.

Table 2. Examples of issues in designing AI computers

| Representation Level |
|---|
| Choosing an appropriate knowledge representation |
| Representing meta-knowledge |
| Acquiring and learning domain knowledge and meta-knowledge |
| Representing knowledge in a distributed fashion |
| Declaring parallelism in AI languages |
| **Control Level** |
| Analyzing data-dependencies |
| Synchronization |
| Maintaining consistency |
| Partitioning AI problems |
| Deciding granularity of parallelism |
| Dynamic scheduling and load balancing |
| Efficient search strategies |
| Trade-offs on using heuristic information |
| Predicting performance and linear scaling |
| **Processor Level** |
| Defining computational models |
| Developing methods to pass information |
| Designing hardware for overhead-intensive operations |
| Designing interconnection structure for load balancing and communication of guiding and pruning information |
| Managing large memory space |

## 2 Representation level

Since 1950, knowledge-representation schemes have been widely discussed in the literature[15-16]. The representation level is an important element in the design process and dictates whether or not the given problem can be solved in a reasonable amount of time. Although various paradigms have been developed, most existing knowledge-representation methods and AI languages were designed for sequential computations, and the requirements of parallel processing were either not taken into account or were only secondary considerations. Moreover, many designers of AI computers start with a given language or knowledge-representation scheme;

hence the representation level is already fixed. Research in designing AI computers has focused on automatic methods to detect parallelism and providing hardware support for time-consuming operations in a given representation but has not provided much to aid users in collecting and organizing knowledge or in designing efficient algorithms.

## 2.1 Domain knowledge representations

Domain knowledge refers to objects, events, and actions. From an implementation point of view, the criteria to evaluate a representation scheme for a multiprocessing system are its declarative power, the degree of knowledge distribution, and its structuralization.

Declarative versus procedural representations: The major knowledge-representation paradigms used today can roughly be classified into declarative and procedural ones, although most practical representation schemes combine features from both. Declarative representations specify static knowledge, while procedural ones specify static knowledge as well as the control information that operates on this static knowledge. Horn clauses (or even first-order logic), semantic networks, and rule-based production systems are examples of declarative representations, while Lisp programs are procedural representations. Frames combine both declarative and procedural information to represent structured knowledge. Attached to each frame is various heuristic information, such as a procedure on using the information in the frame.

A declarative approach allows the hiding of procedural control-flow information, thereby resulting in an easily created, modified, and understood knowledge representation. Declarative representations are referentially transparent; that is, the meaning of a whole can be derived solely from the meaning of its parts and is independent of its historical behavior. This may significantly increase program productivity because of its user orientation and user friendliness.

Declarative representations offer higher potential for parallelism than procedural ones for the same problem, because a declarative representation specifies tasks as a set, while a procedural representation may overconstrain the order of execution by the implicit order of statements. Parallel versions of procedural representations, such as parallel Lisp programs, achieve a limited amount of concurrency, while relying on programmers to specify the parallel tasks[17-18]. However, parallelism in a declarative representation may be restricted by the implementation of the language translators. For example, interpreters for rule-based production systems can be viewed as pattern-directed procedure invocations. Although pattern matching may provide a rich source of parallelism, the match-select-act cycle is a bottleneck and restricts the potential parallelism. Less restrictions are seen in the implementation of logic programming and semantic networks. This is the key reason for the Japanese FGCS project to choose logic as the basic representation. It has also been reported that if 256 000 processing units were used, the Connection Machine, using a semantic network representation, can execute four orders of magnitude faster than a sequential Lisp machine with respect to a number of object-recognition problems[19].

A disadvantage of declarative representations is that their nondeterminism is usually associated with a large search space that may partly counteract the gains of parallel processing; whereas procedural schemes allow the specification and direct interaction of facts and heuristic information, hence eliminating wasteful searches. A trade-off between the degree of parallelism and the size of the search must be made in designing a representation scheme.

Distributed knowledge representations: A second criterion to evaluate a representation scheme is its degree of distribution. In a local representation, each concept is stored in a distinct physical device, and each device may be shared among multiple concepts. Although this simplifies their management, the knowledge will be lost if the device fails. Most current AI systems adopt the local representations.

Recently, distributed representations have been proposed. In this scheme, a piece of knowledge is represented by a large number of units and distributed among multiple physical devices, and each device is shared among multiple knowledge entities. The resulting system is more robust because the failure of one physical device may cause some but not all information to be lost in multiple knowledge entities. Neural networks[20] and the Boltzmann Machine[21] are examples in this class. The proposed Boltzmann Machine consists of a very large network of binary-valued elements that are connected to one another by bidirectional links with real-value weights. The weight on a link represents a weak pairwise constraint between two hypotheses. A positive weight indicates that the two hypotheses tend to support one another, while a negative weight suggests that the two hypotheses should not both be accepted. The quality of a solution is then determined by the total cost of all constraints it violates.

Another interesting distributed knowledge-representation scheme, called Sparse Distributed Memory (SDM), has been proposed by Kanerva[22]. The SDM has a 1 000-bit address to model a random sample of $2^{20}$ physical locations. Given a 1 000-bit read/write address, the locations in the SDM that are within 450 bits of this address are selected associatively. Statistically, nearly 1 000 memory locations will be selected. The word read is a statistical reconstruction by a majority rule. The SDM model was designed with an analogy to the human brain and can perform pattern computations such as looking up patterns similar to a given pattern and generating a pattern that is an abstraction of a given set of similar patterns[23]. Although it is much simplified with respect to the human brain, its concept may lead to a new class of computers suitable for pattern computations.

Distributed representations are generally fault-tolerant in that, within a large parallel network with a few faulty units, the remaining pattern is still usable. This property is very attractive for wafer-scale integration. The disadvantage of distributed representations is that they are hard for an outside observer to understand and modify, so automatic learning schemes must be employed. An open problem at this time is to combine local and distributed representations by decomposing a large knowledge base into partitions and using a local representation for each.

Structuralization of knowledge: A third criterion to evaluate knowledge-representation schemes is their structuralization; this is related to the inference time and the amount of memory space required to store the knowledge. In general, the more structured a knowledge representation is, the less inference time and the more memory space are needed. An experimental comparison of efficiency has been reported for four kinds of knowledge-representation schemes for a pilot expert system, namely, a simple production system, structured production system, frame, and logic[24]. It was found that the volume of knowledge bases for the four schemes were different. In one case, both production systems have 263 rules and 15 000 characters, the frame system has 213 frames and 29 000 characters, and the logic system has 348 clauses and 17 000 characters. The memory space required by the frame system is the largest because some related pieces of knowledge have to be replicated in different frames. Since at most one conclusion is allowed in each Horn clause, the space of the logic system is larger than that of the production systems. The experimental results also show that, with respect to forward and backward reasoning, the frame system is the fastest, while the logic system is the slowest. The efficiency of the frame system is relatively insensitive to the size of the knowledge base because related pieces of knowledge are connected to one another by pointers, thereby limiting searches. The inference time of the simple production system is moderately sensitive to changes in the size of the knowledge base, while that of the logic system is markedly sensitive to changes in size.

Structured knowledge representations are usually desirable as long as the memory space needed is reasonable. To achieve this end, metaknowledge may be included in the knowledge base to reduce the search overhead needed. There are two problems in using metaknowledge. First, it consumes more memory space and may increase the overheads in memory management and communication. Second, metaknowledge in a poorly understood domain may be fallible and may lead the search in the wrong direction, thereby increasing the total search time. Theoretical studies and experimental comparisons are urgently needed to address this space-time trade-off.

## 2.2 Metaknowledge representations

Metaknowledge includes the extent and origin of domain knowledge of a particular object, the reliability of certain information, the possibility that an event will occur, and the precedence constraints. In other words, metaknowledge is knowledge about domain knowledge. Metaknowledge can be considered to exist in a single level or in a hierarchy[25]. In a hierarchical form, metaknowledge is used to decide which domain-dependent actions to perform, while meta-metaknowledge is the control knowledge about metaknowledge. Higher level metaknowledge is common-sense knowledge known to humans.

The use of metaknowledge allows one to express the partial specification of program behavior in a declarative language, hence making programs more aesthetic, simpler to build, and easier to modify. It facilitates incremental system development; that is, one can start from a search--intensive program and incrementally add control information until a possibly search-free program is obtained. Lastly, many knowledge- representation schemes and programming paradigms, such as logic, frames, semantic networks, and object-oriented programming, can be integrated with the aid of metaknowledge[25-26]. Metaknowledge can be classified as deterministic and statistical according to the correctness and efficiency considerations.

Deterministic metaknowledge: Deterministic metaknowledge is related to the correct execution of the algorithm. Metaknowledge about precedence relationships results from a better understanding of the problem; this helps reduce the resource and time complexities. For instance, to solve the problem of sorting a list, it is necessary to analyze the problem, find the appropriate representation, and evaluate the necessary tasks. A list of $n$ elements can be sorted by searching in parallel in $O(\log n!)$ average time ($=O(n\cdot\log n)$) one of the $n!$ permutations that contain the sorted elements; however, an algorithm such as Quick--sort contains functionally dependent subtasks and can sort the list in $O(n\cdot\log n)$ average time using one processor. In general, the deeper we understand the problem to be solved, the larger is the set of necessary precedence constraints and the more efficient is the solution to the problem.

Many AI languages allow programmers to specify the sequence of executions in a serial computer, but the metaknowledge to specify the correct execution in a multiprocessing environment is incomplete or missing. In programs written in pure declarative languages, the static aspects of the represented knowledge are stressed, while the controls are left to the compiler/interpreter. For instance, in a logic program, a clause $a$: $-a_1$, $a_2$, $a_3$, means that $a$ is implied by $a_1$, $a_2$, and $a_3$, but nothing about their functional dependencies is represented. The sequence of executions in a serial computer is correct because a definite search order is imposed, but the precedence relationships among subgoals are unknown to the scheduler in a multiprocessor.

In a number of AI languages such as Prolog, the type and meaning of variables and functions are dynamic and query dependent and cannot be completely specified at compile time. To use metaknowledge in this regard, the semantic meaning of subgoals and operations can be specified, which can be interpreted as precedence relationships by the scheduler at run time. In logic programming, the method to represent semantic information in a general and efficient way is still open.

The metarules used must be sufficient and precise such that all precedence relationships can be derived unambiguously and easily. An important consideration is the scope within which metarules can be applied. Common-sense metarules should be included to operate on more specific metarules specified by the programmers. Using the metarules, the interpreter/complier generates the necessary synchronization primitives.

Several researchers have addressed this problem. Gallaire and Lasserre used metaknowledge expressed as a general or special control strategy in a Prologlike interpreter[27]. In their approach, metaknowledge is made explicit through metarules, each of which describes an action to be undertaken by the interpreter whenever the interpreter focuses its attention on an object involved in the metarule. In LP, a Prolog equation-solver learning system[28], control information is expressed in a declarative representation, and inference is performed at the metalevel. Search at the object level is replaced by search at the “meta” level. Research is necessary to provide a practical method to specify unambiguously the needed synchronization through metaknowledge.

Statistical metaknowledge: Statistical metaknowledge can be used to enhance the computational efficiency of an AI program. Warren used a simple heuristic and reordered only the goals of compound queries written in pure Prolog[29]; even so, he typically obtained query speedups of an order of magnitude. The probability of success of a subgoal and the associated search cost have been found to be useful in guiding the search of logic programs[30-31]. In general, clauses in Prolog with the same head should be ordered such that those likely to succeed with a smaller expected search cost are searched first. In contrast, subgoals within a clause should be ordered such that those likely to fail with a smaller expected search cost are searched first.

In many expert systems, the belief and other measurements of accuracy of the information have been widely used. For example, in MYCIN, the Confidence Factor (CF) is used to decide among alternatives during a consultation session[32], The CF of a rule is a measurement of the association between premises and actions. A positive CF indicates that the evidence confirms the hypothesis, while a negative CF indicates disconfirming evidence.

The representation of metaknowledge about uncertainty is an active topic in AI research. Several methods, such as fuzzy logic and Dempster-Shafer theory, are being studied currently. The proper choice is still unclear.

## 2.3 AI languages and programming

Conventional imperative languages are inefficient and complex to program for symbolic and pattern processing; hence the design of AI programming languages has had a central role in the history of AI research. Frequently, new ideas in AI were accompanied by a new language that was natural for expressing the ideas.

To enhance programmer productivity and take full advantage of parallel processing, declarative languages have been designed for AI programming. Function-, logic-, and object-oriented languages are the major programming paradigms today. Lisp is an early and widely used functional language; it is characterized by symbolic computations, representation of information by lists, and recursion as the only control mechanism. Numerous imperative features have been incorporated into different dialects of Lisp, so most Lisp programs are not actually declarative, but a large enough subset allows declarative programming to be done.

Hybrids of programming paradigms have been developed. One simple approach to combining features from two languages is to provide an interface between the two. Examples include Loglisp[33], Funlog[34], and Oil[35]. The provision of features from multiple languages within a single unified framework, such as Lambda Prolog, has also been proposed. A different approach called narrowing involves replacing pattern matching in functional languages by unification[36]. Logic programs can then be expressed as functions. Recently, three commercial programming tools Kee, Art, and Loops have been introduced, which provide a mechanism to allow multiple paradigms to be used in a program.

New AI languages feature large declarative power, symbolic processing constructs, representing information by lists, and using recursion as the only control mechanism. These languages differ in their expressive power, their ease of implementation, their ability to specify parallelism, and their ability to include heuristic knowledge. A language-oriented AI computer will inherit all the features and limitations of the language it implements. Note that no single paradigm is appropriate for all problems, as one language may be more "natural" than another, depending on the requirements and the personal view. Hence intelligent systems should allow multiple styles, including function, object-, and logic-oriented paradigms.

Expressive power versus ease of implementation: Functional languages, such as pure Lisp[37], Backus' FP[38], Hope[39], and Val[40], share many features with logic languages, including the declarative nature, reliance on recursion, and potential for execution parallelism. Yet they have vital individual features as well, First, in functional programs, input and output variables are fixed, while in logic programs, the modes of variables are query dependent. For example, the statement $z$ = plus($x$, $y$) in a functional program implies that $x$ and $y$ are inputs and $z$ is output. In contrast, in a logic program, the goal sum ($X$, $Y$, $Z$) has eight possible combinations of modes of variables $X$, $Y$, and $Z$. For instance (in, out, in) means that $Y$=$Z$–$X$. Second, in a functional program, only constant and constructor functions can appear in the output; while in a logic program, logic variables can be used as output. Third, pure functional programs are deterministic, and no search is needed, while logic programs are inherently nondeterministic and require searches. Finally, functional programming provides the ability to write high-order functions; that is, a function can be passed as an argument. In contrast, Prolog is a first-order language, although some logic programming languages are not.

The first three properties, especially the nondirectionality, make logic languages more expressive in the sense that a single logic program corresponds to several functional programs. Moreover logic and functional programs are executed using resolution and reduction (or term rewriting), respectively. Note that resolution can use input information implicit in the patterns to cut down the size of the set to be examined. For example, to solve the append subgoal, append ([$P$], [$Q$, $R$], [1, 2, 3]), resolution makes no distinction between inputs and outputs and uses the input information (length of the lists) to select the appropriate clauses and produce bindings for the variables involved. However, in the corresponding functional formulation ([$P$], [$Q$, $R$]) = split([1, 2, 3]), all possible splits of [1, 2, 3] are produced, and the one that splits the list into [$P$] and [$Q$, $R$] will be selected. The previous example illustrates that reduction can lead to overcomputation as compared to resolution.

The crucial disadvantage of functional programming lies in the difficulty to represent the inherent nondeterminism in AI problems. Although the recursive formulation and the leftmost-outermost reduction of functional programs enable depth-first searches naturally, it is difficult to write a heuristic search program by a pure functional language since heuristic searches are inherently history-sensitive. In fact, best-first-search programs written in Lisp include a lot of "setq" and "prog" statements, which are not pure functional primitives[41]. Due to their less expressive power for representing nondeterminism and their inefficiency in dealing with large data structures, pure functional languages are unsuitable for general AI applications.

Although logic languages are more expressive, their implementations, especially in a parallel processing environment, are more difficult due to the nondirectionality of variables. The dynamic nature of modes requires run-time analysis. In contrast, the run-time behavior of functional programs is much simpler to control than that of logic programs, particular in a parallel context. Techniques such as graph reduction and data flow have been developed for the parallel evaluation of functional languages. Further, Lisp has only a few primitive operators and provides unique list structures to compound data objects. These features simplify the implementation of Lisp compilers/interpreters. In fact, Scheme, a dialect of Lisp, has been implemented in a single chip[42]. The implementation, however, may be complicated by the dynamic nature and primitives with side-effects introduced in practical functional languages. Dynamic features, such as random accesses to linked lists, garbage collection, frequent function calls, and dynamic binding of functions, incur extensive run-time overheads.

Obviously, it would be advantageous if the simple controls of functional languages could be implemented in the more expressive logic languages. Considerable efforts have been devoted to combining functional and logic programming[34]. One approach to simplifying logic languages is to introduce directionality of modes of variables[36]. This method degrades its expressive power to that of first-order functional languages. Others attempt to extend functional languages to achieve the expressive power of logic languages but retain most of the underlying functional simplicity. An example is Hope with unification[43]. Unfortunately, up to now, no language exists that has good expressive power while being flexible enough for parallel execution. Efforts are needed in this direction.

Specification of parallelism: Since parallel processing was not a consideration when most existing AI languages, such as Lisp and Prolog, were designed, the precedence restrictions implicit in a sequential execution order cannot be detected easily in a parallel execution. To extend these languages in a parallel processing environment, explicit primitives may have to be included.

In a pure functional language (data-flow language), the meaning of an expression is independent of the history of computations performed prior to the evaluation of this expression. Precedence restrictions occur as a result of function application. Notions such as side effects do not exist, hence all arguments and distinct elements in a dynamically created structure can be evaluated concurrently. For example, to compute the average of numbers in a list *s*, (l.(2.(3.nil))), using the function average(*s*)=div(sum(*s*), count(*s*)), the computations of sum(l.(2.(3.nil))) and count(l.(2.(3.nil))) can proceed concurrently. It has been reported that implementations of functional languages on parallel computers seems easier than that on sequential ones[44].

Note that Lisp and many of its dialects are not pure functional languages. Referential transparency is lost in most Lisp languages due to side effects. The precedence restrictions are represented not only in function calls but also in procedures.

Several parallel Lisp languages have been proposed and implemented. Multilisp, developed by Halstead, has been implemented on a 128-processor Butterfly computer. Concurrency in Multilisp can be specified by means of the pcall and future constructs[45]. pcall embodies an implicit fork join. For example, (pcall ABC) results in the concurrent evaluation of expressions *A*, *B*, and *C*. The form (future *X*) immediately returns future (a pseudo value) for *X* and creates a task to evaluate *X* concurrently, hence allowing concurrency between the computation and the use of *X*. When the evaluation of *X* yields a value, it replaces the future. The future construct is good in expressing mandatory parallelism but is quite expensive in the current Multilisp implementation.

Another parallel Lisp language, Concurrent Lisp[46], is extended from Lisp 1.5 and has three additional primitive functions to specify concurrency: STARTEVAL for process activation, and Critical Region (CR) function, and Conditional Critical Region (CCR) function for mutual exclusion. A multiprocessing program written in Concurrent Lisp is a set of cooperating sequential processes, each of which evaluates its given form. Similar to P/V primitives, CR and CCR have enough power to express process interactions.

In Parlog, a parallel logic programming language[47], every argument has a mode declaration that states whether the argument is input (?) or output (^). For example, in the following statements, mode merge(??^).

merge([$U$ | $X$], $y$, [$U$ | $Z$])←merge($X$, $Y$, $Z$)

the first two lists are merged to form the result. In Concurrent Prolog[48], a read-only annotation (?) is used. For example,

merge([$U$ | $X$], $Y$, [$U$ | $Z$])←merge($X$?, $Y$, $Z$)

indicates that *X* must have a value before merge (*X*?, *Y*, *Z*).can be invoked. Another way to specify the concurrency is to use different symbols to distinguish between "parallel AND" and "sequential AND" such as "," and "&" in Parlog. Guarded clauses are used in Parlog and Concurrent Prolog partly to specify parallelism. A guarded clause has a format, $h$: $-$ $g$ | $b$., where $g$ is the guard of the clause and $b$ is its body. Subgoals in the body can only be evaluated when all subgoals in the guard have succeeded, and values bound have been committed to the body.

Clearly, the previous approach of specifying parallelism by users detracts from the objective of declarative programming, which separates logic from control, or "what" from "how". Both mode declarations in Parlog and read-only annotations in Concurrent Prolog impose a fixed execution order on subgoals, which may be inefficient in parallel processing. On the other hand, distinguishing the guard from the body cannot completely specify the precedence relationships because subgoals in the guard and body may be dependent. The use of guards is also complicated by a lack of general methodology to select subgoals in the guard. Moreover, precedence relationships are a partial order, so the distinction between "sequential AND" and "parallel AND", which are linear orders, is insufficient to specify all precedence relationships. Lastly, owing to the nondeterministic behavior of AI programs, users cannot always specify the parallelism perfectly. A desirable parallel AI language should allow its compiler to detect the parallelism and schedule parallel executions as efficiently as possible.

Object-oriented languages: Object-oriented programming holds promise as a framework for concurrent programming that can be extended to data bases and knowledge bases. It is expected that "object-oriented programming will be in the 1980's what structured programming was in the 1970's"[49]. A variety of object-oriented languages include Smalltalk[50], Loops[51], Actor[52], CommonObjects[53], and many others[54]. Recently, CommonLoops was suggested as a standard for object-oriented extensions to Common Lisp by the Lisp community[55].

Object-oriented programming has been used to express different concepts, but the concept of an object is the common feature in these languages. Objects are entities that combine the properties of procedures and data. Object-oriented programming replaces the conventional operator-operand concept by messages and objects. All actions in an object-oriented program result from sending messages among objects. A selector in the message specifies the operation to be performed. An object responds to messages using its own procedures (called methods) for performing operations. Message sending supports data abstractions, a concept that is necessary but not sufficient for the language to be object-oriented. Object-oriented languages must additionally support the management of data abstractions using abstract data types and the composition of abstract data types through inheritance. Inheritance is used to define objects that are almost like other objects. In fact, object-oriented programming should be characterized by the nature of its type mechanisms rather than the nature of its communication mechanisms; that is, object-oriented programming can be defined as

object-oriented = data abstraction + data types + type inheritance

Object-oriented programming is a paradigm for organizing knowledge domains while allowing communications. Concurrent models, operating systems, and coordination tools are built from low-level objects, such as processes, queues, and semaphores. Hewitt's Actor model is a formalization of the ideas of object-oriented languages: an actor in his model is the analogue of a class or type instance but considers the added effects of parallelism[56]. Computations in the Actor model are partial orders of inherently parallel events having no assignment commands. The language Act3, based on the Actor model, combines the advantages of both object-oriented and functional programming[57]. To support object-oriented programming, appropriate objects representing data structures should exist at the hardware level as objects of "machine data-structure type". This gives birth to the data-type architecture[58], The Apiary network architecture is based on the Actor model[59-60].

## 2.4 Summary

A major problem in the representation level lies in the large amount of knowledge needed to define a good representation and the imprecise nature of this knowledge. Efforts have been directed toward the automatic acquisition of domain knowledge and metaknowledge to lead to a good representation and the design of a language that is more expressive and yet easy to implement in a parallel processing environment. The design of a systematic method to generate alternate representations is particularly desirable. The methodology should start with the problem specification, use automated tools to transform the problem specifications into problem representations, compare alternate representations, and use metaknowledge to guide the generation of different representations.

## 3 Control level

There are four basic issues in the control level of computer-system design. Maintaining consistency of knowledge is important, as incomplete and inconsistent knowledge is often dealt with in AI computations. As multiprocessing is widely used in AI computations, related issues include the decomposition of a problem (or program) into subproblems, the synchronization of cooperating processes, and the scheduling of processes for efficient execution. Although the design issues in the control level are similar to those in traditional multiprocessing systems, AI problems often start with different representations, hence their solutions in the control level may be very different from traditional ones.

### 3.1 Consistency maintenance

Traditional logic is monotonic because new axioms are only added to the list of provable theorems and never cause any to be withdrawn. However, knowledge-based systems on changing real-world domains have to cope with the maintenance of consistent deduction. Classical symbolic logic lacks the tools to deal with inconsistencies caused by new information. Nonmonotonic reasoning has been developed to deal with this problem[61].

Early attempts at consistency maintenance evolved around explicit manipulation of statements. The major system developed was Strips, which dealt with the manipulation of blocks of various sizes, shapers, colors, and locations by a robot[62]. In Strips, the entire data base is searched for inconsistencies when the robot moves a block. System applied inference refers to a system in which the architecture provides a mechanism to automatically maintain the consistency of the data base. The widely publicized system of this nature was Microplanner[63]. In Microplanner, the operators of Strips are replaced by “theorems”. There is no automatic inference mechanism, and the programmer is required to encode all possible implications of a theorem. An improvement to Strips is Doyle’s Truth Maintenance System (TMS) in which the reasons for beliefs are recorded and maintained, and these beliefs can be revised when discoveries contradict assumptions[64]. To attach a justification to a fact, a TMS is designed with a goal that efficiently links consequences and their underlying assumptions. In TMS, each relation has an associated IN and OUT nodes. The statement at this node is true if the statements in the IN list are known to be true and the statements in the OUT list are not true.

A different approach to consistency maintenance was adopted in designing the IBM Yes/MVS expert system that operates on a System 370 computer under the MVS operating system[65]. This expert system is used to schedule a real-time system in which contradiction occurs between the changed facts and the previous consequences. The system removes inconsistent deductions and computes new consequences in accordance with the changed facts. The consistency maintenance mechanism has three parts:

recognition of inconsistencies, modification of the resultant state to remove inconsistencies and rededuce consistent consequence, and hidden control to ensure that all inconsistencies are detected and corrected properly.

Experience on the design of Yes/MVS shows a pitfall in which correcting an inconsistency may cause another inconsistency, which in the process of being corrected reintroduces the first inconsistency. It was also found that knowledge represented in a style for consistency maintenance turned out to be quite modular, and maintaining it has been easier than initially expected.

Nonmonotonic logic has been demonstrated to be feasible but inefficient to implement in a large system. To allow the system to be used in real time, hardware support has to be provided on the time-consuming operations. Fundamental operations such as standard data base functions may have to be implemented in hardware. The management of a virtual memory system to support frequent additions and deletions in a TMS is an important design issue. The maintenance of the appropriate storage organization such that locality is maintained among relations affecting each other is a nontrivial problem. Finally, parallel processing may introduce additional problems of consistency; efficient parallel architectures to process concurrent queries have to be investigated.

## 3.2 Partitioning

In parallel computations, determining the granularity or the minimum size of a subproblem that should be computed by a single processor depends on the inherent parallelism in the problem to be solved. Partitioning can be implemented in different levels. In the higher levels, a complex AI problem is partitioned into several functional tasks, each of which is processed by a functionally distributed computer system. In the lower levels, the control graph of the program is partitioned into atomic operations, each of which can be processed independently.

Partitioning can be performed by users at design time or compilers at compile time or schedulers at run time. In the first method, programmers use a parallel language to specify and partition problems. These languages can define parallel tasks and the associated data communications. Design issues of parallel languages were discussed in Section 2.3. In this section, we discuss static and dynamic partitioning.

Inherent parallelism and granularity: The proper granularity of parallelism should be determined from the inherent parallelism in the problem and the communication overheads involved in synchronization and scheduling. In general, finding the optimal granularity is difficult; however, the degree of parallelism inherent in the problem may provide useful information to guide the design of the architecture.

An example to illustrate the choice of the proper granularity is shown in the design of parallel rule-base systems. Forgy et al. observed that each OPS-5 production, when it fires, manipulates a few (usually two or three) working memory elements and affects only a small number (20-30) of productions[11]. According to this analysis, it appears that only limited speedups are available and that massive parallelism may not be needed. To improve the degree of parallelism, further efforts should be devoted to a) investigating parallel match algorithms, b) designing efficient partitioning strategies, and c) developing techniques to rewrite sequential OPS5 programs into versions more able for parallel processing.

Gupta estimated that the hardware utilization will be around two percent if the Rete match algorithm is mapped directly onto the Dado architecture[66]. He recommended partitioning OPS5 production rules into 32 subsets to exploit the modest amount of production-level parallelism.

Based on Gupta's algorithm, Hillyer and Shaw studied the execution of production systems on the Non-Von computer, a heterogeneous system with 32 Large Processor Elements (LPE's) and 161 000 Small Processor Elements (SPE's)[67]. Each SPE has 64 bytes of RAM to store a condition-element term. The large number of SPE's, which can be viewed as an active memory of LPE's, perform intraproduction tests in a massively associative fashion. The performance is predicted at a rate of more than 850 productions fired per second using hardware comparable in cost to a VAX 11/780. This shows that two orders of magnitude of speedup is achievable by properly partitioning production systems.

The partitioning algorithm used may have significant effects on performance. If a majority of node activations occur within a single partition, then the performance will not be good. Some researchers have reported heuristics for partitioning production systems, such as assigning productions that are sensitive to the same context, goal, or task to different processors in a round-robin fashion. However, preliminary results have shown that these strategies do not bring significant improvement as compared to random partitioning[68], Intelligent partitioning strategies, using knowledge previously known, remain to be developed.

In a multiprocessing system, it is hoped that equal-sized tasks are distributed evenly to all processing units. The above example, however, has shown that this may be impractical because the problems to be solved may have irregularly structured control- and data-flow graphs and data-dependent workloads. In practice, efficient heuristic methods may have to be used to partition the task graph into granules that can be executed in parallel Important related issues to be studied in this case are the design of heterogeneous architectures and the dynamic distribution of workload.

Compiler detection of parallelism: Based on the data dependencies in a program, a compiler may be able to detect the parallel modules in it and partition the program at compile time. An example is the post-compiler of FAIM-1, called an allocator, that performs data-flow analysis on the procedural code and inference connectivity analysis on the logic behavior to statically distribute the fragments to the processing elements[69]. Similar work has been done on partitioning programs for numeric applications[70].

Detection of parallelism in logic programs has centered on detecting AND parallelism and OR parallelism. AND parallelism in logic programs involves the simultaneous execution of subgoals in a clause. Due to shared variables, concurrent execution of two or more subgoals in a clause may result in binding conflicts. The detection of AND parallelism is based on the analysis of input-output modes of arguments in a subgoal. The input and output variables in a logic program denote the direction of binding transfers during unification, in a way similar to the input and output arguments in procedure calls. However, an argument in a logic program can be in the input mode in one instance and in the output mode in another, or may remain unbound. This dynamic behavior prohibits a complete static analysis. Previous research, therefore, developed methods either to provide primitives for users to specify the modes or to assign modes automatically to arguments that can be analyzed at compile time and leave the rest to be resolved at run time. Automatic detection of AND parallelism at compile time can be classified into two types.

(1) Detection of restricted AND-parallelism: DeGroot proposed a typing algorithm to detect restricted AND parallelism[71]. The essential concept is to monitor all potentially executable subgoals and ensure that no two subgoals will share one or more unbound variables if they execute in parallel. A term in a clause can be in one of three types: a) grounded (or constant), b) nongrounded nonvariable (an input variable), or c) variable (an uninstantiated variable). To lower the run-time overhead of checking the contents of terms, a partial check is made at compile time; only terms of type 1 and that of type 3 with different variable names are detected to be independent. All other possibilities remain to be detected at run time. A consequence of this partial check is that a term may occasionally be typed too strongly.

(2) Detection of coupled data-dependencies: Chang, Despain, and DeGroot updated the above typing algorithm by testing for coupled data dependencies at compile time to reduce the run-time overhead[72-73]. In this scheme, variables in a clause are classified into three groups: grounded, coupled, and independent (An independent variable is neither grounded nor coupled with other variables). Two terms are said to be coupled if they share at least a common unbound variable. Two variables are in the same coupled group if the compiler detects that there is a chance for them to be coupled. To find the group a variable belongs to, a programmer has to supply the activation mode of the query and the entry points of the program. The compiler then classifies the variables in the subgoals from left to right and derives the execution graphs and backtracking based on the worse case activation mode of each variable. Multiple execution graphs may be generated at compile time and the appropriate one selected at run time. This scheme has been adopted in the PLM of the Aquarius project[74].

Other heuristic methods of checking types at compile time are also possible. Tung and Moldovan have also investigated a number of heuristics to infer the modes of a given variable and mark all possible input-output modes of arguments in the clauses[75].

Compiler detection of parallelism has the advantages of reduced run-time overhead and programming efforts. Its disadvantage is that it may not be able to detect all the inherent parallelism in a highly expressive AI language and may have to be combined with user declaration and dynamic detection. The restrictions of compiler detection are briefly summarized below.

Special cases: The extraction of parallelism from data-dependency analysis is based on the assumption that if two subgoals do not share any unbound variable, then they can be executed concurrently. This assumption is not true in some special features of the language, such as outputs in Prolog. A solution to this problem is proposed by DeGroot[76].

Procedural dependencies: A procedural dependency exists between two subgoals if their execution order is fixed by their semantics. For example, in the following clause,

$a(X)$: – test_for_ok($X$), work_on($X$)

the subgoal "test_for_ok($X$)" must be executed first. Note that the subgoals in this example cannot be executed concurrently even if $X$ is grounded, because the second subgoal may contain meaningless, inaccurate, or unbound work unless the first subgoal is true. In declarative languages such as Prolog, it is difficult to specify the semantic of subgoals without specifying its explicit control for parallelism. A solution to this problem is proposed by DeGroot[76].

Exponential complexity: It may be difficult to define all possible combinations of modes at compile time as they grow exponentially with the number of potential output variables.

Dynamic detection of parallelism: Many data dependencies in a highly expressive AI language cannot be resolved until run time. For example, a subgoal $p(X, Y)$ in a logic program may be called as $p(X, X)$, which is a coupling dependency on a query with coupled terms introduced at run time. This dependency cannot be detected at compile time. Due to the dynamic nature of AI computations, an AI computer should provide a mechanism to map the program and data onto hardware dynamically.

In general, the computational model can be represented as a token-flow graph with four kinds of nodes: and-decomposition, or- decomposition, and-join, and or-join. The tokens passed along the edges can be demand tokens, data tokens, or control tokens. Conery and Kibler described an AND/OR process system based on a producer-consumer model that dynamically monitors variables and continually develops data-dependency networks to control the order of execution of subgoals, never allowing two potential procedures with the same variable to be executed in parallel[77]. An ordering algorithm, called a connection rule, is used dynamically to determine a generator for each unbound variable. When a subgoal is completed, it is checked to ensure that it did produce all variable bindings it was supposed to; otherwise, the ordering algorithm is evaluated again. Improvements were made to the above scheme to reduce further the run-time overhead and extract more parallelism[78-79].

Since dynamic partitioning must be repeatedly executed at run time, it may reduce the performance gains and could even produce negative gains. The trade-off between static partitioning by an intelligent compiler and dynamic partitioning by a sophisticated operating system is an important issue to be addressed in parallel AI processing. Dynamic partitioning is closely related to dynamic scheduling, and related issues will be discussed in a subsequent section.

Bottleneck analysis: An important issue in partitioning is to decompose the problem evenly, so bottlenecks in performance do not exist. It is easy to see that if a bottleneck requires a fraction of the total computations, then the speedup cannot be more than the reciprocal of this fraction, regardless of how the rest is partitioned. It is well-known that the performance bottleneck of an application executing on a vector computer is its scalar code. Similarly, the performance bottleneck of a parallel AI computation is its sequential part (sequential inference or I/O). An important problem is to find the bottleneck in the problem to be solved.

Experience with designing the Fido vision system at Carnegie-Mellon University has shown that an unbalanced partitioning algorithm can substantially degrade the performance[80]. Adding Warp, a systolic system with peak processing rate of 100 mflops, to a host (a Sun computer and three "standalone processors") seems only to double or triple the speed of the Fido loop. This means that Warp is definitely underutilized; functions on the standalone processors, either in preprocessing or post-processing in using the Warp array, take up a substantial amount of time. It is expected that proper partitioning of vision algorithms will improve its performance significantly.

## 3.3 Synchronization

Synchronization refers to the control of deterministic aspects of computations, while scheduling handles mainly the nondeterministic aspects. The objective of synchronization is to guarantee the correctness of parallel computations such that the results of execution in parallel are the same as those of a sequential execution. That is, the parallel execution is serializable. In some nondeterministic problems, the generation of the same set of results as a sequential execution may not be necessary. For example, a user may wish to obtain a small subset of answers from a large set; the particular answers obtained do not have to be the same in the serial and parallel cases. In this case requirements on synchronization can be relaxed in parallel processing.

Many synchronization primitives used in AI processing are the same as those used in conventional computers. Examples include semaphores, test-and-set, full/empty bits, fetch-and-add, and synchronization-keys. Additionally, new or extended concepts related to synchronization have been introduced by AI researchers, such as the blackboard and actors. In this section, we will survey the synchronization of AI computations in the control and data levels and mechanisms using shared memory and message passing.

Two levels of synchronization: In procedural languages, if a statement precedes another statement in the program, the implication is that this statement should be executed before the second statement if the two statements share common variables; that is, control-level synchronization is implicit when data-level synchronization is needed. This implicit execution order may overspecify the necessary precedence constraints in the problem.

On the other hand, if the tasks are specified as a set using a declarative language, then control-level synchronization is absent, and they can be processed concurrently if they do not share common variables. If they have common variables but are semantically independent, then they can be processed sequentially in an arbitrary order to maintain data-level synchronization.

The difficulty of specifying control-level synchronization when tasks are semantically dependent is a major problem in declarative languages such as Prolog. For example, the decomposition of a set into two subsets in Quicksort must be performed before the subsets are sorted. Hence the tasks for decomposition and for sorting are both semantically and data dependent. To overcome this problem, programmers are provided with additional tools, such as specifying the input/output modes of variables in a Prolog program, to specify control-level synchronization. These primitives may have side effects and may not be able to specify completely all control-level synchronization in all situations. These problems may have to be dealt with at run time until sufficient information is available.

In general, process activations and deactivations can be considered as control-level synchronization, while passing arguments in procedure calls can be considered as data-level synchronization. Both methods can be implemented through a shared memory or by message transfers.

Shared memory: In tightly coupled multiprocessor systems, synchronization is done through a shared memory. Examples of such existing and proposed AI computers include Aquarius[74], Concurrent Lisp machine[81], Concert Multilisp machine[82], and Parallel Inference Engine[83]. In what follows, we will discuss synchronization using blackboards and show methods using shared variables in logic programs.

(1) Blackboard: Historically, the blackboard model was developed for abstracting features of the Hearsay-II speech-understanding system[84]. The model is usually viewed as a problem-solving framework; however, we discuss only its control aspect here. The model consists of three major components: a knowledge source, a blackboard data structure, and control. The knowledge to solve the problem is partitioned into knowledge sources that are kept independently. The data needed to solve the problem concerned include input data, partial solutions, alternatives, and final solutions, which are kept in a global data base, the blackboard. The blackboard can be divided into multiple blackboard panels that correspond to the hierarchy of solution space. Knowledge sources result in changes in the blackboard, which lead to a solution to the problem. Communications and interactions among knowledge sources take place solely through the blackboard. A monitor is needed to ensure that no more than one knowledge source can change the blackboard at one time. There are a set of control modules that monitor changes in the blackboard and decide the appropriate action to take next. The sequence of knowledge-source invocations is dynamic.

The blackboard model provides a useful framework for diverse types of knowledge to cooperate in solving a problem and has been used to many AI applications. Its implementation is similar to that of a critical section in operating systems. In the pure model, the solution is built one step at a time. Currently, extensive research on concurrent access to blackboards is conducted.

Hayes-Roth has proposed a more powerful blackboard control architecture in which control information (metaknowledge) is also stored and updated on a separated control blackboard[85]. This approach adapts to complex control plans as a whole. Operational strategies, heuristic, and scheduling rules can change repeatedly in the course of problem-solving.

(2) Synchronization via shared memory variables: Although Lisp contains a "pure function" subset, it also supports many functions with side effects, such as rplaca, rplacd, set, and input/output functions. These side effects result from procedural dependencies and global (or free) variables and resemble problems in conventional parallel languages. In fact, some shared-memory multiprocessors, such as Concert and Butterfly, support both Multilisp, Simultaneous Pascal, and other parallel languages[82]. Multilisp provides a simple method to wait for values generated in the future. However, as in other languages, procedure activations in Multilisp may not be well nested, and an activation can terminate before another activation it contains. This exception-handing problem has to be addressed in programming the system[86].

Pure Prolog is a single-assignment language. Under this restriction, the distinction between a shared-memory variable and a communication channel vanishes. Since a logic variable is not allowed to be rewritten through side effects, conventional hardware-synchronization mechanisms, such as test-and-set, full/empty-bit method and fetch-and-add, are no longer needed in multiprocessing of pure logic programs[87]. The popular strategy taken now is to provide the programmer with a mechanism to delay process reduction until enough information is available so that a correct decision can be made. Currently, the Concurrent Prolog group is concentrating their efforts on Flat Concurrent Prolog, a subset of Concurrent Prolog. In Guarded Horn Clause (GHC)[88], ICOTs current choice for Kernel Language 1, OR parallelism was eliminated from Concurrent Prolog, and a strict synchronization rule that suspends a subgoal if it tries to write in the parent environment is adhered. This rule made the read-only annotation somewhat superfluous. Although it simplifies the implementation of GHC, some expressive power is lost due to a weaker notion of unification[89].

(3) Joins: As similar to conventional fork-join primitives, static joins can be used for synchronization in parallel AI processing. For example, in multiprocessing of logic programs, a parent node can activate its children in parallel, and each child begins producing all possible answers. The parent waits for each child to complete, collects their answers, computes the "join" of their answers, and passes the entire set of results as its answer. This approach uncovers the greatest AND parallelism in a logic program but is efficient only if the program consists mostly of deterministic procedures and clauses; that is, most variables have only a single binding. For nondeterministic AI problems, joins are impractical because the nondeterminism increases the uncertainty whether a given AND node should be evaluated. Note that if joins are computed dynamically, that is, a parent node collects separate answers from each child as they are produced, then the data-level synchronization employed forms a pipelined computation called dynamic joins. This scheme will be discussed later with respect to synchronization in semantic networks.

Message passing: In passing messages, a communication channel between the sender and receiver processes is required. Synchronization via messages can be achieved through software protocols or specialized hardware. Many existing and proposed AI computers pass around messages of arbitrary complexity and perform complex operations on them. The computing elements are complex, and the communication costs are high. Alternatives to passing messages are discussed in this section.

(1) Massage passing in production systems: Reasoning using forward chaining in production systems has different behavior from reasoning using backward chaining. The behavior in forward chaining is illustrated in OPS5, whose interpreter repeatedly executes a match-select-act cycle. In the match phase, all rules whose conditions are satisfied by the current content of the working memory are selected. This set is called the conflict set. In the select phase, conflict resolution is performed to select one of the productions in the conflict set. In the act phase, the working memory is modified according to the action part of the selected rule. Although the three phases can overlap in a multiprocessing environment, synchronization must be performed to ensure that the result is consistent with that of a sequential execution; that is, all changes in the conflict set must be known prior to the completion of conflict resolution in the next cycle.

Synchronization in the efficient Rete interpreter for OPS5 is based on a data-flow graph, which can be viewed as a collection of tests that progressively determine the productions ready to fire. Inputs to the graph consist of changes to the working memory encoded in tokens. Output tokens specify changes that must be made to the conflict set. Tokens are sent via messages in a multiprocessing system.

(2) Maker passing and value passing: Marker passing has been studied as an alternative to message passing. In such systems, communications among processors are in the form of single-bit markers. An important characteristic is that there is never any contention: if many copies of the same marker arrive at a node at once, they are simply OR'ed together. The order of markers to be passed is determined by an external host.

Marker passing is suitable for systems implementing semantic networks. Nodes in the semantic network are mapped to processors in the system. An example of such a system is NETL[90]. A basic inference operation in semantic networks is set intersection. Analogous to dynamic joins in data bases, set intersections are implemented using data-level synchronization. If an object with $n$ properties is searched, then $n$ commands are sequentially broadcast to all corresponding links, the associated nodes are marked, and the node with $n$ markers reports its identity to the controller. Marker passing is adequate for many recognition problems; however, it may not be sufficient to handle general AI problems. The Connection Machine was originally developed to implement marker passing to retrieve data from semantic networks, but its current version has more powerful processing units that can manipulate address pointers and send arbitrary messages.

In value passing, continuous quantities or numbers are passed around the system, and simple arithmetic operations are performed on these values. Like marker-passing systems, there is no contention in value passing: if several values arrive at a node via different links, they are combined arithmetically, and only one combined value is received. In this sense, value passing systems can be considered as an analog computer. Examples of value-passing system are the Boltzmann machine[91]and other "neural" computation systems[92].

Marker-passing systems do not gracefully handle recognition problems in which the incoming features may be noisy. These problems can be better handled by value-passing system in which each connection has an associated scalar weight that represents the confidence on the incoming values. Many iterative relaxation algorithms that have been proposed for solving low-level vision and speech-understanding problems are ideally suited to value-passing architectures.

(3) Object-oriented and Actor Approaches: In the object-oriented approach, and in particular, the Actor model, an actor is a virtual computing unit defined by its behavior when messages are received. Actors communicate via point-to-point messages that are buffered by a mail system. The behavior of an actor consists of three kinds of actions: a) communicate with specific actors of known mail addresses; b) create new actors; and c) specify a replacement that will accept the next message. Actor languages avoid the assignment command but allow actors to specify a replacement. Replacements can capture history-sensitive information, while allowing concurrent evaluation of data-independent expressions[52]. Message passing in actors, which can be viewed as a parameter-passing mechanism, differs from both call-by-value and call-by- reference.

### 3.4 Scheduling

Scheduling is the assignment of ready tasks to available processors. It is especially important when there is nondeterminism in the algorithm. Scheduling can be static or dynamic. Static scheduling is performed before the tasks are executed, while dynamic scheduling is carried out as the tasks are executed. The actions to be performed in scheduling include a) determination of dependent tasks, b) static reordering of tasks at compile time, c) dynamic selection of tasks at run time when free processors are available, and d) determination of the number of processors to solve a given class of problems cost-effectively. AH schedules can be considered as a search strategy based on a search tree or search graph[93].

Identifying dependencies: Parallel scheduling of AI programs is complicated by their dynamic functional and shared-variable dependencies and the high expressive power of many AI languages. Due to high expressive power, the same program can be used to represent many different dependencies, each of which may be scheduled differently. Identifying dependencies at compile time is also difficult due to the dynamic and nondeterministic nature of executions.

If functional dependencies exist among tasks, then the scheduler must find these dependencies dynamically; if there are no functional dependencies but only shared-variable dependencies, then the scheduler has to compare the merits of all possible schedules. Both cases are not practical because of the high dynamic overhead. As discussed earlier, solutions to detect dependencies are not satisfactory at this time.

A viable approach is to identify the possible dependencies at compile time, statically order all sibling nodes in a search tree for each case, and schedule them according to a parallel depth-first strategy. A simple method was proposed by Warren[29], which orders the subgoals in a clause according to the number of possible solutions generated under the given subgoal. Our experimental simulations indicated that the worst case evaluation time resulting from this method can be worse than the case without reordering, but the best case time can be 2-30 times better. Warren's method does not consider the effects of backtracking, the possible dependencies among subgoals and clauses, and the overhead of finding the solutions. We have proposed a method to represent the effects of backtracking as an absorbing Markov chain[94]. By assuming that sibling nodes are independent, they are reordered to minimize the total expected search cost of the program. Heuristics have been developed to reorder subgoals when they are dependent and have side effects. Our preliminary simulations indicated that the performance is substantially better than that of Warren's method.

Selection strategies: Suppose in the course of evaluating an AI program $n$ active tasks and $m$ processors are available, $1 \leqslant m < n$. The ideal scheduling algorithm should select $m$ active tasks such that this decision will minimize the expected computational time. It is difficult to design such an optimal selection algorithm because a) the metrics to guide the search are estimated heuristically and may be fallible, b) the metrics may be dynamically changing during the search, and c) problem-dependent precedence restrictions may exist that cannot be detected at compile time. As a result, unexpected anomalies: may occur when parallel processing is applied.

The potential parallelism in an AI computation can be classified into two types: deterministic parallelism and nondeterministic parallelism. Deterministic parallelism refers to the concurrent execution of two or more units of computations, all of which are necessary for the completion of the given job. The computational units can be tasks, processes, and/or instructions. Since all units of computation, which are performed concurrently, have AND relations, this kind of parallelism is traditionally called AND-parallelism. Nondeterministic parallelism refers to the search of multiple potential solutions in parallel. Since all potential solutions have OR relations, this kind of parallelism is traditionally called OR parallelism.

Although AND parallelism is treated as deterministic and OR parallelism as nondeterministic in conventional studies, the selection of descendents of an AND task to evaluate is also nondeterministic, as the aim is to select one that fails as soon as possible. Hence scheduling is important for tasks that are nondeterministic but may not be specific with respect to AND or OR parallelism.

In nondeterministic searches, heuristic information to guide the scheduler in selecting nondeterministic tasks is more important than the design of parallel processors, as the number of processors is almost always smaller than the number of processable tasks.

As an example, in selecting nodes to evaluate in a branch-and-bound search tree, which is an OR tree with lower bound values to guide the search, the problem is reduced to finding the $m$ smallest numbers from $n$ numbers. Table 3 shows the results obtained by three architectural approaches. In the first approach, a multistage selection network was designed to perform the selection exactly[95]. In the second approach, a single-stage ring network was used to shuffle the nodes until a complete selection was obtained[96].

In the third approach, a no-wait policy was applied. It was recognized that the heuristic information to guide the search might not always be accurate. Hence the "most promising" task in local memory was always evaluated in each cycle, while the fetch of the "more promising" tasks from other processors was initiated. It was found that, on the average, a minimum of 63 percent of the desirable tasks to be selected were selected by the no-wait policy without any additional overhead on selection, assuming that the $m$ most promising tasks were randomly distributed among the processors[13,96].

Table 3. Selecting the $m$ smallest numbers from $n$ numbers

| Approach | Time complexity in each iteration | Space/hardware complexity for selection | Accuracy of selection |
| --- | --- | --- | --- |
| Multistage selection network | $O(\log m \cdot \log n)$ | $O(n \cdot \log^2 m)$ | 1.0 |
| Single-stage network | $O(m)$ | $O(n)$ | 1.0 |
| No-wait policy | $O(1)$ | $O(m)$ | 0.63 |

The management of the large memory space to store the heuristic information and the large number of intermediate nodes in the search tree is another difficult problem to solve. A trade-off must be made to decide for a given amount of heuristic information and a given architectural model whether the amount of heuristic information should be increased or decreased, and how effective should the new heuristic information be.

The memory space required to store enough heuristic information to avoid backtracking is often prohibitive. For example, assume that all solution trees of a complete binary AND/OR tree with $n$ levels are equally likely. The leaves are assumed to be OR nodes and are at level 0, while the root is an AND node and is a level $n$. We have that $f(n)$, the total number of solution trees, satisfy the following recurrence.

$$f(n)=\begin{cases}1, & n=0 \text{ or } n=1\\ 4f^2(n-2), & n\geqslant 2\end{cases}$$

$$=2^{2(2^{n/2}-1)}$$

For $n = 0$, there is only one node, hence there is one solution tree. For $n$ –1, the root is an AND node with two descendents (see Figure 1(a)). Again, this represents one solution tree. For the general case, each node in level $n$–2 has $f(n–2)$ solution trees (see Figure 1(b)). A solution tree for the root at level $n$ consists of picking two nodes in level $n$–2, a total of four combinations. Each pair of nodes selected in level $n$–2 represent two solution trees, all possible combinations of which will yield a new solution tree. This case is depicted in Figure 1(b).

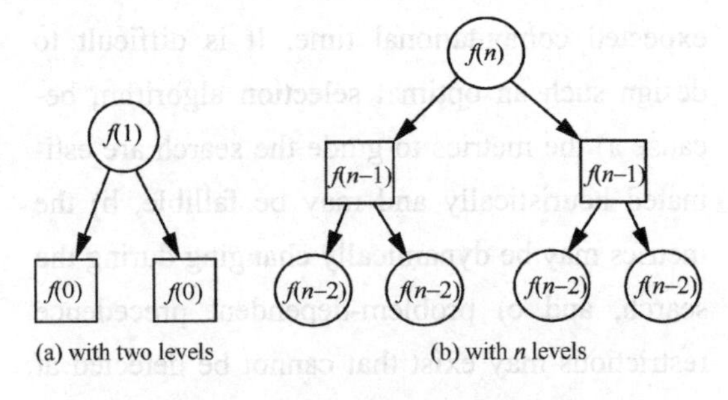


Figure 1. Binary AND/OR trees (Circles represent AND nodes: boxes represent OR nodes.)

Since all solution trees are equally likely, the entropy of the heuristic information to guide the search at the root such that a correct decision is always made without backtracking is

$$I = \sum_{j-1}^{f(n)} \frac{1}{f(n)} \log_2 f(n) = 2(2^{n/2} - 1)$$

which is exponential with respect to the height of the tree.

To manage the large memory space incurred by the storage of intermediate subproblems that may lead to solutions, we have investigated three alternatives to support branch-and-bound algorithms with a best-first search, the results of which are displayed in Table 4. In a direct implementation, the best-first search was implemented on an existing virtual-memory system, a VAX 11/780 computer running 4.2 BSD Unix. In the second approach, a modified virtual memory with specialized fetch and replacement policies was designed to adapt to the characteristics of the search algorithm. In the third approach, the no-wait policy discussed earlier was used to select subproblems in the main memory without waiting for the "most promising" subproblem to be accessed from the secondary memory. Again the no-wait policy is superior in performance[97].

Table 4. Relative times to complete a branch-and-bound algorithm for various memory-management techniques

| Approach | 0/1 integer programming problems | 0/1 knapsack problems |
| --- | --- | --- |
| Direct implementation | 1 | 1 |
| Modified virtual memory | 0.6 | 0.1 |
| No-wait policy | 0.1 | 0.001 |

The nondeterministic nature of computations and the fallibility of heuristic guiding may lead to anomalies of parallelism. When $n$ processors are used to solve the problem, the resulting speedup as compared to a single processor may be less than one, greater than $n$, or between one and $n$. The reasons for this anomalous behavior are due to a) ambiguity in the heuristic information, b) more than one solution node, and c) approximation and dominance tests[98]. As a result, subtrees searched under serial processing will be terminated, and the search will be misled into a different part of the search tree.

In summary, scheduling is important when there is nondeterminism in the problem. Good heuristic metrics to guide the search are usually difficult to design and depend on statistics such as success probabilities, search costs, and problem-dependent parameters. Trade-offs must be made among the dynamic overhead incurred in communicating the heuristic-guiding information, the benefits that would be gained if this information led the search in the right direction, and the granularity of tasks. In practice, the merits of heuristic guiding are not clear, since the heuristic information may be fallible. As a result, some AI architects do not schedule nondeterministic tasks in parallel. The excessive overhead coupled with the fallibility of heuristic information also leads some designers to apply only static scheduling to AI programs.

Pruning: Pruning can be considered as a negative form of heuristic guiding which guides the search to avoid subproblems that will never lead to better or feasible solutions. Pruning is useful in both backward and forward chaining. In backward reasoning, problems are decomposed into smaller subproblems and evaluated independently. There are usually redundant evaluations of the same task in different parts of the search tree when the search trees are recursive. Likewise, in forward reasoning, the more primitive facts are reduced to form more general facts until the query is satisfied. Unnecessary results are generated because it is not clear which reaction will lead to a solution of the problem.

Pruning in search problems can be carried out by dominance relations. When a node $P_i$ dominates another node $P_j$ it implies that the subtree rooted at $P_i$ contains a solution with a value no more (or no less) than the minimum (or maximum) solution value of the subtree rooted at $P_j$.

As an example, consider two assignments $P_1$ and $P_2$ on the same subset of objects to be packed into a knapsack in the 0/1 knapsack problem. If the total profit of the objects assigned to the knapsack for $P_1$ exceeds that of $P_2$ and the total weight of the objects assigned in $P_1$ is less than that of $P_2$, then the best solution expanded from $P_1$ dominates $P_2$.

When parallel, processing is used, it is necessary to keep the set of current dominating nodes (demoted by $N_d$) in memory[13]. These are nodes that have been generated but not yet dominated. In general, $N_d$ can be larger than the set of active nodes. A newly generated node, $P_i$ has to be compared with all nodes in $N_d$ to see whether $P_i$ or any nodes in $N_d$ are dominated.

If $N_d$ is small, it can be stored in a bank of global data registers. However, centralized comparisons are inefficient when $N_d$ is large. A large $N_d$ should then be partitioned into $m$ subsets, $N_d^0, \cdots, N_d^{m-1}$, distributed among the local memories of the $m$ processors. A subproblem generated in processor $i$, is first compared with $N_d^i$; any subproblems in $N_d^i$, dominated by $P_{ij}$ are removed. If $P_{ij}$ is not dominated by a subproblem in $N_d^i$, it is sent to a neighboring processor and the process repeats. If it has not been dominated by any node in $N_d$, $P_{ij}$ eventually returns to processor $i$ and is inserted into $N_d^i$.

There are several problems associated with the use of dominance tests in AI applications. First, dominance relations are very problem-dependent and cannot be derived by a general methodology. Most of the dominance relations have been developed for dynamic programming problems. To derive a dominance relation in a search process, a dominance relation is hypothesized, and a proof is developed to show that the dominance relation is correct. Some progress has been made on using learning-by-experimentation to derive dominance relations for dynamic programming problems[99]. However, automatic proof techniques are largely missing. Moreover, learning-by-experimentation is applicable if there are a very small number of dominance relations that are used frequently in the problem. In many AI applications coded in Prolog, there is a large number of dominance relations, each of which is used infrequently in the program. Some special cases can be solved, such as finding redundant computations in recurrences[100]. For the general case, it is sometimes difficult to find these dominance relations without human ingenuity. Second, many dominance relations are related to the semantics of the applications. A good language to represent semantics is missing at this time. Lastly, the overhead of applying dominance relations is usually very high, and sequential and parallel implementations will incur prohibitive overhead.

Granularity of parallelism: When a parallel computer system with a large number of processors is available, it is necessary to determine the granularity of parallelism, that is, the size of tasks that will be executed as an indivisible unit in a processor. Since many AI problems can be represented by AND/OR trees, some processors have to be idle when nodes close to the root are evaluated. The proper number of processors should be chosen to match the inherent parallelism in the problem to be solved.

The proper granularity is a function of the problem complexity, the shape of the AND/OR tree, and the distribution of processing times of tasks. Many of these parameters are dynamically changing and data-dependent, and only special cases can be analyzed[101]. An important functional requirement for parallel processing of AI programs is the ability to dynamically distribute the workload. For a system with a small granularity, an efficient interconnection network is required to transfer data and control information. In a loosely coupled system with a coarse grain, an effective load balancing mechanism is also needed.

# 4 Processor level

The VLSI technology that has flourished in the past ten years has resulted in the development of many special-purpose computers for AI processing. Architectures for AI processing can be classified into the micro-, macro-, and system-level architectures. Microlevel and macrolevel architectures are discussed in the next two sections. Sections 4.3-4.7 briefly discuss the system-level architectures. A taxonomy of architectures implementing AI systems have also been discussed by Hwang et al.[102].

## 4.1 Microlevel architectures

The microlevel architectures consist of architectural designs that are fundamental to applications in AI. In the design of massively parallel AI machines[91], some of the basic computational problems recognized arc set intersection, transitive closure, contexts and partitions, best-match recognition, Gestalt recognition, and recognition under transformation. These operations may not be unique to AI, and many exist in other applications as well. Due to the simplicity of some of these operations, they are usually implemented directly in hardware, especially in systolic arrays. Many other basic operations can also be implemented in VLSI. Examples include sorting and selection, computing transitive closure, string and pattern matching, selection from secondary memories, dynamic programming evaluations, proximity searches, and unification.

Some AI languages such as Lisp differ from traditional machine languages in that the program/data storage is conceptually an unordered set of linked record structures of various sizes, rather than an ordered indexable vector of numbers or bit fields of a fixed size. The instruction set must be designed according to the storage structure[103]. Additional concepts that are well-suited for list processing are the tagged memory[104] and stack architectures.

## 4.2 Macrolevel architectures

The macrolevel is an intermediate level between the microlevel and the system level. In contrast to the microlevel architectures, macrolevel architectures are (possibly) made up of a variety of microlevel architectures and perform more complex operations. However, they are not considered as a complete AI system but can be taken as more complex supporting mechanisms for the system level. The architectures can be classified into those that manage data, such as dictionary machines, data base machines and structures for garbage collection, and those for searching.

A dictionary machine is an architecture that supports the insertion, deletion, and searching for membership, extremum, and proximity of keys in a data base[105]. Most designs are based on binary-tree architectures; however, designs using radix trees and a small number of processors have been found to be preferable when keys are long and clustered[106].

Data base machines depend on an architectural approach that distributes the search intelligence into the secondary and mass storage and relieves the workload of the central processor. Extensive research has been carried out in the past decade on optical and mass storage, backend storage systems, and data base machines. Earlier data base machines developed were mainly directed toward general-purpose relational data base management systems. Examples include the DBC, Direct, Rap, CASSM, associative array processors, text retrieval systems, and CAFS[107-109] Nearly all current research on data base machines to support knowledge data bases assume that the knowledge data base is relational, hence research is directed toward solving the disk paradox[110] and enhancing previous relational data base machines by extensive parallelism[111-114]. Commercially available data base and backend machines have also been applied in knowledge management[115-116].

Searching is essential to many applications, although unnecessary combinatorial searches should be avoided. The suitability of parallel processing to searching depends on the problem complexity, the problem representation, and the corresponding search algorithms. Parallel algorithms and architectures to support divide-and-conquer, branch-and-bound, and AND/OR-graph search have been developed[13].

Extensive research has been carried out in supporting dynamic data structures in a computer with a limited memory space. Garbage collection is an algorithm that periodically reclaims memory space no longer needed by the users[117]. This is usually transparent to the users and could be implemented in hardware, software, or a combination of both. For efficiency reasons, additional hardware such as stacks and reference counters are usually provided.

### 4.3 Functional-programming-oriented system-level architectures

The objective of writing a functional program is to define a set of (possibly recursive) equations for each function[44]. Data structures are handled by introducing a special class of functions called constructor functions. This view allows functional languages to deal directly with structures that would be termed "abstract" in more conventional languages. Moreover, functions themselves can be passed around as data objects. The design of the necessary computer architecture to support functional languages thus centers around the parallel evaluation of functional programs (function-oriented architectures) and the mechanisms of efficient manipulation of data structures (list-oriented architectures).

In function-oriented architectures, the design issues center on the physical interconnection of processors, the method used to "drive" the computation, the representation of programs and data, the method to invoke and control parallelism, and the optimization techniques[118]. Desirable features of such architectures should include a multiprocessor system with a rich interconnection structure, the representation of list structures by balanced trees, and hardware supports for demand-driven execution, low- overhead process creation, and storage management.

Architectures to support functional-programming languages can be classified as uniprocessor architectures, tree-structured machines, data-driven machines, and demand- driven machines. In a uniprocessor architecture, besides the mechanisms to handle lists, additional stacks to handle function calls and optimization for redundant calls and array operations may be implemented[119-121]. Tree-structured machines usually employ lazy evaluations, but suffer from the bottleneck at the root of the tree[122-124], Data-flow machines are also natural candidates for executing functional programs and have tremendous potential for parallelism. However, the issue of controlling parallelism remains unresolved. A lot of the recent work has concentrated on demand-driven machines which are based on reduction machines on a set of load-balanced (possibly virtual) processors[125-131].

List-oriented architectures are architectures designed to support the manipulation of data structures and objects efficiently. Lisp, a mnemonic for list processing language, is a well-known language to support symbolic processing. There are several reasons why Lisp and list-oriented computers are really needed. First, to relieve the burden on the programmers, Lisp was designed as an untyped language. The computer must be able to identify the types of data, which involves an enormous amount of data-type checking and the use of long strings of instructions at compile and run times. Conventional computers cannot do these efficiently in software. Second, die system must periodically perform garbage collection and reclaim unused memory at run time. This amounts to around 10-30 percent of the total processing time in a conventional computer. Hardware implementation of garbage collection is thus essential. Third, due to the nature of recursion, a stack-oriented architecture is more suitable for list processing. Lastly, list processing usually requires an enormous amount of space, and the data structures are so dynamic that the compiler cannot predict how much space to allocate at compile time. Special hardware to manage the data structures and the large memory space would make the system more efficient[132-133].

The earliest implementation of Lisp machines were the PDP-6 computer and its successors the PDP-10 and PDP-20 made by the Digital Equipment Corporation (DEC)[37]. The half-word instructions and the stack instructions of these machines were developed with Lisp's requirements in mind. Extensive work has been done for the DEC-system 10's and 20's on garbage collection to manage and reclaim the memory space used.

The design of Lisp machines was stalled at MITs AI Laboratory in 1974. Cons, designed in 1976[134], was superseded in 1978 by a second generation Lisp machine, the CADR. This machine was a model for the first commercially available Lisp machines, including the Symbolics LM2, the Xerox 1100 Interlisp workstation, and the Lisp Machine Inc. Series 3 CADR, all of them delivered in 1981. The third-generation machines were based on additional hardware to support data tagging and garbage collection. They are characterized by the Lisp Machines Inc. Lambda supporting Zetalisp and LMLisp, the Symbolics 3600 supporting Zetalisp, Flavors, and Fortran 77, the Xerox 1108 and 1132 supporting Interlisp-D and Smalltalk, and the Fujitsu FACOM Alpha Machine, a backend Lisp processor supporting Maclisp. Most of the Lisp machines support networking using Ethernet. The LMI Lambda has a NuBus developed at MIT to produce a modular, expandable Lisp machine with multiprocessor architecture.

A single-chip processor to support Lisp has been implemented in the MIT Scheme-79 chip[42]. Other experimental computers to support Lisp and list-oriented processing have been reported[135-144]. These machines usually have additional hardware tables, hashing hardware, tag mechanisms, and list processing hardware, or are microprogrammed to provide macroinstructions for list processing. A Lisp chip built by Texas Instruments implements over half a million transistors on a 1-$cm^2$ chip for 60 percent of the functions in a TI Explorer. The implementation on a single chip results in five times improvement in performance[145]. Experimental multiprocessoring systems have been proposed to execute Lisp programs concurrently[46,81,146-150]. Data-flow processing is suitable for Lisp as these programs are generally data driven[151-154]. Other multiprocessing architectures to support list processing have been proposed and developed[59,131,155-157].

Architectures have also been developed to support object-oriented programming languages. Smalltalk, first developed in 1972 by the Xerox Corporation, is recognized as a simple but powerful way of communicating with computers. At MIT, the concept was extended to become the Flavors system. Special hardware and multiprocessors have been proposed to directly support the processing of object-oriented languages[158-161].

Owing to the different motivations and objectives of various functional-programming-oriented architectures, each machine has its own distinct features. For example, the Symbolics 3600[104] was designed for an interactive program development environment where compilation is very frequent and ought to appear instantaneous to the user. This requirement simplified the design of the compiler and results in only a single-address instruction format, no indexed and indirect addressing modes, and other mechanisms to minimize the number of nontrivial choices to be made. On the other hand, the aim in developing Soar[161] was to demonstrate that a Reduced Instruction Set Computer (RISC) could provide high performance in an exploratory programming environment. Instead of microcode, Soar relied on software to provide complicated operations. As a result, more sophisticated software techniques were used.

## 4.4 Logic- and production-oriented system-level architectures

Substantial research has been carried out on parallel computational models of utilizing AND parallelism, OR parallelism, and stream parallelism in logical inference systems, production systems and others. The basic problem on their exponential complexity remains open at this time.

Sequential Prolog machines using software interpretation, emulation, and additional hardware support such as hardware unification and backtracking[162] have been reported. Single-processor systems for production systems using additional data memories[163] and a RISC architecture[11] have been studied.

New logic programming languages suitable for parallel processing have been investigated. In particular, the use of predicate logic[164], extensions of Prolog to become Concurrent Prolog[48], Parlog[165], and Delta-Prolog[166], and parallel production systems[167] have been developed. One interesting parallel language is that of systolic programming, which is useful as an algorithm design and programming methodology for high-level-language parallel computers[168].

Several prototype multiprocessor systems; for processing inference programs and Prolog have been proposed, some of which are currently under construction. These systems include multiprocessors with a shared memory[169], ZMOB, a multiprocessor of Z80's connected by a ring network[170], Aquarius, a heterogeneous multiprocessor with a crossbar switch[74], and Mago, a cellular machine implementing a Prolog compiler that translates a Prolog program into a formal functional program[171]. Techniques for analyzing Prolog programs such that they can be processed on a data-flow architecture have been derived[172-176]. An associative processor has been proposed to carry out propositional and first-order predicate calculus[177].

Dado is a multiprocessor system with a binary-tree interconnection network that implements parallel production systems[178]. Non-Von is another tree architecture used to evaluate production systems at a lower level of granularity[179].

## 4.5 Distributed problem-solving systems

Knowledge in an AI system can sometimes be represented in terms of semantic nets. Several proposed and experimental architectures have been developed. NETL[90] and its generalization to Thistle[91] consist of an array of simple cells with marker-passing capability to perform searches, set-intersections, inheritance of properties and descriptions, and multiple-context operations on semantic nets. Thinking Machine Inc.'s Connection Machine is a cellular machine with 65 536 processing elements. It implements marker passing and virtually reconfigures the processing elements to match the topology of the application semantic nets[7]. Associative processors for processing semantic nets have also been proposed[180].

Some AI architectures are based on frame representations and may be called object-oriented architectures. For example, the Apiary developed at MIT is a multiprocessor actor system[59], An efficient AI architecture may also depend on the problem-solving strategy. A general form of architectures called connectionist architectures evolve from implementing neurons in brains[6]. The basic idea of the Boltzmann machine is the application of statistical mechanics to constrained searches in a parallel network[21]. The most interesting aspect of this machine lies in its domain-independent learning algorithm[181].

With the inclusion of control into stored

knowledge, the resulting system becomes a distributed problem-solving system. These systems are characterized by the relative autonomy of the problem-solving nodes, a direct consequence of the limited communication capability. With the proposed formalism of the contract net, contracts are used to express the control of problem solving in a distributed processor architecture[182]. Related work in this area include Petri-net modeling[183], distributed vehicle-monitoring testbed[184], distributed air-traffic control system[185], and modeling the brain as a distributed system[186-187].

## 4.6 Hybrid systems

It has been suggested that a combination of Lisp, Prolog, and an object-oriented language such as Smalltalk may be a better language for AI applications[188]. This approach can be carried out in two ways. First, multiple AI languages can be implemented using microprogramming on the same computer, so programs written in these languages can be executed independently. For example, Prolog is available as a secondary language on some Lisp machines. A version of a Prolog interpreter with a speed of 4.5 KLIPS (Kilo Logical Inferences Per Second) has been developed for Lisp Machine's Lambda. A second approach is to design a language that combines the desirable features from several AI languages into a new language. Some of the prototype multiprocessors, such as ZMOB and Mago, were developed with a flexible architecture that can implement object-oriented, functional, and logic languages. FAIM-1, a multiprocessor connected in the form of a twisted hex-topology, was designed to implement the features of object-oriented, functional, and logic programming in the Oil programming language[69]. Currently, a parallel version of Scheme similar to MultiLisp is being implemented. Hope, a hybrid functional and logic language, is currently being implemented on Alice[189].

## 4.7 Fifth generation computer projects

The Fifth Generation Computer System (FGCS) project was started in Japan in 1982 to further the research and development of the next generation of computers. It was conjectured that computers of the next decade will be used increasingly for nonnumeric data processing such as symbolic manipulation and applied AI. The goals of the FGCS project are

(1) to implement basic mechanisms for inference, association, and learning in hardware;

(2) to prepare basic AI software to utilize the full power of the basic mechanisms implemented;

(3) to implement the basic mechanisms for retrieving and managing a knowledge base in hardware and software;

(4) to use pattern recognition and AI research achievements in developing user-oriented man-machine interfaces;

(5) to realize supporting environments for resolving the "software crisis" and enhancing software production.

The FGCS project is a marriage between the implementation of a computer system and the requirements specified by applications in AI, such as natural-language understanding and speech recognition. Specific issues studied include the choice of logic programming over functional programming, the design of the basic software systems to support knowledge acquisition, management, learning, and the intelligent interface to users, the design of highly parallel architectures to support inferencing operations, and the design of distributed-function architectures that integrates VLSI technology to support knowledge data bases[190-192].

A first effort in the FGCS project is to implement a sequential inference machine, or Sim[193]. Its first implementation are two medium-performance machines for software development known as Personal Sequential Inference (PSI) machine and Cooperative High-speed Inference (CHI) machine[194]. The PSI and CHI machines have further been implemented in custom LSFs into PS1-II and CHI-II. The PSI-II has been found to have a performance that ranges from 100 to 333 KLIPS for various benchmark programs. Another architectural development is on the knowledge-base machine, Delta[195].

The current efforts in the intermediate stage are on the parallel inference machine, or PIM, and the multi-PSI computers[195]. As an intermediate target, PIM-I is being built now. It consists of about 100 processing elements, with a total speed of 10-20 MLIPS (Mega Logical Inferences Per Second) including overhead caused by Pimos. Eight processing elements with private caches in a cluster are connected through a shared memory, and a switching network is used to connect the clusters. Each processing element will be implemented in standard-cell VLSI chips. The machine language is KL1-B based on GHC[196]. Lastly, the development of the basic software system acts as a bridge to fill the gap between a highly parallel computer architecture and knowledge information processing[197]. The Pimos was designed as a single unified operating system to control the parallel hardware[198], It was built on the multi-PSI (version 2) system. Each PE consists of a PSI-II, 16-MW main memory, and interfaces to the mesh interconnection network. The KL1-B interpreter is implemented in firmware and attains a speed of 100-150 KLIPS[199].

In the final stage, a parallel computer with about 1 000 processing elements and attaining 100 MLIPS to 1 GLIPS (Gega Logical Inferences Per Second) is expected to be built. Although the projects are progressing well, there is a recognition that more research is needed on exploiting intelligence rather than brute-force parallelism. The proposal of the sixth generation computer system project is an indication of efforts in this direction[14].

The Japanese FGCS project has stirred intensive responses from other countries. The British project is a five-year $550 million cooperative program between government and industry that concentrates on software engineering, intelligent knowledge-based systems, VLSI circuitry, and man-machine interfaces. Hardware development has focused on Alice, a Parlog machine using dataflow architectures and implementing both Hope, Prolog, and Lisp[189]. The European Commission has started the $1.5 billion five-year European Strategic Program for Research in Information Technologies (Esprit) in 1984[200]. The program focuses on microelectronics, software technology, advanced information processing, computer-integrated manufacturing, and office automation. In the U.S., the most direct response to the Japanese FGCS project was the establishment of the Microelectronics and Computer Technology Corporation in 1983[4]. The project has an annual budget of $50-$80 million per year.

It has a more evolutionary approach than the revolutionary approach of the Japanese and would yield technology that the corporate sponsors can build into advanced products in the next 10-12 years. Meanwhile, other research organizations have formed to develop future computer technologies of the U. S. in a broader sense. These include Darpa's Strategic Computing and Survivability, the semiconductor industry's Semiconductor Research Corporation, and the Microelectronics Center of North Carolina[4].

## 5 Design decisions of AI-oriented computers

The appropriate methodology to design an AI computer should utilize a top-down design approach: functional requirements should be developed from the problem requirements, which are mapped into hardware based on technological constraints. Similar to the design of conventional computers, a bottom-up design approach is not adequate since special requirements of the applications may not be satisfied. Before a design is made, it is important to understand the applicability of the system to a class of problems and then to strive for high performance in a prototype implementation. Thus knowing that an *m*-processor system gives a *k*-fold increase in performance over a single processor is more important than knowing the maximum instruction rate of a prototype. Proper understanding and analysis of the problem is probably more important than applying brute-force parallelism randomly in the design.

The issues classified in Table 2 provide a view to the sequence of design decisions made in developing a special-purpose computer to support AI processing. The various approaches can be classified as top-down, bottom-up, and middle-out.

Top-down design decisions: This approach starts by defining, specifying, refining, and validating the requirements of the application, devising methods to collect the necessary knowledge and metaknowledge, choosing an appropriate representation for the knowledge and metaknowledge, studying problems related to the control of correct and efficient execution with the given representation scheme, identifying functional requirements of components, and mapping these components into software and microlevel, macrolevel and system-level architectures subject to technological and cost constraints. The process is iterative. For example, the representation of knowledge and the language features may be changed or restricted when it is discovered that the functional requirements found cannot be mapped into a desirable and realizable system with the given technology and cost. In some projects, the requirements may be very loose and span across many different applications. As a result, the languages and knowledge-representation schemes used may be oriented towards general-purpose usage. The Japanese FGCS project is an attempt to use a top-down approach to design an integrated user-oriented intelligent system for a wide spectrum of applications.

Bottom-up design decisions: In this approach, the designers first design the computer system based on a computational model, such as data flow, reduction, and control flow, and the technological and cost limitations. Possible extensions of existing knowledge-representation schemes and languages developed for AI applications are implemented. Finally, AI applications are coded using the representation schemes and languages provided. This is probably the most popular approach to apply a general-purpose or existing system for AI processing. However, it may result in inefficient processing, and the available representation schemes and languages may not satisfy the application requirements completely, ZMOB and Butterfly Multiprocessor are examples in this class.

Middle-out design decisions: This approach is a short cut to the top-down design approach. It starts from a proven and well-established knowledge-representation scheme or AI language (most likely developed for sequential processing) and develops the architecture and the necessary modifications to the language and representation scheme to adapt to the application requirements and the architecture. This is the approach taken by many designers in designing special-purpose computers for AI processing. It may be subdivided into top-first and bottom-first, although both may be iterative. In a top-first middle-out approach, studies are first performed to modify the language and representation scheme to make it more adaptable to the architecture and computational model. Primitives may be added to the language to facilitate parallel processing. Nice features from several languages may be combined. The design of the architecture follows. Alice and FAIM-1 are examples of architectures designed using this approach. In the bottom-first middle-out approach, the chosen language or representation scheme is mapped directly into architecture by providing hardware support for the overhead-intensive operations. Applications are implemented using the language and representation scheme provided. Lisp computers are examples designed with this approach.

## 6 Conclusion

Although many AI computers have been proposed or built, Lisp computers are probably the only architecture that have had widespread use for solving real AI problems. This is probably due to the large investment in software for many applications coded in Lisp. At present, there is no comprehensive methodology for designing parallel AI computers. Research on AI in the past three decades and the recent experience in building AI computers have led to a view that the key issue of an AI system lies in the understanding of the problem rather than efficient software and hardware. In fact, most underlying concepts in AI computers are not new and have been used in conventional systems. For example, hardware stack and tagged memory were proposed before they were used in Lisp computers. However, the above argument does not imply that research on hardware and architectures is not necessary.

To support efficient processing of AI applications, research must be done in developing better AI algorithms, better AI software management methods, and better AI architectures. The development of better algorithms can lead to significant improvements in performance. Many AI algorithms are heuristic in nature, and upper bounds on performance to solve these problems have not been established as in traditional combinatorial problems. As a consequence, the use of better heuristic information, based on common-sense or high-level metaknowledge and better representation of the knowledge, can have far greater improvement in performance than improved computer architecture. Automatic learning methods to aid designers in systematically acquiring and managing new knowledge to be available in the future are very important.

Better AI software management methods are essential in developing more efficient and reliable software for AI processing. AI systems are usually open and cannot be defined based on a closed-world model. The language must be able to support the acquisition of new knowledge and the validation of existing knowledge. Probabilistic reasoning, fuzzy knowledge, and nonmonotonic logic may have to be supported. The verification of the correctness of an AI program is especially difficult due to the imprecise knowledge involved and the disorganized way of managing knowledge in a number of declarative languages and representation schemes. Traditional software engineering design methodologies must be extended to become knowledge engineering to accommodate the characteristics of knowledge in AI applications. Automatic programming is important to aid designers to generate the AI software from specifications.

The role of parallel processing and innovative computer architectures lies in improving the processing time of solving a given AI problem. It is important to realize that parallel processing and better computer architectures cannot be used to overcome the exponential complexity of exhaustive enumeration (unless an exponential amount of hardware is used) and are not very useful to extend the solvable problem space. For a problem with a size that is too large to be solved today by a sequential computer in a reasonable amount of time, it is unlikely that it can be solved by parallel processing alone, even if a linear speedup can be achieved. The decision to implement a given algorithm in hardware depends on the complexity of the problem it solves and its frequency of occurrence. Problems of low complexity can be solved by sequential processing or in hardware if they are frequently encountered; problems of moderate complexity should be solved by parallel processing; and problems of high complexity should be solved by a combination of heuristics and parallel processing. In many AI systems developed today, tasks and operations implemented in hardware are those that are frequently executed and have polynomial complexity. These tasks or operations are identified from the languages or the knowledge-representation schemes supported. The architectural concepts and parallel processing schemes applied may be either well-known conventional concepts or new concepts for nondeterministic and dynamic processing. The role of the computer architects lies in choosing a good representation, recognizing overhead-intensive tasks to maintain and learn metaknowledge, identifying primitive operations in the languages and knowledge- representation schemes, and supporting these tasks in hardware and software.

## References:


[1] WAH B W, LI G J. Tutorial on computers for artificial intelligence applications[M]. New York: IEEE Computer Society Press, 1986.

[2] SIMON H A. Whether software engineering needs to be artificially intelligent[J]. IEEE Transactions on Software Engineering, 1986, SE-12(7): 726-732.

[3] FEIGENBAUME A. Knowledge engineering: the

applied side[M]// HAYESJ E, MICHIE D. Intelligent systems: the unprecedented opportunity. Chichester: Ellis Norwood Ltd., 1983: 37-55.

[4] Special issue on tomorrow's computers[J]. IEEE Spectrum, 1983, 20(11): 51-58, 69.

[5] AMDAHL G M. Tampered expectations in massively parallel processing and semiconductor industry[C]// The 2nd International Conference on Supercomputing. New York: ACM Press, 1987.

[6] FAHLMAN S E, HINTON G E. Connectionist architecture for artificial intelligence[J]. Computer, 1987, 20(1): 100-109.

[7] HILLIS W D. The connection machine[M]. Cambridge: MIT Press, 1985.

[8] WALTZ D L. Applications of the connection machine[J]. Computer, 1987, 20(1): 85-97.

[9] FENNELL R D, LESSER V R. Parallelism in artificial intelligence problem solving: a case study of Hearsay-II[J]. IEEE Transactions on Computers, 1977, C-26(2): 98-111.

[10] LENAT D B. Computer software for intelligent systems[J]. Scientific American, 1984, 251(3): 204-213.

[11] FORGY C L, GUPTA A, NEWELL A, et al. Initial assessment of architectures for production systems[C]// The 4th National Conference on Artificial Intelligence. Palo Alto: AAAI Press, 1984: 116-120.

[12] LI G J, WAH B W. Coping with anomalies in parallel branch-and-bound algorithms[J]. IEEE Transactions on Computers, 1986, C-35(6): 568-573.

[13] WAHB W, LI G J, YU C F. Multiprocessing of combinatorial search problems[J]. Computer, 1985, 18(6): 93-108.

[14] Science and Technology Agency. Promotion of research and development on electronic and information systems that may complement or substitute for human intelligence[Z]. 1985.

[15] BRACHMAN R, LEVESQUE H. Readings in knowledge representation[M]. Los Altos: Morgan Kaufmann, 1985.

[16] DREYFUSH, DREYFUSS. Why expert systems do not exhibit expertise[J]. IEEE Expert, 1986, 1(2): 86-90.

[17] HALSTEAD R H. Implementation of MULTILISP: LISP on multiprocessor[C]// The 1984 ACM Symposium on LISP and Functional Programming. New York: ACM Press, 1984: 9-17.

[18] HALSTEAD R. Design requirements of concurrent Lisp machines[M]// HWANG K, DEGROOT D. Supercomputers and AI machines. New York: McGraw-Hill Companies, 1988.

[19] FLYNN A M, HARRIS J G. Recognition algorithms for the connection machine[C]// The 9th International Joint Conference on Artificial Intelligence. Los Altos: William Kaufmann, 1985: 57-60.

[20] HOPFIELD J J. Neural networks and physical systems with emergent collective computational abilities[J]. Proceedings of the National Academy of Sciences of the United States of America, 1982, 79(8): 2554-2558.

[21] HINTONG E, SEJNOWSKI T J, ACKLEY D H. Boltzmann machine: constraint satisfaction network that learns[Z]. 1984.

[22] KANERVA P. Parallel structures in human and computer memory?[R]. 1986.

[23] DENNING P. A view of Kanerva's sparse distributed memory[R]. 1986.

[24] NIWA K, SASAKI K, IHARA H. An experimental comparison of knowledge representation schemes[J]. AI Magazine, 1984, 5(2): 29-36.

[25] BOWEN K. Meta-level programming and knowledge representation[J]. New Generation Computing, 1985, 3(4): 359-383.

[26] GENESERETH M R. An overview of meta-level architecture[C]// The 3rd National Conference on Artificial Intelligence. Palo Alto: AAAI Press, 1983: 119-124.

[27] GALLAIRE H, LASSERRE C. Metalevel control for logic programs[M]// CLARK K L, TAMLUND A S. Logic programming. New York: Academic Press, 1982: 173-185.

[28] SILVER B. Meta-level inference: representing and learning control information in artificial intelligence[M]. Amsterdam: North-Holland Publishing Company, 1986.

[29] WARREN D H D. Efficient processing of interactive relational database queries expressed in logic[C]// The 7th International Conference on Very Large Data Bases. Piscataway: IEEE Press, 1981: 272-281.

[30] GOOLEY M M, WAH B W. Efficient reordering of Prolog programs[C]// The 4th IEEE International Conference on Data Engineering. Piscataway: IEEE Press, 1988: 71-75.

[31] LI G J, WAH B W. MANIP-2: a multicomputer architecture for evaluating logic programs[C]// 1985 Inter-

national Conference on Parallel Processing. Washington: IEEE Computer Society Press, 1985: 123-130.
[32] BUCHANAN B G, SHORTLIFFE E H. Rule-Based experts programs: the MYCIN experiments of the Stanford heuristic programming project[M]. Boston: Addison-Wesley Longman Publishing, 1984.
[33] ROBINSON J, SIBERT E. LOGLISP: motivation, design, and implementation[M]// CLARK K, TAMLUND S. Logic programming. New York: Academic Press, 1982.
[34] DEGROOT D, LINDSTROM G. Logic programming[M]. Englewood Cliffs: Prentice-Hall, 1985.
[35] DAVIS A L, ROBISON S V. The FAIM-1 symbolic multiprocessing system[C]// The 30th IEEE Computer Society International Conference. Washington: IEEE Computer Society Press, 1985: 370-375.
[36] REDDY U S. On the relationship between logic and functional languages[M]// DEGROOTD, LINDSTRONE G. Logic programming. Englewood Cliffs: Prentice-Hall, 1985.
[37] MCCARTHY J. History of Lisp[J]. ACM SIGPLAN Notices, 1978, 13(8): 217-223.
[38] BACKUS J. Function-level computing[J]. IEEE Spectrum, 1982, 19(8): 22-27.
[39] Bailey R. A hope tutorial[J]. Byte, 1985, 10(8): 235-258.
[40] MCGRAW J R. Data flow computing: Software development[J]. IEEE Transactions on Computers, 1980, C-29(12): 1095-1103.
[41] WINSTONP H, HOM B. Lisp (2nd ed)[M]. Boston: Addison-Wesley Longman Publishing, 1984.
[42] SUSSMAN G L, HOLLOWAY J, STEEL G L, et al. Scheme-79: Lisp on a chip[J]. Computer, 1981, 14(7): 10-21.
[43] DARLINGTON J, FIELD A J, PULL H. The unification of functional and logic languages[R]. 1985.
[44] DARLINGTON J. Functional programming[Z]. 1984.
[45] HALSTEAD R. Parallel symbolic computing[J]. IEEE Computer, 1986, 19(8): 35-43.
[46] SUGIMOTO S, TABAIA K, AGUSA K, et al. Concurrent Lisp on a multi-micro-processor system[C]// The 7th International Joint Conference on Artificial Intelligence. Los Altos: William Kaufmann, 1981: 949-954.
[47] CLARK K, GREGORY S. Note on system programming in PARLOG[C]// 1984 International Conference on Fifth Generation Computer Systems. Amsterdam: North-Holland Publishing Company, 1984: 299-306.
[48] SHAPIRO E, TAKEUCHI A. Object oriented programming in concurrent Prolog[J]. New Generation Computing, 1983, 1(1): 25-48.
[49] RENTSCH T. Object oriented programming[J]. ACM SIGPLAN Notices, 1982, 17(9): 51-57.
[50] GOLDBERG A J, ROBSON D. Smalltalk-80: The language and its implementation[M]. Boston: Addison-Wesley Longman Publishing, 1983.
[51] STEFIK M, BOBROW D G. Object-oriented programming: themes and variations[J]. AI Magazine, 1986, 6(4): 40-62.
[52] AGHA G. Actor: a model of concurrent computation in distributed systems[D]. Cambridge: MIT Press, 1986.
[53] SNYDER A. Object-oriented programming for common Lisp[R]. 1985.
[54] WEGNER P, SHRIVER B. Special issue on object-oriented programming workshop[J]. ACM SIGPLAN Notices, 1986, 21(10).
[55] BOBROW D G, KAHN K, KICZALES G, et al. CommonLoops: Merging common Lisp and object-oriented programming[R]. 1985.
[56] HEWITT C. Viewing control structure as patterns of passing messages[J]. Artificial Intelligence, 1977, 8(3): 323-364.
[57] AGHA G, HEWITT C. Concurrent programming using actors: exploiting large-scale parallelism[C]// International Conference on Foundations of Software Technology and Theoretical Computer Science. Heidelberg: Springer, 1985: 19-41.
[58] GILOI W K. Advanced object oriented architecture[J]. Future Generation Computer Systems, 1985, 2(2): 169-175.
[59] HEWITT C. The apiary network architecture for knowledgeable systems[C]// The 1980 ACM Conference on LISP and Functional Programming. New York: ACM Press, 1980: 107-117.
[60] HEWITT C, LIEBERMAN H. Design issues in parallel architectures for artificial intelligence[C]// The 28th IEEE Computer Society International Conference. Washington: IEEE Computer Society Press, 1984: 418-423.
[61] WINOGRAD T. Extended inference modes in reasoning by computer systems[J]. Artificial Intelligence, 1980, 13(1-2): 5-26.
[62] FIKES R E, NILSSON N J. Strips: a new approach to the application of theorem proving to problem solv-

ing[J]. Artificial Intelligence, 1971, 2(3-4): 189-208.
[63] SUSSMAN G, WINOGRAD T, CHAMIAL E. Micro-planner reference manual[R]. 1970.
[64] DOYLE J. A truth maintenance system[J]. Artificial Intelligence, 1979, 12(3): 231-272.
[65] SCHOR M. Declarative knowledge programming: better than procedural[J]. IEEE Expert, 1986, 1(1): 36-43.
[66] GUPTA A. Implementing OPS5 production systems on DADO[C]// 1984 IEEE Conference on Parallel Processing. Washington: IEEE Computer Society Press, 1984: 83-91.
[67] HILLYER B K, SHAW D E. Execution of OPS5 production systems on a massively parallel machine[J]. Journal of Parallel Distributed Computing, 1986, 3(2): 236-268.
[68] OFLAZER K. Partitioning in parallel processing of production systems[C]// 1984 International Conference on Parallel Processing. Washington: IEEE Computer Society Press, 1984: 92-100.
[69] ANDERSON J M, COATESW S, DAVIS A L, et al. The architecture of FAIM-1[J]. Computer, 1987, 20(1): 55-65.
[70] KUCK D J, DAVIDSON E S, LAWRIE D H, et al. Parallel supercomputing today and the cedar approach[J]. Science, 1986, 231(4741): 967-974.
[71] DEGROOT D. Restricted AND-parallelism[C]// 1984 International Conference on Fifth Generation Computer Systems. Amsterdam: North-Holland Publishing Company, 1984: 471-478.
[72] CHANG J H, DESPAIN A M, DEGROOT D. AND-parallelism of logic programs based on a static data dependency analysis[C]// The 30th IEEE Computer Society International Conference. Washington: IEEE Computer Society Press, 1985: 218-225.
[73] DEGROOT D, CHANG J H. A comparison of two AND-parallel execution models[C]// Proceedings of Hardware and Software Components and Architectures for the 5th Generation. [S.l.:s.n.], 1985: 271-280.
[74] DESPAIN A M, PATT Y N. Aquarius—a high performance computing system for symbolic/numeric application[C]// The 30th IEEE Computer Society International Conference. Washington: IEEE Computer Society Press, 1985: 376-382.
[75] TUNG Y W, MOLDOVAN D. Detection of AND-parallelism in logic programming[C]// 1986 International Conference on Parallel Processing. Washington: IEEE Computer Society Press, 1986: 984-991.
[76] DEGROOT D. Restricted AND-parallelism and side-effects in logic programming[M]// HWANG K, DEGROOT D. Supercomputers and AI machines[M]. New York: McGraw-Hill Companies. 1988.
[77] CONERY J S, KIBLER D F. AND parallelism and nondeterminism in logic programs[J]. New Generation Computing, 1985, 3(1): 43-70.
[78] KIM S, MAENG S, CHO J W. A parallel execution model of logic program based on dependency relationship graph?[C]// 1986 International Conference on Parallel Processing. Washington: IEEE Computer Society Press, 1986: 976-983.
[79] LINY J, KUMAR V. A parallel execution scheme for exploiting AND-parallelism of logic programs[C]// 1986 International Conference on Parallel Processing. Washington: IEEE Computer Society Press, 1986: 972-975.
[80] KLINKER G, CLUNE E, CRISMAN J, et al. The implementation of a complex vision system on systolic array machine[R]. 1986.
[81] SUGIMOTO S, AGUSA K, TABATA K, et al. A multi-microprocessor system for concurrent Lisp[C]// 1983 International Conference on Parallel Processing. Washington: IEEE Computer Society Press, 1983: 135-143.
[82] HALSTEAD R, ANDERSON T, OSBORNE R, et al. Concert: Design of a multiprocessor development system[C]// The 13th Annual Symposium on Computer Architecture. New York: ACM Press, 1986: 40-48.
[83] GOTO A, TANAKA H, MOTO-OKA T. Highly parallel inference engine PIE: goal rewriting model and machine architecture[J]. New Generation Computing, 1984, 2(1): 37-58.
[84] ERMAN L D, HAYES-ROTH F, LESSER V R, et al. The Hearsay-II speech-understanding system: integrating knowledge to resolve uncertainty[J]. ACM Computing Surveys, 1980, 12(2): 213-253.
[85] HAYES-ROTH B. A blackboard architecture for control[J]. Artificial Intelligence, 1985, 26(3): 251-321.
[86] HALSTEAD R, LOAIZA J. Exception handling in multilisp[C]// 1985 International Conference on Parallel Processing. Washington: IEEE Computer Society Press, 1985: 822-830,
[87] LINDSTROM G, PANANGADEN P. Stream-based execution of logic programs[C]// The 1984 Interna-

tional Symposium on Logic Programming. Piscataway: IEEE Press, 1984: 168-176.

[88] UEDA K. Guarded horn clauses[R]. 1985.

[89] TAKEUCHI A, FUKUKAWA K. Parallel logic programming languages[C]// The 3rd Conference on Logic Programming. Berlin: Springer-Verlag, 1986: 242-254.

[90] FAHLMAN S E. NETL: a system for representing and using real-world knowledge[M]. Cambridge: MIT Press, 1979.

[91] FAHLMAN S E, HINTON G E. Massively parallel Architectures for AI: NETL, THISTLE, and Boltzmann Machine[C]// The 3rd National Conference on Artificial Intelligence. Palo Alto: AAAI Press, 1983: 109-113.

[92] HOPFIELD J J, TANK D W. Neural computation of decisions in optimization problems[J]. Biololgical Cybernetics, 1985, 52(3): 1-25.

[93] PEARL J. Heuristics-intelligent search strategies for computer problem solving[M]. Boston: Addison-Wesley Longman Publishing, 1984.

[94] WAH B W, LI G J. How good are parallel and ordered depth-first searches?[C]// 1986 International Conference on Parallel Processing. Washington: IEEE Computer Society Press, 1986: 992-999.

[95] WAH B W, CHEN K L. A partitioning approach to the design of selection networks[J]. IEEE Transactions on Computers, 1984, C-33(3): 261-268.

[96] WAH B W, MA Y W. MANIP-a multicomputer architecture for solving combinatorial extremum problems[J]. IEEE Transactions on Computers, 1984, C-33(5): 377-390.

[97] YU C F, WAH B W. Efficient branch-and-bound algorithms on a two-level memory system[J]. IEEE Transactions on Software Engineering, 1988, SE-14(9): 1342-1356.

[98] LI G J, WAH B W. Computational efficiency of parallel approximate branch-and-bound algorithms[C]// 1984 International Conference on Parallel Processing. Washington: IEEE Computer Society Press, 1984: 473-480.

[99] YU C F, WAH B W. Learning dominance relations in combinatorial search problems[J]. IEEE Transactions on Software Engineering, 1988, SE-14(8): 1155-1175.

[100] CHEN H Y, WAH B W. The RID-REDUNDANT procedure in C-Prolog[C]// International Symposium on Methodologies for Intelligent Systems. [S.l.:s.n.], 1987: 71-75.

[101] LI G J, WAH B W. Optimal granularity of parallel evaluation of AND-trees[C]// 1986 Fall Joint Computer Conference. Washington: IEEE Computer Society Press, 1986: 297-306.

[102] HWANG K, CHOWKWANYUN R, GHOSH J. Computer architectures for implementing AI systems[M]// HWANG K, DEGROOTD. Supercomputers and AI machines. New York: McGraw-Hill Companies, 1988.

[103] STEELE G L, SUSSMAN G J. Design of a Lisp-based microprocessor[J]. Communications of the ACM, 1980, 23(11): 628-645.

[104] MOON D A. Symbolics architecture[J]. Computer, 1987, 20(1): 43-52.

[105] SCHMECK H, SCHRODER H. Dictionary machines for different models of VLSI[J]. IEEE Transactions on Computers,1985, C-34(5): 472-475.

[106] FISHER A L. Dictionary machines with a small number of processors[C]// The 11th Annual Symposium on Computer Architecture. New York: ACM Press, 1984: 151-156.

[107] BABB E. Joined normal form: a storage encoding for relational Databases[J]. ACM Transactions on Database Systems, 1982, 7(4): 588-614.

[108] HSIAOD K. Special issue on database machines[J]. Computer, 1979, 12(3).

[109] LANGDONG G. Special issue on database machines[J]. IEEE Transactions on Computers, 1979, C-28(6).

[110] BORAL H, DEWITT D. Database machine: an idea whose time has passed?[C]// The 2nd International Workshop on Database Machines. [S.l.:s.n.], 1983: 166-167.

[111] MURAKAMI K, KAKUTA T, ONAI R. Architectures and hardware systems: parallel inference machine and knowledge base machine[C]// 1984 International Conference on Fifth Generation Computer Systems. Amsterdam: North-Holland Publishing Company, 1984: 18-36.

[112] SAKAI H, IWATA K, KAMIYA S, et al. Design and implementation of relational database engine[C]// 1984 International Conference on Fifth Generation Computer Systems. Amsterdam: North-Holland Publishing Company, 1984: 419-426.

[113] SHIBAYAMA S, KAKUTA T, MIYAZAKI N, et al. A relational database machine with large semiconductor disk and hardware relational algebra processor[J]. New

Generation Computing, 1984, 2(2): 131-155.

[114] TANAKA Y. MPDC-massive parallel architecture for very large databases[C]// 1984 International Conference on Fifth Generation Computer Systems. Amsterdam: North-Holland Publishing Company, 1984: 113-137.

[115] KELLOGG C. Intelligent assistants for knowledge and information resources management[C]// The 8th International Joint Conference on Artificial Intelligence. Los Altos: William Kaufmann, 1983: 170-172.

[116] NECHES P M. Hardware support for advanced data management systems[J]. Computer, 1984, 17(11): 29-40.

[117] COHEN J. Garbage collection of linked data structures[J]. ACM Computing Surveys, 1981, 13(3): 341-367.

[118] VEGDAHLS R. A survey of proposed architectures for the execution of functional languages[J]. IEEE Transactions on Computers, 1984, C-33(12): 1050-1071.

[119] CASTAN M, ORGANICK E I. M3L: an HLL-RISC processor for parallel execution of FP-language programs[C]// The 9th Annual IEEE/ACM Symposium Computer Architecture. Piscataway: IEEE Press, 1982: 239-247.

[120] STEEL G, SUSSMAN G. Design of Lisp-based processor, or SCHEME: a dielectric Lisp or finite memories considered harmful, or LAMBDA: the ultimate opcode[R]. 1979.

[121] TURNER D A. A new implementation technique for applicative languages[J]. Software Practice and Experience, 1979, 9(1): 31-49.

[122] DAVIS A L. A data flow evaluation system based on the concept of recursive locality[C]// National Computer Conference. Arlington: AFIPS Press, 1979: 1079-1086.

[123] MAGO G. Making parallel computation simple: the FFP machine[C]// The 30th IEEE Computer Society International Conference. Washington: IEEE Computer Society Press, 1985: 424-428.

[124] O'DONNELL J T. A systolic associative Lisp computer architecture with incremental parallel storage management[D]. Iowa City: The University of Iowa, 1981.

[125] CLARKE T, GLADSTONE P, MACLEAN C, et al. SKIM—The S, K, I reduction machine[C]// The 1980 ACM Conference on LISP and Functional Programming. New York: ACM Press, 1980: 128-135.

[126] DARLINGTON J, REEVE M. ALICE and the parallel evaluation of logic programs[Z]. 1983.

[127] KELLER R M, LIN F C H. Simulated performance of a reduction-based multiprocessor[J]. Computer, 1984, 17(7): 70-82.

[128] KELLER R M, LIN F C H, TANAKA J. Rediflow multiprocessing[C]// The 28th IEEE Computer Society International Conference. Washington: IEEE Computer Society Press, 1984: 410-417.

[129] KLUGE W E. Cooperating reduction machines?[J]. IEEE Transactions on Computers, 1983, C-32(11): 1002-1012.

[130] TRELEAVEN P, MOLE G. A multi-processor reduction machine for user-defined reduction languages[C]// The 7th Annual Symposium on Computer Architecture. New York: ACM Press, 1980: 121-130.

[131] TRELEAVEN P C, HOPKINS R P. A recursive computer architecture for VLSI[C]// The 9th Annual Symposium on Computer Architecture. New York: ACM Press, 1982: 229-238.

[132] DEERING M F. Architectures for AI[J]. Byte, 1985, 10(4): 193-206.

[133] FITCH J. Do we really want a Lisp machine?[C]// The ACM SEAS/SMC Annual Meeting. [S.l.:s.n.], 1980.

[134] KNIGHT T. The CONS microprocessor[Z]. 1974.

[135] DEUTSCH P. Experience with a microprogrammed interlisp systems[J]. ACM SIGMICRO Newsletter, 1978, 9(4): 128-129.

[136] GOTO E, IDA T, HIRAKI K, et al. FLATS, a machine for numerical, symbolic and associative computing[C]// The 6th International Joint Conference on Artificial Intelligence. Los Altos: William Kaufmann, 1979: 1058-1066.

[137] GREENFELD N, JERICHO A. A professional's personal computer system[C]// The 8th Annual Symposium on Computer Architecture. New York: ACM Press, 1981: 217-226.

[138] GRISS M, SWANSON M. MBALM/1700: a microprogrammed Lisp machine for the Burroughs B1726[J]. ACM SIGMICRO Newsletter, 1977, 8(3):15-25.

[139] NAGAO M, TSUJII J I, NAKAJIMA K, et al. Lisp machine NK3 and measurement of its performance[C]// The 6th International Joint Conference on Artificial Intelligence. Los Altos: William Kaufmann, 1979: 625-627.

[140] VON PUTTKAMER E. A microprogrammed Lisp machine[J]. Micro-processing Microprogrammings,

1983, 11(1): 9-14.

[141] SANSONNET J, BOTELLA D, PEREZ J. Function distribution in a list-directed architecture[J]. Microprocessing and Microprogrammings, 1982, 9(3): 143-153.

[142] SANSONNET J P, CASTAN M, PERCEBOIS C. M3L: a list-directed architecture[C]// The 7th Annual Symposium on Computer Architecture. New York: ACM Press, 1980: 105-112.

[143] SANSONNET J P, CASTAN M, PERCEBOIS C, et al. Direct execution of Lisp on a list-directed architecture[C]// The Symposium on Architectural Support for Programming Languages and Operating Systems. New York: ACM Press, 1982: 132-139.

[144] TAKI K, KANEDA Y, MAEKAWA S. The experimental Lisp machine[C]// The 6th International Joint Conference on Artificial Intelligence. Los Altos: William Kaufmann, 1979: 865-867.

[145] MATTHEWS G, HEWES R, KRUEGER S. Single-chip processor runs Lisp environments[J]. Computer Design's, 1987, 26(9): 69-76.

[146] GUZMAN A. A heterarchical multi-microprocessor Lisp machine[C]// IEEE Workshop on Computer Architecture for Pattern Analysis and Image Database Management. Piscataway: IEEE Press, 1981: 309-317.

[147] HILL M, EGGERS S, LARUS J, et al. Design decisions in SPUR[J]. Computer, 1986, 19(11): 8-22.

[148] MCKAY D, SHAPIRO S. MULTI: a Lisp based multiprocessing system[C]// The 1980 ACM Conference on LISP and Functional Programming. New York: ACM Press, 1980: 29-37.

[149] MODEL M. Multiprocessing via intercommunicating Lisp systems[C]// The 1980 ACM Conference on LISP and Functional Programming. New York: ACM Press, 1980: 188-195.

[150] WILLIAMS R. A multiprocessing system for the direct execution of Lisp[C]// The 4th Workshop on Computer Architecture for Non-Numeric Processing. New York: ACM Press, 1978: 35-41.

[151] AMAMIYA M, HASEGAWA R, NAKAMURA O, et al. A list-processing-oriented data flow machine architecture[C]// 1982 National Computer Conference. Arlington: AFIPS Press, 1982: 144-151.

[152] AMAMIYA M, HASEGAWA R. Dataflow computing and eager and lazy evaluations[J]. New Generation Computing, 1984, 2(2): 105-129.

[153] YAMAGUCHI Y, TODA K, YUBA T. A performance evaluation of a Lisp-based data-driven machine (EM-3)[C]// The 9th Annual Symposium on Computer Architecture. New York: ACM Press, 1983: 363-369.

[154] YAMAGUCHI Y, TODA K, HERATH J, et al. EM-3: a Lisp-based data-driven machine[C]// 1984 International Conference on Fifth Generation Computer Systems. Amsterdam: North-Holland Publishing Company, 1984: 524-532.

[155] COGHILL G, HARMA K. PLEIADES: a multimicroproccssor interactive knowledge base[J]. Microprocessors Microsystems, 1979, 3(2): 77-82.

[156] DIEL H. Concurrent data access architecture[C]// 1984 International Conference on Fifth Generation Computer Systems. Amsterdam: North-Holland Publishing Company, 1984: 373-388.

[157] GILOI W K, GUETH R. Concepts and realization of a high-performance data type architecture[J]. International Journal of Parallel Programming, 1982, 11(1): 25-54.

[158] ISHIKAWA Y, TOKORO M. The design of an object-oriented architecture[C]// The 11th Annual International Symposium on Computer Architecture. New York: ACM Press, 1984: 178-187.

[159] PLOTKIN A, TABAK D. A tree structured architecture for semantic gap reduction[J]. Computer Architecture News, 1983, 11(4): 30-44.

[160] SUZUKI N, KUBOTA K, AOKI T. SWARD32: a bytecode emulating microprocessor for object-oriented languages[C]// 1984 International Conference on Fifth Generation Computer Systems. Amsterdam: North-Holland Publishing Company, 1984: 389-397.

[161] UNGAR D, BLAU R, FOLEY P, et al. Architecture of SOAR: Smalltalk on RISC[C]// The 11th Annual Symposium on Computer Architecture. New York: ACM Press, 1984: 188-197.

[162] TICK E, WARREN D H D. Towards a pipelined Prolog processor[J]. New Generation Computing, 1984, 2(4): 323-345.

[163] LENAT D B, MCDERMOTT J. Less than general production system architectures[C]// The 5th International Joint Conference on Artificial Intelligence. Los Altos: William Kaufmann, 1977: 923-932.

[164] EMDEN M H, LUCENA-FILHO G J. Predicate logic as a language for parallel programming[M]// TAMLUNDS A, CLARKK. Logic programming. New York: Academic Press, 1982: 189-198.

[165] CLARK K, GREGORY S. PARLOG: parallel pro-

gramming in logic[R]. 1984.

[166] PEREIRA L M, NASR R. Delta-Prolog, a distributed logic programming language[C]// 1984 International Conference on Fifth Generation Computer Systems. Amsterdam: North-Holland Publishing Company, 1984: 283-291.

[167] UHR L M. Parallel-serial production systems[C]// The 6th International Joint Conference on Artificial Intelligence. Los Altos: William Kaufmann, 1979: 911-916.

[168] SHAPIRO E. Systolic programming: a paradigm of parallel processing[C]// 1984 International Conference on Fifth Generation Computer Systems. Amsterdam: North-Holland Publishing Company, 1984: 458-470.

[169] BORGWARDT P. Parallel Prolog using stack segments on shared-memory multiprocessors[C]// IEEE International Symposium on Logic Programming. Piscataway: IEEE Press, 1984: 2-11.

[170] WEISER M, KOGGE S, MCELVANY M C, et al. Status and performance of the ZMOB parallel processing system[C]// The 30th IEEE Computer Society International Conference. Washington: IEEE Computer Society Press, 1985: 71-73.

[171] KOSTER A. Compiling Prolog programs for parallel execution on a cellular machine[C]// 1984 ACM Annual Conference on Computer Science: The Fifth Generation Challenge. New York: ACM Press, 1984: 167-178.

[172] AMAMIYA M, TAKCSUC M, HASEGAWA R, et al. Implementation and evaluation of a list-processing-oriented data flow machine[C]// IEEE 13th International Symposium on Computer Architecture. Piscataway: IEEE Press, 1986: 10-19.

[173] BIC L. Execution of logic programs on a dataflow architecture[C]// The 11th IEEE/ACM Annual International Symposium on Computer Architecture. New York: ACM Press, 1984: 290-296.

[174] HASEGAWA R, AMAMIYA M. Parallel execution of logic programs based on dataflow concept[C]// 1984 International Conference on Fifth Generation Computer Systems. Amsterdam: North-Holland Publishing Company, 1984: 507-516.

[175] IRANI K B, SHIH Y F. Implementation of very large Prolog-based knowledge bases on data flow architectures[C]// The 1st IEEE Conference on Artificial Intelligence Applications. Piscataway: IEEE Press, 1984: 454-459.

[176] ITO N, SHIMIZU H, KISHI M, et al. Data-flow based execution mechanisms of parallel and concurrent Prolog[J]. New Generation Computing, 1985(3): 15-41.

[177] DILGERW, MULLER J. An associative processor for theorem proving[J]. IFAC Proceedings Volumes, 1983 16(20): 489-497.

[178] STOFLO S J. Initial performance of the DADO2 prototype[J]. Computer, 1987, 20(1): 75-84.

[179] SHAW D E. On the range of applicability of an artificial intelligence machine[J]. Artificial Intelligence, 1987, 32(2): 151-172.

[180] MOLDOVAN D I. An associative array architecture intended for semantic network processing[C]// 1984 ACM Annual Conference on Computer Science: The Fifth Generation Challenge. New York: ACM Press, 1984: 212-221.

[181] ACKLEY D H, HINTON G E, SEJNOWSKI T J. A learning algorithm for Boltzmann machines[J]. Cognitive Science, 1985, 9(1): 147-169.

[182] SMITH R G, DAVIS R. Frameworks for cooperation in distributed problem solving[J]. IEEE Transactions on Systems Man & Cybernetics, 1981, SMC-11(l): 61-70.

[183] PAVLIN J. Predicting the performance of distributed knowledge based systems: a modeling approach[C]// The 3rd National Conference on Artificial Intelligence. Palo Alto: AAAI Press, 1983: 314-319.

[184] LESSER V R, CORKILL D D. The distributed vehicle monitoring testbed: a tool for investigating distributed problem solving networks[J]. AI Magazine, 1983, 4(3): 15-33.

[185] CAMMARATA S, MCARTHUR S, STEEB R. Strategies of cooperation in distributed problem solving[C]// The 8th International Joint Conference on Artificial Intelligence. Los Altos: William Kaufmann, 1983: 767-770.

[186] FRITZ W. The intelligent system[J]. ACM SIGART Newsletter, 1984(90): 34-38.

[187] GEVINS A S. Overview of the human brain as a distributed computing network[C]// IEEE International Conference on Computer Design: VLSI in Computers. Piscataway: IEEE Press, 1983: 13-16.

[188] TAKEUCHI I, OKUNO H, OHSATO N. TAO-a harmonic mean of Lisp, Prolog, and Smalltalk[J]. ACM SIGPLAN Notices, 1983, 18(7): 65-74.

[189] SMITH K. New computer breed uses transputers for parallel processing[Z]. 1983.

[190] KAWANOBE K. Current status and future plans of the fifth generation computer system project[C]// 1984 International Conference on Fifth Generation Computer Systems. Amsterdam: North-Holland Publishing Com-

pany, 1984: 3-36.
[191] TRELEAVEN P C, LIMA I G. Japan's fifth-generation computer systems[J]. Computer, 1982, 15(8): 79-88.
[192] UCHIDA S. Inference machines in FGCS project[C]// VLSI'87 International Conference on VLSI. [S.l.:s.n.], 1985.
[193] YOKOI T, UCHIDA S. Sequential inference machine: SIM-its programming and operating system[C]// 1984 International Conference on Fifth Generation Computer Systems. Amsterdam: North-Holland Publishing Company, 1984: 70-81.
[194] TAKI K, YOKOTA M, YAMAMOTO A, et al. Hardware design and implementation of the personal sequential inference machine (PSI)[C]// 1984 International Conference on Fifth Generation Computer Systems. Amsterdam: North-Holland Publishing Company, 1984: 398-409.
[195] MURAKAMI K, KAKUTA T, ONAI R, et al. Research on parallel machine architecture for fifth-generation computer systems[J]. Computer, 1985, 18(6): 76-92.
[196] SATO M, SHIMIZU H, MATSUMOTO A, et al. KLI execution model for PIM cluster with shared memory[C]// The 4th International Conference on Logic Programming. [S.l.:s.n.], 1987: 338-355.
[197] FURUKAWA K, YOKOI T. Basic software system[C]// 1984 International Conference on Fifth Generation Computer Systems. Amsterdam: North-Holland Publishing Company, 1984: 37-57.
[198] TAKI K. The parallel software research and development tool: Multi-PSI system[C]// The 1st Franco-Japanese Symposium on Programming of Future Generation Computers. Amsterdam: Elsevier Science Publishers, 1988: 411-426.
[199] ICHIYOSHI N, MIYAZAKI T, TAKI K. A distributed implementation of flat GHC on the multi-PSI[C]// The 4th International Conference on Logic Programming. [S.l.:s.n], 1987.
[200] ESPRIT: Europe challenges U.S. and Japanese competitors[J]. Future Generation Computer Systems, 1984, 1(1): 61-69.

发表于《模式识别与人工智能》1990年第1期

# 神经网络计算机的体系结构

李国杰

中国科学院计算技术研究所，北京 100190

**摘　要**：本文从计算机体系结构的角度探讨了有关神经网络计算实现的若干问题。首先比较了虚拟实现和光学、电子光学及VLSI神经网络计算机等几种实现方式，然后分析了设计神经网络计算机应考虑的几个问题，如设计方法论、虚拟处理器、并行粒度、连接的局部性与规则性、输入输出以及符号处理与神经网络计算相结合等。

# On neurocomputer architecture

Guo-jie LI

Institute of Computing Technology, Chinese Academy of Sciences, Beijing 100190, China

**Abstract:** By comparing virtual implementations with optical, electro-optical and VLSI implementations of neural computations, some essential issues in designing neurocomputers, such as design methodology, virtual processor, granularity, locality and regularity of connections, input/output, and combination, of symbolic processing with mural computation, are discussed in the viewpoint of computer architecture.

近 3 年来，有关人工神经元网络的模型和学习算法等方面的研究已成为热门课题，但是，在目前的工艺技术基础上，究竟哪种计算机体系结构最适合神经网络计算？在不久的将来，神经网络计算机应如何实现最合算？这些问题至今尚没有明确的答案，本文从计算机体系结构的角度来探讨有关神经网络计算实现的若干问题。

## 1　神经网络计算机的实现途径

一般来讲，神经网络计算机可以用全硬件方式或软件模拟的方式实现。基本的硬件元件有光学器件、光电器件和电子器件[1]。下面对神经网络计算机的目前水平与可能的实现途径做简单的比较。

### 1.1　全硬件实现与虚拟实现

全硬件实现是指物理上的处理单元及通信通道与一个应用问题的神经网络模型中的神经元及连接一一对应，每一个神经元及每一个连接都有与之对应的物理器件。这种器件可以是数字式的，也可以是模拟器件（如运算放大器）。很显然，全硬件实现的优点是速度快。如果用光学器件或光电器件全硬件方式实现神经网络计算，可以满足许多应用的实时要求。但是全硬件实现中各处理单元之间的连接方式一般难以改变，因此，全硬件实现往往在专用神经网络计算机中采用，缺乏通用性、灵活性和可程序性。一种折中的办法是在全硬件中允许连接的权可改变，这样可构成部分通用的神经网络计算机。

如果用 $P$ 个物理单元实现由 $N$ 个神经元组成的神经网络的计算，只要 $P<N$，我们就称它为神经网络的虚拟实现。在虚拟实现中，若干个神经元要映射到一个物理处理单元，软件的采用提高了虚拟实现的通用性与灵活性。从计算机发展的历史来看，虚拟实现的通用机一直是计算机的主流。每当器件的集成度提高一步，求解问题的规模也相应提高。即使将来光学器件或光电器件发展成熟，虚拟的光学（或光电）神经网络计算机也不可缺少。根据目前技术，虚拟电子神经网络计算机是神经网络计算机的主攻方向。

现有的计算机也可以用软件模拟神经网络计算，国外已有几十家公司出售神经网络计算软件包。但是模拟的速度太慢，典型的神经网络计算问题在 VAX11/780 上模拟实现一般要 2～10 h，训练学习甚至要十几天。即使用巨型计算机（如 Cray）或现有的并行计算机，也不能有效地模拟较大的神经网络（$10^8$ 连接）。目前的粗粒度并行机（如 Cray X/MP-2）不能提供充分的通信带宽，而细粒度并行机（如 Connection Machine）又缺乏足够的局部存储器。另一方面，神经网络计算主要的操作是矩阵乘法，现有的计算机指令很多，用于神经网络计算又会浪费大量资源。为了提高虚拟实现的效率，近几年国外推出了一些专门用于神经网络计算的计算机。表 1 的前 6 种机器是目前市场上销售的典型产品。这些产品中多数采用了先进的数字信号处理或浮点流水线 VLSI 器件（如 Weitek XL 或 TM 32020 等），而且大多是作为 PC 机、工作站或小型机的协处理机，用过程调用方式调用神经网络部件。这些产品价格低（几万美元），但性能比 VAX11/780 要高几百到几千倍。MARK-I 是最早投放市场的虚拟神经网络计算机。它是 TRW 公司的产品，采用 VME 总线和 M68020（及 M68881）微处理机（及浮点协理器），处理单元最多可达 15 个，可以支持包含 65 000 个神经元的计算。近几年 TRW 公司又推出 MARK-IV 及 MARK Ⅲ-1 等新产品，前者可支持包含 256 000 个神经元的计算，后者可作为 DEC VAX 计算机的协处理机，它配有完善的软件环境 ANSE。ANZA 和∑-1 都在 PC-AT 机上附加了一块快速浮点计算的加速板。

表 1 的后 5 种计算机是传统并行计算机，用这些计算机模拟神经网络计算往往效率不高。例如 Think Machine（TM）公司的 CM-2 计算能力为 2 500 MIPS，但模拟神经网络计算时只有 13 MIPS。在神经网络计算机中，IPS 表示每秒计算多少条连接（Interconnect），它不完全等于每秒执行多少指令。MX-1/16 是 MIT 林肯实验室研制的多处理机，以一台 LISP 机为主机，另有 16 台每秒 1 000 万次的数字信号协处理机。MX-1/16 的局部存储器和通信带宽都能满足神经网络计算要求，因此效率较高。从各种并行机模拟神经网络计算的效率差别中，我们可发现神经网络计算对计算机体系结构的要求。根据高通信带宽的要求，现有体系结构中的计算机阵列、Systolic 阵列和基于信息传送的多处理机较适合用于神经网络计算。

### 1.2 光学与电子光学神经网络计算机

光学技术的主要特征是高度连接性（扇入扇出数大）和高通信带宽。这些正是实现神经网络计算的主要要求，因此采用光学技术是实现神经网络计算机最理想的途径。需要的光学元件主要是全息存储器，光盘，光滤波器，光

表 1 神经网络计算机与传统并行计算机的性能比较

| | 计算机名称 | 公司 | 计算能力 | 连接数/条 |
|---|---|---|---|---|
| 神经网络计算机 | Mark-III | TRW | 500 KIPS | 1 130 000 |
| | ∑-1 | SAIC | 10 MIPS | 1 000 000 |
| | ANZA Plus | HNC | 6 MIPS | 1 500 000 |
| | NX-16 | HD | 8 MIPS | 1 600 000 |
| | NDS | Nestor | 500 KIPS | 15 000 000 |
| | ODYSSEY | TI | 20 MIPS | 256 000 |
| 传统并行计算机 | WARP | CMU | 10～17 MIPS | 320 000 |
| | Butterfly | BBN | 16 MIPS | 120 000 000 |
| | CM-2 | TM | 13 MIPS | 64 000 000 |
| | MX-1/16 | MIT | 120 MIPS | 50 000 000 |
| | X/MP 2 | Cray | 50 MIPS | 2 000 000 |

调制器（SLM），由光电器件组成的逻辑门阵列，光导纤维及透镜、反射镜片等[2]。光学神经网络计算机可以用模拟量计算（每秒可实现 $10^{12}$ 次乘加运算），也可用双稳器件（SEED）实现数字计算。目前，美国、日本等都在从事光学计算研究。美国从事光学（或电子光学）神经网络计算机研究的主要单位包括 BDM 公司、Caltech、CMU、Hughes 和海军实验室。典型的研究课题有 BDM 公司的 Attentive Associative Memory 和 Caltech 的全息联想存储器。实现光或光电神经网络计算机的主要困难是如何取得输出信号并分配到合适的光处理器件。若采用全息技术连接，每一连接都需要单独的输入头（Spot），$10^9$ 条连接所需接头点的面积近 1 000 $cm^2$。若采用光纤通信，$10^9$ 条连接的连接头总面积超过 3 $m^2$，此外光纤通信的成本也很高。光计算机的成本主要是机械支持和外部接口。对于虚拟光电子神经网络计算机，还必须解决对全息技术编程的控制等问题。由于光（光电）神经网络计算机在性能价格比方面尚不能与电子计算机竞争，近期的研制工作将在实验室进行。多数专家估计要到 21 世纪初（2005 年左右）才有光神经网络计算机上市。

### 1.3 VLSI 神经网络计算机

VLSI 技术的迅速发展为实现神经网络计算提供了一条途径。VLSI 神经网络计算机包括模拟（Anolog）和数字两种。模拟电路速度快，集成度高，但精度低，对噪音、温度敏感，因而制造较困难。按照目前的工艺水平，一片器件只能包含几百个神经元，预计到 20 世纪末，每片最多包括几千个神经元（假定每个神经元有 1 000 条突触）。VLSI 神经网络计算机的成本主要取决于 RAM 的价格。据预测，到 20 世纪 90 年代中期 64 Mbit 的存储器片价格可降至几美元。我们按这种价格估计了构造具有 $10^9$ 个神经元、$10^{10}$ 条连接的实时神经网络（PE 修改频率为 10 Hz）的可能性。假设数字电子计算机周期为 10 ns，每秒可进行 $10^8$ 次连接修改，为了达到每秒 $10^{11}$ 次连接修改，大约需要 500 台处理机。处理机之间的通信可通过光广播实现。若一条连接需用 64 位信息表示，总信息量 $64\times10^{10}$ 位，即需要 $10^4$ 片 RAM（每片 64 Mbit）。总成本约 100 万美元，计算机总体积约 0.36 $m^3$。采用这种 VLSI 神经网络计算机可以满足某些模式识别、声音识别等应用的实时需要。专家们估计大约 1/10 的人工神经网络应用问题可用 VLSI 神经网络计算机解决。

## 2 设计神经网络计算机应考虑的几个问题

如果把人脑看成一台计算机，这种计算机与现有计算机大不相同，它由大量处理机构成（每一人脑神经元至少相当于一台 68000 计算机一样复杂），各处理机之间有大量非规则的全局性的连接，每一处理机异步随机地执行较低速的模拟量计算。而现有的 VLSI 技术适用于局部规则连接的少量处理机同步的高速数字计算。因此设计神经网络计算机面临的关键问题就是：采用不太多的（比如 100 万个以下）比人脑神经元快得多的处理元件，每个元件最多只有几十个扇入扇出，能否制造出一种计算机，使它的功能接近人脑在声音识别、联想记忆等方面的功能？这就要求我们解决人工神经网络模型与现有计算技术之间在处理机数量，连接数量与方式、处理速度等许多方面不匹配的矛盾。下面我们分成几个方面来讨论解决这些矛盾的原则和基本方法。

### 2.1 设计方法论——映射理论

神经网络计算机的设计过程从上到下可分成神经网络应用、神经网络模型、神经网络计算机体系结构与 VLSI 及光电子技术 4 个层次。每一层次都有多种选择，每相邻两层之间存在很多的映射。目前的设计往往是根据设计者的偏好随意选择一种映射，这就很难保证选择到较好的映射。从这种意义上讲，神经网络计算机的设计方法论就是寻找从应用到基础硬件技术之间的 3 次合理映射。限于篇幅，本文只讨论选择从神经网络模型到神经网络计算机体系结构之间的映射应考虑的一些问题。

假设一个神经网络模型由 $N$ 个神经元组成，采用 $P$ 个处理机实现（$P<N$）。我们必须将 $N$ 个神经元分成 $P$ 组，分在同一组的神经元的状态改变由一个处理机模拟实现。两个处理机之间的通信带宽必须满足两组中神经元连接传送信息的速度要求。设处理机 $i$ 中以处理机 $j$ 为邻居的神经元数目为 $V_{ij}$，处理机 $i$、$j$ 之间通信距离为 $D_{ij}$，那么从减少通信量的角度来看，将神经元分割到合适的处理机的原则是 $\mathrm{Min}(\sum_i,\sum_j,V_{ij},D_{ij})$。一般来讲，这一最优映射问题是 NP 完全问题，我们可根据神经网络的具体结构采用启发式的映射方法[3]。

### 2.2 虚拟处理器

随着人工智能应用的推广，计算机的设计由面向 CPU 逐渐朝面向存储器的方向转移。过去为了扩大存储容量采用了辅助存储器（磁盘）但又不至于明显降低存取速度。我们建立了虚拟存储系统这一概念。目前这一概念已被借用到并行处理机系统。一个神经网络模型中包含的大量神经元和连接权等有关信息存于存储器中，实际的处理机数可能远少于神经元数。在计算过程中，存储的神经元信息可自动地调入处理机。当然这些神经元信息也可存于辅存中，按页调入处理机。从用户的角度来看，仿佛神经元计算机有和神经元一样多的处理机，这种技术被称为虚拟处理机技术。应注意商品神经网络计算机说明书和各种文献报道中提到的神经元数目实际上只反映存储器的容量而不表示实际的处理机数目。虚拟处理机效率的高低取决于上面提到的神经元自动映射策略的优化程度高低。

### 2.3 并行粒度

从硬件观点来看，并行粒度可粗略地理解为解决一个问题所使用的处理机数目，处理机数目越多则并行粒度越小。对于神经网络计算，是否在神经元数目范围内，使用的处理机越多，即粒度越小就能越快地完成计算呢？答案是否定的。在这里，我们必须注意现有计算技术的一条重要限制：在一个处理机内存储数据的速度往往比处理机之间的通信速度高一个数量级。因此，当采用过多处理机时，本来可以从内存中取到的数据也必须由其他处理机送来，通信量的增加会大大降低神经网络计算机的实际速度。对于一种神经网络应用问题，根据给定的模型，我们通过分析可以找出最佳的并行粒度。

为确定最佳并行粒度，首先要建立评判粒度选择好坏的标准。常用的标准有两个，一个是并行计算时间 $T_p$，另一个是处理机利用效率 $f=T_1/PT_p$，这里 $T_1$ 是串行计算时间，$P$ 是处理机个数。以 $T_p$ 为标准，这适合某些实时应用，但一般来讲，采用过多处理机，会导致资源浪费。以处理机效率为标准会导致串行，因为串行时 $f$ 最大（$f=1$）。在我们的研究中采用了将两者结合起来的新的标准 $PT_p^2$。因为 $PT_p^2=T_1T_p/f$，所以一般达到较小的 $T_p$ 和较大的 $f$ 时，$PT_p^2$ 最小，这是较折中的选择。我们的研究表明，并行粒度的正确选择取决于神经网络模型中连接的局部性以及处理速度与通信速度之比。

并行粒度研究对神经网络模型提出了新的要求：并行处理机数目与神经网络规模成比例变化（Scalable Problem）。一般来讲，对于一个给定规模的问题，若处理机数目增加，处理机的利用率就会降低。但适当增加问题的规模，处理机的利用率又会恢复到原来水平。在保持处理机利用率不变的条件下，问题规模增加与处理机数目增加的比例关系称为等价效率函数（Iso-Efficient Function），等价效率函数反映了神经网络模型的计算复杂性。必须指出，现有的神经网络模型并没有解决人工智能常有的组合爆炸问题，只是把计算量转移到学习上。因此增加处理机数目一般不能明显增加可解问题的规模。目前已发现有些神经网络模型计算（学习）时间与神经元数量关系不大，但与学习样本的数量有明显的依赖关系。虽然大规模并行处理机是神经网络计算的重要特征，但我们不能把提高性能的希望完全寄托在并行处理上，也不能完全寄托在靠牺牲通用性、灵活性换取来的模拟电路全硬件实现上。根本的出路还是要建立既有容错性、坚韧性、并行性又只有较低计算（学习）复杂性的神经网络计算理论。

## 2.4 连接的局部性与规则性

二维的 VLSI 技术特别适合用于处理具有局部和规则连接的问题，而一般的神经网络模型的连接不一定是局部、规则的。Hopfield、Kohonen 等模型甚至具有类似 Crossbar 的全部连接。因此如何利用 VLSI 技术有效地实现非局部连接神经网络是一个重要的研究课题。现已发现对于某些应用，Boltzmann 计算机模型有较规则的连接，分层的神经网络模型也有相对的局部连接。离散时间离散状态的神经网络模型与格点自动机（CA）的主要区别在于非邻域的连接，研究格点自动机与神经网络计算的关系是解决这一问题的突破口。我们应该看到，由于神经网络计算有很强的容错性，即使神经元在未收到某些远处神经元的信息时就开始计算也可能得到正确的结果，这就大大降低了对同步的要求。我们必须研究在什么条件下舍弃或推迟远程信息仍能保证收敛到正确结果。

## 2.5 输入输出

大量的并行操作可以加速神经网络计算，但是如何把数据输入分送到各处理单元，如何从各处理单元收集结果是必须解决的计算系统问题。这一问题现已成为细粒度并行机（如 CM）的瓶颈，对于大规模并行的神经网络计算机，这一瓶颈将更严重。必须明确不能用神经网络计算机做图像的预处理。应该先用光学或其他预处理压缩维数，抽取图像的特征，然后再用神经网络计算机来处理这些特征。输入输出能力必须与处理能力匹配，神经网络计算机也不能违背这一基本原则。

### 2.6 符号处理与神经网络计算相结合

理论研究已证明，尽管神经网络模型引入了大量连接、分布知识表示、随机计算、任意值的连接权等新内容，Boltzmann 计算机的计算能力还等价于由只具有有限扇入、连接权恒为 1 的组合逻辑电路构成的常规计算机的能力。神经网络计算机不能取代常规计算机已成为学术界的共识。许多专家都认为，未来的第六代计算机应由较粗粒度的巨型计算机与细粒度的神经网络计算机共同组成。神经网络计算机的一个基本特征是全息方式的存储，它适用于低层次的传感处理，其根本的问题是不能抓住结构。而传统的计算机和人工智能技术主要基于算法（包括非确定算法）与数据及知识结构，将这两方面结合起来，充分发挥两者的长处，通过大规模并行处理来实现与人类智能有关的计算，正是结构化连接机制（Structured Connectionist Model）的目标。值得注意的是，神经网络计算解决的问题领域与常规人工智能主要面对的问题领域并没有很多重叠，这为它们的结合提供了推动力。VLSI 技术和 Von Neumann 计算机的迅猛发展使得一切偏离计算机发展主流的计算机产品难以立足。前几年问世的 LISP 计算机纷纷被淘汰，这十分发人深省。神经网络计算机的发展一定要向计算机的主流靠拢，随着主流技术的发展而发展。我们在这方面要保持清醒的头脑，防止跌入体系结构的陷阱。

## 参考文献：


[1] HECHT-NIELSEN R. Performance limits of optical, electro-optical and electronic neurocomputers[C]// Proceedings of SPIE: The International Society for Optical Engineering. [S.l.:s.n.], 1986: 277-306.

[2] TRELEAVEN P. Neurocomputers[Z].

[3] GHOSH J, HWANG K. Mapping neural network onto message-passing multicomputers[J]. Journal of Parallel and Distributed Computing, 1989, 6(2): 291-330.

## 第4章

# 人工智能和计算理论

# 受用一辈子的好奇心

我与人工智能、计算理论的缘分要从小时候讲起。我读初中的时候，似懂非懂地读了一本封面已泛黄的《科学概论》。书中讲了一个有趣的故事：很久以前埃及有个残暴的国王，每天都要杀一个人，杀人之前要被杀者讲一句话，如果讲的是真话就要被砍头，如果说假话就要被绞死。一句话非真即假，因此没人能逃脱被杀的命运。然而有一个聪明人，他讲了一句话："我是被绞死的。"国王砍头也不行绞死也不行，只好把他放了。这个故事讲的是"数理逻辑"，我看了以后觉得太神奇了，当时就想长大了一定要研究这门学问。后来我几经波折终于还是做了计算机与人工智能研究工作，和数理逻辑有很密切的关系。我感受到，小时候的兴趣和好奇真的可能受用一辈子。

人工智能与别的学科不一样，给人高不可攀的感觉，因此人工智能的学术圈滋长了一种崇拜权威的文化。1984 年我参加美国人工智能学会（AAAI）年会，在会上听学术报告，有时会听到："Minsky said"，似乎只要他做的研究成果与 Minsky 说的一样，就一定是正确的。当时在美国的 *Artificial Intelligence* 期刊发表论文，一般在论文的标题下都有一行字，标明此文是某人推荐，推荐者往往是人工智能学界的权威学者。我觉得这种文化对人工智能的发展十分不利。人工智能的发展历史出现 2 次寒冬，与某些权威学者的错误判断有密切关系，其中就有上面提到的人工智能的元老 Minsky。

世界各国大都将人工智能归类于计算机科学，授予人工智能突出贡献者的奖项是计算机学会的图灵奖。但中国的学科分类与其他国家不一样，人工智能划在自动化领域，这可能是受苏联的影响。1981 年成立的中国人工智能学会是挂靠在社会科学院下面的一级学会，不属于中国科学技术协会领导，当时计算机领域的学者很少参加这个学会的活动。为了联合全国人工智能学者，863 计划 306 专家组，特别是戴汝为院士做了不少努力，但由于种种原因，成立统一的中国人工智能学会的目的没有实现。戴汝为在美国普渡大学做访问学者时，我正在那里读博士。回国以后我与他有较多合作。国家智能计算机研究开发中心一成立，就与中国自动化学会合作（戴院士在中国科学院自动化研究所工作），创办了《模式识别与人工智能》期刊，他当主编，我当副主编。这份期刊如今成了我国人工智能领域的主要学术期刊。国家智能计算机研究开发中心后来演变为计算所的一个研究中心，已很少在外面公开活动，但至今仍然是这份期刊的主办单位之一，也算是继续为发展人工智能技术做贡献。

本章收录的文章涉及人工神经网络、模拟退火、SAT 问题、模式识别和图像检索，但我最关注的是人工智能的计算复杂性问题。智能问题多数是 NP 困难问题。如何对付具有指数复杂性的 NP 问题是人工智能研究的核心问题，学术界常常被所谓 NP 问题误导。20 世纪 80 年代有学者发表一篇文章说，神经网络泛化 1/3 就是 NP 完全问题，害得神经网络出现寒冬。其实 NP 问题并没有那么可怕，因为所谓指数复杂性是指最坏情况。一个人工智能问题（例如自然语言理解）如同一片湖水，可能大多数地方很浅，只有个别地方有深沟，但这些深沟的分布很不规则。科学家的责任是识别这些深沟在哪里，剩下的地方可以大胆加速处理。有些理论研究成果像灯塔一样指路，有些理论成果起反作用，人工智能历史上起反作用的理论成果不少，所谓寒冬往往是“大学者”自己吓唬自己造成的。我 1992 年在《模式识别与人工智能》期刊上发表的论文《人工智能的计算复杂性研究》提出了一些有价值的观点，可惜后来没有时间深入做下去。

发表于《计算机学报》1990年第5期

# 基于熵的人工神经网络系统理论初探

李国杰

中国科学院计算技术研究所

**摘　要：**国外目前对人工神经网络的研究多数是从单个状态出发，研究系统的输入输出关系或动态行为。有别于已有的方法，从系统的熵的角度出发对同步并行和异步串行计算进行分析，研究了神经网络系统的吸引性。论证了同步计算中神经网的熵取决于可能状态的总数而与状态分布概率无关。分析了能量函数与熵的关系，推导了Hopfield模型可能状态数目的上界。对于异步随机计算，分析了神经网系统的熵的收敛性以及它与状态迁移矩阵的特征值及迹的关系。

# Entropy based system-theoretical aspect of artificial neural networks

Guo-jie LI

Institute of Computing Technology, Chinese Academy of Sciences

**Abstract:** It is the interest to build up a unifying framework which explains the common nature of all the different artificial neural network (ANN) model. The system-theoretical aspect of ANN was introduced from the stand point of the entry. The convergent behavior of entropy in both synchronous and synchronous neural computation was analyzed, and, with respect to Hopfield CAM model, the upper bound of the number of possible state is derived. The relationship between entropy and state-transition matrix of artificial neural network was studied.

## 1　引言

人工神经网络(以下简称神经网络或神经网)以它的大规模并行性、知识的分布式表示、容错性与自组织性等许多优点吸收着各国学者，近三年来，已形成全球性的神经网络计算研究热潮[1-2]。

神经网计算有多种处理模型和学习模型[3]。本文研究神经网计算过程中宏观不确定性变化的一般规律，即熵的变化规律。文中有些结果基于Hopfield神经网模型[4]，但本文的基本思想和分析方法适用于其他模型。

由 $n$ 个神经元组成的神经网在某一时刻的系统状态可用一个由 $n$ 个元素组成的状态向量

本课题受国家863计划支持

表示，$\boldsymbol{S}=(S_1,S_2,\cdots,S_n)^{\mathrm{T}}$，向量的每一元素 $S_i$ 表示一个神经元的激励状态（若不加说明，下文中“状态”均指系统状态向量而不是指个别神经元的状态）。神经元 $i$ 和 $j$ 的连结强度用权 $W_{ij}$ 表示，所有的连接权构成一个 $n\times n$ 的矩阵 $\boldsymbol{W}$，神经网的状态可根据下述动力学规则演变：

$$\boldsymbol{S}(t+1)=f\left(\boldsymbol{W}\,\boldsymbol{S}(t)-\theta\right) \qquad (1)$$

$$f(x)=\begin{cases}1 & x\geqslant 0\\ -1 & x<0\end{cases} \qquad (2)$$

给定一种计算方式后，一个神经网的行为完全由连接矩阵 $\boldsymbol{W}$ 和非线性阈函数 $f$ 确定。

不论从微观还是从宏观来看，神经网络的行为都与它的计算方式有关，最基本的计算方式有 2 种：同步计算与异步计算。同步计算是指技照式（1）表示的迭代方式计算，即等所有神经元都完成当前迭代计算后才开始下一步迭代，它是一种确定型计算。本文讨论同步计算时只考虑全并行方式。异步计算是指每次对一个或 $m$ 个（$m<n$）随机选取的神经元（或按某种次序选取）进行计算，每次计算得到的该神经元的新值立即在下一次计算中使用。异步串行的计算式如下：

$$S_i(t+1)=f\left(\sum_{j=1}^{t\cdots1}W_{ij}S_j(t+1)+\sum_{j=i}^{n}W_{ij}S_i(t)-\theta_i\right) \qquad (3)$$

异步计算可能带有随机性，神经网系统表现出与同步计算不同的行为。下面我们将分别讨论这 2 种计算方式下熵的变化规律。

## 2 神经网计算中的能量与熵

Hopfield 模型的一个重要贡献是将能量函数概念引入神经网络研究，将神经明的结构（如连接矩阵）与要解决的问题（目标函数）联系起来。在 Hopfield 神经网与 Boltzmann 计算机中，能量函数定义为：

$$E\left(\boldsymbol{S}(t)\right)=-\frac{1}{2}\boldsymbol{S}^{\mathrm{T}}(t)\boldsymbol{W}\boldsymbol{S}(t)+\theta^{\mathrm{T}}\boldsymbol{S}(t) \qquad (4)$$

从自动控制的观点来看，神经网络可以看成是一个非线性动力学系统。它的稳定性信息可以通过关于系统状态的 Liapunov 函数直接得到，而不必求解描述该系统的微分方程。神经网中定义的能量函数就是系统的 Liapunov 函数。应当指出，能量最小化原则并不是对各种神经模型都适用的普遍原则。国外许多学者正在寻求一种适用于各种神经网模型的统一的理论基础。本文提出用熵的理论作为神经网模型的基础，作者认为可能是神经网系统理论研究的一个新方向。

众所周知，人工智能中要处理的问题都带有某种不确定性。从本质上讲，神经网络处理信息的过程就是不确定性减少的过程。信息论告诉我们：衡量一个系统不确定性最本质的参数是系统的熵。对于一个由 $n$ 个神经元组成的神经网来说，每个神经元可取离散的 2 个值之一。假设在某时刻 $t$，某一系统状态 $S_i$ 出现的概率为 $P(i,t)$，我们可定义一个空间量度熵：

$$H(t)=-\sum_{i=1}^{2^n}P(i,t)\log_2 P(i,t) \qquad (5)$$

对于无次序的初始状态集合来说，每个神经元的激励状态可随机地取值，所有 $2^n$ 个可能状态以相等概率出现。因此在时刻 $t=0$，$H(0)=n$，随着系统状态的迁移，有些状态不可能再出现，这些状态称为“禁止状态”。概率 $P(i,t)>0$ 的状态称为 $t$ 时刻的“可能状态”。

本文讨论的熵与能量分别定义在不同层次上，从式（4）与式（5）可以看出，每个状态对应一个能量值，我们可以比较不同状态的能量大小，而熵的大小取决于所有状态的概率分布。熵是用来描述系统不确定性的工具，针对不同应用可采用不同形式的熵。对于模式识别与许多神经网应用来说，我们最关心的是状态概率分布的非均匀性。确切地说，我们注意熵函数的凹性，即要求均匀分布时熵最大而某一状态概率为 1 时熵最小，而且当系统由均匀分布向极不均匀分布（即极少可能状态）演变时，熵逐渐变小。为了描述概率分布的非均匀性，我们可采用下述比对数形式更简单的熵函数：

$$H\left(P_1,P_2,\cdots,P_N\right)=1-\sum_{i=1}^{N}P_i^2=\sum_{i=1}^{N}\sum_{j=1}^{N}P_iP_j,i\neq j \tag{6}$$

其中，$N$ 是系统状态的总数。

上式定义的熵虽然不满足复合性与可加性，但可作为分析神经网系统不确定性的有效工具。如果将第 $t$ 步运算后所有可能状态的概率分布写成向量形式 $\boldsymbol{P}(t)=(P_1(t),P_2(t),\cdots,P_N(t))$，可定义一个状态迁移矩阵 $\boldsymbol{A}$。对于同步并行计算来说，$\boldsymbol{A}$ 是一个布尔矩阵，$A_{ij}=1$ 表示状态 $i$ 经一步并行计算后迁移到状态 $j$。对于异步串行计算来说，选定不同执行次序逐个地对 $n$ 个神经元进行计算，可能将一个系统状态转变到不同的状态。$A_{ij}$ 表示对各种可能计算次序统计得到的状态 $i$ 迁移到 $j$ 的概率。很显然，神经网的状态迁移矩阵是一个马尔可夫转移矩阵。我们可利用马尔可夫链和布尔矩阵的性质来研究神经网。

根据式（6）和马尔可夫转移矩阵的性质

$$\boldsymbol{P}\left(t+1\right)=\boldsymbol{P}\left(t\right)\boldsymbol{A} \tag{7}$$

有

$$\boldsymbol{H}\left(t+1\right)=1-\boldsymbol{P}\left(t\right)\boldsymbol{A}\boldsymbol{A}^{\mathrm{T}}\boldsymbol{P}^{\mathrm{T}}\left(t\right) \tag{8}$$

对照式（8）与式（4）可看出熵与能量函数的对应关系。向量 $\boldsymbol{P}(t)$ 与 $\boldsymbol{S}(t)$ 对应而状态迁移矩阵 $A$ 与连接矩阵 $W$ 对应。本文采用的状态迁移矩阵是研究神经网的重要中间工具，它可以作为一座桥将连接矩阵与神经网的性能联系起来。

如果合理地规定神经元的激励状态可取值的个数与范围，并做合理的解释，也可把神经网的能量函数看成是一种熵，或者反过来，熵也可以看成是一种 Liapunov 函数。Watanabe 论述过熵与 Liapunov 函数的等价性，本文不再赘述。

## 3 同步并行计算方式

同步并行计算方式的优点是能充分发挥神经网模型的并行性。它是一种确定型计算模型，一种状态经过一步并行计算后必然到达某一确定的状态。这种方式不适于全硬件实现的无统一时钟的异步神经网络，但它可以有效地在现有的并行计算机或串行机上模拟实现。

### 3.1 同步计算中熵的单调性

首先，我们证明同步计算中神经网系统的熵取决于可能状态的总数而与可能状态的概率分布无关。由于篇幅限制，本节的定理 1 和定理 2 的证明从略，读者可从参考文献[5]中找到详细证明。

**定理 1**　假设 $M(t)$ 是神经网络第 $t$ 步同步并行计算后可能状态的总数，系统的熵按式（5）或式（6）定义，那么 $H(t+1)\leqslant H(t)$，而且

(a)如果 $M(t)=M(t+1)$ 则 $H(t)=H(t+1)$；

(b)如果 $M(t)>M(t+1)$ 则 $H(t)>H(t+1)$

这一定理说明了可能状态数目在神经网自组织过程中的重要作用。神经网每一步基本操作都是矩阵乘向量，连接矩阵提供的某些信息导致不确定性减少，它通过可能状态数目减少反映出来。现在我们解释神经网中能量函数与熵的关系。先给出一条重要的引理。

**引理 1**　假设计算中出现的所有状态在初始时都是可能状态，如果第 $t$ 步迭代计算后某状态 $S_K$ 是禁止状态，那么第七步以后 $S_K$ 永远是禁止状态。

**定理 2** 假定同步并行计算中出现的所有状态都是初始时的可能状态，对于任意可能状态 $\boldsymbol{S}(t)$，能量函数单调非升或非降，即 $E(\boldsymbol{S}(t))\leqslant E(\boldsymbol{S}(t-1))$ 或 $E(\boldsymbol{S}(t))\geqslant E(\boldsymbol{S}(t-1))$，而且若至少有一个可能状态满足 $E(\boldsymbol{S}(t))<E(\boldsymbol{S}(t-1))$ 或 $E(\boldsymbol{S}(t))>E(\boldsymbol{S}(t-1))$，那么，神经网系统的熵单调下降，即 $H(t+1)<H(t)$。

在定理 2 的证明中，我们没有限制能量函数的具体形式。能量函数单调变化说明系统的状态变化是不可逆的。因而可能状态数必然减少导致熵减少，这一定理与热力学第二定律揭示的封闭系统不可逆过程导致熵增加并不矛盾。因为神经网络是与外界交换物理能量的耗散系统。根据 Prigogine 的耗散结构理论，这种系统的熵向减少方向演变，这就是自组织过程。

### 3.2 可能状态总数的上界

上面的分析已说明同步计算中熵的大小由可能状态总数唯一确定。而一个神经网系统可能状态数目的多少及其减少的速率是由它的连接矩阵和非线性阈函数决定的。下面我们讨论可能状态总数与连接矩阵的关系，并以 Hopfield CAM 模型为例推导可能状态数目的上界。

作为一个简单实例，我们先检查一个 3×3 连接矩阵 $W$。给定一个状态向量$(1,-1,1)^{\mathrm{T}}$，问它是不是一个以 $W$ 为连接矩阵的神经网的可能状态？如果是，有多少输入状态吸引到这一给定状态。这两个问题可以通过解下述线性不等式解决。

$$
\begin{aligned}
W_{11}x_1+W_{12}x_2+W_{13}x_3 &\geqslant 0\\
W_{21}x_1+W_{22}x_2+W_{23}x_3 &< 0\\
W_{31}x_1+W_{32}x_2+W_{33}x_3 &\geqslant 0
\end{aligned}
\qquad (9)
$$

这里假定所有阈值 $\theta_i$ 为零。如果不等式组（9）无解，那么状态(1,-1,1)就不是一个可能状态。相反，如果式（9）有多个解，那么就出现吸引现象。从理论上讲，求解不等式组就是求整数线性规划的满足条件的可行解。这类问题是非确定性多项式（Nondeterministic Polynomial，NP）完全问题。但我们并不需要直接求解这类整数规划问题，而只要弄清楚它的可行解与 $W$ 的关系。

对于神经网来说，弄清楚一步计算的输入输出状态映射十分重要。下面分析 Hopfield CAM 模型的 $2^n$ 个输入状态，经一步并行计算后最多能到达多少个可能状态。为了回答这一问题，先要定义一些术语。

**定义 1** 对于 $m$ 个已存储的模式 $f_1,f_2,\cdots,f_m$，我们称向量 $\boldsymbol{C}_i=(f_1(i),f_2(i),\cdots,f_m(i))^{\mathrm{T}}$ 为第 $i$ 模式列。

例如，如果存储的 3 个模式是

$$
\begin{matrix}
1 & 1 & -1 & -1 & 1 & -1 & -1 & -1\\
1 & 1 & 1 & -1 & -1 & -1 & 1 & -1\\
1 & 1 & 1 & 1 & -1 & -1 & -1 & -1
\end{matrix}
\qquad (10)
$$

那么，第一和第三模式列为 $(1,1,1)^{\mathrm{T}}$ 和 $(-1,1,1)^{\mathrm{T}}$。

**定义 2** 一组与指定的模式列相等的模式列称为该指定模式列的等价类。

**定义 3** 一组与指定的模式列相反的模式列称为该指定模式列的互补类。

例如，上面例子中模式列 2 是模式列 1 的等价类，而模式列 6 和 8 构成模式列 1 的互补类。给定 $m$ 个模式，最多 $m$ 个互相独立的模式列。一组最大的互相独立的模式列称为基本模式列组，其中的每个模式列被称为基本模式列，每个基本模式列可能有它的等价类和互补类。一个基本模式列组所具有等价类和互补类的数目分别记为 $p$ 和 $q$；所有基本模式列对应的等价类和互补类所包含的模式列的总数记为 $r$，在式（10）给出的例子中，选模式列 1、3、4 为基本模式列，则 $p=1$，$q=2$，$r=4$。显然，选取不同的基本模式列组，$p$、$q$、$r$ 的值可能不同。对于一组给定的样本模式来说，我们选择基本模式列的原则是使 $r$ 最大，如对应最大的 $r$ 值有几种可能则选择其中 $q$ 值最小的一种。

Hopfield 最初提出他的模型时，曾做了神经元没有自反馈的限制，即规定 $W_{ii}=0$。这一限制对于某些应用来说是不必要的。有些学者已经研究了带神经元自反馈的模型，即允许 $W_{ii}\neq 0$。他们发现这种神经网易于光学实现并且性能更好[6]。下面的定理给出了有自反馈的神经网一步计算后所剩下的可能状态数目的上界。

**定理 3** 假定一个 Hopfield CAM 神经网由 $n$ 个神经元组成，一共存储了 $m$ 个样本，而且连接矩阵按外积法构成，零输入时输出为 1，共有 $2^n$ 个初始状态，那么，经一步并行计算后最多剩下 $2^{n-r}\cdot\left(\frac{3}{2}\right)^q$ 个可能状态，$r$ 和 $q$ 按上述说明定义。

**证明**：首先，我们要说明按外积法获得的连接矩阵的秩不大于 $m$。假定存储的 $m$ 个模式是 $\boldsymbol{f}_1,\boldsymbol{f}_2,\cdots,\boldsymbol{f}_m$，由外积法规则，有

$$\boldsymbol{W}=\boldsymbol{f}_1^{\mathrm{T}}\boldsymbol{f}_1+\boldsymbol{f}_2^{\mathrm{T}}\boldsymbol{f}_2+\cdots+\boldsymbol{f}_m^{\mathrm{T}}\boldsymbol{f}_m$$

其中，$\boldsymbol{f}_i$ 是行向量，$\boldsymbol{f}_i^{\mathrm{T}}\boldsymbol{f}_i$ 是自相关矩阵。每个自相关矩阵的秩是 1。根据线性代数理论，

$$\mathrm{rank}(\boldsymbol{A}+\boldsymbol{B})\leqslant\mathrm{rank}(\boldsymbol{A})+\mathrm{rank}(\boldsymbol{B})$$

所以 $\mathrm{rank}(\boldsymbol{W})\leqslant m$。

令 $\boldsymbol{W}_i$ 表示连接矩阵 $\boldsymbol{W}$ 的第 $i$ 行，按外积法规则

$$\boldsymbol{W}_i=\sum_{i=1}^{m}\boldsymbol{f}_i(i)\boldsymbol{f}_i$$

这意味着连接矩阵的每一行对应一个模式列。另一方面，连接矩阵的每一行确定一个神经元的输出。因此，每个模式列对应一个神经元。一个基本模式列和它的等价类对应的连接矩阵的各行完全相同，因此，它们所对应的神经元的输出必然相同。这意味着等价类对构成输出状态向量的组合没有贡献。但是，互补类对输出组合可能有影响。假设有 2 个初始状态向量 $\boldsymbol{X}_1$ 和 $\boldsymbol{X}_2$ 使得 $\boldsymbol{W}_i\boldsymbol{X}_1>0$ 而 $\boldsymbol{W}_i\boldsymbol{X}_2=0$，按定理条件规定，这 2 个输入状态都使得神经元 $i$ 输出为 1。再假设连接矩阵还有另一行 $\boldsymbol{W}_j=-\boldsymbol{W}_i$，那么 $\boldsymbol{W}_j\boldsymbol{X}_1<0$ 而 $\boldsymbol{W}_j\boldsymbol{X}_2=0$，因此 1 和−1 都是神经元 $j$ 的可能输出。考虑这一对神经元（即考虑一对互补模式列），3 种输出组合，即(1,1)、(1,−1)、(−1,1)都是可能的。因为存储的样本模式共有 $q$ 个互补类，所以有互补类的基本模式列和它们的互补类所对应的神经元的输出组合最多 $3^q$ 种。没有互补类的基本模式列对应的神经元共有 $2^{m-q}$ 种可能输出组合。除了上述几种情况外，连接矩阵中剩下的行都是与基本模式列组对应的各行的线性组合。由于存在非线性阈函数，它们的输出相关性十分复杂，我们假设在最坏情况下这些行对应的神经元的输出可以是 1 或−1。因此，其对应的输出组合不多于 $2^{n-m-r}$。综合以上各种情况，可推导出一步并行计算后剩下的可能状态数目的上界为

$$2^{(n-m-r)}\cdot 2^{(m-q)}\cdot 3^q=2^{n-r}\left(\frac{3}{2}\right)^q$$

证毕。

从上述定理可知，$r$ 越大，$q$ 越小，则一步计算后剩下的可能状态越少，即熵减小越快。我们选择 $r$ 的最大值（确定 $r$ 后再选 $q$ 的最小值）估计可能状态总数，可得到较精确的上界。

对式（10）给出的例子，由定理 3 得出的可能状态数的上界是 36。我们对所有 256 种输入用计算机做了模拟，发现实际共有 18 种可能状态，这说明理论推导出的上界有较好的精度。应当强调指出，定理 3 给出的上界不适用于无神经元自反馈的神经网。如果所有对角线元素为零，连接矩阵的秩一般大于 $m$，因此可能状

态数往往会增加很多。模拟结果表示，式（10）给出的例子采用零对角线连接矩阵共有 82 种可能输出状态。从可能状态的多少可以定性地解释带神经元自反馈连接的神经网比设有自反馈连接的神经网有更好的吸引性能。但由于前者对应的可能状态一般并不是后者对应的可能状态的子集，难以对上述断言给出严格的证明。

定理 3 规定采用的阈函数是当 $x \geqslant 0$ 时 $f(x)=1$ 否则 $f(x)=-1$。如果采用其他阈函数，可能状态数目的上界将不同。例如，如果采用的阈函数是 $x>0$ 时 $f(x)=1$，$x<0$ 时 $f(x)=-1$，$x=0$ 时状态不变，则可能状态数量会增加很多。它的上界是 $2^{n-p}$，$p$ 是基本模式列的等价类数目。这是因为在这种情况下一对互补的模式列对应的一对神经元有 4 种可能输出组合。模拟结果表明，式（10）给出的例子采用这种阈函数有 54 种可能状态。

定理 3 只给出了一步并行计算后得到的可能状态数的上界，这一上界可作为定性分析神经网系统吸引性的参考。在以后几步计算中，可能状态数目一般会进一步减少，对整个计算过程各步得到的可能状态数的估计还需要做深入的分析。

## 4 异步串行计算方式

上节所证明的熵变化的单调性不能直接推广到异步计算。由于异步计算可能包含随机性，因此熵的变化行为较复杂。本节讨论异步计算过程中熵的变化规律。首先我们着重分析全局稳定的神经网异步计算时熵的变化规律，最后再简单说明非全局稳定的神经网熵的变化特征。

### 4.1 异步计算的状态迁移矩阵

为了便于进行理论分析，本节假定神经网有统一的时钟。每一步从 $n$ 个神经元中随机地选择一个进行计算。假定一步中选择任意神经元的概率都相同（均匀 $1/n$），我们很容易得到异步串行计算的状态迁移矩阵。因为一个状态经一步随机串行计算最多跃迁到 $n$ 种可能状态，这种以串行步为时间单位的状态迁移矩阵是一个稀疏矩阵。我们也可以规定 $n$ 步串行计算为一次“迭代”，一次“迭代”中每个神经元计算一次，但计算顺序可随机选择。以迭代步为时间单位可以得到另一类状态迁移矩阵。对于这两类矩阵来说，吸引子 $i$ 对应的状态迁移概率 $P_{ii}=1$ 而 $P_{ij}=0$，$i \neq j$。下面的分析对这两类矩阵都成立。

在异步计算中，神经网的状态可分为两大状态集合：瞬时状态集合和遍历状态集合。瞬时状态集合中的状态一旦离开这个集合就不会再回到此集合；而一个状态一旦迁移到遍历状态集合就不会再逃出这个集合。在神经网的应用中，人们希望所有的遍历状态都只包含一个状态，这种神经网在计算时不会出现周期振荡，即具有全局稳定性。假设一个全局稳定的神经网的状态总数为 $N$，它对应的 $N$ 维状态迁移矩阵具有下式表示的形式：

$$\boldsymbol{A}=\left(\begin{array}{c|c}\boldsymbol{I} & \boldsymbol{O} \\ \hline \boldsymbol{R} & \boldsymbol{Q}\end{array}\right) \tag{11}$$

其中，$\boldsymbol{I}$ 表示 $r$ 维单位子矩阵，只有对角线元素为 1，它们对应 $r$ 个吸引子。$\boldsymbol{O}$ 表示 $N-r$ 维零子矩阵，它的所有元素为零。$\boldsymbol{Q}$ 表示瞬时状态集合内部的状态迁移概率。$\boldsymbol{R}$ 表示瞬时状态集合中的状态到吸引子的迁移概率。现代控制理论已经揭示，状态迁移矩阵可对角化与系统的稳定有密切关系。全局稳定的神经网具有非周期的状态迁移矩阵，而这种矩阵往往可以对角化。本节的理论分析基于可对角化的矩阵，一个矩阵 $\boldsymbol{A}$ 若能通过相似变换转化成对角矩阵 $\boldsymbol{D}$，即 $\boldsymbol{A}=\boldsymbol{NDN}^{-1}$，则称 $\boldsymbol{A}$ 为可对角化矩阵。一个 $n$ 阶方阵可对角化的充分必要条件是它具有 $n$ 个互相独立的特征向量（不一定要求有 $n$ 个不同的特征值）。事实上，相似变换矩阵 $\boldsymbol{N}$ 的每一列

都是 $A$ 的不同的特征向量，它们互不相关。

## 4.2 异步计算中熵的收敛性

这一小节根据状态迁移矩阵的某些参数推断全局稳定的神经网异步计算中熵的收敛行为。一个矩阵可以按它的特征值，左右特征向量进行分解。假设一个状态迁移矩阵 $A$ 有 $N$ 个特征值 $\lambda_i$，对应的 $N$ 对左、右特征向量分别为 $\pi_i$ 和 $\varphi_i$，即 $\pi_i A=\lambda_i\pi_i$，$A\varphi_i=\lambda_i\varphi_i$。令 $B_i=\varphi_i\pi_i$，我们有

$$A=\sum_{i=1}^{N}\lambda_i B_i, A^t=\sum_{i=1}^{N}\lambda^t{}_i B_i, (A^{\mathrm{T}})^t=\sum_{i=1}^{N}\lambda^t{}_i B_i^{\mathrm{T}}$$

按式（6）熵的定义和 $P(t)=P(0)A^t$，我们得到

$$\begin{aligned}H(t)&=1-P(t)P^{\mathrm{T}}(t)=1-P(0)A^t\left(A^{\mathrm{T}}\right)^t P^{\mathrm{T}}(0)\\&=1-P(0)\left(\sum_{i=0}^{N}\lambda_i^t B_i\right)\left(\sum_{j=1}^{N}\lambda_j^t B_j^{\mathrm{T}}\right)P^{\mathrm{T}}(0)\\&=1-\sum_{i=1}^{N}\sum_{j=1}^{N}c_{ij}\lambda_i^t\lambda_j^t\end{aligned}\quad(12)$$

这里 $c_{ij}=P(0)B_iB_j^{\mathrm{T}}P^{\mathrm{T}}(0)$，与 $t$ 无关。

非周期的马尔可夫转移矩阵的特征值满足 $\lambda_i=1$，$1\leqslant i\leqslant r$；$|\lambda_i|<1$，$r<i\leqslant N$。因此，对于全局稳定的神经网，由式（12）可以看出

$$\lim_{t\to\infty}H(t)=1-P(0)\left(\sum_{i=1}^{r}\sum_{j=1}^{r}B_iB_j^{\mathrm{T}}\right)P^{\mathrm{T}}(0)\quad(13)$$

因为任意状态迁移矩阵对应于 $\lambda_i=1$ 的右特征向量为 $\varphi_i=(1,1,\cdots,1)^{\mathrm{T}}$，所以式（13）可改写成

$$\begin{aligned}\lim_{t\to\infty}H(t)&=1-P(0)\left(\sum_{i=1}^{r}\sum_{j=1}^{r}\varphi_i\pi_i\pi_j^{\mathrm{T}}\varphi_j^{\mathrm{T}}\right)P^{\mathrm{T}}(0)\\&=1-\sum_{i=1}^{r}\sum_{j=1}^{r}\pi_i\pi_j^{\mathrm{T}}\end{aligned}$$

因此，异步串行计算中，其对应的神经系统的熵一定会收敛到一个与初始状态无关的极限值，这个值由状态迁移矩阵对应于 $\lambda=1$ 的左特征向量唯一确定。

必须注意，式（12）中除 $\lambda=1$ 以外的特征值可正可负，特征向量也不是全 1 向量，熵的具体收敛过程与非 1 特征值及初始状态有关。由式（12）可精确描述熵的变化情况，但实际应用中往往不必知道熵变化的详细过程，只要较粗略地估计熵收敛的快慢程度，因而不必知道全部特征值。下面从几个方面讨论熵的收敛性。

（1）与初始状态的关系

常见的神经计算有两大类：一类初始输入的可能状态很多，计算过程中熵逐步减小。另一类计算输入只有少数几种可能状态，但不是所希望的模式，计算时熵开始增大，以后再逐步减小直到输出所希望的模式。计算初期熵增大的阶段称为忘却（Unlearning）阶段。图 1（a）表示第一类计算，图 1（b）表示第二类计算。

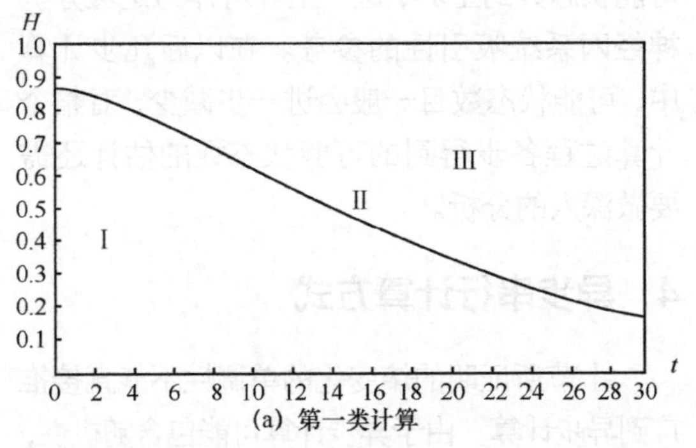

(a) 第一类计算

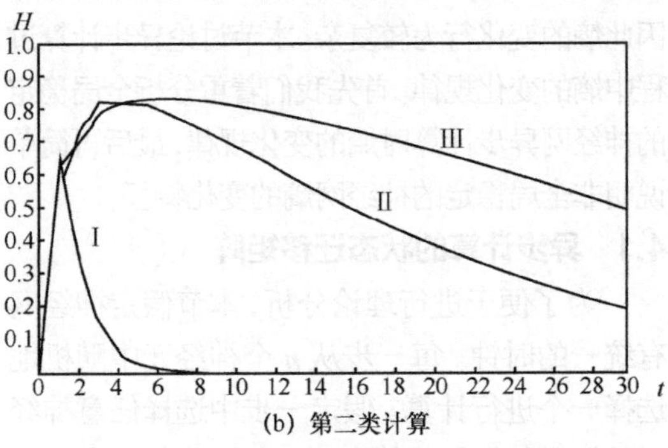

(b) 第二类计算

图 1 神经网系统熵的收敛过程（3 个实例）

（2）最大的非 1 特征值的影响

设状态迁移矩阵除 1 以外最大的特征是 $\lambda_m$，由非负矩阵的性质可知 $1>\lambda_m>|\lambda_i|$，$i>r$，$i\neq m$。熵的收敛速度主要由 $\lambda_m$ 决定。定性地说，全局稳定的神经网系统的熵按几何级数的速率收敛。对于式（11）所示的神经网状态迁移矩阵来说，$\lambda_m$ 的大小与子矩阵 $\boldsymbol{Q}$ 的迹及 $\boldsymbol{Q}$ 的各行元素之和有关。令 $SR_{\max}(\boldsymbol{Q})=\max\limits_{i}\sum\limits_{j=r+1}^{N}\boldsymbol{Q}_{ij}$，$SR_{\max}(\boldsymbol{Q})=\min\limits_{i}\sum\limits_{j=r+1}^{N}\boldsymbol{Q}_{ij}$，那么，子矩阵 $\boldsymbol{Q}$ 的最大特征值 $\lambda_m$ 满足 $SR_{\min}(\boldsymbol{Q})<\lambda_m<SR_{\max}$。对于式（11）所示的矩阵来说，$\boldsymbol{Q}$ 的特征值也就是 $\boldsymbol{A}$ 的特征值。因此，一般来讲迹 $tr(\boldsymbol{Q})$ 大而 $SR_{\min}(\boldsymbol{Q})$ 也较大时，$\lambda_m$ 较大；反之，$tr(\boldsymbol{Q})$ 较小而 $SR_{\max}(\boldsymbol{Q})$ 较小时，$\lambda_m$ 较小。我们也可以启发式地用子矩阵 $\boldsymbol{R}$ 的所有元素 $i$ 和来估计 $\lambda_m$ 的大小，一般这个和值较大时 $\lambda_m$ 较小。图 1 表示 3 种不同状态迁移矩阵所对应的熵收敛过程。曲线 I 对应的状态迁移矩阵的迹 $tr(\boldsymbol{A}_1)=1.1$，$SR_{\max}(\boldsymbol{Q}_1)=0.5$，$SR_{\min}(\boldsymbol{Q}_1)=0.2$，因此，其对应的 $\lambda_m$ 较小，熵收敛很快。曲线 II 对应的 $tr(\boldsymbol{A}_2)=4.5$，$SR_{\max}(\boldsymbol{Q}_2)=1$，$SR_{\min}(\boldsymbol{Q}_2)=0.8$，因此 $\lambda_m$ 较大，熵收敛较慢。

（3）负特征值的影响

如果状态迁移矩阵有负的特征值而且绝对值较大，那么系统的熵将不会单调下降，在开始若干步内熵会出现明显的振荡，时大时小变化。这是因为负的特征值 $\lambda_i(|\lambda_i|<1)$，使 $\lambda_i^t$ 表现出阻尼振荡行为。图 1 中曲线Ⅲ对应的 $tr(\boldsymbol{A}_3)=1.1$，$SR_{\max}(\boldsymbol{Q}_3)=1$，$SR_{\min}(\boldsymbol{Q}_3)=0.8$，其对应的 $\boldsymbol{A}$ 有绝对值较大的负特征值，所以开始几步熵明显振荡且收敛很慢。

## 4.3 异步计算中熵变化的单调性

上面的分析表明从全过程来看异步计算中熵的变化不具有单调性。下面的定理将说明某些全局稳定的神经网系统的熵在若干步异步计算后也会单调减小。

**定理 4** 如果一个神经网系统所有的遍历状态集都只包含一个状态，而且它的状态迁移矩阵可对角化，那么，这系统的熵经若干步异步计算之后一定单调减小。

**证明：**我们采用式（6）定义的熵。设第 $t$ 步迭代后的状态概率分布向量为 $\boldsymbol{P}(t)$，那么

$$H(t+1)=1-\boldsymbol{P}(t+1)\boldsymbol{P}^{\mathrm{T}}(t+1)=$$
$$1-\boldsymbol{P}(t)\boldsymbol{A}\boldsymbol{A}^{\mathrm{T}}\boldsymbol{P}^{\mathrm{T}}(t)$$

$$H(t+1)-H(t)=-\boldsymbol{P}(t)\left(\boldsymbol{A}\boldsymbol{A}^{\mathrm{T}}-I\right)\boldsymbol{P}^{\mathrm{T}}(t)$$

由假设

$$\boldsymbol{A}=\left(\begin{array}{c|c}I & O\\ \hline \boldsymbol{R} & \boldsymbol{Q}\end{array}\right)\quad \boldsymbol{A}^{\mathrm{T}}=\left(\begin{array}{c|c}I & \boldsymbol{R}^{\mathrm{T}}\\ \hline \boldsymbol{O} & \boldsymbol{Q}^{\mathrm{T}}\end{array}\right)$$

$$\boldsymbol{A}\boldsymbol{A}^{\mathrm{T}}-1=\left(\begin{array}{c|c}O & \boldsymbol{R}^{\mathrm{T}}\\ \hline \boldsymbol{R} & \boldsymbol{R}\boldsymbol{R}^{\mathrm{T}}+\boldsymbol{Q}\boldsymbol{Q}^{\mathrm{T}}-I\end{array}\right)$$

因为 $\boldsymbol{R}$ 子矩阵中至少有一个非零项，不妨设 $\boldsymbol{R}_{ij}>0$。根据马尔可夫过程理论，瞬时状态集中的状态出现概率随计算次数增加而趋于零。所以当 $t$ 足够大时 $\boldsymbol{P}_i(t)<\varepsilon$，$r<i\leqslant N$，$\varepsilon$ 是任意小的正数。采用上节使用的可对角化矩阵分解方法，不难证明 2 个瞬时状态的分布概率的比值也趋于常数。令 $\boldsymbol{P}_{\min}(t)=\min\limits_{t<i\leqslant N}\left(\boldsymbol{P}_i(t)\right)$，$\boldsymbol{P}_{\max}(t)=\max\limits_{t<i\leqslant N}\left(\boldsymbol{P}_i(t)\right)$。

适当选取常数 $C_1$ 和时间 $t$ 可使 $\boldsymbol{P}_{\min}(t)>C_1\boldsymbol{P}_{\max}(t)$ 满足。另一方面，如果状态 $j$，$1\leqslant j\leqslant r$，是吸引子，当 $t$ 足够大时，$P_j(t)>C_2$，$C_2$ 是小于 1 的正数。因此，

$$H(t+1)-H(t)=$$
$$-\boldsymbol{P}(t)\left(\begin{array}{c|c}O & \boldsymbol{R}^{\mathrm{T}}\\ \hline \boldsymbol{R} & \boldsymbol{R}\boldsymbol{R}^{\mathrm{T}}+\boldsymbol{Q}\boldsymbol{Q}^{\mathrm{T}}-I\end{array}\right)\boldsymbol{P}^{\mathrm{T}}(t)<$$
$$(N-r)\boldsymbol{P}_{\max}^{2}(t)-\boldsymbol{R}_{ij}\boldsymbol{P}_{\min}(t)\boldsymbol{P}_j(t)\begin{array}{l}r<i\leqslant N\\ 1\leqslant j\leqslant r\end{array}$$

当 $t$ 足够大时，我们有

$$H(t+1)-H(t)<(N-r)\varepsilon-R_{ij}C_1C_2$$

当 $t$ 充分大使得 $\varepsilon<\frac{R_{ij}C_1C_2}{N-r}$ 时，$H(t+1)<H(t)$。　　证毕。

### 4.4 非全局稳定神经网的振荡性

在实际的神经网设计中，一个遍历集内可能有多个状态，系统可能在这几个状态之间周期振荡。这种非全局稳定的神经网的状态迁移矩阵的左上角不是单位子矩阵而是若干个周期子矩阵。这种矩阵有特征值 $\lambda_i=-1$ 或模为 1 的复数特征值。对应的系统的熵将不收敛，熵随时间周期振荡而且其行为与初始状态有关。

## 5 结论

本文提出以熵的理论作为分析各种神经网模型的公共基础，并从熵的观点对同步并行与异步串行计算的神经网系统的吸引性与稳定性做了初步探讨。和目前人工神经网研究中广泛采用的能量函数不同，熵的研究着眼于系统的宏观统计性质，主要关心反映整体性质的可能状态数目、状态迁移矩阵等，而不是个别状态的微观变化过程。本文的研究结果表明无论是同步还是异步计算，在一定条件下神经网系统的熵都单调减小，而且熵减少的速率可以通过对可能数目的估计或通过对状态迁移矩阵的分析加以预测。这些初步结果表明以熵的理论为基础研究各种神经网模型的吸引性、稳定性、自组织性以及学习能力等是一个值得重视的研究方向。它与现有的神经网理论结合起来有可能发展成为更一般的神经网理论。

## 参考文献：


[1] RUMELHART D, MCCLELLAND J L. Parallel distributed processing[M]. Massachusetts: MIT Press, 1986.

[2] HECHT-NIELSEN R. Neurocomputer applications[C]// The Neural Computer Conference. [S.l.:s.n.], 1987: 240-224.

[3] LIPPMANN R P. Introduction to computing with neural nets[J]. IEEE ASSP Magazine, 1987, 4(2): 4-22.

[4] HOPFIELD J J. Neural networks and physical systems with emergent collective computational abilities[J]. Pro. Natl. Acad. Sci., 1982, 79(8): 2554-2558.

[5] 李国杰. 人工神经网络系统的宏观不确定性[R]. 1988.

[6] GINDI G R, GMITRO A F, PARTHASARATHY K. Hopfield model associative memory with nonzero-diagonal terms in memory matrix[J]. Applied Optics, 1988, 27(1): 129-134.

发表于 *International Conference on Circuits and Systems*论文集，1991年

# On the reduction of connections in hopfield associative memories

Zhi-ming TANG[1,2], Guo-jie LI[1,2]

1. Institute of Computing Technology, Chinese Academy of Sciences, Beijing 100080, China

2. National Research Center of Intelligent Computing Systems

**Abstract:** The attempt to reduce the number of interconnections in artificial neural networks has been taken by many researchers to ease the real implementation. Since such reduction usually degrades the performance of ANS, a tradeoff between the ease of implementation and the network performance must be made. This paper proves, for Hopfield neural network, that such tradeoff is impossible, i.e., the reduction is fruitless.

## 1 Introduction

Hopfield neural network[1] provides a powerful model for associative content addressable memory. As stated by Hopfield and others[1-3], a number of stable states can be maintained in the network and each of them can be retrieved with probe vector not far away from it in Hamming distance.

Hopheld's neural net consists of $N$ identical neurons, each being a simple bistable element with two possible states as +l and −l. Every neuron is connected to every other neuron through an weighted link. The operations performed by individual neuron are modeled as summing and thresholding. Information is stored as the interconnection weights which are determined from a set of stored bipolar (+1, −1) library vectors according to Hebbian rules.

The high interconnect tivity of Hopfield network poses problems to both virtual and fully implementation of large networks. For virtual implementation, it causes the rather long iteration cycle; for fully one it causes the difficulty in VLSI layout. And even in optical implementation, any reduction in the number of interconnections will ease the situation.

An immediate observation for solving this problem is to reduce the number of interconnections and to change global interactions to local ones, such as those described in [4-5]. However, the reduction of connection will inevitably alter the performance of the network, the storage capacity for the associative memory case. The chief aim of this paper is to investigate the validity of such reduction, in view of both network

This work is supported by the National High Technology Programme and National Natural Science Foundation of China.

performance and implementation easiness. Our results show that such reduction is fruitless.

Our discussion will be mainly based on the reduction scheme proposed by Somani et al.[4], which is briefly introduced in the next section. The performance or more precisely, the storage capacity of the reduced interconnection neural network (RINN) is analyzed in Section 3. Section 4 presents our basic result that RINN is unjustifiable. The result is generalized to enhanced memory capacity cases in Section 5. And finally, in Section 6, we give the further discussions.

## 2 The reduction of connections and the RINN

The fully connective property of Hopfield neural network cause the difficulty in VLSI implementation, and the long iteration cycle in virtual implementation. Thus, to reduce the number of connections among neurons, i.e., to reduce the number of non-zero weights, and change global interconnections to local ones, is helpful to the fast and effective implementation of neurocomputing.

Somani and Penla studied the possibility of reducing the number of connections in the Hopfield network[4]. They found that, when preserving only weights with relatively larger absolute values and setting others to zero (hence the corresponding connections can be omitted), the performance of the network (e.g. the capacity of associative memory, the ratio of correct convergence of input vectors, etc.) degrades little. On the basis of such RINNs (reduced interconnections neural network), they proposed the concept of CNN (compact neural network) which is fit for the implementation in parallel systems with regular structure.

The reasonability of preserving merely weights with larger absolute values is intuitively obvious. Consider a Hopfield neural network model with $N$ neurons. $\boldsymbol{x}=(x_0,x_1,\cdots,x_{N-1})$ is its state vector (in cases without confusion, $x_i$, $i=0,1,\cdots,N-1$, is also used to represent the $i$-th neuron), $\boldsymbol{W}=(w_{ij})$, $0\leqslant i$, $j\leqslant N-1$, is its weight matrix. Here, we do not assume that the diagonal elements, or $w_{ii}$, are zeros. As proved in [6], when $w_{ii}\neq 0$, the network converges even faster. For a bi-state Hopfield network, i.e., $x_i\in(+1,-1)$ for any $0\leqslant i\leqslant N-1$, the temporal evolution of the network state can be expressed as

$$\boldsymbol{x}(t+1)=\operatorname{sgn}(\boldsymbol{Wx}(t))$$

or

$$x_i(t+1)=\operatorname{sgn}\left(\sum_{j=0}^{N-1}w_{ij}x_j(t)\right)=\begin{cases}+1, & \text{if}\sum_{j=0}^{N-1}w_{ij}x_j(t)\geqslant 0;\\ -1, & \text{if}\sum_{j=0}^{N-1}w_{ij}x_j(t)<0.\end{cases}\tag{1}$$

In Eq. (1), the value of $x_i(t+1)$ is determined by the sign of the $\Sigma$ term, and the influence to this term of $w_{ij}$ with larger absolute value is obviously greater than that of weights with smaller absolute values. However, theoretically, such reasonableness is difficult to prove, because the clipped weight matrix in not necessarily symmetric, and the distribution of non-zero weights follows no definite rules.

## 3 Capacity of RINN

[4] pointed out that the RINN has basically

the same capacity as that of Hopfield networks since it only omits those of weights with relatively smaller absolute values. However, our simulation on RINN showed that the capacity of RINN is affected by the reduction and is less than that of Hopfield nets. That suggests us to restudy the capacity of RINN. Our results revealed that the above conclusion of [4] is not true, and also does not confirm with the practical simulation.

## 3.1 Capacity as the number of library vector stored

If we define the memory capacity of neural network as the number of library vectors which can be perfectly stored and retrieved, then with the same method as used in [3], we have the following result.

**Theorem 1:** The memory capacity of RINNs generated from a Hopfield network with $N$ neurons by preserving the first $C(C \leqslant N)$ weights with maximum absolute values per neuron and setting others to zero is $M \approx \dfrac{C}{2\ln N}$.

**Proof:** Consider the reduced weight matrix $\boldsymbol{T}$. Each row of $\boldsymbol{T}$ has only $C$ non-zero elements. According to the sum-of-outer-products method for constructing the original weight matrix

$$\boldsymbol{W} = \sum_{i=1}^{M} \boldsymbol{W}_i = \sum_{i=1}^{M} \left( \boldsymbol{x}^{(i)} \left( \boldsymbol{x}^{(i)} \right)^T - \boldsymbol{I}_N \right)$$

where, $\boldsymbol{x}^{(i)}, i = 1, \cdots, M$, library vector, $\boldsymbol{I}_N$ denotes the $N \times N$ identity matrix. With the same approach, $\boldsymbol{T}$ can also be expressed as the sum of $M$ matrices $T_1, T_2, \cdots, T_M$, where

$$T_i^{jk} = \begin{cases} W_i^{jk}, T_{jk} \neq 0 \\ 0, \quad T_{jk} = 0 \end{cases}, 1 \leqslant i \leqslant M, 1 \leqslant j、k \leqslant N.$$

Suppose that the $MN$ components of all $M$ memories be i.i.d. $\pm 1$ (probability 1/2 each) random variables. Without loss of generality, when the input vector is $x^{(1)}$, according to the evolution rule(1), we have

$$\begin{aligned} \left( \boldsymbol{x}^{(1)} \right)'_i &= \operatorname{sgn}\left( \left( \boldsymbol{T} \boldsymbol{x}^{(1)} \right)_i \right) \\ &= \operatorname{sgn}\left( \left( T_1 \boldsymbol{x}^{(1)} \right)_i + \sum_{k=2}^{M} \left( T_k \boldsymbol{x}^{(1)} \right)_i \right) \\ &= \operatorname{sgn}\left( s_i + z_i \right) \end{aligned}$$

where $s_i$ is the "signal" and $z_i$ is the "noise". From of the definition of $T_1$, $s_i = C x_i^{(1)}$, which has the same sign as $\boldsymbol{x}^{(1)}$. Our assumptions about the components $\boldsymbol{x}^{(1)}, \cdots, \boldsymbol{x}^{(M)}$ imply that the noise term $z_i$ is a sum of $(m-1)$ i.i.d.random variables, each with mean 0 and variance $C$. Hence if $N$ is fixed and $M$ is large, the normalized noise $z_i \big/ \sqrt{C(M-1)}$ approaches a standard normal random variable. It follows then that the probability that the $i$-th component of $\left( x^{(1)} \right)'$ has the opposite sign of $x_i^{(1)}$ will be approximately

$$\begin{cases} \Phi\left( \dfrac{-C}{\sqrt{C(M-1)}} \right) = \Phi\left( -\sqrt{\dfrac{C}{M-1}} \right), & \text{if } x_i^{(1)} \text{ is positive}, \\ Q\left( \dfrac{C}{\sqrt{C(M-1)}} \right) = Q\left( \sqrt{\dfrac{C}{M-1}} \right), & \text{if } x_i^{(1)} \text{ is negative}, \end{cases}$$

where,

$$\Phi(z) = \frac{1}{\sqrt{2\pi}} \int_{-\infty}^{z} \mathrm{e}^{-t^2/2} dt,$$

$$Q(z)=\frac{1}{\sqrt{2\pi}}\int_z^{\infty}\mathrm{e}^{-t^2/2}dt$$

Obviously, $\Phi(z)=Q(-z)$. Hence no matter what the sign of $x_i^{(1)}$ is, the probability that its sign is different from that of $(x^{(1)})'_i$ is the same. Thus the expected number of error components in $(x^{(1)})'$ is approximately $NQ\left(\sqrt{\frac{C}{M-1}}\right)$.

From [3], with suitable restrictions, the expected number of error elements in $(x^{(1)})'$ is approximately Poisson. Given this, it follows that the probability of the error components, i.e., the probability that $x^{(1)}$ is indeed a fixed point, is given by

$$\beta=\exp\left(-nQ\left(\sqrt{\frac{C}{M-1}}\right)\right)$$

Now suppose we require that this probability be a fixed number very near 1, say $\beta=0.999\,999$. Then converting the preceding expression, we can get

$$Q\left(\sqrt{\frac{C}{M-1}}\right)=\frac{\alpha}{N}$$

where $\alpha=-\ln\beta$. This means that

$$M=\frac{C}{\left(\Phi^{-1}\left(\frac{\alpha}{n}\right)\right)^2}+1$$

However, for small positive values of $x$ we have $\Phi^{-1}(x)\approx\sqrt{2\ln(1/x)}$, and so, since $\alpha$ is fixed,

$$M\approx\frac{C}{2\ln N}$$

Hence the theorem follows.

A direct conclusion from the above theorem is that the capacity of RINN does not simply equal that of Hopfield network, $N/2\ln N$. In fact, when the number of neurons is given, the capacity decreases linearly with the decrease of the number of connections per neuron. Another direct conclusion can be derived from Theorem 1: if the number of connections per neuron is fixed at some constant value, the increase of the number of neurons will cause the slow decrease of network capacity.

**Corollary 1:** If $p(0\leqslant p\leqslant 1)$ is the ratio of the number of non-zero weights per neuron to that of all possible connection, $p=C/N$, then the capacity of the RINN is $M\approx pN/2\ln N$.

Since the capacity of Hopfield network is approximately $N/2\ln N$, as proved in [3], that above corollary means that the capacity of a $p$-connected RINN, is just $p$ times that of the Hopfield network with same neurons.

### 3.2 Capacity as the number of total bits stroed

Another way of defining network capacity is to look at the number of total bits, $B$, stored in the system, if the capacity in the sense of the last subsection is $M$, then $B=MN$, where $N$ is the number of neurons.

With the above definition of $B$, we immediately have the following two results:

- For a Hopfield neural network with $N$ neurons, $B_H=\frac{N^2}{2\ln N}$;
- For a RINN with $N$ neurons and each neuron preserving $C$ connections, $B_R=\frac{CN}{2\ln N}$.

Two direct observations can be drawn from these results:

1. $B$ is proportional to $C$ when $N$ is given;

2. When $C$ is fixed, $B$ monotonically increases as $N$ Increases.

Note that observation 2 is an important complement to connections to the second conclusion derived from Theorem 1 in the last subsection.

Let $p=C/N$, we have $B = p\frac{N^2}{2\ln N}$. This result confirms with Corollary 2.

# 4 RINN is unjustifiable

The objectives of reducing connections of Hopfield network are two-fold: one is the reduction of execution time in virtual implementation, another is the convenience of VLSI implementation. The following theorem shows that the total execution time can not be shortened by reducing connections.

**Theorem 2:** For various $p$-connected RINNs with the same memory capacity, where $0 \leqslant p \leqslant 1$, the fully connected Hopfield network ($p=1$) has the shortest execution time.

**Proof:** We can complete the proof by stating that the total number of connections of a RINN is always greater than that of the corresponding Hopfield net with same memory capacity, no matter which capacity definition is used. Suppose $H$ is a Hopfield network with $N_H$ neurons , hence, in the first kind of meaning, its capacity is $M_H = N_H / 2\ln N_H$ . Let $R$ be any $p(p<1)$ -connected RINN, with $M_R = M_H$ . The number of neurons of $R$ fulfils the following equation:

$$M_R = M_H = pN_R / 2\ln N_R$$

From

$$\frac{N_H}{2\ln N_H} = p\frac{N_R}{2\ln N_R} < \frac{N_R}{2\ln N_R}$$

we can get $N_H < N_R$ , since function $x/\ln x$ is monotonically increasing for $x > \mathrm{e}$. Therefore, $2\ln N_H < 2\ln N_R$, and $N_H < pN_R$. On the other hand, the total number of connections of $R$ is

$$C_R = pN_R \cdot N_R > N_H \cdot N_H = C_H$$

On the other hand, we have

$$B_H = \frac{N_H^2}{2\ln N_H}, \text{ and } B_R = \frac{pN_R^2}{2\ln N_R}$$

To get a RINN with $B_R = B_H$, it is necessary that

$$\frac{N_H^2}{2\ln N_H} = \frac{pN_R^2}{2\ln N_R} \tag{2}$$

Since $p<1$, (2) means

$$\frac{N_H^2}{\ln N_H} < \frac{N_R^2}{\ln N_R} \tag{3}$$

It is easy to see that function $f(x) = x^2 / \ln x$ is monotonically increasing for large positive $x$. Therefore, from (3), we have $N_R > N_H$ for any large $N_H$ , $N_R$ . And under the same condition, we have $\ln N_R > \ln N_H$ . So form (2), we can get $pN_R^2 > N_H^2$, which directly means $C_R > C_H$. Since the computational amount to complete a single iteration is proportional to the total number of connections, the computational amount of $R$ is greater than that of $H$. Hence the theorem follows.

Therefore, it is impossible to shorten the total computation time of the Hopfield network by reducing the network connections. However, is it possible to reduce the communication amount in the virtual implementation with connection reduction?

At first, since the network connection can

not be localized merely with reducing the number of connections, the amount of communication saved from less connections per neuron is difficult to offset the increased one due to the adoption of more neurons. In fact, according to the analysis on virtual implementation in [7], for fully connected networks, the communication amount of each processor is on the order of $N_H$. If the distribution of non-zero weights is still even in the reduced network, the communication amount will be $pN_R$. From the proof of Theorem 2, it is obvious that $pN_R>N_H$. Secondly, even with proper permutation, the reduced network have its connection localized, the reduced communication amount has little effect in improving the performance. Actually, as shown in the appendix, in virtually implemented networks, the large amount of communication is relatively smaller compared with that of computation. Hence the total execution time cannot be greatly reduced even there is no communication. On the other hand, the increase of computation amount caused by reducing connections, as shown in Theorem 2, will probably lengthen the total execution time.

From the point of view of full implementation with VLSI technology, if with some appropriate permutation, the remain connections can be localized, the implementation will become easier. However, such easiness does not help in real situation.

There is a limit posed by the technology level on the magnitude of $N$, where $N$ is the number of neurons in a fully interconnected Hopfield network which can be embedded in a single chip. Let it be $N_{\max}$. For a Hopfield network with $N \leqslant N_{\max}$, connection reduction is unnecessary, since the corresponding RINN with same capacity will consume more silicon area. In fact, the total length of wire in the Hopfield case is $N_H^2 \cdot \sqrt{N_H}$, which is less than $pN_R^2 \cdot \sqrt{pN_R}$, that of RINN. In the case where $N>N_{\max}$ virtual implementation is necessary, hence RINN is also helpless.

Concluding the above discussion, it is obvious that, for Hopfield networks, the reduction of connections is fruitless.

## 5 More general cases

The storage capacity of Hopfield neural networks as derived in [3] is based on the assumption that the components of all library vectors stored are all i.i.d. random $\pm 1$ variables. Such a capacity is relatively small, and much work has been done to enhance the network capacity[8-9]. For enhanced storage capacity, e.g., the capacity on the order of $N$ or $N^2$, is RINN still unjustifiable?

**Theorem 3:** Given a fully interconnected artificial neural network with $N$ neuros. If its capacity in on the order of $N^{\alpha+1}$, and the capacity of the corresponding RINN obtained from preserving $C(C<N)$ weights with relatively larger absolute values and setting all others to zerois on the order of $CN^{\alpha}$, then to attain the same memory capacity,

(1) fully connected one needs less connections if $\alpha<1$;

(2) both networks need the same connections if $\alpha=1$;

(3) RINN needs less connection if $\alpha>1$.

**Proof:** We use the same notations as those

in the proof of Theorem 2. According to the conditions of the theorem,

$$M_H = N_H^{\alpha+1}, \text{ and } M_R = CN_R^{\alpha}$$

From $N_H^{\alpha+1} = CN_R^{\alpha}$, and $C < N_R$, we can get $N_H < N_R$, for otherwise,

$$CN_R^{\alpha} < N_R \cdot N_R^{\alpha} \leqslant N_H^{\alpha+1}$$

Since $CN_R \cdot N_R^{\alpha-1} = N_H^2 \cdot N_H^{\alpha-1}$, and $N_R > N_H$, we have

$$\begin{cases} CN_R > N_H^2, & \text{if } \alpha < 1, \\ CN_R = N_H^2, & \text{if } \alpha = 1, \\ CN_R < N_H^2, & \text{if } \alpha > 1. \end{cases}$$

Hence the theorem follows.

If capacity is counted as the number of total bits stored in the network, using the previous notations, we can immediately get the following corollary.

**Corollary 2:** If $B_H = N_H M_H$ and $B_R = N_R M_R$, $M_H$, $M_R$ as defined above, then to attain same capacity, or $B_H = B_R$,

(1) fully connected one needs less connections if $\alpha < 0$;

(2) both networks need the same connections if $\alpha = 0$;

(3) RINN needs less connections if $\alpha > 0$.

## 6 Concluding remarks

Ease of implementation is an important criterion for evaluating the performance of artificial neural networks. In view of this criterion, Hopfield neural network is not a good candidate since its complexity can not be simplified even with connection reduction. Nevertheless, our above results on the Hopfield model are all based on the assumptions that the library vectors are pairwise orthogonal, and all components of these vectors are i.i.d. random ±1 variables. The suitability of such results on other kinds of library vectors is still an open problem.

The discussion on more general cases is based on the assumption that for networks with higher capacities, the capacity decrease is still proportional to the reduction ratio of connections per neuron. Although this is a more or less subjective hypothesis, the corresponding result reveals the potentiality of the reduction approach on other kinds of neural networks. Seeking for such networks and finding their appropriate implementation strategy will become the theme of further study.

## References:

[1] HOPFIELD J J. Neural networks and physical systems with emergent collective computational abilities[J]. Proceedings of the National Academy of Sciences, 1982, 79: 2554-2558.

[2] YASER S, ABU-MOSTAFA Y, ST JACQUES J. Information capacity of the hopfield model[J]. IEEE Transactions on Information Theory, 1985, 31(4): 461-464.

[3] MCELIECE R, POSNER E, RODEMICH E, et al. The capacity of the hopfield associative memory[J]. IEEE Transactions on Information Theory, 1987, 33(4): 461-482.

[4] SOMANI A K, PENLA. Compact neural network[C]// The 1987 International Conference on Neural Networks. [S.l.:s.n.], 1987: 191-198.

[5] AKERS L A, WALKER M R, FERRY D K, et al. Limited interconnectivity in synthetic neural systems[J]. Neural Computers, 1988: 407-416.

[6] GINDI G R, GMITRO A F, PARTHASARATHY K. Hopfield model associative memory with non-zero-diagonal terms in memory matrix[J]. Applied Optics, 1988, 27(1): 129-134.

[7] TANG Z M. Granularity and performance ofmessage-passing multicomputer[D]. Beijing: Institute of Computing Technology, Chinese Academy of Sciences,

1990.
[8] LEE Y C, DOOLEN G, CHEN H H, et al. Machine learning using a higher-order correlation network[J]. Physica D: Nonlinear Phenomena, 1986, 276-306.
[9] LITTLE M J, BAK C S. Enhanced memory capacity of a Hopfield neural network[C]// Real-Time Signal Processing IX. International Society for Optics and Photonics. [S.l.:s.n.], 1986: 150-156.

发表于 *Journal of Computer Science and Technology*, 1991年第4期

# General simulated annealing

Xin YAO[1], Guo-jie LI[2]

1. Dept. of Computer Science, University of Science and Technology of China, Anhui 230026

2. National Research Center for Intelligent Computing Systems, Beijing 100080

**Abstract:** Simulated annealing is a new kind of random search methods developed in recent years. It can also be considered as an extension to the classical hill-climbing method in Al, probabilistic hill-cimbing. One of its most important features is its global convergence. The convergence of simulated annealing algorithm is determined by state generating probability, state accepting probability, and temperature decreasing rate. This paper gives a generalized simulated annealing algorithm with dynamic generating and accepting probabilities. The paper also shows that the generating and accepting probabilities can adopt many different kinds of distributions while the global convergence is guaranteed.

## 1 Introduction

Simulated Annealing (SA in short ) is a powerful way to solve hard combinatorial optimization problems. It was first used by S. Kirkpatrik et al.[1] and V . Cerny[2] for computer design and combinatorial optimization. It was also used in image processing[3] and neural network computing[4]. It not only has the merit of global convergence, but also needs little domain knowledge in searching. This is very helpful in solving many practical problems, e.g., VLSI design, image processing, and neural network computing.

The idea of SA algorithm came from the analogy to the cooling process of solid materials[5]. When we want to crytallize some solid, we first melt it and then slowly cool it down. If the cooling rate is sufficiently slow around the freezing point, the solid will be crytallized, which corresponds to the minimum energy state of the solid. When solving combinatorial optimization problems, SA algorithm does the same way. It starts with an arbitrary initial state. At each step, it selects the next state from neighboring states at random. As its control parameter "temperature" goes down, SA algorithm will converge to the global optimum state, which is analogous to the minimum energy state in physical cooling process.

In the classical hill-climbing method, the algorithm only accepts the better state at each step. This often leads to the local optimum. SA algorithm overcomes the drawback, and converges to the global optimum state. At each step, it accepts not only better state but also worse

This research is supported by the National 863 Project of China.

state with some probability. This accepting probability for worse state gradually goes to zero as the temperature goes down.

In this paper, we simply call the problem instances as problems whenever the meaning is clear from the context. A combinatorial optimization problem can be informally defined as the problem of finding the optimum state in the finite problem state (solution ) space $S$, where each state (solution) $i$ has a value $c_i$ called state value, the optimum state is the state with minimum (minimization problems) or maximum (maximization problems) value. This paper only deals with the minimization problems.

For a finite state space, a neighborhood system can be defined as $\mathcal{N} = \{N_i | i \in S, N_i \subseteq S\}$, it satisfies: (a) $i \notin N_i$; (b) $j \in N_i \langle = \rangle i \in N_j$. Obviously, $N_i$ denotes the set of neighboring states of state $i$. It is generated by perturbations from state $i$. The perturbation method is different for different problems.

The main steps of SA algorithm can be described below.

PROCEDURE SA-algorithm ($i_0$,$T_0$);

{$i_0$ is an initial state, $T_0$ is the initial temperature.}

(1) $X= i_0$ ;$n$:=0 ;  {$X$ stands for the current state.}

(2) REPEAT

(2.1 ) REPEAT

(2.1.1) $j :=$ generate $(X, T_n)$ ;

(2.1.2) IF accept $(X, j, T_n)$ THEN $X:=j$

(2.2) UNTIL 'inner-loop stop criterion'

(2.3) $T_{n+1}$: = update $(T_n)$ ; $n := n+ i$

(3) UNTIL 'final stop criterion'

{Finally, $X$ is the optimum or near optimum state found.}

There are three functions which determine the convergence of SA algorithm, i.e., generate $(X, T_n)$, accept $(X, j, T_n)$ and update $(T_n)$. In practice, initial temperature $T_0$, 'inner-loop stop criterion', and 'final stop criterion' are also important parameters.

In order to describe the generate $(X,T)$ and accept $(X,j,T)$ functions, let $g_{ij}(T)$ be the probability of generating $j$ from $i$ (Note: $g_{ij}(T)$=0 if $j \notin N$) and $a_{ij}(T)$ be the probability of accepting $j$ as new current state provided the old one is $i$.

The behavior of SA algorithm can be described by nonstationary Markov chains[6]. Denote the directed graph underlying the Markov chain as $G = (V,E)$, the node set as $V=S$, and the edge set as $E = \{(i, j) | i, j \in S, j \in N_i\}$. The one-step transition probabilities of the Markov chain associated with SA algorithm are:

$$p_{ij}(T) = \begin{cases} g_{ij}(T) a_{ij}(T), & j \in N_i, j \neq i \\ 0, & j \notin N_i, j \neq i \\ 1 - \sum\limits_{k \in N_i} p_{ik}(T), & j \notin N_i, j = i \end{cases} \quad (1)$$

All previous SA theories suppose $g_{ij}(T)$ is independent of temperature $T$. This method is inefficient in practice because the algorithm can run faster if it goes further in one step. H. Szu[7] has recently proposed a new SA algorithm, Fast Simulated Annealing (FSA). He used Cauchy distribution to generate next state. In this paper, we show that not only can Cauchy distribution be used as generating probability, but a set of other distributions can also be used.

As to the accepting probability $a_{ij}(T)$, a

common form is

$$a_{ij}(T)=\min\left\{1,\exp\left[-\left(c_j-c_i\right)/T\right]\right\} \qquad (2)$$

where $c_i$, and $c_j$ are state values. We also give a weaker restriction on $a_{ij}(T)$ in this paper, i.e. $a_{ij}(T)$ can adopt other forms as long as it satisfies some conditions. S. Anily and A. Federgruen[8] have discussed the SA with general accepting probability, but supposed $g_{ij}(T)$ is independent of temperature $T$. We have considered the SA with both general generating and accepting probabilities in this paper.

The theoretical analysis of SA algorithm is very important to understand the essential features which make the algorithm work well and to suggest techniques for controlling its operation. The theory can also be very helpful in practice to guide the selection of various parameters of the algorithm.

Section 2 of this paper gives some necessary notations and preliminaries; Section 3 proves our main results, i.e., the convergence of generalized SA; and Section 4 is the conclusion.

## 2 Noptions and preliminaries

Before proving our main results, it is necessary to introduce some definitions and theorems[6,8-11].

**Definition 1:** Let $\{P_k\}_{k=1}^{\infty}$ be transition matrices for a nonstationary Markov chain with starting vector $f(0)$. Define

$$f(k)=f(0)P_1P_2\cdots P_k$$

$$f(m,k)=f(0)P_{m+1}P_{m+2}\cdots P_k$$

$$P(m,k)=P_{m+1}P_{m+2}\cdots P_k$$

**Definition 2:** If $f=(f_1,f_2,\cdots)$ is a vector, define the norm of $f$ by $\|f\|=\sum_{i=1}^{\infty}|f_i|$

If $A=(a_{ij})$ is a square matrix, define the norm of $A$ by $\|A\|=\sup_i\sum_{j=1}^{\infty}|a_{ij}|$.

**Definition 3:** A nonstationary Markov chain is called weakly ergodic if for all $m$,

$$\lim_{k\to\infty}\sup_{f(0),g(0)}\left\|f(m,k)-g(m,k)\right\|=0$$

where $f(0)$ and $g(0)$ are starting vectors.

**Definition 4:** A nonstationary Markov chain is called strongly ergodic if there exists a vector $q=(q_1,q_2,\cdots)$, with $\|q\|=1$ and $q_i\geqslant 0$ for $i=1,2,\cdots$, such that for all $m$

$$\lim_{k\to\infty}\sup_{f(0)}\left\|f(m,k)-q\right\|=0$$

where $f(0)$ is a starting vector.

**Definition 5:** Let $P$ be a stochastic matrix. The ergodic coefficient of $P$, denoted by $\alpha(P)$, is defined by

$$\alpha(P)=1-\sup_{i,k}\sum_{j=1}^{\infty}\max\left(0,p_{ij}-p_{kj}\right)$$

and the delta coefficient of $P$, denoted by $\delta(P)$, is defined by

$$\delta(P)=1-\alpha(P)$$

**Lemma 1:** Let $P$ be a stochastic matrix. Then

(1) $0\leqslant\alpha(P)\leqslant 1$

(2) $\alpha(P)=1-(1/2)\sup_{i,k}\sum_{j=1}^{\infty}\left|p_{ij}-p_{kj}\right|$

(3) $\alpha(P)=\inf_{i,k}\sum_{j=1}^{\infty}\min\left(p_{ij},p_{kj}\right)$

**Theorem 1:** Let $\{X_n\}$ be a nonstationary Markov chain with transition matrices $\{P_n\}_{n=1}^{\infty}$. The chain $\{X_n\}$ is weakly ergodic if and only if there exists a subdivision of $P_1P_2P_3\cdots$ into blocks of matrices $[P_1P_2\cdots P_{n_1}][P_{n_1+1}P_{n_1+2}\cdots P_{n_2}]\cdots[P_{n_j+1}P_{n_j+2}\cdots P_{n_{j+1}}]\cdots$ such that $\sum_{j=0}^{\infty}\alpha\left(P\left(n_j,n_{j+1}\right)\right)=\infty$, where $n_0=0$.

**Definition 6:** A class $F\subset C^1$ of functions defined on (0, 1] is a closed class of asymptotically monotone functions (CAM) if

(a) $f\in F\Rightarrow f'\in F$ and $-f\in F$;

(b) $f,g\in F\Rightarrow(f+g)$ and $(f*g)\in F$; and

(c) all $f\in F$ change signs finitely often on $(0,1]$.

**Definition 7:** A class $F$ of functions defined on (0,1] is a rationally closed class of bounded variation (RCBV) if

(a) $f\in F\Rightarrow f$ is of bounded variation on $(0,1]$;

(b) $f\in F\Rightarrow -f\in F$;

(c) $f,g\in F\Rightarrow(f+g)$ and $(f*g)\in F$; and

(d) $f,g\in F$ with $(f/g)$ bounded on $(0,1]\Rightarrow f/g$ is of bonded variation.

**Definition 8:** A real-valued function $f$ is an exponential sum in $1/c$ if it is of the form $\sum_{i=1}^{\infty}Q_i(1/c)e^{\lambda_i/c}$ with $\lambda_i$, a given real number, and $Q_i(\cdot)$, a given polynomial $(i=1,2,\cdots,n)$.

**Definition 9:** A real-valued function $f$ is an exponential rational in $1/c$ if it is the ratio of two exponential sums in $1/c$. Note that the exponential rationals contain the rational functions as well as the exponential sums as subclasses.

**Proposition 1:** The following classes of functions defined on (0,1] are RCBV and CAM:

(a) the polynomials (RCBV and CAM);

(b) the rational functions (RCBV and CAM);

(c) the piecewise rationals (RCBV); and

(d) the exponential rationals (CAM).

**Definition 10:** Let $\{a(n)\}_{n=1}^{\infty}$ be a sequence with $\{a(n)\}\in\mathbb{R}^m$ for some $m\geqslant 1$. The (vector) function $\tilde{a}(c):(0,1]\to\mathbb{R}^m$ is an extension of the sequence if $\tilde{a}(c_n)=a(n)$ for some sequence $\{c(n)\}_{n=1}^{\infty}$, with $\lim_{n\to\infty}c_n=0$.

**Definition 11:** A nonstationary Markov chain $\{P(n)\}_{n=1}^{\infty}$ is said to have the regular extension $P(\cdot)$ if a real number $c^*>0$ exists such that the collection of subchains of $P(c)$ is identical for all $c<c^*$.

**Theorem 2:** Let $\{P(n)\}_{n=1}^{\infty}$ be a weakly ergodic nonstationary Markov chain and $P(c)$ a regular extension such that all entry functions $P_{ij}(c)(1\leqslant i,j\leqslant N)$ belong to

(a) a closed class of asymptotically monotone functions, or

(b) a rationally closed class of functions of bounded variation.

Then $\{P(n)\}_{n=1}^{\infty}$ is strongly ergodic. Moreover, for $n$ sufficiently large, each $P(n)$ has a sufficiently unique steady-state distribution $\psi(n)$ with $\lim_{n\to\infty}\psi(n)=q^*$ and $\lim_{n\to\infty}p_{ij}(m,n)=q_j^*$ for all $1\leqslant i,j\leqslant N$ and $m\geqslant 1$.

Without loss of generality, suppose that the inner-loop of SA algorithm iterates only once at each temperature, that is,

$$T_0>T_1>T_2>\cdots,\ \text{and}\ \lim_{n\to\infty}T_n=0 \quad (3)$$

We have shown a result on the optimal iterations for the inner-loop in another paper[10].

## 3 SA with general generating and accepting probabilities

The convergence of generalized SA can be

proved by the strong ergodicity of the Markov chain associated with it, i.e. $\lim_{k\to\infty}\sup_{f(0)}\left\|\boldsymbol{f}(m,k)-\boldsymbol{q}\right\|=0$ for all $m$. Then, we need only to verify that $\boldsymbol{q}=\boldsymbol{q}^*$, where $\boldsymbol{q}^*$ is the optimum (minimum ) probability vector of SA. Obviously,

$$q_i^*=0 \text{ for } i\notin S^* \tag{4}$$

$$S^*=\left\{i \middle| i\in S, c_i\leqslant c_j, j\in S\right\} \tag{5}$$

## 3.1 Weak ergodicity of the Markov chain associated with SA

We first prove the weak ergodicity of the Markov chain associated with SA. In order to use Theorem 1, it is necessary to analyze the ergodic coefficient of the transition matrices.

For convenience sake, abbreviate $P(T_n)$, $g_{ij}(T_n)$, and $a_{ij}(T_n)$, etc. as $P(n)$, $g_{ij}(n)$, and $a_{ij}(n)$ respectively, where $P(T_n)=(p_{ij}(T_n))_{|s|\times|s|}$ is the transition matrices of the Markov chain, and $P_{ij}(T_n)$ is defined by (1). Let

$$g^-(n)=\min_{i\in S, j\in N_i}\left\{g_{ij}(n)\right\} \tag{6}$$

$$a^-(n)=\min_{i\in S, j\in N_i}\left\{a_{ij}(n)\right\} \tag{7}$$

The radius of graph $G$ underlying the Markov chain is defined by:

$$r=\min_{i\in S-S_{lm}}\left\{\max_{j\in S}\left\{d_{ij}\right\}\right\} \tag{8}$$

where $S_{lm}=\left\{i \middle| i\in S, c_i\geqslant c_j \forall j\in N_i\right\}$ (9)

$d_{ij}$ is the distance from $i$ to $j$ measured by the minimum path length between them in $G$. Denote the state attaining the minimum value in (8) as $i_c$,which is called the center of graph $G$. $i_c$ can be reached with positive probability from any state in $S$ in no more than $r$ transitions.

**Theorem 3**: Suppose the generating probability $g_{ij}(n)$ of generalized SA satisfies:

(gl) $g_{ij}(n)$ is a non-increasing function of $n$ for sufficiently large $n$ and $i\in S$, $j\in N_i$;

(g2) $\lim_{n\to\infty} g_{ij}(n)$ exists for $i\in S$ and $j\in N_i$;

and the accepting probability $a_{ij}(n)$ satisfies:

(al) If $c_i<c_j$, $i,j\in S$, then $a_{ij}(n)$ is a decreasing function of $n$ for sufficiently large $n$, and $\lim_{n\to\infty} a_{ij}(n)=0$;

(a2) If $c_i\geqslant c_j$, $i,j\in S$, then $a_{ij}(n)=1$ for all $n$.

Then the Markov chain associated with generalized SA is weakly ergodic if

$$\sum_{k=n_1}^{\infty}\left[g^-(kr)a^-(kr)\right]^r=\infty \tag{10}$$

where $n_1$ is a constant, $g^-(n)$ and $a^-(n)$ are defined by (6) and (7).

**Proof**: For $i\in S$, $j\in N_i$, and sufficiently large $n$,

$$p_{ij}(n)=g_{ij}(n)a_{ij}(n)\geqslant g^-(n)a^-(n) \tag{11}$$

For $i\in S-S_{lm}$, there exists $j\in N_i$ such that $c_i<c_j$.Thus, $p_{ij}(n)$ is a decreasing function of $n$ for sufficiently large $n$. This, together with (1) and conditions (g) and (a), leads to the conclusion that $p_{ij}(n)$ is an increasing function of $n$ for sufficiently large $n$. Hence,

$\forall\, i\in S-S_{lm}, n\geqslant n_1 r$, where $n_1$ *is a* constant, $r$ is defined by (8), and

$$p_{ii}(n)\geqslant g^-(n)a^-(n) \tag{12}$$

Here we have used the fact that $g^-(n)a^-(n)$ decreases with increasing $n$ for sufficiently large $n$.

Remember that $i_c$ is the center of graph $G$ underlying the Markov chain. For $n\geqslant n_1 r$ and $i\in S$

$$p_{ii_c}(n-r,n) \geqslant \prod_{k=n-r+1}^{n}\left[g^-(k)a^-(k)\right] \geqslant \left[g^-(n)a^-(n)\right]^r \quad (13)$$

Using (13) and Lemma 1 (3) we get, for $n \geqslant n_1 r$,

$$\alpha(P(n-r,n)) = \min_{i,k}\left\{\sum_{j=1}^{\infty}(p_{ij}, p_{kj})\right\} \geqslant \min_{i,k}\left\{\min\left(p_{ii_c}, p_{ki_c}\right)\right\} \geqslant \left[g^-(n)a^-(n)\right]^r \quad (14)$$

From (14) and (10), we have

$$\sum_{k=n_1}^{\infty}\alpha(P(kr-r,kr)) \geqslant \sum_{k=n_1}^{\infty}\left[g^-(kr)a^-(kr)\right]^r = \infty \quad (15)$$

The weak ergodicity of the Markov chain is proved by (15) and Theorem 1. Q.E.D.

If a set of $g_{ij}(n)$ and $a_{ij}(n)$ is chosen according to the conditions specified in Theorem 3, we get a weakly ergodic Markov chain. This means that the asymptotic behavior of the corresponding SA algorithm is independent of initial states.

**Corollary 1**: Let

$$g_{ij}(T_n) = \begin{cases} w_{ij}/w_i, & j \in N_i \\ 0, & j \notin N_i \end{cases} \quad (16a)$$

where $w_i = \sum_{j \in N_i} w_{ij}$, $w_i$ is independent of $n$, (16b)

and $a_{ij}(T_n)$ defined by (2). The Markov chain associated with SA is weakly ergodic if

$$T_n = \lambda/\ln(n+n_0), \quad n = 0,1,2,\cdots \quad (17a)$$

where $n_0 (\geqslant 2)$ is a constant, $\lambda \geqslant (\Delta c^+)r$ (17b)

$$\Delta c^+ = \max_{i \in S, j \in N_i}\{|c_j - c_i|\} \quad (17c)$$

**Proof**: From (6) and (7) we get

$$g^-(kr) = \min_{i \in S, j \in N_i}\{g_{ij}(kr)\} = \min_{i \in S, j \in N_i}\{w_{ij}/w_i\} = g^-$$

$$a^-(kr) = \min_{i \in S, j \in N_i}\{a_{ij}(kr)\} \geqslant \exp\left[-\Delta c^+/T_{kr}\right] = (kr+n_0)^{-(\Delta c^+)/\lambda} = r^{-(\Delta c^+)/\lambda}(k+n_0/r)^{-(\Delta c^+)/\lambda}$$

Hence, by using (17b),

$$\sum_{k=0}^{\infty}\left[g^-(kr)a^-(kr)\right]^r \geqslant (g^-)^r r^{-(\Delta c^+)r/\lambda}\sum_{k=0}^{\infty}(k+n_0/r)^{-(\Delta c^+)r/\lambda} = \infty$$

Q.E.D.

Corollary 1 is the result proved before[8-10]. It is only a special case of our general theorem. In Corollary 1, we suppose $g_{ij}(T_n)$ is independent of $T_n$. Each state can only generate its neighbors. This often leads to the low efficiency of the SA algorithm. One way to accelerate the annealing process is to jump occasionally a long distance at one annealing step. Certain probability distributions can be used to generate the distance. Because the state space $S$ is discrete, the distance between two states $i$ and $j$ is also discrete. It, denoted by $d_{ij}$, can be considered as the distance from point $i$ to point $j$ in $\mathbb{R}^m$ if every state in $S$ has $m$ components.

Let $f(x)$ denote the continuous probability distribution used to generate the distance between current state $i$ and next state $j$. The generating probability $g_{ij}(T_n)$ of SA algorithm depends on the probability of the random number generated by $f(x)$ falling into the small range around $d_{ij}$. Let $\xi$ denote the random number generated by $f(x)$. A simple way to define $g_{ij}(T_n)$ is as follows.

$$g_{ij}(T_n)=Pr\{d_{ij}-1/2<\xi<d_{ij}+1/2\}=$$
$$\int_{d_{ij}-1/2}^{d_{ij}+1/2} f(x)\,dx \approx f(d_{ij}) \tag{18}$$

In practice, $g_{ij}(T_n)$ can be multiplied by a constant factor to improve the performance of the SA algorithm. This does not affect our theoretical results. Sometimes, there are perhaps several states which have the same distance from current state. The case can be solved by simply dividing $g_{ij}(T_n)$ among those states. Note that $f(x)$ usually has a parameter corresponding to the temperature of SA algorithm. Hence, (18) can be written as

$$g_{ij}(T_n)=f(d_{ij},T_n) \tag{19}$$

**Corollary 2**: Let $f(d_{ij},T_n)$ be normal distribution and $a_{ij}(T_n)$ defined by (2). The Markov chain associated with SA is weakly ergodic if

$$T_n=\lambda_G/\ln(n+n_0),\quad n=0,1,2,\cdots \tag{20a}$$

where $n_0\,(\geqslant 2)$ is a constant,

$$\lambda_G \geqslant \left[(d^+)^2+\Delta c^+\right]r \tag{20b}$$

$$d^+=\max_{i\in S,j\in N_i}\{d_{ij}\} \tag{20c}$$

$\Delta c^+$ is defined by (17c).

**Proof**: $f(d_{ij},T_n)=1/(2\pi T_n)^{1/2}\cdot\exp\left[-d_{ij}^2/(2T_n)\right]$

To make things simple, constants in the above formula can be omitted without loss of generality. Hence,

$$g_{ij}(T_n)=1/T_n^{1/2}\cdot\exp\left(-d_{ij}^2/T_n\right) \tag{21}$$

Using (20) we get

$$g^-(kr)=1/T_{kr}^{1/2}\cdot\exp\left[-(d^+)^2/T_{kr}\right]$$
$$=1/\lambda_G^{1/2}\cdot\left[\ln(kr+n_0)\right]^{1/2}\cdot(kr+n_0)^{-(d^+)^2/\lambda_G}$$

$$a^-(kr)\geqslant\exp\left(-\Delta c^+/T_{kr}\right)=$$
$$(kr+n_0)^{-(\Delta c^+)/\lambda_G}\sum_{k=0}^{\infty}\left[g^-(kr)a^-(kr)\right]^r$$
$$\geqslant 1/\lambda_G^{r/2}\cdot r^{-\left[(d^+)^2+\Delta c^+\right]r/\lambda_G}\cdot\sum_{k=0}^{\infty}\left[\ln(kr+n_0)\right]^{r/2}\cdot$$
$$(k+n_0/r)^{-\left[(d^+)^2+\Delta c^+\right]r/\lambda_G}=\infty$$

Noting that $\lambda_G$ satisfies (20b), we prove the corollary.

Q.E.D.

**Corollary 3**: Let $f(d_{ij},T_n)$ be Cauchy distribution and $a_{ij}(T_n)$ defined by (2). The Markov chain associated with SA is weakly ergodic if

$$T_n=\lambda_c/\ln(n+n_0),\quad n=0,1,2,\cdots \tag{22a}$$

where $n_0\,(\geqslant 2)$ is a constants,

$$\lambda_c\geqslant(\Delta c^+)r \tag{22b}$$

**Proof:** $g_{ij}(T_n)=T_n/\left(T_n^2+d_{ij}^2\right) \quad (23)$

$$g^-(kr)=T_{kr}/\left[T_{kr}^2+(d^+)^2\right]=$$
$$\lambda_c/\left[\lambda_c^2/\ln(kr+n_0)+(d^+)^2\ln(kr+n_0)\right]$$
$$a^-(kr)\geqslant\exp\left(-\Delta c^+/T_{kr}\right)=(kr+n_0)^{-\Delta c^+/\lambda_c}$$

The remaining steps of the proof are the same as Corollary 2's.

Q.E.D.

**Corollary 4**: Let $f(d_{ij},T_n)$ be exponential distribution and $a_{ij}(T_n)$ defined by (2). The Markov chain associated with SA is weakly ergodic if

$$T_n=\lambda_e/\ln(n+n_0),\quad n=0,1,2,\cdots \tag{24a}$$

where $n_0\,(\geqslant 2)$ is a constant,

$$\lambda_e\geqslant(d^++\Delta c^+)r \tag{24b}$$

**Proof**:

$$g_{ij}(T_n)=1/T_n\cdot\exp\left(-d_{ij}/T_n\right)$$
$$g^-(kr)=1/\lambda_e\cdot\ln(kr+n_0)\cdot(kr+n_0)^{-d^+/\lambda_e}$$

$$a^{-}(kr) \geqslant (kr+n_0)^{-\Delta c^{+}/\lambda_e} \quad (25)$$

The remaining steps are the same as Corollary 2's. Q.E.D.

**Corollary 5:** Let $f(d_{ij}, T_n)$ be Weibull distribution and $a_{ij}(T_n)$ defined by (2). The Markov chain associated with SA is weakly ergodic if

$$T_n = \lambda_w / \ln(n+n_0), \quad n=0,1,2,\cdots \quad (26a)$$

where $n_0 (\geqslant 2)$ is a constant,

$$\lambda_w \geqslant \left[(d^{+})^2 + \Delta c^{+}\right] r \quad (26b)$$

**Proof:**

$$g_{ij}(T_n) = d_{ij}/T_n \cdot \exp(-d_{ij}^2/T_n) \quad (27)$$

$$g^{-}(kr) = d^{-}/\lambda_w \cdot \ln(kr+n_0) \cdot (kr+n_0)^{-(d^{+})^2/\lambda_w}$$

where $d^{-} = \min_{i \in S, j \in N_i} \{d_{ij}\}$

$$a^{-}(kr) \geqslant (kr+n_0)^{-\Delta c^{+}/\lambda_w}$$

The remaining steps are trivial. Q.E.D.

So far some corollaries of Theorem 3 have been given. More can be added to this list. In the above corollaries, we suppose $a_{ij}(T_n)$ is defined by (2). We can also easily prove that the results are true for

$$a_{ij}(T_n) = \min\left\{1, \left[1+\exp\left((c_j - c_i)/T_n\right)\right]^{-1}\right\} \quad (28)$$

and some other kinds of $a_{ij}(T_n)$.

## 3.2 Strong ergodicity and convergence of the Markov chain associated with SA

The weak ergodicity of the Markov chain associated with SA guarantees that the asymptotic behavior of SA is independent of the initial state. It says nothing about the convergence of SA. But under certain conditions, strong ergodicity can be derived and the convergence is proved. This section shows the strong ergodicity and the convergence of SA under Theorem 3's conditions by means of Theorem 2.

**Theorem 4:** Under the assumptions of Corollary 1, the Markov chain associated with SA is strongly ergodic and converges to optimum vector $q^{*}$.

**Proof:** From (16) and (2) we get

$$p_{ij}(T_n) = \begin{cases} w_{ij}/w_i \cdot \min\{1, \exp[-(c_j - c_i)/T_n]\}, & j \in N_i \\ 0, & j \notin N_i, j \neq i \\ 1 - \sum_{k \in N_i} p_{ki}(T_n), & j = i \end{cases} \quad (29)$$

First we should prove that $\{p_{ij}(T_n)\}_{n=0}^{\infty}$ has a regular extension $P(c), c \in (0,1]$, which is obtained by substituting $c$ for $T_n$ in $p_{ij}(T_n)$.

Note that $\{T_n\}_{n=0}^{\infty}$ satisfies (3), hence we know $P(c)$ is an extension of $\{p_{ij}(T_n)\}_{n=0}^{\infty}$ by Definition 10. The regularity of $P(c)$ can easily be shown by a known result[9].

If there exists a real number $c_0 > 0$ such that $\{(i,j) \mid p_{ij}(c) > 0\}$ is identical for all $c < c_0$, $P(c)$ is a regular extension.

Next we know from Proposition 1 (d) that $P_{ij}(c)$ is CAM. Thus, the Markov chain is strongly ergodic because of Corollary 1 and Theorem 2. Moreover, for sufficiently large $n$, each $P(n)$ has a unique steady-state distribution $\psi(n)$ with $\lim_{n\to\infty} \psi_i(n) = q_i^{*}$ and $\lim_{n\to\infty} p_{ij}(m,n) = q_j^{*}$ for all $i, j \in S$ and $m \geqslant 1$.

It is easy to verify that $\psi(n) = (\psi_i(n))$ is

$$\psi_i(T_n)= \left[w_i\cdot\exp(-c_i/T_n)\right]\Big/\left[\sum_{i\in S}w_i\cdot\exp(-c_i/T_n)\right] \tag{30}$$

Hence,

$$q_i^*=\lim_{n\to\infty}\psi_i(T_n)=\begin{cases} w_i\Big/\sum_{j\in S^*}w_j, & i\in S^*\\ 0, & i\notin S^*\end{cases} \tag{31}$$

This means that SA converges to the optimum state vector. Q.E.D.

Following the same proving steps of Theorem 4, we can show other useful results from Corollaries 2~5.

**Theorem 5:** Under the assumptions of Corollaries 2~5, the Markov chains associated with SA are strongly ergodic and converge to the optimum vectors.

**Proof:** Here we show the conclusion is true for Corollary 4's case (exponential distribution). The other three cases are the same.

For (25) and (2) we get

$$p_{ij}(T_n)= \begin{cases} 1/T_n\cdot\exp(-d_{ij}/T_n)\cdot\\ \min\{1,\exp[-(c_j-c_i)/T_n]\}, & j\in N_i\\ 0, & j\notin N_i, j\neq i\\ 1-\sum_{k\in N_i}p_{ik}(T_n), & j=i\end{cases} \tag{32}$$

First, a regular extension $P(c)$ of $\{p_{ij}(T_n)\}_{n=0}^{\infty}$ is obtained by substituting $c$ for $T_n$ in (32), $c\in(0,1]$.

Next, we know $P_{ij}(c)$ is CAM for $i,j\in S$ from Proposition 1 (d).

Hence the strong ergodicity is proved. Because

$$\psi_i(T_n)p_{ij}(T_n)=\psi_j(T_n)p_{ji}(T_n)$$

we get $\psi_i(T_n)/\psi_j(T_n)= p_{ji}(T_n)/p_{ij}(T_n)= \exp[(c_j-c_i)/T_n]$

Thus, for $i\in S$

$$\psi_i(T_n)=\exp(-c_i/T_n)\Big/\sum_{j\in S}\exp(-c_i/T_n) \tag{33}$$

$$q_{e1}^*=\lim_{n\to\infty}\psi_i(T_n)=\begin{cases} 1/|S^*|, & i\in S^*\\ 0, & i\notin S^*\end{cases} \tag{34}$$

Q.E.D.

Theorem 5 shows that many probability distributions can be used to generate the displacement between two states in a combinatorial state space. But it is worth noting that Theorems 3-5 only give a set of sufficient conditions for the convergence of SA.

The results shown in this section have a common point, that is, the cooling rate $T_n$ has the form of $D/\ln(n+n_0)$, where $D$ is a constant. In fact, some distributions are better than others. The comparisons can be made either experimentally or theoretically. The latter case needs more sophisticated mathematical proof to show a faster cooling rate than $D/\ln(n+n_0)$. H.H. Szu and R.L. Hartley[7] first proposed fast SA using Cauchy distribution and claimed that they had proved its convergence for $T_n=1/n$. Although the theoretical proof has not been seen, the result is very encouraging. We think more empirical studies are needed to evaluate its practical advantages.

## 4 Conclusions

This paper proves the convergence of SA algorithms with general generating and accepting probabilities. It shows that not only can normal and Cauchy distributions be used as generating probabilities, but others, e g., exponential and

Weibull distributions, can also be used. The paper discusses different accepting probabilities as well. We think it is necessary to consider the two probabilities together because each of them cannot produce satisfactory results independently. We need to find a good pair of generating and accepting probabilities in order to form a satisfactory SA which can solve a wide range of practical combinatorial optimization problems. This paper is a first step towards the goal. To make the algorithm practically useful, much more work should be done, especially the empirical study of SA algorithms with different generating and accepting probabilities.

**Acknowledgement**

we gratefully acknowledge the help and support Profs. Yan-Min Yang and Guo-Liang Chen in the research.

## References:

[1] KIRKPATRICK S, GELATT C D, VECCHI M P. Optimization by simulated annealing[J]. Science, 1983, 220(4598): 671-680.

[2] CERNY V. Thermodynamical approach to the traveling salesman problem: an efficient simulation algorithm[J]. Journal of Optimization Theory & Applications, 1985, 45(1): 41-51.

[3] GEMEN S, GEMEN D. Stochastic relaxation, gibbs distributions, and the bayes restoration of images[J]. IEEE Transactions on Pattern Analysis & Machine Intelligence, 1984: 721-741.

[4] ACKLEY D H, HINTON G E, SEJNOWSKI T J. A learning algorithm for boltzmann machines[J]. Cognitive Science, 1985, 9(1): 147-169.

[5] METROPOLOS N, POSENBLUTH A W, POSEBLUTH MN, et al. Equations of state calculations by fast computing machines[J]. Journal of Chemical Physics, 1983, 21: 1087-1091.

[6] ISAACSON D L, MADSEN R W. Markov chains, theory and applications[M]. New Jersey: John Wiley & Sons, Inc., 1976.

[7] SZU H H, HARTLEY R L . Nonconvex optimization by fast simulated annealing[J]. Proceedings of the IEEE, 1987, 75(11): 1538-1540.

[8] ANILY A, FEDERGRUEN A. Simulated annealing methods with general acceptance probabilities[J]. Journal of Applied Probability, 1987, 24(3): 657-667.

[9] ANILY A, FEDERGRUEN A. Ergodicity in parametric nonstationary markov chains: an application to simulated annealing methods[J]. Operations Research, 1987, 35(6): 867-874.

[10] YAO X, LI G J. Performance analysis of simulated annealing[C]// The 3rd Pan-Pacific Computer Conference. [S.l.:s.n.], 1989.

[11] ROYDEN H L. Real analysis, 2nd edn[M]. New York: Macmillan Publishing Co., Inc., 1968.

发表于《模式识别与人工智能》1992年第3期

# 人工智能的计算复杂性研究

李国杰

国家智能计算机研究开发中心，北京

**摘　要：** 首先澄清一些关于人工智能计算复杂性的模糊认识，强调计算复杂性依赖于问题描述而独立于算法。为了说明这一领域的研究方向，对机器学习与人工神经网络方面的计算复杂性研究的历史与成果做了简要的介绍与分析。作者强调人工智能研究应集中于易解问题，通过对智能问题本身的深入理解，找到合适的问题描述、对输入输出加适当限制，或发现新的“计算”模型将难解问题变为易解问题。最后指出，计算效率研究可以看成是一种反向的复杂性研究，应十分重视提高有限计算机资源下的实际解题效率。

**关键词：** 人工智能；计算复杂性；计算效率

# Computational complexity of artificial intelligence

Guo-jie LI

National Research Center for Intelligent Computing Systems, Beijing

**Abstract:** Some concepts on computational complexity of artificial intelligence are first clarified. It is reminded that the complexity is dependent on the problem description and independent to the algorithms used. To show the research direction, the research results on complexity of machine learning and neural computations are briefly surveyed. The author points out that, for artificial intelligence researchers, the efforts should be focused on the tractable problems rather than intractable problems. What need to be attacked are to understand deeply the problems to be solved, to find the proper descriptions of the problems, to add some restrains on the input and output of the problem descriptions, and to discover the novel computational models. Finally, it is claimed that the study of computational efficiency is the reverse study of complexity, and attention should be paid to the improvement of computational efficiency under limited resources (time, space).

**Key words:** artificial intelligence, computational complexity, machine learning, neural computation, computational efficiency

## 1　前言

人工智能是所有科学技术问题中最困难最复杂的问题之一。现已证明在人工智能领域遇到的许多问题都具有指数甚至更高的计算复杂性。著名学者 Rich 关于人工智能的定义已被广泛引用：人工智能就是关于利用问题领域知识在多项式时间内解决具有指数复杂性问题的技术的研究[1]。这一定义叙述欠严谨，因而引起异议[2]，但这一定义非常清楚地指明如何对付指数复杂性是人工智能要解决的根本问题。在目前的人工智能研究中，有不少偏重理论的学者致力于证明某种推理或学习问题是非确定性多项式（Nondeterministic Polynomially，NP）困难问题。他们的工作是重要的，如同“永动机是不可能的”这一类定律对力学系统设计有指导意义一样，揭示智能问题的本质复杂性可以使设计智能系统的人减少盲目性，少走弯路。但是从事实现工作的人更需要指明出路的理论，呼唤卡诺式的理论结果。自从卡诺推导出热机效率的极限是 $1-T_1/T_2$（$T_1$，$T_2$ 分别是卡诺循环的低温与高温）以后，热机的研究才得到突飞猛进的发展。许多致力于模式识别与人工智能研究的学者满足于提出一些颇为有效的启发式方法，但如果缺乏对计算复杂性的清醒认识，不了解所采用方法的极限能力，往往要花费许多精力。本文下面几节将澄清关于计算复杂性的某些模糊认识，综述一些重要理论结果并阐明作者关于复杂性研究的立场与观点。

## 2 可计算性与计算复杂性

可计算性与计算复杂性是两个既有区别又有一定联系的概念。有些学者在研究过程中不自觉地混淆了这两个概念。比如有些论文以人脑处理模式识别（包括识别拓扑结构）很快而现有计算机效率低为根据，得出人工神经网络的能力超过图灵机及认知不等价于计算的结论，这就是将计算复杂性问题当成可计算问题了。请注意 Minsky 在批评 Perceptron 时，只是说拓扑识别具有阶数较高的多项式复杂性，并没有讲它是一个不可计算问题。根据图灵机模型，什么是“计算”已有十分严格的定义。原则上讲，图灵机是讨论可计算性的模型，它比讨论计算复杂性时用到的各种模型（如判断树、共享存储等）更一般、更抽象。但不可计算或不可判定问题也可以看成是计算复杂性的最高层次，例如停机问题（一个不可判定问题）也属于 NP 困难问题（但不是 NP 完全问题）。因此寻求新的模型有效地解决 NP 困难问题也是对图灵机模型的冲击。

在讨论计算复杂性时，我们经常提到 NP 困难问题与 NP 完全问题。这两类问题有联系也有区别。通俗地讲，所谓 NP 困难问题就是目前还没有或不可能找到多项式时间内求解算法的问题（包括某些不存在解题算法的问题）。理论上严格定义 NP 困难问题需引入归约性，即一个问题 L 是 NP 困难。当且仅当可满足性问题（即判定谓词公式真假）可归约到 L。NP 完全问题是 NP 困难问题的一个子集。所谓“完全”的意思是这个集合中任意一个问题可在多项式时间内求解，当且仅当此集合中其他所有问题在多项式时间内可解。NP 完全问题一定是判定问题，而 NP 困难问题可以是优化问题或不可计算问题。一个问题是 NP 完全问题除了满足归约性之外，还一定要属于 NP，即此问题在非确定的图灵机上多项式时间内可解。粗略地说，属于 P 的问题是容易计算的问题，而属于 NP 的问题是容易检查的问题，即只要给出一个答案，就能在多项式时间内判断这答案是不是该问题的解。例如给出图上一条路径，我们很容易判断它是不是 Hamiton 圈。目前的研究工作大都是基于大家熟知的模型分析某个抽象出来的数学问题是不是 NP 完全问题，如一些图论问题与人工神经网络问题等。对于人类的认知、推理、学习来说，我们无法直接讨论其复杂性，必须先有描述这些智能过程的模型。每种问题模型有其固有的复杂性。如同物理学中的能量守恒一样，计算复杂性是计算机科学中最重要的守恒量。对于人工智能学者来说，我们最需要的复杂性理论应能揭示问题模型与计算复杂性的关系，并能指导研究者建立合适的问题模型，使其固有的复杂性低到可接受的程度。

传统的人工智能基于 Church-Turing 假设，即承认任何可以有效计算的函数（Effectively Computable Function）能在图灵机上实现。Newell 和 Simon 提出的物理符号系统假设与 Church-Turing 假设思想是一致的，只是更加明确地表示他们相信认知过程可以用物理符号系统实现。必须指出，所谓 Church-Turing 假设或者更简洁的“认知等价于计算”假设只是一种假设，无法给出证明也不容易找到反例，因为这一等价关系的一边有严格的形式化定义（即图灵可计算性），而另一边（认知或 Effectively Computable Function）却只有直观的定性描述。人们常以形象思维、创造性思维为例说明认知不同于计算，但究竟什么是形象思维并没有明确定义。常识告诉我们要说明事物 $A$ 不属于集合 $B$，一个起码的条件是必须知道 $A$ 是什么，或者至少要知道 $A$ 的特征是什么，如果连 $A$ 的特征都讲不清，怎么能确定 $A$ 是否属于 $B$。笔者认为，由于我们对于人脑功能的了解太少，至少在今后几十年里从可计算性的角度争论认知是否等价于计算不会有结果，也许从计算复杂性的角度开展研究更有意义，更有可能取得突破。

## 3 问题描述与计算复杂性

计算复杂性反映一种问题的本质性的困难程度，常用对应问题规模 $n$ 的下界函数 $g(n)$ 表示，它表示求解给定问题的任何算法必须花费的时间（空间）的下界。在这里，所谓“问题”是指包含各种问题实例（Problem Instance）的一类问题，如 $n$ 皇后、TSP 等。对于一个给定的问题来说，它的计算复杂性是个不变量，即问题的计算复杂性与采用的算法无关。算法的复杂性与问题的复杂性常容易混淆，我们在文献中常看到这样的提法：通过获取知识采用某种启发式算法，将某人工智能问题的复杂性降低。这是一种错误的观念，因为改进算法不能改变问题本身的困难程度。

那么，问题的困难程度与什么有关？有没有化繁为简解决问题的出路？不难理解，一个问题的困难程度与问题的描述有关，特别是与它的输入表示有关。改变了问题描述，实际上是把原来的问题变成另一个问题，其困难程度可能大不一样。人们常常发问：人脑是如何克服或绕过组合爆炸？这种提问的前提假设是诸如从人群中识别一个熟悉面孔之类的问题一定是一个具有指数复杂性的问题。这种前提恰恰把事情搞颠倒了。其实对于人脑来说，识别面孔本来就是容易的事，采用计算复杂性的术语来说，只有 $O(1)$ 复杂性，不存在绕过指数爆炸的问题。而我们目前采用的对识别面孔这一问题的描述没有反映人脑的功能，我们的描述造成了指数爆炸。因此，归根结底人工智能的出路在于对问题本身的深入了解。

在问题描述中，输入的表示很关键。在图灵机模型中，输入变量都是整数。近几年来，国外在连续量（实数）计算模型的研究方面已取得一些突破性进展。Blum 等学者 1989 年提出的实数环上的计算模型已引起较广泛的重视[3]。在这一模型中，实数不是看成无限位的二进制数，而是看成一个数学实体（Entity），因而基本操作是有理数操作而不是位操作。根据这一模型建立的“通用计算机”可将图灵机看成整数计算的特例。Blum 等人还建立了一整套与已有计算复杂性对应的实数环上的复杂性理论，包括 NP 完全性等。在这一新的计算模型中，停机问题不再是不可判定问题，这说明其“计算”能力超过图灵机。由于任何物理系统的取值只有有限的测控能力，如何在“计算机”中表示实数这一种数学抽象已成为新模型是否有用的关键。与上述工作相呼应的是对 Church-Turing 假设的扩充：任何物理上可实现的系统（或过程）都可用一通用“计算机”完全模拟（Perfectly Simulated），即有效可计算。计算复杂性与可计算性研究已经不限于计算机科学本身，数学家、物理学家、脑科学家等纷纷介入预示着新的突破可能更早到来。但是应当指出这条路十分艰难。最近有些学者已证明对于神经网络计算来说，连续量信息等价于 $n^3 \log n$ 位信息，并指出模拟元件只有具有相当于指数多位有效位时，才有更强的计算能力。

## 4 不同方面的复杂性

传统的人工智能与计算机科学中讨论的复杂性主要是组合复杂性，即讨论计算时间（空间）与问题规模的函数关系。这种复杂性与计算模型及输入的表示有关但计算中需要的信息是完全的、严格的、无代价的。很多人工智能问题只有部分的、不精确的信息，获取与求解问题有关的信息需要付出代价，因而人们往往只能近似求解或定性求解。对于这一类问题来说，Traub 等学者建立了基于信息的复杂性理论，讨论为了将解的不确定性限制在某一给定范围，究竟需要获得多少信息[4]。这种复杂性理论与计算模型无关。

在过去几十年内，人们大量使用的是串行计算机，因此研究计算复杂性大多是以求解一个问题所需的逻辑操作的总步数来衡量。随着并行与分布式计算机的逐步普及，通信开销已上升为不可忽视甚至主要的时间开销，因而研究通信复杂性已成为当前的热门课题。由于人工智能问题本质上涉及非确定计算，研究人工智能问题的通信复杂性不同于 VLSI 计算的 $AT^2$ 理论。在确定型计算中只需研究完成一个计算至少需要多少信息交换。对于并行搜索这类人工智能问题来说，增加通信往往能减少因全局信息不足引起的不必要计算，大大缩短计算时间。因此人工智能问题、计算复杂性与通信复杂性交织在一起时，需要做全面地分析。对于人工神经网络来说，通信复杂性还与是否采用模拟电路有关。传统的组合逻辑电路的计算能力正比于输入端数目；而以模拟量运算的单个神经元的计算能力与输入端数目的平方成正比，因而模拟量运算倾向于采用大量输入端。神经网络计算中互连数目的多少与性能密切相关，我们已证明对于联想记忆模型来说，减少连接数必然导致记忆容量减少，以存储容量/计算量衡量，全连接的 Hopfield 网是最优方案[5]。

运行智能程序慢往往是由于主存耗尽导致过多地与磁盘交换信息。空间复杂性反映对存储容量的需求，它在实际运行中往往比时间复杂性更重要但容易被忽视。理论上 $A^*$ 是最优的搜索算法，但由于其指数级空间复杂性实际上几乎不可行。近年来人们更加重视只有线性空间复杂性的 $IDA^*$ 和 $IRA^*$ 算法[6]。

## 5 关于人工智能计算复杂性的某些结果

在这篇短文中不可能列举有关人工智能问题计算复杂性的大量研究成果，只能通过几个例子说明这方面研究的发展方向。

从 20 世纪 70 年代后期开始，在人工智能领域中有关复杂性的研究已逐步展开，其中比较重要的工作有，关于受限满足问题[7]、启发式搜索[8]、机器学习[9]和自然语言[10]的复杂性分析。近几年来，关于计算复杂性的研究已深入到外推（Abduction）[11]、非单调推理[12]和概率推理[13]。在这一节中，我们以机器学习和人工神经网络的复杂性研究为例说明研究的重点已从对一般问题的研究转移到限制较强的问题。

关于机器学习的形式化模型最早的工作要追溯到 1967 年 Gold 的工作，他研究非常一般的学习理论，即从无限的字符串序列中寻找对应的语法规则，他的模型对特征空间、输入输出的数目及数据结构都没有限制。这实际上是可计算性研究，还谈不上计算复杂性，实用性较差。1984 年 Valiant 提出的可学习理论[9]，不再讨论绝对的可学习性，而是对固定长度的字符串研究在有限时间内可行的可学习性。他提出 Probably Approximate Correct（PAC）模型，用概率衡量学习的成功率。PAC 有较强适应性，其样本复杂性与时间复杂性对误差与可信度不很敏感。所谓学习的样本复杂性是指达到一定的学习正确率与可信度，至少需要多少训练样本。对于机器学习来说，这是与时间复杂性一样重要的判据。到了 20 世纪 90 年代，Judd 做出了引人注目的关于机器学习复杂性方面的研究成果[14]，Judd 研究对于具有特定结构的机器来说什么样的任务可以有效地学习。他集中精力于神经网络的训练问题，即 load 问题，他的结论是 loading 是 NP 完全问题，泛化（Generalization）1/3，即训练完只改变 1/3 输入已是 NP 完全问题。Judd 认为神经网络原则上不能泛化，现在谈论预估未学习过的测试样本的识别率还为时过早。

对于一般较成熟的学科来说，往往是逐步扩大研究的领域范围，减少对问题的限制，从具体的研究逐步抽象出普遍适用的规律与方法。但是机器学习的复杂性研究却走一条相反的路。这种逐步后退的现象说明有关人工智能的计算复杂性研究难度很大，直接研究具有一般性的大问题难以获得很有意义的成果。正如 Simon 教授所说，人工智能目前需要上千个小的突破。

当 Hopfield 用人工神经网络求解一个旅行推销员问题，在很短时间内就找到十分接近最优解的近似解时，曾预言神经网络计算可能是求解组合优化问题的有效途径。不少人认为神经网络计算将如一夜东风吹散笼罩在人工智能上空指数爆炸的乌云。但是最近两年的复杂性研究结果表明神经网络计算也不是灵丹妙药。B.Goodman 已证明存在一个多项式规模的神经网络能解决任何一个 NP 困难问题就意味着 NP=co-NP。NP=co-NP 意思是若一个问题属于 NP，则它的补问题也属于 NP。所谓补问题是从相反的角度提问，例如问一个图是否不存在 Himilton 圈。即使对于近似计算来说，神经网络计算也不一定能保证精度。Goodman 还证明了对于旅行推销员问题来说，除非 P=NP，否则不存在一个多项式规模的神经网络可以保证近似解精度。我国年青学者姚新推广了 Goodman 的结果，他证明用多项式规模神经网络找一类 Continously Reducible NP 完全问题也是不可能的，除非 NP=co-NP。大量研究结果使多数学者相信 NP=co-NP 及 P=NP 都是不成立的。所以上述结果表明人工智能中的许多问题对于神经网络计算模型来说具有与传统计算一样的计算复杂性。

光学计算可以改变许多数值计算问题的计算复杂性，例如 FFT 和矩阵乘法在光学计算中均可以一步完成。但是目前的研究结果初步表明图灵机模型下不可判定与具有指数复杂性的难解问题不大可能在光学计算机上变成一个简单问题。

## 6　难解（Intractable）与易解（Tractable）问题

计算机科学中难解问题与易解问题的分界线在于是否已有多项式复杂性的解题算法。尽管复杂性为高阶多项式（$O(n^3)$以上）的算法计算机实际处理时几乎不可忍受，从理论上还是归于易解问题。十多年来关于人工智能问题计算复杂性的研究结果几乎全都是针对难解问题。很少人认真研究人工智能中的易解问题，这实在是极不正常的现象。有些明智的学者也开始呼吁重视易解性（Tractability）研究[2]。所谓难解问题的前提假设往往是求正确的解或最优解，它的问题描述导致了最坏情况下出现指数复杂性。如果我们的出发点不是求完全正确的解或最优解，而是较快地得到一个解（不一定完全正确或最优），则需要另一种问题描述。这就是说，要把易解性作为讨论人工智能的基础和归宿。有了这样的观念更新，人工智能才能真正在实际生产与人类生活中发挥大的作用。目前的做法是先面对一个难解问题，然后在此基础上做一些近似处理或启发式处理，试图减少一些计算量。这种迂回的做法不能保证得到易解问题。许多 NP 困难问题求近似解仍然是指数复杂性。

一个 NP 困难问题中往往只有很少的实例具有指数复杂性，而且这些“最坏”的实例的分布有某种规律可寻。Taylor 等人在 1991 年 IJCAI 大会发表的一篇论文中指出 NPC 问题可以用少至一个参数排序，真正难解的问题实例只分布在这些参数的某些临界值上，这些临界值将 NPC 问题分成两个特性不同的区域[15]。例如判断一个图是否存在 Hamilton 圈，只有当平均连接性为 $\ln N+\ln\ln N$ 时难以判断，小于这个临界值几乎没有 Hamilton 圈，大于它几乎一定有解。这两种情形用搜索方法都能较快出判断。这两个不同区域性质不同，类似物理上固体与液体的相变，相变点对应某一序列参数的临界点。有了这种分析结果，我们就可以预先估计一个问题实例易解还是难解。从易解性的角度来看，所谓 NP 困难问题并不可怕，对这种困难问题的绝大多数实例集合，即将输入（或输出）做某种限制的问题实例子集，往往能找到非常有效的算法。例如对于非对称的一百万个城市旅行推销员问题来说，采用赋值算法几个小时计算就能找出最优解，若只求一百万个皇后问题的一个解（不是全部解）半个小时就能找出来。这些事例均说明对输入输出加一点限制，问题的描述稍稍改变，原来的难解问题往往能变成易解问题。我们的任务就是找导致易解问题的合适的问题描述。

20 世纪 90 年代人工智能的特点之一是将智能技术推向真正实用，可扩展性（Scalability）成了各国学者注意的焦点。对不同应用，实际需要解决的问题规模有差别，但一般来讲总要到“千”这一数量级。对于这样的大问题，不管什么“计算”系统都不可能接受指数复杂性，也就是说，能在人工系统或人机协作系统上求解的实际应用问题一定是易解问题。人工智能要走出玩小孩游戏（Toy Problem）的圈子只有两条出路，要么承认 Church-Turing 假设，以现有计算机能力为基础（与图灵机能力只差多项式倍），寻找合适的问题描述，发现人工智能中的易解问题，要么不承认 Church-Turing 假设，寻求新的“计算”模型，使得对人脑易解的问题在新的模型中也易解。总之，NP 困难问题不应成为人工智能的禁区或拦路虎。在功能更强的计算模型找到之前可以先把一个 NP 困难问题分解成若干易解子问题（排除临界区），逐个解决。

## 7　提高有限资源条件下的解题效率

计算的时间、空间开销不只是与问题规模 $n$ 有关，而且取决于解的精度（优化问题）或解的正确率、成功率（判定问题）。设 $\varepsilon$ 表示解的误差（与最优解相比），$p$ 表示获得解的成功率或正确率，则计算复杂性可以表示为函数 $f(1/\varepsilon,n)$ 或 $f\left(p,n\right)$。从这种意义来理解，所谓计算复杂性也反映计算中资源耗费与所获得结果好坏的关系。换言之，研究计算复杂性的本质就是研究计算的效率。从事理论研究的学者一般讨论当 $n$ 增大或 $\varepsilon$ 减小时。$f(1/\varepsilon,n)$ 如何变化，尤其热衷于讨论 $n$ 趋于无穷大时的渐近特征，即使得到资源消耗很惊人的结论，理论上也是允许的。但从实际出发，无论是允许的计算时间和可用内存都是有限制的，所以更有价值的研究应当是做反方向的复杂性研究，即对给定的计算时间或空间，可能达到的最小 $\varepsilon$（最大 $p$）和最大 $n$ 是多少。笔者过去做的研究工作大都是这种面向实用的复杂性分析，即计算效率分析[16-17]近几年来国外开始重视有限资源下计算效率的研究，已经有一些成果。对于计算机资源相对缺乏的发展中国家来说，更应重视这一研究方向。

还应当指出，通常的计算复杂性研究都是以某种基本操作时间为时间单元，这样得到的计算时间只是虚拟时间。虚拟时间与实际运行时间开销并不是简单的线性关系，要想建立真正有实际价值的效率估计理论，还要与机器的结构、通信联系起来考虑。本文一开始就提到目前人工智能的复杂性理论结果与实际智能系统的实现距离太大，不能起到像卡诺的热机效率理论那样的作用；而从事模式识别与智能程序设计的工程技术人员又只是从经验出发不自觉地修改问题描述、选择易解问题和控制问题规模。今后的努力方向应当是使这两部分工作更有效地结合，以面向实际的复杂性理论为指导设计出更有效的、能解决大的应用问题的智能程序与智能系统。

## 参考文献：

[1] RICH E. Artificial intelligence[M]. New York: McGraw-Hill, 1983.

[2] BYLANDER T. Tractability and artificial intelligence[J]. Journal of Experimental & Theoretical Artificial Intelligence, 1991, 3(3): 171-178.

[3] BLUM L, SHUB M, SMALE S. On a theory of computation and complexity over the real number, np-completeness[J]. Bulletin of the American Mathematical Society, 1989, 21(1): 1-16.

[4] WOZNIAKOWASKI H. A survey of information-based complexity[J]. Journal of Complexity, 1935, 1(1): 11-44.

[5] TANG Z, LI G J. On the reduction of connections in Hopfield associative memories[C]// China 1991 International Conference on Circuits & Systems. Piscataway: IEEE Press, 1991: 774-779.

[6] LI G J, WAH B. Parallel iterative refining a* search: an efficient search scheme for solving combinatorial optimization problems[C]// International Conference on Parallel Processing. Piscataway: IEEE Press, 1921: 608-615.

[7] DAVIS E . Constraint propagation with interval labels[J]. Artificial Intelligence, 1987, 32(3): 281-331.

[8] PEARL J. Hcuristics: intelligent search strategies for computer problem solving[M]. New Jersey: Addsion-Wesley, 1984.

[9] VALIANT G. A theory of the learnable[J]. Communication of the ACM, 1984, 27(11): 1134-1142.

[10] BERWICK R, BARTON G, RISTAD E. Computational complexity and natural language[M]. Massachusetts: MIT press, 1987.

[11] BYLANDER T, ALLEMANG D, TANNER M C, et al. The computational complexity of abduction[J]. Artificial Intelligence, 1991, 49(1-3): 25-60.

[12] KAUTZ H A, SELMAN B. Hard problems for simple default logics[J]. Artificial Intelligence, 1991, 49(1-3): 243-279.

[13] COOPER G F. The computational complexity of probabilistic inference using bayesian belief networks[J]. Artificial Intelligence, 1990, 42(2-3): 393-405.

[14] JUDD J S. Neural network design and the complexity of learning[M]. Massachusetts: MIT press, 1990.

[15] TAYLOR W M, CHEESEMAN P, KANEFSKY B. Where the really hard problems are[C]// International Joint Conference on Artificial Intelligence [S.l.:s.n.], 1991: 331-337.

[16] LI G J, WAH B W. Computational efficiency of parallel combinatorial OR-tree searches[J]. IEEE Transactions on Software Engineering, 1991, 16(1): 13-31.

[17] WAH B W, LI G J. How good are parallel and ordered depth-first searches[C]// International Conference on Parallel Processing. Piscataway: IEEE Press, 1986: 992-999.

发表于《模式识别与人工智能》1993年第2期

# 求解可满足性（SAT）问题的算法综述

顾钧[1,2]，李国杰[2]

1. 卡尔加里大学电机与计算机工程系，加拿大；

2. 国家智能计算机研究开发中心，北京 100080

**摘　要：** 可满足性（SAT）问题是人工智能、数理逻辑和计算理论中的一个最基本的问题。传统的方法将可满足性问题当成离散的有约束的判定问题处理。在这篇文章中，综述了求解SAT问题的几类主要算法，重点讨论用优化方法求解SAT问题。

**关键词：** 人工智能；局部搜索；全局搜索；组合优化；运筹学；可满足性问题

# Algorithms for the satisfiability problem: a survey

Jun GU[1,2], Guo-jie LI[2]

1. Department of Electrical and Computer Engineering, University of Calgary, Canada T2N 1N4

2. National Research Center for Intelligent Computing Systems, Beijing 100080

**Abstract:** The satisfiability (SAT) problem is a fundamental problem in artificial intelligence, mathematical logic, constraint satisfaction, VLSI engineering, and computing theory. Methods to solve the satisfiability problem play an important role in the development of computing theory and systems. Traditional methods treat the SAT problem as a discrete, constrained decision problem. In this paper, we survey several major categories of SAT algorithms with the emphasis on introducing efficient optimization approaches to the SAT problem.

**Key words:** artificial intelligence, local search, global search, combinatorial optimization, operations research, satisfiability (SAT) problem

## 1 引言

问题求解是人工智能的核心问题和主要应用。演绎、推理、规划等都属于问题求解技术，自然语言理解、图像理解、专家系统、定理证明和机器人规划等应用都采用了问题求解技术。一般来讲，用人工智能技术求解的问题包括两大类：判定问题与优化问题，多数是求满足某些约束条件的可行解，即带约束的判定问题。传统的人工智能采用状态空间与问题归约的搜索方法、几乎处处碰到指数爆炸，难以解决大规模的实际智能应用问题。近年来国外一些学者从运筹学的角度出发，将离散量与连续量的计算方法结合起来，借用运筹学计算中非常成熟的优化技术来解决人工智能中传统的判定问题，为实际上有效地解决大规模的困难问题提供了一条途径、可满足性（SAT）问题是最典型的 NP 完全问题，本文通过介绍解决 SAT 问题的各种有效算法说明在符号处理中引入运筹学的优化技术和已知的数学计算方法将大大扩展可解的智能应用问题的规模，促进智能技术真正实用化。

## 2 可满足性问题

可满足性问题包含 3 个部分。

- $m$ 个变量的集合：$x_1, x_2, x_m$。
- 基本式 W(literals)的集合：一个基本式是一个变量或它的非。
- $n$ 个不同子句的集合；$C_1, C_2, C_n$ 每一子句由用逻辑或操作（V）连接的基本式组成。

所谓可满足性问题就是确定是否存在一组变量的逻辑赋值（真或假）使得合取范式（CNF）$C_1 \wedge C_2 \wedge \ldots \wedge C_n$ 的值为真，若 CNF 公式可为真，则称此 SAT 问题有解，否则称为无解。

可满足性问题是解决数理逻辑、推理、机器学习、约束满足等许多理论与实际问题的基础性问题。SAT 问题是一大类 NP 完全问题的核心[1-2]。求解 SAT 问题在研究人工智能和计算理论与系统中起很重要的作用，因此，设计有效的算法求解SAT问题一直受到人们相当的重视。

在过去的 20 年里，已经提出了许多求解 SAT 问题的算法。根据变量域（离散或连续）和有无约束。SAT 算法大致可分成 4 类：离散约束算法、离散无约束算法、约束规划算法和无约束规划算法。实际上，大多数串行 SAT 算法可以映射到并行计算机从而得到 4 类对应的并行 SAT 算法。大多数已有的 SAT 算法属于上述 4 类之一或是它们的某种组合[3-5]，从另一角度来看，基本上只有两类 SAT 算法，即完备的算法与不完备的算法。完备的 SAT 算法既能检验其可满足性又可检验其不可满足性，即对于一个给定的 CNF 公式来说，可以找到适当的变量赋值使其为真或者证明不存在这样的赋值。而不完备的 SAT 算法只能对某些 CNF 公式找到一个随机的解，验证可满足性，若一 CNF 公式无解，则不能给出任何回答。

## 3 基本的 SAT 问题实例模型

在人工智能与计算理论中，所谓“问题”是有相同描述的各种各样具体问题实例（Instance）的总称。问题实例通常用来研究计算复杂性和测试算法性能。下述几个问题模型在 SAT 算法研究中经常用到，前面两个模型产生随机 CNF 公式。随机产生问题实例是算法设计的一种标准技术[6-9]。

① 匀称 $l$-SAT 模型。在此模型中，一个随机产生的 CNF 公式由独立的随机产生的子句组成。每个子句都包含某些变量或它的非，且每一基本式在一个子句中只出现一次，即对于 $x_i \in X$、$\Pr[Q = x_i] = 1/|X|$，这里 $|X|$ 表示变量集合 $X$ 中元素的个数。基本式 $Q$ 取非的概率为 $p$。这一模型在文献［3-5, 8］中采用。

② 平均 $l$-SAT 模型。在此模型中，一个随机产生的 CNF 公式由独立的随机产生的子句组成。子句中每一变量和它的非均以 $p/2$ 的概率出现，一个变量在一个子句中不出现的概率为 $1-p$。一个子句中基本式的平均数近似为 $l$。这一模型在文献［3-5, 7-11］中采用。

③ 有一定结构的 SAT 模型。此模型用于研究 SAT 算法解有一定结构的问题的性能，例如图着色问题和 $n$ 皇后问题[3-5,8]。

④ 实际问题实例模型。许多学者对实际

应用领域中提出的问题实例进行研究。

为了理解 SAT 问题实例的计算难度，需要用到两个基本的相互联系的概念。一个是 SAT 问题实例的困难程度（Hardness），另一个是 SAT 问题实例的难易分布(Hard-and-easy Distribution)[5]，一个 SAT 问题的问题空间可分成两部分：一部分包含全部有解的实例，另一部分包含全部无解的实例。对于随机产生的 CNF 范式来说,一个问题实例的困难程度与它离这两部分的分界的远近有关，离分界面越近的实例越困难。与一个问题的计算复杂性类似,一个问题实例的困难程度是与算法无关的本质特征,它取决于 SAT 问题模型。用平均 $l$-SAT 模型产生的实例中包含的有解实例要比匀称 $l$-SAT 模型产生的有解实例少得多。一般来讲，对于一个随机产生的具有 $n$ 个子句、$m$ 个变量、每一个子句平均 $l$ 个基本式的 CNF 公式来说，当 $n/m$ 增加或 $l$ 减少（$l \geqslant 3$）时，其困难程度增加。也就是说，较少基本式与较多子句使得所有子句都为真的可能性减少。3-SAT 问题被公认为是很困难的问题之一。目前已有大量研究致力于设计有效的算法解决 $l$ 小而 $n/m$ 大的 SAT 问题。

SAT 问题实例的难易分布指的是 SAT 问题实例的分布情况。这一分布不仅与 SAT 问题模型有关，而且与解决问题的算法有关。从某一算法得出的难易分布一般不能推广到另一种算法。一般的计算复杂性理论主要关心最难的实例的困难程度，但当我们研究算法的效率时，应当注意实例的难易分布，弄清楚每一算法适用于哪一部分问题实例。

## 4　离散约束算法

这一类算法将 SAT 问题当作有约束条件的判定问题，采用带约束条件的离散搜索与推理来求解 SAT 问题。一种最直接的方法是枚举所有可能的变量赋值，检查是否有一组赋值使给出的 CNF 公式为真。这种方式是不可接受的，因为其计算时间随问题规模增大而指数增长。在各种已提出的求解技术中，例如相容性算法、回溯算法[12]、项重写、产生式系统、多值逻辑、归结[13]和独立算法等，归结是最广泛使用的方法。

Davis-Putnam 算法(以下简称 D-P 算法)、简化的D-P算法及变量严格有序的D-P算法都是归结原则的特殊形式。D-P 算法已经成为实际求解 SAT 问题的主要方法。在某些限制条件下，对于随机产生的具有 $n$ 个子句、$m$ 个变量、每个子句含 $l$ 个基本式的 CNF 公式来说，已证明 D-P 算法的平均时间复杂性为 $O(n^{o(m/l)})$[7]。如此高的计算复杂性及其限制条件使 D-P 算法的应用范围较小。一般来讲，基于归结的方法不能处理中等规摸以上的推理问题。

## 5　局部搜索算法

对于局部搜索来说，若将一个 CNF 公式中值为假的子句的数目作为目标函数的值。SAT 问题就变为无约束的离散优化（极小化）问题[4-5,8,10,14]。局部搜索（即局部优化）是 20 世纪 60 年代中期就已提出的一种用于对付组合优化问题 NP 困难性的技术。它是处理组合优化问题少数几种成功的方法之一。局部搜索是在离散的搜索空间中运用连续量优化技术的一种初级形式。

在局部搜索中，任意一组变量赋值都看成是可行解域中的一个解点。给定一个具有目标函数 $f$ 和可行解域 $R$ 的优化问题[1]，对于每一个解点 $x_k \in R$ 来说，一个典型的局部搜索算法需要一个预先定义的邻域 $N(x_k) \subset R$。每当得到一个当前的解点 $x_k$，就在集合 $N(x_k)$ 中搜索一

[1] 不失一般性，本文只讨论求极小值问题

个满足 $f(x_{k+1})<f(x_k)$ 的新解点。如果这样的解点存在，它就变成一个新的当前解点，再继续作局部搜索；否则 $x_k$ 就保留作为对于 $N(x_k)$ 的局部最优解点。局部搜索可能产生一组可行解点，它们中每一个相对其邻域都是局部最优解点。将局部搜索方法运用于一个具体问题，我们只需要规定邻域和执行一个随机过程去产生一个初始的可行解。

## 5.1 运用局部搜索求解 SAT 问题

算法 1 表示了求解 SAT 问题的典型的局部搜索算法。它包含一个初始化阶段与一个搜索阶段。算法一开始产生一个 SAT 问题实例并选择一个初始解点，以值为假的子句数目作目标函数值。在每一次迭代中，执行 test-swap() 函数检查将 $x_i$ 改变赋值其目标函数值能否进一步变小，如果能，则执行 perform-swap()过程并通过 evaluate-object- function()改进目标函数值。当找 SAT 问题实例的解到，则算法终结。实际上，在搜索过程中算法可能陷入某一局部最优点。为了提高算法的收敛性能，算法中需要增加局部“陷阱”处理程序 local-handler()。至今已提出许多局部搜索算法[4-5,8,10,14]。如果我们将局部搜索技术嵌入各种搜索框架，可以有效地检验相当多的 CNF 公式的可满足性和不可满足性。

表 1 给出了 D-P 算法及一个 SAT 1.7 算法[4-5,8,14]求解匀称 *l*- SAT 问题实例的有代表性的执行结果。从此表可以看出对这些随机给出的实例，SAT1 算法的执行速度要比 D-P 算法快几百到几千倍。在过去几年里，我们比较了各种 SAT 算法的性能，发现这些新 SAT 算法的执行速度比归结法、GSAT 算法[15]及内点法[11]高出几个数量级。

算法 SAT 1.0 如下。

```
procedure SAT 1.0 ( )
begin
    /* initialization */
    get_a _SAT_instance( );
    x0:= select _an_initial_point( );
    f:= compute_conflicts( x0 );

    /* search */
    k:= 0;
    while f ≠ 0  do
    begin
        for each variable i := 1 to m do
            /* if swap (xi, x̄i) does not increase f*/
            if test_swap (xi, x̄i) then
            begin
                xk+1:=perform_swap (xi, x̄i);
                f:=compute _conflicts (xk+1);
            end;
        if local then local_handler( );
        k:=k+1;
    end;
end;
```

我们也研究过某些局部搜索算法的平均时间复杂性[4-5,14]。对于随机产生的具有 $n$ 个子句、$m$ 个变量、每个子句 $l$ 个基本式的 CNF 公式来说，当 $l \geqslant \log m-\log\log m-C$ 及 $n/m \leqslant 2^{l-2}/l$ 时（C 为常数），SAT 1.1[2]算法检验可满足性的平均复杂性为 $O(m^{o(1)}n)$。当 $l \geqslant 3$ 且 $n/m \leqslant \alpha 2^l/l$（$\alpha$ 为小于 $l$ 的常数），SAT 1.2 算法检验可满足性的平均复杂性为 $O(m^{o(1)}n^2)$。当 $l \geqslant 3$ 且 $n/m \leqslant \alpha 2^l/l$ 时，SAT 1.3 算法的平均复杂性为 $O\left(\ln(m\log m)^2\right)$。这些结果表明，就验证可满足性而言局部搜索的

[2] 文中提的 SAT 1.1，SAT 1.7 等各种算法请参见[3-5,8].

表 1 D-P 算法及一个 SAT 1.7 算法求解匀称 *l*- SAT 问题实例的有代表性的执行结果

| Problems<br>Clause，Variable，Literal | | | Execution Time<br>Davin-Putnam Algorithm | | | | Execution Time<br>SAT 1.7 Algorithm | | | |
|---|---|---|---|---|---|---|---|---|---|---|
| *n* | *m* | *l* | S/F | Min | Mean | Max | S/F | Min | Mean | Max |
| 500 | 500 | 3 | 10/0 | 2.066 | 2.159 | 2.249 | 10/0 | 0.010 | 0.013 | 0.020 |
| 750 | 500 | 3 | 10/0 | 2.699 | 2.016 | 3.183 | 10/0 | 0.010 | 0.015 | 0.020 |
| 1 000 | 500 | 3 | 10/0 | 3.149 | 3.657 | 6.316 | 10/0 | 0.020 | 0.035 | 0.050 |
| 1 250 | 500 | 3 | 1/9 | 3.966 | 5.797 | 10.02 | 10/0 | 0.030 | 0.049 | 0.080 |
| 1 500 | 500 | 3 | 6/4 | 7.499 | 9.147 | 11.60 | 10/0 | 0.050 | 0.108 | 0.220 |
| 1 000 | 500 | 4 | 10/0 | 4.533 | 4.684 | 4.933 | 10/0 | 0.020 | 0.026 | 0.030 |
| 1 500 | 500 | 4 | 10/0 | 6.766 | 7.960 | 16.90 | 10/0 | 0.030 | 0.040 | 0.060 |
| 2 000 | 500 | 4 | 8/2 | 8.999 | 10.27 | 14.25 | 10/0 | 0.040 | 0.066 | 0.100 |
| 2 500 | 500 | 4 | 2/8 | 13.63 | 15.96 | 18.28 | 10/0 | 0.060 | 0.074 | 0.090 |
| 3 000 | 500 | 4 | 1/9 | 46.33 | 46.33 | 46.33 | 10/0 | 0.070 | 0.118 | 0.160 |
| 3 000 | 500 | 5 | 10/0 | 16.23 | 16.90 | 18.82 | 10/0 | 0.070 | 0.094 | 0.140 |
| 4 000 | 500 | 5 | 5/5 | 21.72 | 28.39 | 44.56 | 10/0 | 0.090 | 0.119 | 0.160 |
| 5 000 | 500 | 5 | 0/10 | >1 200 | >1 200 | >1 200 | 10/0 | 0.110 | 0.180 | 0.290 |
| 6 000 | 500 | 5 | 0/10 | >1 440 | >1 440 | >1 440 | 10/0 | 0.210 | 0.313 | 0.450 |
| 7 000 | 500 | 5 | 0/10 | >1 680 | >1 680 | >1 680 | 10/0 | 0.380 | 0.591 | 0.980 |
| 10 000 | 1 000 | 10 | 10/0 | 99.71 | 1018 | 103.1 | 10/0 | 0.030 | 0.047 | 0.070 |
| 12 000 | 1 000 | 10 | 10/0 | 122.1 | 124.3 | 126.7 | 10/0 | 0.030 | 0.073 | 0.100 |
| 14 000 | 1 000 | 10 | 10/0 | 140.7 | 145.2 | 148.7 | 10/0 | 0.040 | 0.077 | 0.150 |
| 16 000 | 1 000 | 10 | 10/0 | 165.1 | 167.1 | 168.7 | 10/0 | 0.060 | 0.098 | 0.140 |
| 18 000 | 1 000 | 10 | 10/0 | 185.5 | 188.6 | 189.9 | 10/0 | 0.060 | 0.136 | 0.210 |

注：表 1 在 SUN SPARC 2 工作站运行 D-P 算法和 SAT 算法的实际性能比较（10 个实例的平均时间，时间单位是秒，符号 S/F 表示在 120 ×*n*/*m* 秒时间内算法是否成功地找到解，S 为成功，F 为失败）

平均复杂性大大优于具有指数平均复杂性 $O(n^{o(m/l)})$ 的 D - P 算法。

## 5.2 从 SAT 到 CSP

SAT 问题是典型的约束满足问题（CSP）。传统的相容性算法通过局部一致性检查处理 CSP 问题，能力有限而且效率低。事实上这种方法连小规模的 CSP 问题都解决不了。如果我们将 CSP 问题中不一致的总量表示为一个目标函数的值，那么，与 SAT 问题一样，CSP 问题也就变为无约束的优化问题。因而许多已有的优化技术都可用来有效地解决 CSP 问题。迄今我们已经尝试解决了数百个困难的搜索问题。

*n* 皇后问题是一个典型的 CSP 问题。根据 Johnson 和 Kumar 的综述[16-17]，Sosic 和 Gu 最早采用局部优化算法求解 *n* 皇后问题。他们提出这一算法的 4 种基本版本。QS1 算法在 Utah 大学 AI 课程中第一次采用[18]。这种优化技术的求解效率大大高于传统的回溯技术。优化方法求解 *n* 皇后问题的规模已超过百万个皇后。而已报道的结果表明，在一定时间内采用回溯方法最多只能得到 97 个皇后的解。QS1 算法的修改版本 QS2、QS3、QS4 在 1989 年就已

提出[19-20]。1990 年 Minton 等人[21]报告了类似的用局部搜索求解 *n* 皇后的结果。

最近 Selman、Levesque 和 Mitchell 报告了 GSAT 算法的实验结果。根据 Selman[15,22]，用局部搜索求解 *n* 皇后问题的方法是他们的 GSAT 算法的思想源泉（原始动力），他们给出了一些用困难程度较低的 SAT 问题模型产生的易解问题实例的解答，这些问题模型中没有包含较难的问题实例，因而对 SAT 问题实例的困难程度做出了不正确的结论，从而导致对 GSAT 算法的性能也做了不恰当的评价。

## 6 全局搜索算法

在全局搜索或全局优化中，采用特殊的变换模型将离散的布尔空间 $\{0,1\}^m$ 上的 SAT 问题转换成实空间 $R^m$ 上的"连续量 SAT 问题"。因而具有离散判定问题形式的 SAT 问题转换成了连续量的无约束优化问题，可以采用许多已知的全局优化方法求解[3-4,8,10,23-24]。

### 6.1 全局优化

当人们处理一个需要选择一组相互关联的变量值的复杂判定问题时，一般应集中考虑一个或少数几个考核性能和评价决策质量的目标函数。通过目标函数可将搜索、判定与优化统一起来，优化过程的核心就是在满足对决策变量的限制条件下求目标函数的极值。

计算机实现重复的操作十分有效，大多数已有的优化算法都设计成一个逐步求精过程。寻找一个向量作为优化问题的解的典型过程如下：首先选择一个初始向量 $\boldsymbol{x}_0$，然后执行算法产生一个改进的向量 $\boldsymbol{x}_1$。继续这一过程，找到一系列逐步改善的向量 $\boldsymbol{x}_0,\boldsymbol{x}_1,\cdots,\boldsymbol{x}_k,\cdots$，直到达到最优解 $\boldsymbol{x}^*$。这是用局部搜索和全局搜索求解 SAT 问题共同的基本思想。

局部搜索和全局搜索都采用局部邻域搜索方法求解极小化问题，一个重要区别仅仅在于其目标函数减小的方式。在局部搜索中，只要目标函数值减小，新的解点就被接受，即类似于贪婪算法（Greedy Algorithm）；而在全局搜索中只接受能导致达到全局最优值的解点。在全局搜索的每一迭代步中都要利用已知的全局信息寻找这一步中目标函数的最优值，即沿这一迭代步中最优的解点指出的方向下降，从这一意义上说，局部搜索是无方向的定性的全局搜索，而全局搜索是有方向的定量的局部搜索，大多数全局搜索都要做连续量的定量计算。

本文讨论的搜索实质上是运筹学中的全局优化和组合优化的结合。人工智能中讨论的许多搜索及有关工作可以包含在传统的优化范围之内。本文论述的搜索是指问题空间中解点的迁移与"进化（Evolution）"。这一思想同样运用于模拟退火和遗传算法。而在经典的状态空间及问题归约搜索，即所谓与或树的搜索中，每一节点实质上代表问题空间的某个区域，可包含若干解点[25]。对优化技术不太熟悉的读者可参阅有关运筹学与非线性规划方面的教科书。

许多搜索问题本质上是离散的，怎样用连续量的优化技术解决离散的搜索问题？这是我们关心的焦点，目前已提出了几种方法[4]。第一种方法是直接将离散值当成问题的连续变量的约束条件。第二种方法是将离散值编码成一种成本函数，加到目标函数之中。第三种方法是给出大多数离散值作为对其值作特殊解释的函数的一个单独的实例。第四种方法是运用"载波原理（Carrier Principle）"实现一种通用搜索的混合模型，其方法如下。在远距离通信中，语音信号不能直接传送，在传送之前通过一个调制器将语音信号加到高频载波信号上；到了接收端，一个解调器将语音信号从载波中分离出来。这一载波原理启发了优化的混合模型，即把一个离散搜索问题放入一个具有连续变量的搜索空间，然后采用全局搜索的方法求解，最后将连续变量的结果“译码”变成对应的离散解，即原来的离散搜索问题的解。

## 6.2 用全局搜索求解 SAT 问题

最近几年来，不少学者采用约束规划技术求解 SAT 问题，发现这种方法比归结方法快[9,26-27]，但一般不具有鲁棒的收敛性，常常难以解决难的 SAT 问题。Hooker[9]报告了割平面法求解小规模 SAT 问题的计算结果，对于大规模的问题他的方法的计算行为很难预测。Kamath 等人采用内点法改进了 Hooker 的结果，他们证明某些大规模的 SAT 问题实例可以求解，但对于难的 SAT 问题实例来说，这一方法的收敛性较差。

在 1987 年和 1988 年两年内，几个通用的可满足性（Universal SAT）问题求解模型，UniSAT，已提出来，即把在$\{0,1\}^m$布尔空间的离散约束 SAT 问题转换成实空间 $R^m$ 上的连续无约束 UniSAT 问题。各种非线性规划算法都可用来求解 UniSAT 问题，并已导出许多系列连续 SAT 算法[3-4,8]。我们已在实际应用中测试了这些全局搜索算法，发现其性能大大优于现有的各种 SAT 算法与技术[3-4,8]。如果将全局搜索技术嵌入到各种搜索模式中，大量 CNF 公式都能较快地判定其有解或无解。

算法 2 给出了一个求解 SAT 问题的通用的优化算法 SAT 6.0[3-4,8,10,23-24]。大多数求解 SAT 问题的全局搜索算法可以从这一通用算法导出，这一算法包括以下几个步骤。过程 get_a_SAT_ instance()产生一个具有 $n$ 个子句、$m$ 个变量、每一子句平均 $l$ 个基本式的 SAT 问题实例，建立表示给定 SAT 问题的目标函数，一开始选择一个初始解点并计算其目标函数值，在每一步迭代的搜索过程中，用 test_min ()函数检查选择一些变量做优化目标函数能否达到这一步迭代的最小值（有时这一步可省去）。每一步迭代中，$y_{i(k)}s$ 可以是一个变量，即进行线性搜索（Line Search），也可以是几个变量，如多维搜索或所有变量，即最速下降搜索。如果选择的变量目标函数能达到这一步的最小值，则通过 perform_min()实现极小化操作并用 evaluate_object_function()计算 $f$ 的值，即得到这一步迭代 $f$ 的最小值。当搜索接近于解点时，调用 approximate()函数做连续量 $Y_{k+1}$ 到离散量 $X^*$ 的转换。由于每一步迭代只考虑部分变量并且在一个邻域内搜索，全局搜索也有可能陷入局部最优点，为了提高算法的收敛性，需要采用某种跳出局部最优的机制。当找到 SAT 问题的一个解，算法结束。我们已研究了包括基于最速下降法（SAT 7 系列）、改进的最速下降法（SAT 8 系列）、牛顿法（SAT 10 系列）、准牛顿法（SAT 11 系列）、下降法（SAT 14 系列）、割平面法（SAT 15 系列）、共轭方向法（SAT 16 系列）、椭球体法（SAT 17 系列）、同伦法（SAT 18 系列）、布尔差分法（SAT 20 系列）及其他技术的各种新的 SAT 算法[8]。实践表明这些全局搜索算法可以较快的计算速度和满意的收敛性能求解规模非常大或较难的 SAT 问题。算法 SAT 6.0 如下。

```
procedure SAT 6.0( )
begin
    /* initialization */
    get_a _SAT_instance( );
    Y0 := select _an_initial_solution( );
    f(Y0 ):= evaluate_object_function( );

    /* search */
    k := 0;
    while not(solution - found( )) do
        for some  y_i(k)s ∈ Y_k
        begin
            /* minimizer */
```

```
                if, test_min(f(y_{i(k)}s))  then
                begin
                        Y_{k+1}:=perform_
min(f(y_{i(k)}s));
                        f(Y_{k+1}):=evaluate_  ob-
ject_function ();
                end
                if close_to_solution( ) then X*:
= approximate(Y_{k+1});
            end
            /*local handler*/
            if local then local_handler( );
            k:=k+1;
        end;
    end;
```

我们已研究了一些求解 SAT 问题的全局搜索算法的收敛性能和平均时间复杂性，并且证明了在某些条件下，最速下降法具有线性收敛率 $\beta<1$，牛顿法有二阶收敛率。对于 $m$ 个变量的 SAT 问题来说，协调下降法的收敛率近似为 $1-\beta/m$。我们还证明了对于随机产生的CNF公式来说，若 $t\geqslant 3$，$n/m\leqslant\alpha 2^t/t$，协调下降的平均时间复杂性为 $O(\ln(m\log m)^2)$ [14,23]。

## 7 求解 SAT 问题的并行算法

并行处理技术已用于加速 SAT 问题的求解。并行的离散约束算法包括并行相容性算法[28-31]，布尔差分法[32]以及并行 PROLOG 语言等。一些并行局部搜索算法也已实现[5,8,28,32]。Kamath 等人[11]在 KORBX(R)并行向量机上实现了用内点算法求解 SAT 问题。

## 8 结论

SAT 问题是 NP 困难问题的核心，传统的方法把 SAT 问题当作离散的约束判定问题处理，近年来采用优化技术求解 SAT 问题已成为一个很活跃的领域，这方面的研究促进了运筹学技术、组合优化技术与有效的搜索算法设计相互结合与集成。由于线性与非线性规划的理论与方法都相当成熟，从工程应用角度来看，将优化技术与基本的搜索框架集成起来，是一种解决搜索问题的通用而且自然的方法。人工智能的主要任务是问题求解，我们不应当作茧自缚，将问题求解技术局限于离散的非数值的处理。连续量计算与离散符号处理的结合、现有数学成果与智能技术的结合将大大拓宽人工智能应用的范围。达尔文进化论的精髓就是“杂种优势”，如何在问题求解中发挥杂种优势是我们需要重视的新课题。

## 参考文献：


[1] COOK S A. The complexity of theorem-proving procedures[C]// The 3rd Annual ACM Symposium on Theory of Computing. New York: ACM Press, 1971: 151-158.

[2] GAREY M R, JOHNSON D S. Computers and intractability: a guide to the theory of NP-completeness[M]. New York: W. H. Freeman and Company, 1979.

[3] GU J. Global search for satisfiability (SAT) problem[J]. Submitted for Publication, 1989.

[4] GU J. On optimizing a search problem[J]. World Scientific, 1992, 1: 63-105.

[5] GU J. Local search for satisfiability (SAT) problem[J]. IEEE Transactions on Systems, Man and Cybernetics, 1993, 23(4): 1108-1129.

[6] CORMEN T H, LEISERSON C E, RIVEST R D. Introduction to algorithms[M]. Massachusetts: MIT Press, 1990.

[7] GOLDBERG A, PURDOM P, BROWN C. Average time analysis of simplified davis-putnam procedures[J]. Information Processing Letters, 1982, 15(2): 72-75.

[8] GU J. How to solve very large-scale satisfiability (vlss) problems[R]. [S.l.]: Present in part in Gu J, Benchmarking SAT Algorithms, Technical Report UCECE-TR-90-002,1990.

[9] HOOKER J N. Resolution vs. cutting plane solution of inference problems: some computational experience[J].

Operations Research Letters, 1988, 7(1): 1-7.

[10] GU J. Efficient local search for very large-scale satisfiability problems[J]. ACM SIGART Bulletin, 1992, 3(1): 8-12.

[11] KAMATH A P, KARMARKER N K, PAMAKRISHNAN K G, et al. Computational experience with an interior pointalgorithm on satisfiability problem[J]. Annals of Operations Research, 1990, 25: 43-58.

[12] BROWN C A, PURDOM P W. An average time analysis of backtracking[J]. SIAM Journal on Computing, 1981, 10(3): 583-593.

[13] GENESERETH M R, NILSSON N J. Logical foundations of artificial intelligence[M]. San Francisco: Morgan Kaufmann Publishers, 1987.

[14] GU J, GU Q P. Average time complexities of several local search algorithms for the satisfiability Problem[R]. [S.l.]: Technical Report UCECE-TR-91-004, Submit for publication, 1991.

[15] SELMAN B, LEVESQUE H, MITCHELL D. A new method for solving hard satisfiability problems[C]// The 10th National Conference on Artificial Intelligence. [S.l.:s.n.], 1992: 440-446.

[16] JOHNSON W L. Letter from the edior[J]. ACM SIGART Bulletin, 1991, 2(2): 1.

[17] KUMAR V. Algorithms for constraint-satisfaction problems: a survey[J]. American Association for Artificial Intelligence, 1992, 13(1): 32-44.

[18] SOSIC R, GU J. How to search for millions of queens[R]. University of Utah: Technical Report UUCS-TR-88-008, Dept. Of Computer Science, 1988.

[19] SOSIC R, GU J. Fast search algorithms for the n-queens problem[J]. IEEE Transactions on Systems Man & Cybernetics, 1991, 21(6): 1572-1576.

[20] SOSIC R, GU J. Efficient local search with conflict minimization: a case study of the n-queens problem[J]. IEEE Transactions on Knowledge and Data Engineering, 1994, 6(5): 661-668.

[21] MINTON S, JOHNSTON M D , PHILIPS A B, et al. Solving large-scale constraint-satisfaction and scheduling problems using a heuristic repair method[C]// The of the 8th National Conference on Artificial Intelligence. [S.l.:s.n.], 1990.

[22] SELMAN B. Private Communication[Z]. 1992.

[23] GU J, DU D Z. Mathematical property of the unisat problem[J]. Ongoing Research, 1990, present.

[24] GU J. The unisat problem models (appendix)[J]. IEEE Transactions on Pattern Analysis and Machine Intelligence, 1992, 14: 857 -865.

[25] LI G J, WAH B W. Computational efficiency of parallel combinatorial or-tree searches[J]. IEEE Transactions on Software Engineering, 1990, 16(1): 13-31.

[26] HOOKER J N. A quantitative approach to logical inference[J]. Decision Support Systems, 1988, 4(1): 45-69.

[27] JEROSLOW R G. Computation-oriented reductions of predicate to propositional logic[J]. Decision Support Systems, 1988, 4(2): 183-197.

[28] GU J. Parallel SAT algorithms and architectures[Z]. 1985.

[29] GU J, WANG W, HENDERSON T C. A parallel architecture for discrete relaxation algorithm[J]. IEEE Transactions on Pattern Analysis and Machine Intelligence, 1987, PAMI-9(6): 816-831.

[30] GU J, WANG W. VLSI architecture for discrete relaxation algorithm[M]. New York: Oxford University Press, 1990: 111-136.

[31] GU J, WANG W. A novel discrete relaxation architecture[J]. IEEE Transactions on Pattern Analysis and Machine Intelligence, 1992, 14(8): 857-865.

[32] SOSIC R, GU J, JOHNSON R R. The unison algorithm: fast evaluation of boolean expressions[J]. ACM Transactions on Design Automation of Electronic Systems, 1996, 1(4): 456-477.

发表于《计算机学报》1995年第9期

# 技术综合集成在模式识别中的应用

张永慧[1,2]，刘昌平[1,2]，罗公[1,2]，李国杰[1,2]

1. 中国科学院计算技术研究所；

2. 智能计算机研究开发中心

**摘　要**：介绍了应用综合技术集成的方法解决印刷体汉字识别系统误识率太高的重大难题，并通过集成系统的实践证实了其技术集成优势。由于识别方法的互补效应，不但提高了识别的正确率，而且使误识率得到大幅度的降低。采用该集成办法研制的系统，经过100万字的实际文章的测试，系统的识别率超过98%，误识率小于0.3%，尤其是汉字的误识率小于0.1%。

**关键词**：图像处理；特征抽取；汉字识别

# Integrating comprehensive techniques in pattern recognition

Yong-hui ZHANG[1,2], Chang-ping LIU[1,2], Gong LUO[1,2], Guo-jie LI[1,2]

1. National Research Center for Intelligent Computing Systems, Beijing 100080

2. Institute of Computing Technology, The Chinese Academy of Sciences, Beijing 100080

**Abstract**: High error rate is an essential problem for an OCR system. It is difficult to develop an OCR system with both high correct rare and low error rate. This paper introduces an integrated system for printed Chinese character recognition. It integrates three recognition methods into an OCR system. Because the three methods are complementary each other, the integrated system achieved high correct rate and very low error rate. A rest about 1 000 000 printed Chinese characters shows that the correct rate is above 98% and the error rate is less than 0.3% especially the Chinese character error rate is less than 0.1%.

**Key words:** image processing, feature extraction, Chinese character recognition

## 1　问题的提出

“文字识别”是模式识别研究的一个重要的应用领域，是当今世界上非常活跃的研究课题，汉字识别又是文字识别研究中最难的研究课题，其字符集是西文符号量的几百倍。由于世界上使用汉字的人数占总人数的四分之一，所以在未来信息化时代，汉字信息的自动录入

本课题得到 883 高技术基金资助

具有重要的实用价值。我国的“印刷汉字识别技术”已从纯理论和方法的研究，发展到面向实用的技术研究。有几种识别系统已面世，但在实用中尚还存在着诸多问题，除需要方便、省时、友好的交互界面外，其关键仍是自动录入的质量，尽管对质量好的印刷汉字其识别率已达到 97%~98%的高水平。然而，面对大量的不同字体、字号、数字、符号，不同的印刷厂家，印刷在不同的纸质上的各种各样的文章，有的识别率可能在 95%~96%，甚至可能还低。录入效果最好的一篇 5 000 字的文章也需要查出100个错字，再经过人工输入计算机，这就提出一个校对、编改的问题，它直接影响到自动录入系统的总体效率，影响到高技术产品的推广应用。

解决系统的应用问题，一是提高系统的技术指标，最好达到 100%的正确识别率；另一是采取一些补救措施，节省校对、编改的时间。前者就目前计算机硬、软件的技术水平以及录入文章的印刷质量（就多数印刷质量而言）难以实现，后者就是本文中采用“高新技术集成”的观点，在保证进一步提高识别率的基础上，大幅度地将误识字变成拒识字，节省了人工校改的时间，提高系统的实用水平。这是通过两次评测的统计分析[1-2]提出的，并通过“集成式印刷体汉字识别系统”（简称集成系统）的实践得到了证实。新系统提高了识别正确率、使误识率大幅度地降低，实现了识别技术的重大突破。在此基础上为用户提供一个“拒识符”引导、不用原稿的校对编改环境，大大缩短了校编时间。

## 2 技术综合集成的方案设计

作为在微机上开发的应用软件，其系统的硬件配置和集成环境应有一个基本规定，其原则一方面从当前用户能接受并熟悉的机型，并有超前的选型考虑，另一方面从软件运行的条件、环境以及要达到的适当速度考虑资源的配置，量后确定的基本机型配置如下。

- 机型。微机系统，以 PC-486/33 的微机为基本机型，内存资源至少 4 MB，适应常用的几种汉字系统。

- 配套设备。输入设备：HP ScanJet II p 图像扫描仪 300DPI，16 级灰度，256 级亮度可调，除键盘外，配备鼠标；输出设备：显示器 VGA 显示方式，EPSON LQ-1600K 打印机。系统结构的设计：采用开放式的环境，模块化的结构。

其方案特点如下。

（1）影响关键技术指标的核心模块：切分和识别模块都是通过严格的测试、综合评比选出的，它们是国内多年研究成果中的精华，切分的优势和识别模块的互补，使集成系统具备了单个系统无法实现的技术指标：在短期内，使系统进一步提高识别率，特别是使误识率大幅度降低。

（2）影响系统使用效率的人机交互界面，是以鼠标驱动的自动、半自动、手动相配合交互方式，尤以图像处理方便为特点，配有文字行倾斜自动矫正，文章自动分栏、画框，多文字框自动连接、切分和识别一次完成，还备有一批量图像的自动识别，以尽量简化人工操作。

（3）集成后处理对识别子模块的判决是采用综合评分的办法，利用上下文信息和语言学知识，利用对“半词”的特殊处理，制定的一套综合体现“高频优先”和“整体优先”的评分原则，确保误识率再度降低。

（4）为方便用户最后的文字处理，系统还为输出正文设计了校改环境，它是图、文并存的一种人机交互的编校软件，缩短了编改、校审的时间。

（5）模块化的图形窗口控制界面，开放式体系结构为系统各模块的集成提供了方便的接口环境，利于模块的更新、产品的换代。

以下分别介绍各部分功能和实现的办法。

## 3 交互式界直设计

交互式界面是人机交互的图形窗口，它主要完成系统环境的设置以及识别前的各项处理。这部分是直接与用户见面，为用户服务的多个模块。比如系统的设置，根据系统所配置的不同扫描输入设备，以配备相应的驱动模块，同样还需要选择用户系统中使用的中文系统等。

这部分主要功能是图像处理。其中还包括扫描图像的灰度、分辨率的选择。这是设备配有的功能，需要提供用户来选择和进行参数的调整。主要有 4 个模块实现识别的前处理，如图 1 所示。

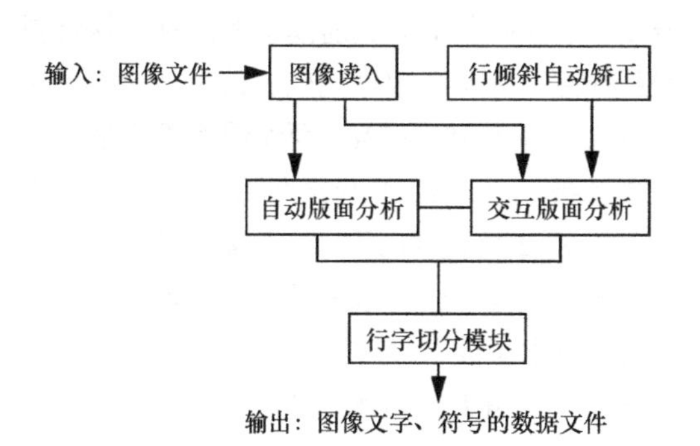


**图 1 识别前的图像处理**

（1）图像读入模块：通过扫描输入设备直接读取图像文件，或将存盘的图像文件读取，为加速处理提供扩展内存空间作为图像存储空间，并同时在显示器上显示压缩、非压缩两种图像窗口相对应的画面。

（2）版面倾斜的自动校正模块：它是对整幅画面中的文字行倾斜，通过计算其倾斜角来移动整个画面，根据识别的技术要求，其行倾斜自动矫正的最大角度为 1.5 度，太大角度的矫正会影响系统的识别率。

（3）版面分析模块：由于扫描的图像文件不单纯是文字图像，还包括插图、分栏线、表格线、文稿的横、竖排版，有些需要录入的文章可能是报纸上的某段，受设备幅面的限制只能扫描一部分，为了满足各种各样文章的自动录入，这个功能模块设计得比较灵活，既可实现全自动，也可为用户选择手动、半自动。

- 自动：适用简单版面分析，不需人工干预，自动分栏，去掉画面，将文字块按文章的顺序连接起来。
- 半自动：对较复杂的版面或报纸上的文章，可先采用手动选择整篇或部分文章的外框，再交由自动版面分析。
- 手动：手动版面分析可以使用户任意选择文字框进行识别，这种手动录入的功能，对办公人员选择性的摘录是非常方便的。

除版面分析的功能外，还配备人机交互的调整功能，对自动版面分析的错误可以方便地修改、删除等。

（4）文字块的行、字切分模块：行、字切分模块是识别模块的前处理模块，它是一种特殊的图像处理模块，直接影响识别系统的技术指标。由于印刷排版的复杂性、多变性，汉字的分离字以及数字、英文字母、标点符号混排在汉字文件中，加之字体、字号的不同，印刷质量、油墨、纸张的不同，如此复杂的文字框内把每个文字、每个符号切分的正确无误也是尚未攻克的难题。这个模块的选择是通过 100 多篇各式各样的文字块的实测评比，选出最好的切分模块，可实现对横、竖排版的不同风格，不同行字间距、不同字号的各个文字框的切分，并能对“二、八、儿”等二分字、“三、川”等三分字做上标记，对汉字、标点符号、数字、英文字母混排做到正确切分。

## 4 识别技术的优选与综合集成

模式识别的方法在汉字识别的应用研究中已取得可喜的成果。在样本集的模式识别试验中，通过学习可以达到百分之百的识别正确率。在应用研究中已有多个产品上市，从单一字体到多字体混排，从识别铅印的书刊到各种报纸，甚至打印的文章，逐渐在增加功能，更新版本，但在技术上的突破均遇到困难。由于模式样板与实际的印刷文字总有着千差万别，包括扫描的文字周边，内部的噪声，文字的大、小变化，印刷造成的断笔、连笔以及扫描灰度、扫描分辨率选择不当都会造成文字模糊不清，使这些文字的实际特征与这些文字的样本特征显露出差别。这些差别对不同的特征选择也不相同，有些特征对这种差别不敏感，而有些特征对这种差别将会出现判别错误。

多年来的实验表明，任何一种特征的模式识别方法在实际应用中，都需要不断地学习、修补特征模式。然而通过学习、修正总不是解决问题的根本办法。如何突破这个技术难点，如何解决误识字的修改，是识别技术在应用中遇到的一大难题。

通过评测后[1-2]的分析，我们发现现有的几个识别系统对同样的文章其识别率相差不大，但误识字却极少有相同的字，即甲系统的误识字，而乙系统对该字识别是正确的，丙系统的误识字，甲、乙系统识别均正确，这就说明这几个系统的识别方法具有互补性。

本系统的识别子模块是采取评测选优的方法，通过100万字的实际文章评测，选出互补性最佳组合的3个子模块，实施综合集成。

参加评测的是采用不同识别方法的 4 个模块，单个识别模块（不计切分错）的识别结果比较见表1。

利用简单的三取二表决将 4 个识别模块组成 A、B、C、D 4 组组合如下。

- A 组：由 1#、2#、3#识别模块的组合。
- B 组：由 2#、3#、4#识别模块的组合。
- C 组：由 3#、4#、1#识别模块的组合。
- D 组：由 4#、1#、2#识别模块的组合。

4 种组合采用三取二的简单表决结果见表 2，可以看出 4 个识别模块的识别方法都具有互补性。从 4 组数据比较，其识别率都比单个模块提高了 1%～2%，而误识率都降低了一个数量级。从 4 组中选出最好的一组优化组合（见表 2 中的 D 组）作为集成系统的优选识别模块。表 2 还可以说明，采用多种识别方法的“技术集成”，对汉字识别系统向实用化发展是一个重大的技术突破。

**表 1 单个识别模块（不计切分错）的识别结果比较**

| | 1#识别模块 | 2#识别模块 | 3#识别模块 | 4#识别模块 |
|---|---|---|---|---|
| 总体识别率 | 97.83% | 97.02% | 96.57% | 96.45% |
| 误识率 | 2.04% | 2.98% | 3.42% | 3.48% |
| 汉字识别率 | 97.84% | 97.61% | 96.68% | 96.18% |
| 误识率 | 2.02% | 2.39% | 3.32% | 3.80% |
| 符号识别率 | 97.78% | 94.82% | 98.16% | 37.45% |
| 误识率 | 2.11% | 5.17% | 3.18% | 2.25% |

表 2　4 种组合采用三取二的简单表决结果

| | A 组表决结果 | B 组表决结果 | C 组表决结果 | D 组表决结果 |
|---|---|---|---|---|
| 总体识别率 | 98.66% | 98.50% | 98.64% | 98.74% |
| 误识率 | 0.33% | 0.35% | 0.23% | 0.23% |
| 汉字识别率 | 98.91% | 98.76% | 98.77% | 98.90% |
| 误识率 | 0.23% | 0.25% | 0.14% | 0.13% |
| 符号识别率 | 97.73% | 97.55% | 98.12% | 98.17% |
| 误识率 | 0.70% | 0.76% | 0.57% | 0.61% |

集成系统的识别模块是由 D 组的 3 个子模块和集成判别模块组成，见图 2。1#、2#识别模块是串行操作完成对切分后的每个图像文字、符号进行识别，包括抽取特征，与特征库比较、分类、判决、后处理等，先后产生两个识别结果文件。3#识别模块是利用 1#、2#识别模块的识别结果文件进行代码比校，对编码不同的文字及一些特殊的文字集进行再识别，产生 3#识别模块的识别结果文件。

图 2 展示了汉字识别子模块的集成框图。

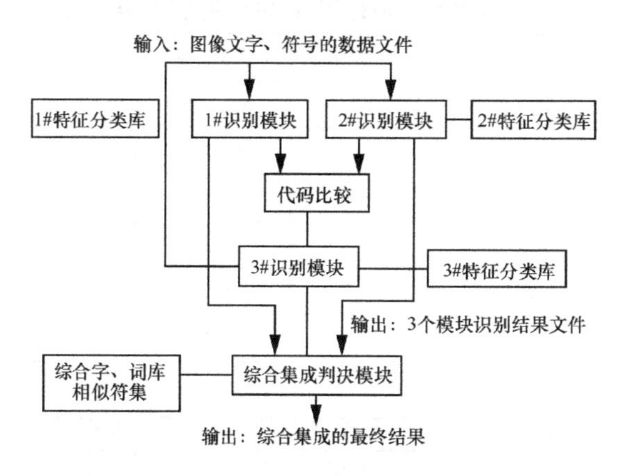


图 2　汉字识别子模块的集成框图示意

这里采用的 3 种特征分类的方法是北京信息工程学院的“点特征法”[3]，中国科学院沈阳自动化研究所的“改进的汉明网”及“文字周边链码”的方法[4]，和清华大学的“微结构分析法”[5]，这些方法都是经过实际的产品考验已成熟的分类方法，这 3 个采用不同方法分类的识别模块都是多年从事模式识别研究的结晶，此次集成的各识别模块在技术上又做了较大的改进，尤其对楷体的识别得到较好的解决。此外，对标点符号、英文字母的识别相对于集成前的水平，有了较大的提高，并加强了对易错的相似字的特殊处理。

集成判决模块不是采用简单的三取二表决，而是采用上下文信息关系和语言学知识，确保系统的识别正确率，最大限度地降低误识率。对 3 种识别结果参照可疑字符集和模糊分词进行综合评判，同时考虑了二字词的前后联词关系，还考虑到多字词、高频词、半独立词在上下文当中的特殊性质，并以一定的方式和强度体现在综合评分之中，在模糊分词算法中利用了动态规划技术把整体性质好的候选字优选出来，把整体性质差的候选字存疑，给出可疑字（或称拒识字）标记，确保识别的正确率，避免了误识字的遗漏。

对标点符号的识别结果，用了十几个小模块作分别处理。综合各种标点符号的一系列语言学知识、特殊的用法做出判决。如单、双引号的用法，小数点与人名姓间圆点的不同用法、破折号、删节号的用法等。

## 5　输出正文的校改环境

集成系统从图像输入到正文输出，中间生

成许多临时文件，其中包括切分生成的图像文字、符号的数据文件及索引文件，3 个识别模块识别的中间结果等，这些不可多得的数据应当充分被利用，为校对、编改服务，系统正是利用这些有利条件设计了一个查阅、编辑模块，它提供了一个带有图像窗口的正文编辑器。正文窗口提供识别结果的正文输出，图像窗口是提供图像文字、符号的显示窗口，此外又提供中间结果的文字框，显示对应文字集成前的 3 个识别结果。

这个编辑模块与一般的正文编辑不同，它有如下的一些特点。

（1）校对不必查原稿：利用切分后的文字图像，根据正文的游标位置，将图像文字显示在图像窗口内，并将前、后的各两个文字图像同时显示，节省了翻阅原稿的时间。

（2）“标记”引导修改：误识字变成拒识字，在正文中做上标记符，利用定义的特殊键，使游标跟踪“标记符”，不必查找，一键到位。

（3）文字框的一键输入：由于文字框跟踪游标位置显示 3 个识别结果作为候选字，根据图像窗口显示的文字图像，即可对照选键，将文字框正确的文字选进正文。

这种人机交互、图文并存的编辑功能，使用户的校对、编改省时、省力、方便适用，加快了对录入正文的校对、修改。

系统还为用户提供了文件的管理功能。在校对、编改完成后，可以将所有的中间文件，包括扫描的图像文件统一删除，节省用户的盘空间。

## 6 系统测试结果与比较

### 6.1 测试结果

简、繁体的集成系统分别通过 100 万字、30 万字的各种印刷文章的测试，简、繁两种字体集成系统的主要性能指标见表 3。

表 3 简、繁两种字体集成系统的主要性能指标

| | 简体集成系统 | 繁体集成系统 |
|---|---|---|
| 正确切分率 | 99.77% | 99.76% |
| 中间结果（不计噪音、切分错的统计结果） | | |
| 总计识别率 | 98.27% | 95.59% |
| 汉字 | 98.39% | 96.55% |
| 符号 | 97.81% | 89.08% |
| 总计误识率 | 0.16% | 0.31% |
| 汉字 | 0.09% | 0.20% |
| 符号 | 0.44% | 1.03% |
| 最后结果(包括噪音、切分错在内的统计结果) | | |
| 系统识别率 | 98.04% | 95.53% |
| 误识率 | 0.22% | 0.36% |

### 6.2 比较

我们对参加集成的 3 个识别模块在不计噪音、切分错的条件下，进行了识别的各项技术指标测试，将其最好结果与集成后的系统指标进行了比较，简体单个最好的识别模块和集成识别结果比较见表 4，繁体单个最好的识别模块和集成识别结果比较见表 5。

表 4 简体单个最好的识别模块和集成识别结果比较

| | 单个识别模式 | 简单表决集成 | 效果 |
|---|---|---|---|
| 总体识别率 | 97.83% | 98.74% | +0.94% |
| 误识率 | 2.4% | 0.33% | −1.81% |
| 汉字识别率 | 97.84% | 98.90% | +1.06% |
| 误识率 | 2.02% | 0.13% | −1.89% |
| 符号识别率 | 97.78% | 98.16% | +0.38% |
| 误识率 | 2.11% | 0.61% | −1.95% |

表 5 繁体单个最好的识别模块和集成识别结果比较

| | 单个识别模式 | 简单表决集成 | 效果 |
|---|---|---|---|
| 总体识别率 | 94.09% | 95.59% | +1.50% |
| 误识率 | 5.27% | 0.31% | -5.46% |
| 汉字识别率 | 95.30% | 96.55% | +1.25% |
| 误识率 | 4.70% | 0.20% | -4.50% |
| 符号识别率 | 85.88% | 89.08% | +3.20% |
| 误识率 | 13.12% | 1.08% | -12.09% |

从表 4、表 5 中的效果栏可以看出，识别率均以增加而误识率均以减少的百分比出现，集成系统无论是简体还是繁体，其简单表决的效果都比单个最好的识别模块的识别率高出一个百分点，而误识率几乎都降低一个数量级，尤其简、繁体的汉字误识率实现千分之一到二的重大的突破。实践证明了采用技术集成可以实现基本不需要校对的汉字自动录入，根据我们掌握的材料，简体的集成系统在识别率与误识率上已达到世界领先水平。

## 7 结束语

印刷体汉字识别集成系统并非是完美无缺的，由于为提高识别正确率，确保误识率降到最低，甚至要求基本无误识，在各识别模块都增加了特征库，相应地增加了程序的运行时间，造成识别速度比预计的降低些。这也是系统集成的缺点，若能进一步在特征级集成，会提高整体的识别速度，同时由于微处理机的速度不断提高，使识别速度的矛盾日趋缓和，特别是批处理功能的应用使自动化录入更加方便。

印刷体汉字识别集成系统是一个开放式系统，采用的模块化体系结构，利于增加功能，更新版本。未来系统会从技术集成向多功能、综合集成进一步发展，实现各种识别功能互补，实现汉字智能输入的综合集成，这将会为办公自动化提供更方便的输入手段。

目前集成系统实现了多个文字框的多字体、多字号混排的自动识别。但对中英文混排的科技文章识别，尚在切分上存在一定困难，多文种识别、表格的识别也是在实际应用中亟须解决的研究课题。

## 参考文献：

[1] 张永蕙, 刘昌平. 从评测看汉字识别研究中存在的问题[C]// 第四届全国汉字及汉语语音识别学术会议论文集. 杭州: 出版者不详, 1992.

[2] 张永慧, 刘昌平. 汉字识别系统评测结果分析和建议[C]// 第三届中文信息处理国际会议. 北京: 出版者不详, 1992: 407-412.

[3] 张忻中, 沈兰生, 等. 印刷体汉字文本识别系统[C]// 第四届全国汉字及汉语语音识别学术会议论文集. 杭州: 出版者不详, 1992.

[4] 杨力, 张佩芬, 武玉朴, 等. 多功能实用汉字识别系统[C]// 第四届全国汉字及汉语语音识别学术会议论文集. 杭州:出版者不详, 1992: 174-179.

[5] 吴佑寿, 丁晓青. 基于微结构的多字体多字号汉字识别系统[C]// 汉语与东方语言国际会议论文集. 出版地不详: 出版者不详, 1990.

第 5 章

# 并行算法、网格和生物信息学

# 并行算法研究从物理学家开始

算法（Algorithm）是计算机科学的核心内容，但在相当长的时间内，国内没有真正开展计算机算法研究。我在国内读大学和研究生时，书店里只有“计算方法”之类的教科书，讲的都是科学计算的精度和收敛性等，国外称为“数值分析（Numerical Analysis）”，大学里真正教“Algorithm”的老师寥寥无几。1988 年我参加国内第一次算法大会，发现参加会议的几乎都是数学界的学者，与我在美国学习研究的算法不是一回事。改革开放以后，一些访问学者回国，开始提倡研究“非数值计算”，才逐步重视真正的计算机算法研究。姚期智先生到清华大学任教以后，培养了一批算法研究的顶尖学者，为我国的算法研究做出了重大贡献。

中国学术界对计算机算法的忽视可能有历史原因。计算机理论的发展有两大基础，一个是数理逻辑，另一个是算法。欧洲擅长前者，美国擅长后者。我国早期从事计算机理论研究的学者，如胡世华、王湘浩等，都是做数理逻辑研究。后来从事计算机理论研究的主将，如周巢尘、李未等，也是跟随欧洲这条以逻辑为主的技术路线。姚期智先生回国以前，国内缺乏计算机算法研究的旗手。

我国的并行算法研究起步更晚。在推动并行算法研究的过程中，物理学家起了不小的作用。20 世纪 80 年代末，美国哥伦比亚大学的物理学家着手自己研制适合理论物理研究的专用超级计算机，李政道先生把并行计算这股风带到中国。他在北京建立了以理论物理为主要研究方向的中国高等科学技术中心，破格吸收我加入。应他的邀请，1987 年我在理论物理所专门讲授了一门并行计算课程，彭恒武、郝柏林等老科学家每堂课都坐在台下听课，我深深感受到老物理学家对并行计算技术的渴求。后来李政道先生和夏培肃、郝柏林教授合作，申请到国家自然科学基金重大项目，研制适合混沌计算的 BJ-01、BJ-02 并行计算机。

国内最早在大规模并行计算机上调试并行算法的科研人员中也有物理学家。1995 年“曙光 1000”大规模并行计算机做出来后，没有人会用，有些学者将“曙光 1000”大规模并行计算机比喻成一匹长了 32 条腿的马（它有 32 颗 CPU），难以驾驭。当时，中国科学院物理研究所的王鼎盛、生物物理研究所的陈润生、中国科学技术大学的陈国良、中国科学院软件研究所的孙家昶、中国科学院计算技术研究所的孙凝晖等科研人员成立了一个研究并行算法和并行软件的小组，构成一部“三套马车”。他们经常在计算所北楼 200 房间讨论怎么驯服这匹 32 条腿的“烈马”。应用、算法、软件和系统结构的核心骨干这么密切的合作，在国外也很难见到。这种合作产生了深远的影响，引领了国内并行算法和并行软件研究，为后来斩获超级计算机应用 Gordon Bell 大奖奠定了基础。若干年后，这个不到10人的跨学科小组出了4位院士。

算法研究与人工智能的发展相伴而行，自20世纪80年代以来，一批在人工智能学术圈内不受待见的学者另立门户，亮出“计算智能”的旗号，在人工智能的寒冬时期坚持研究以概率统计推理为主的算法，如人工神经网络、遗传算法、模拟退火、演化计算等，基于概率统计的算法在机器学习等领域显示了巨大的威力，为今天人工智能的复兴做出了重大贡献。我回国后指导学生做的研究大多也是基于概率统计的算法，我的学生姚新后来成为国际上演化计算的佼佼者。

我担任计算所所长后不久，通过与中国科学院生物物理研究所陈润生研究员（后来成为中国科学院院士）合作，开始招收生物信息学方向的研究生，后来发展成国内生物信息学领域颇有影响的一支团队，连续在 *Nature* 上发表多篇论文，他们开发的 Pfind 软件已成蛋白质结构分析的主流工具软件。跨学科研究大家都在讲，但真正做起来难度很大。在计算所做的生物信息学研究，主要是算法研究和计算机模拟，生物学界称为“干实验”，必须等生物学家真正做完对应的“湿实验”验证了模拟结果，才能发表论文。因此，开始几年只能做配角，求生物学家做实验。我的学生卜东波曾得到加拿大皇家科学院院士、滑铁卢大学著名计算机科学家李明教授的指导，在算法研究上颇有造诣，他们做的蛋白质结构预测几乎完全正确，难怪他在 *Nucleic Acids Research* 上发表的论文有700多次引用。

本世纪初，中国科学院计算技术研究所率先开展了网格（Grid）研究，学术带头人是徐志伟。本章收录了一篇关于网格的论文，反映了计算所当时的研究成果。网格计算走的是学院派的路子，在标准规范上花费了大量的功夫，设定的目标很宏伟，要在跨平台、跨组织的复杂异构环境中共享资源和协同解决问题，但实际上能做到的主要是高性能科学计算。后来兴起的云计算，吸收了网格计算的一些概念和可推广的技术，由企业牵头走了一条更现实的路子，按用户的需求提供共享资源服务，现在已成为主要的信息基础设施。

发表于 Computer Software and Application Conference论文集, 1985年

# Parallel processing of serial dynamic programming problems

Guo-jie LI, Benjamin W. WAH

Department of Electrical and Computer Engineering and the Coordinated Science Laboratory,

University of Illinois at Urbana-Champaign, Urbana, IL 61801

**Abstract:** Dynamic programming problems are classified into monadic-serial, polyadic-serial, monadic-nonserial, and polyadic-nonserial formulations. Problems in serial formulation can be implemented easily in systolic arrays, while problems in nonserial formulations may have to be transformed into serial ones before efficient implementations can be found. This paper presents parallel architectures for problems formulated in (monadic- or polyadic-) serial dynamic programming formulations. A monadic-serial dynamic programming problem can be solved as the search of an optimal path in a multistage graph and can be computed as a string of matrix multiplications. Two efficient systolic array designs are presented. A polyadic-serial dynamic programming problem can be solved by either a divide-and-conquer algorithm or the search of optimal solutions in a serial AND/OR graph. The optimal granularity of parallel divide-and-bound algorithms is analyzed and simulated. Some transformational techniques to convert nonserial problems into serial ones are briefly discussed.

**Key words:** AND/OR graph, dynamic programming, matrix multiplication, monadic, multistage graph, nonserial, parallel processing, polyadic, serial, systolic arrays

## 1 Introduction

Dynamic Programming (DP) a powerful optimization methodology that is widely applied to a large number of areas including optimal control, industrial engineering, economics, and artificial intelligence[1-5]. Many practical problems involving a sequence of interrelated decisions can be solved by DP efficiently. Bellman has characterized DP through the Principle of Optimality, which states that an optimal sequence of decisions has the property that whatever the initial stale and decision are, the remaining decisions must constitute an optimal decision sequence with regard to the state resulting from the first decision[1]. Subsequently, numerous efforts have been devoted to the rigorous mathematic framework and effective evaluation of DP problems[5-7].

In general, DP is an approach that yields a

Research supported partially by the National Science Foundation Grant ECS-80-16580 and by CIDMAC, a research unit of Purdue University, sponsored by Purdue, Cincinnati Milicron Corporation, Control Data Corporation, Cummins Engine Company, Ransburg Corporation and TRW.

transformation of the problem into a more suitable form for optimization, but is not an algorithm for optimizing the objective function. Moreover, DP can be interpreted differently depending on the computational approach. Bellman, Dreyfus, White, and many others viewed DP as a multistage optimization technique, that is, reducing a single $N$-dimensional problem to a sequence of $N$ one-dimensional problems[1,5]. The decisions that transform an initial state into a final state must be ordered in terms of stages, and functional equations relate state values in successive stages. The use of monotones sequential processes has been proved by Karp and Held to correspond naturally to DP[7] and has been further developed by Ibaraki[6] and Kumar[8]. On the other band, Gensi and Montanari have shown that formulating a DP problem in terms of polyadic functional equations is equivalent to searching for a minimum-cost solution tree in an AND/OR graph with monotone cost function[9]. DP can also be formulated as a special case of the branch-and-bound algorithm, which is a general top-down OR-tree search procedure with dominance tests[10-12]. Lastly, nonserial DP has been shown to be optimal among all nonoverlapping comparison algorithms[2,13].

Although DP has long been recognized as a powerful approach to solving a wide spectrum of optimization problems, its applicability has been somewhat limited due to the large computational requirements. Recent advances in Very-Large-Scale Integration (VLSI) and multiprocessor technologies have provided feasible means of implementation. Casti, et al., have studied parallelism in DP[14]. Guibas, Kung and Thompson have proposed a VLSI algorithm for solving the optimal parenthesization problem[15]. Linear pipelines for DP have been described recently[16]. Clarke and Dyer have designed a systolic array for curve and line detection in terms of nonserial DP[3]. Wah, et al., have proposed parallel processing for branch-and-bound algorithms with dominance tests[17]. However, these studies were directed towards the implementation of a few special cases of DP formulations.

In this paper, we classify DP problems into monadic-serial, polyadic-serial, monadic-nonserial, and polyadic-nonserial formulations. Potential parallelism and the corresponding parallel architectures for solving serial DP problems are investigated. Generally, a problem can be expressed in both monadic and polyadic formulations, and the efficiency and costs of implementations must be compared. It has been shown that unrestricted nonserial optimization problems are NP-hard, but that problems with a favorable pattern of term interactions may be solved efficiently[2]. An approach to solve a nonserial DP problem with some structural properties is to first convert it into a serial DP problem, and to map the serial formulation into an appropriate parallel architecture. Some methods for the conversion are proposed in this paper.

## 2 Classification of dynamic programming formulations

A DP formulation is represented in a recursive functional equation whose left-hand side identifies a function name and whose right-hand side is an expression involving the maximization (or minimization) of values of some cost functions. Note that the cost functions are neither restricted to be monadic nor additive; however, they must be monotone in order for the Principle of Optimality to hold. DP formulations are clarified according to the form of the functional equations and the nature of recursion.

DP problems can be solved as either the search of an optimal path in a multistage graph or as the search for an optimal solution in an AND/OR graph. We will adopt the graph search as a paradigm to illustrate the various approaches of DP. For DP problems in serial formulations, the corresponding graph representations have serial structures, and hence can be implemented easily in systolic arrays. A type of serial graphs of special interest is the multistage graph in which nodes are decomposed into stages, and nodes in one stage are connected to nodes in adjacent stages only. Figure 1 depicts two examples of multistage graphs.

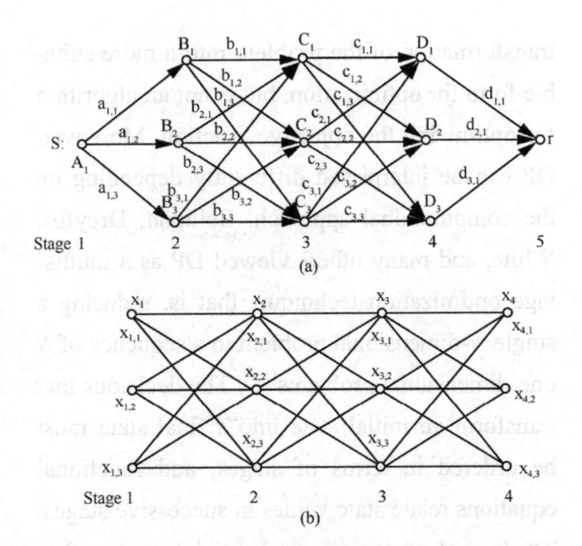

Figure 1. (a) A multistage graph with five stages and three nodes in each intermediate stage. (The cost on each edge is constant.) (b) A multistage graph with four stages (variables) and three vertices (quantized values) in each stage (The cost on each edge is a function of the nodes connected.)

## 2.1 Monadic versus polyadic formulation

A DP formulation is called monadic if its cost function involves only one recursive term, otherwise it is called polyadic. The distinction is illustrated by an example of finding the minimum-cost path in a multistage graph. For a multistage graph, let $e_{i,j}$ be the cost of an edge. The cost of a path from source, *s*, to sink, *t*, is the sum or costs on the edges of the path. Define $f_1(i)$ as the minimum cost of a path from *i* to *t*. Thus the cost of a path from *i* to *t* via a neighbor *j* is $c_{i,j}+f_1(j)$. To find $f_1(i)$, paths through all possible neighbors must be compared. Hence

$$f_1(i)=\min_j[c_{i,j}+f_1(j)] \tag{1}$$

This equation is termed a forward function al equation. Similarly, if $f_2(i)$ is defined as the minimum cost of a path from *s* to *i*, then the functional equation becomes

$$f_2(i)=\min_j[f_2(j)+c_{j,i}] \tag{2}$$

This equation is termed a backward functional equation. The formulations in Eq's (1) and (2) are monadic since each cost function involved one recursive term only.

Eq's (1) and (2) can be generalized to find the optimal path form any vertex *i* to any other vertex *j*. The functional equation is

$$f_3(i,j)=\min_{k\in V_m}[f_3(i,k)+f_3(k,j)] \tag{3}$$

where $f_3(i,j)$ is the minimum cost of a path traversing from *i* to *j* and passing through a node in stage $V_m$. This cost function is polyadic because it involves more than one recursive term.

Examples of this kind of problems include finding the optimal binary search tree and computing the minimum-cost order of multiplying a string of matrices.

For polyadic DP formulations, Bellman's Principle of Optimality must be generalized to include the statement that "all subsequences of an optimal policy are also optima." For instance, according to Eq. (3), if it is found that the minimum-cost path from *i* to *j* passes through *k*, then the subpath from *i* to *k* of this optimal path must be optimal over all subpaths from *i* to *k*; so is the subpath from *k* to *j*.

## 2.2 Serial versus nonserial formulations

The distinction between serial and nonserial optimization problems is based on both the form of their objective functions and the nature of recursion. From the objective function, an optimization problem is said to be serial if all terms of its objective function share one variable with its predecessor term (except for the first term); and another one with its successor term (except for the last term); otherwise, it is said to be nonserial. The name "serial" refers to the interaction graph to represent the problem, in which vertices stand for variables, and an edge exist between two vertices if and only if two variables belong to a term of the objective function[2]. It is obvious that a serial optimization problem has a corresponding interaction graph with a serial structure.

An example of a serial optimization problem is depicted in Figure 1(b). In this multistage graph, each stage, $X_i$, $1 \leqslant i \leqslant N = 4$, stands for a discrete variable, and Node $x_{i,j}$ stands for the *j*'th value taken by variable $X_i$. Bold characters are used to denote vectors and matrices, and variables here can be considered as vectors of defined values. If the cost of edge $(x_{i,j}, x_{i+1,j_{i+1}})$ is $g_1(x_{i,j_i}, x_{i+1,j_{i+1}})$, then the minimum-cost path from any node in stage 1 to any node in stage *N* is

$$\min_X f(X) = \min_X \sum_{i=1}^{N-t} g_i(X_i, X_{i+1}) \tag{4}$$

where $X$ is the set of discrete variables $\{X_i, \cdots, X_N\}$. In Eq. (4), every term of the objective function has two variables that only interact with variables in the neighboring terms. Therefore, Eq. (4) is a serial optimization problem.

Many practical DP problems can be represented in a serial formulation. For a traffic-control problem, $X_i$ can be the possible times for the traffic light to be in state *i*, and the cost on an edge of the graph representation is the difference in timing. For a circuit-design problem, $X_i$ can be the possible voltages at poiot *i*, and the cost of an edge of the graph representation may be the corresponding power dissipation. For a fluid-flow problem, $X_i$ can be the possible pressure values in the *i*'th pump, and function *f* may be the flow rate for a given pressure. For a scheduling problem, $X_i$ can be the possible task service times for the *i*'th task, and the edge cost reflects the delay. Note that the optimal-path problem in multistage graphs is a special case of serial optimization problems.

In contrast, the objective function of a general nonserial optimization problem has the following form.

$$f(X) = \mathop{\Phi}_{i=1}^{N} g_i(X^i) \tag{5}$$

where $X$=($X_1$,…,$X_N$) is a set of discrete variables, $X^i \subseteq X_i$, and $\Phi$ is a monotone function relating the $g_i s$ together. For example, the following equation is a nonserial optimization problem.

$$\min_X \{g_1(X_1,X_2,X_4)+g_2(X_3,X_4)+g_3(X_2,X_5)\} \quad (6)$$

where $X=\{X_1,\cdots,X_5\}$.

From the viewpoint of recursion, a DP problem can be represented as a folded AND/OR tree (or AND/OR graph) in which the nodes are classified into levels or stages[18]. If this AND/OR graph has a serial structure such that arcs only exist between adjacent levels, then the corresponding DP problem has a serial formulation. For nonserial DP problems, the dependency between states is not restricted to successive stages, but may exist between states in arbitrary stages. In the corresponding AND/OR graphs, the arcs are not restricted to successive levels, but may run between any two arbitrary levels.

As an example, consider the problem of finding the optimal order of multiplying a string of matrices. For simplicity, consider the evaluation of the product of four matrice.

$$\boldsymbol{M}=\boldsymbol{M}_1\times\boldsymbol{M}_2\times\boldsymbol{M}_3\times\boldsymbol{M}_4$$

where $\boldsymbol{M}_i$, $1\leqslant i\leqslant 4$, is a matrix with $r_{i-1}$ rows and $r_i$ columns. Let $m_{i,j}$ be the minimum cost of computing $\boldsymbol{M}_i\times\cdots\times\boldsymbol{M}_j$. Clearly,

$$m_{i,j}=\min_{i\leqslant k<j}\left(m_{i,k}+m_{k-1,j}+r_{i-1}\cdot r_k\cdot r_j\right) \quad (7)$$

The solution to be found is $m_{1,4}$. This formulation is polyadic-nonserial and can be represented as the search of an AND/OR graph as shown in Figure 2, where the AND-nodes[1] denote multiplications and the OR-nodes denote comparisons. In Figure 2, the topmost node represents the original problem of multiplying four matrices. There can be achieved in three ways: ① $(\boldsymbol{M}_1\times\boldsymbol{M}_2\times\boldsymbol{M}_3)\times\boldsymbol{M}_4$; ② $(\boldsymbol{M}_1\times\boldsymbol{M}_2)\times(\boldsymbol{M}_3\times\boldsymbol{M}_4)$; or ③ $\boldsymbol{M}_1\times(\boldsymbol{M}_2\times\boldsymbol{M}_3\times\boldsymbol{M}_4)$. These three alternatives are represented by the three AND-nodes in the second level. Note that the first AND-nodes in the second level is connected to the node representing $m_{4,4}$ in the bottommost level. Similarly, the third node is connected to the node representing $m_{1,1}$ in the bottommost level. These arcs do not connect nodes in adjacent levels, hence formulation in Eq. (7) is polyadic-nonserial.

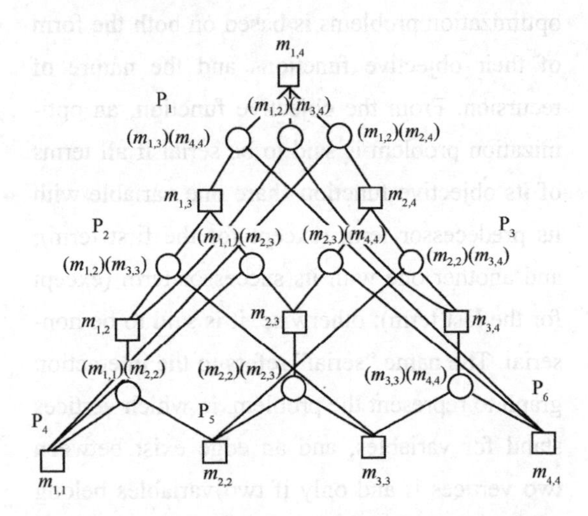


Figure 2. An AND/OR graph representation of finding the optimal order of multiplying a string of four matrices. (AND nodes are represented as circle s and indicate multiplications; and OR nodes are represented as squares and indicate comparisons.)

---

1 The definitions of AND and OR nodes used are due to Martelli and Montanari[18]. The roles of the AND and OR nodes are reversed in Nilsson's definition[19].

We have classified DP problems in terms of their recursive functional equations and objective functions. Monadic and polyadic DP formulations are distinct approaches to representing various optimization problems, while serial and nonserial optimization problems are problems solvable by the corresponding DP formulations.

# 3 Systolic arrays for monadic-serial formulations

Monadic-serial DP problems can be conveniently solved as the multiplication of a string of matrices. In this section, two efficient systolic designs are presented. The proposed designs do not exploit all potential parallelism of solving a given problem, especially when the number of stages is large. Other parallel designs using different formulations may allow a higher degree of parallelism and will be discussed later.

## 3.1 Solving monadic-serial DP problems as strings of matrix multiplications

Recall that the search for a solution of a monadic-serial problem can be viewed as finding a path in a multistage graph. For the multistage graph in Figure 1(a) and from Eq. (2), $f(\boldsymbol{C}_1)$, the minimum cost from $\boldsymbol{C}_1$ to $t$, is

$$f(\boldsymbol{C}_1)=\min\{c_{1,1}+d_{1,1},c_{1,2}+d_{2,1},c_{1,3}+d_{3,1}\} \quad (8)$$

$f(\boldsymbol{C}_2)$ and $f(\boldsymbol{C}_3)$ are obtained similarly.

Eq. (8) is similar to an inner-product operation. If we define matrix multiplication in terms of a closed semi-ring (R, MIN, +, $+\infty$, 0) in which "MIN" corresponds addition and "+" corresponds to multiplication in conventional matrix multiplications[20], then Eq. (8) becomes

$$f(\boldsymbol{C})=\boldsymbol{C}\cdot\boldsymbol{D}=\begin{pmatrix} f(\boldsymbol{C}_1) \\ f(\boldsymbol{C}_2) \\ f(\boldsymbol{C}_3)\end{pmatrix}= \begin{pmatrix} c_{1,1} & c_{1,2} & c_{1,3} \\ c_{2,1} & c_{2,2} & c_{2,3} \\ c_{3,1} & c_{3,2} & c_{3,3}\end{pmatrix}\begin{pmatrix} d_{1,1} \\ d_{2,1} \\ d_{3,1}\end{pmatrix} \quad (9)$$

Likewise, we have

$$f(\boldsymbol{B})=\boldsymbol{B}\cdot(\boldsymbol{C}\cdot\boldsymbol{D})$$

$$f(\boldsymbol{A})=\boldsymbol{A}\cdot(\boldsymbol{B}\cdot(\boldsymbol{C}\cdot\boldsymbol{D}))$$

Thus solving the multistage-graph problem with a backward monadic DP formulation is equivalent to multiplying a string of matrices. The order of multiplications is reversed in forward monadic DP formulations.

For a multistage graph with $N$ stages and $m$ vertices in each stage, the computational complexity is $O(m^2N)$. For single-source and single-sink problems, the first and last matrices degenerate into row and column vectors, respectively.

## 3.2 Systolic array for string of matrix multiplication with broadcasting

A linear systolic array with parallel inputs and broadcasting for evaluating monadic-serial DP problems is described in this section. The following scheme is based on multiplying a matrix with a vector. Figure 3(a) depicts a scheme for computing $(\boldsymbol{A}\cdot(\boldsymbol{B}\cdot(\boldsymbol{C}\cdot\boldsymbol{D})))$ for the multistage graph in Figure 1(a). An iteration is defined as a time unit during which data are shifted or broadcast into the Processing Elements (PEs) and a multiply-accumulate operation is carried out in each. The iteration numbers are indicated in Figure 3(a). All input matrices are fed into the systolic array in the same format. In the first

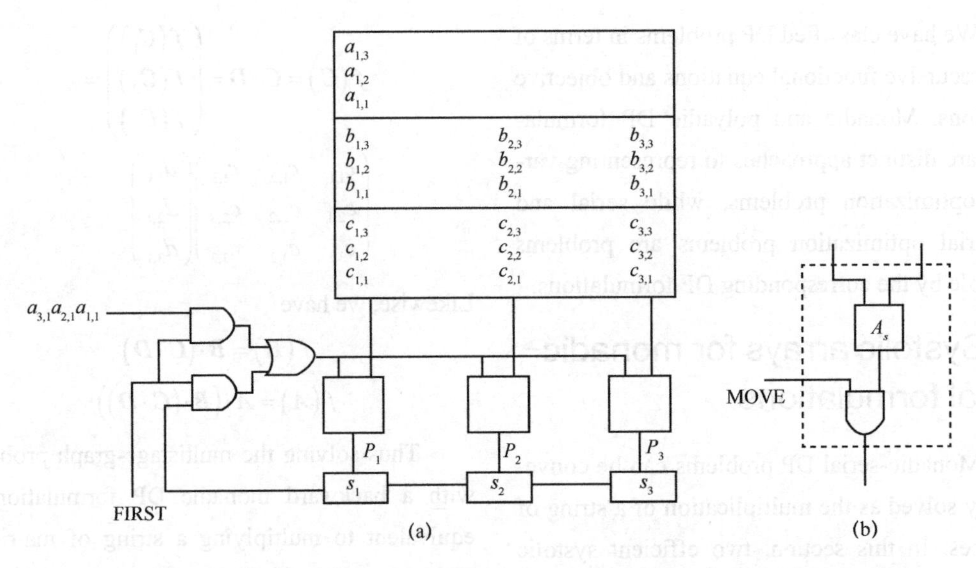


Figure 3. (a) A systolic array with broadcasts for computing a string of matrix multiplications.
(b) Structure of PE $P_i$

three iterations, $\boldsymbol{C}\cdot\boldsymbol{D}$ is evaluated. The control signal FIRST is one, $\boldsymbol{D}$, the input vector, is broadcast to all PEs; and the intermediate results of $f(\boldsymbol{C}_i)$, $i=1,2,3$, remain stationary.

At the end of the third iteration, the result vector is gated into registers $s_1,s_2,s_3$, by the control signal MOVE (see Figure 3(b)), and FIRST is set to zero. Since FIRST is zero, $f(\boldsymbol{C}_i)$, $i=1,2,3$ are fed back and broadcast as new inputs. In the following three iterations, $\boldsymbol{B}\cdot f(\boldsymbol{C})=\boldsymbol{B}\cdot(\boldsymbol{C}\cdot\boldsymbol{D})$ is computed. At the end of the sixth iteration, the output vector $\{f(\boldsymbol{B}_i),i=1,2,3\}$ is formed. In the last three iteration, input vectors $A$ and $\{f(\boldsymbol{B}_i),i=1,2,3\}$ enter PE $P_1$ to form the final result.

To search a multistage graph with $(N+1)$ stages and $m$ nodes in each intermediate stage (the first and last stages have one node each), it takes $N\cdot m$ iterations with $m$ PEs. There is no delay between feeding successive input matrices into the systolic array, and the PEs are kept busy most of the time. In contrast, it takes $(N-2)m^2+m$ iterations to solve the problem with a single PE. Define PU, the processor utilization, as the ratio of the number of serial iterations to the product of the number of parallel iterations and the number of PEs. PU for the above systolic array is

$$\mathrm{PU}=\frac{(N-2)m^2+m}{N\cdot m\cdot m}=\frac{N-2}{N}+\frac{1}{N\cdot m} \tag{10}$$

when $N$ and $m$ are large, PU is very close to 1. A non-broadcast version of the above design is shown elsewhere[21].

Although the proposed systolic array is designed for matrices in which each element is a single constant, it can be extended to many practical sequentially-controlled systems such as Kalman filtering, inventory systems, and multistage production processes in which each matrix element is a vector with many quantised values.

In this case, the potential parallelism could be very large.

### 3.3 Systolic arrays for string of matrix multiplications with serial inputs

The decree of parallelism of the proposed scheme in the last section is restricted by the limited number of I/O ports in a VLSI chip and the fact that the radio of the computational overhead to the I/O overhead is relatively low in matrix-vector multiplication. The I/O bottleneck is due to the large number of edge costa that must be fed into the systolic array. For the serial problem formulated by Eq. (4) and illustrated in Figure 1(b), the edge costs are expressed as functions of the nodes connected, and hence only the values of the nodes have to be input. This results in an order-of-magnitude reduction in the input overhead. In this section, we develop an efficient and practical design for this type of problems.

The search for an optimal assignment of $X_i s$ in Eq. (4) corresponds to the search for the shortest path in a multistage graph, where nodes in each stage represent values that can be assigned to a variable. An example graph with four variables, each of which can take on three quantized values, is shown in Figure l(b). There are multiple sources and sinks, and all possible paths from any vertex in stage 1 to any vertex in stage $N$ must be compared. Systolic processing is suitable when the number of quantized values in each stage is constant, and the $f_i s$, the functions to compute edges costs, are independent of $i$.

To solve Eq. (4), the variables can be eliminated one by one. First, $X_1$ is considered. Since only one term, $f(X_1,X_2)$, is affected by $X_1$, it is sufficient to compute

$$h(X_2)=\min_{X_1} f(X_1,X_2) \tag{11}$$

In other words,

$$h\left(x_{2,j_2}\right)=\min_{x_{1,j_1}\in X_1} f\left(x_{1,j_1},x_{2,j_2}\right) \quad x_{2,j_2}\in X_2$$

The optimization problem then becomes

$$\min_X f(X)=\min_{X-(X_1)}\left\{h(X_2)+\sum_{i=2}^{N-1} f(X_i,X_{i+1})\right\} \tag{12}$$

If $h(X_k)$ is defined as

$$h(X_k)=\min_{X_{k-1}}\left\{h(X_{k-1})+f(X_{k-1},X_k)\right\} \tag{13}$$

$$2\leqslant k\leqslant N$$

or

$$h\left(x_{k,j_k}\right)=\min_{x_{k-1,j_{k-1}}\in X_{k-1}}\left\{h\left(x_{k-1,j_{k-1}}\right)+f\left(x_{k-1,j_{k-1}},x_{k,j_k}\right)\right\}$$

$$x_{k,j_k}\in X_k,\quad 2\leqslant k\leqslant N$$

then $h(x_{k,j_k})$ represents the shortest path from any vertex in stage 1 to $x_{k,j_k}$. After eliminating $k-1$ variables, $X_1,\cdots,X_{k-1}$, the remaining optimization problem becomes

$$\min_X f(X)=\min_{X-(X_1,\cdots,X_{k-1})}\left\{h(X_k)+\sum_{i=k}^{N-1} f(X_i,X_{i+1})\right\} \tag{14}$$

Finally, we get $h(X_N)$, each element of which represents the shortest path from any vertex in stage 1 to a node of stage $N$. The problem is solved by comparing the $m$ elements of $h(X_N)$.

Figure 4 shows a systolic array with three PEs that performs the search of the graph in Figure 1(b). PE $P_i$ consists of three registers, $R_i,K_i,H_i$, and three operation components, $F_i$, $A_i$, $C_i$. Input data pass through $R_i$ in a pipelined fashion. Feedback data are maintained in $K_i$, and $H_i$ until new data replace them. The

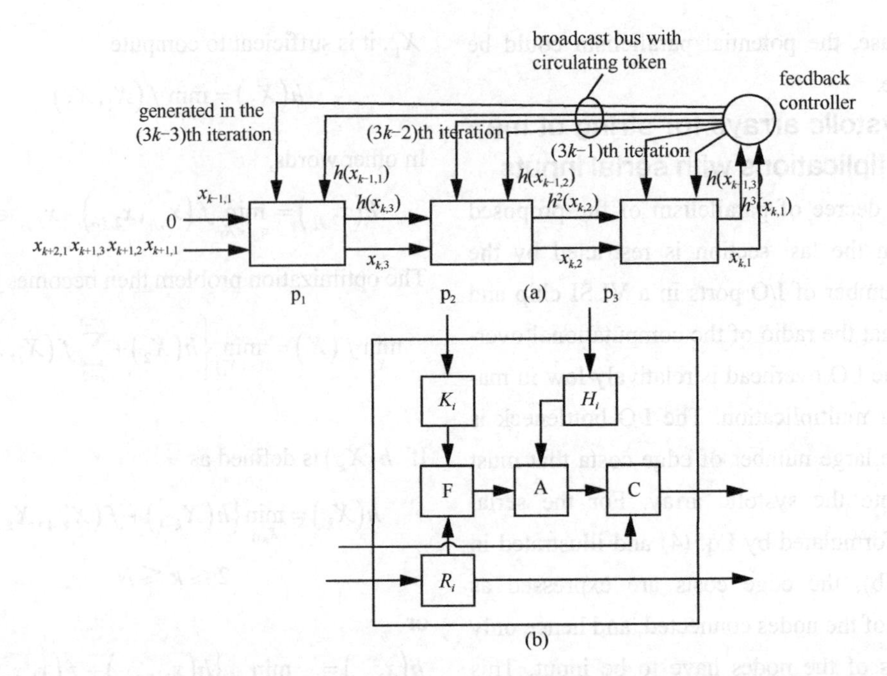


Figure 4. A systolic array with serial inputs and outputs to solve a monadic-serial DP problem: (a) State of outputs at the end of the (3*k*)'th iteration, $2 \leqslant k \leqslant N$ (outputs at the end of the feedback controller are generated in the (3*k*−1)'th, (3*k*−2)'th, and (3*k*−3)'th iterations); (b) Registers and operation units in the *i*'th PE

operation components, $F_i$, $A_i$, $C_i$, are used to compute function *f*, and perform additions and comparisons, respectively. For simplicity, function *f* is assumed to be independent of *i*, and hence the subscripts in $F_i, A_i$, and $C_i$ will be dropped. The connections of the registers and operation components are shown in Figure 4(b).

The systolic array is initialized by zeroing all registers, $H_i$s and $K_i$s, and by sequentially loading input data in $X_1$, $x_{1,m}, \cdots, x_{1,1}$, to PEs $P_1, \cdots, P_m$. As the intermediate results are shifted out from $P_3$, the feedback controller feeds them back in a round-robin fashion. Referring to Figure 4(a), when $x_{2,1}$ enters $P_1$, $x_{1,1}$ and $h(x_{1,1})$ (equals 0) leave $P_3$ and are fed back to $P_1$ through the feedback controller. $f(x_{1,1}, x_{2,1})$ and $h^1(x_{2,1}) = \min(0, h(x_{1,1}) + f(x_{1,1}, x_{2,1}))$ are then computed in $P_1$. In the next iteration, $x_{2,2}$ enter $P_1$, $x_{2,1}$ and $h^1(x_{2,1})$ are shifted to $P_2$, and $x_{1,2}$ and $h(x_{1,2})$ (equals 0) are fed back by the feedback controller to $P_2$. In $P_2$, $f(x_{1,2}, x_{2,1})$ and $h^2(x_{2,1}) = \min(h^1(x_{2,1}), h(x_{1,2}) + f(x_{1,2}, x_{2,1}))$ are computed. In $P_1$, $f(x_{1,1}, x_{2,2})$ and $h^1(x_{2,2}) = \min(0, h(x_{1,1}) + f(x_{1,1}, x_{2,2}))$ are computed. When $x_{2,1}$ and $h^2(x_{2,1})$ arrive at $P_3$, $h(x_{2,1}) = h^3(x_{2,1})$ is evaluated, and $x_{2,1}$, and $h(x_{2,1})$ are fed back to $P_1$ at the end of this iteration. Input data are continuously shifted into the pipeline, and the process is repeated. For the graph in Figure 1(b), the process is completed in fifteen iterations.

In general, to evaluate the optimal path for an $N$-stage graph, each with $m$ quantized values, a pipeline with $m$ PEs is needed. Between the $((k-1)m+1)$' st and $(k\cdot m)$' th iterations, $2\leqslant k\leqslant N$, $x_{k,1},\cdots,x_{k,m}$ enter the $R$-pipeline; $x_{k-1,i}$ and $h(x_{k-1,i}),1\leqslant i\leqslant m$, are fed back to registers $K_i$ and $H_i$, in the $((k-1)m+i)$' th iteration; and $h^m(x_{k,1}),h^{m-1}(x_{k,2}),\cdots,h^1(x_{k,m})$ are obtained at the end of the $(k\cdot m)$' th iteration in $P_m,\cdots,P_1$. After $N\cdot m$ iteration, $h^m(x_{N,1}),\cdots,h^1(x_{N,m})$ are obtained in $P_m,\cdots,P_1$, and the final solution is obtained in $P_m$ by comparing $h(x_{N,1}),\cdots,h(x_{N,m})$. This is done by setting $F=0$ in the last $m$ iterations and circulating the values of $h(x_{N,i})$, $1\leqslant i\leqslant m$, through the pipeline. Therefore, the total computational time is $(N+1)m$ iterations, each of which includes the time for the computation of function $f$, one addition, and one comparison. PU for this scheme is $((N-1)m^2+m)/((N+1)m\cdot m)=1$.

Although distinct feedback lines are shown in Figure 4(a), only one of the feedback lines is used in any iteration. Hence a single broadcast bus suffices, and the station to pick up the data from the bus is controlled by a circulating token.

If the optimal path in addition to the optimal cost value is desired, $N$ path registers, each of which can store $m$ indices, are needed in $P_m$. In the computation of $h^m(x_{k+1,i})=\min_j\{h(x_{k,i})+f(x_{k,j'},x_{k+1,i})\}$, $1\leqslant k\leqslant N-1$, index $j'$, $1\leqslant j'\leqslant m$, of the edge $(x_{k,j'},x_{k+1,i})$ belonging to the optimal path from any vertex in stage 1 to Vertex $x_{k+1,i}$ must be propagated in the pipeline and is known to $P_m$. Index $j'$ is stored in the $i$'th word of the $k$'th path register. The pointers stored in the path registers are used to trace the optimal path at the end of the computation.

# 4 Parallel processing of polyadic-serial DP problems

We have shown that a serial optimization problem can be solved as the multiplication of a string of matrices. However, a problem expressed in a monadic-serial formulation does not exploit all the potential parallelism because the order of matrix multiplication is fixed. On the other hand, there is more flexibility for parallelism when the problem is formulated in a polyadic equation because the matrices can be multiplied recursively in a more flexible order.

A polyadic-serial DP formulation can be solved by a divide-and-conquer algorithm, the multiplication of a string of matrices, or the search of an AND/OR graph. The last two alternatives are related since the evaluation of a set of AND-nodes and their common parent is equivalent to computing an item in a matrix multiplication, that is, an inter-product of a row vector and a column vector. In general, if the number of states in each stage is large and constant, then the matrix-multiplication method is preferable to an AND/OR-graph search, as more potential parallelism can be exploited. When the AND/OR graph is nonserial and irregular, the graph-search method is more beneficial.

## 4.1 Solving polyadic-serial DP problems by divide-and-conquer algorithms

Consider the polyadic-serial DP formulation in Eq. (3) for the multistage-path problem in Figure 1(a).

$$f_3(s,t) = \min_{1\in\{c1,c2,c3\}} \{f_3(s,k) + f_3(k,t)\} \quad (15)$$

where $f_3(i,j)$ is the cost of the optimal path from $i$ to $j$, and $k$ is a node in stage 2 of the graph. In matrix notations, let $f_3(V_i,V_j)$ be a cost matrix, each element of which denotes the cost of the optimal path from a vertex in stage $i$ to a vertex in stage $j$. It is easy to see, for an intermediate stage $k$ between $i$ and $j$, that

$$f_3(V_i,V_j) = f_3(V_i,V_k)\cdot f_3(V_k,V_j) \quad (16)$$

This formulation allows a string of matrix multiplication to be reduced to two smaller strings of matrix multiplications. The substring of matrices can partitioned further until each substring contains only one matrix.

The fastest way to multiply $N$ $m$-by-$m$ matrices is to locale the matrices in the leaves of a complete binary tree of height $\lceil \log_2 N \rceil$. The $N$-stage graph problem can be solved in $O(m\lceil \log_2 N \rceil)$ time units with $\lceil N/2 \rceil$ processors[2] or matrix- multiplication systolic arrays[22]. PU[3] for this approach is relatively low due to the large number of idle processors.

One important issue in parallel divide-and-conquer algorithms is the granularity of parallelism[17]. This is the minimum site of a subproblem that is evaluated by a processor in order to achieve the optimal performance, as measured by the PU, the $AT^2$, or the $KT^2$ criteria, where $A$, $K$, $T$ is the area of a VLSI implementation, the number of processors, and the computational time, respectively.

Parallel divide-and-conquer algorithms is a parallel AND-tree search that can roughly be divided into three phases; start-up, computation, and wind-down. In the start-up phase, the problem is split, and the tasks diffuse through the network. During the computation phase, all processors are kept busy until the number of tasks in the system is less than the number of processors. In the wind-down phase, the results are combined together, and some processors may be idle. PU depends on the ratio between the amount of time spent in the computation phase and that of the other phases. The time complexity of searching a binary AND-tree of $N$ leaves can be formulated in the following recursive equation:

$$T(N) = \begin{cases} S(N) + 2T\left(\dfrac{N}{2}\right) + C(N) & N>1 \\ O(1) & N=1 \end{cases} \quad (17)$$

where $S(N)$ and $C(N)$ are the time complexities of the start-up and wind-down phases. The granularity that results in the optimal PU is related to the complexity of $S(N)$ and $C(N)$. In finding the sum or the maximum of $N$ numbers, $S(N)+C(N)=O(1)$, and using $O(N/(\log_2 N))$ processors will achieve the maximum PU[23-25]. In sorting $N$ numbers, $S(N)+C(N)=O(N)$, and $\log_2 N$ processors should be used to maximize the PU. We have studied the asymptotic PUand found that $N/\log_2 N$ is a threshold when $S(N)+C(N)=O(1)$ [21]. For $k$ (a function of $N$) processors, PU is one, between zero and one, or zero when the limiting ratio of $k$ and $N/\log_2 N$ is

[2] Processors and systdic arrays are synonyrnous here.

[3] PU refers to the utilization of the all the processors or matrixmultiplication systolic arrays and does not refer to the utilization of the Pes within a systolic array as in section 3.

zero, greater than zero, or approaching infinity, respectively.

Since PU increases monotonically with decreasing number of processors, it is not adequate to measure the effects of parallel processing. Another appropriate measure is the $KT^2$ criterion, which considers both PU and computational time. The following theorem proves the lower-bound $KT^2$ complexity of divide-and-conquer algorithms for solving polyadic-serial DP problems. This lower bound is attained when $k(N)$ is $\theta(N/(\log_2 N))$.

**Theorem 1:** Suppose that a string of $N$ $m$-by-$m$ matrices are multiplied by $K(N)$ processors in time $T(N)$ using a parallel divide-and-conquer algorithm, and that each processor performs a multiplication of a pair of $m$-by-$m$ matrices in $T_1$ time units. Then $K(N)T^2(N) \geqslant \theta(N \cdot \log_2 N)T_1^2$, and equality holds when $K(N)=\theta(N/\log_2 N)$ ($\theta$ indicates the set of functions of the same order).

**Proof:** The multiplication of a string of $N$ matrices by a divide-and-conquer algorithm can be represented in a complete binary tree with $N$ terminal. The number of matrix-multiplications, or the number of nonterminal, is $N-1$. The operations are roughly divided into two phases: computation and wind-down (no start-up phase in this case). During the computation phase, all processors are kept busy until half of the number of intermediate matrices to be multiplied is less than the number of processors. There are $(N-1)-(K(N)-1)=N-K(N)$ matrix multiplications to be evaluated, and at least $(N/K(N)-1)\cdot T_1$ time units are needed. In the wind-down phase, the results are combined, together, and some processors would be idle. According to the data dependence, at least $\log_2 K(N)\cdot T_1$ time units are required in this phase. Therefore, the following lower bound of time complexity holds.

$$T(N) \geqslant \left(\frac{N}{K(N)}-1+\log_2 K(N)\right)\cdot T_1 \quad (18)$$

where $1 \leqslant K(N) \leqslant N$. For simplicity, the constant term in Eq. (18) can be ignored without affecting the validity of the following proof. The $KT^2$ lower bound is derived as

$$K(N)T^2(N) \geqslant \left(\frac{N^2}{K(N)}+2N\cdot\log_2 K(N)+K(N)\cdot\log_2^2 K(N)\right)\cdot T_1^2 \quad (19)$$

To find the order-of-magnitude minimum of Eq. (19), it is necessary to compare the following three cases. When $K(N)=\theta(N/\log_2 N)$, $K(N)T^2(N)=\theta((N\cdot\log_2 N)T_1^2)$. In contrast, when $K(N)<\theta(N/\log_2 N)$, the first term on the right-hand side of Eq. (19) is

$$\frac{N^2}{K(N)} > \theta(N\cdot\log_2 N) \quad (20)$$

when $K(N) > \theta(N/\log_2 N)$, the third term on the right-hand side of Eq. (19) is

$$K(N)\log_2^2 K(N) > \theta(N\cdot\log_2 N) \quad (21)$$

since $\log_2^2 K(N) \geqslant \theta(\log_2^2 N)$. The above analysis shows that the $KT^2$ complexity is $\Omega((N\cdot\log_2 N)T_1^2)$, and that $\theta(N/\log_2 N)$ is the optimal granularity to achieve this lower bound.

To investigate the relationship between $K$ and $KT^2$, the exact time required to multiply $N$ $m$-by-$m$ matrices using $K$ processors is derived.

The total time required is

$$T = T_c + T_w = \left\lfloor \frac{N-1}{K} \right\rfloor \cdot T_1 + \left\lceil \log_2 \left( N + K - 1 - K \cdot \left\lfloor \frac{N-1}{K} \right\rfloor \right) \right\rceil \cdot T_1 \tag{22}$$

Where $T_c$ and $T_w$ represent the times in the computation and wind-down phases, respectively. The numerical evaluations of Eq. (22) for $N = 4\,096$ is shown in Figure 5, in which $KT^2$ is minimum when 431 or 465 processors are used. Notice that the curve is not smooth because the time needed in the wind-down phase is decreased by 1 whenever $N$ is divisible by $K$, and this affects $KT^2$ significantly, especially when $K$ is large. The simulation results for different values of $N$ verify that the optimal granularity is close to $N/\log_2 N$.

When $N$ is large and $\lfloor (N-1)/K \rfloor$ is approximately equal to $(N-1)/K$, $T_c = (N/K)-1$, and $T_w \simeq \log_2 K$.

$$T \simeq \frac{N}{K} - 1 + \log_2 K \tag{23}$$

$KT^2$ will achieve the minimum value when $T_c = T_w$. This can be shown by differentiating $KT^2(K)$ with respect to $K$ and relaxing the constraint that $K$ is an integer.

$$\frac{\partial KT^2(K)}{\partial K} = T^2 + 2KT\frac{\partial T}{\partial K} \tag{24}$$

From Eq's (23) and (24), we get $\frac{\partial KT^2(K)}{\partial K} = 0$ if $(N/K-1) = \log_2 K$, i.e., $T_c = T_w$. This means $KT^2$ for solving polyadic serial DP problems by parallel divide-and-conquer algorithms approaches minimum if $N$ is large and the times needed in the computation and wind-down phases are approximately equal.

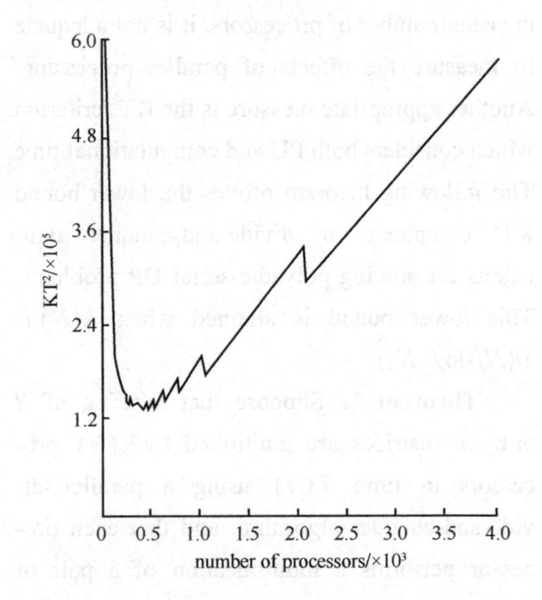


Figure 5. Simulation results of finding the optimal granularity of parallel divide-and-conquer algorithms ($N$=4 096)

So far, the matrices are assumed to have identical dimensions, when this is not true, the order in which the matrices are multiplied together has a significant effect on the total number of operations. Finding the optimal order of multiplying a string of matrices with different dimensions is itself a polyadic-nonserial DP problem, the so-called secondary optimization problem[26-27]. Guibas, Kung, and Thompson have proposed a systolic array to solve the optimal parenthesization problem, which can be used to compute the minimum-cost order of multiplying a string of matrices[15]. Once the optimal order is found, the processors can be assigned to evaluate the matrix multiplication in the defined order and in an asynchronous fashion. In this sense, the tree of matrix multiplications can be treated as a dataflow graph.

## 4.2 Solving polyadic-serial DP problems by AND/OR graph searches

In this section, we discuss the evaluation of polyadic-serial DP problems as AND/OR graph searches. AND/OR graphs are naturally obtained by using a problem-reduction method to represent the DP problem. The mapping of a regular and serial AND/OR graph to a systolic array is straightforward and will not be illustrated here[21].

Polyadic-serial problems are discussed with respect to the search of a multistage graph as formulated by Eq. (4). Suppose an $(N+1)$-stage graph, with stages from 0 to $N$ and $m$ nodes in each stage, is divided into $p$ subgraphs, each of which contains $(N/p)+1$ consecutive stages. For simplity, assume that $N=p'$, where $l$ is a non-negative integer. The minimum cost path has to pass through one and only one vertex in stages 0, $N/p, 2N/p, \cdots, N$ in the segmented graph. The cost of a path is equal to the sum of coats of the $p$-subpaths. There are $m^{p+1}$ possible combinations of subpaths from stage 0 to stage $N$ that must be considered for the optimal path. If all the $m^2$ subpaths from the $m$ vertices in stage $i\,N/p$ to the $m$ vertices is stage $(i+1)N/p$, $0 \leqslant i \leqslant p-1$, have been optimized, then the $(N+1)$-stage graph is reduced to a $(p+1)$-stage graph. By using a divide-and-conquer algorithm, each subgraph with $(N/p+1)$ stages is further divided into $p$ smaller subgraphs. This partitioning process continues until each subgraph has only two consecutive stages.

The partitioning process can be conveniently represented in an AND/OR graph, is which an AND-node corresponds to a summation, and an OR-node corrcsponds to alternative selections or comparisons. In this case, we have a regular and serial AND/OR graph of height $2 \cdot \log_p N$, whose AND-nodes have $p$ branches ($p$-arc nodes) and whose OR-nodes have $m^{p-1}$ branches ($m^{p-1}$-arc nodes). Figure 6 shows an AND/OR graph that represents the reduction of the multistage-graph problem with $m=2$ and $p=2$ from three stages to one stage. The four nodes at the top of the AND/OR graph represent the four possible alternate paths in the reduced two-stage graph. The shortest path is obtained by comparing the coats of these paths.

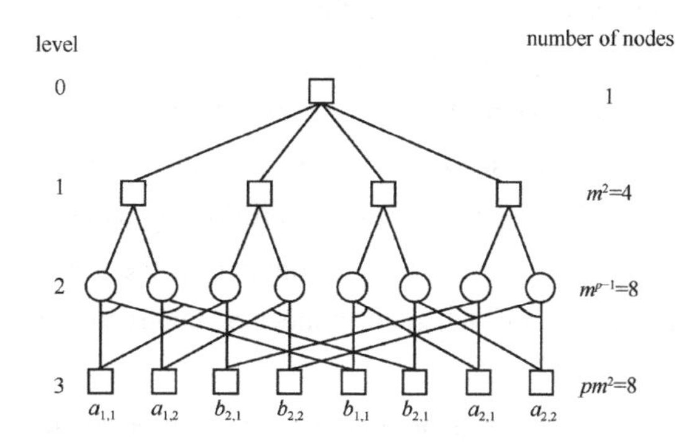


Figure 6. An AND/OR graph representation of the reduction in finding an optimal path in a 3-stage graph to a 1-stage graph. ($m$=2; $p$=2; AND nodes are represented as circles and indicate summations; and OR nodes are represented as squares and indicate comparisons. The values in the terminal nodes are edge costs: costs between stages 1 and 2 are $a_{i,j}$, $i,j \in \{1,2\}$; costs between stages 2 and 3 are $b_{i,j}$)

The relationship between DP and graph search was investigated by Martelli and Montanari[18] who showed that, in the case of polyadic cost functions, the solution of a DP problem can be obtained by finding a minimal-cost solution tree in an AND/OR graph. This equivalence allows various graph searching techniques to be translated into techniques for solving DP problems[28]. For those acyclic AND/OR graphs with positive arc costs. Martelli and Montanari have named them as additive[18] and have proposed top-down and button-up search algorithms. A similar algorithm, called AO*, for searching hypergraphs was discussed by Nilsson[19].

The above AND/OR graph representation of a polyadic DP problem can be considered as a folded AND/OR tree. It is easy to see that the efficiency of solving DP problem by searching an AND/OR graph depends on the graph structure (parameter $p$). The following theorem analyses the optimal structure.

**Theorem 2**: If a serial DP problem is solved by searching a serial and regular AND/OR graph, then binary partitioning, namely, using 2-arc AND nodes, is optimal in the sense of minimizing the total number of nodes in the AND/OR graph.

The proof is omitted here and can be found elsewhere[21]. The reasonableness of this theorem can be interpreted intuitively. For an AND/OR graph, the larger the value of $p$ is, the less the Principle of Optimality is applied. In the extreme case, $p=N$, the corresponding AND/OR-graph search becomes a brute-force search, and the Principle of Optimality is never used. In contrast, in the case of binary partitioning unnecessary computation are pruned by comparisons *m* the AND/OR graph.

For multistage-graph problems represented in an irregular but serial AND/OR graph, the number of nodes in the graph depends on the ordering of stage reduction. However, it is not difficult to demonstrate that binary partitioning is also optimal in this case. Assume that stages $i_1,\cdots,i_4$ with $m_1,\cdots,m_4$ nodes are to be reduced to two stages $i_1$ and $i_4$. If 3-arc AND-nodes are used, then $m_1\cdot m_2\cdot m_3\cdot m_4$ comparisons are needed to eliminate stage 2 and 3. However, when 2-arc AND-nodes are used, $m_1\cdot m_3\cdot(m_2+m_4)$ comparisons are needed if stage 2 is eliminated first, and $m_2\cdot m_4\cdot(m_1+m_3)$ comparisons are needed if stage 3 is eliminated first. It is easy to see that using 3-arc AND-nodes requires more comparisons as long as $m_i\geqslant 2$, $1\leqslant i\leqslant 4$. Furthermore, binary partitioning requires less additions since only one addition is needed for each AND-node.

## 5 Solving nonserial DP problems

The key of DP is to break a complex optimization problem into a sequence of easier subproblems. In serial optimization into problems, variables are shared by successive terms in the objective function and hence can be dealt with one by one. This serial structure allows efficient parallel processing, especially systolic processing. On the other hand, to implement nonserial DP problems by systolic processing, they may have to be transformed into the corresponding serial formulations before they are implemented. This transformation is possible if the nonserial problems have some special structures.

One way to convert a nonserial formulation into a serial one is to combine several primary variables into a new variable. The method is illustrated by the following example on monadic-nonserial problems. For instance, let $V=\{V_1,\cdots,V_N\}$ be a set of discrete variables and the objective function be

$$f(V)=g_1(V_1,V_2,V_3)+g_2(V_2,V_3,V_4)+\cdots+g_{N-2}(V_{N-2},V_{N-1},V_N) \quad (25)$$

To transform this into a serial formulation, the variables are eliminated one by one as follows. Let $h_1(V_2,V_3)=\min_{V_1} g_1(V_1,V_2,V_3)$. Then

$$\min_V f(V)=\min_{V-\{V_1\}}\left\{h_1(V_2,V_3)+\sum_{i=2}^{N-2} g_i(V_i,V_{i+1},V_{i+2})\right\} \quad (26)$$

If $h_k(V_{k+1},V_{k+2})$ is defined as

$$h_k(V_{k+1},V_{k+2})=\min_{V_k}\{h_{k-1}(V_k,V_{k+1})+g_k(V_k,V_{k+1},V_{k+2})\} \quad (27)$$

then it represents the minimum of the summation of the first $k$ terms of $f(V)$. After eliminating $k$ variables, $V_1,\cdots,V_k$ the remaining optimization problem becomes

$$\min_V f(V)=\min_{V-(V_1,\cdots,V_k)}\left\{h_k(V_{k+1},V_{k+2})+\sum_{i=k+1}^{N-2} g_1(V_i,V_{i+1},V_{i+2})\right\} \quad (28)$$

The variables are, therefore, eliminated in the order $V_1,\cdots,V_N$. Here, variables $V_1$ and $V_{i+1}$ are treated as s single variable in a combined stage $V_1$'. If $m_k$, $1\leqslant k\leqslant n$ quantized values are allowed for $V_k$, then there would be $m_i\cdot m_{i+1}$ quantized values in the combined stage $V_1$', and $m_{i-1}\cdot m_i\cdot m_{i+1}$ steps are required to eliminate $V_{i-1}$ , in which a step consists of a computation of function $f$, an addition, and a comparison operation. The process of elimination the remaining variables is repealed until $V_{N-1}$ and $V_N$ remain. The optimal solution is obtained by comparing all values of $h_{N-2}(V_{N-1},V_N)$. The total number of steps required to compute Eq. (25) is

$$\sum_{k=1}^{N-2}(m_k\cdot m_{k+1}\cdot m_{k+2})+m_{N-1}\cdot m_N \quad (29)$$

In short, the monadic-nonserial problem in Eq. (25) is solved from the following serial problem.

$$\min_V f(V)=\min_V\{g'_1(V'_1,V'_2)+g'_2(V'_2,V'_3)+\cdots+g'_{N-2}(V'_{N-2},V'_{N-1})\} \quad (30)$$

where the new variable $V_i$' is combined from $V_i$ and $V_{i+1}$. From this example, it is observed that more operations are needed for evaluating monadic-nonserial DP problems than that of monadic-serial DP problems. However, the potential parallelism is higher, and there is no increase in delay in processing the transformed problem.

Another way to solve nonserial problems is to convert the nonserial AND/OR graph into a serial graph by adding dummy nodes such that all arcs connect nodes in adjacent stages. In this case, the nonserial DP problems are characterized by high-order recursive functions. The method of adding dummy nodes essentially adds new functions to the problem, which converts the high-order recursion to a linear recursive function. For instance, in Figure 2, the bottommost OR-nodes can be connected to their parents via other intermediate AND-nodes in adjacent levels. This transformed AND/OR graph is suitable for VLSI implementation since the interconnections can be mapped into a planar structure. However, the transformation may introduce additional delay and redundant hardware. A systolic array for finding the optimal order of multiplying a systematic of matrices is designed by the systematic method mentioned above[21].

## 6 Conclusions

Dynamic programming formulations have been classified according to the structure of the functional equations. A given problem can usually be formulated in multiple ways, hence it is important to compare the alternative implementation. The applicability of systolic processing is most suitable when the formulation is serial.

Many sequential decision problems have serial formulation that can be considered as searching multistage graphs. If there are a large number of states AND/OR quantized values in each stage, then a monadic formulation is more appropriate, and the problem is efficiently solved as a serial string of matrix multiplications. Two efficient and practical systolic arrays have been developed. On the other hand. If the number of stages is large, then the problem should be put into a polyadic formulation. The matrices are grouped into a binary tree and multiplied by a divide-and-conquer algorithm. We have found the $KT^2$ lower bound for multiplying a tiring of $N$ $m$-by-$m$ matrices, where $K$ is the number of processors. It is shown that dividing the string into $\theta(N/\log_2 N)$ groups is optimal in the sense of achieving the lower bound.

When the formulation is nonserial, it might have to be transformed into a serial one before an efficient design can be found. A monadic-nonserial formulation can be transformed into a monadic-serial one by grouping state variables. A polyadic-nonserial problem can be represented as the search of an optimal solution in an AND/OR graph, which can be transformed AND/OR graph for a serial problem by adding dummy nodes. The transformed into an AND/OR graph can be mapped directly into a planar systolic array by using appropriate control signals. The additional hardware are and delay introduced is problem dependent.

## Reference:


[1] BELLMAN R, DREYFUS S. Applied dynamic programming[M]. Princeton: Princeton University Press, 1962.

[2] BERTELE U, BRIOSCHI F. Nonserial dynamic programming[M]. New York: Academic Press, 1972.

[3] CLARKE M, DYER C. Systolic array for s dynamic programming application[C]// The 12th Workshop on Applied Imagery Pattern Recognition. Piscataway: IEEE Press, 1983.

[4] NEY H. Dynamic programming as a technique for pattern recognition[C]// The 6th International Conference on Pattern Recognition. 1982: 1119-1125.

[5] WHITE D. Dynamic programming[M]. Edinburgh: Oliver &Boyd, 1969.

[6] IBARAKI T. Solvable classes of discrete dynamic programming[J]. Journal of Mathematical Analysis & Applications, 1973, 43(3): 642-693.

[7] KARP R, HELD K M. Finite state processes and dynamic programming[J]. Siam Journal on Applied Mathematics, 1967, 15(3): 693-718.

[8] KUMAR V. A general bottom-up procedure for searching and/or graphs[C]// National Conference on Artificial Intelligence. Texas: [s.n.], 1984: 182-187.

[9] GNESI S, MONTANARI U, MARTELLI A. Dynamic programming as graph searching: an algebraic approach[J]. Journal of the ACM, 1981, 28(4): 737-751.

[10] IBARAKI T. The power of dominance relations in branch-and-bound algorithms[J]. Journal of the ACM, 1977, 24(2): 264-279.

[11] LI G J, WAH B W. Computational efficiency of parallel approximate branch-and-bound algorithms[C]// The International Conference on Parallel Processing. [S.l.:s.n.], 1984: 473-480.

[12] MORIN T L, MARSTEN R E. Branch-and-bound

strategies for dynamic programming[J]. Operations Research, 1976, 24(4): 611-627.

[13] ROSENTHAL A. Dynamic programming is optimal for nonserial optimization problems[J]. Siam Journal on Computing, 1982, 11(1):47-59.

[14] CASTI J, RICHARDSON M, LARSON R. Dynamic programming and parallel computers[J]. Journal of Optimization Theory & Applications, 1973, 12(4): 423-438.

[15] GUIBAS L, KUNG H T, THOMPSON C. Direct VLSI implementations of combinatorial algorithms[C]//The1st Caltech Conference on VLSI. [S.l.:s.n.], 1979: 509-525.

[16] VWIMAN P, RAMAKRISHNAN V. Dynamic programming and transitive closure on linear pipelines[C]// The International Conference on Parallel Processing. [S.l.:s.n.], 1984: 359-364.

[17] WAN B H, LI G J, YU C F. Multiprocessing of combinatorial search problems[J]. IEEE Computer, 1985, 18(6):93-108.

[18] MARTELLI A, MONTANARI U. Additive and/or graphs[C]// The International Joint Conference on Artificial Intelligence. [S.l.: s.n.], 1973: 1-11.

[19] NILSSON N J. Principle of artificial intelligence[M]. Palo Alto: Tioga Publishing Co., 1980.

[20] AHO A V, HOPCROFT J E, ULLMAN J D, et al. The design and analysis of computer algorithm[M]. Massachusetts: Addison-Wesley, 1974.

[21] LI G J, WAH B W. Parallel processing for dynamic programming problems[C]// The International Conference on Parallel Processing. [S.l.:s.n.], 1985.

[22] LI G J, WAH B W. The design of optimal systolic algorithms[J]. IEEE Transactions on Computers, 1985, C-34(1): 66-77.

[23] BAUSET G, STEVENSON D. Optimal sorting algorithm for parallel computers[J]. IEEE Transactions on Computers, 1978, C-27(1): 84-87.

[24] KUCK D J. A survey of parallel machine organization and programming[J]. ACM Computing Surveys, 1977, 9(1): 29-59.

[25] SAVAGE C D. Parallel algorithms for graph theoretic programming problems[D]. Urbana-Champaign: University of Illinois at Urbana-Champaign, 1978.

[26] BERTELE U, BRIOSCHI F. A new algorithm for the solution of the secondary optimization problem in non-serial dynamic programming[J]. Journal of Mathematical Analysis & Applications, 1969, 27(3): 565-574.

[27] BRIOSCHI F, EVEN S. Minimizing the number of operations in certain discrete-variable optimization problems[J]. Operations Research, 1970, 18(1): 66-81.

[28] PEARL J. Heuristics: intelligent search strategy for computer problems solving[M]. Massachusetts: Addison- Wesley, 1984.

发表于 IEEE International Conference on Tools for AI论文集, 1991年

# New crossover operators in genetic algorithms

Shang YI, Guo-jie LI

National Research Center for Intelligent Computing Systems (NCIC),

Beijing 100080, China

**Abstract:** This paper presents two new crossover operators in genetic algorithms for solving some combinatorial prob¬lems with ordering. One is enhanced order crossover (EOX). The other, GREE, is a heuristic crossover for a class of combinatorial optimization problems, such as traveling salesman problems (TSP). Genetic algorithms using GREE as unique crossover run very fast and get good solutions. Combining GREE with EOX, genetic algorithms can find optimal or very near optimal solutions in rather short time.

## 1 Introduction

The ordering problem is a kind of combinatorial optimization problems, such as Traveling Salesman Problems (TSP) and scheduling problems, in which the goodness of a solution depends strongly on the order of elements. It is quiet difficult to solve ordering problems by Genetic Algorithms (GAs)[1]. By analyzing survival probability of the schema under Order Crossover (OX) and comparing with that under traditional simple crossover, we design the Enhanced Order Crossover (EOX), which augments the survival probability of ordering schema. Experiments performed on TSP go some way to support the theoretical analysis.

Heuristic crossover GREE is based on schema analysis of GAs. Although GREE is designed for TSP, its idea is applicable to other ordering-only problems. The famous 318 city TSP, LIN318[2], is attacked by GAs with GREE, and best tours obtained are only 5% longer than the optimum.

## 2 Enhanced order crossover(EOX)

Recombination, in which crossovers are the key, ought to create new individuals with minimum disruption of the allocation strategy dictated by reproduction alone. Traditional simple crossover satisfies the requirement rather well, but OX does not. The survival probability of ordering schema (o-schema) under OX is relatively small, which leads to the design of EOX.

### 2.1 Order crossover (OX)

In ordering problems, solutions are represented by permutations of a list and order-

This project is supported by China 863 program.

ing schema (o-schema)[3] is used to analyze the performance of GAs. In o-schema, only relative positions of elements are considered. Schema has been normalized by taking the first element to be the successor of the longest string of don't care symbols "!". Assume $K$ is the length of a cut, $D$ is the length of a schema $H$, $L$ *is* the length of the string and $P(H, t)$ is the expected proportion of $H$ represented in population at time $t$. The survival probability $p(s)$ of o-schema $H$ under OX is:

$$p(s) = p(s/w)p(w) + p(s/o)p(o) + p(s/1c)p(1c) + p(s/2c)p(2c) \tag{1}$$

where

$w$ (within) $H$ is entirely within the cut section;

$c$ (outside) $H$ is entirely outside the cut section;

l$c$ (1 cut) $H$ contains one cut point;

2$c$ (2 cuts) $H$ contains two cut points.

We define schemata with $O = D$ as continuous schemata in the following part of this paper. For simplicity, we only consider survival probability of continuous schemata. Schemata whose $D$'s are close to $O$'s have similar behavior.

The expected survival probability $A_1$ of o-schema $H$ under OX is:

$$\begin{aligned} A_1 &= P_1\left(1 - P(H,t)\right) + P_2 \cdot P(H,t) \\ &= P_1 + (P_2 - P_1)P(H,t) \end{aligned} \tag{2}$$

where, $P_1$ is the survival probability of $H$ under OX when only one parent contains $H$; $P_2$ is the survival probability of $H$ under OX when both parents contain $H$. From [3], we have

$$P_1 \approx \frac{(L-D)(L-D+1)}{2L(L-1)} + \frac{1}{L-1}\sum_{K=1}^{L-1}\left(1-\frac{K}{L}\right)^D \tag{3}$$

$$P_2 = \frac{(L-D)(L-D+1)}{L(L-1)}\left(1+\frac{L-D+1}{L}\right) + \frac{2(D-1)(L-D+1)}{L(L-1)}\frac{D+2}{2L} + \frac{(D-1)(D-2)(L+2)}{2L^2(L-1)} \tag{4}$$

Due to space limitation, the complicated mathematical derivation is eliminated in this paper and it can be found in NCIC technical report.

## 2.2 EOX

In OX, when the section of the first parent between the first and second cut points is copied to the offspring, the remaining places are filled with elements not occurring in this crossover section, using the order the elements are found in the second parent after the second cut points. Therefore, when both parents contain continuous o-schema $H$, the survival probability of $H$ under crossover is sensitive to the position of the second cut point in the second parent.

EOX proceeds almost the same as OX. The difference is only that after two cut points are chosen at random in the first parent, the second parent is rotated until the element just before the second cut point is the same as the element just before the second cut point in the first parent. This procedure is shown in the following example.

| | | 1st cut \|point | 2nd cut \|point |
|---|---|---|---|
| parent1 | h k c e | f d b l a | i g j |
| parent2 | a b c d | e f g h i | j k l |
| rotated parent2 | e f g h | i j k l a | b c d |
| offspring | h i j k | f d b l a | c e g |

The expected survival probability $A_2$ of o-schema $H$ under EOX is:

$$A_2 = P_1'(1-P(H,t)) + P_2' \cdot P(H,t) = P_1' + (P_2' - P_1')P(H,t) \quad (5)$$

$P_1'$ equals $P_1$ of OX, but $P_2'$ is different from $P_2$ and is calculated as follows.

$$P_2' \approx E[p(w)] + E[p(o)] + \frac{E[p(1c)]}{2} + E[p(2c)] = \frac{(D-1)^2 + L(L-D)}{L(L-1)} \quad (6)$$

When $L \gg 1$,

$$P_2' \approx 1 - \frac{D}{L} + \left(\frac{D}{L}\right)^2 \quad (7)$$

When $D \approx L/2$, $P_2$ reaches its approximate minimum 3/4.

In case of $L \gg 1$, expected survival probabilities of schema $H$ under OX and EOX are shown in Table 1. When $D$ is close to $L$, $A_2 \approx P(H,t)$. This implies if $P(H,t)$ is large, even though $H$ is long, the survival probabilities of $H$ under EOX is high. But because $A_1 \approx P(H,t)/2$, even though the proportion of $H$ is high, only half of $H$ can survive under OX. Because long, high-order and high-performance schemata are difficult to survive and so cannot construct better, longer schemata further, GAs with OX are difficult to get good results.

Table 1. Expected survival probability $A_1$ and $A_2$

| $D$ | 1 | … | $L/2$ | … | $L$ |
|---|---|---|---|---|---|
| $A_1$ | 1 | | $\frac{1}{8}+\frac{5}{16}P(H,t)$ | | $\frac{1}{2}P(H,t)$ |
| $A_2$ | 1 | | $\frac{1}{8}+\frac{5}{8}P(H,t)$ | | $P(H,t)$ |

In traditional bit representation with $k$-ary alphabet where only value schema (v-schema) is involved, each particular string of length $L$ is a representative of $2^L$ v-schemata. The total number of v-schemata in the representation is $(k+1)^L$. In the permutation representation where o-schema is involved, each particular string of length $L$ is still a representative of $2^L$ o-schemata. But the total number of o-schemata represented is much larger than $(L-1)!$. In the same number of trials, the proportion of o-schemata explored by GAs in ordering problems is far less than the proportion of v-schemata explored in value problems. In contrast to OX, EOX is good at exploitation and weak at exploration. These result in the fact that GAs using EOX are easier to trap in local optima. Therefore it is necessary for EOX to cooperate with other operators that can enhance exploration ability to enable GAs to jump out local optima.

## 2.3 Experimental results

The efficiency of EOX is tested by solving TSP. A 30 cities Euclidean TSP[3] is used for experiments. The length of optimal tour is 423.7. Crossover is always applied to 80% of the population. The termination condition of GAs is that the rate of min, the minimum tour length in population, to exp, the average tour length of population, is larger than 0.95, i.e. $\min/\exp > 0.95$.

For different population sizes, typical results of GAs using OX and EOX are listed in Table 2, respectively.

Although GAs using EOX are converged quickly, they are easily trapped in local optima. However, when the operator INVE, which randomly inverses a part of the string, is incorporated, and elitist strategy (the best individual obtained so far is reserved and inserted into following generation) is used, the result is greatly improved. INVE is applied to 10% of population. Typical results of GAs using INVE are listed in Table 3, where time is measured by number of generations, solution means the length of shortest tour found.

Table 2. Comparison between OX and EOX

| Popu. size | OX | | EOX | |
|---|---|---|---|---|
| | Time | Solu. | Time | Solu. |
| 30 | 290 | 508 | 113 | 575 |
| | 217 | 488 | 67 | 616 |
| | 201 | 537 | 60 | 684 |
| | 263 | 467 | 79 | 538 |
| | 464 | 506 | 41 | 759 |
| 60 | 1 554 | 425.5 | 90 | 572 |
| | 1 512 | 424.9 | 156 | 530 |
| | 915 | 429.5 | 86 | 565 |
| | 1 783 | 447.9 | 96 | 519 |
| | 1 518 | 423.7 | 100 | 574 |

Table 3. Solving TSP by GAs using EOX and INVE

| Popu.size | Time | Solu. |
|---|---|---|
| 60 | 330 | 425.3 |
| | 249 | 476.4 |
| | 343 | 428.8 |
| 90 | 349 | 428.6 |
| | 402 | 423.7 |
| | 481 | 425.2 |
| | 574 | 425.2 |
| | 531 | 425.5 |

# 3 Heuristic crossover operator GREE

## 3.1 Mechanics of GREE

We define o-schemata with $D=O=2$ as 2-order continuous o-schemata. In problems where compact schemata are important (such as TSP), 2-order continuous o-schemata of high performance are basic building blocks. GREE is able to proliferate high-performance 2-order continuous schemata in population.

First, we calculate the performance of 2-order continuous schemata in TSP. Suppose city number is *n*, the length of edge *ij* between city $i$ and $j$ is $w_{ij}$, $(i,j=1,\cdots,n)$, $w_{ii}=0$. The performance of o-schema $H=lm!\cdots!$ is the average length of all tours passing edge *lm*.

There are $(n-2)!$ different tours passing *lm*, among which each tour is composed of 3 parts: ① edge *lm*; ② edges *li* and *mj*, ( $i,j\in P_{n-2}$, $i\neq j$, $P_{n-2}$ is the set of all cities (nodes) except *l* and *m*); ③ path $i\cdots j$, which consists of $n-3$ edges. Totally, for all $i\in P_{n-2}$, $(n-3)!$ tours pass edge *li*, for all $j\in P_{n-2}$, $(n-3)!$ tours pass edge *mj*, and for all $i,j\in P_{n-2}, i\neq j$, tours pass edge *ij*.

So the average length $\overline{L_{lm}}$ of the $(n-2)!$ tours passing *lm* is:

$$
\begin{aligned}
\overline{L_{lm}} &= \frac{(n-2)!w_{lm}+(n-3)!\sum_{i\in P_{n-2}}(w_{li}+w_{mi})}{(n-2)!} \\
&\quad + \frac{2(n-3)!\sum_{i,j\in P_{n-2}}w_{ij}}{(n-2)!} \\
&= w_{lm}+\frac{1}{n-2}\sum_{i\in P_{n-2}}(w_{li}+w_{mi}) \\
&\quad +\frac{2}{n-2}\sum_{i,j\in P_{n-2}}w_{ij} \\
&= \frac{n-4}{n-2}w_{lm}-\frac{1}{n-2}\sum_{i\in P_{n-2}}(w_{li}+w_{mi}) \\
&\quad +\frac{2}{n-2}\sum_{i,j\in P_n}w_{ij}
\end{aligned}
\tag{8}
$$

where $\sum_{i,j\in P_{n-2}}w_{ij}$ represents the total length of different edges construted by cities in $P_{n-2}$. Since the third item in the last equation is constant to all

2-order continuous schemata, we choose the first two items as discriminant item $\Theta$ :

$$\Theta_{lm} = \sum_{i\in P_{n-2}} (w_{li} + w_{mi}) - (n-4) w_{lm} \qquad (9)$$

2-order continuous schemata with bigger $\Theta$ is better.

Now it is ready to describe the mechanics of heuristic crossover GREE, which is just like Grefenstette's heuristic crossover[4], called GREFEN in this paper. GREE constructs an offspring from two parent tours as follows. Pick a random city as the star ting point for the child's tour. Compare the four edges leaving the starting city in the parents and choose the edge with the biggest $\Theta$ (instead of the shortest edge). If the parental edge with biggest $\Theta$ would introduce a cycle into the partial tour, then choose the edge with the second biggest $\Theta$, and so on. If all the four edges introduce cycles, then extend the tour by a random edge. Continue to extend the partial tour using these steps, until a complete tour is generated. The discriminant items $\Theta_{ij}$, $(i, j = 1, \cdots, n)$ can be calculated before GAs run. So the computational complexity of GREE's constructing an offspring is $O(n)$, similar to that of GREFEN.

## 3.2 Experimental results

To examine the performance of heuristic crossover operators, a number of instances of TSP with 3 060 100 and 318 cities are solved, and the results on 30 and 318 cities TSP are presented here. Experimental results of GAs using GREE and GREFEN on 30 city TSP, respectively, are listed in Table 4. Crossover rates are 80%. Elitist strategy is used. GAs with GREE run faster and reaches better result than GAs with GREFEN, and the former always find tours within 5% from optimal tour when population is not too small.

Table 4. Comparing GREE and GREFEN

| Popu. size | GREE | | GREFEN | |
|---|---|---|---|---|
| | Time | Solu. | Time | Solu. |
| 30 | 16 | 432.7 | 16 | 441.4 |
| | 12 | 440.2 | 27 | 437.7 |
| | 24 | 452.3 | 25 | 440.3 |
| | 20 | 434.0 | 15 | 440.7 |
| | 12 | 432.2 | 22 | 459.8 |
| 60 | 20 | 430.5 | 30 | 437.9 |
| | 15 | 436.8 | 20 | 440.7 |
| | 19 | 428.8 | 33 | 433.0 |
| | 20 | 431.8 | 21 | 442.7 |
| | 18 | 430.5 | 28 | 436.6 |

A 318 city TSP, LIN318, is also solved by GAs with GREE. If combining GAs with other local search methods, results can be improved further. In the experiments, when GAs ended (no more than 500 generations), a fine tuning method, 2-opt[2] is introduced. Errors are reduced from 10% to 5% with respect to optimal tour. Typical results are listed in Table 5. The optimal tour length of the problem is 41.3[5].

Table 5. Using GREE to solve LIN318

| Popu. size | GREE | | Final |
|---|---|---|---|
| | Time | Solu. | Solu. |
| 600 | 396 | 45.1 | 43.3 |
| | 440 | 45.6 | 44.6 |
| | 500 | 44.3 | 42.9 |
| 900 | 500 | 45.2 | 43.6 |
| | 500 | 44.3 | 43.1 |

## 3.3 Combining GREE with EOX

EOX is able to search and preserve long, high-order, high-performance schemata. But using it directly to initial population is inefficient. GREE runs fast but cannot get very good results. We use GREE to initial population first, and when 2-order continuous schemata of high performance are rich in population, which are basic building blocks of high-order, high-performance schemata, then the EOX is adopted. Experiments on 30 cities TSP get excellent results. In experiments, GREE is used until $\min/\exp > 0.9$. Typical results are list in Table 6. Optimal or very near optimal tours are found in rather short time.

Table 6. GAs using GREE→ EOX

| Popu. size | Time | Solu. |
|---|---|---|
| 60 | 99 | 423.9 |
| | 46 | 424.9 |
| | 55 | 424.4 |
| | 47 | 428.8 |
| | 104 | 423.7 |
| 90 | 124 | 423.9 |
| | 87 | 423.7 |
| | 151 | 423.9 |
| | 76 | 423.9 |
| | 105 | 424.1 |

## References:


[1] GOLDBERG D E. Genetic algorithms in search, optimization and machine learning[M]. Massachusetts: Addison-Wesley, 1989.

[2] LIN S, KERNIGHAN B W. An effective heuristic algorithm for the traveling-salesman problem[J]. Operations Research, 1973, 21(2): 498-516.

[3] OLIVER I M, SMITH D J, HOLLAND J R C. A study of permutation crossover operators on the traveling-salesman problem[C]// The 2nd International Conference on Genetic Algorithms and Their Application. 1987: 224-230.

[4] GREFENSTETTE J J, GOPAL R, ROSMAITA B J, et al. Genetic algorithms for the traveling salesman problem[C]// The International Conference on Genetic Algorithmsand Their Application. [S.l.:s.n.], 1985: 160-168.

[5] LAELER E L, LENSTRA J K, KAN A H G R, et al. The traveling salesman problem[M]. [S.l.]: John Wiley & Sons, 1985.

发表于《计算机学报》1997年第1期

# 求解SAT问题的局部搜索算法及其平均时间复杂性分析

刘涛[1,2]，李国杰[1,2]

1. 中国科学院计算技术研究所，北京 100080；

2. 国家智能计算机研究开发中心，北京 100080

**摘　要：**SAT问题在人工智能、VLSI设计和计算理论等领域有着广泛的应用背景。近年来，局部搜索算法在求解SAT问题时得到了巨大的成功。除提出了多种改进策略之外，还对一般局部搜索算法进行了平均时间复杂性分析。假设SAT问题中CNF公式的子句数为$n$，变量数为$m$，每个子句长度为$L$，则局部搜索算法的平均时间复杂性分析结果为：当$n<2^L m$时，求解随机SAT问题的算法复杂性为$O(n^{O(1)}L(1+n^{1+\varepsilon}/m)),\varepsilon>0$，特别地，对于$n/m\leqslant 1$时的3-SAT问题，算法复杂性为$O(n^2(1+n^{1+\varepsilon}/m))$。当$n\ll m$时，算法复杂性为$O(nL)$。

**关键词：**SAT问题；局部搜索；回溯算法；平均时间复杂性

**中图分类号：**TP301

# Local search for solving sat problems and its average time complexity

Tao LIU[1,2], Guo-jie LI[1,2]

1. Institute of Computing Technology, Chinese Academy of Sciences, Beijing 100080;

2. National Research Center for intelligent Computing Systems, Beijing 100080

**Abstract:** This paper develops several efficient search strategies for the simple local search algorithm, and analyzes the average time complexity of the simple local search algorithm. For the randomly generated CNF formulae with $n$ clauses, $m$ variables, and $L$ laterals in each clause, the average run time of the local search algorithm for verifying the satisfiability is $O(n^{O(1)}L(1+n^{1+\varepsilon}/m)), \varepsilon > 0$, where $n < 2^L m$. Especially, for 3-SAT problems, the average time complexity is $O(n^2(1+n^{1+\varepsilon}/m))$, where $n/m \leqslant 1$. When $n \ll m$, the local search algorithm has $O(nL)$ average time complexity.

**Key words:** SAT problem, local search, backtracking algorithm, average time complexity

## 1 引言

对于一个随机生成的 CNF 公式来说，我们假定它有 $n$ 个子句，$m$ 个变量，而其中每个子句有 $L$ 个变量（或文字）SAT 问题就是要确定这个 CNF 公式是否可满足。有关求解 SAT 问题的局部搜索算法平均时间复杂性分析，目前已有一些研究成果。如文献[1]就给出了至少 3 种局部搜索算法的平均时间复杂性分析结果，其中最好的算法复杂性为 $O(\ln(m\log m)^2)$，其中 $L>3$，$n/m\leqslant\alpha\cdot 2^L/L,\alpha<L$。然而，我们在实验结果的分析中发现，当固定 $n$，增大 $m$ 时，局部搜索算法求解 SAT 问题所需时间明显减少，这主要是由于其算法分析时所做的假设条件与我们所实现的算法中所遇问题例有不同之处，因此我们将结合文献[1]中的算法分析给出局部搜索算法平均时间复杂性分析。

一般说来，算法分析就是确定算法所需时间的上界，并以此从理论上评价算法的效率。早期的算法分析一般只研究算法在最坏情况下的时间复杂性。后来，人们发展了平均时间复杂性分析方法。

客观地讲，上述两种分析方法各有利弊。由于本文中所涉及的研究对象主要是随机 SAT 问题，因此只讨论算法的平均时间复杂性，如不特别说明，算法复杂性即是指算法的平均时间复杂性。

## 2 求解 SAT 问题的局部搜索算法

J.Gu[1-3]和 B.Selman 等人的工作已经证实了优化方法求解 SAT 问题的有效性。我们在这里将算法的适用范围限定在约束较弱（足以使问题有解）的问题例上，然后给出几种新的求解策略。

运用优化方法的首要一步是建立目标函数。如前所述，我们将冲突数（不满足子句个数）作为目标函数，由于每个变量 $x_i(i=1,2,\cdots,m)$ 只能取二值（0 或 1），其可行解空间大小为 $2^m$。尽管保证找到全局最优解不是一件容易的事情，而对 SAT 问题，其目标函数的下界至少是零，一旦某个局部最优解使冲突数变为零，算法就可以停止并输出一个解。

求解 SAT 问题的局部搜索过程描述如下。

```
Procedure Find solution
Input: g(x) /*布尔函数*/
Output: True or Local Minimum Conflicts/*“真”或局部最小冲突数*/
begin
    x= initial_solution ()
    for K=1 to Numberofloops
        if Conflicts(x)==0 then return true;
        for j=l to n
            if Mj≠0 then
                w(Mj)=v(Mj)+1; /*Mj 的权加 1*/
                with probability P(j)=kw(Mj)
                    x'=testswap(xi);/ * xi∈ Mj*/
                    if conflicts (x')≤ Conflicts(x)
                    then r= x'
            end if
        endfor
    endfor
end procedure
```

其中 $x$ 代表一个可行解的形式，$x_i$ 是单项式 $M_j$ 中的一个变量。$P(j)$ 的大小由 $M_j$ 的权值来决定，权值越大，$P(j)$ 越大。

函数 testswap 的功能是将所有 $\boldsymbol{x}_i$ 代之以 $\overline{x}_i$ (=1−$x_i$)，并返回一个新的解矢量 $\boldsymbol{x}'$。函数 Conflicts 计算当前变量赋值的冲突数，注意实

际上只需计算由于 $x_i$ 的“翻转”而引起的冲突数变化。

上述 Find solution 过程与现有局部搜索算法有两个重要区别。首先，我们在这里使用的不等式为“≤”而不是“<”。这样做的一个直观解释是要与目标函数的选取协调起来。我们知道,冲突数的最大值只能是子句的个数 $n$。在大多数情况下，冲突值都分布在值较小的范围里，算法中使用的不等式“≤”可以扩大算法的搜索空间，避免算法经常落入局部最优点。此外，我们在算法中使用“权”来避免算法陷入局部最优并加快算法的收敛速度。经验还告诉我们，子句的权值越大，表明该子句“经常”得不到满足，实质上是因为这些子句中包含了某些强约束变量。对于原始的随机（均匀分布）SAT 问题而言，各变量的约束强弱比较接近，所以常常是不满足子句的个数较小，但它们又不是固定的几个子句。在这种情况下，该问题例多半是无解的，然而，要证明它无解却是相当困难的，当问题例的规模稍大时甚至不能在容许的时间内证明它的无解性。我们把这一个困难归因于 Co-NP 问题的难解性。

上述局部搜索算法的收敛速度较快，但在另一方面这也是其缺陷之所在，因为对于困难的（有解）问题例来说有时会找不到解。为此，我们提供了如下几种附加策略。

（1）当冲突数在有限步之内不再降低时，重新初始化矢量 $\boldsymbol{x}$，再进行一次寻解过程。

（2）当已得到一个局部最优解时，强行改变部分变量的赋值，再进行寻解过程。

（3）当所求解的问题例难度较大时，可以将布尔函数 $g(x)$ 连续化，并增加一个罚函数使目标函数变为如下形式（非线性规划问题）：

$$G(x)=\sum_{i=1}^{n}\prod_{J=1}^{L}y_{ij}^{2}+\prod_{j=1}^{m}\beta_j x_j^2(1-x_j)^2$$

其中

$$y_{ij}=\begin{cases}1-x_j, x_j\in C_i\\ x_j, \bar{x}_j\in C_i\end{cases}$$

$\beta_j$（>0）可由子句长度决定，一般说来，$L$ 越大，$\beta_j$ 值越大。由于 $x_j$ 已变为 0 到 1 之间的连续变量，罚函数（$G(x)$中的第二个和式）的作用是对 $x_j$ 的非整数取值进行惩罚，以尽量避开非整数解。

（4）由于 SAT 问题可转化为一整数线性规划问题（Integer Linear Programming，ILP），可以使用隐枚举法或割平面法来求解变换后的问题。

一个 SAT 问题例的具体变换规则如下。首先将每个子句 $C_i$ 变为一个等价的不等式方程：

$$\sum_{x\in C_i}x+\sum_{\bar{x}\in C_i}(1-x)\geqslant 1$$

然后将第一个不等式方程变为：

$$\sum_{x\in C_i}x+\sum_{\bar{x}\in C_i}(1-x)\geqslant 1$$

目标函数选为 $y$，问题变为求 $y$ 的最大值。原布尔表达式有解当且仅当上述优化问题的整数最优解使 $y\geqslant 1$。

在此要特别强调的是，在一般情况下，整数线性规划问题的解不可能用可接受的方式舍入对应线性规则问题的解而得到（这同样适用非线性整数规划）。如在 3-SAT 问题中，若取所有 $x_i=1/2$，则它可以使所有约束方程得以满足，并使目标函数 $y\geqslant 1$。然而，这种最优解使我们无法通过舍入方法来取得相应 SAT 问题例的 0-1 解。

线性规划的对偶性允许人们成功地运用对偶单纯形算法，而整数线性规划的对偶理论却并不那么成熟。问题很可能就在于整数线性规划的难解性，因为即使整数线性规划及其对偶问题都存在一个整数解，也不能保证这两个解相等。

虽然有人会怀疑局部搜索法只适用于容易

的问题例，但事实是它同样适用于难解的问题例（只要所遇到的问题例有解）。我们所面对的一个主要困难是难以分析局部搜索算法在求解困难问题例时的算法复杂法。尽管上面我们已给出了各种求解策略来改进局部搜索算法的性能，为了抓住问题的本质，本文将只对一般的局部搜索算法进行算法平均时间复杂性分析。

下面是局部搜索算法的一般性描述。

```
Procedure Local Search
Input: one SAT problem instances
Output: one solution of false
begin
        xi= initial solution;
        while there exists a feasible solution
xi∈N(xi)
        /*其中 N(xs)表示 xs 的邻域内可行
解集合*/
        let  c(xt) < c(xs)  do
        xs = xt
        if  c(xs) == 0  return (0);/*返回一
个解*/
        end while
end Procedure
```

## 3 求解 SAT 问题的局部搜索算法复杂性分析

如前所述，随机生成的 SAT 问题满足均匀分布。如果以 $x=\{x_1,x_2,\cdots,x_m\}$ 表示所有变量的集合，$w$ 表示子句中的一个文字，则

$$P_r(w=x_i)=P_r(w=\bar{x}_i)=\frac{1}{2m}$$

此外，我们还设定所有问题例的子句长度相同并且在任何一个子句中都不允许有重复变量。

给定一个矢量 $\boldsymbol{v}\in\{0,1\}^m$，$v_k$ 表示 $\boldsymbol{v}$ 中的第 $k$ 个分量，$F(\boldsymbol{v})$、$C_i(\boldsymbol{v})$ 和 $w(\boldsymbol{v})$ 分别表示布尔表达式 $F$，子句 $C_t$ 和文字 $w$ 的值。如不产生混淆，有时我们也用 $C_i$ 表示子句 $C_i$ 中出现的文字集合。

假设 $S_0(\boldsymbol{v})=\{C_i|C_t(\boldsymbol{v})=0\}$，$S_1(\boldsymbol{v})=\{C_t|C_i(\boldsymbol{v})=1\}$，则显然有 $|S_0(\boldsymbol{v})|+|S_1(\boldsymbol{v})|=n$，$n$ 是布尔表达式中的子句数，其中 $|S_0(\boldsymbol{v})|$ 就是局部搜索法选定的目标函数。

**引理 1**[1]：$\left(1-\frac{1}{n}\right)^n$ 是一个单调增函数，

$$\lim_{n\to\infty}\left(1-\frac{1}{n}\right)^n=1/e,$$

当 $n\geqslant 2$ 时

$$\left(1-\frac{1}{n}\right)^n\geqslant 1/4。$$

**引理 2**[1]：设 $h_1,h_2>1$ 和正整数 $t$，$G(t,h_1,h_2)$ $=\min\{r|$在 $r$ 次独立的伯努利试验（每次成功的可能性至少 $1/h_1$）中，至少有 $t$ 次成功的可能性不小于 $1-1/h_2\}$，那么

$$G(t,h_1,h_2)\leqslant 2h_1(t+\ln h_2)$$

**引理 3**[1]：假设局部搜索法的初始解矢量为 $\boldsymbol{v}_0$，而且随机生成的布尔表达式有 $n$ 个子句，$m$ 个变量，子句长度为 $L$，那么

$$P_r(|S_0(\boldsymbol{v}_0)|\leqslant n/2^L)\geqslant 1/2$$

**定义 1**：设 $k\geqslant 1$，矢量 $\boldsymbol{v}_{k-1}$ 使 $|S_0(\boldsymbol{v}_{k-1})|>0$，事件 $E$ 被定义为：

$$\exists C_i\in S_0(\boldsymbol{v}_{k-1}),\ \exists \boldsymbol{v}_k\in N(\boldsymbol{v}_{k-1})$$

s.t.

$$C_i(\boldsymbol{v}_k)=1\ \text{and}\ (\forall C_j\in S_1(\boldsymbol{v}_{k-1}),C_j(\boldsymbol{v}_k)=1),$$

其中，

$$\mathrm{N}(v)=\{v'\,|\,v'\in\{0,1\}^m\ 且\ H(v,v')\leqslant 1\}$$

而 $H(\boldsymbol{v},\boldsymbol{v}')$ 为 $\boldsymbol{v}$ 与 $\boldsymbol{v}'$ 之间的 Hamming 距离。

**引理 4**[1]：

$$P_r(E)\geqslant 1-\left[1-\left(1-\frac{k-1}{m}\right)\left(1-\frac{1}{2^{L-1}m}\right)^n\right]^{L\cdot|S_0(v_{k-1})|}$$

**引理 5**[4]**：** 当$n<2^L m$时，局部搜索法所求得的第一个局部最优解就是 SAT 问题解的可能性，

$$p=\frac{1}{2}\prod_{i=1}^{n/2^L}\left\{1-\left[1-\left(1-\frac{n}{2^L m}\right)\left(1-\frac{n}{2^{L-1}m}\right)^n\right]^{L\times i}\right\}$$

可以从引理 5 看出 $p$ 随 $m$ 或 $L$ 的增大而增大但随 $n$ 的增大而减小。注意引理中要求$n<2^L m$，这一限制条件的合理性可以通过实验得到证实，因为当$n\geqslant 2^L m$时随机 SAT 问题例一般是不可满足的。

**定理 1：** 当$n<2^L m$时，局部搜索算法的平均时间复杂性为$O(n^{O(1)}L(1+n^{1+\varepsilon}/m))$，$\varepsilon$ 是一个充分小的正数。特别地，当$\beta\geqslant(1-\beta)^L$、$\beta=(1-\alpha)\left(1-\frac{2L\alpha}{n}\right)^n$、$\alpha=\frac{n}{2^L m}$时，局部搜索算法复杂性为$O(n^2L(1+n^{1+\varepsilon}/m))$。

**证明**：设$n, m$是充分大的正整数，$L\geqslant 3$，由引理 5 知：

$$p=\frac{1}{2}\prod_{i=1}^{n/2^L}\left\{1-\left[1-\left(1-\frac{n}{2^L m}\right)\left(1-\frac{n}{2^{L-1}m}\right)^n\right]^{L\times i}\right\}$$

设$\alpha=\frac{n}{2^L m}<1, \beta=(1-\alpha)\left(1-\frac{2L\alpha}{n}\right)^n\geqslant(1-\alpha)(1/4)^{2L\alpha}$，$c\log n<n/2^L$，且$(1-\beta)^{Lc}\leqslant 1/2$则

$$p\geqslant\frac{1}{2}\times\prod_{i=1}^{c\log n}\left\{1-\left[1-(1-\alpha)\left(1-\frac{2L\alpha}{n}\right)^n\right]^{L\times i}\right\}\times$$

$$\prod_{i=c\log n+1}^{n/2^L}\left\{1-\left[1-(1-\alpha)\left(1-\frac{2L\alpha}{n}\right)^n\right]^{L\times i}\right\}$$

$$\geqslant\frac{1}{2}\times\beta^{c\log n}[1-(1-\beta)^{Lc\log n}]^{n/2^L}$$

$$\geqslant\frac{1}{2}\times\beta^{c\log n}(1-1/n)^{n/2^L}$$

考虑到$\beta<1$，由引理 1

$$p\geqslant\frac{1}{n^r} \quad (1)$$

其中 $r$ 是一个正常数（通常情况下，$r$ 随 $m$ 和 $L$ 的增大而减小）。

当$\alpha$足够小以使$\beta\geqslant(1-\beta)^L$时，总可以找到一个常数$c>0$使

$$\beta^c\geqslant 1/2 \text{和} (1-\beta)^{Lc}\leqslant 1/2$$

同时成立，此时

$$p\geqslant\frac{1}{2n} \quad (2)$$

由式（1）及引理 2，取$h_1=n^r$，$t$=1 和$h_2=\mathrm{e}^k$，则

$$G(1,n^r,n^k)\leqslant 2n^r(1+k)$$

所以局部搜索法在求得第$2n^r(1+k)$个局部最优解之前就找到 SAT 问题解的概率至少是$1-1/h_2=1-\mathrm{e}^{-k}$。

注意每次初始化解的时间复杂性是$O(nL)$。另外算法找到一个局部最优解平均需要进行$O(n^{1+\varepsilon})$次“翻转（Swap）”操作，而一次“翻转”操作的平均时间复杂性为$\Theta(nL/m)$）。因此，算法复杂性为$O(n^{O(1)}L(1+n^{1+\varepsilon}/m))$。

由式（2）及引理 2，取$h_1=2n$，$t$=1 和$h_2=\mathrm{e}^k$，则当$\alpha$足够小以使$\beta\geqslant(1-\beta)$成立时，

$$G(1,2n,\mathrm{e}^k)\leqslant 4n(1+k)$$

所以局部搜索法此时在求得第$4n(1+k)$个局部最优解之前就找到 SAT 问题解的概率至少是$1-1/h_2=1-\mathrm{e}^{-k}$。

由此可以看出，在这种情况下，算法复杂性为$O(n^2L(1+n^{1+\varepsilon}/m))$。　　证毕。

注：本文中将用$\log x$来表示$\log_2 x$。

图 1~图 3 分别说明了算法所需时间与 $n$、$m$、$L$ 之间的关系。注意本文图中的时间单位皆为“秒”。

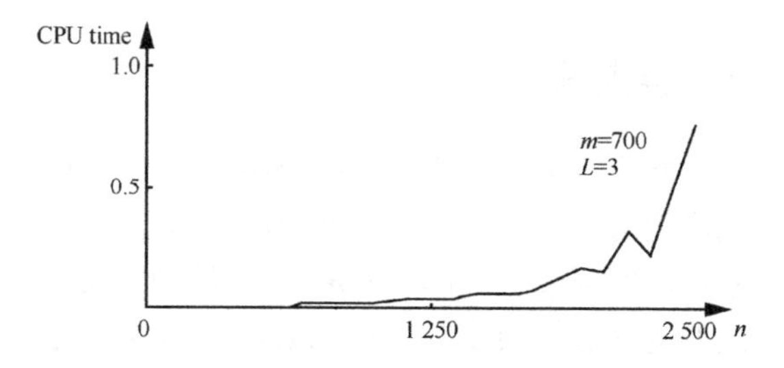


**图 1　局部搜索算法所需时间与 $n$ 之间的关系**

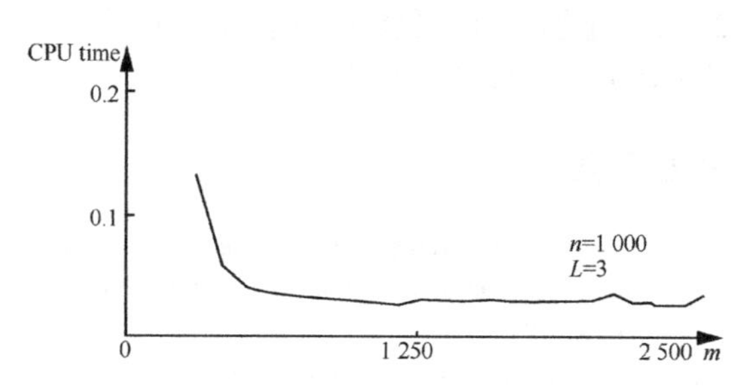


**图 2　局部搜索算法所需时间与 $m$ 之间的关系**

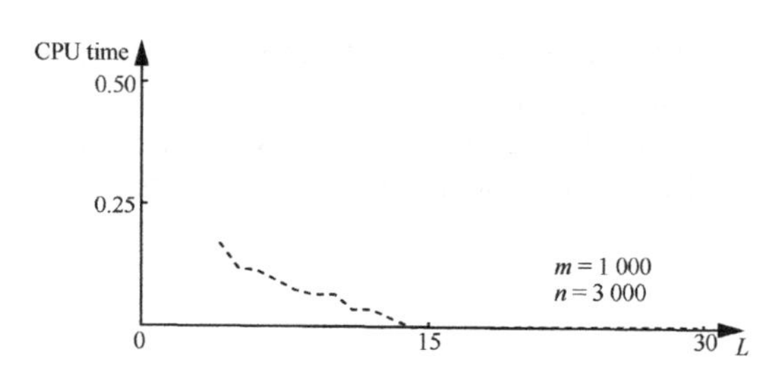


**图 3　局部搜索算法所需时间与 $L$ 之间的关系**

说明：定理 1 只是说明了局部搜索算法复杂性的一种上界，由于其中忽略了 $L$ 的某些作用，所以它与实验结果有一定出入。

图 4 描述了局部搜索算法复杂性与 $n/m$ 之间的关系。

**推论 1:** 设有一个 $n/m \leqslant 1$ 的 3-SAT 问题，局部搜索算法求解此问题的算法复杂性为 $O(n^2(1+n^{1+\varepsilon}/m))$ 。

**证明:**　设 $n \leqslant m$ ，$L=3$，则

$$\alpha = \frac{n}{2^L m} \leqslant 1/8$$

$$\beta = (1-\alpha)\left(1-\frac{2L\alpha}{n}\right) \geqslant (7/8)\left(1-\frac{3}{4n}\right)^n$$

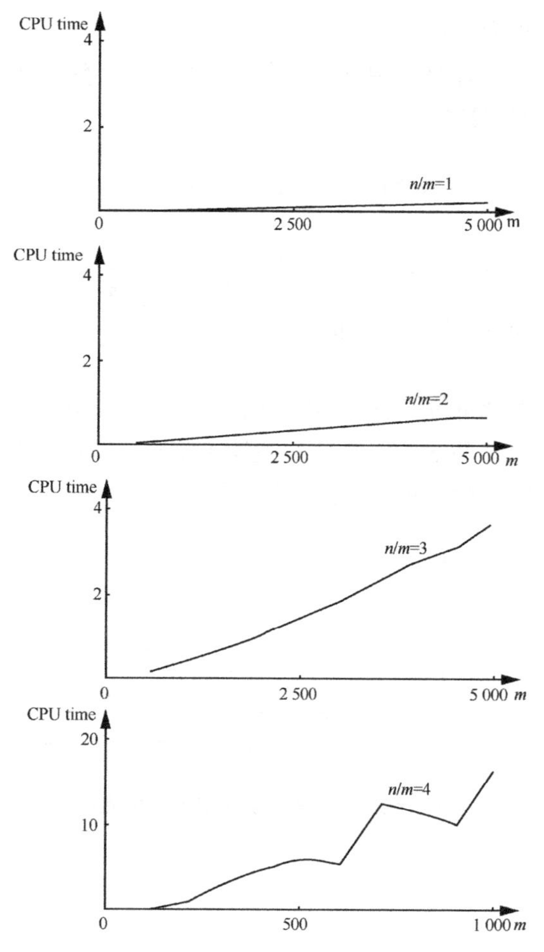


**图 4　局部搜索算法所需时间曲线**

当 $n$ 足够大时，$\beta \geqslant 1/3$，因此

$$\beta \geqslant (1-\beta)^L$$

由定理 1 可知算法复杂性为 $O(n^2(1+n^{1+\varepsilon}/m))$。

证毕。

为保证每个变量在布尔表达式中皆会出现，一般需假设 $Ln \geqslant m$ 即

$$\frac{1}{L2^L} \leqslant \alpha < 1$$

然而，对于随机问题而言，这种限制并没有太大必要。在上述分析中实际上也是允许 $m \gg n$ 这种情况的，因此定理 1 和推论 1 在 $m \gg n$ 时同样成立。

**推论 2:** 当 $m \gg i$ 时，局部搜索算法复杂

性为 $O(nL)$ 。

**证明：** 在定理 1 的证明过程中，我们曾经得到如下不等式：

$$p \geqslant (1/2)\cdot \beta c \log n \cdot \left(1-\frac{1}{n}\right)^{n/2^L}$$

其中 $\beta=(1-\alpha)\left(1-\frac{2La}{n}\right)^n, \alpha=\frac{n}{2^L m}$，当 $m \gg n$ 时，$\alpha \to 0$，$\beta \to 1$，

$$p \geqslant 1/4$$

由此可得算法复杂性为 $O(nL)$ 。 证毕。

图 5 说明了当 $m$ 逐渐变大时，算法所需时间渐渐变小并最终稳定下来。

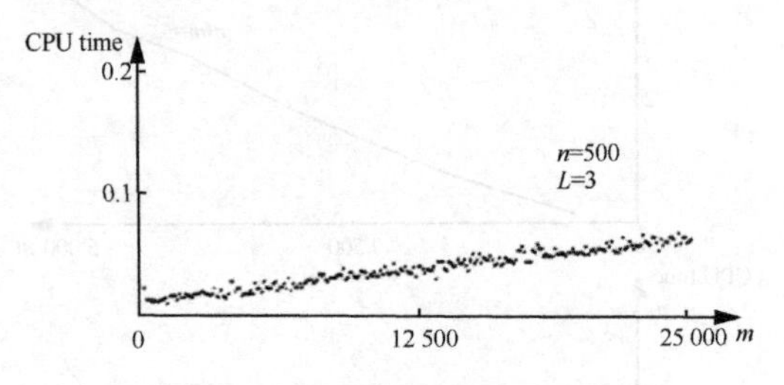


**图 5 $m \gg n$ 时局部搜索算法所需时间曲线**

## 4 结束语

由以上分析可以得到如下结论：当 SAT 问题的约束较弱时，局部搜索算法具有多项式平均时间复杂性。当 $n < 2^L m$ 时，求解随机 SAT 问题的算法复杂性为 $O(n^{O(1)}L(1+n^{1+\varepsilon}/m))$ ，是充分小的正数，特别地，对于 $n/m \leqslant 1$ 时的 3-SAT 问题，算法复杂性为 $O[n^2(1+n^{1+\varepsilon}/m)]$ 。当 $n \ll m$ 时，算法复杂性为 $O(nL)$ 。

## 参考文献：

[1] GU J, GU Q P. Average time complexities of several local search algorithms for the satisfiability problem (SAT)[R]. 1991.

[2] GU J. Efficient local search for very large-scale satisfiability problems[J]. Acm Sigart Bulletin, 1992, 3(1): 8-12.

[3] GU J. Constraint-based search[M]. New York: Cambridge University Press, 1993.

[4] AHRONI R, ERDOS P, LINAL N. Optima of dual integer linear programs[J]. Combinatorica, 1988, 8(1): 13-20.

发表于 *Nucleic Acids Research*, 2003年第9期

# Topological structure analysis of the protein-protein interaction network in budding yeast

Dong-bo BU[1], Yi ZHAO[1], Lun CAI[1], Hong XUE[2], Xiao-peng ZHU[2], Hong-chao LU[1], Jing-fen ZHANG[1], Shi-wei SUN[1], Lun-jiang LING[2], Nan ZHANG[2], Guo-jie LI[1], Run-sheng CHEN[1,2*]

1. Bioinformatics Research Group, Key Laboratory of Intelligent Information Processing, Institute of Computing Technology

2. Bioinformatics Laboratory, Institute of Biophysics, Chinese Academy of Sciences, Beijing, China

**Abstract:** Interaction detection methods have led to the discovery of thousands of interactions between proteins, and discerning relevance within large-scale data sets is important to present-day biology. Here, a spectral method derived from graph theory was introduced to uncover hidden topological structures (i.e. quasi-cliques and quasi-bipartites) of complicated protein-protein interaction networks. Our analyses suggest that these hidden topological structures consist of biologically relevant functional groups. This result motivates a new method to predict the function of uncharacterized proteins based on the classification of known proteins within topological structures. Using this spectral analysis method, 48 quasi-cliques and six quasi-bipartites were isolated from a network involving 11 855 interactions among 2 617 proteins in budding yeast, and 76 uncharacterized proteins were assigned functions.

## 1 Introduction

With the availability of complete DNA sequence data for many prokaryotic and eukaryotic genomes, a formidable challenge of post-genomic biology is to understand how genetic information results in the concerted action of gene products both temporally and spatially to achieve biological function, as well as how they interact with each other to create an organism. It is important to develop reliable proteome-wide approaches for a better understanding of protein functions[1-2]. Genomic approaches have been used to predict functions of a large number of genes based on their sequences. However, as we know, proteins rarely act alone at the biochemical level; rather,

* To whom correspondence should be addressed at Bioinformatics Laboratory, Institute of Biophysics, Chinese Academy of Sciences, Beijing, China.

Tel: +8610 64888543; Fax +8610 64871293; Email: crs@sum5.ibp.ac.cn

Correspondence may aslo be addressed to Guojie Li. Tel: +8610 62565533; Fax: +8610 62567724; Email: lig@ict.ac.cn

The authors wish it to known that, in their opinion, the first two authors should be regraded as joint First Authors

they interact with other proteins as an assembly to perform particular cellular tasks. Having systematic functions, these assemblies represent more than the sum of their parts[3]. Traditionally, protein interactions were studied individually by genetic, biochemical and biophysical techniques focusing on a few proteins at a time[4]. It is increasingly realized that dissecting the genetic and biochemical circuitry of a cell prevents us from further understanding the biological processes as a whole. Basic constituents of cellular protein complexes and pathways, protein-protein interactions are key determinants of protein function. It is believed that all biological processes are essentially and accurately carried out through protein-protein interactions.

In the last 3 years, high-throughput interaction detection approaches, such as yeast two-hybrid systems[5-6], protein complex purification techniques using mass spectrometry[3,7], correlated messenger RNA expression profiles[8-9], genetic interaction data[10-11] and 'in silico' interaction predictions derived from gene context analysis gene fusion[12-13], gene neighborhood[14-15] and gene co-occurrences or phylogenetic profiles[16-17], have been developed and they have created a number of datasets regarding protein-protein interactions for several model organisms (Saccharomyces cerevisiae, Caenorhabditis elegans and Helicobacter pylori). These large-scale datasets open a door to comprehensive understanding of the genetic and biochemical phenomena in a cell. Subsequently, several promising methods have been successfully applied to this field. For instance, Schwikowski et al.[18] and Hishigaki et al.[19] predicted uncharacterized proteins based on interacting partners; Maslov and Sneppen[20] analyzed the stable topological properties of interaction networks; Ge et al.[21] provided the first global evidence that genes with similar expression profiles are more likely to encode interacting proteins; and Fraser et al.[22] revealed that the connectivity of well-conserved proteins in the network is negatively correlated with their rate of evolution. These studies revealed that the available data from protein-protein interaction networks in S.cerevisiae share some unexpected features with other complex networks.

The topological pattern of interactions is a rich source of biological functional information, and therefore we need to develop methods to mine and to understand the interaction networks. Here, we applied the spectral analysis method, which has been successful used in other fields[23], to proteomics to identify topological structures of interaction networks, i.e. quasi-cliques and quasi-bipartites. Interestingly, we found that the proteins within same group share similar biological functions. Moreover, for one-third of proteins that are still uncharacterized in S.cerevisiae, this method provides a new approach to predict their functions based on topological structures.

## 2 Materials and methods

### 2.1 Spectral analysis

Spectral analysis is a powerful tool to reveal high-level structures underlying enormous and complicated relationships. As a famous paradigm, David Gibson, Jon Kleinberg and Prabhakar Raghavan did excellent work on extracting information from link structure of the Web[23-24]. The World Wide Web is known to be composed of an increasing number of pages with hyperlinks pointing to other pages. Despite high complexity of the Web structure, spectral analysis was successfully used to discover 'authoritative' information sources and 'hub' pages joining authoritative ones together.

We applied the spectral analysis method to complicated protein-protein interaction networks and identified interesting topological structures. In this method, a network is represented by a bi-directed graph $G(V,E)$, i.e. vertex set including each protein as a vertex $V=\{P_1,P_2\cdots P_n\}$ , and the edge set $E=\{(P_i,P_j|$ there is an interaction between protein $P_i$ and $P_i$ $\}$. The symmetric $n\times n$ adjacent matrix is defined as $\boldsymbol{A}=(a_{ij})$ , where $a_{ij}=1$ if $(P_i,P_j)\in E$ , and $a_{ij}=0$ if $(P_i,P_j)\notin E$ .

Spectrum of the adjacency matrix $\boldsymbol{A}$ is essentially a reasonable measurement of properties of nodes that could be propagated across the interactions. Let us consider assigning a score to each node to represent their intensity, say $X$. A node with a high score would increase its neighbors' score through their interactions. In other words, two nodes are mutually reinforcing, which is in nature a cyclic definition of scores:

$$\Delta X_i=\sum_{j=1}^{n}a_{ij}\times X_j$$

The iteration method derived from Gibson et al.[23] and Kleinberg[24] is introduced to break such a cycle. It is interesting that $X_i$ converges to a fixed point from any initializing assignment, and it can be proved that the fixed point is one of the eigenvectors of matrix $\boldsymbol{A}$, which means it is an intrinsic characteristic of interactions. Moreover, since matrix $\boldsymbol{A}$ is symmetric, all of its eigenvectors are mutually orthogonal, which means that the corresponding properties are also mutually independent. In other words, each eigenvector represents a special property that none of the others could represent.

## 2.2 Identification of topological structures

From a topological point of view, the spectrum helps to uncover the hidden topological structures of a complex interaction network. We found that for each eigenvector with a positive eigenvalue, the proteins corresponding to absolutely larger components tend to form a quasi-clique (i.e. every two of them tend to interact with each other) (Figure 1(a)), whereas for each eigenvector with a negative eigenvalue, such proteins tend to form a quasi-bipartite (i.e. the proteins in which two disjoint subsets express high level connectivity between sets rather than within sets) (Figure 1(b)).

This observation can be explained as follows. The maximal eigenvalue of an adjacent matrix is the maximal value of

$$Q(\vec{x})=\sum_{i=1}^{m}A_{ij}x_ix_j=$$

$$\sum_{v_iv_j\in E}x_ix_j \text{ subjected to} \sum_{i=1}^{m}x_ix_j=1$$

(where $x_i$ is the $i$th component of the eigenvector). Other positive eigenvalues can also be described as the maximal value $Q$ with orthogonal condition. Since $Q$ is the summary of $x_ix_j$ corresponding to edge $v_iv_j$, it would be maximal when the nodes with more edges are assigned a larger value with the same signal, which form a quasi-clique intuitively. Similar quasi-bipartites would be obtained eigenvectors with negative

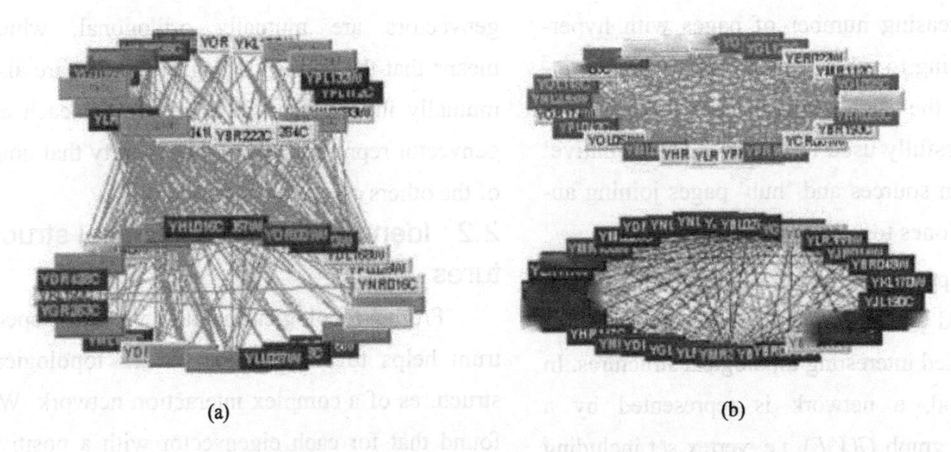

Figure 1. The topological structures of protein-protein interaction networks. In a quasi-clique, proteins tend to interact with each other (a), while in a quasi-bipartite, proteins between sets have denser interactions than those within sets (b)

eigenvalues.

We applied the clustering coefficient (CC)[25-26] in our analysis to quantify a quasi-clique's tendency to form a cluster. The ratio between the number of edges that actually exist between these $N$ nodes and the total number $N(N-1)/2$ gives the CC-value of a quasi-clique, i.e. $\mathrm{CC}=E/[N*(N-1)/2]*100\%$, where $E$ is the number of interactions within the clique and $N$ is the number of proteins in it. CC is greater than 0 and less than 1. A value close to 1 represents a clique close to a complete graph.

## 2.3 Assignment of annotation and P-values to quasi-cliques

As an isolated quasi-clique may involve different functional categories, P-values[27-28] are used as criteria to assign each quasi-clique a main function. Hypergeometric distribution was applied to model the probability of observing at least $k$, proteins from a quasi-clique size $n$ by chance in a category containing $C$ proteins from a total genome size of $G$ proteins, such that the $P$-value is given by

$$P=1-\sum_{i=0}^{k-1}\frac{\binom{C}{i}\binom{G-C}{n-i}}{\binom{G}{n}}$$

The above test measures whether a quasi-clique is enriched with proteins from a particular category more than would be expected by chance. If the $P$-value of a category is near 0, the proteins of the category in a quasi-clique will have a low probability of being chosen by chance. Here, we assigned each quasi-clique the main function with the lowest $P$-value in all categories.

# 3 Results

## 3.1 Data source and analysis

Among the interactions produced by high-throughput methods there could be many false positives. To measure their accuracy and to identify the biases, von Mering et al. [4] assessed a total of 80 000 interactions among 5 400 yeast proteins reported previously and assigned each interaction a confidence value. In order to reduce the interference by false positives, we focused on 11 855 interactions with high and medium confidence among 2 617 proteins.

To analyze the interaction dataset, first we applied the spectral method to calculate all eigenvalues and eigenvectors of the adjacency matrix corresponding to the network. The following criteria were then used to generate quasi-cliques based on eigenvectors with larger and positive eigenvalues. (i) All the proteins were sorted by their absolute weight value in an eigenvector, and the top 10% were selected. (ii) Every protein must interact with at least 20% of the members. Here, we used CC-value to measure the degree of the interconnectivity between nodes and tuned the parameter to guarantee the quality of those cliques. (iii) A quasi-clique must contain at least 10 proteins. As a result, we yielded 48 quasi-cliques, among which the largest one contains 109 proteins (quasi-clique 1 in Table 1) and the smallest one contains 10 proteins (quasi-clique 45 in Table 1); on average, a quasi-clique contains 26.6 proteins (a protein may appear in different quasi--cliques). Similar analysis based on eigenvector with negative eigenvalue produced six quasi-bipartites.

The two topological structures show different interaction patterns. In a quasi-clique proteins tend to interact with each other (Figure 1(a), while in a quasi-bipartite, proteins between sets have denser interactions than those within sets (Figure 1(b). Identification of the above topological structures could not only represent the complicated interaction network in order, but also make the complicated network more convenient to analyze.

Table 1. Annotation of all quasi-cliques

| No. | Protein No. | Interaction No. | CC/% | Distribution /% | | | *P*-value (log10) | Function |
|---|---|---|---|---|---|---|---|---|
| | | | | M | U | D | | |
| 1 | 109 | 2 978 | 50 | 72 | 6 | 22 | 75 | Ribosome biogenesis |
| 2 | 97 | 2 327 | 49 | 72 | 5 | 23 | 67 | Ribosome biogenesis |
| 3 | 68 | 955 | 41 | 19 | 50 | 31 | 10 | rRNA processing |
| 4 | 44 | 570 | 60 | 18 | 50 | 32 | 6 | rRNA processing |
| 5 | 37 | 423 | 63 | 19 | 49 | 32 | 6 | rRNA processing |
| 6 | 34 | 239 | 42 | 76 | 15 | 9 | 25 | Ribosome biogenesis |
| 7 | 24 | 230 | 83 | 50 | 0 | 50 | 15 | General transcription activities |
| 8 | 40 | 333 | 42 | 80 | 10 | 10 | 47 | Splicing |
| 9 | 34 | 329 | 58 | 9 | 50 | 41 | 5 | Other tRNA-transcription activities |
| 10 | 54 | 1 018 | 71 | 78 | 9 | 13 | 41 | Ribosome biogenesis |
| 11 | 24 | 200 | 72 | 79 | 4 | 17 | 28 | Cytoplasmic and nuclear degradation |
| 12 | 44 | 410 | 43 | 59 | 11 | 30 | 20 | Ribosome biogenesis |
| 13 | 32 | 267 | 53 | 34 | 3 | 63 | 15 | rRNA synthesis |
| 14 | 34 | 214 | 38 | 18 | 21 | 61 | 7 | Amino acid degradation (catabolism) |
| 15 | 21 | 153 | 72 | 52 | 14 | 34 | 14 | rRNA processing |
| 16 | 31 | 189 | 40 | 39 | 19 | 42 | 7 | Oxidation of fatty acids; lipid, fatty-acid and isoprenoid biosynthesis |

(continued)

| No. | Protein No. | Interaction No. | CC/% | Distribution /% | | | P-value (log10) | Function |
|---|---|---|---|---|---|---|---|---|
| | | | | M | U | D | | |
| 17 | 27 | 251 | 71 | 11 | 67 | 22 | 2 | tRNA transcription; other control of cellular organization; other transcription activities |
| 18 | 16 | 106 | 88 | 63 | 19 | 18 | 16 | mRNA processing (splicing, 5'-, 3'-end processing) |
| 19 | 21 | 119 | 56 | 57 | 24 | 19 | 9 | Ribosome biogenesis |
| 20 | 35 | 281 | 47 | 60 | 3 | 37 | 17 | Ribosome biogenesis |
| 21 | 24 | 119 | 43 | 25 | 21 | 54 | 5 | Lipid, fatty-acid and isoprenoid biosynthesis |
| 22 | 16 | 62 | 51 | 19 | 13 | 68 | 7 | Osmosensing; protein binding |
| 23 | 13 | 78 | 100 | 69 | 31 | 0 | 8 | Ribosome biogenesis |
| 24 | 27 | 157 | 44 | 56 | 4 | 40 | 25 | rRNA synthesis |
| 25 | 14 | 46 | 50 | 36 | 7 | 57 | 6 | Respiration |
| 26 | 28 | 142 | 37 | 50 | 0 | 50 | 23 | rRNA synthesis |
| 27 | 21 | 134 | 63 | 71 | 0 | 29 | 20 | Splicing |
| 28 | 23 | 153 | 60 | 65 | 17 | 18 | 19 | Splicing |
| 29 | 17 | 86 | 63 | 82 | 0 | 18 | 27 | rRNA synthesis |
| 30 | 18 | 92 | 60 | 89 | 0 | 11 | 32 | Transport ATPases |
| 31 | 19 | 84 | 49 | 37 | 5 | 58 | 11 | Tricarboxylic-acid pathway (citrate cycle, Krebs cycle, TCA cycle) |
| 32 | 15 | 70 | 66 | 40 | 7 | 53 | 6 | Lipid, fatty-acid and isoprenoid biosynthesis |
| 33 | 11 | 51 | 92 | 27 | 9 | 64 | 4 | Homeostasis of metal ions (Na, K, Ca, etc.) |
| 34 | 12 | 57 | 86 | 33 | 8 | 59 | 6 | Homeostasis of metal ions (Na, K, Ca, etc.) |
| 35 | 20 | 96 | 50 | 35 | 5 | 60 | 11 | Tricarboxylic-acid pathway (citrate cycle, Krebs cycle, TCA cycle) |
| 36 | 11 | 45 | 82 | 55 | 27 | 18 | 2 | Assembly of protein complexes; lipid, fatty-acid and isoprenoid biosynthesis; cell wall |
| 37 | 11 | 51 | 93 | 27 | 9 | 64 | 4 | Homeostasis of metal ions (Na, K, Ca, etc.) |
| 38 | 19 | 136 | 79 | 79 | 5 | 16 | 22 | Cytoplasmic and nuclear degradation |
| 39 | 13 | 72 | 92 | 100 | 0 | 0 | 22 | Cytoplasmic and nuclear degradation |
| 40 | 14 | 59 | 64 | 29 | 21 | 50 | 2 | Aminoacyl-tRNA-synthetases; cell wall |
| 41 | 13 | 47 | 60 | 54 | 8 | 38 | 12 | Tricarboxylic-acid pathway (citrate cycle, Krebs cycle, TCA cycle) |
| 42 | 12 | 58 | 88 | 42 | 8 | 50 | 5 | Lipid, fatty-acid and isoprenoid biosynthesis |
| 43 | 19 | 91 | 53 | 21 | 37 | 42 | 4 | rRNA processing |
| 44 | 11 | 52 | 94 | 64 | 9 | 27 | 6 | Ribosome biogenesis |
| 45 | 10 | 24 | 53 | 50 | 0 | 50 | 5 | Amino acid biosynthesis |
| 46 | 11 | 52 | 94 | 64 | 9 | 27 | 6 | Ribosome biogenesis |
| 47 | 15 | 40 | 38 | 20 | 0 | 80 | 6 | Other proteolytic degradation |
| 48 | 15 | 51 | 48 | 40 | 20 | 40 | 6 | Lipid, fatty-acid and isoprenoid biosynthesis |

No., quasi-clique No. Protein No., the number of proteins in the quasi-clique. Interaction No., the number of interactions within the quasi-clique. CC, represents the CC-value of a quasi-clique (see Materials and Methods). Function, the assigned function of the quasi-clique. Distribution, percentages of the following three classes in the quasi-clique: M, the percentage of proteins which have the main function; U, the percentage of uncharacterized proteins and D, the percentage of proteins which are discordant with the quasi-clique's function.

### 3.2 Annotation of quasi-cliques

For each of the 48 quasi-cliques, we calculated its *P*-value and annotated it based on the Munich Information Center (MIPS) hierarchical functional categories. MIPS allows a protein to appear in more than one category, which was taken into account in the calculation of *P*-value. As a result, 43 quasi-cliques were annotated with one functional category and the other five quasi-cliques were assigned to a set of functional categories (Table 1; see Supplementary Material for complete data sets).

We investigated the functions of individual proteins in quasi-cliques and found that most of them usually share common functions, including ribosome biogenesis, rRNA and tRNA synthesis, processing, transcription control and mRNA splicing, etc. (Figure 2 and Table 1). Only a small fraction of the proteins turn out uncharacterized or have functions conflicting with the common function of the quasi-clique, as shown in Figure 2. This could be explained by either unavoidable false positive interactions under the current experimental conditions or that the proteins really share this kind of function but it is yet not proved.

To visualize protein interactions and functional annotations, we have developed a software package that, along with the complete set of data generated by our algorithm, is publicly available. Using this software, users can view topological structures and find annotations of proteins and their interactions conveniently.

### 3.3 Functional prediction for uncharacterized proteins in quasi-cliques

The isolated quasi-cliques give a good clue to predict functions of the uncharacterized proteins. Among the 2 617 proteins in the raw dataset, 555 were uncharacterized according to MIPS hierarchical functional categories[4]. For the 76

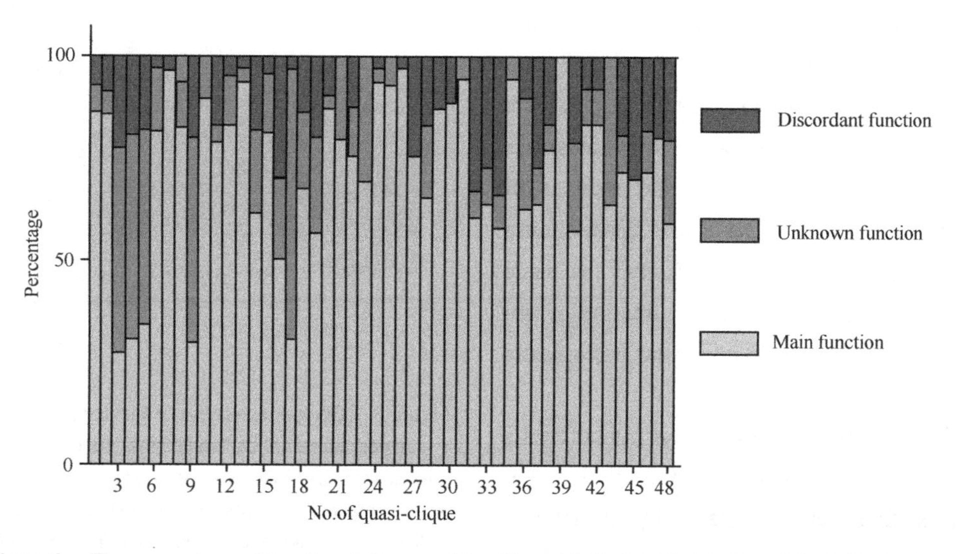


Figure 2. The percentage of functional classes of the 48 quasi-cliques. Distribution of the following three classes: main function, percentage of the proteins that have the main function; unknown function, percentage of the uncharacterized proteins; and discordant function, percentage of the proteins that have discordant functions

uncharacterized proteins in the 48 quasi-cliques, we assigned for each one a function according to the main function of its hosting quasi-clique. If a protein falls into more than one quasi-clique, the main function of the quasi-clique with the lowest *P*-value was assigned to it. If multiple hosting quasi-cliques have the lowest *P*-value, or a quasi-clique has multiple main functions, a set of functions would be assigned to the protein. The 76 unknown proteins and their predicted functions with the corresponding *P*-values are listed in Table 2. There are 43 rRNA processing proteins, seven proteins related to pre-RNA processing, 11 proteins related to ribosome biogenesis, and the other 15 proteins related to energy, metabolism, cytoskeleton and transcription-regulating (See Table 2 for complete data).

We assessed the ability of the *P*-value to annotate and assign functions using the same approach as Wu et al.[28]. As a control, we created and analyzed random networks with the same interaction distribution as the original network. The results show that among the 48 quasi-cliques of our experimental data, >87.5% were significant in one or more annotation categories at $P \leqslant 0.01/N_c$ (here $N_c$ is the number of categories), whereas <2.1% of quasi-cliques identified from random network met the same criteria. This means a substantial fraction of isolated quasi-cliques are likely to be biologically meaningful.

Some of our predictions were supported by recent experimental evidence. Of all the quasi-cliques, five were dominated by uncharacterized proteins (functions are unknown for at least 50% of proteins, Figure 2), which imply that those unknown proteins in a same quasi-clique may form a large complex relating to a certain cellular process. For quasi-cliques 3 and 4, most of the proteins were predicted to mediate rRNA processing, which is partly consistent with the results from recent experiments[29-31] (Figure 3).

The ORF name of proteins is listed in the 'Protein' column, corresponding *P*-value is listed in the middle column and predicted function for each protein is listed in the 'Predicted function' column.

# 4 Discussion

The yeast large-scale protein-protein interaction data have broadened our view of protein functions in this proteomics era. The biological processes of a cell are controlled by interacting proteins in metabolic and signaling pathways and in complexes such as the molecular machines that synthesize and use adenosine triphosphate,

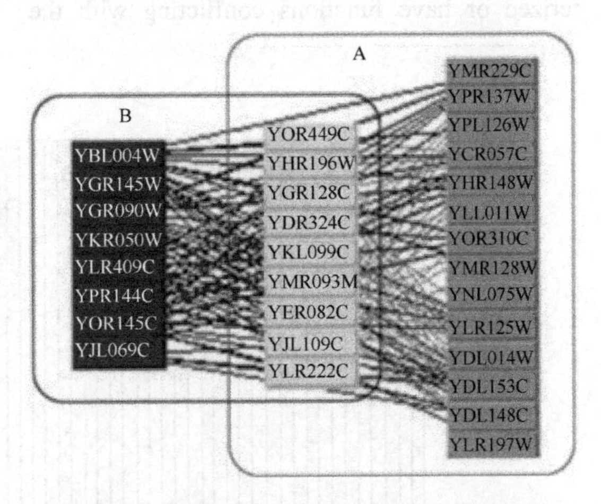


Figure 3. Comparison between function prediction and experimental annotation for small subunit (SSU) processome. (A) SSU processome that is supported by experimental evidence (the yellow and the green proteins); (B) our predictions based on quasi-clique 3 (the red proteins and the yellow proteins). The yellow ones are the overlap of (A) and (B). This suggests that our prediction is partly consistent with the experimental evidence[31]

Table 2. Prediction for uncharacterized proteins involved in 48 quasi-cliques

| Protein | $P$-value($\log_{10}$) | Predicted function |
|---|---|---|
| YLR421C | 28 | Cytoplasmic and nuclear degradation |
| YNL168C | 7 | Amino acid degradation (catabolism) |
| YDL193W | 6 | Lipid, fatty-acid and isoprenoid biosynthesis |
| YDR018C | 6 | Lipid, fatty-acid and isoprenoid biosynthesis |
| YNL026W | 6 | Lipid, fatty-acid and isoprenoid biosynthesis |
| YJL046W | 12 | Tricarboxylic-acid pathway (citrate cycle, Krebs cycle, TCA cycle) |
| YPL252C | 6 | Homeostasis of metal ions (Na, K, Ca, etc.) |
| YLR074C | 2 | tRNA transcription; other control of cellular organization; other transcription activities |
| YNL123W | 7 | Osmosensing; protein binding |
| YGL211W | 2 | Aminoacyl-tRNA-synthetases; cell wall |
| YGL211W | 2 | Assembly of protein complexes; lipid, fatty-acid and isoprenoid biosynthesis; cell wall |
| YBL055C | 7 | Oxidation of fatty acids; lipid, fatty-acid and isoprenoid biosynthesis |
| YDR428C | 7 | Oxidation of fatty acids; lipid, fatty-acid and isoprenoid biosynthesis |
| YGR263C | 7 | Oxidation of fatty acids; lipid, fatty-acid and isoprenoid biosynthesis |
| YOR093C | 7 | Oxidation of fatty acids; lipid, fatty-acid and isoprenoid biosynthesis |
| YGL059W | 7 | Oxidation of fatty acids; lipid, fatty-acid and isoprenoid biosynthesis |
| YDR428C | 7 | Amino acid degradation (catabolism) |
| YGR263C | 7 | Amino acid degradation (catabolism) |
| YOR093C | 7 | Amino acid degradation (catabolism) |
| YGL059W | 7 | Amino acid degradation (catabolism) |
| YGL059W | 7 | Osmosensing; protein binding |
| YJR008W | 25 | rRNA synthesis |
| YKL099C | 6 | rRNA processing |
| YBL004W | 10 | rRNA processing |
| YDL213C | 4 | rRNA processing |
| YDR324C | 10 | rRNA processing |
| YDR449C | 10 | rRNA processing |
| YDR496C | 10 | rRNA processing |
| YER082C | 10 | rRNA processing |
| YER126C | 10 | rRNA processing |
| YFR001W | 6 | rRNA processing |
| YGL111W | 10 | rRNA processing |
| YGR090W | 10 | rRNA processing |
| YGR103W | 10 | rRNA processing |
| YGR128C | 10 | rRNA processing |
| YGR145W | 10 | rRNA processing |
| YHR052W | 10 | rRNA processing |
| YHR088W | 14 | rRNA processing |
| YHR196W | 6 | rRNA processing |
| YHR197W | 10 | rRNA processing |
| YJL069C | 6 | rRNA processing |

(continued)

| Protein | $P$-value($log_{10}$) | Predicted function |
|---|---|---|
| YJL109C | 10 | rRNA processing |
| YKL014C | 10 | rRNA processing |
| YKR060W | 10 | rRNA processing |
| YKR081C | 10 | rRNA processing |
| YLR022C | 14 | rRNA processing |
| YLR106C | 6 | rRNA processing |
| YLR186W | 10 | rRNA processing |
| YLR222C | 10 | rRNA processing |
| YLR276C | 10 | rRNA processing |
| YLR409C | 10 | rRNA processing |
| YMR049C | 10 | rRNA processing |
| YMR093W | 10 | rRNA processing |
| YNL002C | 10 | rRNA processing |
| YNL110C | 10 | rRNA processing |
| YNL182C | 10 | rRNA processing |
| YNR053C | 10 | rRNA processing |
| YOL041C | 10 | rRNA processing |
| YOL077C | 10 | rRNA processing |
| YOR001W | 14 | rRNA processing |
| YOR145C | 6 | rRNA processing |
| YPL012W | 10 | rRNA processing |
| YPL093W | 10 | rRNA processing |
| YPR144C | 10 | rRNA processing |
| YDL209C | 47 | Splicing |
| YGR278W | 47 | Splicing |
| YLR424W | 47 | Splicing |
| YPL151C | 47 | Splicing |
| YGR156W | 16 | mRNA processing (splicing, 5'-, 3'-end processing) |
| YKL018W | 16 | mRNA processing (splicing, 5'-, 3'-end processing) |
| YKL059C | 16 | mRNA processing (splicing, 5'-, 3'-end processing) |
| YDR036C | 20 | Ribosome biogenesis |
| YDR101C | 75 | Ribosome biogenesis |
| YGL129C | 20 | Ribosome biogenesis |
| YGR283C | 75 | Ribosome biogenesis |
| YIL093C | 20 | Ribosome biogenesis |
| YJR014W | 75 | Ribosome biogenesis |
| YKL155C | 20 | Ribosome biogenesis |
| YMR074C | 6 | Ribosome biogenesis |
| YMR158W | 75 | Ribosome biogenesis |
| YMR310C | 75 | Ribosome biogenesis |
| YNL177C | 75 | Ribosome biogenesis |

replicate and transcribe genes, or build up the cytoskeletal infrastructure[32-33]. The knowledge regarding protein-protein interactions has been accumulated by biochemical and genetic experiments, including the widely used high-throughput interaction detection methods, such as the yeast two-hybrid system and protein complex purification techniques using mass spectrometry. Now, a challenging task is to decipher the relationships between individual proteins and to understand the molecular organization of cellular networks. Here, for the first time, we analyzed the complicated protein interaction networks using the spectral analysis method. This approach is useful in revealing hidden topological structures, including quasi-cliques and quasi-bipartites, which exhibit meaningful information of a complex network. Figure 4(a) shows a part of the original interaction network, which contains 109 proteins. It looks confusing and difficult to assimilate before analysis. In contrast, a tightly interacting quasi-clique including 68 proteins was found from this part of network by spectral analysis. This suggests that a network actually is not random as it appears (Figure 4(b)).

As part of these studies, we first offered a flexible and promising large-scale protein function prediction system based on spectral analysis. Compared with the previous approaches, what we presented here has a number of practical advantages. Previous methods used partners or neighbors alone to perform the prediction, whereas our method utilized the more informative topological structure of the whole network, and produced some results that were not covered by the previous predictions. The 76 proteins contain 43 rRNA processing proteins, seven proteins related to pre-RNA processing, 11 proteins related to ribosome biogenesis and another 15 proteins related to energy, metabolism, cytoskeleton and transcription regulation. As a control, we created and analyzed random networks with the same interaction distribution as the original network. The results show that among the 48 quasi-cliques of our experimental data, $>87.5\%$ were significant in one or more annotation categories at $P \leqslant 0.01/N_C$ (here $N_C$ is the number of categories), whereas $<2.1\%$ of quasi-cliques identified from a random network met the same

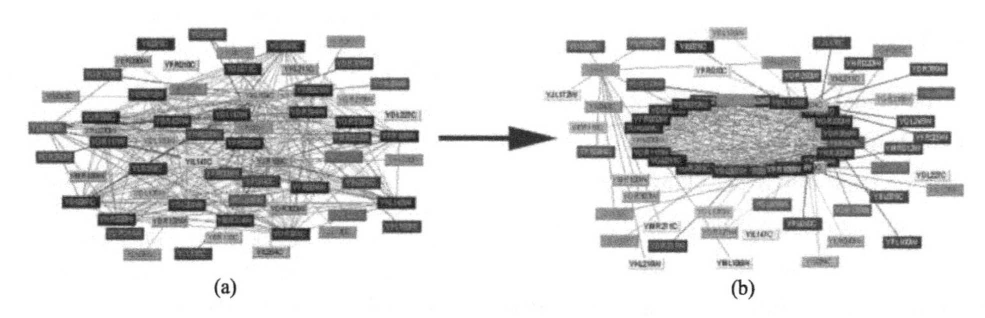

(a) (b)

Figure 4. Comparison of different visual representations with and without topological structure. The original protein-protein interaction network is rather miscellaneous and difficult to assimilate (a). The spectral analysis revealed a hidden topological structure underlying the miscellaneous network (b)

criteria. Some of our predictions have been proved by experiments published recently. This suggests that our prediction method is accurate. Furthermore, this method is a universal one that could be used to predict protein function in other organisms.

Although the initial results are promising, the current method is still far from perfect. We have not yet fully explored all quasi-cliques, for that the problem has been proved to be NP-Complete. Therefore new methods should be developed to reveal more sophisticated topological features. It should be pointed out that prediction accuracy is affected by knowledge of known annotations and false positive interactions. It is well known that so far annotations of proteins in databases are incomplete, i.e. a number of proteins with well-characterized function, or at least well-supported functional prediction, are annotated as 'unknown function' in MIPS. This introduces additional uncertainties into our prediction. We believe that our prediction would be better if a more accurate interaction and annotation dataset was applied.

Supplementary Material is available at NAR Online.

## 5 Acknowledgements

We would like to acknowledge with deep appreciation Professor Soren Norby for his examining and revising this paper. This work was supported by the Chinese Academy of Sciences Grant No.KSCX2-2-07, National Sciences Foundation of China Grant No.39890070, the National High Technology Development Program of China under Grant No.2002AA231031, National Key Basic Research & Development Program (973) under Grant No.2002CB713805, the National Grand Fundamental Research 973 Program of China under Grant No. G1998030510 and Beijing Science and Technology Commission Grant No. H010210010113.

## References:

[1] FIELDS S. The future is function[J]. Nature Genetics, 1997, 15(4): 325-327.

[2] RAIN J C, SELIG L, DE REUSE H, et al. The protein-protein interaction map of Helicobacter pylori[J]. Nature, 2001, 409(6817): 211-215.

[3] GAVIN A C, BOSCHE M, KRAUSE R, et al. Functional organization of the yeast proteome by systematic analysis of protein complexes[J]. Nature, 2002, 415(6868): 141-147.

[4] VON MERING C, KRAUSE R, SNEL B, et al. Comparative assessment of large-scale data sets of protein-protein interactions[J]. Nature, 2002, 417: 399-403.

[5] UETZ P, GIOT L, CAGNEY G, et al. A comprehensive analysis of protein-protein interactions in saccharomyces cerevisiae[J]. Nature, 2000, 403(6770): 623-627.

[6] ITO T, CHIBA T, OZAWA R, et al. A comprehensive two-hybrid analysis to explore the yeast protein interactome[J]. Pro. Natl. Acad. Sci. USA, 2001, 98(8): 4569-4574.

[7] HO Y, GRUHLER A, HEILBUT A, et al. Systematic identification of protein complexes in saccharomyces cerevisiae by mass spectrometry[J]. Nature, 2002, 415(6868): 180-183.

[8] CHO R J, CAMPBELL M J, WINZELER E A, et al. A genome-wide transcriptional analysis of the mitotic cell cycle[J]. Molecular Cell, 1998, 2(1): c65-73.

[9] HUGHES T R, MARTON M J, JONES A R, et al. Functional discovery via a compendium of expression profiles[J]. Cell, 2000, 102(1): 109-126.

[10] TONG A H, DREES B, NARDELLI G, et al. Systematic genetic analysis with ordered arrays of yeast deletion mutants[J]. Science, 2001, 294(5550): 2364-2368.

[11] MEWES H W, FRISHMAN D, GILDENER U, et al. MIPS: a database for genomes and protein sequences[J]. Nucleic Acids Research, 2002, 30(1): 31-34.

[12] ENRIGHT A J, ILIOPOULOS I, KYRPIDES N C, et al. Protein interaction maps for complete genomes based on gene fusion events[J]. Nature, 1999, 402: 86-90.

[13] MARCOTTE E M, PELLEGRINI M, NG H L, et al. Detecting protein function and protein-protein interactions from genome sequences[J]. Science, 1999, 285(5428):751-753.

[14] OVERBEEK R, FONSTEIN M, D'SOUZA M, et al. The use of gene clusters to infer functional coupling[J]. Pro. Natl. Acad. Sci. USA, 1999, 96: 2896-2901.

[15] DANDEKAR T, SNEL B, HUYNEN M, et al. Conservation of gene order: a fingerprint of proteins that physically interact[J]. Trends in Biochemical Sciences, 1998, 23(9): 324-328.

[16] PELLEGRINI M, MARCOTTE M, THOMPSON M J, et al. Assigning protein functions by comparative genome analysis: protein phylogenetic profiles[J]. Pro. Natl. Acad. Sci. USA, 1999, 96: 4285-4288.

[17] HUYNEN M A, BORK P. Measuring genome evolution[J]. Pro. Natl. Acad. Sci. USA, 1998, 95(11): 5849-5856.

[18] SCHWIKOWSKI B, UETA P, FIELDS S. A network of protein–protein interactions in yeast[J]. Nature Biotechnology, 2000, 18: 1257-1261.

[19] HISHIGAKI H, NAKAI K, ONO T, et al. Assessment of prediction accuracy of protein function from protein–protein interaction data[J]. Yeast, 2001, 18(6): 523-53.

[20] MASLOV S, SNEPPEN K. Specificity and stability in topology of protein networks[J]. Science, 2002, 296(5569): 910-913.

[21] GE H, LIU Z, CHURCH G M, et al. Correlation between transcriptome and interactome mapping data from saccharomyces cerevisiae[J]. Nature Genetics, 2002, 29(4): 482-486.

[22] FRASER H B, HIRSH A E, STEINMETZ L M, et al. Evolutionary rate in the protein interaction network[J]. Science, 2002, 296(5568): 750-752.

[23] GIBSON Q, KLEINBERG J, RAGHAVAN P. Inferring web communities from link topology[C]// The 9th ACM Conference on Hypertext and Hypermedia. New York: ACM Press, 1998.

[24] KLEINBERG J M. Authoritative sources in a hyper-linked environment[C]// The 9th ACM Conference on Hypertext and Hypermedia. New York: ACM Press, 1998.

[25] BOLLOBAS B. Modern graph theory[M]. New York: Springer-Verlag, Inc, 1998.

[26] WATTS D J, STROGATZ S H. Collective dynamics of "small-world" networks[J]. Nature, 1998, 393: 440-442.

[27] TAVAZOIE S, HUGHES J, CAMPBELL M, et al. Systematic determination of genetic network architecture[J]. Nature Genetics, 1999, 22: 281-285.

[28] WU L F, HUGHES T R, DAVIERWALA A P, et al. Large-scale prediction of Saccharomyces cerevisiae gene function using overlapping transcriptional clusters[J]. Nature Genetics, 2002, 31(3): 255-265.

[29] HARNPICHARNCHAI P, JAKOVLJEVIC J, HORSEY E, et al. Composition and functional characterization of yeast 66s ribosome assembly intermediates[J]. Molecular Cell, 2001, 8(3): 505-515.

[30] BASSLER J, GRANDI P, GADAL O, et al. Identification of a 60s preribosomal particle that is closely linked to nuclear export[J]. Molecular Cell, 2001, 8(3): 517-529.

[31] DRAGON F, GALLAGHER J E G, COMPAGNONE-POST P A, et al. A large nucleolar U3 ribonucleoprotein required for 18S ribosomal RNA biogenesis[J]. Nature, 2002, 417(6892): 967-970.

[32] ALBERTS B, BRAY D, LEWIS J, et al. Molecular biology of the cell, 3rd edn[M]. New York: Garland Publishing Inc., 1983.

[33] LODISH H, BERK A, ZIPURSKY S L, et al. Molecular cell biology, 3rd edn[M]. New York: Scientific American Books, 1995.

发表于 *Journal of Computer Science and Technology*, 2006年第5期

# Usability issues of grid system software

Zhi-wei XU[1], Hao-jie ZHOU[1,2], Guo-jie LI[1]

1. Institute of Computing Technology, Chinese Academy of Sciences, Beijing 100080, P.R. China

2. Graduate University of the Chinese Academy of Sciences, Beijing 100039, P.R. China

**Abstract:** This paper addresses the problem why grid technology has not spread as fast as the Web technology of the 1990's. In the past 10 years, considerable efforts have been put into grid computing. Much progress has been made and more importantly, fundamental challenges and essential issues of this field are emerging. This paper focuses on the area of grid system software research, and argues that usability of grid system software must be enhanced. It identifies four usability issues, drawing from international grid research experiences. It also presents advances by the Vega Grid team in addressing these challenges.

**Key words:** grid computing, system software, usability, distributed system architecture

## 1 Introduction

The idea of grid computing can be traced back to the utility computing concept of the 1960's. It was revived in the 1990's with the introduction of metacomputing[1], and later the concept of computational grids[2], analogous to electrical power grids. In the early 2000's, grid computing research flourished worldwide, with a lot of efforts put into building wide-area, decentralized IT infrastructure, or cyberinfrastructure[3], whereby various resources and services can be provisioned on demand. Viewed as "the next phase of distributed computing", grid computing has grown into a rich field, embracing distributed application solutions, business services, distributed middleware, distributed system software, and even servers and storage hardware.

Grid system software is an important research area in the grid computing arena. Its current landscape can be summarized as follows: 1) the architecture of grid system software has gone through a number of transitions, and is converging towards into a style of high-performance service-oriented architecture, as exemplified by the Open Grid Service Architecture (OGSA) proposed by the Global Grid Forum (GGF); 2) with grid computing embraced by both academia and the industry, many products and services are available, either as open source or as proprietary offerings; 3) many grid applications and services are available now, utilizing various grid system software technologies, ranging from scientific research to financial applications and online

This work is supported in part by the National Natural Science Foundation of China (Grant Nos.60573102,90412010) and the National Grand Foundamental Research 973 Program of China (Grant Nos.2003CB317000, 2005CB321800).

games.

However, grid technology spreads much slower than Web technology of the 1990's. There are many reasons for this. Coordinated resource sharing and collaboration, in a wide area and crossing multiple administrative domains, is a far more complex problem than sharing static Web pages, thus must overcome more technical hurdles. There are more standards efforts needed by grids than the HTML, URL, HTTP needed by Web. Grid system software is still a young field, lacking both theoretical understanding and mature, stable implementation.

Another main reason is that it is still difficult to use grid technology, from grid construction, resource deployment, application development, system management, to end user experiences. Much research efforts in the past were devoted to the basic services of OGSA. With these services gradually in place, there is a growing need to address the usability issues.

In this paper, we examine the status and trends of grid system software, from the user's viewpoint. The emphasis is on identifying the usability gap. Different grid environments, usage modalities and usability issues are analyzed, from representative international grid research projects. We also discuss research advances from the Vega Grid team in addressing usability issues.

The rest of this paper is organized as follows. In Section 2, we analyze grid system environment and summarize usage modalities from representative grid system software efforts in the international community (e.g., OGSA, TeraGrid, Globus, NAREGI, and China National Grid). Section 3 discusses usability issues. Section 4 presents research advances in the Vega grid software suite. Section 5 draws conclusion remarks.

## 2 Usage modalities in grid environments

In this paper, usability issues refer to those requirements of grid system software which affect ease of use. To better understand these issues, we need to carefully examine the grid environments, the users involved, and the usage modalities.

### 2.1 Environment for grid system software

Figure 1 illustrates a conceptual grid environment for the discussions in this paper. A grid consists of four types of elements: users, applications, system software, and sites.

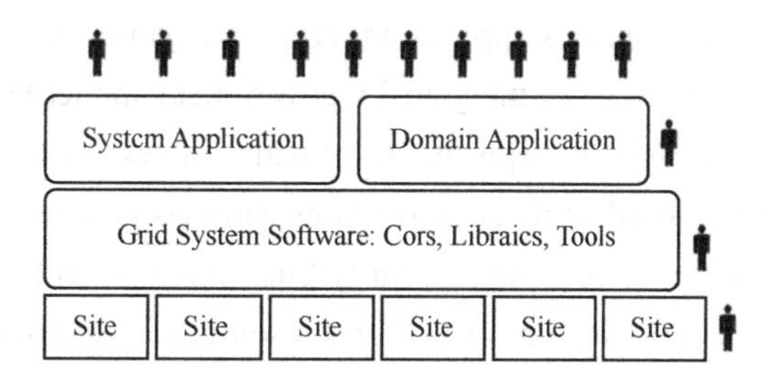


Figure 1. Four types of elements in a grid: users, applications, system software, and sites

Physically, a grid consists of multiple sites, which are distributed in a wide area (in different cities or even different continents) and connected by a public network (e.g., the Internet) or a private network (e.g., a VPN). Each site is a computing center or data center, providing physical resources such as computing/storage hardware, local operating systems, application software packages, licenses, data, instruments and sensors.

The sites in a grid are usually owned and managed by multiple administrative domains. This paper makes the simple assumption that each site corresponds to a domain. In reality, a domain could manage multiple sites, and a site could host multiple domains.

On top of the interconnected sites runs a layer of grid system software, which is responsible for integrating the sites into a cohesive grid system, thus eliminating silos. The grid system software consists of three types of components: one or more always-on runtime subsystems (cores), libraries and services, and user operable utility tools.

The grid system software supports two types of applications. Domain applications are those that provide domain-specific application functions, such as a bioinformatics application, an enterprise information dashboard, a manufacturing logistics application, an online games application, or an integrated e-government application. System applications provide values common to all application domains, such as grid resource accounting, security/trust management, searching and mining of grid resources information. The distinction is not absolute between system applications and utility tools of the system software.

In real world usage patterns, there are many types of grid users, and many ways to classify them. Figure 1 illustrates one scheme to classify users into four basic types:

• End users use a domain (or system) application to attain some business (common) value from the grid. They normally are aware of the grid system software only through the utility tools provided, but are not necessarily concerned with API or other details.

• Developers (application developers) use the grid system software to develop domain or system applications. They may be fully aware of all grid system software interfaces.

• Grid administrators manage the grid as a whole. They are most concerned with the management tools of the grid system software, from user management to security policies, service provisioning, configuration management, monitoring and accounting, and management of virtual organizations.

• Site administrators each manage a site. They need to interact with the grid system software by specifying physical resources/services and the mapping between grid and site users.

## 2.2 Classification of Grids

There are hundreds of on-going grid projects worldwide (see [4] for a sample of operational and planned grids). From the users' viewpoint, these grids can be projected into a two-dimensional space. We select a few representative grid systems to illustrate the classification scheme in Figure 2.

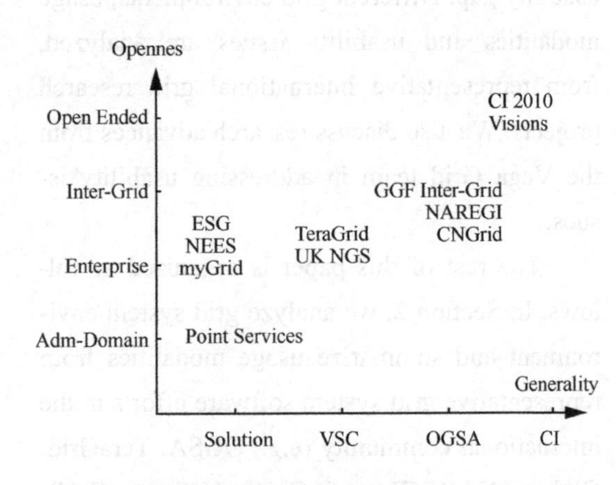


Figure 2. Classification of grids by openness and generality

The vertical axis in Figure 2 shows the openness of ad-ministration and controlled accesses for a grid. The most closed grids have one administrative domain, and all ad-ministration policies and rules are dictated by the domain manager.

An enterprise grid, having multiple administrative domains, is often governed by a committee of stakeholders, who establish a cohesive set of policies and rules for user management, resource/service accesses, security, scheduling, accounting, etc. However, such a grid does allow access and sharing of resources previously owned and controlled by individual domains.

The next level of openness comes with inter-grid systems, where multiple interoperable grids must open their resources and policies for a wider scope of sharing. Users of one grid can construct, deploy, or utilize resources in another grid, although the two grids may have different sets of policies and rules. GGF is now advocating a cooperation project called Grid Interoperation (GIN) to interconnect main national grids.

Finally, a long term goal is to have open ended grid systems, similar to the Web today. Any individual, being a scientist or a school kid, can become a legitimate user in a matter of a few seconds, without explicit committee approval. The same individual can also contribute to the grids as easily as putting an HTML page on the Web. He could autonomously create and deploy a resource, a service, even a grid, without going through a lengthy apply-approve process.

The horizontal axis of Figure 2 shows the scope of generality visible to the users. The most focused (thus least general-purpose) grids are those that provide a solution for a specific application domain. Three examples are shown in Figure 2. The Earth System Grid (ESG)[5] aims to providing a grid solution for modeling and simulation of Earth's climate. NEES is an application grid for a Network for Earthquake Engineering and Simulation. The my Grid provides a grid solution for bioinformatics applications.

Such solution grids are the most numerous today, especially in the USA and Europe. They directly provide domain-specific application values to the users, and encourage cross-discipline cooperation between business domain specialists and computer professionals. Users mostly see business-level interfaces, not the grid system software, as the latter is often hidden in the application solution, even encapsulated in the application codes (see Figure 3).

A virtual supercomputer (VSC) grid provides a general-purpose platform, not specific to any application domain. Two typical examples are the US TeraGrid and UK National Grid Service (NGS). Like a traditional supercomputer, such a grid provides uniform user management (such as identification and authentication), a home directory for each user, and scratch spaces for temporary files. A user can upload his own application codes and data, and execute interactive commands as well as batch jobs.

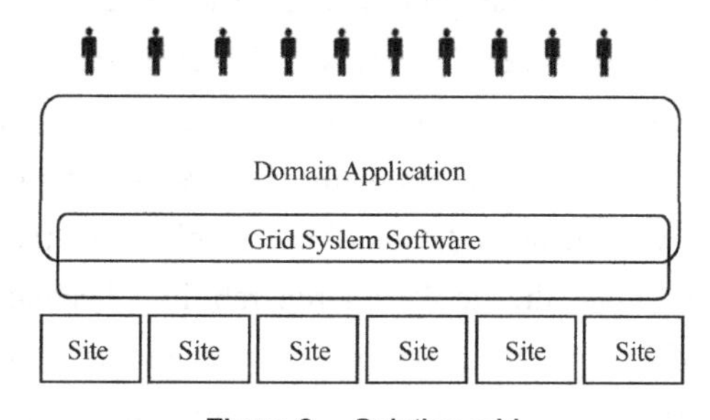

Figure 3. Solution grid

Unlike a single supercomputer, a VSC grid integrates several sites offering high- performance computing, storage, and visualization functions. The grid system software provides a uniform environment, such that users can expect, by default, the availability of the same interoperable grid software tools and services on every site. This is illustrated in Figure 4.

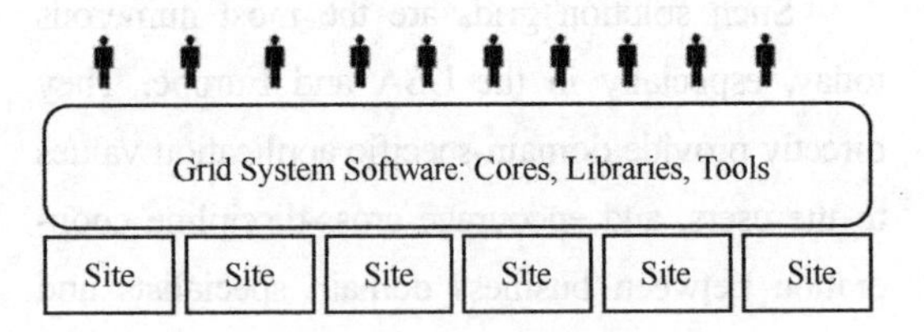


Figure 4. Virtual supercomputer (VSC) grid

For instance, on TeraGrid, each site provides a set of grid software tools known as the Common TeraGrid Software Stack (CTSS). The current CTSS includes the Globus 2.4 software for common grid runtime, GridFTP for cross-site data, transfer, MyProxy for user credentials, GSI for grid security, Condor-G for job submission, and MPICH-G2 for grid-enabled message passing. Additional values are added by a Storage Resource Broker (SRB) service, accounting information service, visualization tools, etc.

The OGSA mode is more general than the VSC mode in the following three senses. First, it employs a service-oriented architecture, and attempts to provide grid system software functions uniformly as services. This is different from the previous approach of a bag of ad hoc tools, and could alleviate the problem of users needing to know the tedious details of each tool and worrying about the tools interoperability. Second, OGSA aims to providing more than the bare essentials of a supercomputer environment such as user management, files storage and transfer, interactive and batch jobs management. The on-going OGSA standards efforts have made advances in sophisticated data, management, information utilization, workflow, and semantic grid services. It enables grids to transition from a scientific computing platform to a more general IT platform, delivering services for computing, data management, information management, and knowledge processing. Third, an OGSA grid supports the concept of virtual organization (VO), a dynamic partition of grid resources/services and associated policies. OGSA grids are illustrated in Figure 5.

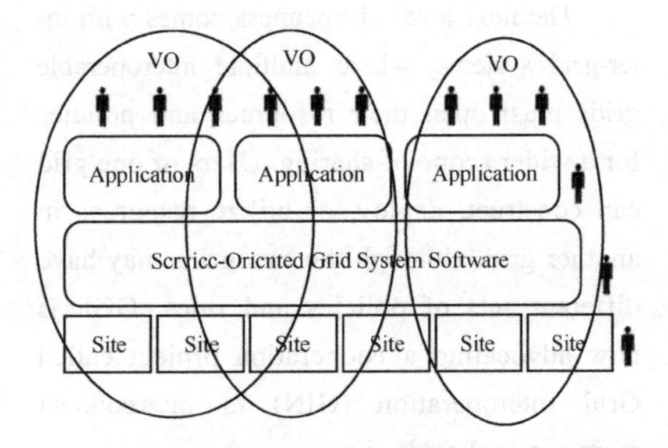


Figure 5. OGSA grid

Some recent grid projects adopted the OGSA mode, such as China National Grid (CNGrid)[6], the Japanese National Research Grid Initiative (NAREGI)[7], and several semantic grid efforts[8].The next phase of TeraGrid will incorporate the Globus Toolkit 4 grid system software[9], moving towards the OGSA mode. In this regard, the NAREGI grid software is the most aggressive. At its release in late 2007, it will implement dozens of OGSA related GGF standards or drafts.

Cyberinfrastructure (CI), as illustrated in

Figure 6, is still a set of ongoing researchgoals[3]. Cyberinfrastructure differs from the previous three modes in several important ways.

Multiple grids. In addition to multiple VOs, multiple grids are supported on the same infrastructure, including solution grids, VSC grids, and OGSA grids. This implies that 1) the grids can share the same infrastructure platform, and 2) the grids can access one another, thus sharing and collaboration can cross boundaries of grids. These grids use and become part of the infrastructure.

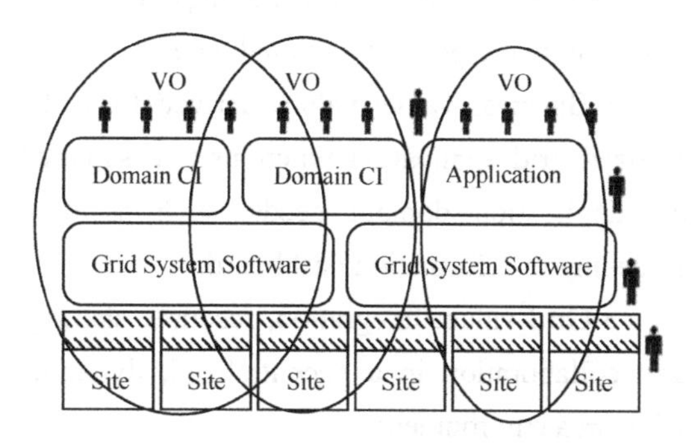


Figure 6. Illustration of cyberinfrastructure

Shared infrastructure. Cyberinfrastructure emphasizes an interoperable infrastructure shared by all application domains, on top of which domain-specific CIs can be built. This shared infrastructure offers common supports for collaboration and sharing of “HEC (high-end computing) resources, massive data stores, broad and deep databases, high-bandwidth networks, middleware, shared software libraries and tools, services, people, and domain-specific software”[3].

Virtualization. To support the dynamic, diverse environment of multiple grids, multiple application domains, and multiple VOs, the sites should be “grid-oriented” or “grid enabling”. A common approach is to utilize virtualization technology. For instance, IBM has selected virtualization as the main theme for its next generation servers, to support its On Demand initiative. In the open source community, various virtual machines are used, such as Java virtual machines, XEN, grid virtual machine[7], and service virtual machine[10]. In ideal cases, each sites can configure a suitable software stack and environment, including local operating system image, application software, libraries, and provide proper bindings and policy mappings for the grid in need. These resources plus the bare metal hardware are properly configured and provisioned on demand.

Empowering individuals. Cyberinfrastructure is not geared towards big organizations only, where all resources, services, policies are well defined/deployed and managed by committees and IT professionals. Instead, cyberinfrastructure, by providing all necessary plumbing, could drastically lower the barriers. It enables individuals, illustrated by the bigger human figures in Figure 6 who do not belong to any team, to access and consume resources as easily as big teams. Furthermore, it also empowers individuals *to* produce: to construct a grid, initiate a VO, deploy an application service, or contribute a resource. Peter Freeman et al.[3] believe that: “in the long run, the empowerment of individuals and small teams may be where the most significant changes in S&E (science and engineering) take place.”

## 2.3 Usage modalities

From a single user’s viewpoint, we can project the usage modalities into a two-dimensional space depicted in Figure 7, where a sample of such modalities are shown. The vertical axis of Figure 7 measures the giving factor, which also indirectly indicates how much the user is aware of other users. While a lot of users just consume grid resources, contribution-based usage cannot be ignored. Many users, including end users, are contributing to the grid, by creating, producing, and deploying resources, services, applications, and VO's. Collaboration among users can be viewed as containing both consumption and production.

The horizontal axis of Figure 7 measures the interactivity factor, which also indirectly indicates how much the user is aware of the responsiveness of grid system.

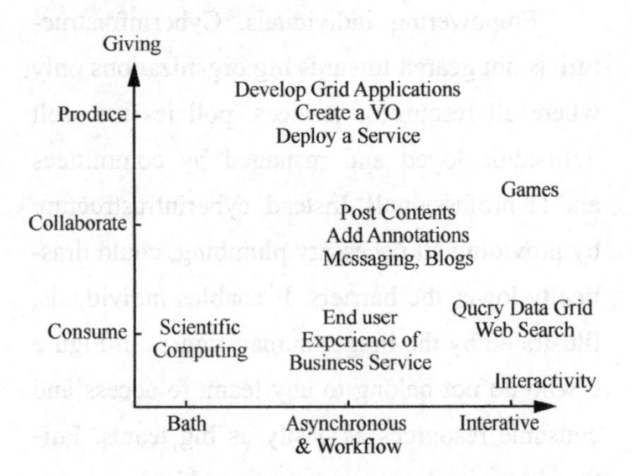


Figure 7. Single user usage modalities in a grid

In an online game usage mode, a user sends the game grid interactive commands or events, and expects immediate responses from the game grid and from other players. Studies for interactive user experience show that "crisp" responses within 150 milliseconds by the system are preferred[11].

Users have bigger latency tolerance with other usage modalities. In scientific computing on a virtual supercomputer grid, a user submits a job to the grid and only expects the batch scheduler timely responses that the submission is indeed successfully accepted by a batch queue. In a business process workflow scenario, the user can also tolerate a reasonable delay after a user input.

## 3 Grid values and usability issues

From an end-user's viewpoint, a grid system must provide both functional and nonfunctional values, for the grid technology to spread. There are two types of functional values:

• Business value: values derived from the realized grid application functions that satisfied the application domain needs, such as drug screening or information dashboard.

• Grid value: integration, resource sharing, and collaboration in a decentralized, dynamic, wide area environment.

There are three types of nonfunctional values:

• Security: authentication, authorization, availability (and reliability), trust, privacy, etc.

• Efficiency: low cost, low overhead, high utilization, and high quality of services.

• Usability: ease of use issues, such as the four issues of single system image, portability, agility, and managed code discussed below.

These values must all present for the grid technology to become widely deployed. In the past, not enough attention is paid to usability issues. Usability is more than just designing a graphic user interface, grid portal or problem solving environment, it must be considered when designing the architecture and standards of grids.

In addition to supporting various usage modalities, a deeper requirement of grid usability is to liberate users. A grid system, with its system software core, libraries, configuration files, environment variables, tools, and system applications, should take care of tedious, routine details, and prevent these overheads from appearing in user actions or application codes.

Furthermore, the grid system software should strive to satisfy the usability requirements without too much overhead. Currently, calling a trivial grid service (e.g., an Apache Tomcat+Axis service or a WSRF service) needs a few dozens milliseconds. Adding security features increases this overhead to hundreds of milliseconds[12]. Calling a GRAM service on a grid based on Globus Toolkit 4 could take 0.4~2 seconds. In the current implementations of the Japanese NAREGI grid[7] and the China National Grid[6], with their virtualized services and other features to support usability and security, calling an end-user level service need to call several Axis or WSRF type of services. Consequently, calling a NAREGI service needs tens of seconds, and calling a CNGrid service could exert an overhead of 2~7 seconds.

There already exist grids that are fast and have good usability. However, these are mostly solution grids, where the efficiency and usability issues are resolved by the application developer and embedded in the design and codes of the application software. The challenge for the next few years is to develop general-purpose grid system software that is both efficient and easy to use.

Four usability requirements are identified below that the grid system software should support. By "support", we mean that the grid system software should enable the user to construct a grid exhibiting these features. While grids and service-oriented architecture have the potential to enhance usability, the current implementations still have a large gap from the requirements. Grids today are far behind traditional computer systems and the Web technology, with respect to the four usability issues of single system image, portability, agility, and managed code.

Single System Image. The grid system software should offer simple, uniform rules for the user to design and use the proper level of single system image. One analogy is the Web technology, at the heart of which is a single space of URL links, which theoretically is the entire Web. Such simplicity and clarity of single system image is one reason why the Web spreads so fast.

This is not the case with current grids. The usual case is that single system image issue is not explicitly discussed. A bag of techniques are implicitly offered, such as URL, XML namespace, WSDL, Web addressing endpoints, file directories, without a coherent set of guidelines for constructing and using a grid. In an application grid, a user sees application level services. In TeraGrid, a user sees his home directory, scratch spaces, and multiple supercomputers. In CNGrid, a user sees a number of VO's called agoras, where user-level services crossing multiple sites can be accessed.

Portability. The grid system software should itself be portable. But more importantly, it should support application portability and service portability among different VO's and grids. An application program, written in some high-level language, should be portable on different grids, just as it is portable on different computers, with the support of compilers. A service should be portable on different grids, just as a hardware device is portable on different computers, with the same system software and hardware platforms.

We do not yet have any system software like a grid compiler. It is not easy to port an application grid, for instance myGrid, to the TeraGrid or the UK NGS platforms. We still cannot do it in the way we port a scientific application program, for instance NASTRAN, from a Unix platform to a Linux or Windows platform.

Agility. The grid system software should support changes in application requirements, security and resource use policies, grid services, grid system software, and the hosting environment, without unnecessary human involvement. One of the key advantages of grids and SOA is loose coupling and support for agility. Changes in one part do not require an overhaul of the entire system.

While some solution grids already provide for business agility, the current implementations of grid system software still lack good support for agility. Often the application codes need to be modified to accommodate changes in resources or policies. Many grid applications need to access a service via its URL. When the URL changes, often the application codes have to be modified.

Managed Code. The grid system software should support the "do no harm" philosophy, by enforcing complexity boundaries, lest one user's grid application should affect other users or the grid system. Again, this is not well supported in today's grid system software implementations.

## 4 Vega suite of grid software

The Vega Suite of grid software aims to supporting system software requirements for the cyberinfrastructure of Figure 6. Usability issues are addressed through a loosely coupled architecture and a set of systems abstractions.

### 4.1 Architecture of the Vega Suite

Figure 8 depicts the Vega Suite architecture. The suite consists of four components. The grid operating system core and utilities (Vega GOS) provide common supports. For those CI's that focus on information sharing, integration, and management, the Vega information grid software (VIG) provides additional functions. The VINCA software provides functions for those CI's that require supports for workflows and business processes. GSML is a new XML-based language that supports application integration and end-user interaction. Users who prefer not to learn a new language can use traditional Java plus APIs from Vega GOS, VIG, and VINCA.

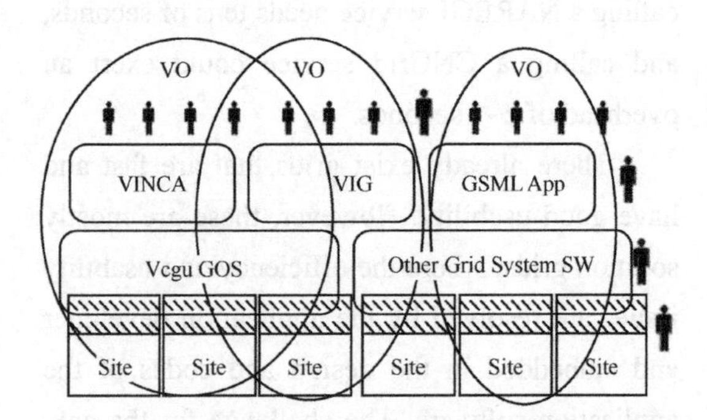


Figure 8. Illustration of the Vega Suite of grid software

These software packages are loosely coupled, meaning they can be used separately and evolved independently. This is analogous to the LAMP suite of open source software (Linux,

Apache, MySQL, PHP/Perl/Python). GSML, VINCA, and VIG run on top of the Vega GOS, but they can also run on other grid system software platforms, such as Apache Tomcat/Axis or GT4. In addition, the Vega Suite is designed to support interoperability with other grid system software. A grid built with the Vega Suite can access resources in another grid built on top of different grid system software.

## 4.2 System abstractions

The Vega Suite employ four abstractions called resource space, grip, agora, and funnel. The last one is to aid grid programming and discussed elsewhere[13]. Below we discuss how the first three abstractions help address the usability issues.

To support the cyberinfrastructure with multiple VO's and multiple grids, the application developer and the site manager face the challenge of dealing with multiple interfaces. The Vega strategy is to separate the development of applications and resources from grid system software.

When a resource (e.g., a BLAST program for gene sequence) is developed and then deployed as a physical service, it is done following the WSDL standard, independent of any grid system software concerns. Then the system software can link the physical service (e.g., named grid.net/blast2006) to a virtual service in a grid.

When a domain application such as a bioinformatics workflow is developed, it is done again independent of any grid system software concerns. The application codes only use effective services, which are business (application-level) services, instead of physical services. When and after the application is deployed on a grid, these business services are bound to virtual services, which in turn are bound to physical services.

This service virtualization scheme, utilizing the abstraction of effective-virtual-physical resource spaces, help achieve portability and agility. For instance, when the physical service blast2006 is changed to blast2007, or its URL is moved, there is no need to modify applications using this service. When an application is ported to another grid, we only need to establish the necessary mappings among the resource spaces.

To support a proper level of single system image and managed code, the Vega Suite employs the abstractions of agora and grip, and separate application logic from context and policies. An agora is a realization of the VO concept. It maintains two sets of members, the members of users and the members of services. In addition, it maintains context and policies shared by the members. Current agora implementation supports context and policies related to security (authentication, authorization, access control) and resource use (e.g., load balancing, accounting). The important point is that the application developer does not have to write codes to take care of such concerns.

When a grid application is evoked to run, a grip is created for the lifecycle of the application. The codes of the grip include two parts. One part corresponds to the application logic and associated data, structure, which is provided by the application developer. The other part corresponds to the common support for virtualization, context, policies, accessing physical services, and other low-level details (e.g., XML documents translation, SOAP messaging, exception handling), which is provided via the Vega Suite interface.

A grip is to grid what a process is to a traditional operating system. However, unlike a process, a grip can run on multiple nodes of a site, or even on multiple sites, while still maintain a single system image of coordinated resources and services.

The grip and the agora are system abstractions, and can provide automatic checking against bad effects such as invalid or unauthorized access to grid resources and services. However, the current Vega Suite design cannot realize fully managed codes. If a user writes a malicious code in the application logic without using the Vega Suite interface, the Vega grid system software cannot catch it. This exists in other grid and SOA systems, too. For instance, a grid allowing a user to submit a batch job has limited protection if the job contains malicious codes. An approach offering better supports of managed code is proposed in [10], which uses a service virtual machine to dynamically check all application codes.

### 4.3 Implement at ion and evaluation

Part of the work in Vega GOS presented above have been implemented and deployed in China National GridI[6]. The application experience with CNGrid has been invaluable to the new design of the Vega Suite. Most ideas presented above have been implemented in Vega Suite version 1 (alpha). Preliminary tests show that the Vega approach is feasible in enhancing supports for grid usability.

For instance, although the current Vega Suite implementation is based on J2SE and Apache Tomcat/Axis, we have demonstrated that the Vega Suite can work with international grid system software platforms such as GT4[9]. The Vega Suite can use GT4 as its kernel. Alternatively, a grid application based on the Vega Suite can access a GT4 grid service via a simple gateway, utilizing the grip and the agora constructs.

The Vega Suite also has improved performance than the China National Grid system software. With all features supporting service virtualization, security and context enabled, the overhead for calling a business service has been reduced to around 0.4 seconds. This improvement is mainly due to a more efficient implementation of the grip construct. There is still much room for efficiency improvement, as we have not utilized locality well yet. In addition to the obvious case of service locality, we are also exploring the cases of context locality and policy locality.

## 5 Conclusions

Good usability is a must for grid technology to be widely used. By reviewing international grid research experiences and usage modalities, we have identified four usability issues: single system image, portability, agility, and managed code.

The Vega Suite addresses the usability issues with a loosely coupled architecture and four system abstractions. Preliminary evaluations show that this is a feasible approach. It can enhance usability, while at the same time maintain security and improve efficiency.

## References:

[1] CATLELL C E, SMARR L. Metacomputing[J]. Com-

munications of the ACM, 1992, 36(6): 44-52.
[2] FOSTER I, KESSELMAN C. The grid: blueprint for a new computing infrastructure[M]. San Francisco: Morgan Kaufmann, 1998.
[3] FREEMAN P A, CRAWFORD D L, KIM S T, et al. Cyberinfrastructure for science and engineering: promises and challenges[J]. Proceedings of the IEEE, 2005, 93(3): 682-691.
[4] PARASHAR M, LEE C A. Special issue on grid computing[J]. Proceedings of the IEEE, 2005, 93(3): 479-484.
[5] BERNHOLDT D, BHARATHI S, BROWN D, et al. The earth system grid: supporting the next generation of climate modeling research[J]. Proceedings of the IEEE, 2005, 93(3): 485-495.
[6] ZHA L, LI W, YU H , et al. System software for China national grid[C]// IFIP International Conference on Network & Parallel Computing. Heidelberg: Springer Press, 2005: 14-21.
[7] MATSUOKA S, HATANAKA M, NAKANO Y, et al. Design and implementation of naregi super schedule based on the OGSA architecture[J]. Journal of Computer Science and Technology, 2006, 21(4): 521-528.
[8] ROURE D D, JENNINGS N R, SHADBOLT N R. The semantic grid: past, present and future[C]// The 2nd European Conference on the Semantic Web: Research and Applications. Piscataway: IEEE Press, 2005.
[9] FOSTER I. Globus toolkit version 4: software for service-oriented systems[J]. Journal of Computer Science & Technology, 2006, 21(4): 513-520.
[10] WANG X, XIAO L, LI W, et al. Abacus: a service-oriented programming language for grid applications[C]// The 2nd IEEE International Conference on Services Computing. Piscataway: IEEE Press, 2005: 225-232.
[11] TOLIA N, ANDERSEN D G, SATYANARAYANAN M. Quantifying interactive user experience on thin clients[J]. Computer, 2006, 39(3): 46-52.
[12] HUMPHREY M, WASSON G, JACKSON K, et al. State and events for web services: a comparison of five WS-resource framework and WS-notification implementations[C]// The 14th IEEE International Symposium on High Performance Distributed Computing. Piscataway: IEEE, 2005: 24-27.
[13] SHU C C, YU H Y, LIU H Z. Beap: an end-user agile programming paradigm for business applications[J]. Journal of Computer Science and Technology, 2006, 21(4): 609-619.

第6章

# 高性能计算机

# 曙光高性能计算机的艰苦历程

为了更广泛地听取国内外专家的意见，确定国家智能计算机研究开发中心的研究方向，我们于1990年5月在北京饭店召开了智能计算机发展战略国际研讨会。我们邀请了时任美国总统科学顾问的许瓦尔兹教授、人工神经网络理论的奠基者之一霍普菲尔德教授、日本第五代计算机的负责人之一田中英彦教授、美国伊利诺伊大学的华云生教授、南加利福尼亚大学的黄铠教授、波音公司的德格鲁特研究员等参加会议并发表意见。我国吴文俊教授等100多名学者到会。参加会议的多数外国专家不赞成我们走第五代计算机的路，建议根据中国国情，先研制比个人计算机性能高一档的工作站。这次会议对国家智能计算机研究开发中心选择以通用的并行计算机为主攻方向起到了重要的推动作用。

研制"曙光一号"是国家智能计算机研究开发中心历史上精彩的一幕。当时我们决定派一支小分队到美国进行研发。小分队在硅谷租间房子安顿下来后，需要什么软件和零部件，打个电话就有人送来，有些软件还让我们免费试用。这种"借树开花""借腹生子"的做法大大缩短了机器的研制周期。樊建平等几名派出的开发人员创造了一个中国计算机研制历史上的奇迹，不到一年时间就完成了"曙光一号"的研制，载誉归来，实现了他们在"人生能有几回搏"誓师大会上讲的"不做成机器回来就无脸见江东父老"的诺言。与现在的十亿亿次超级计算机相比，"曙光一号"真是"小巫见大巫"。但"曙光一号"的研制成功开辟了一条在开放和市场竞争条件下发展高技术的新路。

"曙光一号"研制成功以后，国家智能计算机研究开发中心开始研制"曙光1000"大规模并行机。大规模并行机的关键技术是把大量处理机有效连接起来的高速互连网络和每个处理单元的核心操作系统。国家智能计算机研究开发中心率先在国内突破了"蛀洞路由（Wormhole Routing）"关键技术，为我国发展大规模并行机开拓了一条道路。这款芯片的研制者是当时刚进中国科学院计算技术研究所的小伙子曾嵘。但其实他在硕士期间研究的是计算机下围棋的软件，没有碰过集成电路。1997年我访问麻省理工学院时告诉Bill Dally教授（Wormhole Routing技术的发明者），我们已研制成功异步蛀洞路由芯片，他很惊讶，因为异步电路的调试很困难，他做异步路由芯片也曾失败过。这件事给我们的启发是，要信任有潜力的年轻人，他们可能做出意想不到的出色成果。"曙光1000"是国内研制成功的第一个实际运算速度超过每秒10亿次浮点运算的并行机（Linpack速度超过每秒15亿次）。1997年我们获得国家科技进步奖一等奖。

在研制“曙光1000”（当时称为“东方一号”）的过程中，我应邀到在日本举行的1994年并行体系结构、算法和网络国际会议（ISPAN）上做大会特邀报告，介绍国家智能计算机研究开发中心在并行处理方面开展的研究工作。在会上我遇到并行处理领域的权威H. Siegel教授（我在普渡大学读博士时修过他讲的计算机系统结构课，但他可能并不认识我）。1996年，ISPAN在北京召开，他第一次来中国，在会上他专门组织了一次小型茶话会，对中国大加赞赏。2001年，我获得普渡大学颁发给杰出校友的Outstanding Engineer奖。当时我的导师华云生等华人教授都已离开普渡大学，我一直纳闷是谁提名我获这个奖的，后来有人告诉我是H. Siegel教授，真是没有想到。

曙光高性能计算机是我国“发展高科技，实现产业化”的成功案例。20世纪90年代，国内高性能计算机被IBM等外国公司垄断，长期被西方禁运。如今曙光高性能计算机在国内的市场占有率远远超过IBM等外国公司，国内市场份额多年位居第一，2020年曙光公司的市值已超过500亿元。2019年美国政府将曙光公司列入限制技术出口的实体清单，《华尔街日报》发表文章解释，这是因为曙光公司掌握了“通往技术王国的钥匙（The Key to The Kingdom）”，这件事从反面凸显了曙光公司在实现我国高技术自主可控中的地位。

中国科学院计算技术研究所和曙光公司不慕虚名，研制高性能计算机的目的就是为了市场应用，因此特别重视可市场化的关键技术，重视可规模化、标准化生产的技术。本章收录的有关集群文件系统的论文是漫长研发过程的缩影，从孙凝晖开始，经过熊劲等几代博士生连续攻关，再将人和技术转移到曙光公司，公司又严格按市场要求开发了3～4年，才形成曙光公司今天的主打产品——Parastor云存储系统。在几代曙光超级计算机的研制中，有许多令人难忘的场面。2019年，在完成“曙光7000”调试的最后冲刺中，一天中竟有50多名员工晕倒送医院医治，第二天这些员工又赶到机房上班。没有如此顽强的“曙光人”，曙光计算机决不会有今天的辉煌。

发表于《计算物理》1992 年第 4 期

# 高速科学计算与大规模并行机

李国杰

中国科学院计算技术研究所国家智能计算机研究开发中心，北京 100080

**摘　要：**本文分析了高速科学计算中的“巨大挑战（Grand Challenges）”问题对大规模并行处理（MPP）系统的需求，比较了向量机与MPP，介绍了MPP的发展现状及关键技术，最后扼要地论述了并行计算模型、并行算法与并行程序环境。本文的目的是促使计算物理学者重视并行算法和并行科学计算软件的研究。

**关键词：**大规模并行处理；大规模并行机

# High speed scientific computations and massive parallel computer

Guo-jie LI

Institute of Computing Technology, Academia Sinica, Beijing 100080, China

**Abstract:** In this article, we analyze the requirements of “Grand challenges” problems in high speed scientific computations for massive parallel processing (MPP) systems, compare vector supercomputers with MPP, introduce the present state of the art and key technology of MPP, and discuss the issues of parallel computational models, parallel algorithms and parallel programming environment. The purpose of this article is to call attention of researchers in computional physics area to the MPP.

**Key words:** massive parallel processing (MPP), massive parallel computer (MPC)

## 1　引言

20 世纪 90 年代是计算机突飞猛进的时代，计算机的发展将从根本上改变人们生产、生活及科研的方式。大致来讲，计算机的总的发展趋势是小型化、并行化、网络化与智能化，20 世纪 90 年代最受重视的计算机技术是基于精简指令集计算机（RISC）的微处理机芯片技术与并行分布处理技术。以高速计算、高速通信以及多媒质智能化接口技术为标志的计算机高技术受到各国政府、产业及科研部门的普遍重视。限于篇幅，本文不能对计算机的发展趋势做全面论述，只能着重就与高速科学计算有密切关系的大规模并行计算机谈一点个人看法。

## 2 高速科学计算是发展超级计算机[1]（Supercomputer）的主要牵引力

众所周知，计算机的速度总是满足不了科学计算的实际需要，理论物理、量子化学等科学研究要求计算机的速度有几个甚至十几个数量级的提高。用目前最快的向量机做 50 个原子的计算模拟需要 50~100 小时，因其计算量与原子个数的立方成正比，目前的计算资源无法模拟 500 个以上的原子行为。自从发现高温超导现象以后，许多物理学家致力于探索超导机理，他们普遍认为只有根据相对论量子电动力学计算出材料的超导性质才能真正理解高温超导，但由于自由度太多，目前的计算机不可能求解固体的 Schroedinger 方程。基础理论研究呼唤高性能的计算机，这一领域对计算能力的要求无止境。

为了满足工业部门和科学研究的需要，并进一步提高在高速计算领域的竞争能力，美国政府（包括国防部、自然科学基金会、能源部等 8 个部门）制订了为期 5 年（1992—1996 年）的高性能计算与通信发展计划（HPCC），这一计划的主要目标是在相当宽的重要领域内计算能力达到每秒 $10^{12}$ 个操作（Teraops），国家研究与教育计算机网的通信能力达到每秒 $10^9$ 位（Giga bits/s）。1992 年和 1993 年两财政年度 HPCC 的投资分别为 6.5 亿和 8 亿美元. 投资的重点（占 43%）是发展先进的软件技术与并行算法。HPCC 计划被称为“巨大挑战”（Grand Challenges）计划。美国许多有名的大学与国家计算中心都投入了大量人力物力攻克各种“Grand Challenges”问题，即大型的复杂的科学工程计算问题。HPCC 计划要求到 1996 年实际计算速度比 1992 年提高几百倍。要实现这一目标，关键技术是可扩展的大规模并行计算（Scalable Massive Parallel Computing），即系统的性能随着处理机个数增加而接近正比地增加。欧洲各国与日本也十分重视这一研究方向。我国这方面的研究刚刚起步，但越来越被重视，估计在“八五”计划期间我国也会研制出自己的大规模并行机。

## 3 向量机与并行机

迄今为止，解决高速科学计算问题主要还是依靠向量机。十几年来，向量机的代表是 CRAY 公司的产品，最新推出的产品是 CRAY-YMP/C90，整个系统由 16 台 PE 组成，每个 PE 有两条向量流水线，主频 250 MHz，每 PE 可达峰值速度 1 GFlops，参考价格 3 000 万美元（16PE）。日本研制生产的向量机的硬件水平已超过美国，如 NEC 公司推出的 SX-3 向量机，主频周期只有 2.9 ns，比 C90 还快，而且采用多个向量功能单元，一条流水线每个时钟期可产生 4 个浮点结果，有资料报道，SX-3 峰值可达 22 GFlops（4PE），实际测试对于高度向量化而且向量很长的 Benchmark 程序 SX-3 运行速度是 CRAY YMP 的 8 倍（比较单个 PE）。向量机已使用了十多年，开始使用时人们把它看成专用机，经过十多年努力，向量化编译技术日益成熟，适用范围逐步扩大，因而现在的用户已经几乎把向量机看成科学工程计算的通用机了。

向量机再向前发展有什么障碍呢？首先，向量机的效率取决于应用程序中可向量化语句占的比例。CRAY 公司检查了三十多个常用科学计算程序，发现平均向量化率为 60%～80%，这意味着如果仅仅采用向量流水技术，加速比不超过 5。另外，提高向量机性能的主要途径是加快主频，目前时钟周期已接近 1 ns，再提高潜力不大。CRAY 公司一直在攻砷化镓芯片技术，争取实现 1 ns 时钟，但基于砷化

[1] 作者不赞成把计算机分成巨、大、中、小、微几类，将 Supercomputer 译成巨型机容易使人误解为只是体积大，译成超级计算机似乎更恰当。

镓技术的 CRAY-3 迟迟没有推出，连唯一的用户劳伦斯利弗莫尔国家实验室也取消了订货，可见难度超过预料。第一台砷化镓计算机产品 C3800 出自 Convex 公司，但其重点不是攻主频而是攻集成度，每个 PE 峰值速度只有 120 MFlops，还不如 ECL 和 CMOS 技术。向量机主要开发低层次的并行性，即 Loop 语句级的并行。低层次并行有一定限度，并行计算领域的权威学者 Kuck 教授的研究结果表明，从统计平均观点来看，开发 Loop 级的并行性加速比一般不超过 20。因此，最近几年更多的学者转向研究高层次的并行，即从模型与算法级上开发并行，这方面的潜力更大。

虽然向量机也可看成一种类型的并行机，但今天人们讲并行机或程序并行化通常不是指向量机或程序向量化，而指 SIMD 阵列机，共享存储多处理机和松耦合多计算机系统。可向量化的程序不一定能有效地并行化；相反，可并行化的程序不一定能有效地向量化。公正地说，并行机与向量机各有自己的应用领域，并不能相互取代。当然，有些领域两种机器都适用，MPP 已成为传统向量机的市场竞争对手。关于大规模并行机究竟能不能做到向量机的通用程度，最近在 *Supercomputing Review* 期刊上有一场争论，有人认为大规模并行机只能是专用机，但也有些计算中心特别热衷于 MPC，例如美国 Sandia 国家实验室，他们专门成立了一个大规模并行计算研究实验室（MPCRL），已在 MPC 上出色地进行了电子结构、分子动力学、催化剂设计等方面的研究。笔者认为现在谈论 MPC 是否只能专用还为时过早，就像 15 年前谈论向量机是否专用一样. 随着大规模并行处理技术的日益成熟，其应用面必将逐步扩大。

为了促进 MPP 技术的发展，美国 1987 年专门设立了 Gorden Bell 奖，分别奖励计算机实际速度、性能价格比和编译的加速比方面的世界冠军，表 1 列出了获奖者的世界纪录。

从表 1 看出,这 3 项纪录每年都大幅度提高。1991 年宣布的 1990 年最高实际计算速度是在 CM-2 机器上的地震模型程序（14.1 GFlops）。性能价格比最高的程序是 Oak Ridge 国家实验室开发的计算超导特性的 Fortran 程序，在 Intel/i860（1 28PE）实现 2.53 GFlops，机器价格为 300 万美元，这一程序移植到 IBM RS/ 6000 可运行速度为 16.7 MFlops，价格 8 500 美元，达到 2.1 GFlops/ $ 1M 最高纪录，但该程序在 IBM RS/ 6000 工作站上要运行一个月之久。编译获得的最高加速比是在 CM-2 机（相当 2 048FPU）上实现椭圆格点产生程序，先用预编译 KAP/F77-F90 得到 F90 并行程序，再用 CM Fortran 编译。从这些例子来看，并行处理（包括编译）技术在学校和科研单位已达到相当满意的水平。但是，多数专家估计，要在工业界和大量应用部门普及并行机还需要 10 年以上时间。

**表 1 获奖者的世界纪录**

| 年份 | 性能/(GFlops) | 性能价格比/(GFlops/ $ 1M) | 加速比 |
|---|---|---|---|
| 1987 年 | 0.45 | 0.03 | 600 |
| 1988 年 | 1.0 | 0.05 | 800 |
| 1989 年 | 6.0 | 0.5 | 1 100 |
| 1990 年 | 14.0 | 2.0 | 1 800 |

## 4 大规模并行处理机

经过几年准备，大规模并行机已从实验室走向市场。去年好几家公司陆续宣布推出 MPC，在这一领域目前走在前列的有 Intel、Connection Machine、nCUBE 等公司。最近，nCUBE 公司宣布 1994 年将推出峰值速度达

6.5 TFlops（6 万 5 千亿次）的 nCUBE-3，由 64 000 个 PE 组成。美国《商业周刊》报道，日本政府 支持一项被称为“Real World Computing Project”的计划，打算用 100 万个处理机构成一大规模并行机，峰值速度高达 125 TIPS！笔者认为要实现这样的系统，除了解决并行技术外，还必须从根本上降低器件功耗，解决散热问题。即使一个处理机功耗按 10 W 估计（目前水平一般超过 100 W），整个系统功耗将高达 1 万千瓦！超导计算机是降低功耗的一条重要出路。其他公司也将在 1995 年前后推出万亿次并行机。美国许多专家预测 1994 年将是 MPC 摊牌的一年，以 MPC 为标志的计算机高技术竞争正在激烈进行。

究竟一个系统具有多少处理机才算大规模并行并没有明确的定义，实际上人们在称呼大规模并行机时隐含了对单个 PE 性能的折衷考虑。如采用最高档微处理芯片做 PE，目前几百台处理机构成的系统已可以认为是 MPP 了。一个 MPP 系统究竟是用功能较强而台数相对较少的 PE 构成，还是由大量的较弱的处理单元（1 位或 4 位处理单元）构成，几年前曾有过争论。现在看来，各大公司采取的技术路线已基本接近，这就是以先进的 RISC 芯片（10~100 MFlops）为基础，采用几千甚至几万个处理机构成可扩展的并行机系统。Connetion Machine 公司曾经致力于 64 000 个位片机组成系统，目前已放弃原来的路线，改用 SPARC 芯片，每个 PE 再增加 4 条流水线算术部件，使浮点处理能力提高到 128 MFlops，该公司宣布的 CM-5 大规模并行机目前最大节点数为 1 000，价格为 2 500 万美元。CRAY 公司到目前为止生产的各种向量机最多只有 16 个 PE，1991 年 CRAY 公司收购了以研制 MPP 闻名的 FPS 公司，并考虑采用 DEC 公司的 ALPHA 芯片（200 MIPS）做 MPC（最近有报导可能不选用 ALPHA 芯片）。美国国防部投资 1 270 万美元支持 CRAY 公司做 MPC，1993 年打算推出 100 GFlops 的并行机，1995 年推出 1 TFlops 并行机。IBM 公司计划 1993 年将推出 512 台 RS/6000 组成的并行机。DEC 公司已决定放弃 VAX 系列，采用他们设计的最高档 RISC 芯片（ALPHA）生产并行机。这些大公司的转向和殊途同归说明 MPP 是发展高性能计算机的必由之路。

为了说明 MPP 的当前水平和发展方向，下面我们较详细地介绍一下 Intel 公司即将推出的大规模并行机 Paragon X/PS。Intel 公司得到美国国防部 DARPA 的支持，开展了研制世界上最高速计算机的试金石（Touchstone）计划。1992 年已完成该计划的阶段成果 Delta 机，该机包含 528 个节点，峰值速度达 320 亿次。预计 1993 年中将完成该计划的最终产品 Sigma 机，对外型号叫 Paragon。该系统包含 4 000 个 i860XP 处理机，峰值速度高达每秒 3 000 亿次浮点计算。经进一步扩大规模，1995 年将达到每秒万亿次水平。Paragon 是一个典型的分布存储 MIMD 系统，每一处理机通过一个专门设计 Routing Chip 连到二维 Mesh 网络的节点上，称为节点机。过去几年内，分布存储（或称松耦合）并行机多数采用超立体互连结构，由于通信速度的提高和采用蛀孔（Wormhole）技术，节点之间的距离已不再是影响传输延迟的主要因素，因此这两年研制的 MPC 大多采用两维 Mesh 结构。所谓蛀孔技术，通俗地讲是把打电话（Circuit Switch）和打电报（Packet Switch）两种文式结合起来，并把要传送的信息分成一个字节一段（Flit），像虫子蛀洞一样，报文跟着报文头一个节点一个节点地传送。如传输中受到阻塞，不必把全部报文存入某一节点的缓冲存储，而是停在通信链路上。这种方式使通信速度大大加快。Paragon 每一节点用一个 i860 处理机专管通信，并采用 Wormhole 技术，每条链路通信带宽高达 200 MB/s。系统软件方面 Paragon 采用可望成为工业标准的 OSF/1.1 分布式操作系统，它以 Mach 3.0 为基础。UNIX 操作系统本来的特点是精练，后来越搞越庞大，不适于 MPP。20 世纪 90 年代操作系统的主要方向是发展微核（Micro Kernel）技术，即把 UNIX 的最基本部分（如通信、存储管理等）纳入微核中，而把文件系统等非基本部分作为一个 UNIX 服务器放到 Kernel 外面，这样既简化了操作系统，又使并行操作系统及用户的界面与 UNIX 兼容。Paragon 采用的另一项关键技术是共享虚存技术。由于并行机的可扩展性（Scalability）和易于编程相互矛盾，过去几年内研制生产的并行机只能满足一方面的要求，要么是共享主存的多处理机，编程较容易，但扩充受限制；要么是分布存储并行机，容易扩充，但用户编程困难。共享虚存技术较满意地解决了这一矛盾，使得并行机的存储器物理上分布在各处理机中（便于扩充系统），而逻辑上有统一的地址空间，用户编程就如同在共享主存多处理机上一样方便，现有的串行程序也有可能通过自动并行识别和编译直接在 MPC 上使用。这一技术虽然不完全成熟，但必将受到用户欢迎。

MPC 的一个重要优点是与相近档次的机器比较其性能价格比高。当然，如果光看花一元钱能买多少次计算，应当说工作站比 MPC 更合算。工作站是做串行标量计算，尤其是做 CAD 的理想工具，图形功能强，通用性强，用户使用十分方便，今后 10 年还会有大的发展，但用工作站做稀疏矩阵计算就会慢得难以忍受，不能解决大型科学计算问题。有人怀疑 MPC 的性能价格比未必会高于向量机，他们的理由是由于 VLSI 技术发展，CPU 芯片已不再是计算机系统的主要成本，计算机的主要成本是存储器和通信部件，如果 MPC 的存储容量与通信能力与向量机一样，则成本应基本相同。这是一种似是而非的看法。存储容量相同不一定成本一样，关键是看用什么手段达到足够快的存取带宽。MPC 的存储器分布在各个节点中，每个节点类似一个工作站母板，实现较简单；而要保证向量机的高速流水线的集中供数取数，必须采取非常高速的电路技术（如 ECL 技术）和昂贵的冷却技术。必须指出，衡量一个系统的整体性能，不能只看 CPU 芯片的峰值速度，通信能力或者说存储带宽是非常重要的指标。一般来讲，只有保证每进行一次浮点运算能提供 2 ~3 个浮点数，实际计算速度才能接近峰值速度，由于芯片集成度限制，芯片内的高速缓存容量总是不够用（i860XR 才 8 KB），而且 RISC 技术强调寄存器运算尽量减少存储器存取，因此编译优化技术显得格外重要。一台并行机的实际计算能力与通用性很大程度上取决于通信能力。思维机器公司的 CM-5 每个节点峰值速度高达 128 MFlops, 但每条链路通信带宽只有 5~20 MB/s（Cluster 内部 20 MB/s, 外部 5 MB/s），不到 Paragon X/PS 通信带宽的 1/10。可以断言，CM-5 的通用性较差。欧洲一些国家采用 Inmos 公司生产 Transputer 也研制成功一些 MPC，例如英国研制成功 QCD-20 并行机，每个节点用 i860 计算，另用两个 Transputer T800 做通信，因 T800 的每条链路带宽只有 20 Mbit/s（相当 2.5 MByte/s），通信能力很差，只能专门用于量子色动力学这类计算量很大而通信少的研究，该机峰值速度可达 17 GFlops。

限于篇幅，我们不能对 MPC 做更多介绍。下面用一张表（表 2）列出最近推出的几种有代表性 MPC 和向量机，读者可自己比较它们的性能。

表 2 最近推出的大规模并行机与向量机

| 公司 | 机器 | 处理机数 | 峰值速度/GFlops | 存储容量/GB | 参考价格/万美元 |
| --- | --- | --- | --- | --- | --- |
| Intel | Paragon XP / S | 4 000 | 300 | 512 | 55 |
| TMC | CM-5 | 1 000 | 128 | 32 | 25 |
| Alliant* | Compus 800 | 800 | 32 | 128 | 20 |
| nCUBE | nCUBE-2S | 8 192 | 34 | 512 | ≤34 |
| CRAY | C90 | 16 | 16 | 2 | 30 |
| NEC | SX-3 | 4 | 22 | 2 | 24 |

*Alliant 公司生产共享存储大规模并行机，已于 1992 年 6 月宣布倒闭。

## 5 并行算法与程序并行化

影响并行机推广使用的主要障碍是缺乏软件支持。如果现有的串行计算机的速度已满足要求而价钱又可以接受，没有一个用户会主动要求用并行机。使用大规模并行机要求用户从计算模型和算法的层次重新考虑如何计算，从较高的层次开发问题中包含的并行性。下面我们从计算模型、算法和并行编程环境三方面简单介绍一下这方面的进展。

并行性包括控制级和数据级两个层次。从控制这一层次很难做到大规模并行，目前能有效运行的大规模并行应用程序大多采用 SIMD（单指令流多数据流）、SPMD（单程序多数据流）或 ASIMD（Autonomous SIMD，自治的 SIMD），即主要是开发数据层次的并行。所谓 SPMD 是指各个节点上运行同样的程序，只是各自的数据不一样，各个节点之间异步，不必像SIMD机器那样每一处理机同一时刻执行同样的指令，节点之间可通过相互通信实现同步和协作。用区域分裂法求解大型偏微分方程就是这种模式。如果一台机器分成几个 Cluster，每个 Cluster 运行 SPMD 模式，但每个 Cluster 运行的程序不同，则称为 MPMD 模式。ASIMD 是对传统 SIMD 的一个扩充与改进，增加了系统适应能力。所谓自治性（Autonomy）包括执行、寻址、连接和 I/O 等方面的灵活性，但又基本上保持 SIMD 机的可程序性。以 Marpar 公司推出的 MP-1 型机（16 000 个 4 位机组成）是 ASIMD 模式的代表，这种 SIMD 与 MIMD 的折衷模式正逐步得到推广。

对于串行计算机不同的算法解同一问题可能在效率上有差别，但差别一般不超过几倍或几十倍。但是对于大规模并行计算，好的算法可能使效率成千上万倍地提高。将最佳串行算法并行化有时也能得到较好甚至最佳并行算法，但在多数情况下好的并行算法往往要以非最佳甚至不好的串行算法为基础。有些串行算法早已被人抛弃，但加以改造后可能得到很有效的并行算法，例如曾获得超级计算机奥林匹克奖的并行油藏模拟算法就是基于 1938 年的一项早已被废弃的研究成果。研究并行算法的学者需要解放思想，打破传统的串行计算的思想束缚。长期以来，人们承认 Amdahl 定律的假设，即程序中串行部分所占的比例是固定的，与求解问题的规模无关。因此总是用一固定规模的问题计算加速比，强调只要串行部分占 1/10，不管用多少处理机，最大加速比不会超过 10。最近几年大规模并行计算的实践证明，在相当多的科学计算程序中，串行部分的绝对计算量基本上不变，当问题规模增大时，串行部分的比例大大下降，因而可获得千倍以上的加速。究竟具有这种性质的问题是较普遍存在的还是只在少数领域出现，目前还无定论，看法上还有分歧。但至少我们可以说，很多人默认的 Amdahl 定律的前提假设并不普遍成立，或者说，我们不应该只用同一规模的问题来比较串行并行的效率。并行机应该用来计算大问题。这也就是讲并行计算的粒度应足够大。还应该指出，不是问题规模大了，加速比肯定就能上去，影响并行计算速度的另一个重要因素是是否具有足够的局部主存。如果每个处理机局部存储器太小，计算时频繁访问磁盘，计算速度一定慢得不可忍受。关于并行存储系统与性能的关系已超出本文范围，此处不再赘述。

用户最关心的还是程序语言和编程及调试环境，应当说这是并行机的薄弱环节。前些年这方面的重点研究方向是并行语言与并行编译，人们试图制定并行 Fortran、并行 C 语言的国际标准。许多学校和公司已投入了大量人力物力研究并行编译技术与并行性自动识别，希望把串行程序自动地或半自动地（加一定用户干预）转变为并行程序. 这方面的工作已取得不少进展，本文前面提到的获 Gorden Bell 奖的一项成果就用到 Kuck and Associate 公司的 KAP 并行编译器。但总体来讲，通用的并行编译技术离实用化还有一段距离，也许再过 10 年，局面会有较大改观。目前大多数并行机上的编程环境采用并行库程序的方法，例如 MOMDizer、Express、Linda 等，这些环境基本上不需要改变原来的串行程序的风格，用户只需在程序中调用并行库函数，因而对程序员要求低，较易于推广使用。这一类环境可移植性强，目前已移植到 IBM、CRAY、TMC、nCUBE、Intel 等大公司的并行机上。一般来讲，移植性强的软件实现效率较低，如何提高这些并行环境的实现效率还有待进一步努力。至于并行语言的标准化看来还需要一段时间，目前有一定影响的并行 Fortran 语言是 F90 和 Fortran D。

## 6 结束语

美国的 Grand Challenges 计划中列举的大规模并行计算包括磁记录技术、药物设计、催化、燃烧、海洋模型、臭氧洞、空气污染等应用领域，其中不少与计算物理有关。近几年关于 Supercomputing 的会议文集与学报论文中，计算物理占相当大比重。我国的高技术计划与其他部门的计划已在考虑研制大规模并行机。因此大规模并行对于我国学者已不是遥不可及的事。从现在起，切实加强并行物理模型、并行算法与并行程序的研究已刻不容缓，趁计算物理学会召开年会之际，献上此文，希望引起计算物理学界对大规模并行计算的重视，使我国在这一高技术领域有所作为，尽量缩小与国外的差距。

发表于《中国科学院院刊》1994 年第 4 期

# "曙光一号"并行计算机

李国杰，陈鸿安，樊建平，刘金水

中国科学院计算技术研究所国家智能计算机研究开发中心，北京 100080

## 1 背景和目标

近几年来，国际上计算机与半导体工业发展飞速，在不断提高单机或单处理器性能的同时，多处理机系统和并行计算机已大量进入市场。这是因为并行机具有比单机更高的速度、更高的作业吞吐量、更高的性能价格比、更高的可扩充性和可靠性。因此，并行机已成为 20 世纪 90 年代计算机发展的主流。

并行机的发展建立在高性能的 RISC 处理机芯片基础上，采用当前流行的操作系统和一系列工业标准，形成了"开放的"体系结构。这种开放系统打破了少数公司对计算机的垄断，给后来者和小公司提供了绝好的机会，一批名不见经传的小公司已挤入国际著名计算机生产商的行列。国内的改革开放和国际趋于缓和的形势为我们学习和利用国外先进技术、建立国际合作提供了机会。并行机的研制在国外也是起步不久的新领域，关键在软件上，技术并不很成熟。发展并行机是我们面临的一次新的机遇。"曙光一号"并行计算机的研制就是在这样背景下的一次成功的尝试。

在 863 计划智能计算机专家组的正确指导下，设在中国科学院计算技术研究所的国家智能计算机研究开发中心通过分析并行机的发展现状和趋势，选定了以研制全对称共享存储多处理机（SMP）系统的"曙光一号"和分布存储的大规模多处理机（MPP）的"东方一号"为主攻方向。从用户使用的角度看，目前最成熟的还是共享存储多处理机。从国内外市场需求看，目前需要量最大的也是共享存储的多处理器服务器。国内已研制成功的几种共享多存储多处理系统不是全对称的，而是靠在一台主机上串行地运行操作系统的核心来支持用户程序。操作系统核心本身并没有并行化。根据国外对典型的服务器运行进程的有关统计，平均一个进程花费 40%～50%的时间运行系统核心代码。因此，采用非对称方式，许多应用只适用于两个 CPU 的多机系统，即使每个用户进程花费 10%的时间执行核心代码，这种多处理机最多只能供 10 个用户同时使用。所谓全对称的多处理机则是指每一个CPU 对主存和外设的访问以及接收中断请求等都具有相等的能力。每个 CPU 都可以同时运行系统核心程序，响应中断，也可以运行用户程序，这样就大大扩充了同时运行的用户数。"曙光一号"瞄准全对称多处理机的方向，把攻克和掌握 20 世纪 90 年代全对称多处理机技术作为自己的重要目标。

同样是多处理机，并行计算的粒度（Grain）差别很大。国内已研制的多处理机最细粒度只能到并行循环这一级，采用进程切换，这样的进程切换的开销在毫秒级。但是在大量科学工程计算、智能应用和实时处理中需要细粒度并行。这就要求引入多线程（Multi-Thread）技术，大大降低上下文切换（Conext Switch）的开销，这也是20世纪90年代多处理机的一项关键技术。“曙光一号”也把具有多线程机制、支持细粒度并行作为技术突破的重要方向。

我们以邓小平同志“发展高科技，实现产业化”的题词作为研制“曙光一号”的指导思想。从设计伊始就定下一个目标：“曙光一号”不能是一个原理性样机，而应是具有市场竞争力的高科技产品。为了在国内批量生产，不能选用国外禁运的 CPU 芯片，而要选择在国内能买到的、支持细粒度并行、实时处理能力强的芯片，在当时的条件下，只有 Motorola 的88K 芯片具备这些条件。要攻克全对称多处理和支持细粒度并行的多线程技术，目的是增强产品的竞争力而不是纯学术研究。在硬件和软件设计时特别强调采用公认的工业标准，要在遵从工业标准的基础上积极创新，以利于应用软件的移植和推广应用以及设备的更换和扩充，系统总体强调开放性和通用性。设计从高起点起步，采用最先进的设计工具，通过加强国际合作，尽量缩短产品开发周期。

## 2 特点和性能

“曙光一号”并行计算机系统是我国自行研制的第一台用微处理机芯片构成的全对称多处理机（SMP），又是一个通用的并行计算平台。它采用20世纪90年代的最新技术，如精简指令系统 RISC 技术、多种大规模集成电路芯片和大容量动态存储器；设计了内部高速总线，实现了各主要部件之间的高速数据通信；选用最新的现场可编程门阵列（FPGA），自行设计并实现了多处理机中断控制器 CD-92，处理多处理机的中断请求，保证每个处理机平等访问内部总线上的资源，形成了一个全对称的紧耦合的共享存储的体系结构；在输入/输出端口上采用了 VMEbus、SCSI、Ethernet 和 RS-232C 等标准的总线技术，可灵活地连接网络，配置系统。

“曙光一号”的操作系统 SNIX（Symmetric UNIX）是通过对 UNIX 版本细粒度加锁以及动态分配 I/O 中断向量方法，实现多处理机系统的对称式处理，通过在 UNIX 操作系统核心中增加共享资源进程以及成群调度策略，在用户空间以库函数的方式实现线程（Threads）概念，支持中细粒度的并行计算。在操作系统之上，提供了大量 UNIX 实用程序、标准化的串行语言编程环境与调试环境（包括串行 Fortran 语言的自动识别器）、X-Window/Motif 图形用户界面、基于 TCP/IP 协调的网络软件、第四代语言、数据库管理系统、智能应用开发环境以及分布式程序设计 Express 等。

“曙光一号”在预先研究的基础上，经过一年多的艰苦努力终于研制成功，实现了20世纪90年代并行计算机的主流技术：全对称多处理机系统。自行设计的 CD92 多处理机中断控断控制器（两块万级门规模的大规模集成电路芯片）和局部总线支持了全对称多处理机的硬件实现，操作系统核心并行化的改造和并行编译支持其软件实现。此外还有以下几个特点。

- 支持细粒度并行的多线程技术。
- 不仅支持多用户多任务的多处理，而且真正加速单任务的并行。
- 提供具有世界先进水平的并行编译。
- 硬件和软件都遵循国际公认的工业标准，是开放性系统，扩充和更换设备灵活。软件界面友好的移植十分方便，应用软件丰富。
- 中断处理速度快，适于做事务处理服务器、网络通信服务器。
- 系统扩展性强，支持的用户多。

• 实时处理能力强，适于实时仿真。

• 系统稳定可靠。

经过严格测试表明，“曙光一号”具有很高的性能，全部达到或超过设计指标。这里简列几项。

• 8CPU 实测定点速度为 305.25 MIPS，浮点速度为 35.10 MFlops。

• 并列加速比：4CPU 是单 CPU 的 3.4~5.7 倍。

• 操作系统核心代码并行加速比：4CPU 下为 3.5。

• 中断响应系统开销小于 17. 6 μs。

• 多用户吐量：严格同时运行 128 个用户，系统响应时间只增加 12%。

“曙光一号”性能优于国外 20 世纪 90 年代初同类产品水平，一家外国公司承认我们的系统软件水平比他们的同类机器好。根据我们最近获得的机能测试报告，“曙光一号”上下文切换（以空系统调用测）远远小于 SUN 公司和 IBM 公司最近推出的同类型机器。

## 3 市场与应用

“曙光一号”于 1993 年 10 月通过了由国家科委组织的技术鉴定，受到鉴定委员会的高度评价。鉴定委员会认为：“曙光一号”是 863 高科技计划信息领域的一项重大成果，达到了 20世纪90 年代初同类计算机产品的国际先进水平，不但是一项科研成果，而且是一个具有市场竞争力的、性能价格比高的高技术产品，具有广阔的应用前景。

1994 年 10 月，国务院总理李鹏在八届人大二次会议的政府工作报告中指出：我国科技事业在深化改革中取得新的进步，银河 II 型计算机、“曙光一号”并行计算机等一批科研成果达到世界先进水平。国家科委和中国科学院领导十分重视将科技成果转化为生产力，要把“曙光一号”推向市场，促进高性能计算机的产业化。国家科委党组做出决定：将曙光工程列为 1994 年优先实施的重大科技产业工程，并作为首选项目，拨款 200 万元作为紧急启动费用。中国科学院工作会议上把“曙光一号”列为五大产业化战役之一。

国家智能中心十分重视市场和推广应用。几乎在研制“曙光一号”的同时，组建了成果推广部，随后又正式注册成立了“北京市曙光计算机公司”，开展了市场和销售工作，取得初步的成绩。

1993 年 12 月初，第一台“曙光一号”安装在武汉大学软件基地，用作校园联网服务器；中国科学技术大学计算机系也安装了一台“曙光一号”并行计算机，用于国际联网服务器；全国高新技术开发区联网、航天部、内贸部和铁道部等单位正在洽谈合同；清华大学在 CIMS 中也选用并安装了一台“曙光一号”作为大型企业 MIS 系统服务器；作为援外项目，一台“曙光一号”即将出口，装备埃及穆巴拉克国家实验室；南京新技术开发区也将安装一台，用于信息演示中心和开发区联网；国家科委办公室自动化系统也采用一台“曙光一号”作为服务器，即将安装；邮电部业务管理系统也将安装曙光机作为服务器；湖南省邮电业务系统准备采用多台“曙光一号”作为地市级服务器；湖南省计算机所已签订合同，购置“曙光一号”做软件开发平台。

由于计算机发展异常迅速，每一、二年就会有新一代的机种问世。要形成产业就必须不断升级换代，以满足市场需求和保护用户资源，在升级换代中不断扩大市场，增加市场占有率。“曙光一号”的升级和形成曙光系列的工作已在进行，曙光系列机的市场将配合国家经济信息化和三金工程，争取直接为三金工程服务。曙光系列主要面向通信，通过研制通信需要的硬件和软件，使曙光系列在通信服务上形成特色；同时重点开辟邮电领域与商业自动化的管理市场，使曙光系列计算机在大型商场、企业和政府机关实现管理计算机化中做出贡献。

为了推广曙光系列计算机，形成高性能计算机产业，在国家科委有关部门推动下，1994 年 7 月已正式签订成立中外合资曙光计算机有限公司的协议，注册资本 7 812 万元（合 880 万美元），其中“曙光一号”知识产权占 240 万美元。该公司的成立受到国内外各大公司的关注，许多公司主动前来洽谈合作。Motorola 公司将智能机中心与曙光公司选作重要的合作伙伴，打算今后在曙光系列产品升级与推广中全面合作。“曙光二号”“曙光三号”产品 1995 年将推向市场。我们相信，曙光计算机将在我国高性能计算机市场上占有一席之地。

## 4 几点体会

外国计算机大量进入中国市场，我国计算机产业面临严重的挑战。我们是一退到底，甘于做代理商，还是知难而进，力争一席之地？对此必须作出回答。曙光计算机研制成功与初步走向市场给我们许多启示。我们认为，并行计算机的发展为我们提供了难得的机遇，我们有可能后来居上，这一机遇一定要抓住，千万不可错过。要想后来居上，一定要有所赶。在曙光计算机研制中，我们体会到要改变过去研制成果难以推向市场的局面，应该“三不做”：即完全仿制，没有自己知识产权的产品不做；仅仅为了填补国内空白，没有市场竞争力的产品不做；不符合国际工业标准，与主流技术脱节的产品不做。我们集中力量突破从国外买不到的或附加值高的技术，开发有自己特色知识产权的技术，设计与生产都从全世界范围内择优。在研制过程中大胆重用年轻科技人才，用开放的思路研制开放的计算机系统，坚持以应用带动整机研制和生产，以我们诚心诚意的服务和负责到底的精神取得用户信任，战胜国外对手。国产高性能计算机的曙光已经出现在地平线上，让我们以千百倍的热情去迎接光明的未来。

发表于《高技术通讯》1994 年第 7 期

# “曙光一号”并行计算机的系统软件与特点

樊建平，李国杰

国家智能计算机研究开发中心，北京 100080

“曙光一号”共享存储多处理器系统提供一个通用的并行计算平台，以支持人工智能应用、科学计算以及一般信息处理系统的开发。系统设计的主要目标一方面是攻克全对称多处理机的设计与实现技术，另一方面要求它成为具有市场竞争力的产品。为实现上述的设计目标，在“曙光一号”的实现过程中采用了对称式多处理机技术与细粒度并行处理技术，实现策略是采取尽量多的工业标准，使系统成为一个开放式的可扩展性好的并行计算环境。

“曙光一号”的操作系统 SNIX（Symmetric UNIX）是通过对 UNIX 单机版本细粒度加锁以及动态分配 I/O 中断向量的方法，实现多机系统的对称式处理。通过在 UNIX 操作系统核心中增加共享资源进程以及成群调度（Gang Scheduling）策略，在用户空间以库函数的方式实现线程（Threads）概念，支持中微粒度的并行计算。在 SNIX 操作系统之上，包括了大量的 UNIX 实用程序、标准化的串并行语言编程环境与调试环境（包括串行 Fortran 语言的自动识别器）、X Window/Motjf 图形用户界面、基于 TCP/IP 的网络软件、第四代语言以及数据库管理系统、智能应用开发环境以及分布式程序设计环境 Express 等。

图 1 按层次方式给出了“曙光一号”系统的整个系统构成，包括硬件、操作系统核心、实用程序、程序设计环境、用户界面、应用系统开发环境等诸方面。图 2 则更加详细地以图形方式进行了说明。以下将按它们在图中从低到高的顺序分别进行描述。

| 智能应用开发环境 |
| --- |
| 分布式程序设计环境 |
| 数据库管理系统 |
| X Window/Motif 图形用户界面 |
| 网络通信软件 |
| 程序设计语言开发环境 |
| 操作系统实用程元 |
| 操作系统SNIX核心界面库函数 |
| 监控器、操作系统引导程序Sysboot、SNIX核心 |

**图 1 “曙光一号”系统软件构成示意图**

## 1 各个构件介绍

“曙光一号”是基于总线式共享主存的多处理器系统。基本系统包括 4 个 Motorola RISC 芯片 MC88100 以及 8 个高速缓存及虚实地址转换部件 MC88200，系统主存存储器为 64 MB DRAM（可扩展到 768 MB），CD92 多处理机中断控制器提供多处理器之间相互中断与多个时钟发生器以及 CPU 标识寄存器等功能。系统有两个 SCSI 接口，一个 Ethernet，4 个 RS232 串行口。VME 底板总线给整个系统提

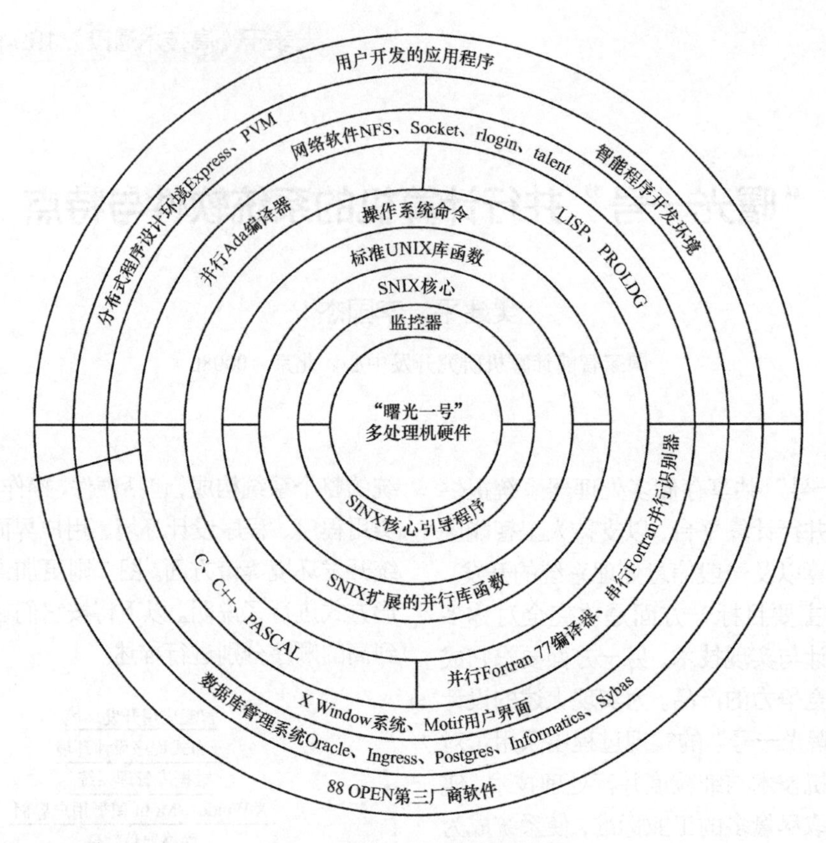


**图 2 "曙光一号"系统软件图示**

供了较强的扩展能力，可支持存储器扩展板、通信板、SCSI 接口板、FDDI 接口板等。

Monitor 作为系统加电后运行的最初软件，主要功能包括硬件配制的设置、硬件详细测试以及引导"sysboot"程序等。sysboot 程序完成 SNIX 操作系统核心加载的主要工作。系统库函数包括传统的 UNIX 标准库函数与 SNIX 扩展的库函数，扩展的函数库包括提供 Pthreads 并行线程库函数，它是用户编写并行程序的基础。SNIX 操作系统实用程序包括一般用户实用程序、程度员实用程序以及系统管理员实用程序。SNIX 操作系统提供了较丰富的高级语言编译系统，包括并行与串行两大类：串行语言包括过程式（C. Fortran、C++、Pascal）与函数式（LISP、Prolog 等）两类。并行程序设计语言大约可分为 3 类：已有串行语言+并行库函数、已有串行语言+并行语法结构、设计新的并行程序语言。在"曙光一号"中，并行语言采用前两种形式，即 C+并行库函数（Pthreads）、并行 Fortran 与并行 Ada。其中并行 Fortran 还提供一个串行 Fortran 程序的自动并行识别器，它通过数据相关性分析将用户的串行 Fortran 程序自动转换为并行 Fortran。"曙光一号"系统实现 UNIX 系统中的标准的 TCP/IP，可方便地与其他计算机互联。图形用户界面采取标准的 X Window/ Motif。在操作系统之上，提供 Oracle、Ingres、Postgres 数据库管理系统及第四代语言 RDB。通过 Express 与 PVM 分布式计算环境，系统可扩展到 16 处理器用于科学计算。

“曙光一号”系统有多个智能程度开发环境，其中包括由国家智能计算机研究开发中心自行开发的“曙光智能应用开发环境”，它在功能上类似于国外流行的智能开发环境 Prokappa、Nexpert、由浙江大学开发的“知识级的问题求解建模环境 KMEPS”、由西安交通大学开发的“对象函数型人工智能语言系统”、由石油大学与国家智能机中心共同开发的“Powerl-油气田布井决策系统”、由中国科学院系统科学研究所研制的基于吴方法的“几何定理证明器”软件系统、由北京信息工程学院开发的“全文信息检索系统”（中英文）等。

88OPEN 组织成立于 1988 年，主要是针对 88K 处理器构成的硬件与 UNIX 操作系统软件定义的基于二进制兼容的标准。它有几百个成员，上千个应用软件包符合该标准。这些软件分属以下领域：人工智能、数据库管理、一般事务处理（记账、商务管理、财务模型等）、图象图形处理、生产管理、网络与通信、办公自动化系统、科学与工程计算、软件工程管理等。“曙光一号”系统满足 88OPEN 标准。

## 2 “曙光一号”系统软件的特点

“曙光一号”并行计算机系统的设计以及开发环境的设置目标可归纳为以下 4 个。

- 采取国际标准，提供一个开放式的计算环境。
- 通过并行化操作系统核心及核心之上的各种实用程序，提高系统的响应及任务吞吐能力，充分发挥多机系统在网络系统、事务处理、信息处理方面的应用。
- 通过增加各种并行机制，支持中微粒度的并行计算问题。
- 在系统软件的各个层次上提供灵活的界面，支持不同用户的特殊需求（如实时处理、图形图象处理等）。

在软硬件方面采取封闭的做法已经被证明是一条死路，开发出的系统在市场上没有任何竞争能力。“曙光一号”系统软件在设计初期就将该点作为第一目标。采取的标准包括：SNIX（UNIX 的并行扩展）是操作系统中事实上的标准 AT&.T UNIX V 的扩展，遵从 POSIX 及 SVID 标准。系统通过提供流行的多种高级语言编译系统（如 C、C++、Fortran 等）方便了应用系统的升级以及较好地解决了已有应用软件的移植问题。图形用户界面采用 X 窗口软件（Motif 界面），用户可直观自然地使用系统。系统支持较流行的数据库管理系统（如 Oracle、Ingres 等），增强了其在信息处理领域的应用能力。函数式语言编程环境以及多个智能应用的开发系统为智能应用程序的开发提供了灵活与高效的环境，是“曙光一号”在智能方面的集中体现。由于“曙光一号”系统遵从国际标准化组织 88BOPEN 所规定的标准 BCS（Binary Compatibility Standard），因而“曙光一号”系统可运行几百个由第三厂商开发的满足 BCS 要求的应用软件（领域包括商业、金融、保险、建筑、出版、教育、办公自动化、科学计算等）。多个“曙光一号”系统板还可以通过 VMEbus 或者 Ethernet 相连，在分布式程序设计环境 Express 的支持下形成功能更加强大的系统。

“曙光一号”系统通过对称式的硬件设计以及并行化的操作系统提供一个真正的多任务处理环境，在传统的支持多任务的单处理机系统中，其操作系统在模拟一个多机系统（通过频繁切换不同用户程序提供多道程序运行环境）上要花费大量时间。SNIX 操作系统提供真正的多道程序运行环境，当可运行的用户程序数目小于或者等于处理器个数时，它们将真正并行地执行。这就意味着“曙光一号”系统较单机系统在进程切换方面的开销少许多，随 CPU 个数的增加，该方面得到的好处将更加明显，整个系统的任务吞吐量较单机系统有明显的提高。我们在实现过程中主要采取两种技术措施，包括基于全局数据依赖性分析并行化 UNIX 操作系统核心代码、采取动态均匀中断向量分配算法实现高效的 I/O 操作。

为了加速单任务的执行速度和支持中微粒度的并行计算，我们对传统 UNIX 进行了大量的改造，包括通过对核心数据结构枷锁，使核心本身可多次再入，BP 每一个处理器可并行地执行核心代码，向用户程序提供服务（包括系统调用、中断服务等人为在用户空间支持当今流行的线程概念），加速中微粒度并行任务的执行速度，在原 UNIX 进程基础上增加了共享进程资源的机制（包括共享进程地址空间及文件指针等）。操作系统核心还提供 gang CPU 调度机制，增强用户程序对 CPU 的控制能力，在操作系统核心之上提供并行库函数 Pthreads、并行程序设计语言 Fortran. Ada 以及并行调试工具等，大大提高了用户编制并行程序的能力。

为支持特殊的应用需求（如实时处理、多操作环境等），操作系统核心还提供各式系统调用，方便那些需要直接控制硬件的用户。这包括将操作系统核心地址空间直接映射到用户空间的能力，用户程序直接处理例外及中断的能力等。

## 3 软件的进一步发展计划

“曙光一号”系统软件的发展主要针对实际应用，包括：根据用户需要增加更多的语言编译系统，完善用户的编程环境，移植更多的数据库管理系统（如 Sysbase）以及基于不同数据库系统的应用系统（如银行处理系统、期货交易系统、邮电部门管理系统、股市交易系统、商场管理、一个较大部门的管理系统等），移植较流行的商务办公室自动化软件，为特定部门开发应用系统（如石油勘探系统等），增加更多的支持人工智能研究与开发的支撑环境。“曙光一号”作为高性能的网络服务器，希望其能在国民经济的主战场上发挥更大的作用。

发表于 International Symposium on Parallel Architectures, Algorithms and Network 论文集, 1994 年

# NCIC's research and development in parallel processing

Guo-jie LI

National Research Center for Intelligent Computing Systems (NCIC), Beijing 100080, China

**Abstract:** National Research Center for Intelligent Computing Systems(NCIC for short) is the unique national hi-tech R/D center for advanced computing technology in China. In this overview, we first introduce China's Hi-Tech R&D Programme (863 programme) and NCIC, then we reported the state of the art of parallel processing at NCIC. This article discussed the key technologies being exploited by the representative Chinese R&D teams and the wide applications of parallel computers in China. The key technologies in parallel processing we are attacking and reported in this article include wormhole routing and other efficient switching techniques, the Easter series MPP systems, the Dawning series symmetric and multi-thread multiprocessor, parallel operating systems and parallel file systems, parallel compiler and efficient programming tool. The future research directions at NCIC are also mentioned.

## 1 China's Hi-Tech Programme (863 Programme) and NCIC

The National Hi-Tech R&D Programme was officially launched by the Chinese government in March 1986, so the China's Hi-Tech R&D Programme is often called 863 Programme. Concentrated on limited targets, seven areas were selected in this programme, which include biotechnology, space technology, information technology, laser technology, automation technology, energy technology and advanced materials. The total budget is about 1.2 billion dollars. Advanced computing system is the most important subject in the selected information area.

The 863 Programme is a long term programme lasting for 15 years, from 1986 to 2000. It is not a rigid or fixed programme, which will be modified to adapt for the demands of industries and markets and to keep abreast of the latest trends. Under the influence of the Japan's Fifth Generation Computer Systems, one of the original objective in 863 programme is to build intelligent computing systems and our center is named by National Research Center for Intelligent Computing Systems (NCIC for short). Now we change our research direction into advanced computing technology for intelligent applications, especially parallel and distributed computing technology and natural I/O.

Every subject in 863 programme is supervised by a steering (experts) committee, set up by the State Science and Technology Commission. The steering (experts) committee is responsible for working out the R&D plan and strategy, selecting the research teams, managing the research funds, and dealing with technology transfer. During the past 8 years, more than 200 research teams and about 1 500 top-level researchers in the computer area from more than 100 universities and research institutes have been involved in our subject of 863 programme.

National Research Center for Intelligent Computing Systems (NCIC) was formally set up in 1990 and it is the heavy-weight in R&D on advanced computing technology in China. About 70 high level researchers are now working at the center, which is equipped with advanced facilities. The major tasks of NCIC include:

- to address the fundamental theoretical researches on advanced computing systems and artificial intelligence;
- to design and implement the prototypes of advanced SMP and MPP parallel computers, which are the key objectives in each stage stipulated in 863 programme and can be converted into competitive commercial products;
- to integrate the applicable R&D results obtained by many selected teams over China involved in this programme and make the integrated products marketable;
- to carry out academic exchange and international cooperation, and keep in contact with foreign researchers.

In the remaining part of this article we will introduce some achievements of NCIC on parallel processing and discuss some projects undertaken by our center.

## 2 Wormhole routing and MPP system

Efficient routing of messages is critical to design of massive parallel processors(MPP). In 1986, Dally and Seitz proposed Wormhole routing mechanism which is widely adopted in many successful commercial and research machine like Paragon and J-machine[1-2]. If a large-scale wormhole routing network is to be constructed, it is unfeasible to provide a common clock for all computer nodes, since the clock-skew is a serious problem for long distance signal propagation. Therefore, self-time circuit design has become an important issue. We have successfully designed and implemented an asynchronous wormhole routing chip, called AWRC-1, in which flits passing between two adjacent nodes must use hand-shaking protocol, and all control blocks are implemented as asynchronous finite state machine. Due to the self timed implementation, the ASIC design is much more difficult than the conventional synchronous chip design. To simplify and modularize the wormhole chip design, we adopted the partition approach, that is, the AWRC-1 has been designed such that the exact same function is performed in each dimension. The modular design allows two or high dimension wormhole routing chip to be implemented with ease. AWRC-1 is a two dimension wormhole routing chip which contains 10 physical data channels, the communication bandwidth of each channel is 50 MB/s, so the total bandwidth of an AWRC-1 chip can reach 500

MB/s. Although some authors have proposed adaptive routing, the specious claims for the advantages of adaptive routing are critiqued. Some simulation results show that dimension-order routing achieves both higher throughput and lower latency than adaptive routing. In our design, the simple X-Y routing is used. We found X-Y routing is efficient for our applications.

The AWRC-1 chips are used as the connector for each node in Eastern-1 MPP system built by our center. Eastern series MPP system is the key project of 863 programme. The major application areas of the developed MPP systems include weather forecast, oil exploration and oil deposit simulation, and other grand challenge scientific and engineering problems in China. Since low dimensional networks are more scalable than high dimensional ones, and it is shown that for network of constant wire bisection, low dimensional networks have a lower latency and higher hot-spot throughput than high dimensional networks. Our simulations also support this claim. Hence, our Easter series MPP systems are designed under the mesh architecture. Easter-1 with 36 PE and 2.5 GFlops will delivered in this year, Easter-2, which is consisted of more than 100 PE and will achieve 20 GFlops, is under construction and planed to be completed in March 1996. The latter will be designed and implemented in cooperation with US computer companies. The system software and programming environment of the Easter MPP systems will follow the international standard. PVM, Express, parallel Fortran and a rich parallel math library will be provided. Efficient parallel algorithms, especially in the area of computational intelligence, are also studied at our center.

To find efficient switch and flow control mechanism in the networks, we have paid much efforts to study new routing mechanisms and performance analysis of mesh networks. Most of routing schemes proposed in the past either sacrifice adaptability for deadlock freedom or at the expense of adding a large amount of virtual channels. One of my former student proposed a new adaptive message routing model, based on which only two or three virtual channels to share one physical channel for the mesh or Tori topology respectively are needed to construct fully adaptable routing scheme[3]. The key idea is to divide channels into two separate groups: basic channels that responsible for deadlock freedom, and extended channels that are in charge of adaptability. It is well known that the wormhole routing model has an advantage that the latency is independent of the distance. Our simulation results, however, shown that when wormhole routing is used, the network can operate at maximum of the 40 to 50 percent of network capacity. Furthermore, the larger the network size, the lower the available capacity. In general, when the number of nodes in a network is larger than 64, the performance of virtual-cut-through is better than wormhole routing. From this observation, we proposed VCT-like switching model, which compromises wormhole routing and virtual-cut-through. Our simulation promises that the latency for VCT-like model is as small as that for wormhole routing and the available capacity is close to that for virtual-cut-through. For virtual-cut-through and store-and-forward techniques, the accurate empirical formulas can be derived based on queuing theory, but for wormhole routing, it is difficult to apply queuing theory into analysis of network latency. Therefore we give a method called wait-time- recurrence model with the message-flow model. By using these models, we can derive an empirical formula for analysis of network latency.

## 3 Dawning series SMP and its applications

In October of 1993, NCIC delivered its first multiprocessor product, called Dawning-1 computer, which is a high performance multiprocessor system consisting of 4 to 16 Motorola 88K RISC processors and up to 768 MB shared memory[4]. Dawning-1 is the first symmetrical multiprocessor made in China. It supports not only multitask and multi-user applications as most commercial SMP products do but also fine grain parallel computation and real time applications. Even for the problem with very small size, such as 32 point FFT, the near linear speedup can be obtained in Dawning-1. Comparing with IBM, SUN and many other commercial products, Dawning-1 achieves the shortest context switch time. Differing from the computers developed in China, in the past decades, the Dawning-1 is a real open system which adopts various industrial or de facto standards, such as VME bus, SCSI bus, Ethernet, X-window and Motif.

The major features of Dawning-1 are in the two fold aspect. In the hardware design, we have designed and implemented an ASIC chips for interrupt control, called CD92, which is the critical component for multiprocessing and real time applications. With the aid of the ASIC chips, the system overhead of an interrupt is less than 17. In the software aspect, we have designed a fully symmetrical multithreading UNIX operating system, called SNIX (Symmetrical UNIX), in which the UNIX kernel is parallelized, the efficient multi-thread mechanism and gang scheduling police are provided. The key issues we have addressed are the strategy on adding the lock to critical sections and the method to solve the deadlock problems. We have also designed and implemented an efficient tool, called PORT, for automatically translating sequential programs into parallel ones. In the next section, the tool PORT will be discussed.

Since Dawning-1 is a cost-effective general purpose multiprocessor, it is hopeful to be a competitive high performance computer product in China's market. A large amount of venture capital from foreign banks and companies have been invested in NCIC and the first high performance computer company based on the intellectual property of Chinese researchers, the Dawning International Co. Ltd has been set up and started the business. Motorola has chosen NCIC and Dawning company as its strategic partner in China and signed contract with us to cooperatively design and implement the upgrade products of Dawning-1. We are planning to deliver Dawning-3 and Dawning-4 in 1995 and 1996, respectively. In Dawning-3, two or four power-PC processors will be used to build an advanced SMP system. Differing from other power-PC products, Dawning-3 will be focused on communication applications and capable of extending to MPP system. That is, the mother-board of Dawning-3 can be viewed as a node of MPP system. In the system design, we will consider the architecture requirements mentioned above. Dawning-4 will be based on power-PC 620 and the performance is expected to be much higher than the most existing SMP in the world.

China is a developing country but the potential computer market is very large. In 1993, the shipment of high performance computers in China was 1.18 times that of 1992 and it is forecased that this year's shipment will be double that of last year. The Dawning series multiprocessors can be applied to very wide application areas, such as the network servers for communication, database servers for banks and other OLTP service, office automation, and management information systems. Chinese government is now devoting much efforts to develop Chinese information highway (Golden Bridge project), credit cards (Golden Card project), and EDI systems (Golden Custom project), so called Three-Golden Programme. Recently, many other global computer networks have been involved in the national project and may be called Eight Golden Programme. Certainly, Dawning series computers will take an important role in the reconstructure of China's economy.

## 4 Parallel program restructuring tools

Although several kind of multiprocessors are commercially available in the present time, the parallel efficiency and program productivity are not satisfactory. We have tested a lot of parallel programs and found that, for quite complex programs, some commercial compiler fails to generate correct execution code. We have studied optimizing parallel compiler, called Parallel Optimizing Restructuring Tool, PORT for short. PORT is a set of automatic source to source restructuring tools that translates FORTRAN 77 into extended FORTRAN 77 with parallel directives and assertions. The ultimate goal of program reconstruction is to exploit parallelism in programs and to maximize the speed of code execution[5]. The major features of PORT are efficient flow analysis, data dependence test, and DO loop transformation. Data analysis includes local, intraprocedural and interprocedural analysis. Intraprocedural control flow analysis generates a control flow graph (CFG), while interprocedural analysis generates a call graph (CG) and converts control dependence (IF statements) into data dependence. Data dependence tests contain separability tests, extended GCD tests, extended Benerjee tests, and other data dependence tests. Do loop transformation package performs loop interchange, loop distribution, loop index set splitting and loop fusion. It also generates synchronization FOR loops that contain loop-carried dependence. Hence, PORT is able to parallelize all nested loops rather than only the innermost ones. PORT also provides static and dynamic analysis tools.

The total amount of PORT'S source code is more than 60 000 lines. We have passed the Perfect Benchmarks Tests, which contain 13 application programs, and shown the perfect correctness. We have tested the SGFs parallel compiler pfa (parallel Fortran Accelerator) and found that it cannot correctly reconstruct our test programs but PORT has passed all tests. For standard Livermore Benchmark, more test programs can be parallelized by using PORT than PFA does, and the execution speed of the programs compiled by

PORT is higher than other available parallel compilers. The compiling speed is 2 000 to 2 500 lines source code per minute.

Based on PORT, we also developed a graphical environment, called PORTGraph to debug FORTRAN 77 programs and assist the transformation of sequential FORTRAN 77 programs into parallel ones that run on massive parallel processor systems. PORTGraph provides a collection of efficient and powerful graphical tools for static analysis, dynamic analysis, program debug and monitor, and system status report. PORTGraph is easy to be ported in a wide range of architecture such as supercomputers, mainframes, and workstations. This tool displays not only the static information such as the calling relationships and data dependence relationships, but also a lot of dynamic information, for example, the current execution path on the call graph, the current control flow on the CFG. The rich debug information is available when running PORTGraph tool.

# 5 Further researches at NCIC

NCIC has done many other researches related to parallel processing, such as constructing distributed operating system and parallel file systems, designing the parallel hardware algorithm on Chinese character and speech recognition. All these activities are omitted in this article. Only the topics of the further researches at NCIC concerning with parallel processing are listed below.

- Downsized MPP system. In our next MPP system, the basic component is the printed circuit board containing nine high performance CPU chips (PowerPC 604 is the first choice). The separated PCB may be used as an accelerator for workstation, which is affordable for Chinese customers.
- Chinese Information Platform. Based on our Dawning series multiprocessors, we will develop a framework, a kind of midware, to provide a unified platform, for building various MIS, CIMS, and multimedia systems.
- High availability multiprocessor. Most users of OLTP systems in China require high availability function.
- Enhancing superscalar CPU by embedding or attaching efficient vector processor.
- Optical crossbar. Cooperated with the selected university, we plan to build a prototype of parallel processor with optical crossbar.

# References:


[1] DALLY J, SEITZ C L. The torus routing chip[J]. Journal of Distributed Computing,1986, 1(3): 187-196.

[2] DALLY W J, CHIEN A A, FISKE S, et al. The message-driven processor: a multicomputer processor node with efficient mechanisms[C]//The 1991 IEEE International Conference on Computer Design on VLSI in Computer & Processors. Piscataway: IEEE Press, 1992.

[3] SHEN W, CHEUNG Y S. Basic-and extended: a new adaptable message routing model[J]. To appear in Journal of Computer Science and Technology.

[4] LI J, et al. The Dawning-1 parallel computer[J]. To appear in the special issue of Journal of Computers, in Chinese.

[5] ZHANG Q, QIAO R L. The parallelized and optimized program reconstructuring tool[J]. To appear in the special issue of Journal of Computers, in Chinese.

发表于《计算机学报》1994 年第 12 期

# 共享存储多处理机系统中的多级高速缓存

熊劲[1,2]，李国杰[1,2]

1. 国家智能计算机研究开发中心，北京 100080;

2. 中国科学院计算技术研究所，北京 100080

**摘　要**：共享存储多处理机系统中，存储子系统的性能是影响整个系统性能的关键之一。我们通过基于访存地址流的模拟，从缺失率、平均访存时间和总线占用3个方面，对共享存储多处理机系统中的两种两级缓存方案做了性能比较，并将它们同没有第二级缓存的情形做了性能比较。

**关键词**：高速缓存；缓存一致性；基于访存地址流的模拟

# Multi-level cache in shared-memory multiprocessors

Jin XIONG[1,2], Guo-jie LI[1,2]

1. National Research Center for Intelligent Computing Systems, Beijing 100080, China

2. Institute of Computing Technology, Chinese Academy of Sciences, Beijing 100080, China

**Abstract:** In shared-memory multiprocessor systems, the performance of the memory subsystem is very important to the whole system's performance. By trace-driven simulation, this paper compares the miss ratio, the average access time and the bus utilization of two two-level cache coherence schemes. The performance of the two schemes is also compared with that of a one-level cache system.

**Key words:** cache, cache coherence, trace-driven simulation

## 1 引言

在单处理机系统中，为了缓解快速的处理机周期与慢速的存储器存取之间的矛盾，而将高速缓存（cache，简称缓存）引入了计算机系统[1]。在有缓存的系统中，影响系统效率的主要因素是缓存的缺失率。影响缓存缺失率的因素有缓存容量、缓存存储块大小及相关系数。虽然缓存越大，缓存的缺失率就越低，但是缓存随容植的增大，其速度会急剧下降。对块大小及相关系数，情况也如此。Przybylski发现采用单级缓存可以获得的性能提高是有

本课题得到 863 计划与国家自然科学基金项目资助。

限的，当性能达到一定程度后，改变任何缓存参数，都几乎不能使性能得到更进一步的改进。但是如果在 CPU 与主存之间采用多级缓存，则可以使性能得到更多的改进[2]。

在如图 1 所示的这种结构的多处理机系统中，共享数据都在存储器中，不进入缓存，只利用总线将各模块的存储器连通，变成共享存储器。总线是该系统的瓶颈，如果在这种系统中增加一级缓存（如图 2 所示），则可以利用时空局部性来改进访存性能，而且也减轻了总线的负担。不过，增加一级缓存使缓存一致性的维护变得困难和复杂。如何维护这样一个两级缓存系统的一致性，以及这种两级缓存系统的性能如何，这正是本文要研究的。

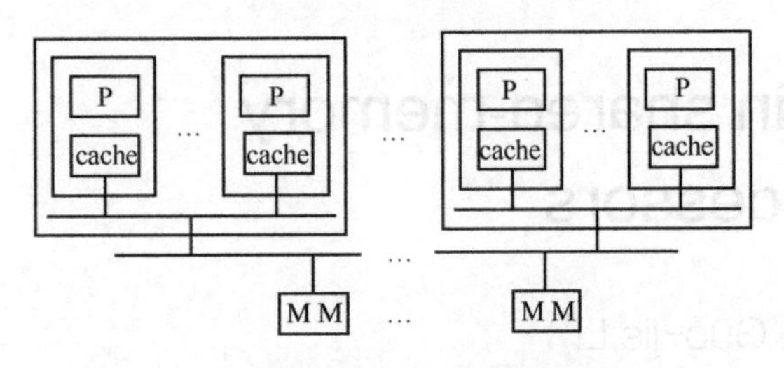


图 1　一种单级缓存的共享存储多处理机系统

在研制“曙光一号”多处理机的过程中，我们研究了如图 2 所示的两级缓存系统，给出了两种维护缓存一致性的方案，并采用基于访存地址流的模拟（Trace-Driven Simulation），对它们的性能做了比较，并同没有第二级缓存的情形做了比较。

## 2　维护两级缓存一致性的方案

对于如图 2 所示的两级缓存系统，为了便于维护一致性和减轻总线的负担，两线缓存的容量、块大小和关联数要满足 MLI 性质[3]。因此我们只考虑 SLC 的容量是其对应的所有 FLC 容量之和的正整数倍，而且两级缓存的块大小是相同的这种情形。为了减轻第二级总线的负担，第二级缓存采用 write-back 策略，因此当第二级缓存中的某块被替换回主存时，需先将对应所有 FLC 中该块复本都作废（若有 Dirty 块，还需先写回再作废）。当作废一个 SLC 块时，必须同时作废与之对应的所有 FLC 中的该块复本。当某个 SLC 块的状态变为 Shared 时，与其对应的所有 FLC 中对应块的状态也要相应地变为 Shared。第一级缓存可以采用 write-back 策略，也可以采用 write-through 策略。由此有两种方案，前者对应于 WB-WB 方案，后者对应于 WT-WB 方案。一个写操作可以直接写状态为 Exclusive/Dirty 的块，而对状态为 Shared 的块，则应先作废所有 FLC 和所有 SLC 中的对应复本之后才能写。当发生缺失时，先看同一模块上的其他 FLC 中有无复本。若没有，则从对应 SLC 中读，若对应

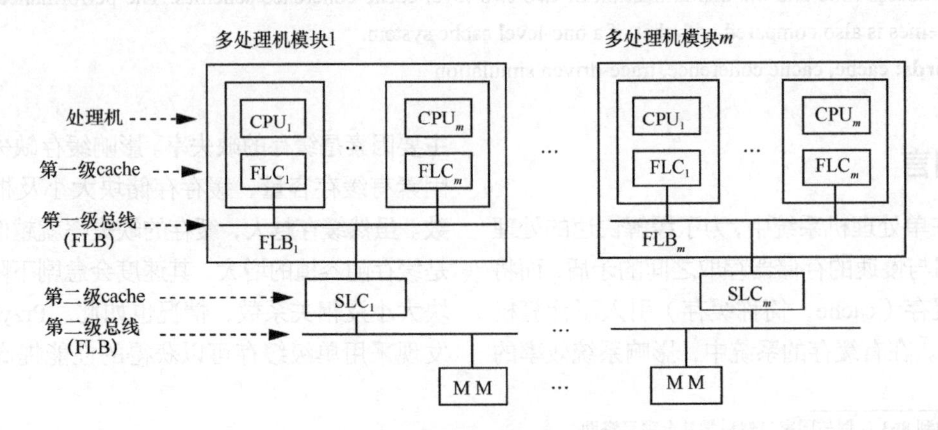


图 2　一种两级缓存共享存储多处理机结构

SLC中也没有，则先让该SLC从其他SLC(它可能需从其对应 FLC 中获得该块的最新复本)或其他主存中读得该块。

WB-WB 方案下两级缓存都采用基于MESI 协议的方案来维护一致性。两级缓存中存储块都可以有4种状态：Invalid、Exclusive、Dirty和Shared。

| FLC中块状态 | 对应SLC中块状态 | 真实状态 |
| --- | --- | --- |
| Invalid | Invalid/Exclusive/Dirty/Shared | Invalid |
| Exclu-sive | Exclusive/Dirty | Exclusive |
| Dirty | Exclusive/Dirty | Dirty |
| Shared | Exclusive/Dirty | Shared-within-the-Module |
| | Shared | Shared-among-Modules/Shared-within-the-Module-and-among-Modules |

WT-WB 方案下第一级缓存采用 write-through 写回策略，因此第一级缓存采用基于ESI 协议的方案维护一致性。FLC 中块状态只有3种：Invalid、Exclusive 和 Shared。

| FLC中块状态 | 对应SLC中块状态 | 真实状态 |
| --- | --- | --- |
| Invalid | Invalid/Exclusive/Dirty/Shared | Invalid |
| Exclusive | Exclusive | Exclusive |
| | Dirty | Dirty |
| Shared | Exclusive/Dirty | Shared-within-the-Module |
| | Shared | Shared-among-Modules/Shared-within-the-Module-and-amongs-Modules |

## 3 基于访存地址流的模拟模型

我们模拟的模型是由两个多处理机模块构成的，每个多处理机模块上有4个CPU。每个 FLC 是将指令缓存和数据缓存分离的，并且需要由 TLB 把虚地址转换成实地址。对于SLC，不再分指令缓存和数据缓存，而统一于一个缓存，假设每个CPU类似于MC88100[4]，是精减指令集计算机。一个CPU周期为40 ns。假设缓存控制器都类似于 MC88200[5]，读/写第一级缓存只需要1个周期，而第二级缓存比第一级缓存慢一倍。假设读存储器需要6个周期，写存储器需要3个周期，则存储器的恢复时间需要3个周期。

我们采用软件模拟上一节中的两个方案。模拟程序是在 SUN SPARC II Station 上用C语言实现的。模拟程序的输入包括对缓存层次的详细描述和若干个访存地址流。我们所用的访存地址流是由美国Illinois大学的Kent Fuchs 教授提供的，包括3个程序的访存地址流：Gravsim、Tgen 和 Fsim。Gravsim 是采用 Garnes 和 Hut 算法对 Gravitational N-body 的模拟器，Tgen 用并行搜索算法来产生顺序电路和组合电路的测试模式，Fsim 是数字电路的并行错误模拟器。所有的访存地址流都是在 Encore-91 机器下使用7个CPU 运行时得到的。文献[6]中有这些地址的详细说明。表1中给出了它们的有关信息。地址流中B所定义的访存操作有6种：读/写局部数据、读/写全局数据、取指令和切换。缓存层次的描述包括每一级缓存的组织，即指令缓存同数据缓存分离还是统一、缓存的容量、关联数以及缓存的块大小。模拟程序的运行结果送到一个输出文件，该文件中有用于计算缺失率、总线占用和平均访存时间的各种统计数据。

## 4 模拟结果

我们将第 2 节中的两个两级缓存一致性方案做了性能比较，并将它们同如图1所示的没有第二级缓存的结构(称为1-Level 方案)进行了比较。

### 4.1 缺失率

模块 $i$ 上FLC的平均缺失率 $AVGR_i$，$SLC_i$

**表 1　访存地址流有关信息**

| | totalref. | instrnref. | dataread | datawrite | shared read | shared write | context switch |
|---|---|---|---|---|---|---|---|
| Tgen | 4 606 308 | 2 577 103 | 1 819 235 | 209 941 | 898 472 | 30 267 | 29 |
| Gravsim | 4 487 687 | 2 405 651 | 1 651 142 | 430 882 | 579 212 | 14 841 | 12 |
| Fsim | 4 906 401 | 3 009 644 | 1 808 344 | 88 404 | 1 343 515 | 15 594 | 9 |

的全局缺失率 $GR_i$ 和局部缺失率 $LR_i$ 分别定义如下：

$$AVGR_i = \frac{\sum_j 在FLC_{ij}中缺失的访问个数}{\sum_j 处理机P_{ij}所发出的访问个数}$$

$$GR_i = \frac{在SLC_i中缺失的访问个数}{模块i上所有处理机所发出的访问个数之和}$$

$$LR_i = \frac{在SLC_i中缺失的访问个数}{模块i上所有处理机传到SLC_i的访问个数之和}$$

表 2 给出了以 Tgen 程序的地址流作为输入时的模拟结果。

1-Level 方案下数据在 FLC 中的缺失率远远高于其他两种方案。这是因为在 1-Level 方案下，所有共享数据都不进入缓存；指令在 FLC 中的缺失率比其他两种方案都略低一点，这是因为指令不存在共享的情形。对于两级缓存方案，当第二级缓存中的块被替换时，其相应第一级缓存中的块都要作废，因此这就使得第一级缓存的指令缺失率稍稍增加。

**表 2　Tgen 下两级缓存的缺失率**

| | | | 1-Level | WB-WB | WT-WB |
|---|---|---|---|---|---|
| 第一级 cache 的平均缺失率 | 模块 1 | 指令 | 0.119 576 | 0.119 638 | 0.119 622 |
| | | 读 | 0.650 676 | 0.083 755 | 0.083 601 |
| | | 写 | 0.381 608 | 0.358 008 | 0.356 981 |
| | 模块 2 | 指令 | 0.407 148 | 0.407 208 | 0.407 188 |
| | | 读 | 0.639 728 | 0.352 934 | 0.352 360 |
| | | 写 | 0.379 658 | 0.360 969 | 0.359 089 |
| 第二级 cache 的全局缺失率 | SLC1 | 指令 | | 0.119 612 | 0.119 611 |
| | | 读 | | 0.082 846 | 0.082 816 |
| | | 写 | | 0.343 841 | 0.343 99 |
| | SLC2 | 指令 | | 0.407 194 | 0.407 184 |
| | | 读 | | 0.350 739 | 0.350 542 |
| | | 写 | | 0.351 807 | 0.350 969 |
| 第二级 cache 的局部缺失率 | SLC1 | 指令 | | 0.999 789 | 0.999 909 |
| | | 读 | | 0.991 213 | 0.991 983 |
| | | 写 | | 0.960 931 | 0.962 738 |
| | SLC2 | 指令 | | 0.999 964 | 0.999 990 |
| | | 读 | | 0.993 916 | 0.994 977 |
| | | 写 | | 0.974 636 | 0.977 408 |

注：其中，FLC 的 I/Dcache 各 2K，SLC 为 128K，块大小均为 4 个字，关联数均为 40

WB-WB 方案与 WT-WB 方案相比，无论是取指令还是读写数据，第一级缓存的缺失率和第二级缓存的全局缺失率都是 WB-WB 方案的比 WT-WB 方案的要略高一点；而第二级缓存的局部缺失率反而是 WB-WB 方案的比 WT-WB 方案的略低。

值得指出的是，在我们采用的 3 个地址流中，绝大多数数据只使用一次，因而第二级缓存局部缺失率极高。表面上看，这种极端的例子似乎说明第二级缓存不必使用，但事实上，采用两级缓存的作用在于大大降低第一级缓存中的读缺失率（从 0.65 降到 0.08），因而明显缩短了平均访存时间。

## 4.2 平均访存时间

平均访存时间定义为平均一个访存要用多少个 CPU 周期才能完成。对于两级缓存系统，平均访存时间 $T$ 是这样来计算的：

$$T = h_1 \times t_1 + (1-h_1) \times h_2 \times t_2 + (1-h_1-(1-h_1) \times h_2) \times t_m$$

其中，$h_i$ 是第 $i$ 级缓存的命中率。$t_i$ 是第 $i$ 级缓存的访问时间，即从处理机发出访问请求到第 $i$ 级缓存把所请求的数据提供给处理机所用的时间。$t_m$ 是存储器的访问时间，即从处理机发出请求到存储器把被请求的数据提供给处理机所用的时间。

表 3 给出了 3 种方案在 3 种访存地址流下的平均访存时间（以 CPU 周期为单位）。

表 3　平均访存时间

| 方案 | 1-Level | WB-WB | WT-WB |
|---|---|---|---|
| Tgen | 8.905 676 | 6.696 181 | 6.709 753 |
| Gravsim | 14.913 963 | 12.310 973 | 12.059 739 |
| Fsim | 5.469 253 | 3.482 434 | 3.500 529 |

1-Level 的平均访存时间比两级缓存方案长 20%~60%，这是因为 1-Level 方案中，一级缓存方案的读缺失率明显高于两级缓存方案。对于 Tgen 和 Fsim 这两个程序的地址流，WB-WB 方案的平均访存时间比 WT-WB 方案的略少；而对于 Gravsim 程序的地址流，WB-WB 方案的平均访存时间却比 WT-WB 方案的略多。这是由于 Tgen 程序和 Fsim 程序下，缺失较少；而 Gravsim 下缺失较多。

## 4.3 总线占用情况

表 4 给出以 Fsim 程序的地址流作为输入时的结果。

1-Level 方案下，第二级总线占用次数远远比其他两种方案低，因为这时仅当写共享数据时才需占用第二级总线；其第一级总线的占用在多数情况下比其他两种方案多，有时（我们这里仅 Tgen 的 FLB2) 却比其他两种方案高。因为，1-Level 方案下，第一级总线的占用完全由第一级缓存的缺失情况决定，此时没有维护一致性的总线开销。因此 1-Level 方案同其他两种方案之间第一级总线占用的差异实际上反映了维护缓存一致性的总线开销与避开一致性所带来的缺失增长引起的总线开销之间的 trade-off。

无论第一级总线还是第二级总线，WT-WB 方案下总的总线占用都比 WB-WB 方案下多。

对于第一级总线，发作废或查询信号以及读第二级缓存的总线开销，都是 WB-WB 方案下的多一些，而写第二级缓存的总线开销却是 WT-WB 方案下的远比 WB-WB 方案下的多得多。这说明 WT-WB 方案以增加写访问的第一级总线开销为代价换来的维护一致性和读访问时第一级总线开销的减少不足以抵消它为此付出的代价。

对于第二级总线，两种方案发作废或查询信号的总线开销相差很少，读主存的总线开销是 WB-WB 方案的更多，但写主存和替换的总线开销都是 WT-WB 方案的更多。

我们还发现，对于第一级总线，其开销主要来自于发作废信号（占 40%左右）和读第二级缓存（占 50%左右），它们占了总开销的 90%左右。对于第二级总线，读/写主存的总线开销占了总开销的 90%左右，而维护一致性的开销只占 10%左右。

## 5 结束语

我们通过基于访存地址流的模拟，从缺失率、总线占用情况和平均访存时间 3 个方面全面地比较了没有第二级缓存和两种两级缓存方案的存储系统的性能。我们发现，增加一级缓存能明显地改进存储系统的性能，平均访存时间缩短 20%~60%。就两种两级缓存方案相比，WB-WB 方案下两级缓存的缺失率都比 WT-WB 方案下的高一点；而 WB-WB 方案下的总线开销却比 WT-WB 方案下少许多。当访存中缺失较少时，WB-WB 方案的平均访存时间更少。总体说来，两种方案的性能差不多。本文研究结果为解决曙光系列多处理机系统的可扩展问题提供了实验依据。

## 参考文献：


[1] SMITH A J. Cache memories[J]. ACM Computing Survey, 1982, 14(3).


**表 4 Fsim 下各级总线的占用情况**

| 对比项 | | | 1-Level | WB-WB | WT-WB |
|---|---|---|---|---|---|
| 第一级总线 | 总的占用次数 | FLB1 | 957 586 | 371 417 | 402 023 |
| | | FLB2 | 722 702 | 320 505 | 348 873 |
| | 发作废信号的次数 | FLB1 | 155 94 | 157 614 (0.424 359) | 157 502 (0.391 774) |
| | | FLB2 | \ | 134 146 (0.418 546) | 134 012 (0.384 128) |
| | 发查询信号的次数 | FLB1 | \ | 13 033 (0.035 090) | 12 910 (0. 032 113) |
| | | FLB2 | \ | 11 595 (0. 036 177) | 11 520 (0.033 021) |
| | 返回作废 FLC | | \ | 53 | 55 |
| | 返回查询 FLC | | \ | 55 | 55 |
| | 查询或作废时总线传块次数 | FLB1 | \ | 107 (0.000 228) | 84 (0.000 209) |
| | | FLB2 | \ | 62 (0.000 193) | 45 (0.000 129) |
| | 占用总线读第二级 cache | FLB1 | \ | 185 434(0.499 261) | 184 967 (0. 460 091) |
| | | FLB2 | \ | 161 693 (0.504 494) | 161 308 (0.462 369) |
| | 占用总线写第二级 cache | FLB1 | \ | 15 229 (0.041 002) | 46 560 (0. 115 814) |
| | | FLB2 | \ | 13 009 (0.040 589) | 41 988 (0.056 200) |
| 第二级总线 | 总的占用次数 | | \ | 403 502 | 403 971 |
| | 发作废信号次数 | | \ | 12 717（0.031 517) | 12 718 (0.031 482) |
| | 发查询信号次数 | | \ | 24 235 (0.060 062) | 24 242 (0.060 009) |
| | 作废时写回主存 | | \ | 3 | 5 |
| | 查询时写回主存 | | \ | 48 | 50 |
| | 查询或作废时总线传块 | | \ | 77 (0.000 191) | 84 (0.000 208) |
| | 占用总线读主存 | | \ | 344 476 (0.853 716) | 344 224 (0.852 101) |
| | 占用总线写主存 | | \ | 21 997 (0.054 515) | 22703 (0.056 200) |
| | SLC 替换写回主存 | | \ | 21 946 | 22 648 |

注：括号中的数据为该占用次数在总的占用次数中所占的百分比

[2] PRZYBYLSKI S A. Cache and memory hierarchy design[M]//A performance-directed approach. San Francisco: Morgan Kaufmann Publishers, 1990.
[3] BAER J L, WANG W H. On the inclusion properties for multi-level cache hierarchies[C]//The 15th Annual International Symposium on Computer Architecture. Piscataway: IEEE Press, 1988.
[4] MC88100 RSIC microprocessor user's manual (Second Edition)[Z]. Englewood Cliffs: Prentice -Hall, 1990.
[5] MC88200 cache/memory management unit user's manual (Second edition) [Z]. Englewood Cliffs, Prentice Hall, 1990.
[6] JANSSENS B J. Generation of multiprocessor address traces and their use in the performance analysis of cache-based error recovery methods[D]. Urbana: University of Illinois at Urbana-Champaign, 1991.

发表于《计算机研究与发展》1995 年第 1 期

# 并行操作系统的现状与发展趋势

樊建平，李国杰

中国科学院计算技术研究所国家智能计算机研究开发中心，北京 100080

**摘　要**：本文从操作系统的发展史、当前的发展状态、未来的发展趋势以及技术研究的几个热点等方面进行了阐述。

**关键词**：操作系统；操作系统的发展；并行机

# Overview of multiprocessor operating systems

Jian-ping FAN, Guo-jie LI

National Research Center for Intelligent Computing Systems,

Institute of Computing Technology, Chinese Academy of Sciences, Beijing 100080, China

**Abstract:** In this paper, the history, current status and future trends of the multiprocessor operating systems(MOS) are described.

**Key words:** MOS history, current status, future trends of MOS

## 1　引言

计算机硬件技术在系统体系结构（并行、分布）、处理器的 RISC 化、大容量存储器、大容量高速磁盘等外部设备、高速网络器件以及联网技术等方面取得了飞跃发展。计算机软件技术在标准制定（POSIX、COSE 标准）、软件设计方法（面向对象、模块化等）、软件工程开发与维护工具等方面日趋成熟。用户对计算机的需求也在不断提高与变化（主要包括文字、图象、声音等输入输出技术，大量的数据处理与通信，超高速计算，实时控制等）。作为管理系统硬件资源以及提供用户界面的操作系统本身也有了长足的发展与进步。涉及的研究领域在不断扩大。

操作系统从无到有，大致与硬件技术的发展相对应，归纳为 4 个阶段：（1）（1945—1955 年）电子管计算机与操作系统的早期形式监控器（monitor）；（2）（1955—1965 年）晶体管计算机与批处理系统；（3）（1965—1980 年）中小规模集成电路构成的计算机与支持分时多道程序设计的操作系统；（4）（1980—1994 年）大规模集成电路构成的单机、多机、网络硬件环境与并行操作系统。各个时期的操作系统与硬件环境相适应，功能上不断完善以满足用户的新需求。

20 世纪 80 年代初中期开始，随着并行分布硬件体系结构的成熟、并行与分布算法的发展，并行操作系统从发展初期逐步走向成熟。操作系统由于涉及面广以及内部构成较复杂等原因，一个系统从设计到最终形成产品，并且被广泛推广使用需经 10~15 年的发展周期，这也决定了并行操作系统从改造已有的单机版本（并行化）、增加模块（网络模块、网络文件系统、分布式服务器等）逐渐向全新的设计（如微核心化、面向对象的操作系统设计等）方向发展。在改造已有的操作系统与设计新的操作系统过程中，形成了多种产品，出现了多个研究热点。

## 2 商品化并行操作系统的发展状况

操作系统是一种开发周期长、用户使用的惯性大、界面定义逐渐标准化的系统。个人计算机（PC）领域， DOS、Microsoft Window 已经成为事实上的标准。PC 以外的计算机，UNIX 是事实上的标准。基于 UNIX 操作系统定义了一系列的工业与国际标准（POSIX、SVID、COSE 等）。并行操作系统的发展也是在满足这一标准的前提下从两个方向展开的：一是改造满足标准的单处理机操作系统，增加新的功能模块；二是设计全新的操作系统，其界面满足现有的标准。现已商品化程度较高的并行操作系统基本上属于第一类。第二类操作系统逐渐走向成熟，部分已商品化。

目前市场上已有不同厂商推出多个并行操作系统的版本，包括：SUN 公司的 Solaris、AT&T UNIX SYSTEM V 4.0 MP 版本、DG 公司的 DG/ UX、SGI 公司的 IRIX、MIPS 公司的 RISC/OS 5.0（现属于 SGI）、Cray 公司的 Unicos、Concurrent 公司的 RTU、Sequent 公司的 DYNIX、Encore 公司的 UMAX、Motorola 公司的 Motorola OS、Alliant 公司的 Concentrix、IBM 支持多机结构的 AIX 以及 OSF 组织推出的早期 UNIX 版本 OSF/1.0 等。

早期的并行操作系统由于考虑到硬件系统的非对称性以及实现的简单性，采取主从方式并行化单机 UNIX 操作系统，即任何时刻只允许一个处理器（可固定或者浮动）执行操作系统核心程序（I/O 中断、文件操作等），其他处理器可执行用户程序。这种实现方法不能很好地发挥对称式硬件的特点。目前这种实现方法在商品化系统中被完全放弃，代之以完全对称式的操作系统核心并行化的方法。为充分发挥多处理机硬件环境下的多任务处理能力，所有公司的并行 UNIX 版本都对 UNIX 核心代码进行了并行化改造，使核心本身是可再入式的，即多个 CPU 可并行执行核心代码程序。每一个公司还针对其目标市场对 UNIX 进行了特殊的改造（包括在核心内与核心外），包括支持细粒度的并行计算、实时处理、分布式计算、多用户操作环境等。操作系统核心的并行化技术主要涉及的问题包括：加锁的策略、锁类型的选择、加锁位置的选择、死锁的预防、性能统计分析改进工具。加锁的策略可根据多机硬件特征（CPU 的个数）与应用领域（数据处理、并行计算、实时处理等）进行粗粒度与细粒度加锁的选择，也可以对使用频繁的子系统进行细粒度加锁，使用频度小的子系统进行粗粒度的加锁等。锁类型有简单的 spin-lock、信号灯机制、read-write 锁等，涉及的问题包括如何高效地在不同的硬件中进行实现、在不同的加锁位置选择哪一种锁。加锁位置的选择主要针对共享的数据结构来进行。死锁问题的解决方法包括完全预防死锁的发生以及死锁发生时如何解锁两种方案，在系统进行初步的并行化处理后，还需要通过一些专门的性能调试工具进行并行化后性能的调整，以期达到最好的效果。

为支持细粒度的并行计算，还需要对UNIX本身进行较大的改造。核心方面除要进行细粒度加锁外，还需要实现较传统UNIX进程代价小的调度单位（轻进程、线程、共享资源进 程等）与适合并行处理的调度算法（gang调度）。在核心之外，还需要增加并行库函数以及并行程序设计语言、并行调试器、并行程序性能评价工具等。

为支持实时处理，除核心要进行细粒度的并行化外，还需要在进程调度界面、调度算法（SCHED_FIFO、SCHED_RR）、CPU屏蔽与接收中断的处理、虚拟地址空间的管理（用户与核心空间合二为一、锁定用户程序内等）以及使用界面等诸方面进行改造。

为支持分布式计算，要增加使用灵活的消息传递界面与高效的低层协议。比如不通过核心直接在用户空间进行消息的传递；使用RISC CPU中的 Register窗口加速消息传递的速度等。在核心之外还需要有分布式程序设计环境等，现流行的并行分布计算环境包括Express、PYM、Linda等。

## 3 基于微核心的未来操作系统的研究与发展

微核心化操作系统，顾名思义，就是尽量缩小操作系统核心的规模，将传统操作系统核心实现的内容或称核心提供的服务提高到核心外的用户模块中进行实现，发展微核心化操作系统的主要动因是解决可移植性（包括操作系统自身与用户程序）、可扩展性（方便地增加各种服务模块）与可靠性。UNIX操作系统在发展初期以精小、易移植著称，随着计算机硬件体系结构的不断变化（多处理机结构——共享与分布存储器）与用户所需功能的不断扩展，UNIX逐渐庞大起来。可移植性、可扩展性与可维护性越来越差。究其原因，UNIX原本是为单处理机开发的支持分时多道程序设计的一种环境，在设计初期并没有考虑到今后要管理网络、多机等硬件与支持并行、分布与面向对象的程序设计。微核心化思想就是将传统操作系统核心的功能进行大量的裁减，核心中只保留必需的模块（内存管理、处理器低层的管理、IPC、外部设备驱动程序等），其他的部分（文件系统、网络系统、进程管理）提到核心外进行实现。

为什么说微核心化操作系统将领导未来操作系统的发展潮流？其原因可归纳为以下 4 个方面。

- 功能上已实现现有的操作系统标准，解决了用户程序的继承性。
- 从操作系统的构成技术角度考虑更加合理，在模块化、可移植性与扩展性方面具有明显的优势。
- 主要的商品化操作系统供应商已经在其下一代产品中采纳了微核心化的思想或者其产品就是一个微核心化的操作系统。
- 支持不同的硬件体系结构（并行、分布、传统单机、嵌入式系统等）与应用需求（并行计算、实时处理、面向目标的程序设计、分布式计算等）。

操作系统经过几十年的发展，在研究单位、厂商与用户的共同努力下已形成被广泛采纳的标准（各种层次：包括系统调用、库函数、实用程序等），如 DOS、MSC Window、POSIX、COSE、X Window等，为通过其他技术途径实现这些标准提供了可能，同时也保证了用户对新系统接受的程度。有关操作系统微核心化的研究从20世纪80年代初就开始了，最早的系统包括卡内基梅隆大学（CMU）的 RIG、

ACCENT 以及 1984 年推出的 MACH 操作系统。法国从 20 世纪 80 年代初期研制的 Chorus 操作系统、荷兰从 20 世纪 80 年代中期开始研制的 Amoeba 等都是具有代表性的工作。经过十几年的发展，操作系统的微核心化技术已经进入成熟与商品化的阶段。现在市场上已经销售的商品包括 Microsoft 公司的 Window NT、OSF 组织的 OSF/1.3（基于 MACH 3.0）、NEXT 公司的 NEXTSTEP、USL 与 Chorus System 联合推出的 Chorus/MiX、Unisys 公司的 CTOS 以及 QNX 提供的 QNX 等。正在研制的产品包括 SUN 公司在 Solaris 操作系统之后将要推出的基于微核心化技术的 SpringOS（支持面向对象的程序设计）、IBM 公司为支持 Taligent 环境研制的 Workplace OS、Apple 公司将于 1996 年推出的具有微核心化的操作系统 MAC System 7.5 等。还有一些仅在研究单位研制并且使用的系统，如：AT&T 公司的 Plan 9、Standford 公司的 V kernel、以色列 Hebrew 大学开发的 MOS、DEC 公司的 Topaz 等。所有这一切对微核心化操作系统替代传统的操作系统打下了坚实的基础。

采用微核心化思想实现新一代操作系统已经是学术界与商业界的共识。现在的主要争论点是实现的技术方面，包括微核心应该提供哪些服务；I/O 驱动程序在进行高度抽象尽量减少与硬件的交互界面后，同时如何提高其效率；从提高效率的角度考虑，哪些模块应该在核心空间运行；是否要保留传统操作系统（UNIX）子系统的代码，并且在微核心化操作系统中继续使用或者完全重写等。

图 1 以 MACH 3.0 微核心为蓝本给出了微核心化操作系统的构成层次。MACH 核心只包括进程间通信、虚存管理、进程与线程管理、处理器管理、I/O 设备操作低层模块 5 个部分。基于 MACH 3.0 建立的 UNIX 操作系统（如 OSF/1.3）的其他部分都以 Server 形式在用户空间运行。这是一个较理想的微核心化操作系统。出于市场目标与效率的考虑，各个商家对哪一部分放入核心运行有不同的选择。Microsoft 公司在设计其 Window NT 时，将图 1 中 1～3 个层次的内容均放入核心态运行，主要支持 OS/ 2、Window 32、Posix 与 DOS 4 种操作系统界面。IBM 公司在其 Workplace OS 中，将第二层次的 Server 放入核心空间运行。在 Chours/MiX 中核心部分只有 4 个部分：虚存管理、进程与线程管理、处理器管理、I/O 设备操作低层模块。进程间通信模块也被放入核心外。

另一个研究热点是如何提高基于消息传递机制的微核心化操作系统在效率上与传统无模块化结构的操作系统（基于子程序调用）效率上的差别。最省时的做法就是将各类核心外的服务模块放入核心内执行（MSC Window NT），

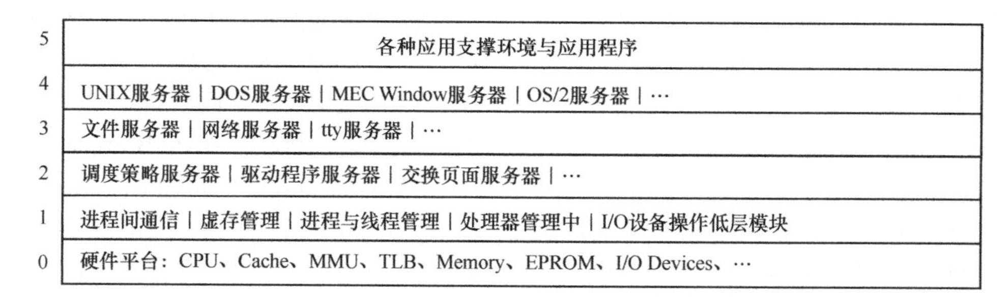


图 1　微核心化操作系统的构成

改消息传递为程序调用或者在核心内仍然采取消息传递方式(省去两次空间切换的开销)。这种方法虽然可提高效率，但对消息传递机制的效率如何提高本身并没有进行研究。有关提高消息传送效率的研究包括：如何不经过操作系统核心在两个用户空间直接进行消息的传递；启用开销较少的线程来进行消息的传递(切换在CPU寄存器中进行，不通过存储器)；如何增加一定的硬件设备，使CPU寄存器之间直接通信。

在微核心化操作系统之上支持多种流行的操作系统界面环境也是多个厂商研究的一个焦点。例如：Microsoft公司在Window NT之上实现DOS、Window 3.1、Win 32、OS/2与POSIX界面；IBM公司的Workplace OS将支持DOS、Windows、OS/2、AIX等；Sunsoft公司的Solaris与Spring OS支持Window 3.1等。实现方法有指令的模拟与解释、机器指令的直接翻译等。

其他的研究热点包括如何高效地并行化已有的子系统或者设计新的子系统问题，包括文件子系统、tty子系统、I/O驱动子系统、进程管理子系统、网络通信子系统以及安全子系统等。

## 4 操作系统领域的一些研究热点

(1)效率问题

主要的方法是通过设计各类跟踪统计工具找出系统中的效率瓶颈，然后再设法解决。从观察的范围看，有宏观与微观两类。前者将整个计算机系统看成一个不可分裂的整体，从大范围内找瓶颈；后者集中于某一点上，如与CPU相关的程序(如中断与例外处理程序)的效率、具体一个操作系统子系统等。从宏观角度看，CPU以每年50%~100%的速度增加，DRAM的容量(单位面积)每3年增加4倍，存取速度增加较缓慢，磁盘容量年增长30%，存取速度增长也很缓慢，如何解决CPU与DRAM速度之间的差别以及DRAM与磁盘存取速度的差别是硬件设计与操作系统实现需要考虑的问题。前一个问题通过增加Cache来解决，涉及的问题包括Cache大小(数据与指令)、Cache数据一致性等。

(2)调度单元的研究

该方面的研究主要集中于线程界面的定义与实现两个方面。

(3)共享虚拟存储

共享虚拟存储机制一个主要特点就是在分布存储的多处理机系统中为各结点机提供一个可共享的“虚拟地址空间”，各个结点机可直接对该虚空间上的任何地址进行读写与执行指令等操作。这一概念实际上是传统操作系统的“虚拟存储空间”在分布式环境下的进一步扩展。它不仅实现了传统“虚存”中的页表到当前结点物理存储器与磁盘块的映射，同时还完成了“Page”虚拟页表到其他结点局部物理存储器的映射操作。

采用共享虚拟存储技术可带来以下好处：①在分布式环境下向用户提供一种“共享存储”的编程模型，否则用户需通过显式的消息传递(Message Passing)方式来编制程序；②用户程序可并行执行；③单个结点上运行的进程可利用其他结点上的局部存储器作为缓存器；④非常自然地支持进程的迁移。

该方面研究的主要领域：①保证“Page”到不同局部存储器中的数据一致性协议与算法。算法包括集中式管理算法、分布式算法、动态算法等。协议有Write-Broadcast、Page Ownership、Multiple Consistency Protocols。②在具体实现环境下的效率考虑(页大小、申请空间的策略等)。③如何以硬件或者软硬结合的方式进行实现。现已经实现的共享虚拟存储系统有：Dash、Ivy、Linda、Memnet、Mermaid、Mirage Munin、Plus、Shiva等。

（4）文件系统

该研究主要集中在解决高效性、网络透明性、可靠性、可扩展性以及易使用性等方面。主要的发展是因为CPU的运算速度在不断加快、DRAM容量的增大、网络普及需要资源共享以及I/O存取速度进展缓慢。研究的领域包括：文件数据的组织与存放策略、用户使用界面的定义、文件数据分布存储的一致性问题；从系统效率以及容错性方面出发研究数据冗余存储的策略；物理存储设备的组织与构成（DRAM、DISK、CDROM）等。另外，由于存储器芯片价格下降很快，如何建立基于内存（DRAM、NVRAM）的文件系统也将在不远的将来成为一个热点。

（5）新概念的操作系统

面向对象的操作系统的研究也是学术单位与计算机厂商的重点之一。一个方向是设计操作系统核心时采用面向对象的程序设计方法，另一方向是在操作系统之上建立面向对象的程序开发支撑环境。Microsoft NT、Chorus、Amoeba、Elmwood等操作系统采用了面向对象的程序设计语言或思想进行了构造。在核心之上构造的面向对象的环境包括：Chorus核心上的COOL（Chorus Object-Oriented Layer）、Microsoft公司建立在Window NT之上的OLE、Apple公司的OpenDoc、IBM等公司联合开发的Taligent Frameworks、IBM公司的DSOM等。

另一个兴起的操作系统研究领域就是单地址空间操作系统。如果说CPU芯片从16位虚地址空间过渡到32位虚地址空间曾经引起OS设计革新的话（单用户到多用户），从32位到64位可能会引发OS设计的另一场革命。生产支持64位虚拟空间的CPU芯片已经是硬件发展的大趋势，64位虚拟空间已经完全有能力存放现代操作系统中的所有用户进程与操作系统本身。现有操作系统中跨地址空间进行通信与同步等操作的效率低下问题将迎刃而解。目前该方面的研究刚刚起步，研究的重点集中在虚拟地址保护与映射机制的分离技术，支持单地址空间OS的硬件新机制。

## 5 结束语

本文从操作系统的发展史、当前的发展状态、未来的发展趋势以及技术研究的热点等几方面进行了阐述。阐述主要集中于传统单机UNIX的并行改造、微核心化操作系统的发展走向，以及在提高效率、调度单元、共享虚拟存储机制、文件系统、新概念操作系统等几方面。操作系统是计算机系统的核心软件，其涉及的研究范围很大，本文阐述的内容是其主流发展的一部分。

发表于 *Journal of Computer Science and Technology*, 2001 年第 6 期

# BCL-3: a high performance basic communication protocol for commodity superserver DAWNING-3000

Jie MA, Jin HE, Dan MENG, Guo-jie LI

Institute of Computing Technology, Chinese Academy of Sciences, Beijing 100080, China

**Abstract:** This paper introduces the design and implementation of BCL-3, a high performance low-level communication software running on a cluster of SMPs (CLUMPS) called DAWNING-3000. BCL-3 provides flexible and sufficient functionality to fulfill the communication requirements of fundamental system software developed for DAWNING-3000 while guaranteeing security, scalability, and reliability. Important features of BCL-3 are presented in the paper, including special support for SMP and heterogeneous network environment, semi-user-level communication, reliable and ordered data transfer and scalable flow control. The performance evaluation of BCL-3 over Myrinet is also given.

**Key words:** cluster communication system, user-level communication, scalable flow control, CLUMPS

## 1 Introduction

Clusters have been popular platforms for high performance computing in recent years. They are widely used in scientific and engineering computing, business computing, and Internet information services. Communication performance is one of the most critical factors determining the performance of a whole cluster system. So how to improve the performance of communication is a hot research topic in cluster computing. Meanwhile, building a cluster with commodity SMPs (CLUMPS) is becoming a trend since this approach can increase the packing density of system while reducing cost. CLUMPS provides different hardware for intra- and inter-node communication, that is, shared memory and message-passing network. Communication software should take the difference into account and provide corresponding protocols so that the capacity of underlying communication hardware can be fully used for various applications[1].

BCL-3 is a high-performance low-level communication protocol designed for and implemented on DAWNING-3000. It provides reliable and ordered message passing with flow control and error correction. DAWNING-3000 is

This work was supported by the National 863 High-Tech Programme of China (No.863-306-ZD01-01).

a CLUMPS consisting of 70 nodes. Each node is a 4-way 375 MHz Power3 SMP, running IBM AIX4.3.3. All nodes are connected with Fast Ethernet, Myrinet and Dnet. Dnet is high performance interconnect developed on our own. Both Myrinet and Dnet are supported by BCL-3. It consists of a library in the user space, a device driver in the kernel space, and a control program called MCP running on the network interface. A few important features of BCL-3 are described as follows.

(1) Flexible and sufficient functionality

BCL-3 provides a channel-based asynchronous communication mechanism, which supports point-to-point message passing and remote memory access (RMA) operations. It provides efficient communication support for the implementations of high-level communication software and other system software on DAWNING-3000, including PVM (parallel virtual machine), MPI (message-passing interface), JIAJIA (a software DSM) and a cluster file system called COSMOS. BCL-3 also implements automatic node status detection, node isolating/rejoining functionality, and application program exception handling to guarantee system availability and stability. Although more functionality can decrease communication performance, especially latency, it is necessary for a commodity system.

(2) Support for SMPs

BCL-3 provides different protocols for intra- and inter-node communication respectively. Intra-node communication is implemented via shared memory while inter-node via message-passing network. This improves intra-node communication performance dramatically.

(3) Zero memory copy and pin-down cache

BCL-3 implements direct user-to-user data transfer within inter-node communication. Sending and receiving data buffers provided by users can be pinned down in physical memory and therefore become DMA-able during communication process. The pin-down Cache is implemented to reuse the pinned down area. All these make contributions to bandwidth improvement. BCL-3 provides almost all hardware bandwidth to its users.

(4) Semi-user-level communication

BCL-3 implements user-level communication on receiving side. But on sending side, there is a need for senders to trap into OS kernel and invoke message-sending operations. Compared with user-level communication, only one kernel trapping is incorporated into the communication path. Although it can increase communication latency a little bit, this combination has a few advantages, which will be discussed in later section.

(5) Heterogeneous network environment support

Because the BCL-3 library in user space is independent of underlying network interface, binary code written in it or in a high-level communication library such as PVM, MPI and JIAJIA on top of it can run on any combination of networks. Applications written in BCL-3 need not to be recompiled. This feature is especially useful for applications running over a cluster of clusters.

(6) Scalability and reliability

BCL-3 adopts ACK/NAK mechanism to guarantee reliable and ordered data communication. ACK/NAK-based flow control improves the scalability in comparison with credit-based flow control. Memory consumption is no longer proportional to system scale. Currently, ACK/NAK mechanism is implemented on the network interfaces of Myrinet and Dnet.

The communication performance of BCL-3 has been measured on DAWNING-3000 over Myrinet. As a reliable and ordered message-passing protocol, it provides 2.7 μs one-way latency and 391 MB/s bandwidth of intra-node communication while 18.3 μs and 146 MB/s of inter-node's. Performance of MPI on top of BCL-3 is also presented.

Sections 2 and 3 introduce the design and implementation of BCL-3 over Myrinet in detail. Then the evaluation of BCL-3 on DAWNING-3000 platform is shown in Section 4. Finally, we present our conclusions and discuss ongoing research in Section 5.

## 2 Design issues

(1) Communication services

Services provided by a communication system can greatly impact its performance. Many communication protocols only provide limited services and show excellent performance. However, the limited services will make other system software on high layers more difficult to implement. BCL-3 provides reliable and ordered message passing with flow control and error correction. Although reliable and ordered delivery reduces the performance of BCL-3, it will reduce more performance when it is implemented on higher layers.

Furthermore, RMA and special message arrival informing mechanism are also provided to support DSMs and distributed file systems. RMA makes it possible to deliver messages without cooperation of the other side. It is essential for software DSMs and distributed file systems. Also, BCL-3 provides "select" mechanism to inform the arrival of messages so that the application can use BCL-3 message passing as a socket. It makes the applications using TCP / IP easy to be ported to using BCL-3.

(2) Semi-user-level communication

In traditional communication protocols, such as TCP /IP, the main communication path involves OS kernel trapping and interrupts handling. All these overheads cause high end-to-end message latencies. User-level communication allows applications to directly access the network interface cards (NIC) without operating system intervention on both sending and receiving sides. Messages are transferred directly to and from user-space by the network interface card while observing the traditional protection boundaries between processes. It reduces the communication overhead by avoiding kernel traps on the data path.

However, in commodity systems, security is the most important requirement. User level communication protocol exposes all the control data structures to user space so that any mistake or malice operation will cause the system to break down. Kernel-level communication protocol protects the important data structures in kernel space.

To provide a low-latency and secure communication, BCL-3 uses a combined method which ensures the system security by protecting the most important data structures in kernel space and reduces communication overhead by diminishing some unnecessary kernel traps. Furthermore, it can make the application binary portable, suitable in heterogeneous network environments and more efficient. BCL-3 is separated into two parts, the user-level library and the kernel extension. All the system-related operations are hidden in the kernel extension, so that the user-level library can be the same on different platforms. Application, which links the same user-level library, can be ported to different platforms with binary codes. Also, the differences between heterogeneous network environments are hidden in the kernel extension.

(3) Communication mechanism and semantics

There are three commonly used message-passing modes, synchronous message passing, blocking message passing and non-blocking message passing. It is clear that synchronous and blocking message passing can be implemented by non-blocking message passing. So BCL-3 only provides non-blocking sending and receiving semantics.

Channels are used to identify the message transfer paths in BCL-3. The system channel is used to transfer small messages. The rendezvous semantic is used to transfer large messages. Rendezvous means the send is guaranteed to complete only if a corresponding receive has been posted. RMA operations are presented in BCL-3 by using a special type of channel. A message arrival informing mechanism, which is compatible with the TCP /IP socket, is also provided so that applications can use the "select" system call to wait for the incoming messages.

(4) Reliability and scalability

Although the reliable protocol can be implemented at higher levels, it is more efficient to implement the reliable protocol in this communication layer. BCL-3 provides a reliable and ordered message passing data path. It uses an ACK/NAK based reliable protocol. By using CRC verification and sequential number checking on each packet transferred on the net, BCL-3 can ensure the correctness and order of packets. Since the communication hardware is highly reliable, this ACK/NAK based protocol is very efficient on our platform. Compared with credit-based reliable protocol, the ACK/NAK based protocol is scalable. It is easy for the protocol to be used in large-scale cluster.

(5) Zero memory copy and buffer management

To implement zero memory copy, the user buffer should be pinned in physical memory and the virtual address should be translated into physical address before starting communication. This operation will increase the communication overhead. In order to reduce the communication overhead, the pin-down cache[2] and three types of buffers are introduced in BCL-3. They can reduce the overhead and improve the system performance. These two techniques will be discussed in detail in the next section.

(6) Special support for SMPs

In the context of SMPs, communication protocol needs to support several processes on one node. When the memory copy bandwidth is lower than the network bandwidth, it will provide a high performance to transfer messages via NIC even if the two processes are in the same node. But the memory copy bandwidth is improved heavily now. It is costly and has no meaning to transfer intra-node messages as the same manner to transfer inter-node messages. BCL-3 uses direct memory copy instead of transfer via network. Shared memory and direct copy from another process's memory space can be used to implement intra-node message passing. BCL-3 uses the former mechanism, which will be discussed later[3-5].

## 3 BCL-3 Protocol

BCL-3 is a low-level communication software used on DAWNING-3000. It has two versions. One is implemented on the Myrinet and the other is implemented on the Dnet. The upper levels can make full use of the hardware performance via BCL-3. BCL-3 supports point-to-point message passing. All other collective message passing should be implemented in the higher level software.

### 3.1 Architecture

Figure 1 is the protocol stack from the angle of the BCL-3 applications. Applications can directly access the BCL-3 level or use other functionality levels. BCL-3 is the lowest level software in DAWNING-3000's communication software. MPI (Message-Passing Interface) is implemented directly on BCL-3. PVM (Parallel Virtual Machine) is implemented on ADI-2 (Abstract Device Interface) of MPI. Notice that this may be changed according to different implementations that are done. In order to keep high performance, we can also port PVM directly on top of BCL-3. TCP /IP was designed to port onto our BCL-3 in the next step. A prototype will be completed early next year. The other two components in the stack are JIAJIA and COSMOS. The former is a DSM (Distributed Shared Memory) software, and the latter is a distributed file system of DAWNING-3000.

| Application | | | | |
|---|---|---|---|---|
| PVM | MPI | TCP/UDP | JIAJIA(DSA) | COSMOS(FS) |
| ADI-2 | | IP | | |
| BCL | | | | |

Figure 1. Protocol stack of DAWNING-3000 software

BCL-3 is divided into three parts, the on-card control program (MCP, Myrinet Control Program), the device driver (DD) and the BCL library. Myrinet has an on-card processor, which can control the card to transfer packets by using three DMA engines. In BCL-3, MCP controls all the inter-node packet transfers. MCP completes a sending operation by reading send request in the card's local memory, sending/receiving messages with DMA engines and informing the user process of completion. Device driver is a kernel extension, which provides several ioctl() calls to control the hardware. It posts operation requests to the on-card memory. These requests include the sending requests and other requests, such as initialize/close communication port requests and initialize channel requests. Device driver also implements some functional operations, which need to be executed in the kernel environment. Such operations include the memory pin/unpin operation and physical memory address conversion. BCL library includes all implementations of BCL API.

### 3.2 Communication mechanism

Communication is occurred between ports. Each process can create only one port to communicate with others. A process is labeled with its node number and port number. The pair of node number and port number is the unique identifier of the process within the application. Each port has a sending request queue, a receiving buffer pool and the corresponding event queues (sending event queue and receiving event queue). The receiving buffer pool is composed of several channels. There are three types of channels, system channel, normal channel and open channel.

When a process (sender) wants to send a message to another process (receiver), the sender should compose a send request and put it into the send request queue. Destination is specified by its node number, port number and channel number in the request. The receiving channel should be ready before the message arrives at the receive side. A receive event will be generated when a complete message arrives. On the send side, a send event will be put into the send event queue when a sending operation is over.

As shown in Figure 2, the system channel is designed to transfer small messages. Each process has one system channel. Every system channel has a buffer pool, which is initialized when the process starts. It is organized as an FIFO queue. When a short message arrives, it will automatically be put to the first free buffer in the buffer pool. The incoming message will be discarded if there is no free buffer in the pool. After the receiver gets the message, the buffer will be returned to the buffer pool.

The normal channel is designed to transfer messages with the rendezvous semantics. Each process has several normal channels. A normal channel has a user-specified buffer. The receiving process needs to prepare the receive channel before the sending process starts transmission. A receive event will be generated when a message completed. A normal channel should be initialized before each receiving operation.

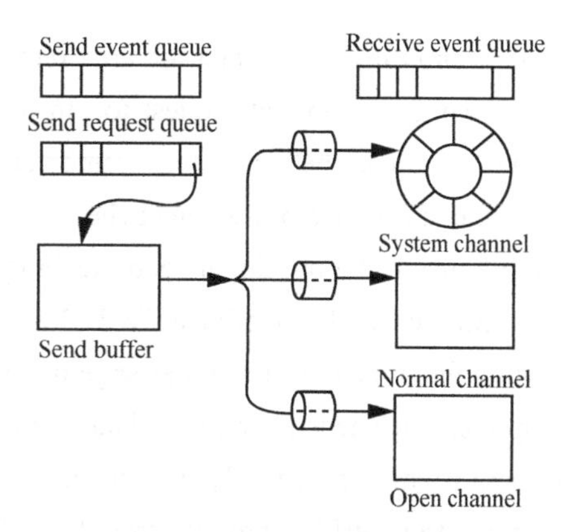


Figure 2. Communication mechanism of BCL-3

The open channel is designed to perform RMA. Each process has several open channels. An open channel has a user-specified buffer. Unlike the normal channel, only once does it need to be initialized before receiving operations. A receiving event will be generated when an RMA operation is completed.

### 3.3 Buffer management

A zero copy message transfer is realized with the help of the Myrinet DMA capability. The user program accesses only virtual addresses while the DMA engines access only physical memory addresses. To realize the DMA-issued message transfer, both the sending and receiving buffers must be pinned in the physical memory and the physical memory addresses must be known. The pin-down operation is costly in most operating systems. This overhead can be reduced if an application repeatedly transfers data from the same memory area without releasing the pinned-down area. When a request to release a pinned-down memory area is issued, the actual release operation is postponed until total physical memory reserved for the pinned-down area is used up. If the release primitive for pinned-down area has been issued but in fact the area remains pinned down, a subsequent request to pin down the same area can be answered immediately. This technique is called pin-down cache.

While using the pin-down cache technique, there is still a cache-hit overhead. BCL-3 avoids it by using the pinned buffer in message passing. If an application needs to transfer data from the same memory area repeatedly, it can request to pin down the buffer before transfer. The pinned-down buffer should be released after all transmissions. BCL-3 defines three types of buffer: normal buffer, enabled buffer and DMA buffer. A normal buffer is a user buffer that is not pinned. The communication primitives need to perform pin/unpin operation when using this type of buffer. An enabled buffer is a user buffer that is already pinned in the physical memory. A DMA buffer is a system buffer allocated by a special call, which is pinned in the physical memory. There is no need to check the pin-down cache before transmission when using the last two types of buffers. It reduces the overhead when using the pin-down cache.

## 3.4 Intra-node communication

There are several ways to move data from one process to another within one node (Figure 3). The traditional way is to move data as the same way as between nodes. Process A first transfers the data to NIC (Network Interface Circuit) by DMA. Then NIC transfers them back to process B. While the memory copy bandwidth is much higher than the DMA bandwidth, a good solution is to use shared memory to implement intra-node communication. Process A first copies the data to a shared memory area. Then process B copies them out. Another way is to move data directly from user space to user space.

Shared memory based intra-node communication can make the system more reliable. Any mistake or malice operation during a directly inter-process memory access can cause the target process to crash. By using shared memory, the sender process can only destroy the shared area. It will not affect other processes' space. This

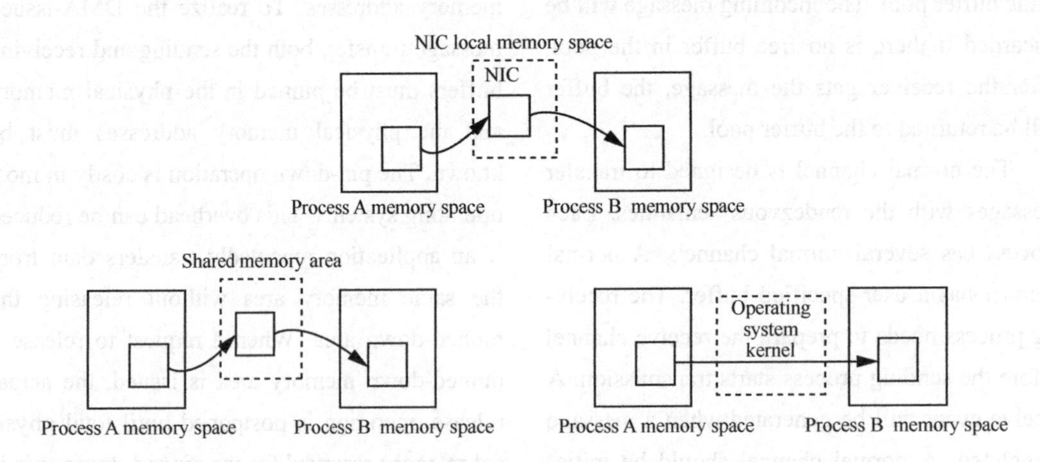


Figure 3. Several ways to move data between processes within one node

guarantees the independence of each process.

BCL-3 uses shared memory based intra-node communication. The internal buffer queue is used to transfer messages from one process to another within a node. This queue consists of a list of buffers. Each pair of processes has two queues (Figure 4). To ensure the message sequence, BCL-3 uses the sequential number to decide whether the operation should continue or not. Shared memory based intra-node communication needs an extra memory copy other than the direct memory copy solution. BCL-3 reduces the extra overhead by using the pipeline message passing technique.

### 3.5 Flow control

BCL-3 provides reliable and ordered delivery. The receiving process will simply discard the packet when it detects an error. The sending process will retransmit it when timeout. When a source node sends two messages to the same destination, a receiving operation executed firstly at the destination will receive the first message. A sequential number based ACK/NAK protocol is used in BCL-3 to implement reliable and ordered delivery. Considering the high reliability of the network hardware, this algorithm is efficient.

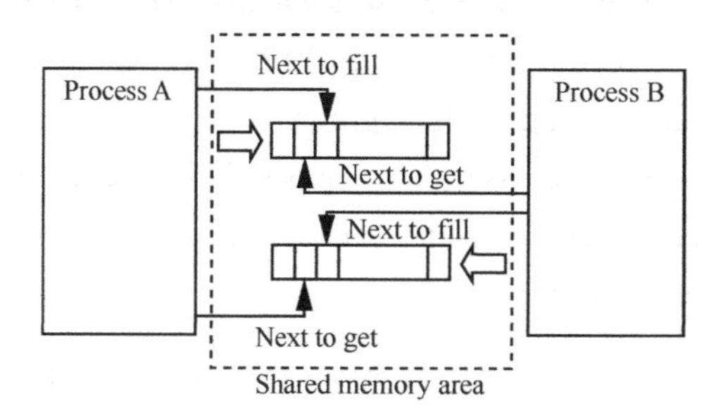


Figure 4. Intra-node communication in BCL-3

Flow control is used to avoid buffer overflow and system deadlock. In BCL-3, the receiving process simply discards messages when there is no enough buffers. The sending process will re-send them when timeout. An overflowed packet is handled as an error packet.

## 4 Performance and analysis

All the tests[6] are done on DAWNING-3000, which consists of 70 IBM270 workstations. Each node is a 4-way 375 MHz Power3 SMP, which is running IBM AIX4.3.3. Myrinet[7] M2M-PCI64A NICs are used on each node. All nodes are interconnected by M2M-OCT-SW8 switches.

The first tests are raw point-to-point communications. Latency and bandwidth are measured on DAWNING-3000. Both the inter-node and the intra-node communications are tested. The result is shown in Figure 5 and Figure 6. The minimal latency is 2.7 μs within one node and 18.3 μs between nodes. The bandwidth is 391 MB/s within one node (with the effect of cache) and 146 MB/s between nodes. And the half-bandwidth is reached with less than 4 KB message.

Table 1 shows the comparison of three protocols. GM is a message-based communication system for the Myrinet, which is designed and implemented by Myricom. GM does not provide special support for SMP. Only inter-node communication performance data are given in the table. The range for GM's one-way short-message latency on a wide variety of hosts is from 13.37 to 21 μs. And the peak bandwidth is over 140 MB/s. BCL-3 reaches almost the same performance and provides a more reliable (using the kernel-level communication) and more complex (special support for SMP) protocol.

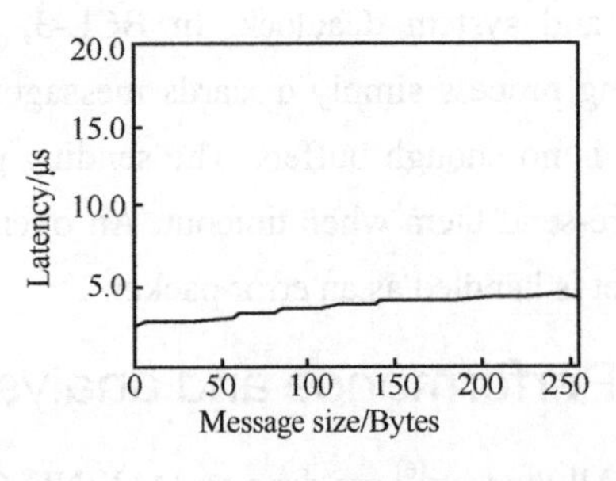

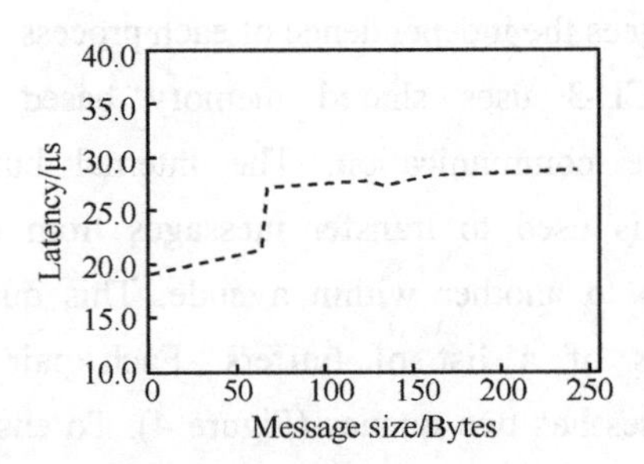


Figure 5. Intra-node (left) and inter-node (right) latency of BCL-3

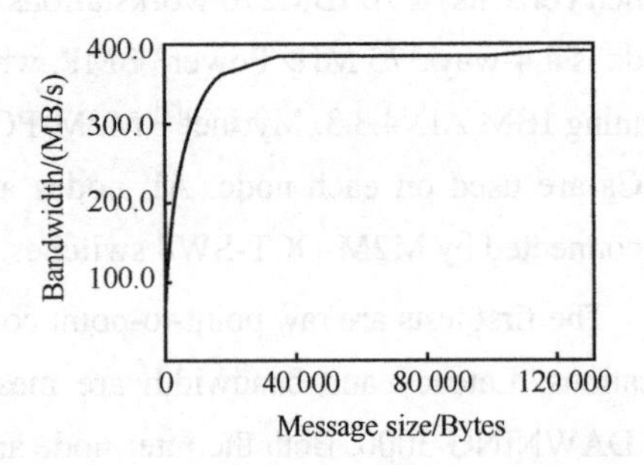

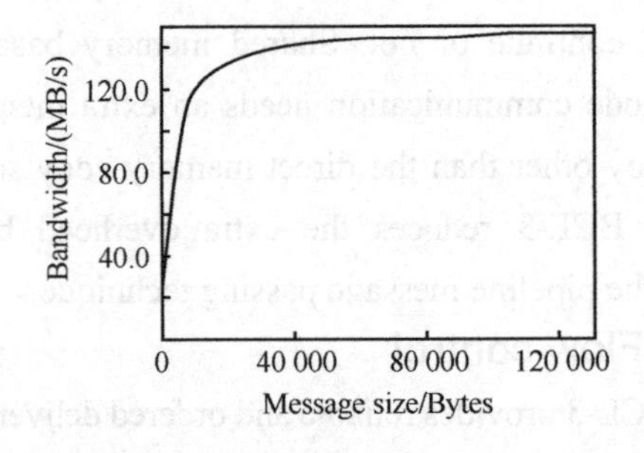


Figure 6. Intra-node (left) and inter-node (right) bandwidth of BCL-3

Table 1. Comparison of different communication protocols

| Protocol | Latency/μs | | Bandwidth/(MB/s) | |
|---|---|---|---|---|
| | Intra-node | Inter-node | Intra-node | Inter-node |
| GM | - | 13.37~21 | | >140 |
| AM- II | 3.6 | 27.5 | 160 | 32.8 |
| BIP-SMP | 1.8 | 5.7 | 160 | 126 |
| BCL-3 | 2.7 | 18.3 | 391 | 146 |

AM-II[8-11](Active Messages) is similar to remote procedure call (RPC) mechanism. Compared with AM-II, BCL-3 has a better latency in both the intra-node and the inter-node communications. It is meaningless to compare the bandwidths of these two protocols since AM-II needs an extra memory copy when transferring a message while BCL-3 does not. BCL-3 reaches a much higher bandwidth.

BIP[12-14] (Basic Interface for Parallelism) is a low level message passing system developed by the Laboratory for High Performance Computing in Lyon, France. It has a very low latency. But it does not provide the functionality of flow control and error correction. Its bandwidth is lower than that of BCL-3.

Figure 7 and Figure 8 show the performance of MPI over BCL-3. The minimal latency is 6.3 μs within one node and 23.7 μs between nodes. The bandwidth is 328 MB/s within one node (with the effect of cache) and 131 MB/s between nodes. If the message length grows to 8 MB, the intra-node decreases. This shows the influence of cache. The bandwidth can be very high without cache replacement.

## 5 Conclusion and future work

BCL-3 is available to easily build clusters of commodity SMPs. This communication layer combined with MPI over BCL-3 provides an

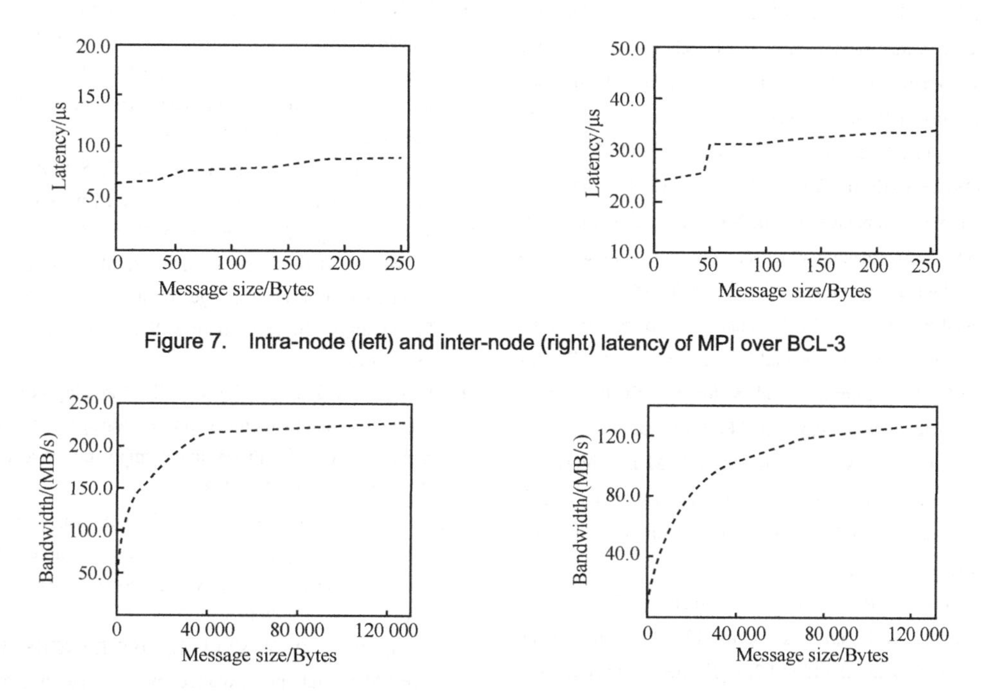


Figure 7. Intra-node (left) and inter-node (right) latency of MPI over BCL-3

Figure 8. Intra-node (left) and inter-node (right) bandwidth of MPI over BCL-3

efficient message passing interface for a cluster of commodity SMPs. BCL-3 achieves 2.7 μs of latency and 391 MB/s of bandwidth for intra-node message-passing. The performance of inter-node communication is 18.3 μs of latency and 146 MB/s of bandwidth. MPI over BCL-3 can achieve 6.3 μs of latency and 328 MB/s of bandwidth for intra-node message passing, and 23.7 μs of latency and 131 MB/s of bandwidth for inter-node message passing. BCL-3 is a flexible and reliable protocol. Point-to-point message passing and RMA are both presented in BCL-3. The combination of the kernel and user-level communications makes the system reliable and portable.

Although BCL-3 is at this time implemented on AIX, it can be ported to other platforms easily because it does not need to modify the source codes of the operating system kernel.

The bandwidth provided by Myrinet is approximately 160 MB/s (1.28 Gbit/s). BCL-3's bandwidth almost reaches the hardware limitation. To break the limitation, we plan to design a double-card system that uses two Myrinet cards to perform communication. It ought to get a higher bandwidth by this means.

Finally, we have tested BCL-3 on a large-scale cluster of SMP nodes, DAWNING-3000. There are 70 nodes and each node has four processors. This shows the scalability of BCL-3.

## References:

[1] HWANG K, XU Z W. Scalable parallel computing: technology, architecture, programming[M]. New York: McGraw- Hill, 1998.

[2] TEZUKA H, O'CARROLL F, HORI A, et al. Pin-down cache: a virtual memory management technique for zero-copy communication[C]//The 12th International Parallel Processing Symposium & 9th Symposium on Parallel and Distributed Processing. Piscataway: IEEE Press, 1998: 308-314.

[3] FOSTER I, GEISLER J, KESSELMAN C, et al. Managing multiple communication methods in high-performance networked computing systems[J]. Journal of Parallel and Distributed Computing, 1997, 40(1): 35-48.

[4] GROPP W W, LUSK E L. A taxonomy of programming models for symmetric multiprocessors and SMP clusters[C]//Programming Models for Massively Parallel Computers. Piscataway: IEEE Press, 1995: 2-7.

[5] MELLOR-CRUMMEY J M, SCOTT M L. Algorithms for scalable synchronization on shared-memory multiprocessors[J]. ACM Transactions on Computer Systems, 1991, 9(1): 21-65.

[6] LUECKE G R, COYLE J J. Comparing the performance of MPI on the Cray T3E-900, the Cray origin 2000 and the IBM P2SC[J]. Electronic Journal of Performance Evaluation and Modelling for Computer Systems (PEMCS), 1998.

[7] BODEN N J, COHEN D, FELDERMAN R E, et al. Myrinet-a gigabit-per-second local-area network[J]. IEEE Micro, 1995, 15(1): 29-38.

[8] VON EICKEN T, CULLER D E, GOLDSTEIN S C, et al. Active messages: a mechanism for integrated communication and computation[C]//The 19th Annual International Symposium on Computer Architecture. Piscataway: IEEE Press, 1992: 256-266.

[9] VON EICKEN T, AVULA V, BASU A, et al. Low-latency communication over ATM networks using active messages[J]. IEEE Micro, 1995, 15(1): 46-53.

[10] LUMETTA S S, MAINWARING A M, CULLER D E. Multi-protocol active messages on a cluster of SMP's[C]//Proceedings Supercomputing 97. Piscataway: IEEE Press, 1997.

[11] LUMETTA S S, CULLER D E. Managing concurrent access for shared memory active messages[C]//The International Parallel Processing Symposium. Piscataway: IEEE Press, 1998: 272-278.

[12] PRYLLI L, TOURANCHEAU B. BIP: a new protocol designed for high-performance networking on Myrinet[C]//Workshop PC-NOW, IPPS/SPDP98. Heidelberg: Springer, 1998: 472-485.

[13] GEOFFRAY P, PRYLLI L, TOURANCHEAU B. BIP-SMP: high performance message passing over a cluster of commodity SMP's[C]//Proceedings Supercomputing 99. Piscataway: IEEE Press, 1999.

[14] PRYLLI L, TOURANCHEAU B, WESTRELIN R. An improved NIC program for high-performance MPI[C]//Proceedings ACM99. New York: ACM Press, 1999.

发表于 *Science*，2002 年第 5565 期

# A draft sequence of the rice genome ( Oryza sativa L. ssp. indica )

Jun YU[1,2,3,4], Song-nian HU[1], Jun WANG[1,2,5]
Gane Ka-Shu WONG[1,2,4], Song-gang LI[1,5], Bin LIU[1], Ya-jun DENG[1,6], Li DAI[1], Yan ZHOU[2,7], Xiu-qing ZHANG[1,3], Meng-liang CAO[8], Jing LIU[2], Jian-dong SUN[1], Jia-bin TANG[1,3], Yan-jiong CHEN[1,6], Xiao-bing HUANG[1], Wei LIN[2], Chen YE[1], Wei TONG[1], Li-juan CONG[1], Jia-ning GENG[1], Yu-jun HAN[1], Lin LI[1], Wei LI[1,9], Guang-qiang HU[1], Xian-gang HUANG[1], Wen-jie LI[1], Jian LI[1], Zhan-wei LIU[1], Long LI[1], Jian-ping LIU[1], Qiu-hui QI[1], Jin-song LIU[1], Li LI[1], Tao LI[1], Xue-gang WANG[1], Hong LU[1], Ting-ting WU[1], Miao ZHU[1], Pei-xiang NI[1], Hua HAN[1], Wei DONG[1,3], Xiao-yu REN[1], Xiao-li FENG[1,3], Peng CUI[1], Xian-ran LI[1], Hao WANG[1], Xin XU[1], Wen-xue ZHAI[3], Zhao XU[1], Jin-song ZHANG[3], Si-jie HE[3], Jian-guo ZHANG[1], Ji-chen XU[3], Kun-lin ZHANG[1,5], Xian-wu ZHENG[3], Jian-hai DONG[2], Wan-yong ZENG[3], Lin TAO[2], Jia YE[2], Jun TAN[2], Xi-de REN[1], Xue-wei CHEN[3], Jun HE[2], Dao-feng LIU[3], Wei TIAN[2,6], Chao-guang TIAN[1], Hong-ai XIA[1], Qi-yu BAO[1], Gang LI[1], Hui GAO[1], Ting CAO[1], Juan WANG[1], Wen-ming ZHAO[1], Ping LI[3], Wei CHEN[1], Xu-dong WANG[3], Yong ZHANG[1,5], Jian-fei HU[1,5], Jing WANG[1,5], Song LIU[1], Jian YANG[1], Guang-yu ZHANG[1], Yu-qing XIONG[1], Zhi-jie LI[1], Long MAO[3], Cheng-shu ZHOU[8], Zhen ZHU[3], Run-sheng CHEN[1,9], Bai-lin HAO[2,10], Wei-mou ZHENG[1,10], Shou-yi CHEN[3], Wei GUO[11], Guo-jie LI[12], Si-qi LIU[1,2], Ming TAO[1,2], Jian WANG[1,2], Li-huang ZHU[3], Long-ping YUAN[8], Huan-ming YANG[1,2,3]

1. Beijing Genomics Institute/Center of Genomics and Bioinformatics, Chinese Academy of Sciences, Beijing 101300, China

2. Hangzhou Genomics Institute–Institute of Bioinformatics of Zhejiang University–Key Laboratory of Bioinformatics of Zhejiang Province, Hangzhou 310007, China

3. Institute of Genetics, Chinese Academy of Sciences, Beijing 100101, China

4. University of Washington Genome Center, Department of Medicine, Seattle, WA 98195, USA

5. College of Life Sciences, Peking University, Beijing 100871, China

6. Medical College, Xi'an Jiaotong University, Xi'an 710061, China
7. Fudan University, Shanghai 200433, China
8. National Hybrid Rice R&D Center, Changsha 410125, China
9. Laboratory of Bioinformatics, Institute of Biophysics, Chinese Academy of Sciences, Beijing 100101, China
10. Institute of Theoretical Physics, Chinese Academy of Sciences, Beijing 100080, China
11. Digital China Ltd., Beijing 100080, China
12. Institute of Computing Technology, Chinese Academy of Sciences, Beijing 100080, China

**Abstract:** We have produced a draft sequence of the rice genome for the most widely cultivated subspecies in China, Oryza sativa L. ssp. indica, by whole-genome shotgun sequencing. The genome was 466 megabases in size, with an estimated 46 022 to 55 615 genes. Functional coverage in the assembled sequences was 92.0%. About 42.2% of the genome was in exact 20-nucleotide oligomer repeats, and most of the transposons were in the intergenic regions between genes. Although 80.6% of predicted Arabidopsis thaliana genes had a homolog in rice, only 49.4% of predicted rice genes had a homolog in A. thaliana. The large proportion of rice genes with no recognizable homologs is due to a gradient in the GC content of rice coding sequences.

Rice is the most important crop for human consumption, providing staple food for more than half the world's population. The euchromatic portion of the rice genome is estimated to be 430 MB in size[1-3], which is the smallest of the cereal crops. It is 3.7 times larger than that of A. thaliana[4-6], and 6.7 times smaller than that of the human[7-8]. The well-established protocols for high-efficiency genetic transformation, widespread availability of high-density genetic and physical maps[9-10], and high degrees of synteny among cereal genomes[11-15] combine to make rice a unique organism for studying the physiology, developmental biology, genetics, and evolution of plants. The International Rice Genome Sequencing Project (IRGSP)[16] has already delivered a substantial amount of sequence for the japonica (Nip-ponbare) subspecies, in bacterial artificial chromosome (BAC) and P1-derived artificial chromosome (PAC)–sized contigs. Working independently, Monsanto and Syngenta[17-18] established proprietary working drafts for japonica, in April 2000 and February 2001, respectively. The Monsanto sequence has been used to assist in the efforts of the IRGSP.

We are releasing a draft genome sequence for rice from 93-11[19], which is a cultivar of Oryza sativa L. ssp. indica, the major rice subspecies grown in China and many other Asia-Pacific regions. It is the paternal cultivar of a super-hybrid rice, Liang-You-Pei-Jiu (LYP9), which has 20% to 30% more yield per hectare than the other rice crops in cultivation[20]. The maternal cultivar of LYP9 is Pei-Ai 64s (PA64s), which has a major background of indica and a minor background of japonica and javanica, two other commonly cultivated subspecies. We have also produced a low-coverage draft sequence for PA64s. A preliminary assembly and analysis on a subset of this sequence was published in the Chinese Science Bulletin[21]. Our discussion will focus largely on the genome landscape of rice, how it differs from that of the other sequenced plant, A. thaliana, and how both plant genomes differ from that of the human. We will show that rice genes exhibit a gradient in GC content, codon usage, and amino acid usage. This compositional gradient reflects a unique phenomenon in the evolutionary history of rice, and perhaps all monocot plants, but not eudicot plants. As a result, about one-half of the predicted rice genes have no obvious homolog in A. thaliana, whereas the other half is almost a replica of the A. thaliana gene set.

The entire rice genome sequence can be downloaded from our Web site at http://btn.genomics.org.cn/rice. Following our announcement of the rice genome sequence at the annual Plant, Animal and Microbe Genomes (PAG X) conference, in San Diego, during the ensuing period from 14 January to 2 March 2002, this sequence was downloaded 556 times, and the BLAST search facilities were used 7008 times by 343 individuals. This sequence has also been deposited at the DNA Data Bank of Japan/European Molecular Biology Laboratory/GenBank under the project accession number AAAA00000000. The version described in this paper is AAAA01000000.

**Experimental design**. The rice genome project at the Beijing Genomics Institute has been designed in two stages. This is a report on stage I, the primary objective of which was to generate a draft sequence of rice at ~4X coverage for 93-11. A similar amount of data will eventually be generated for…

发表于 IEEE International Conference on Cluster Computing 论文集, 2003 年

# Design and performance of the dawning cluster file system

Jin XIONG, Si-ning WU, Dan MENG, Ning-hui SUN, Guo-jie LI

National Research Center for Intelligent Computing Systems,

Institute of Computing Technology, Chinese Academy of Sciences

**Abstract:** Cluster file system is a key component of system software of clusters. It attracts more and more attention in recent years. In this paper, we introduce the design and implementation of DCFS(the Dawning Cluster File System) -a cluster file system developed for Dawning4000-L. DCFS is a global file system sharing among all cluster nodes. Applications see a single uniform name space, and can use system calls to access DCFS files. The features of DCFS include its scalable architecture, metadata policy, server-side optimization, flexible communication mechanism and easy management. Performance tests of DCFS on Dawning4000-L show that DCFS can provide high aggregate bandwidth and throughput.

## 1 Introduction

The cluster computing technology[1] has matured as a mainstream method for building high performance computers. A shared global file system, which is also called cluster file system, is required by cluster systems. A cluster file system is marked by three characteristics: a shared file system, a global name space and a system call interface.

Early distributed file systems[2], such as NFS and AFS, are aimed at sharing files among network computers, with limitations when used as a cluster file system. With its single-server architecture and non-UNIX semantics, NFS can hardly support a cluster with large numbers of nodes.

The focal research issues of cluster file system are:

- High performance: including bandwidth and throughput. According to the US DOE&DOD, the requirement of aggregate bandwidth was about 20~60 GB/s in 2002, and will be increased to 50~200 GB/s in 2005.
- High scalability: the ability to support large clusters with hundreds even thousands of nodes. In recent years, there has been the requirement of large file system whose volume is 100 TB~1 PB servicing more than one thousand nodes.

This work was supported by the National High-Technology Research and Development Program of China (863 Program) under the Grant No.2002AA1Z2102 and the Grant No.2002AA104410.

- High availability: that is, a cluster file system is accessible without stopping the file services in the presence of component failures.
- Manageability: that is, a cluster file system should provide friendly utilities for deployment, configuration and steering.

Generally, there are two approaches to implement cluster file systems all through the ages. One is based on file servers and another is base on shared storage[3]. In order to improve the bandwidth, the disk striping technology in RAID is used in some cluster file systems in which file data are striped across disks of multiple file servers, such as UCB's xFS[4], IBM's GPFS[5], Clemson University's PVFS[6], etc. Some shared storage file systems are symmetric, that is, there is no server, such as Sistina's GFS[3,7], while others are asymmetric, that is, there are some kind of servers. For example, IBM's Tivoli SANergy[8] and SGI's CXFS[9] need a metadata server, and IBM's GPFS need a lock manager. However, SANergy, CXFS and PVFS do not support multiple metadata servers.

The Dawning Cluster File System (DCFS) is a file-server based cluster file system designed for the Dawning 4000-L Linux cluster with the following features:

- Single system image: DCFS is a global file system sharing among all nodes. Applications see a single uniform name space, and can use system calls to access DCFS files.
- Scalable architecture: The architecture of multiple storage servers and multiple metadata servers makes DCFS more scalable.
- High bandwidth: With file data striping and server-side optimization, DCFS can provide high aggregate bandwidth.
- Powerful metadata processing: DCFS is able to make use of the processing power of multiple metadata servers.
- Flexible communication mechanism: DCFS encapsulates its inside communication into an abstraction layer supporting either TCP or VIA.
- Easy management: DCFS provides management utilities to aid DCFS administration.

The remainder of this paper is organized as follows. The design and implementation of DCFS are described in Section 2. The performance evaluation of DCFS on Dawning 4000-L machine is presented in Section 3. Concluding remarks are presented in Section 4. Finally, we outline future work in Section 5.

# 2 DCFS architecture and design

## 2.1 Architecture

The architecture of DCFS is depicted in Figure 1 DCFS splits the file service into two orthogonal parts: operations on file contents and operations on metadata of files and the file system. There are multiple metadata servers providing storage and maintenance of file attributes, directories and the superblock. There are multiple storage servers for the storage and access of actual file contents. The architecture of multiple storage and metadata servers makes DCFS more scalable. The workload can be dispatched to a group of collaborative servers. In DCFS, metadata are stored in regular files of the native file system (such as EXT2) on metadata servers. And

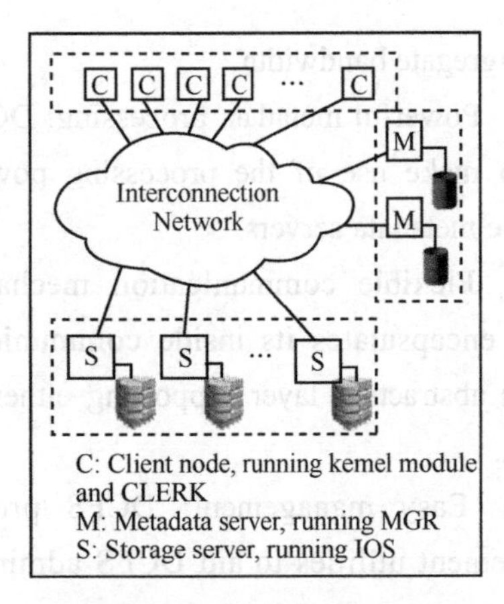


Figure 1. DCFS architecture

file data are stored on multiple storage servers in RAID 0 style of striping.

On client nodes which mount DCFS, the DCFS kernel module implements the interface with VFS and a user-level process called CLERK takes charge of communication with metadata servers and storage servers. So that applications can use standard system calls provided by Linux to access DCFS files, just like the way they access local files.

There is a user-level process, called CND, which acts as an agent to maintain the configuration status of DCFS. It can run on whichever nodes of the system.

## 2.2 Metadata handling

Metadata processing is very important for file systems. According to SFS3.0[10], metadata requests account for over 60% of all requests in NFSv3. The approach of a single metadata server is simple, but limits the performance and scalability. The approach of multiple metadata servers is certainly more complex and difficult to recovery from failures, but is good for performance and scalability.

Unlike NFS, PVFS, and SANergy, which use the structure of local file system on the server to maintain their file system metadata, DCFS maintains and interprets metadata by itself for performance considerations.

In DCFS, there is a metadata server called super-manager who maintains the superblock, root directory and root inode of DCFS. An approach called first-level subtree distribution shown in Figure 2 is used to distribute the metadata among MGRs. Each metadata server stores and maintains a subtree of DCFS file system's root directory. When creating a new object in the root directory, the super-manager makes the decision that which metadata server is going to store and maintain the new object's metadata. The advantage of this distribution policy is that it keeps the parent-child relationship of the objects in DCFS file system. So that majority of metadata operations can be accomplished by requesting only one metadata server. In order to further improve performance, DCFS implements directory cache and inode cache on metadata servers.

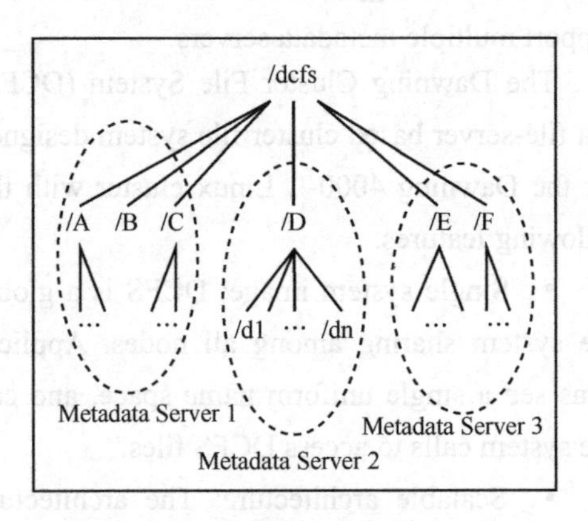


Figure 2. The distribution policy of metadata

## 2.3 Server-side optimization

Storage servers are optimized by server-side data caching, multi-threading, and overlapping of disk access with network transfer. There are four kinds of threads on each storage server, as shown in Figure 3 The heartbeat thread is used for state monitoring. The flushing thread runs in the background and periodically flushes dirty cache blocks to disks. The actual file read and write operations are performed by the working threads. The scheduler thread decides whether to dispatch a thread pair or a single thread to serve this request according to the requested data size. A concept of thread pair is proposed, one thread is in charge of transferring data through network and another is responsible for writing data into or reading data from the storage. The thread pair improves the concurrency of disk access and network transfer.

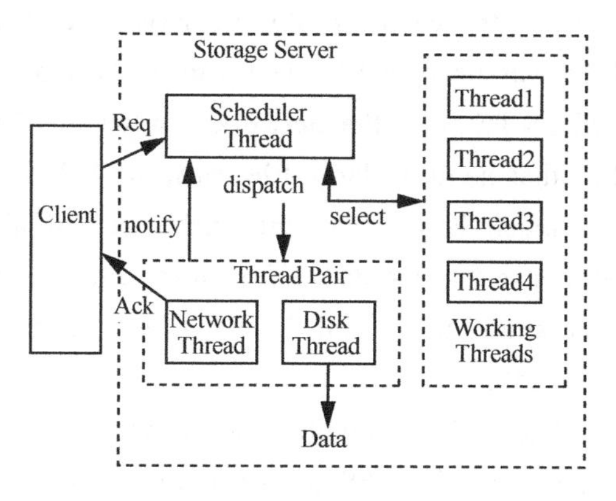


Figure 3. Multi-threaded storage server

## 2.4 State monitoring

In order to discover component failures as soon as possible, a status monitoring mechanism is designed. DCFS organizes the servers and clients into two rings: one for all DCFS clients, another for all DCFS servers. Each one on the rings periodically sends heartbeat messages to its left and right neighbors. If some one couldn't get its neighbor's heartbeat for a specific time, then it will send a message to a special HA-master daemon to inform the abnormity. There is a backup of HA-master running on another node. The backup will take over the master when the HA-master is unreachable.

## 2.5 Communication layer

Cluster file systems are heavily influenced by interconnection technology. In recent years, there have been many efforts in developing high-speed interconnection networks, such as Gigabit Ethernet, Fibre Channel, Myrinet, SCI, etc. And the developing of high performance communication protocols is also in progress, including VIA and IBA. We make an effort to encapsulate inside communication of DCFS into a flexible communication layer, which can not only use stream type communication protocols such as TCP and UDP, but also exploit high-performance communication protocols, such as VIA, IBA, etc.

## 2.6 Management

An agent and a set of utilities are designed to complete the management work, including system configuration, starting and stopping DCFS service, creating a new file system, and reporting the status. The agent process holds all the configuration and status information of DCFS and is in charge of receiving management requests from the administration utilities, passing the requests to relevant DCFS client/server daemons and waiting for the answers.

# 3 Performance evaluation

The performance of DCFS is evaluated as follows:

- Peak performance: It is the highest performance with fixed number of servers;

• Server speedup: It is defined as the peak performance with N servers divided by the peak performance with one server. It shows the performance enhancement with the increase of the number of servers;

• Efficiency: It is defined as the peak bandwidth divided by the total disk bandwidth of N storage servers when disk performance is dominant. Then the cost of DCFS protocol is approximately 1 minus the efficiency. More formally,

$$E = \frac{BW_{Peak}}{BW_{Disk} \times N_{IOS}} \quad (1)$$

$$C = 1 - E \quad (2)$$

where E is efficiency, C is DCFS protocol cost, $N_{IOS}$ is the number of storage servers, $BW_{Disk}$ is the disk I/O bandwidth, and $BW_{Peak}$ is the peak bandwidth with $N_{IOS}$ storage servers.

• Sustainability: It is the maximum number of clients that can be supported simultaneously by a specific number of servers.

Besides, the performance of DCFS and that of PVFS are compared.

### 3.1 Test platform

The performance results of DCFS are gotten on the Dawning 4000-L. The machine consists of 322 nodes, with 2 login nodes, 256 compute nodes and 64 database nodes. The nodes are interconnected through four networks: Fast Ethernet, Gigabit Ethernet, Myrinet and a serial network for diagnosis. Our performance tests were done on 32 compute nodes. Each compute node is a Dawning Tiankuo® R220XP server running Linux 2.4.18-3smp, with 2 2.4 GHz Intel® Xeon™ Processors, 2 Gbyte memory and 3 73 Gbyte SCSI disks. Both DCFS and PVFS ran on the Gigabit Ethernet on which the TCP/IP communication bandwidth is about 106.2 MB/s (measured by netperf for 16 KB messages). The disks are Seagate Ultra320 with model number ST373307LC (with 8 MB data buffer and 2.99 ms average latency). Both sequential read and write bandwidth of the disks are about 60 MB/s (measured by iozone on EXT2).

### 3.2 Aggregate bandwidth

The bandwidth of DCFS is determined by three factors: the network bandwidth, the disk I/O rate and the cost of DCFS protocol. We designed two sets of tests: small-file tests and large-file tests. In small-file tests, all files can be fit in the cache of storage servers so that there was no disk I/O. These tests show how well DCFS uses the network bandwidth. On the other hand, in large-file tests, the I/O size was much larger than the cache size of all servers so that the disk I/O is dominant. These tests show how well DCFS uses the disk bandwidth.

Figure 4 and Figure 5 show the aggregate bandwidths of DCFS for small files and large files respectively. In these tests, DCFS was configured with one metadata server, and each client node ran a test process iozone that read or wrote a DCFS file of specific size, which is $\frac{256 \times N_{IOS}}{N_{Client}}$ Mbytes for small-file tests and $\frac{6 \times N_{IOS}}{N_{Client}}$ Gbytes for large-file tests, where $N_{IOS}$ is the number of storage servers in use and $N_{Client}$ is the number of client nodes in use. The processes access different files.

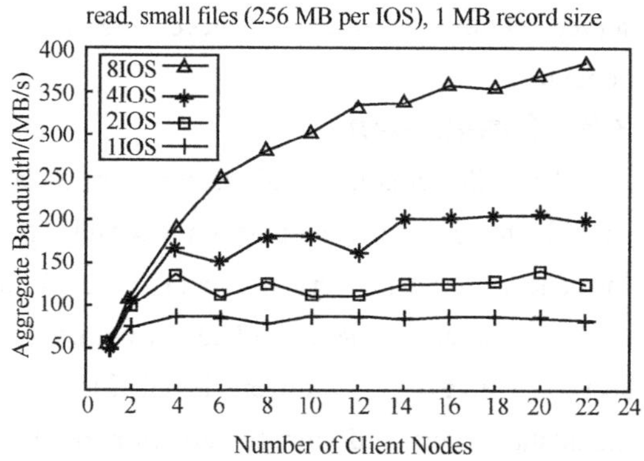

Figure 4. (a) DCFS read bandwidth for small files

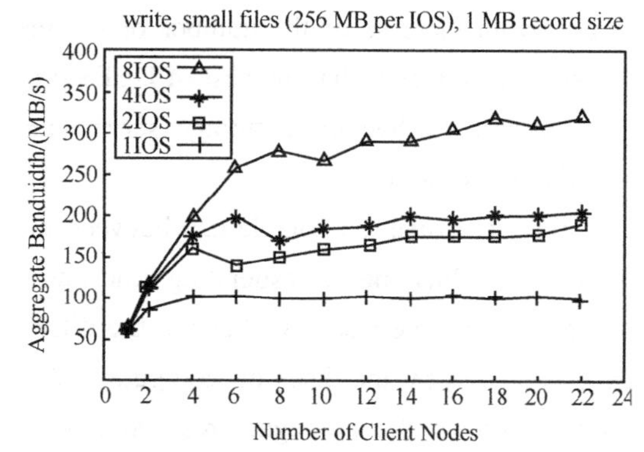

Figure 4. (b) DCFS write bandwidth for small files

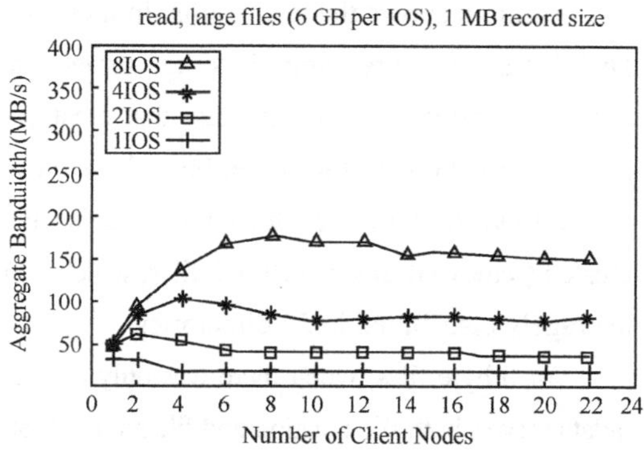

Figure 5. (a) DCFS read bandwidth for large files

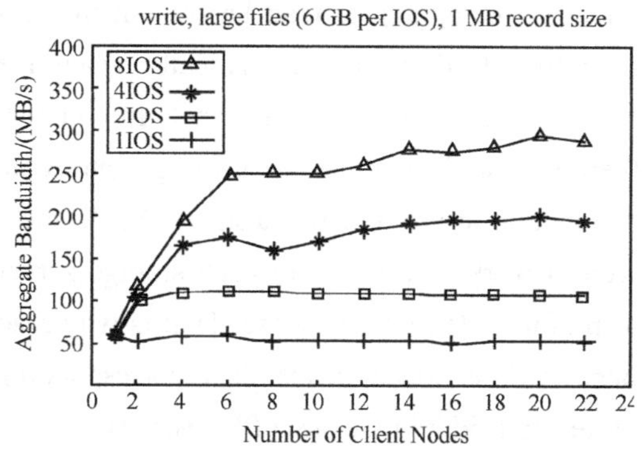

Figure 5. (b) DCFS write bandwidth for large files

Figure 4 is largely similar to Figure 5 except with higher values. And in both situations, one storage server can support more than 22 clients simultaneously without distinct performance decline. The peak bandwidths, speedups and efficiencies in large-file situation are summarized in Table 1. We noted that the performance enhancement is almost proportional to the increase of the number of storage servers in large-file situation. However, the efficiency de-

Table 1. Peak bandwidths, speedups and efficiencies corresponding to Figure 5

| Num. of IOSes | Total RW Size /MB | Read | | | Write | | |
|---|---|---|---|---|---|---|---|
| | | Peak Bandwidth / (MB/sec) | Speedup | Efficiency | Peak Bandwidth / (MB/sec) | Speedup | Efficiency |
| 1 | 6 144 | 31.192 | 1 | 89.28% | 59.578 | 1 | 94.13% |
| 2 | 12 288 | 60.790 | 1.95 | 86.98% | 111.725 | 1.88 | 88.26% |
| 4 | 24 576 | 100.347 | 3.22 | 71.81% | 198.84 | 3.34 | 78.54% |
| 8 | 49 152 | 178.361 | 5.72 | 64.82% | 292.576 | 4.91 | 57.78% |

Disk bandwidth for 4 threads to access 6 Gbytes is about 34.943 MB/sec for read, and about 63.292 MB/sec for write

creases with increase of the number of storage servers, which means that the cost of DCFS protocol increases when more storage servers are added to the system.

We also noted the notable gap between the write bandwidths and corresponding read bandwidths in large-file tests. We believe the difference is caused by the disks on compute nodes. Table 2 shows the aggregate read and write bandwidths of the disk for various numbers of threads. The aggregate read bandwidth of multiple threads decreased dramatically when the number of threads was more than 1. In contrast, there was no distinct decline for aggregate write bandwidth. In performance tests of DCFS, there were 4 working threads for each storage server to perform file read and write. That is why read bandwidth is much lower than corresponding write bandwidth in the large file tests. This result suggests that the number of threads on storage servers should be adjustable according to disk behavior.

### 3.3 Throughput

The file creation and removal rates are suitable to show the metadata processing performance. In DCFS, creation and removal for 0-byte files are done completely by metadata servers. However, for nonzero-length files, both metadata servers and storage servers participate in the operations.

Figure 6 shows the test results. In these tests, DCFS was configured with 4 storage servers, and each client node ran a test process that created or removed 50 000 byte files or 50 004 Kbyte files in a directory. And for 22 client nodes, these tests created (removed) totally 110 000 files, which were averagely distributed in 22 directories.

For 0 byte files, when there was only one metadata server, both file creation and file removal rates reached the peaks at 10 client nodes, 545.07 files/sec

Table 2. Disk bandwidths with various numbers of threads

| Num. of Threads | Total Size /(MB) | Read Bandwidth /(MB/sec) | Write Bandwidth /(MB/sec) |
|---|---|---|---|
| 1 | 6 144 | 66.800 | 63.193 |
| 2 | 6 144 | 41.476 | 63.902 |
| 4 | 6 144 | 34.943 | 63.292 |
| 6 | 6 144 | 33.860 | 62.789 |
| 8 | 6 144 | 33.880 | 62.709 |
| 10 | 6 144 | 33.850 | 61.661 |
| 12 | 6 144 | 33.399 | 61.809 |
| 14 | 6 144 | 33.155 | 60.922 |
| 16 | 6 144 | 34.107 | 61.355 |

The bandwidths are sequential read/write bandwidths measured by iozone on EXT2. Total access size is 6 Gbytes, and record size is 16 Kbytes

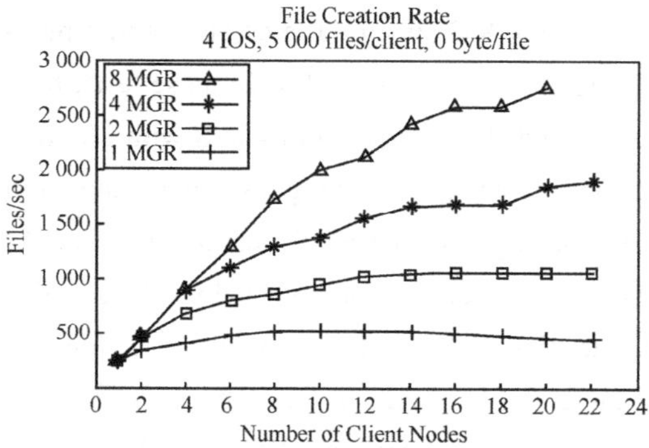


Figure 6. (a) DCFS file creation rate for 0 byte files

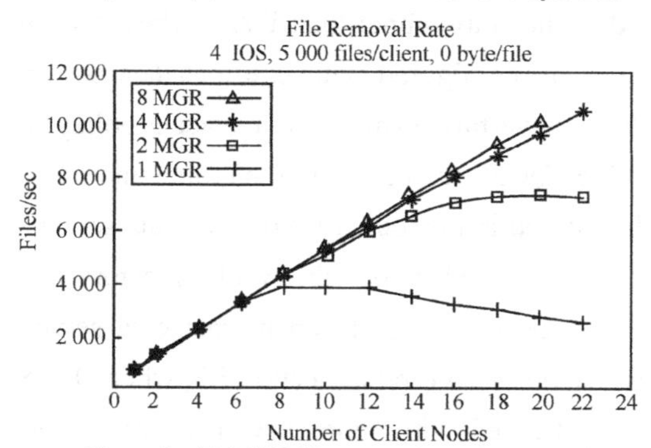


Figure 6. (b) DCFS file removal rate for 0 byte files

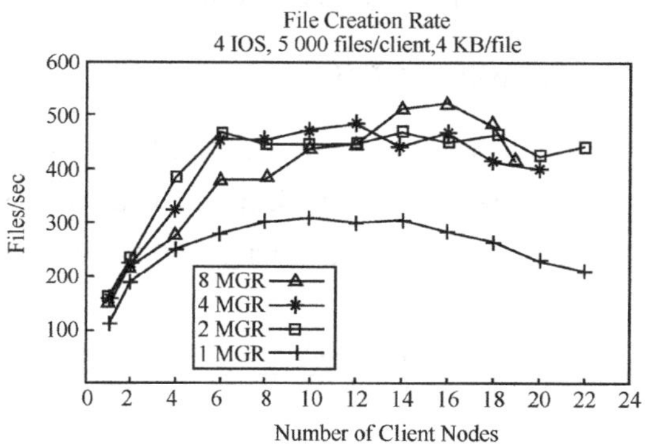


Figure 6. (c) DCFS creation throughput for 4 096 Kbyte files

for creation and 3 903.97 files/sec for removal. When the number of metadata server was more than 2, the tests did not reach the peak rates within 22 client nodes.

The file creation rate for 4 Kbyte files was much lower than that for 0 byte files by comparing Figure 6(c) with Figure 6(a) The file creation rate for 4 Kbyte files enhanced approximately 50% when the number of metadata servers increased from 1 to 2, but the file creation rates with 4 or 8 metadata servers were not higher than that with 2 metadata servers. We believe it is because the performance bottleneck in this situation is file write on storage servers. And the flat organization of files on storage servers, that is, all the files on storage servers are put into a single directory, is another reason. The figure of file removal rates for 4 Kbyte files is not presented here, because it is almost same as Figure 6(b).

## 3.4 Comparison

We compared the performance of DCFS with that of PVFS version 1.5.4. There are many differences between DCFS and PVFS. There are multiple metadata servers in DCFS, while there is only one metadata server in PVFS. DCFS implements multi-threaded storage server. However, PVFS's storage servers (IODs in term of PVFS) are single-threaded.

Figure 7 shows I/O bandwidth of DCFS and PVFS. In these tests, both DCFS and PVFS were configured with 8 storage servers and one metadata server. And both of them used the Gigabit Ethernet for communication. In Figure 7(a), each client node ran a test process that read/wrote a file of 2 $048/N_{\text{Client}}$ Mbytes, where $N_{\text{Client}}$ is the number of client nodes. Similarly, in Figure 7(b), each test process read/wrote a file of 49 $152/N_{\text{Client}}$ Mbytes.

For small files, read bandwidth of PVFS is higher than that of DCFS. However, for large files, read bandwidth of PVFS is lower than that of DCFS. Moreover, write bandwidth of DCFS in either situation is higher than that of PVFS. The higher bandwidth of DCFS for large files indicates DCFS exploits disk I/O better than PVFS does, because storage servers in DCFS are multi-threaded so that disk accesses are overlapped with network transfers.

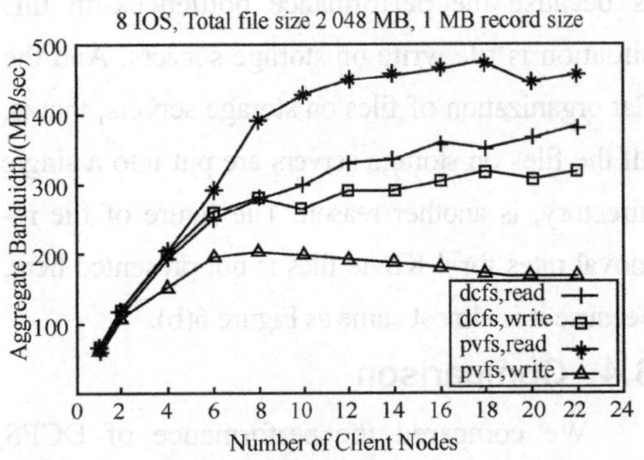


Figure 7. (a) Bandwidth of DCFS and PVFS for small files

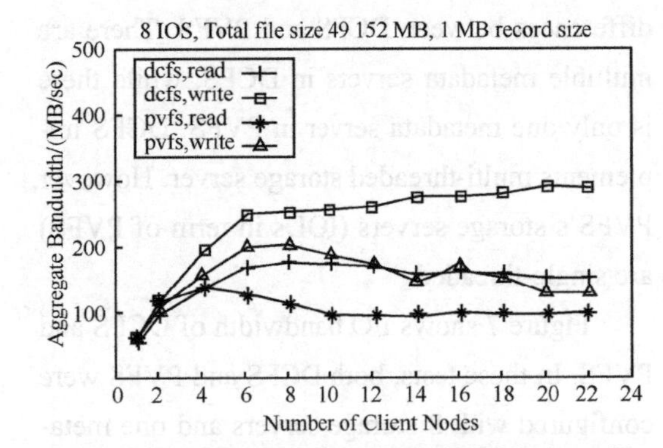


Figure 7. (b) Bandwidth of DCFS and PVFS for large files

Figure 8 shows the file creation and removal rates of DCFS and PVFS. In these tests, both DCFS and PVFS were configured with one metadata server and 4 storage servers. And both used the Gigabit Ethernet for communication. As shown in Figure 8, for both 0 byte files and 4 Kbyte files, DCFS exhibits much higher file creation and file removal rates than PVFS does. These results indicate that DCSF is able to provide better throughput than PVFS when both of them are configured with one metadata server. We believe this owns to the metadata policy of DCFS. DCFS maintains and interprets the metadata stored in the local files on metadata servers. However, PVFS builds the whole directory tree in the local file system on its metadata server. Moreover, according to Section 3.3, when DCFS is configured with more metadata servers, its throughput can be enhanced greatly.

# 4 Conclusions

In this paper, we have described the design and implementation of the Dawning Cluster File System (DCFS) for Dawning 4000-L Linux

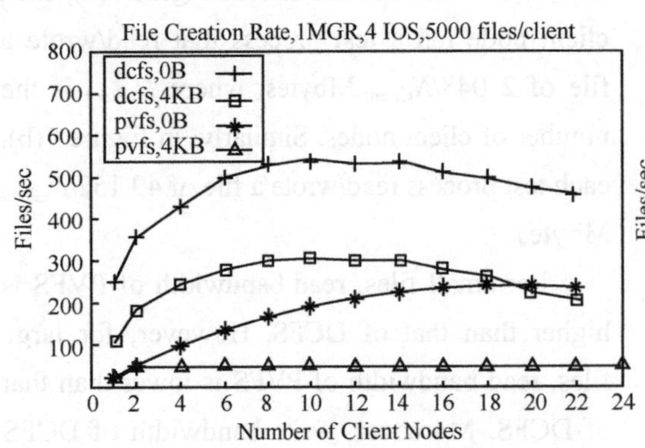


Figure 8. (a) File creation rate of DCFS and PVFS for 5 000 files per client

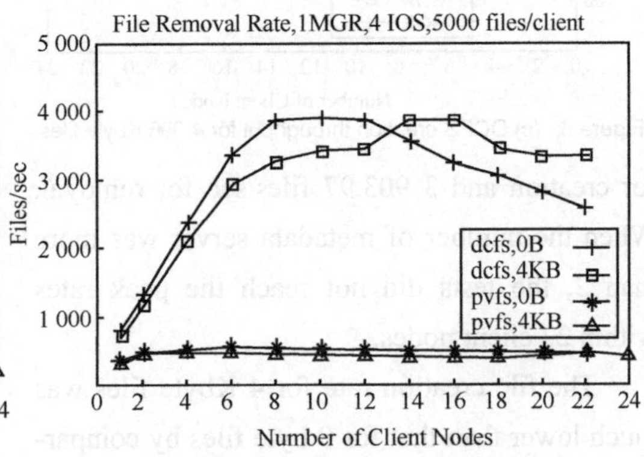


Figure 8. (b) File removal rate of DCFS and PVFS for 5 000 files per client

cluster. In order to achieve scalability and high performance, DCFS uses different policies for file data access and metadata access. File data are stored on multiple storage servers in a RAID-0 style of striping. Metadata are distributed among multiple metadata servers in a first-level subtree approach. Storage servers are multi-threaded to overlap disk access with network transfer.

The performance of DCFS on Dawning 4000-L is encouraging. DCFS exhibits satisfactory speedup for both bandwidth and throughput. And compared with PVFS, DCFS exhibits higher bandwidth for large files, which owns to the multi-threaded storage servers of DCFS, and much higher file creation and removal rates in all situations, which owns to the metadata policy of DCFS.

Compared with the previous cluster file system COSMOS[11] for Dawning 2000 and Dawning 3000, DCFS has similar architecture. However, DCFS has very different implementation policies, including the metadata policy, the file striping policy, the communication mechanism, disk access mechanism, the caching policy and the management policy.

## 5 Future work

More performance analysis of DCFS is needed, especially under heavy-load conditions of large numbers of clients, and under conditions of 8 or more storage servers.

As the performance tests shown, the efficiency of DCFS is not satisfactory when the number of storage servers is 8 or more. We believe this is because of the cost of DCFS protocol, which needs to be further improved.

Although multiple metadata servers in DCFS improve the metadata processing performance, they complicated the failure recovery. It is difficult to ensure the metadata on all metadata servers consistent when failures occur. We are working on this problem.

The metadata distribution and workload may unbalance according to the current metadata distribution policy in DCFS, so better metadata distribution policy is needed.

In order to support high-performance scientific computing applications, we plan to implement a user level MPI-IO library on top of DCFS. This will greatly improve the bandwidth for MPI-IO based applications.

## Acknowledgements

We would like to thanks Guo-zhong Sun, Rong-feng Tang and Zhi-hua Fan for helping us to do the performance tests.

## References:

[1] HWANG K, XU Z W. Scalable parallel computing: technology, architecture, programming[R]. The McGraw-Hill Companies, 1998.

[2] VAHALIA U. UNIX internals: the new frontiers[R]. Prentice-Hall, Inc., 1995.

[3] SOLTIS S R, RUWART T M, O'KEEFE M T. The global file system[C]//The 5th NASA Goddard Conference on Mass Storage Systems and Technologies. [S.l.:s.n.], 1996: 319-342.

[4] ANDERSON T E, DAHLIN M D, NEEFE J M, et al. Serverless network file systems[C]//The 15th ACM Symposium on Operating Systems Principles. [S.l.:s.n.], 1995: 109-126.

[5] SCHMUCK F, HASKIN R. GPFS: a shared-disk file system for large computing clusters[C]//The 1st USENIX Conference on File and Storage. New York: ACM Press, 2002: 231-244.

[6] CARNS P H, LIGON III W B, ROSS R B, et al. PVFS:

a parallel file system for Linux clusters[C]//The 4th Annual Linux Showcase and Conference. New York: ACM Press, 2000: 317-327.

[7] O'KEEFE M T. Shared file systems and fibre channel[C]//The 6th Goddard Conference on Mass Storage Systems and Technologies in Cooperation with the 15th IEEE Symposium on Mass Storage Systems. [S.l.:s.n.], 1998: 1-16.

[8] IBM Tivoli SANergy administrator's guide version 3 release 2[R]. IBM Corporation, 2002.

[9] CXFSTM: a clustered SAN file system from SGI[Z]. Silicon Graphics, Inc., 2001.

[10] SPEC. SFS 3.0 Documentation Version 1.0[R]. Standard Performance Evaluation Corporation, 2001.

[11] DU C, XU Z W. Performance analysis of a cluster file system[C]//The 2000 International Conference on Parallel and Distributed Technologies and Applications (PDPTA'2000). [S.l.:s.n.], 2000: 1933-1938.

发表于 *IEEE Transactions on Parallel and Distributed Systems*, 2011 年第 5 期

# Metadata distribution and consistency techniques for large-scale cluster file systems

Jin XIONG, *Member, IEEE Computer Society*, Yi-ming HU, *Senior Member, IEEE*, Guo-jie LI, Rong-feng TANG, Zhi-hua FAN

**Abstract:** Most supercomputers nowadays are based on large clusters, which call for sophisticated, scalable, and decentralized metadata processing techniques. From the perspective of maximizing metadata throughput, an ideal metadata distribution policy should automatically balance the namespace locality and even distribution without manual intervention. None of existing metadata distribution schemes is designed to make such a balance. We propose a novel metadata distribution policy, Dynamic Dir-Grain (DDG), which seeks to balance the requirements of keeping namespace locality and even distribution of the load by dynamic partitioning of the namespace into size-adjustable hierarchical units. Extensive simulation and measurement results show that DDG policies with a proper granularity significantly outperform traditional techniques such as the Random policy and the Subtree policy by 40 percent to 62 times. In addition, from the perspective of file system reliability, metadata consistency is an equally important issue. However, it is complicated by dynamic metadata distribution. Metadata consistency of cross-metadata server operations cannot be solved by traditional metadata journaling on each server. While traditional two-phase commit (2PC) algorithm can be used, it is too costly for distributed file systems. We proposed a consistent metadata processing protocol, S2PC-MP, which combines the two-phase commit algorithm with metadata processing to reduce overheads. Our measurement results show that S2PC-MP not only ensures fast recovery, but also greatly reduces fail-free execution overheads.

**Key words:** distributed file systems, metadata management

## 1 Introduction

Metadata processing is a key issue for large-scale cluster file systems. Metadata are the data that describe the organization and structure of a file system, usually including directory contents, file attributes, file block pointers, organization and state information of physical space, etc. Metadata processing includes the maintenance of not only the namespace, but also file attributes and locations of file blocks. Although the amount of metadata is small, metadata operations account for over 60 percent of the operations in typical workloads[1]. Therefore, the efficiency of metadata processing will ultimately affect the overall file system performance.

Large clusters nowadays are at a scale of thousands to tens of thousands of nodes per cluster, with an order of magnitude more CPU cores. As their capacity and user requirements increase, so is the demand for high-performance distributed file systems on these clusters. According to Dayal's statistical data[2], which cover over on 13 file systems from five supercomputing sites and two file systems from department file servers, the total number of files as well as directories of each file system is several millions to tens of millions. Moreover, the processes running on each CPU core issue I/O accesses independently, resulting in large numbers of concurrent accesses to the file systems. Therefore, file systems on large-scale clusters are required to handle tens of thousands of files' metadata per second. These requirements indicate that centralized metadata processing that relies on a single metadata server is impractical and it is necessary to use decentralized metadata processing[3-7] that utilizes a group of metadata servers (MDS).

Namespace partitioning and metadata distribution are the key issues in decentralized metadata processing that determine metadata processing throughput. Many existing systems use static partitioning[8-10] which relies on administrators to manually partition the namespace and distribute the metadata. This static method is obviously very difficult to maintain for large-scale clusters. Although a lot of systems use dynamic partitioning, they either emphasize on keeping namespace locality[6,11-12] or focus on equal distribution[5,13-15], and hence, neglect the tradeoff between the two aspects. As a result, these systems cannot make full use of all metadata servers' processing and storage resources. In contrast to these systems, we propose a metadata distribution policy that dynamically partitions the namespace into size-adjustable hierarchical units and achieves a good balance between the namespace locality and equal distribution.

In addition, metadata consistency (the atomicity of a metadata operation) is complicated by cross-metadata server operations (known as distributed operations, see detailed discussions in Section 2.1.1). Individual journaling metadata modifications on each server, as in journaling file systems[16], cannot ensure consistency of distributed operations. To address this problem, existing systems either ensure metadata consistency by using the high-overhead two-phase commit (2PC) protocol at the sacrifice of performance[17-18], or pursue high performance at the cost of relaxing the consistency model by allowing orphan objects[1] in the namespace[5,19].

This paper makes two contributions. First, we present a dynamic metadata distribution policy called dynamic dir-grain policy (DDG) to improve the metadata throughput by using a triple-defined distribution granularity to dynamically partition the namespace and balance metadata distribution and the namespace locality. Our experimental results show that DDG policies with proper granularity setting outperform two traditional policies-the Random policy and the Subtree policy by 40 percent to 62 times under metadata intensive workloads.

Second, we propose a distributed metadata processing policy called S2PC-MP that deals with metadata consistency for distributed operations. By combining the two-phase commit protocol with metadata operations, S2PC-MP reduces the overhead of failure-free execution and ensures quick recovery in the presence of metadata server failures or client failures. Our performance results show that only the creation and deletion throughputs decreased about 10 percent,

[1] We refer to files or directories in a namespace as objects in this paper.

and the read/write bandwidth and transaction throughputs were almost unaffected. Recovery can be completed within tens of seconds.

The rest of this paper is organized as follows: In Section 2, we summarize previous research work and techniques. We introduce the new metadata distribution policy in Section 3 and the S2PC-MP technique in Section 4. In Section 5, we evaluate the performances of the proposed solutions and compare them with those of two well-known policies. We conclude the paper in Section 6.

# 2 Related work

Decentralized metadata processing (DMP) is necessary for large-scale distributed file systems to provide scalable performance. No matter metadata are stored on each metadata server's private storage[5,8-9,11,17] or on a storage system shared by all metadata servers[4,6-7], DMP partitions the namespace among the servers. Metadata distribution and metadata consistency are two key issues that not only affect the metadata performance, but also determine file system reliability.

## 2.1 Metadata distribution policies

In essence, a metadata distribution policy first partitions the namespace into distribution units of smaller size, and then maps these units to the metadata server set.

### 2.1.1 Static partitioning versus dynamic partitioning

Metadata distribution policies can be divided into two types according to how the namespace is partitioned.

In static partitioning, the namespace is manually partitioned by administrators according to their experiences and target application requirements. Early distributed file systems, such as NFS, Andrew[8], Sprite[9], and some recent distributed file systems, such as CXFS[10] and StorageTank[4], use static partitioning.

In dynamic partitioning, the namespace is automatically partitioned and mapped, usually at the time of object creation in the namespace. So in dynamic partitioning, there are distributed metadata operations in addition to ordinary metadata operations. A distributed metadata operation (or distributed operation for short) must be performed by collaboration of two or more metadata servers, while an ordinary metadata operation (or ordinary operation for short) can be performed by just one metadata server. For example, when creating a new file *f*1 under a directory *d*1 which is managed by server1, a dynamic partitioning policy may decide that *f*1 is managed by server2, and hence, the create operation needs to request both server1 and server2. File systems using dynamic partitioning include xFS[13], Slice[17], PVFS2[5], Kosha[11], DCFS[12], GIGA+[20], etc.

Obviously, dynamic partitioning greatly reduces the maintenance cost in these file systems. Static partitioning requires administrators to perform too much maintenance work, including namespace partitioning and mapping in the beginning, and constant readjusting during the whole life of the file systems.

Our DDG policy belongs to dynamic partitioning, and it does not require human intervention in namespace management.

### 2.1.2 Partition granularity

Metadata distribution policies can be divided into three types according to their partition granularity.

In subtree partitioning, the namespace is partitioned into several subtrees, which are then mapped to the metadata servers. The file systems employing static partitioning of namespace usually use subtree partitioning. Examples include NFS, Andrew[8], Sprite[9], CXFS[10], and StorageTank[4]. This type of partitioning is known as static subtree partitioning. Some file systems, such as Kosha[11], IFS[18], DCFS[12], and Ceph[6], also use dynamic subtree partitioning. Although dynamic subtree partitioning overcomes the management problem of the static subtree partitioning, both of them suffer from the problem that their distribution granularity is too large to balance both the resource usage and the workload among the servers.

In single object partitioning, the namespace is partitioned into individual objects (files or directories), which are then mapped to metadata servers through a random algorithm. xFS[13], HBA[14], G-HBA[15], PVFS2[5], etc. use this method. This approach can achieve better load balance than subtree partitioning because of the smallest partition granularity, but the major limitation of this method is that there are too many branch points (to be discussed in Section 3.1), and hence, too many distributed metadata operations to achieve high metadata throughput.

In directory segment partitioning, each directory in the namespace is partitioned into fixed-sized segments through extendible hashing, and these segments are mapped to metadata servers. GIGA+[20] and SkyFS[21] use this method. The idea of partitioning a single directory is motivated by current requirements of very large directories with millions of entries in each directory. Directory segment partitioning is targeted to solve the problem of a very large single directory. However, for many file systems in super-computing centers, 90 percent directories have less than 64 entries[2]. For these cases, this method is less effective. An even bigger issue is that it is difficult to support POSIX readdir operation, which is widely used in supercomputing centers.

Different from the above methods, our DDG policy dynamically partitions the namespace into smaller hierarchical units. The novelty of DDG policy is that it provides a way to control distribution granularity, by which it can keep namespace locality for the majority of objects after namespace partitioning. Therefore, it can achieve better metadata throughput than subtree partitioning and single object partitioning. Moreover, DDG policy supports most POSIX file operations including readdir, and also partitions directories, hence can deal with both large and small directories. Therefore, it is a more general methodology than directory segment partitioning.

#### 2.1.3 Mapping method

In general, there are three methods to map each unit to a metadata server.

Manual mapping (NFS, Andrew[8], Sprite[9], etc.) relies on administrators to perform the mapping and uses a static table to maintain the mapping. Apparently, too much manpower and lack of flexibility are the main drawbacks of such a method.

Random mapping (xFS[13], PVFS2[5], HBA[14], ANU randomization[22], G-HBA[15], etc.) uses a random algorithm to automatically map units to servers. Additionally, xFS uses a table to maintain the mapping, while HBA uses a Bloom Filter-based method and significantly reduced the memory overhead. However, in the former three systems, the mapping cannot adjust automatically with the change of access loads or server configuration. On the contrary, ANU randomization also remaps units to servers according to observed request latencies, and G-HBA maintains additional information to support the server configuration changes.

Hashing-based mapping (LazyHybrid[23], GIGA+[20] etc.) uses a simple hash algorithm to automatically map units to servers. Such a simple hashing makes them less adaptable to the changes of server configurations. LazyHybrid eliminates hierarchical relationships of objects, resulting in a severe performance decline for some operations, such as readdir, renaming a directory, and changing the mode or owner of a directory.

DDG policy uses random mapping. Our system retains the simplicity and even distribution of random mapping and avoids the difficulty of maintaining the mapping by encoding each object's location (the metadata server ID) into its unique ID (inode number).

## 2.2 Metadata consistency

A file system operation consists of a sequence of indivisible suboperations, each of which modifies a single piece of metadata. Therefore, a file system operation is similar to a transaction and has ACID properties[24]. Metadata consistency refers to a metadata operation's result is either it successfully completed or never happened. This issue is important since a metadata operation usually modifies multiple data structures. In native file systems, inconsistency is mainly caused by metadata caching if different types of metadata are cached in different memory locations, and written back individually. In distributed metadata processing, inconsistency is caused not only by metadata caching but also by distributed operations. For example, the create operation creates a new file *f*1 under a directory *d*1. If *d*1 and *f*1 are maintained by two distinct servers ($S_1$ and $S_2$, respectively), $S_1$ and $S_2$ modify different data structures, respectively. If either $S_1$ or $S_2$ crashes, the result on this server may be lost, thus may lead to inconsistent metadata on these two servers. The metadata consistency issue that this paper intends to address is to make an entire metadata operation atomic across multiple metadata servers.

Journaling file systems[16] ensure metadata consistency by logging metadata updates. Many distributed file systems, such as NFS, Calypso[25], CXFS[10], and Lustre[26], have a single metadata server, and hence, can rely on server-side journaling file systems to ensure metadata consistency. In addition, some distributed file systems, such as Sprite[27], StorageTank[4], GPFS[28], and GFS[29], have no distributed metadata operations, so they can ensure metadata consistency by logging metadata updates on individual servers.

Individual logging on each server cannot ensure consistency of distributed operations each of which requires collaboration of two or more servers. The 2PC[30-32] in distributed transaction processing is used by some distributed file systems, such as Slice[17] and IFS[18], in their metadata processing. However, the main concern of using 2PC in metadata processing is its high fail-free execution overhead introduced by two rounds of messages as well as multiple disk writes in a single commit. It is unclear from the literature that any optimization efforts were undertaken in these systems to reduce the high overheads of 2PC.

To avoid 2PC's high overhead and complexity, neither DiFFS[19] nor PVFS2[5] uses 2PC, although they have distributed metadata operations. However, both of them may have "orphan" objects after server failures and they ensure "orphan" objects will not be seen or accessed by users. They achieve high performance by relaxing the consistency model, thus allowing orphan objects.

Ursa Minor avoids 2PC by migrating the required metadata to a single metadata server and executing the operation on that server[7]. This approach simplified the implementation of multiserver operations. However, the performance overheads for multiserver operations are high due to metadata migration. Since its target environments have much few multiserver operations (less than 1 percent), the performance loss is acceptable.

We propose a novel scheme called S2PC-MP to deal with the high failure-free execution overhead of 2PC in metadata processing. By combining 2PC with each metadata operation instead of starting 2PC at the end of each operation, our scheme significantly reduces failure-free execution overhead by reducing the number of messages between collaborative servers and by reducing the number of disk writes.

Some systems, such as the GFS[33], use metadata replication to ensure metadata reliability and availability, and the primary copy protocol to make multiple replicas consistent. Both 2PC and primary copy serialize metadata at a single node for modification operations. However, metadata replication does not partition the namespace, and GFS uses a centralized metadata management scheme. Through replication, GFS allows multiple servers to service read requests, but all write requests must be serviced by the master node, which is a potential bottleneck for scalability. We believe that future applications need both metadata replication and metadata distribution. The former improves availability, while the latter improves scalability. We plan to integrate both methods in metadata management in our future work.

## 3 Dynamic dir-grain policy

We designed a dynamic metadata distribution policy to fully exploit all metadata servers' processing and storage resources.

### 3.1 Namespace locality

The hierarchical structure of file system namespace means that there is some kind of relationship among the objects in it. Many metadata operations, such as lookup, create, remove, mkdir, rmdir, and rename, access or modify not only the target object itself, but also its parent directory. If the target and its parent are maintained by the same metadata server, then those operations can be performed by just this server. Otherwise, two or more servers are needed. We call an object in the namespace a branch point if its parent and it are maintained by two different metadata servers. A branch point is either a directory or a file. And we call the former a branch directory and the latter a branch file. Branch points break up the relationships of the objects and weaken the namespace locality.

Branch points are inevitable in decentralized metadata processing, otherwise all files and directories in the namespace have to be managed by the same metadata server. Inducing branch points is an intrinsic nature of metadata distribution. However, branch points weaken the name-space locality, causing more processing overheads.

Metadata throughput is the total number of operations processed by all the metadata servers per second. It is affected by the following two factors:

- The degree of balance of object distribution. Very imbalanced distribution causes some metadata servers to become too busy to process a large number of requests, while leaving some other servers mostly idle.
- The number of branch points. More branch points result in more distributed metadata operations, which are much more expensive than ordinary operations. Each distributed operation needs to interact with other metadata servers to complete the required operation. Moreover, distributed operations (such as file creation and deletion) which modify the namespace need 2PC to ensure metadata consistency. As discussed in Section 4.2, each distributed modification operation needs three network messages and four synchronous disk writes of metadata log, while an ordinary modification operation needs no network message and only one asynchronous disk write of metadata log. A distributed lookup needs an inter MDS message, while an ordinary lookup does not. Therefore, more branch points not only lead to longer latencies of the operations, but also reduce metadata throughput by causing heavier network traffic and more disk writes.

These two factors are often incompatible. For example, a round-robin policy that tries to evenly distribute objects across multiple servers is likely to generate a large number of branch points. To obtain optimal performance, it is important to seek a balance between them.

### 3.2 Distribution granularity

The size of a distribution unit is known as distribution granularity. A smaller granularity results in more even distribution but more branch points, while a larger granularity leads to fewer branch points but less even distribution. The distribution unit of single object partitioning (discussed in Section 2.1.2) is a single object, so its granularity is 1, the smallest possible value. These policies emphasize on even distribution, at the cost of a large number of branch points. On the other hand, the distribution unit of subtree partitioning is a whole subtree, so its granularity is much larger. Subtree policies emphasize on keeping the namespace locality, which results in imbalanced distribution.

In order to find a compromise between the two factors, we introduce D-D-F granularity, which defines a hierarchical unit in the namespace by a triple <DirDep, DirWid, FileWid>. DirDep stands for directory depth, which defines the maximum number of levels of directories in a distribution unit. DirWid stands for directory width, which defines the maximum number of child directories[2] of each directory in a distribution unit. FileWid stands for file width, which defines the maximum number of child files of each directory in a distribution unit. Figure 1

[2] Child directories of a directory "P" are those directories whose parent is P. And child files of a directory "P" are those files whose parent is P.

shows an example of a distribution unit with a granularity *G*=<3,2,4>.

Uppercase stands for directory, while lowercase stands for file. Directory A1 is the beginning of unit *Ω*1. All files or directories inside the dashed circle belong to unit *Ω*1. Files and directories outside the dashed circle belong to other units.

The existing partitioning policies described in Section 2.1.2 can also be viewed as special cases of certain D-D-F granularities. For example, the granularity of single object partitioning is <1,1,1>. And the granularity of directory segment partitioning is <1,*m*,*n*>, where *m*+*n* equals to segment size. The root subtree policy uses two different granularities: <1,1,1> for root and <∞,∞,∞> for other directories. Kosha's subtree policy also uses two granularities: <1,1,1> for objects in levels higher than or equal to the distribution level, and <∞,∞,∞> otherwise.

## 3.3 Distribution algorithm

Our distribution policy (DDG) assigns a metadata server to an object when the object is created. Suppose *D* is a directory in unit *Ω*. Suppose that *D*' is a directory created most recently under *D*, and *D*' belongs to unit *Ω*'. And suppose that *f* is a file created most recently under *D*, and *f* belongs to unit *Ω*''. The values related with *D* include:

- dir_depth: the depth of *D* in unit *Ω*,
- dir_mds: the metadata server that maintains unit *Ω*',
- dir_num: the number of child directories of *D* in unit *Ω*',
- file_mds: the metadata server that maintains unit *Ω*'',
- file_num: the number of child files of *D* in unit *Ω*''.

For example, for A1 in Figure 1, its dir_depth=1, its dir_mds=MDS(*Ω*2), its dir_num=1, its file_mds=MDS(*Ω*2), and its file_num=1. For C1 in Figure 1, dir_depth=3, its dir_mds=MDS(*Ω*3), its dir_num=1, its file_mds=MDS(*Ω*3), and its file_num=1.

Suppose the distribution granularity is <DirDep, DirWid, FileWid>, and a user wants to create a new object *X* under directory *P*. Five conditions should be taken into account when assigning a metadata server for object *X*.

1. If *X* is a file, and its parent *P*'s file_

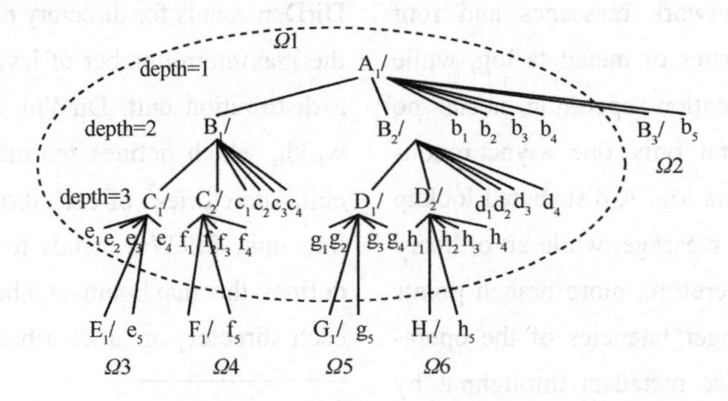


Figure 1. The largest distribution unit with a granularity *G*= <3,2,4>

num+1 ≤ FileWid, then assign *P*'s file_mds to X.

2. If *X* is a file, and its parent *P*'s file_num+1>FileWid, then choose a metadata server $MDS_i$ according to certain random function, and assign $MDS_i$ to *X*.

3. If *X* is a directory, and its parent *P*'s dir_depth+1 ≤ DirDep and *P*'s dir_num +1 ≤ DirWid, then assign *P*'s dir_mds to *X*.

4. If *X* is a directory, and its parent *P*'s dir_depth+1 ≤ DirDep and *P*'s dir_num+1> DirWid, then choose a metadata server $MDS_j$ according to certain random function, and assign $MDS_j$ to *X*.

5. If *X* is a directory, and its parent *P*'s dir_depth+1>DirDep, then choose a metadata server $MDS_k$ according to certain random function, and assign $MDS_k$ to *X*.

After assigning the metadata server for the new object *X*, DDG also modifies the five values of *P* and *X*, accordingly. The pseudocode of the DDG algorithm is depicted in Figure 2.

```
If (X is a file) then
    If (P.file_num+ 1<= GFileWid) then
        Choose P.file_mds to be X's server
        P.file_num←P.file_num+1
    else
        randomly select a server MDS_i to be X's server
        P.file mds ← MDS_i
        P.filc_num ←1
    endif
else if (X is a directory) then
    if (P.dir_depth +1 > G DirDep) then
        randomly select a server MDS_j to be X's server
        P.dir_mds ←MDS
        P.dir_num ←1
        X.dir_depth ←1
        X.dir_mds←MDS_j
        X.dir_num←0
        X.file_mds ←MDS_j
        X.file_num←0
    else if (P.dir_num+1<=G DirWid) then
        Choose P.dir_mds to be X's server
        P.dir_num ← P.dir_num+1
        X.dir_depth←P.dir_depth+l
        Xdir_mds←P.dir_mds
        X.dir_num←0
        X.file_mds←P.file_mds
        X.file_num ←0;
    else
        randomly select a server MDS_k to be X's server
        P.dir_mds ←MDS_k
        Pdir_mum←1
        X.dir_depth ←1
        X.dir_mds←MDS_k
        X.dir_num←0
        X.file_mds←MDS_k
        X.file_num←0
    endif
endif
```

Figure 2. The DDG algorithm

The performance of DDG policy depends on the values of D-D-F, and the proper values of D-D-F depend on the namespace characteristics of the target file system, access workloads, and the metadata servers' processing ability. Our evaluation in Section 5 arrived at a set of reasonable values for DDF based on our data sets and system configuration. These values are applicable for similar environments. Since in different systems, the namespace structures, the access workloads, and metadata servers are very different, there are no such values of D-D-F that are suitable for all environments. Therefore, a preferable way is to adjust the values of D-D-F adaptively by DDG policy itself with the growth of the namespace, and changes of access workloads and server configuration. However, current DDG does not support such adaptation and the values of D-D-F can only be readjusted by administrators.

## 4 Reliable distributed metadata processing protocol

Since 2PC is the basis of our protocol, we first give a brief introduction to 2PC in this section.

### 4.1 Two-phase commit protocol (2PC)

Atomic commitment protocols are proposed

to ensure consistency of distributed transactions, especially in the presence of failures[31]. And 2PC protocol[30-32] is the simplest and most popular atomic commitment protocol. The sites involved in a distributed transaction usually play two different roles. The site that initiates the transaction is called the coordinator, while the remaining sites are called participants. Both the coordinator and all participants write down their progress by writing log records to stable storage in the course of commitment.

The first phase of 2PC is called voting phase, in which the coordinator sends a VOTE-REQ message to all participants to ask them if they agree to commit. Meanwhile, it writes the "Start-2PC" record in its log. And a participant writes the "Yes" record in its log if it votes Yes and sends a VOTE-Yes message back to the coordinator. Otherwise, it writes the "Abort" record and sends a VOTE-No message back. The participants that vote Yes then wait for the next message from the coordinator.

The second phase is called commitment phase. After the coordinator has collected the reply messages (either VOTE-No or VOTE-Yes) from all participants, it then decides Commit by writing the "Commit" record in its log if all votes are VOTE-Yes, and then sends COMMIT-REQ messages to all participants. Otherwise, the coordinator decides Abort by writing the "Abort" record in its log and sends ABORT-REQ messages to all participants that vote yes. A participant writes the "Commit" record in its log if it receives "COMMIT-REQ". Otherwise, it writes the "Abort" record in its log. And the participant sends an acknowledgement message to the coordinator.

Suppose there is just one participant besides of the coordinator, in normal case when no failure occurs in the course of transaction commitment, 2PC needs four messages and five disk writes of log records. In 2PC protocol, both the coordinator and participants that vote yes wait messages. The coordinator waits votes and acknowledgement messages from all participants, while the participants wait the VOTE-REQ and the final decision from the coordinator. The waiting message will never arrive if failures occur in the course of commitment. To address this problem, 2PC protocol uses timeout mechanism and the cooperative termination protocol[31].

## 4.2 S2PC-MP protocol

Since operations that need more than two metadata servers are very scarce, S2PC-MP only deals with operations that need exactly two metadata servers[3]. For convenience, any distributed operation can be denoted by executing a suboperation (Sub-op1) on the first metadata server, then executing another suboperation (Sub-op2) on the second metadata server. The former is called a "coordinator", which receives the metadata request from a client, while the latter is called a "participant," which receives the metadata request from the coordinator. In S2PC-MP, create, remove, mkdir, and rmdir operations are split among the two servers in the way showed in Table 1.

The S2PC-MP protocol is depicted in Figure 3(a). The key idea here is to combine request processing with commitment to reduce the processing latency. It is unnecessary for the

[3]Operations that may require more than two metadata servers are link and rename. However, both of them seldom occur in typical workloads. So they are handled in a different way.

Table 1. Split of operations

| Operations | Sub-op1 (Coordinator) | Sub-op2 (Participant) |
|---|---|---|
| create | Insert a new entry in the parent directory, and update parent inode | Allocate an inode and fill it, set the flag that indicates it is a regular file |
| remove | Remove the entry in the parent directory, and update parent inode | Decrease the nlink of the file, if nlink reaches 0 and open-count is 1, then free the inode of the file |
| mkdir | Insert a new entry in the parent directory, and update parent inode | Allocate an inode and fill it, set the flag that indicates it is directory, and create a native file to store the entries for the new dir |
| rmdir | Delete the entry in the parent directory, and update parent inode | Decrease the nlink of the dir, if nlink reaches 0 and open-count is 1, then free the inode of the dir, and remove the corresponding native file |

coordinator to send the request to the participant if Sup-op1 fails. The coordinator will send the request to the participant only when Sup-op1 succeeds, and this request indicates that the coordinator has voted "Yes". And as a comparison, the metadata processing directly based on 2PC without any improvement is depicted in Figure 3(b).

In S2PC-MP, the coordinator first completes Sup-op1 and writes the result record into its log. If the Sub-op1 succeeds, it sends the operation request (OP-REQ) to the participant. This request not only tells what operation the participant should perform, but also implies that the coordinator has voted "Yes." Therefore, the participant can decide whether the whole operation should be "Committed" or "Aborted" according to the results of Sub-op2. Then it sends its decision ("Commit" or "Abort") as well as the result through the message "Commit-REQ" or "Abort-REQ" to the coordinator. The coordinator writes "Commit" or "Abort" record in its log according to this message, then sends an "ACK" message to the participant. After receiving the "ACK" message, the participant writes an "End" record in its log, and the whole processing is now completed.

For a distributed operation, S2PC-MP needs only three messages between the two servers, compared with five messages needed by Figure 3(b). Although the two suboperations can be processed in parallel in Figure 3(b), the gain of such parallel processing is very limited because both Sub-op1 and Sub-op2 typically cost only little processing time. Therefore, by reducing messages of each distributed operation, S2PC-MP not only reduces the response time of them, but also improves metadata throughput by greatly reducing network traffic. Moreover, the message waiting mechanism in S2PC-MP is different from that in the 2PC. In 2PC, the server does nothing when it is waiting for a specific message, and it ends the waiting if the message arrives or it timeouts. In order to better utilize the server's processing and storage resources, in S2PC-MP, the server will process other requests during waiting by putting the half-completed request into a waiting queue. When the message which it

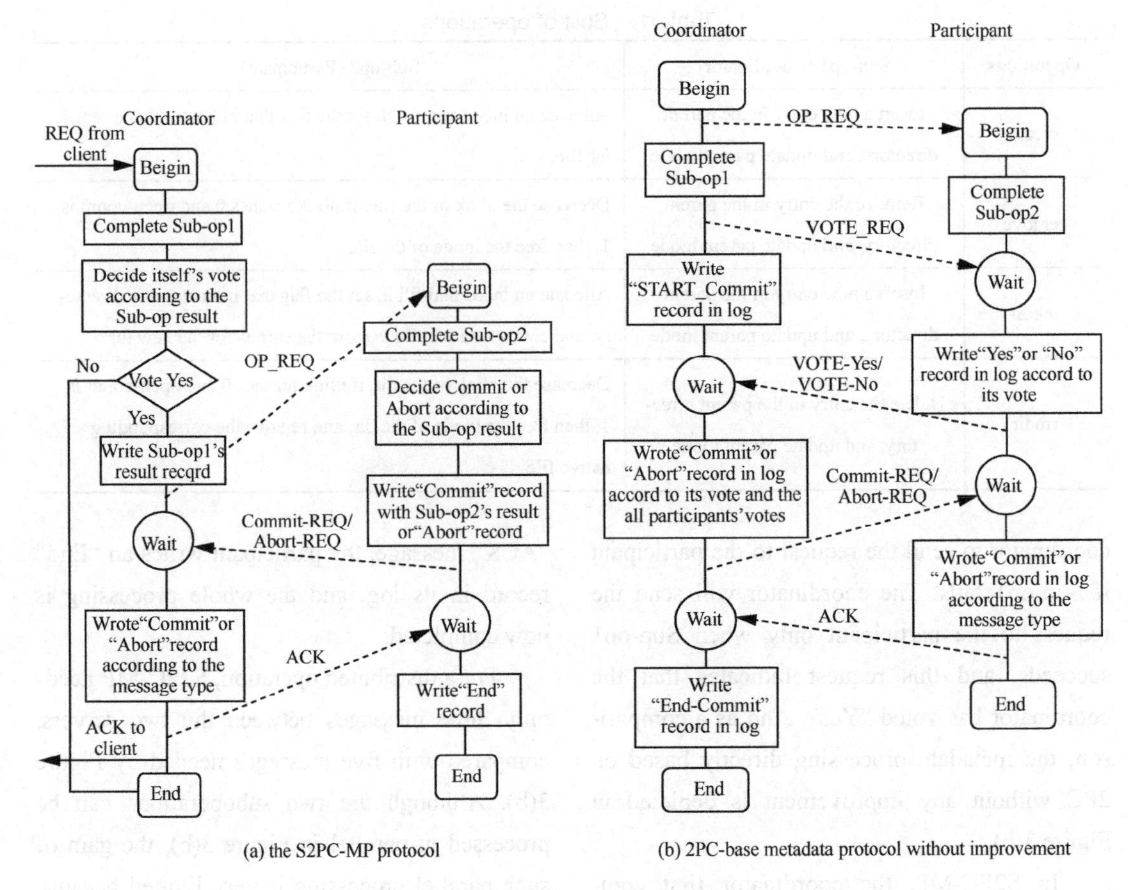

(a) the S2PC-MP protocol (b) 2PC-base metadata protocol without improvement

Figure 3. Metadata processing protocols

has been waiting arrives, it will take the half-completed request out of the waiting queue, and continue processing it. This optimization is shown in Figure 4.

If some metadata server crashes, some waiting messages may never arrive, resulting in some requests staying in the waiting queue forever. Therefore, S2PC-MP needs to handle these waiting requests in its failure recovery protocol, which will be discussed in Section 4.4.

## 4.3 Metadata log

### 4.3.1 Log records

By combining operation result records with commitment progress records, there is a total of four types of records in S2PC-MP protocol. They are as follows:

- Operation result record. It not only contains the result of corresponding operation, but also indicates the beginning of a distributed operation.
- Commit record. For the coordinator, it also indicates the end of the operation.
- Abort record. For the coordinator, it also indicates the end of the operation.
- End record. It is the end of the operation and only appears in the participant.

For failure-free execution of a distributed

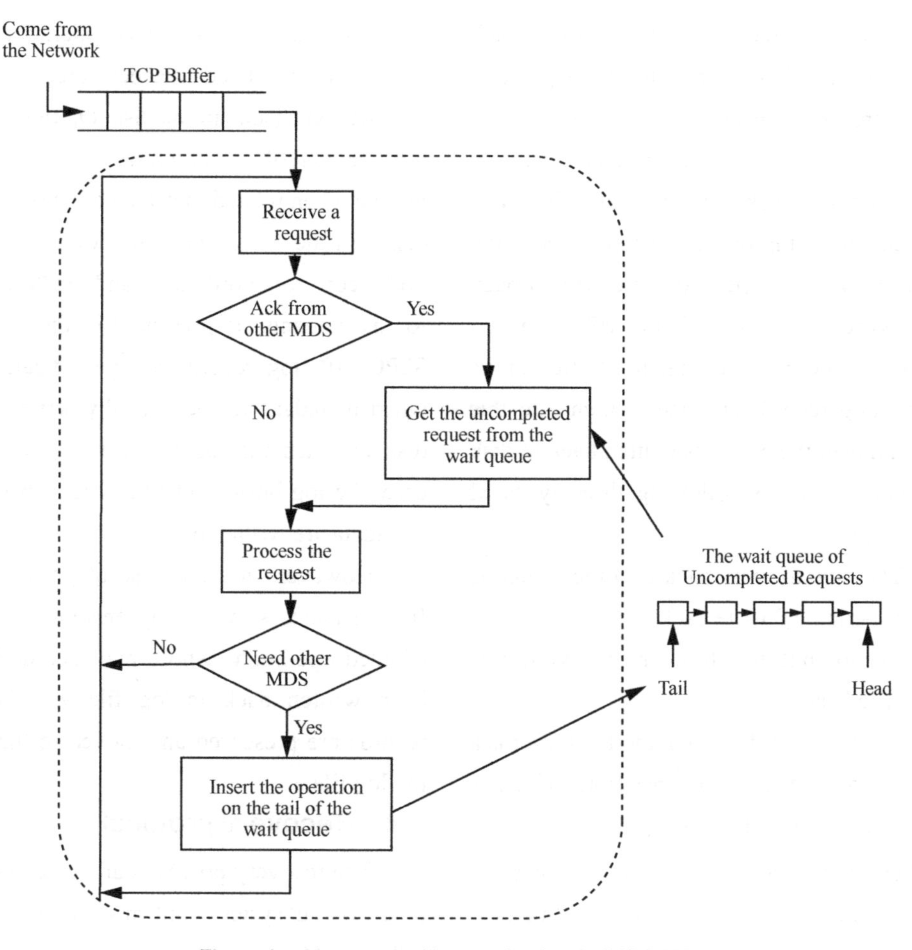


Figure 4. Message waiting mechanism in S2PC-MP

operation, there are two records for each distributed operation in the coordinator's log file. They are 1) Sub-op1's result record, and 2) Commit or Abort record. And there are also two records in the participant's log file. They are 1) Commit record with Sub-op2's result or Abort record, and 2) End record. Each of the above records should be written through to disk.

During recovery of a distributed operation, the system needs to know the state of the operation on the collaborated metadata servers. As a result, we need to carefully design the format of the log records such that the collaborated metadata servers can recognize the related log records of the same operation easily. In our system, each operation is uniquely identified by an ID which is the coalescence of the coordinator's ID and the operation's sequence number assigned by the coordinator. Therefore, we can easily match the log records on different metadata servers by operation IDs.

### 4.3.2 Write-back of log records

Every modification, either an ordinary operation or a distributed operation, has its log records. And these records must be written back to a log file on disk before the corresponding metadata are written back to disk.

Each log record for a distributed operation must be written to disk synchronously. However, motivated by improving performance, log records for ordinary operations are written back to disk asynchronously. In S2PC-MP, each metadata server keeps a log record buffer in its memory. Log records are first written into this buffer, and then the buffer is written back to disk as a whole when the following three types of events occur:

1. The time to write back is come, which is known as periodical write back.

2. The log buffer is full, which is known as forced write back.

3. The last record written into the buffer is a log record of a distributed operation, which is known as protocol write back.

Thus, if a metadata server crashes or powers off, the records in its log buffer will be lost. However, these records correspond to ordinary operations. Their loss will not affect the recovery of metadata consistency. Therefore, the metadata on all servers will recover to a consistent state. Although results of some latest ordinary operations may be lost, the effect is the same as that of other file systems using metadata journaling technology. One solution to avoid data loss is to keep the log buffer on a nonvolatile storage device, such as NVRAM.

#### 4.3.3 Pruning of log records

Like other file systems using metadata journaling, the size of the log file is fixed, 50 MB by default. And the log file is written sequentially. Log records of an ordinary operation can be pruned when related metadata are written from the metadata cache to the disk. On the other hand, log records of a distributed operation can be pruned when related metadata are written back to disk and the end record (the "Commit" or "Abort" record for coordinator, and the "End" record for participants) is presented in the log file. In S2PC-MP, log records are periodically pruned when metadata are periodically written back to disk or when the log file is full. For the latter case, the log buffer must be written back before metadata are written back.

However, in the course of pruning the log file, metadata servers may encounter some distributed operations whose end records have not been written back in log file yet. These log records are preserved and moved to the head of the log file.

### 4.4 Recovery protocol

The recovery process starts when the failure detection subsystem confirms a crash. After the failed metadata server reboots, it informs all other metadata servers to go into the recovery state, in which they do not accept any new coming requests and write their log buffer to disk. The server then reads its log file and recovers according to log records. The result of any ordinary operation is recovered by its records, while the result of any distributed operation is recovered according to steps described in Sections 4.4.1 and 4.4.2 accordingly.

Although all metadata servers must stop processing client request when a failed metadata server restarts, the period of pause processing is very brief, typically tens of seconds, as showed in Section 5.2.2. The S2PC-MP recovery protocol enables fast failure recovery.

### 4.4.1 Recovery for the coordinator

If the current record in the log is an operation result record, and the rebooted server is the coordinator of the distributed operation, then the server tries to search for other records of this operation in the log file.

If it finds the Commit record of this operation, this means that the operation was committed finally. It then writes the operation result to disk and sends an "ACK" message to the participant. Receiving the "ACK" message, the participant will look up the related request in the waiting queue and delete it from the queue if finding it, and write the End record in the log file.

If it finds the Abort record of this operation, this means that the operation was abort finally. So it will not write back the result. And it also sends an "ACK" message to the participant.

If it finds neither of the above records, it does not know what the final decision is. So it sends a "Decision-REQ" message to the participant to ask for the final decision and waits for the answer by blocking itself. After receiving the "Decision-REQ," the participant will look up the records of this operation and send a reply message back. If the participant answers that it has the Commit or Abort record, then the rebooted server writes the Sup-op1's result to disk or ignores Sup-op1's result, respectively, and sends an "ACK" message to the participant. If the participant answers that it has only Sub-op2's record and no any other records, then the rebooted server ignores Sub-op1's result because neither results has been written to disk. If the participant answers that it has no records of this operation, then the rebooted server ignores Sub-op1's result because the participant never handled related request.

### 4.4.2 Recovery for the participant

If current record is an operation result record, and the rebooted server is the participant of the distributed operation, it tries to search for other records of this operation in its log file.

If it finds the End record, this means the operation is committed or aborted and there must be a Commit or Abort record too. So it writes Sub-op2's result to disk if it has the Commit record or ignores Sub-op2's result if it has the Abort record.

If it finds the Commit record, but does not find the End record, then it sends a "Commit-REQ" to the coordinator, writes Sub-op2's result to disk, and waits for the coordinator's "ACK" message by blocking itself. After receiving the "Commit-REQ," the coordinator looks up the request related to the operation in the waiting queue and deletes it if finding it, writes the Commit record in the log file and sends an "ACK" message to the rebooted server. When the rebooted server receives the "ACK" message, it completes the recovery of this operation and goes on to process the next record.

If it finds the Abort record but does not find the End record, or it finds only Sup-op2's record but does not find the other records, then it sends an "Abort-REQ" to the coordinator, and waits for the coordinator's "ACK" message by blocking itself. Afterwards, the coordinator and the reboot server do the same as the above case.

### 4.4.3 Additional steps

After all the records of the rebooted server's

log file are properly handled according to Sections 4.4.1 and 4.4.2, additional steps are required to recover other metadata servers' state.

The rebooted server will send a "Recover-END" message to all other servers to indicate that it has completed recovery. And it waits for response messages from all other servers. When it receives them, it goes into normal state and can handle client requests now.

When receiving the "Recover-END" message, the other servers should first do the following two things. 1) They parse their log files, and undo the result of those operation result records which has not any other records, and the peer server of which is the failed server. In this case, the failed MDS is the participant of these operations. However, the participant crashed before handling the requests about these operations. So its log file has no records about these operations, while the other MDSs have only Sup-op1 result records in their log files. This step is to undo these Sub-op1s' results. 2) They search their waiting queues for the requests that are sent to the failed server and delete them from the queue.

After completing the above two steps, these servers send a response message to the rebooted server. And then they exit the recovery state and go back to the normal state, and continue to process requests from clients.

### 4.5 Validation

In S2PC-MP, every distributed operation is processed by a sequence of steps in a time order. As shown in Figure 3(a), these steps are:

1. The coordinator performs Sub-op1. The completion of this step is shown by Sub-op1's result record in the coordinator's log file;

2. The participant performs Sub-op2. The completion of this step is shown by Sub-op2's result record in the participant's log file;

3. The participant makes the final decision. The completion of this step is shown by either the "Commit" record or the "Abort" record in the participant's log file;

4. The coordinator commits or aborts according to final decision. The completion of this step is shown by either the "Commit" record or the "Abort" record in the coordinator's log file;

5. The participant completes the operation. The completion of this step is shown by the "End" record in the participant's log file.

It is easy to prove that if a distributed operation follows S2PC-MP, shown in Figure 3(a), the two servers can reach to consistent results through the recovery protocol in Section 4.4, whichever server crashes in the middle of the operation. Therefore, the S2PC-MP as well as the recovery protocol can ensure metadata consistency. We omit the proof of this conclusion because it is beyond the scope of this paper.

## 5 Performance evaluation

In this section, we evaluate the performance of the DDG policy and the S2PC-MP protocol using detailed simulation and actual measurements.

### 5.1 Performance of DDG

We evaluate the metadata processing performance of DDG from three aspects. We first conducted extensive simulation study to investigate its performance under different granularities and compare them with two baseline policies, the Random policy and the Subtree policy. We then evaluate its scalability as the number of metadata servers increases. Finally, to validate our simulation results, we implemented DDG in a real system, and conducted detailed and extensive measurements.

The DDG and the two baseline policies are implemented in the same simulator (simu_mds, see Section 5.1.3) and the real system (DCFS2, see Section 5.1.5). The only difference between them is the way to select a metadata server for a new object. In the Random policy, each MDS chooses an MDS for a newly created object in a round-robin fashion. And in the Subtree policy, the MDS that maintains the root directory chooses a MDS for each object created in the root directory in a round-robin fashion, and objects in directories other than the root directory are assigned to the MDS that maintains their parent directory.

### 5.1.1 The test namespaces

We used two namespaces in our performance tests. One of them, denoted by Namespace1, is a snapshot of the home directory on the NFS server of our laboratory. The snapshot, taken in early 2005, contains 698 320 files and 25 864 directories. Because Namespace1 is from a real environment, it can represent the file systems of the similar environments. However, Namespace1 is too small compared with current and future file systems. Therefore, we generated a larger namespace, denoted by Namespace2, with similar characteristics as Namespace1, by an automatic tool. It has 1 010 000 files and 34 721 directories. The characteristics of these two namespaces are shown in Figure 5.

The two namespaces have similar characteristics, including: 1) over 70 percent directories

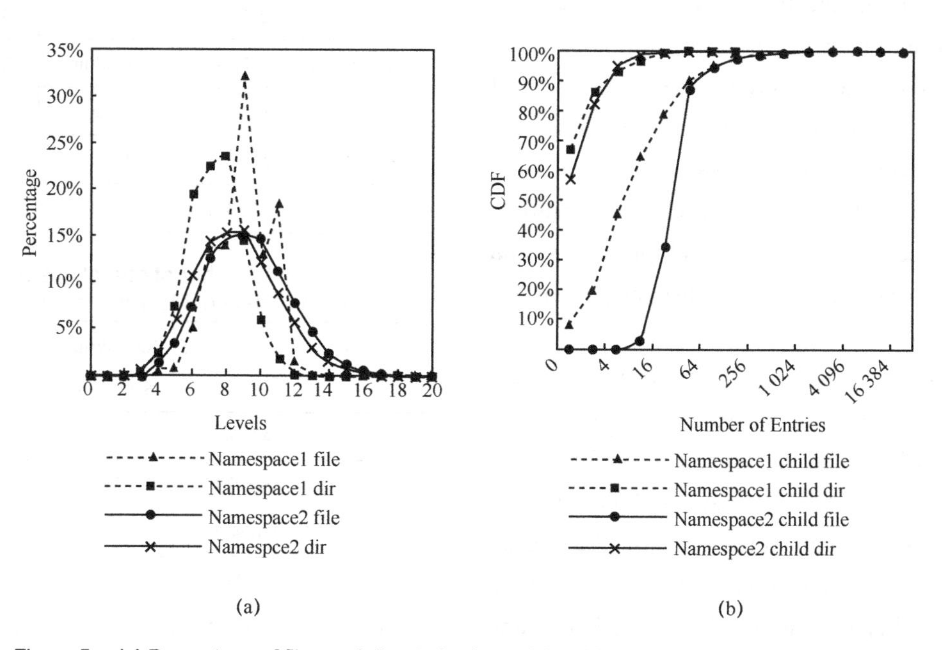


Figure 5. (a) Percentage of files and directories in each level for Namespace1 and Namespace2
(b) Cumulative distribution function of number of child directories and child files of each directory

and files are located in several continuous levels; 2) over 92 percent directories have 0~4 child directories; 3) over 86 percent directories have 0~32 child files, except that the numbers of child files in Namespace1 are distributed over a much wider range. Moreover, the distribution of directory sizes in these two namespaces matches with recently published data[32] over 13 file systems from five supercomputing sites and two file systems from department file servers. Dayal[2] found that 90 percent directories have less than 32 entries (including both child files and child directories) for 11 file systems[2].

### 5.1.2 Benchmark programs

We used two benchmark programs to evaluate our designs. Macrobenchmark1 is a metadata-intensive benchmark, which takes a namespace as the input. It first recreates the hierarchy and all the files according to the input namespace. It then checks the attributes of all files and directories through the stat system call. Finally, it deletes all the files as well as the whole hierarchy. The time spent in each phase is measured. Macrobenchmark1 allows multiple test processes running on multiple nodes, and each process deals with a disjoint part of the namespace.

Macrobenchmark2 is another multiprocess program simulating a real environment with plenty of concurrent processes performing a large number of I/O operations. Each of these processes randomly selects files in the namespace and performs certain operations on these files, so they may access the same file or same directory. The benchmark is both metadata-intensive and data-intensive, generating a lot of file read and write operations. The percentages of each type of operations issued by Macrobenchmark2 are given in Table 2, which reflects workloads similar to SPECsfs3.0[34]. Macrobenchmark2 also takes a namespace as the input and consists of three phases: the warm-up phase, the testing phase, and the postprocessing phase. In the warm-up phase, it rebuilds the hierarchy according to the input namespace. In the test phase, it issues a set of test operations specified in a description file. In the postprocessing phase, it calculates and outputs analysis statistics. The number of operations issued by all client nodes as well as the total amount of data read or written by all client nodes is listed in Table 3.

Table 2. Operations composition of Macrobenchmark2

| Operations | Percentage | Operations | Percent |
|---|---|---|---|
| Create | 5.55% | Stat | 9.98% |
| Mkdir | 1.71% | Readdir | 7.70% |
| Open | 35.45% | Remove | 1.07% |
| Read | 25.67% | Rmdir | 0.04% |
| Write | 12.83% | | |

Table 3. Total number of operations and the amount of data of Macrobenchmark2

| Number of Clients | Total number of Operations | Total Write Size /MB | Total Read Size/MB |
|---|---|---|---|
| 1 | 243 398 | 1 205 | 3 437 |
| 2 | 486 933 | 2 344 | 6 364 |
| 3 | 730 842 | 3 803 | 9 309 |
| 4 | 974 430 | 4 964 | 10 992 |
| 5 | 1 218 173 | 5 651 | 15 010 |
| 6 | 1 461 699 | 6 959 | 18 610 |
| 8 | 1 949 044 | 13 931 | 23 049 |

### 5.1.3 Simulation results

In order to compare the performance of different distribution policies, we implemented the two baseline policies (Random, Subtree) as well as DDG in a metadata server simulator (simu_mds), which is a user-level process. We use n processes to simulate n metadata servers. The simulation experiments ran Macrobenchmark1, which took both Namespace1 and Namespace2 as the data sets. We simulated a system with 16 metadata servers and 32 client nodes. In these tests, each simu_mds has an inode cache that keeps 128k inodes and a dentry cache that keeps 128k directory entries. Both of them use the LRU algorithm for cache replacement.

The results in Figures 6~12 were obtained by running simu_mds on two nodes, with all clients on one node and all metadata servers on the other node. Each node has two quad-core processors (Intel Xeon E5335 @2.0 GHz), 8 GB memory, and three 146 GB SCSI disk (Seagate ST3146707LC). The operating system is CentOS4.4. The two nodes are interconnected by a Gigabit Ethernet network.

The normalized throughputs (which use the throughput of Subtree as the baseline) of file

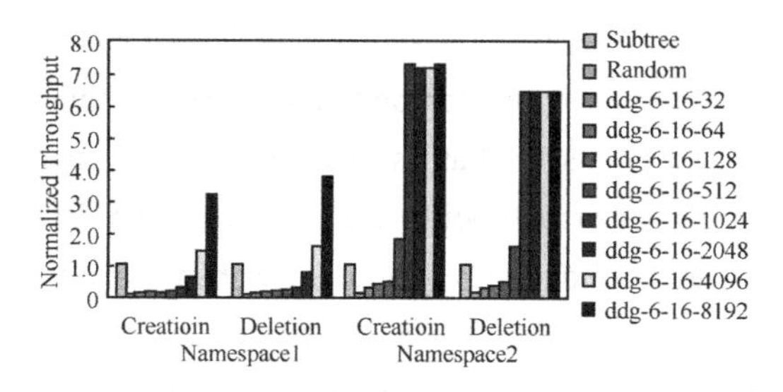


Figure 6. Normalized metadata throughput for file creation and deletion with different FileWid (Macrobenchmark1)

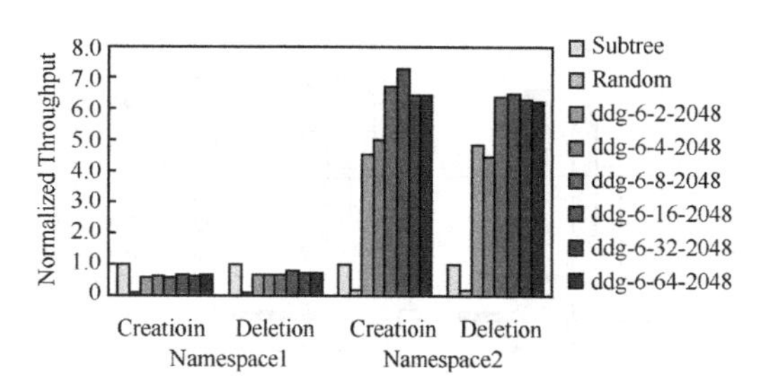


Figure 7. Normalized metadata throughput for file creation and deletion with different DirWid (Macrobenchmark1)

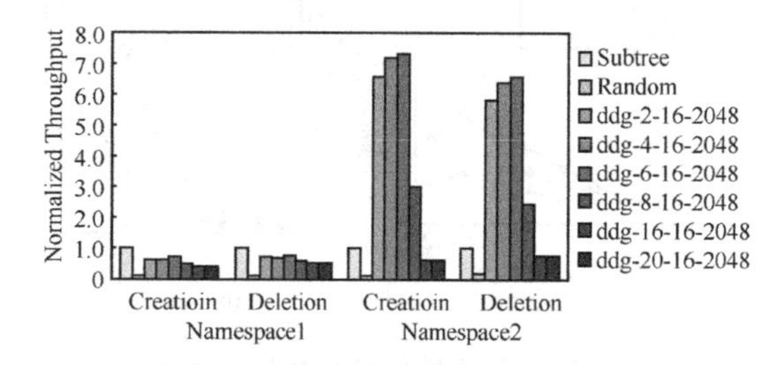


Figure 8. Normalized metadata throughput for file creation and deletion with different DirDep (Macrobenchmark1)

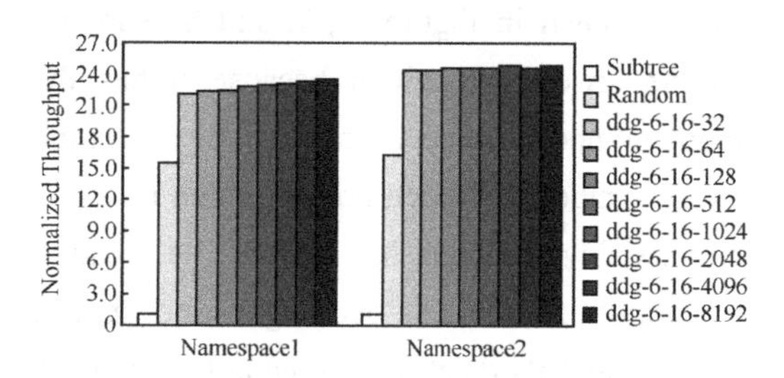


Figure 9. Normalized throughput for file stat operation with different FileWid (Macrobenchmark1)

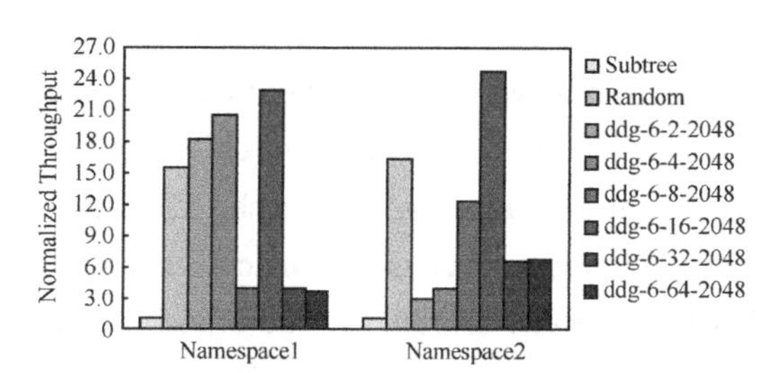


Figure 10. Normalized throughput for file stat operation with different DirWid (Macrobenchmark1)

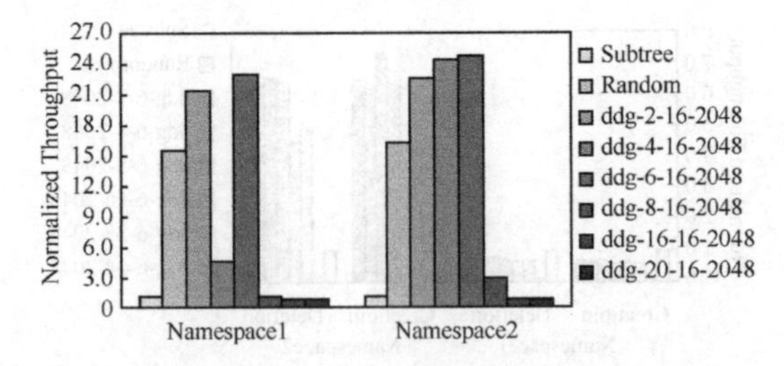


Figure 11. Normalized throughput for file stat operation with different DirDep (Macrobenchmark1)

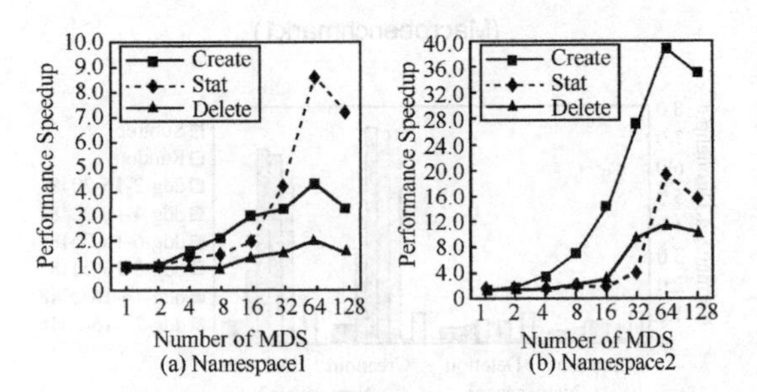


Figure 12. Scalability of metadata servers (Macrobenchmark1)

creation and deletion under different granularities are shown in Figures 6, 7, and 8. Random has the lowest throughput because it has the largest number of branch points. Subtree has the fewest number of branch points, which means it has very few metadata log writes. The creation throughput of Subtree is over eight times that of Random, and its deletion throughput is over six times that of Random. However, since the metadata is very unevenly distributed among the metadata servers in Subtree, some individual metadata servers have a large number of disk reads and writes because the number of objects on it is much larger than the metadata cache. DDG policies with proper granularity can achieve a better balance between distribution and branch points, and hence, their creation and deletion throughputs are 1.4~7 times that of Subtree, and 1.5~63 times that of Random.

As Figure 6 shows, for DDG policies, the throughputs of both creation and deletion increase with FileWid. Smaller FileWid results in more branch points, and hence, more disk writes for metadata log and more messages among metadata servers. DDG outperforms Subtree until FileWid is larger than or equal to 4 096 for Namespace1 and 512 for Namespace2, respectively. From the profiling data, we found that in Namespace1, about 93.45 percent of files are in directories with less than 4 096 child files. Similarly, in Namespace2, about 89.25 percent of files are in directories with less than 512 child files. These results indicate that from the perspective of file creation and deletion throughput, it is preferable to set FileWid to the value such that over 90 percent of files that are in the directories each of which has no more than FileWid number of files.

In Figure 7, we can see that for Namespace1, DirWid has less effect on throughputs than FileWid. Since branch points consist of branch files and branch directories, the proportions of branch directories to total branch points are affected by both FileWid (which determines the number of branch files) and DirWid (which determines the number of branch directories). Because FileWid was set to 2 048, the branch directories account for only 1/4 to 1/8 of total branch points in Namespace1, and hence, the number of branch points changes little with different DirWid values. On the other hand, due to setting FileWid to 2 048, the branch directories account for 100 percent of branch points in Namespace2. So DirWid has some obvious effect on throughputs for Name-space2. Throughputs increase with the increase of DirWid, but decline slightly when DirWid is larger than or equal to 32, because uneven distribution causes more disk reads and writes. These results indicate that from the perspective of file creation and deletion throughput, DirWid is not as important as FileWid, and a value around 16 is preferable for the namespaces in which 90 percent directories are in the directories that have no more than 32 child directories.

As shown in Figure 8, throughputs first increase slightly until DirDep gets 6, and then decline steeply. This is because that a larger DirDep value, like Subtree, may cause much skewed metadata distribution among metadata servers, resulting in extensive disk reads and writes on individual metadata servers. These results indicate that small DirDep values such as 2~6 are preferable.

The throughputs of file stat are showed in Figures 9, 10, and 11. These figures show very different results with that of file creation and deletion. Since file stat is a read only operation, there is no 2PC log, as a result, Subtree and DDG with large DirDep perform worst. And benefited from even metadata distribution, Random's throughput is over 15 times that of Subtree. DDG policies with proper granularity outperform Random by about 10~50 percent because they balance the distribution and namespace locality, and hence, have 3~10 times less inter MDS messages than Random. And DDG policies with proper granularity can achieve 3~24 times performance improvement for file stat over Subtree.

Figure 9 shows that stat throughputs are not sensitive to FileWid. This is very different from file creation and deletion. We believe the reason is that each metadata server has a large metadata cache, which can hold 128k objects, and the distribution skew caused by FileWid is not beyond the limit of the cache.

As Figures 10 and 11 show, DDG has very poor stat performance for both large DirWid and DirDep because of uneven metadata distribution. This is similar with the file creation and deletion. When choosing DDG values, it is very important to keep this in mind.

#### 5.1.4 Scalability analysis

The scalability of metadata processing is an important consideration for a cluster file system designed for large systems. For distributed metadata processing, scalability lies in two aspects: 1) Server scalability, which is how the system performance under fixed workloads changes as the number of metadata servers increases; 2) Client scalability, which is how the system performance changes as the number of clients, and hence, the workload increases.

In this section, we discuss the server scalability of DDG found in the simulation experiments. Figure 12 shows the throughput speedups of DDG-2-16-2048 with 128 client processes and 1~128 metadata servers. Note that the 128 client processes were constantly feeding requests to servers at a very high rate, so the aggregated request rate was equivalent to that generated by a much larger group of client nodes in a real system, where clients do not perform file I/O constantly. As shown in the Figure 12, performance speeds up well for Namespace2, but does not speed up well for Namespace1. Through profiling, we found that Namespace1 has over 42 thousand branch files, while Namespace2 has no branch files. As a result, Namespace1 has a large number of synchronous log writes than Namespace2, which counteracts some of the gains by adding more MDS. Through profiling, we also found that for Namespace1, metadata distribution is more uneven with the increase of the number of metadata servers, which also impairs its speedup.

### 5.1.5 Results on the real system

In addition to simulation experiments, we also implemented the three policies in DCFS2 and measured their performance on a real system. DCFS2 is a cluster file system we designed for very large scale, supercomputer- class clusters. It consists of a storage space manager, a group of metadata servers, a group of IP-addressable storage servers (called SuperNBD servers) for file data storage, and a large number of client nodes.

The tests were done on 17 nodes of a very large cluster, each of which has two AMD Opteron 242 processors, 2 GB of memory, and a 37 GB Seagate ST373307LC SCSI disk, running Turbolinux 3.2.2. The nodes were interconnected by a Gigabit Ethernet. Because of the space limitation, we only present the results of DDG-4-8-128 and compared them with those of Random and Subtree.

Our goal is to measure the performance, interactions, and data distributions between multiple metadata servers. As a result, while this is a relatively small test platform, we configured DCFS2 to run six metadata servers, one storage space manager, two SuperNBD servers, and eight client nodes. While we did not run the experiments on a full production system with thousands of nodes, we believe that our results are still valid for large systems. We are only interested in metadata server performance, and the six metadata servers (which can be further scaled up, as indicated in the above discussion) can easily support a large number of client nodes and disks. In these tests, each MDS used about 800 MB memory as its metadata cache.

Figure 13(a), showing the results of Macrobenchmark1, confirms that DDG significantly outperforms the baseline systems. For file creation, DDG is 66 and 15 percent faster than Subtree and Random, respectively. For file deletion throughput, DDG is 91 and 24 percent higher than those of Subtree and Random, respectively.

Figure 13(b) compares the aggregate throughputs of all clients running Macrobenchmark2 on Namespace1. While this benchmark is both metadata-intensive and data-intensive, DDG still shows the highest throughputs, about 100 percent higher than that of Subtree, and 10 percent higher than that of the Random.

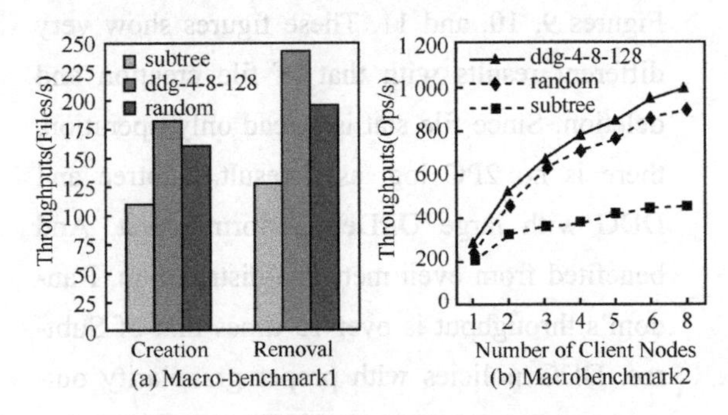


Figure 13. Throughputs on the real system (Namespace1)

Through careful analyses[35], we found that although Subtree issued the fewest number of requests, the distribution of these requests was very unbalanced, resulted in the lowest aggregate throughput. Random, on the other hand, produced the most balanced metadata distribution. However, it also generated 50 percent more branch points and 27 percent more requests than

DDG did. DDG achieved the best performance by striking a right balance between the number of requests and the balance of distribution.

## 5.2 Performance of the S2PC-MP

The S2PC-MP was implemented in DCFS2. We evaluated the performance of the protocol by measuring the overhead of writing log records required by 2PC as well as the efficiency of recovery from a metadata server crash.

The test platform consisted of 20 nodes, each of which is the same as those described in Section 5.1.5. DCFS2 was configured with two metadata servers, one storage space manager, one SuperNBD server, and 16 client nodes.

### 5.2.1 Overhead of metadata logging

We measured the cost of metadata logging of S2PC-MP by comparing the read/write bandwidth, the file creation throughput, the file deletion throughput, and the transaction throughput with and without writing logging.

Iozone[10], running in the cluster mode and the stonewall mode, was used to measure the read and write performance. The total amount of data read/written was 528 MB, and the data size of each read/write operation was 1 MB. As shown in Figure 14, except in a few cases, there was no obvious difference between the aggregate read/write bandwidth with and without writing logging. DCFS2 successfully minimized the logging overheads through many optimization methods, such as asynchronous writing log records, asynchronous updating of file size and mtime in write operations, and no updating of atime in read operations.

Postmark[29] was used to measure the file creation throughput, the file deletion throughput, and the transaction throughput, and results are shown in Figure 15. Each client node ran a postmark process. They created or deleted 48 000 files in 48 directories and performed 96 000 transactions. In the cases of file creation and deletion throughputs, turning on write logging decreased performance by only 10 percent. For the transaction throughput, there was no obvious difference between turning the write logging on and off.

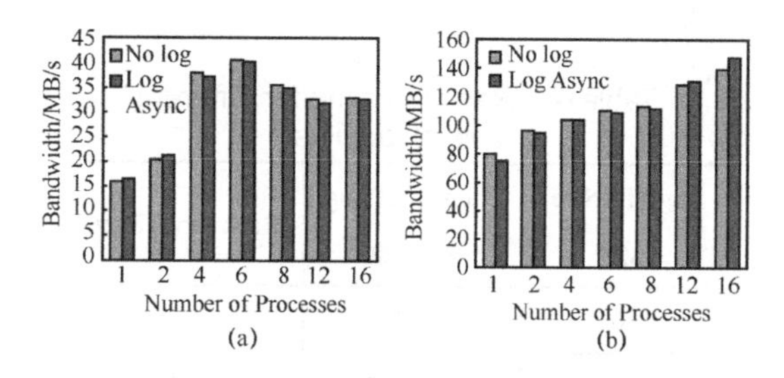


Figure 14. (a) Read (b) write bandwidths

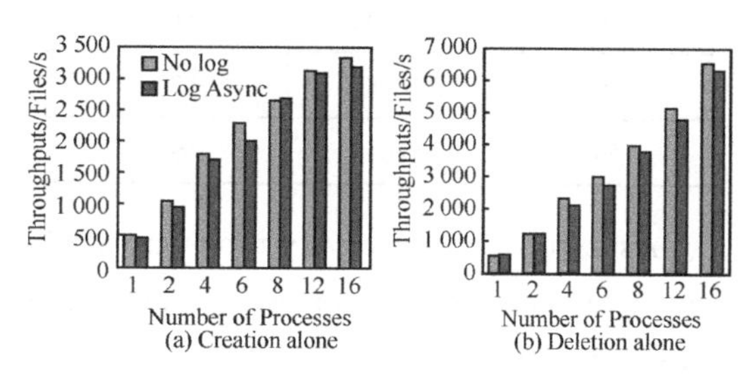


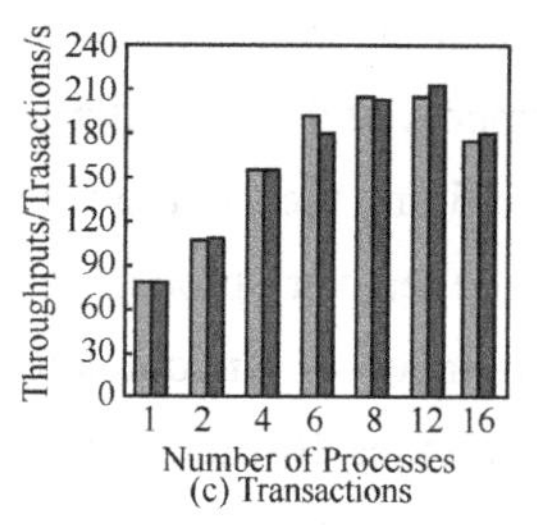


Figure 15. Throughputs by Postmark

### 5.2.2 Efficiency for recovery

For S2PC-MP, the time to recover from a metadata server failure is independent of the file system size, which may be large and in PB-scale. The recovery time is determined by the size of the log file and by the number of uncompleted distributed operations in the log file, whose results need to be restored by communicating with other metadata servers.

Table 4 shows the recovery time for different log file sizes. In all these tests, each log file had 721 uncompleted distributed operations, regardless of its size. As a result, we can isolate the effect of the log file sizes to the recovery time.

Table 4. Recovery time for different size of log files

| Log File Size /MB | Number of Log Records | Recovery Time/s |
|---|---|---|
| 10 | 71 998 | 4.14 |
| 20 | 143 607 | 7.87 |
| 30 | 215 332 | 11.65 |
| 40 | 287 087 | 15.73 |
| 50 | 359 044 | 19.26 |

As expected, the recovery time was within tens of seconds and was proportional to the log file's size.

Table 5 compares the recovery times of different numbers of distributed operations in the log file while fixing the log file size at 10 MB. It is clear that the recovery process is very efficient. When the number of uncompleted distributed operations increases 10 times, the recovery time increased only one time.

# 6 Conclusion

Most supercomputers nowadays are based on large clusters. As their capacity and user requirement increase, so is the demand for high-performance distributed file systems on these clusters. While traditional metadata management techniques such as single metadata server or simple distribution algorithm may be adequate for small to medium systems, large-scale high-performance cluster systems call for more sophisticated decentralized metadata processing technology.

Table 5. Recovery time for different number of uncompleted distributed operations

| Number of Log Records | Number of Uncompleted Distributed Op. | Percentage | Recovery Time/s |
|---|---|---|---|
| 71 763 | 0 | 0 | 4.12 |
| 71 998 | 720 | 1% | 4.14 |
| 72 262 | 1 446 | 2% | 4.39 |
| 72 508 | 2 178 | 3% | 4.64 |
| 72 760 | 2 912 | 4% | 4.89 |
| 73 019 | 3 655 | 5% | 5.33 |
| 74 338 | 7 440 | 10% | 9.26 |

This paper presents a set of solutions to two fundamentally important issues in decentralized metadata processing: 1) how to distribute the objects in the namespace across metadata servers, which affects the metadata throughput; and 2) how to keep the cross-metadata server operations consistent in the presence of server failures, which determines the reliability and availability of the file system.

Our DDG policy dynamically partitions the namespace into hierarchical units according to a triple-defined distribution granularity, and achieves a good balance between maintaining namespace locality and equal distribution, which is important to achieve high metadata throughput. Extensive simulation and measurement results show that DDG policies with proper granularity significantly outperform the Random policy and the Subtree policy by 40 percent to 62 times.

We proposed a technique called S2PC-MP for consistency of cross-server operations. It reduces overhead in normal processing by combining commit operations with metadata operations, and can quickly recover metadata consistency after any server crashes. Our performance results show that the creation and deletion throughputs decrease only about 10 percent and recovery can be completed within tens of seconds.

We implemented DDG algorithm and S2PC-MP protocol in DCFS2, a large-scale cluster file system for supercomputers. However, these methods can also be used in other cluster file systems which provide a hierarchical namespace to users.

## Acknowledgments

The authors wish to thank Prof. Ning-hui Sun, Prof. Dan Meng, Dr. Si-ning Wu, Dr. Jie Ma, and many other colleagues for their discussion and helpful suggestions. The authors wish to thank the Scientific Computing Platform of ICT for providing its namespaces to us. This work was supported in part by the Natural Science Foundation of China under grant no. 60970025, and the National High-Technology Research and Development Program of China under grants nos. 2006AA01A102, 2009AA01Z139, and 2009AA01A129. Yi-ming Hu is supported by National Science Foundation under grant no. CCF-0541103. The fifth author (Zhihua Fan) participated in this work when he was a graduate student at the Institute of Computing Technology, CAS.

## References:

[1] ROSELLI D, LORCH J, ANDERSON T. A comparison of file system workloads[C]//USENIX Annual Technical Conference. Berkeley: USENIX Association, 2000: 41-45.

[2] DAYAL S. Characterizing HEC storage systems at rest[R]. Pittsburgh: Carnegie Mellon University, 2008.

[3] KARAMANOLIS C, LIU L, MAHOLINGAM M, et al. An architecture for scalable and manageable file services[R]. Palo Alto: HP Labs, 2001.

[4] MENON J, PEASE D A, REES R, et al. IBM storage tank-a heterogeneous scalable SAN file system[J]. IBM Systems Journal, 2003, 42(2): 250-267.

[5] PVFS2 Development Team. Parallel virtue file system, Version 2[Z]. 2003.

[6] WEIL S A, POLLACK K T, BRANDT S A, et al. Dynamic metadata management for petabyte-scale file systems[C]//IEEE/ACM High Performance Computing, Networking and Storage Conference (SC '04). Piscataway: IEEE Press, 2004.

[7] SINNAMOHIDEEN S, SAMBASIVAN R, HENDRICKS J, et al. A transparently-scalable metadata service for the ursa minor storage system[C]//USENIX Annual Technical Conference(USENIX ATC '10). Berkeley: USENIX Association, 2010.

[8] SATYANARAYANAN M, HOWARD J H, NICHOLS D A, et al. The ITC distributed file system: principles and design[J]. ACM SIGOPS Operating Systems Review, 1985, 19(5): 35-50.

[9] OUSTERHOUT J K, CHERENSON A R, DOUGLIS F, et al. The sprite network operating system[J]. Computer, 1988, 21(2): 23-36.

[10] SHEPARD L, EPPE E. SGI infinite storage shared filesystem CXFS: a high-performance, multi-OS SAN file system from SGI[Z]. San Francisco: Silicon Graphics, Inc., 2004.

[11] BUTT A R, JOHNSON T A, ZHENG Y, et al. Kosha: a peer-to-peer enhancement for the network file system[C]//IEEE/ ACM High Performance Computing, Networking and Storage Conference. Piscataway: IEEE Press, 2004.

[12] XIONG J, WU S, MENG D, et al. Design and performance of the dawning cluster file system[C]//IEEE International Conference on Cluster Computing. Piscataway: IEEE Press, 2003: 232-239.

[13] ANDERSON T E, DAHLIN M D, NEEFE J M, et al.

Serverless network file systems[C]// The 15th ACM Symposium on Operating Systems Principles. New York: ACM Press, 1995: 109-126.

[14] ZHU Y, JIANG H, WANG J, et al. HBA: distributed metadata management for large cluster-based storage systems[J]. IEEE Transactions on Parallel and Distributed Systems, 2008, 19(6): 750-763.

[15] HUA Y, ZHU Y, JIANG H, et al. Scalable and adaptive metadata management in ultra large-scale file systems[C]//The 28th International Conference on Distributed Computing Systems(ICDCS'08). Piscataway: IEEE Press, 2008: 403-410.

[16] VAHALIA U. UNIX Internals: the new frontiers[M]. New Jersey: Prentice-Hall, Inc., 1996.

[17] ANDERSON D C, CHASE J S, VAHDAT A M. Interposed request routing for scalable network storage[J]. ACM Transactions on Computer Systems, 2002, 20(1): 25-48.

[18] JI M, FELTEN E W, WANG R, et al. Archipelago: an island-based file system for highly available and scalable internet services[C]//The 4th USENIX Windows Systems Symposium. Berkeley: USENIX Association, 2000.

[19] ZHANG Z, KARAMANOLIS C. Designing a Robust namespace for distributed file services[C]//The 20th IEEE Symposium on Reliable Distributed Systems. Piscataway: IEEE Press, 2001: 162-171.

[20] PATIL S, GIBSON G. GIGA+: scalable directories for shared file systems[R]. Pittsburgh: Carnegie Mellon University, 2008.

[21] XING J, XIONG J, SUN N, et al. Adaptive and scalable metadata management to support a trillion files[C]//The Conference on High Performance Computing Networking, Storage and Analysis. Piscataway: IEEE Press, 2009.

[22] WU C, BURNS R. Handling heterogeneity in shared-disk file systems[C]// The International Conference on High Performance Computing and Comm. Piscataway: IEEE Press, 2003.

[23] BRANDT S A, XUE L, MILLER E L, et al. Efficient metadata management in large distributed file systems[C]//The 20th IEEE/11th NASA Goddard Conference on Mass Storage Systems and Technologies MSST'03). Piscataway: IEEE Press, 2003.

[24] HAERDER T, REUTER A. Principles of transaction-oriented database recovery[J]. ACM Computing Surveys, 1983, 15(4): 287-317.

[25] DEVARAKONDA M, KISH B, MOHINDRA A. Recovery in the Calypso file system[J]. ACM Transactions on Computer Systems, 1996, 14(3): 287-310.

[26] BRAAM P J. The lustre storage architecture[Z]. [S.l.]: Cluster File Systems, Inc., 2003.

[27] BAKER M, OUSTERHOUT J. Availability in the sprite distributed file system[J]. ACM SIGOPS Operating Systems Review, 1991, 25(2): 95-98.

[28] SCHMUCK F, HASKIN R. GPFS: a shared-disk file system for large computing clusters[C]//The 1st USENIX Conference on File and Storage Technologies (FAST'02). Berkeley: USENIX Association, 2002.

[29] PRESLAN K W, BARRY A, BRASSOW J, et al. Implementing journaling in a linux shared disk file system[C]//The 8th NASA Goddard Conference on Mass Storage Systems and Technologies in Cooperation with the 17th IEEE Symposium on Mass Storage Systems. Piscataway: IEEE Press, 2000: 351-378.

[30] GRAY J. Notes on data base operating systems[M]. Operating Systems. Heidelberg: Springer-Verlag, 1978: 393-481.

[31] BERNSTEIN P A, HADZILACOS V, GOODMAN N. Concurrency control and recovery in database systems[M]. New Jersey: Addison-Wesley, 1987.

[32] ÖZSU M T, VALDURIEZ P. Principles of distributed database systems(2nd Edition)[M]. New Jersey: Prentice-Hall, Inc., 1999.

[33] GHEMAWAT S, GOBIOFF H, LEUNG S T. The Google file system[C]//The 19th ACM Symposium on Operating Systems Principles (SOSP'03). New York: ACM Press, 2003: 29-43.

[34] SPEC. SFS 3.0 Documentation Version 1.0, Standard Performance Evaluation Corp[Z]. 2001.

[35] XIONG J, TANG R, WU S, et al. An efficient metadata distribution policy for cluster file systems[C]//The IEEE International Conference on Cluster Computing (CLUSTER'05). Piscataway: IEEE Press, 2005.

发表于《中国科学院院刊》2019 年第 6 期

# 发展高性能计算需要思考的几个战略性问题

李国杰

中国科学院计算技术研究所

高性能计算机是我国科学技术快速发展的标志性成果，已成为继高铁之后的又一张“中国名片”。一个发展中国家在尖端计算技术上能迅速走到世界前列，这是一件了不起的事情。对于我国高性能计算机的现状，有人极力赞美——“中国超级计算机技术实力碾压美国”的醒目标题曾在网上刷屏；也有人表示疑虑，认为国产的超级计算机是“用航母运载沙丁鱼”。对发展高性能计算的目标和策略选择，学术界也有不同的看法。正确的战略决策来源于对国情和技术发展趋势实事求是的分析，而不是玩弄技术新名词的“纸上谈兵”。习近平总书记指出：“坚持实事求是，最基础的工作在于搞清楚‘实事’，就是了解实际、掌握实情。这就要求我们必须不断对实际情况做深入系统而不是笼统的调查研究，使思想、行动、决策符合客观实际。”在攀登计算机领域“珠穆朗玛峰”的关键时刻，我们需要遵循习近平总书记的指示，对我国高性能计算机的这件“实事”做深入系统的调查研究，做出符合客观实际的决策。

《中国科学院院刊》作为“国家科学思想库核心媒体”，是中国科学院建设国家高端智库的核心传播平台。《中国科学院院刊》2019 年第 6 期推出“中国高性能计算发展战略”专题，邀请院内外工作在第一线的专家，对涉及高性能计算发展的战略性问题进行深入探讨，旨在凝聚科技界、产业界及社会各界的共识，推动中国高性能计算更理性、更健康地向更高的目标发展。

在讨论与高性能计算有关的战略问题之前，先要明确高性能计算机究竟是指什么。高性能计算机并没有严格的定义，人们在不同的场合讲的“高性能计算机”的含义可能也不一样。①国际组织定义。国际上有一个为世界上最高性能的 500 台计算机排名的组织[1]，最近一次排名是 2019 年 6 月，第 500 名的峰值性能是 2.1 PFlops（2 100 万亿次浮点计算每秒）。在这个组织的网站上，高性能计算机（High Performance Computer，HPC）和超级计算机（Supercomputer）是混用的，不加区分地当成一种计算机类型。也就是说，目前他们把超级计算机（高性能计算机）圈定在 PFlops 级（千万亿次浮点计算每秒）计算机水平。本专题讨论的重点也是 PFlops 级以上的超级计算机。②市场的定义。企业在销售计算机时，高性能计算机是指区别于个人电脑（PC）与低档服务器的计算机，往往认为价格在 10 万元以上的

[1]TOP500 榜单始于 1993 年，是对全球已安装的超级计算机“排座次”的知名排行榜，由美国和德国超算专家联合编制。TOP500 榜单每半年发布一次。排行榜主要编撰人为美国田纳西大学计算机学教授杰克·唐加拉。

就是高性能计算机，而把超级计算机看作最高档的几百台高性能计算机。

请注意，本专题讨论的“高性能计算”，包括硬件、软件、算法、应用、产业生态环境等，不仅仅限于构建“高性能计算机”——一字之差反映不同的战略思维。高性能计算本身就是国家的战略重器，涉及的战略性问题很多，由于篇幅有限，下面列出几个社会各界较为关心的战略性问题，稍做说明，供读者参考。

## 1 发展高性能计算的目的究竟是什么？

高性能计算可以应用于核模拟、密码破译、气候模拟、宇宙探索、基因研究、灾害预报、工业设计、新药研制、材料研究、动漫渲染等众多领域，对国防、国民经济建设和民生福祉都有不可替代的重大作用，发展高性能计算就是要让这巨大的作用发挥出来。同时，高性能计算也是中美大国博弈的重要领域，每一次较量的胜利都会给国人极大的激励，有力增强了民族自豪感和凝聚力。因此，发展高性能计算意义重大。

我国发展高性能计算需要正确处理世界排名与实际效用这两者的关系。其实，我国与美国在高性能计算领域的博弈主要是因为该领域研究对国防、经济和民生的实际效益，而不是某一次排名是否第一。只要认清楚这一点，两者就统一了。如果不重视实际应用绩效，而只把排名第一作为“政治正确”的标志，可能会产生误导。

## 2 如何全面部署计算机科研与产业的发展？

如果把高性能计算机理解成超级计算机，其在整个计算机产业中占比并不大。超级计算机主要是用来解决其他计算机解决不了的挑战性问题，采用几万个甚至百万个以上的处理器并行协同解决一个大问题。在实际应用中，更多的场合是需要同时响应大量的任务请求，即不是要算得快，而是要算得多。这一类应用需要高通量计算机，主要由云计算中心和大数据中心部署。目前银行等金融行业还在大量采购美国 IBM 的主机系统（Mainframe），他们买的主要不是计算速度，而是可靠性和软件的兼容性，业界称为高可靠或高可用系统。我国的计算机产业要从中低端向高端发展，因此我们的任务不仅是发展超级计算机，而且包含发展高端计算机。

美国政府 2015 年发布的“国家战略性计算计划”（NSCI）就是一个较全面的顶层规划，而 2016 年启动的 E 级计算机研制计划（ECP）只是美国能源部对 NSCI 计划的响应。我国国家重点研发计划中有“E 级计算机关键技术验证系统”重点专项，但没有包括其他高端计算级的顶层规划。在我国，高通量计算机至今没有重大项目支持，几大网络服务商需要的云计算和数据中心设备基本上是自行设计的，并委托其他公司组装。如果长期缺乏全国科技力量的支持，我国网络服务企业将难以形成全球竞争优势。

## 3 我国应重点发展什么类型的高性能计算机？

高性能计算机有两种基本类型：① 能力（Capability）型，强调解决单一复杂问题的最高计算速度，尽量缩短求解一个最大最难问题的时间；② 容量（Capacity）型，强调同时处理多个大任务，每一个任务只用到计算机的一部分能力。全球 TOP 500 超级计算机大多数属于容量型。科学研究对计算能力的需求是无止境的，E 级（$10^{18}$）计算机做出来后，还会提出 Z 级（$10^{21}$）计算的需求。研制能力型超级计算机必须突破现有计算机的技术瓶颈，以引领计算机技术的发展，因此，美国的 ECP 计划的目标是研制能力型计算机。

世界上最高水平的超级计算机主要用于科学研究，而科学计算在高性能计算机应用中占的比例已不到 10%。近几年大数据分析和机器学习等人工智能应用已成为高性能计算机的主要负载，2017 年智能应用在中国高性能计算机应用中的占比已提升到 56%，估计这个比例今后还将继续扩大。美国、日本等国纷纷将正在研制的超级计算机称为智能计算机。

长期以来，评测超级计算机的性能都采用 Linpack 测试程序，这是一个求解线性方程组的程序。这个程序的优点是可扩展性特别好，现在还没找到比它扩展性更好的测试程序。而且，Linpack 是 CPU 密集应用的程序，可以测出几乎满负荷、满功耗下的计算机浮点计算性能。从这个意义上讲，Linpack 是测试超级计算机可靠性和稳定性的理想程序。但是，求解线性方程组终究只是一种应用，全面衡量超级计算机的性能需要更合适的基准（Benchmark）测试程序，可惜现在还没有。由于功耗的限制，发展通用超级计算机已遇到极大的困难，近年来领域专用超级计算机成为热门研究方向，Linpack 显然不适合作为领域专用计算机的测试标准。

我国应重点发展什么类型的高性能计算机，这不是一个学术问题，而是一个科技需求问题，只有通过对我国国防、经济、科研和民生的潜在需求的认真调研才能回答。但有一点可以肯定，容量性超级计算机、智能计算机、领域专用超级计算机与能力型超级计算机一样重要，在做科技决策时应统筹兼顾。把研制 E 级高性能计算机的全部人力物力都投在争取 Linpack 指标世界第一可能是不明智的决策。

## 4　中国到底有没有对高性能计算的迫切需求，现在的应用水平怎么样？

从理论上讲，我国对超级计算机肯定有强烈需求；但从实际上讲，这一需求还与一个国家的科研水平、经济水平有关。2018 年，中国气象局安装了派-曙光超级计算机，峰值计算性能是 8 PFlops，计算能力已跃居气象领域世界第 3 位。众所周知，气象领域是使用超级计算的大户，目前能正常发挥作用的超级计算机离 E 级计算还有两个数量级的差距。气象部门要把 E 级超级计算机用起来，必须在基础研究、算法、软件和人才培养上做出巨大的努力。有人说，先有 E 级机，才会有 E 级计算的需求。这是对的，我们需要在 E 级计算机上培养 E 级用户。但一台超级计算机的平均有效寿命只有 5 年，5 年内哪些应用领域的用户可以培养出来也需要通过调研做出判断。

我国现有的超算中心究竟应用效益高不高是一个颇有争议的问题。有些超算中心宣称效益非常好，支持了上千项国家重大科技项目，产生了近百亿元经济效益；而媒体上也有文章说超算中心核心应用拓展不够，没有产生预期功效。造成这种局面的原因是缺乏第三方的公正评估。国家应组织有公信力的评测机构或学会对全国的超算中心做一次评估，了解清楚超算中心究竟完成了多少事关国家重大战略需求的计算任务？借助超算做出了哪些重大科学发现？对经济发展作出了哪些不可替代的贡献？P 级（$10^{15}$）以上的计算任务究竟占多大比例？……只有按照习近平总书记讲的搞清楚“实事”，掌握了实情，才能对我国超算的实际应用水平做出正确判断。

## 5　发展高性能计算要强调应用牵引还是技术驱动？

人们常说，发展科技既要需求牵引，又要

技术驱动。但在实际过程中，往往有所偏重。比较而言，美国发展超级计算机主要是应用牵引，而我国侧重于技术驱动。从一个例子可以看出美国应用牵引的倾向。美国最近开展的ECP计划的负责人Paul Messina是美国阿贡实验室的计算机应用科学家，ECP计划则是由阿贡实验室（超级计算机的应用方）主导的科研项目。在Summit计算机交付之前，美国能源部已经成立了25个应用软件研发小组，设计能够利用E级计算机的软件。ECP计划是否成功的指标不是Linpack性能，而是这25个应用性能的“几何平均值”，这意味着其中任何一个应用的性能都不能很差。美国是先有挑战性应用问题，为解决应用问题造新的计算机；我国的做法则是先造出世界领先的机器，再来找应用。发展超级计算机一定要国家战略导向，以及战略中的挑战问题导向。在研制新的超级计算机之前，应用部门一定要先把急需解决的挑战问题明明白白地提出来，用可考核的应用性能指标来评价正在研制的计算机。在应用牵引上我们应虚心地向美国同行学习。

强调应用牵引不是说技术驱动不重要。由于摩尔定律临近极限，学术界普遍认为现在是系统结构研究的黄金时代，但系统结构研究的困难超出人们的预期。

在ECP计划刚启动时，Paul Messina强调E级计算机研制要走所谓高架轨道（Higher Trajectory），两年以后描述E级计算的用语已经从“新型（Novel）”转向比较乏味的“先进（Advanced）”，Summit计算机的重大技术突破也不多。对于E级计算机和以后更高性能的超级计算机研制者，能耗、访存、通信、可靠性、应用性能这几道“高墙”必须越过。没有关键技术的重大突破，超级计算机不可能再上一个大台阶。中国计算机学者应当在这一征程中做出载入史册的贡献。国家在安排高性能计算重大科研任务时，不能只盯住工程任务，应更加重视颠覆性器件（如新型存储器件、超导、量子、光子器件等，以及几种器件的跨界协同设计）和变革性系统结构的基础研究。降低功耗的技术突破要摆在最优先的位置。

## 6 如何建立发展高性能计算的生态环境?

所谓科研和产业生态环境是指围绕着一个目标形成的，从基础研究、技术突破、产品研发到应用推广的协作共同体，不是简单的链条，而是相互关联的社会网络。

对于我国高性能计算生态环境而言，最薄弱的环节是软件。目前，我国大型科学计算的应用软件大部分依靠进口。我国的超算经费用于应用软件开发的还不到10%，美国相应的投入资金约为中国的6倍。振兴软件的关键是人才，目前能培养高性能计算软件人才的大学很少，因此建议应扩大该领域的招生名额。美国参与ECP计划软件开发的大学并不都是一流大学，一般的大学也承担了开发任务。

我国高性能计算生态环境的另一个薄弱环节是企业应用。美国公司的超算系统规模是中国公司的10倍多。例如，汽车行业的通用、克莱斯勒等公司，每家都有10多个超算系统，英国的BP石油公司也有世界上最大的工业用超级计算机。我国使用高性能计算机较多的是BAT（百度、阿里巴巴、腾讯）等网络服务公司，而在制造业的应用则规模较小。只有企业较普遍地用上了高性能计算机，才能真正走上高质量发展道路。构建高性能计算生态环境时，还要重视发挥骨干企业的作用。高性能计算机研发的一次性工程（即非经常性工程，NRE）费用很高，只有通过企业的工业化设计，采用标准化组件和Scale Down技术，才能使小规模的高性能计算机具有很高的性能价格比，通过批量销售收回 NRE 成本，才能使国家科研投入获得较高的回报。

本期专题由中国科学院计算技术研究所孙凝晖研究员、中科曙光公司历军总裁指导推进，文章作者还包括谭光明、金钟、迟学斌、孙家昶、李根国、冯圣中、范东睿、詹剑锋等，他们都是第一线的科研人员，有些已在高性能计算领域耕耘了二三十年。上述几个战略性问题在他们的文章中都有较详细的论述。一线科研人员的战略思考是基于常年的实践体会的，既有“顶天”的技术眼光，又很接地气，值得决策部门重视。

# 第 7 章

# 微处理器

# 中国科学院计算技术研究所研制微处理器的不懈努力

中国信息产业缺“芯”少“魂”是全国人民的心病，但1986年启动的863计划没有设立研制微处理器（CPU）的项目，理由是863计划民口15年的总投入只有50亿元，全部投给微处理器一个项目都不够。2000年，我开始担任中国科学院计算技术研究所（以下简称中科院计算所）所长，一上任就强调中科院计算所是国立研究所，应主动做与国家命运相关的大事。CPU就成了我首先考虑的大事。当时美国冒出一家叫Transmeta的新公司，采用“Code Morphing”的二进制翻译技术，推出可运行x86程序的超长指令芯片Crusoe。据我们了解，二进制编译的源头技术出自俄罗斯的Elbrus公司。中科院计算所有较强的编译队伍，采用二进制翻译做与x86应用程序兼容的芯片可能是一条奇径。2000年，我率团去莫斯科与俄罗斯科学家谈合作，在国家没有项目支持的情况下，我斗胆答应预付300万美元的知识产权费，但他们的开价是3亿美元，根本谈不拢。与俄罗斯合作失败后，我就下决心自己拉队伍做CPU。中国科学院每年给中科院计算所的知识创新工程经费为2 000万元，我计划用其中的一半研制CPU芯片。

龙芯CPU的起步很艰难。2001年“十五”期间863计划IC设计重大专项预启动时，全国批准了20多个项目，中科院计算所申请20万元都没有评上。这是因为我国一直把发展集成电路看成微电子专业的事，长期忽视计算机系统专业对发展CPU可能做出的贡献。龙芯一号问世和后来的不断升级才逐步改变了这种看法。2004年，科技部组织国家中长期科技发展规划（2006—2020年）战略研究，我担任战略高技术专家组副组长（组长是中国科学院的路甬祥院长）和执笔组负责人。在战略研究快结束时，负责遴选重大专项的主管领导江上舟希望CPU能被立为重大专项，我写了一份设立高性能通用CPU重大专项的立项报告，通过政策组评审后由组长马俊如上报。后来在论证中，增加了基础软件和核心电子器件，形成了核高基重大专项。龙芯CPU在“十一五”期间得到核高基重大专项的支持，但到“十二五”就停止了。2010年以后，龙芯公司是在民间资本的支撑下，通过自己在市场上磨炼拼搏成长壮大起来的。胡伟武领导的龙芯团队20年来目标如一，不放弃、不动摇，现在已经成为国防应用和党政办公系统的主要国产CPU供应商。在龙芯发展的过程中，中科院计算所对龙芯给予了全力以赴的支持，在核高基经费到位之前，中科院计算所曾经为龙芯垫付了8 000万元之多的研发费用。

龙芯在发展CPU的技术路线选择上也走过弯路。“十一五”期间，国内3家承担核高基重大专项CPU研制任务的单位都在研制用于超级计算机的CPU（俗称大CPU），过度追求多核浮点计算峰值性能单一指标，放松了单核性能的提高。龙芯CPU的单核性能从2006年的龙芯2E到2013年的龙芯3B1500只提高了50%左右，而此期间市场主流x86处理器的单核性能提高了5倍以上。也就是说，“十一五”期间我国CPU通用处理性能与国外的差距从一两倍扩大到一个数量级。从2013年开始，龙芯公司对CPU的研发路线进行了重大调整，重点致力于提高单核性能，龙芯3A3000和龙芯3A4000的单核性能提高了几倍，大大缩小了与国外的差距。在高主频、低功耗、强通用性等方面，龙芯与x86芯片还有不小差距。要在广阔的民口市场上打造有竞争力的生态环境，龙芯还有很长的路要走，任重道远。

中科院计算所研制的微处理器不只是龙芯CPU。从2006年到2015年，我作为首席科学家连续承担了两次973项目，第一个项目是“延长摩尔定律的微处理芯片新原理、新结构与新方法研究”，第二个项目是“高通量计算系统的构建原理、支撑技术及云服务应用”，这两个项目都探索了新的微处理器，为研制万亿次级的众核芯片和适合高通量计算的芯片开辟了一条新路。在973项目成果的基础上，中科院计算所创办了一家以视频流处理为主要应用方向的芯片设计公司——中科睿芯科技有限公司。高通量计算是我们在973计划项目研究中提出的新概念，反映了云计算和移动互联网兴起后出现的新计算模式，不是追求算得快，而是追求算得多。这个概念已被学术界和企业界普遍接受。在此基础上，中科院计算所又提出了“信息高铁”新概念，为构建信息基础设施提出了新的目标。

研制CPU有几条不同的道路，需要采用“两条腿走路”的方针。龙芯走的是在x86体系之外另外开辟产业生态的路，先在国防等对安全可控有较高要求的部门建立根据地，再逐步向民口扩展。但在金融、通信、电力等众多民口部门，x86已经是主流的产业生态，积累了数万亿元的应用软件，特别是在服务器市场上，x86 CPU已占到90%以上的份额。因此，在x86生态中取得发展的主动权是国内产业界必须考虑的大事。通过与AMD公司合作，曙光公司为第一大股东的海光公司已经推出国际先进水平的32核服务器CPU，在引进消化再创新的道路上迈出了新的步伐。近10年来，中科院计算所率先研制成功神经网络加速芯片——寒武纪芯片，引领了国际上发展人工智能加速芯片的热潮。中科院计算所还在大力发展RISC-V芯片和开源硬件，以期为物联网等未来的应用打造新的产业生态环境。海光服务器CPU、寒武纪芯片和RISC-V等新型芯片的研究成果没有反映在这本选集中。

发表于*IEEE Micro*，2009年第2期

# Godson-3: a scalable multicore RISC processor with x86 emulation

Wei-wu HU, Jian WANG, Xiang GAO, Yun-ji CHEN, Qi LIU, Guo-jie LI

Institute of Computing Technology, Chinese Academy of Sciences

**Abstract:** The Godson-3 microprocessor aims at high-throughput server applications, high-performance scientific computing, and high-end embedded applications. It offers a scalable network on chip, hardware support for x86 emulation, and a reconfigurable architecture. The four-core Godson-3 chip is fabricated with 65 nm CMOS technology. 8- and 16-core Godson-3 chips are in development.

Godson-3 is the third generation of the Godson microprocessor series, a project of the Institute of Computing Technology at the Chinese Academy of Sciences. As a multicore processor, Godson-3 targets high-throughput server applications, high-performance scientific computing, and high-end embedded applications.

Godson-3's scalable and distributed on-chip network connects processor cores and globally addressed level-two (L2) cache modules. A directory-based cache-coherence protocol maintains multiple level-one (LI) copies of the same L2 block. Godson-3's MIPS64-compatible superscalar reduced-instruction-set-computing (RISC) processor core is designed for high performance and low power dissipation. It also supports efficient x86 to MIPS binary translation through dedicated hardware support.

Godson-3 adopts the scalable mesh of crossbar (SMOC) on-chip network topology. Using the SMOC architecture, a 2×2 mesh network can support a 16-core processor, and a 4×4 mesh network can support a 64-core processor. We've already defined the four-core, eight-core, and 16-core product chips, and we've designed and fabricated four-core Godson-3 based on 65 nm CMOS technology. The eight-core and 16-core Godson-3 chips are still in physical implementation.

## 1 CPU core features

The GS464 is a general-purpose processor core, upgraded from the Godson-2 microprocessor. The GS464's four-way superscalar execution mechanism has extremely high requirements for resolving inter-instruction dependency and providing instructions and data. It therefore uses out-of-order execution and aggressive cache design to improve pipeline efficiency, as other modern microprocessors do[1-6].

The GS464 out-of-order execution scheme combines register renaming, dynamic scheduling[7], and branch prediction. GS464 has a 64-entry physical register file for fixed-point and floating-point register mapping. In GS464, the 16-entry fixed-point reservation station and the 16-entry floating-point reservation station issue instructions out of order, whereas the 64-entry reorder queue commits out-of-order executed instructions in program order. GS464 also implements a 16-entry branch target buffer, an 8 KB entry branch history table, a 9 bit global history register, and a 4-entry return address stack for branch prediction.

GS464 has two fixed-point functional units and two floating-point functional units. Both fixed-point units execute addition, subtraction, logical, shift, and comparison instructions. In addition, the first fixed-point arithmetic logic unit (ALU1) executes trap, conditional move, and branch instructions; the second (ALU2) executes multiplication and division instructions. The first floating-point unit (FALU1) can execute all floating-point instructions; the second floating-point unit (FALU2) can execute floating-point addition, subtraction, and multiplication instructions.

The GS464 memory system supports 64 bit virtual addresses and 48 bit physical addresses, and can access a 128 bit quad word in one cycle. GS464 has a 64 KB L1 instruction cache and a 64 KB L1 data cache; both are four-way set associative. GS464's fully associative translation lookaside buffer (TLB) has 64 entries, each of which maps an odd page and an even page. The 24-entry memory-access queue, which contains a content-addressable memory for dynamic memory disambiguation, supports out-of-order memory access, nonblocking cache, and load speculation. GS464's memory-access pipeline includes four cycles: address calculation, TLB and cache reading, tag comparison, and write back.

GS464 implements the Enhanced JTAG (EJTAG) standard for debugging and performance tuning. We implemented error-correcting code for the data cache and parity check for the instruction cache. We optimized the CPU core for low power dissipation in the architectural, logical, and physical design stages.

Figure 1 shows the architecture of the GS464 CPU core.

We designed several CPU chips with different process technologies based on the GS464 architecture. The Godson-2F[8]—which integrates the GS464 core, a 512 KB L2 cache, a 333 MHz DDR2 controller, and a PCI/PCIX controller—is based on 90 nm CMOS technology. It achieves 1 GHz with both SPECint2000 and SPECfp2000 scores of more than 500. The chip includes 51 million transistors and has a die size of 42 square millimeters. It consumes from 3 to 5 watts depending on the application.

## 2 Hardware support for x86 to MIPS binary translation

To support x86 emulation, Godson-3 provides hardware support for binary translation from x86 to MIPS in its GS464 core. Although the Crusoe processor supports translation from x86 to VLIW[9], no commercial RISC processor provides dedicated support for x86 emulation because of the difference between x86 and RISC[1-5]. Because some x86-related features aren't present in MIPS(EFlags, the floating-point register stack, segment addressing mode, and so

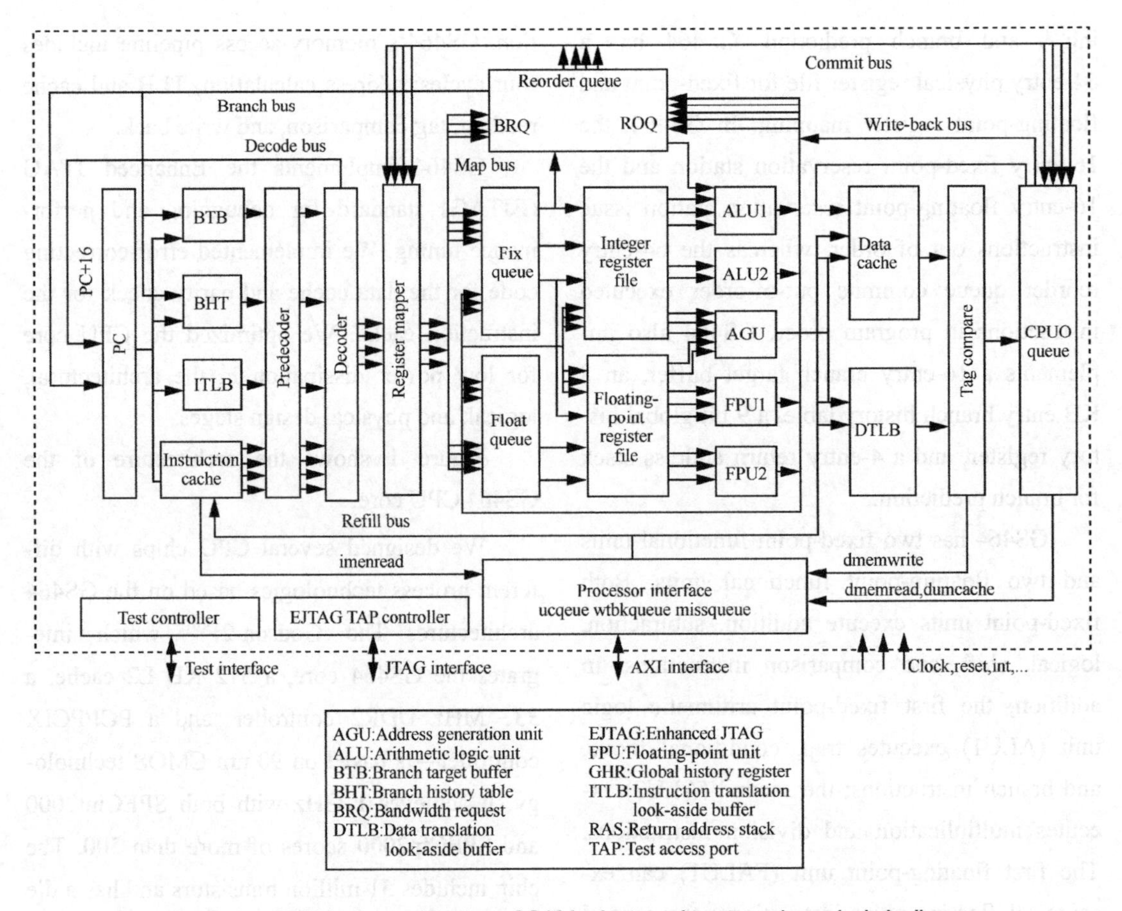


Figure 1. GS464 microarchitecture. GS464 adopts a nine-stage dynamical pipeline

on), software-based translation from x86 binary to MIPS binary is inefficient[10-11]. In many cases, translating an x86 instruction requires tens of MIPS instructions because of the difference between the x86 and MIPS ISAs. Godson-3's x86 binary translator smoothes translation from x86 binary to MIPS binary with minimal hardware support. To achieve this, GS464 defines new instructions and runtime environments through the MIPS64 user-defined interface (UDI) to bridge the gap between the x86 and MIPS64 ISAs. It defines and implements new instructions in MIPS format for functions that are in x86 ISA but not in MIPS64 ISA.

The Godson-3 virtual machine will be compatible with x86 at both the ISA level and Linux application binary interface (ABI) level. We'll build the system-level and process-level virtual machines accordingly. The ISA-level compatibility is for low-cost PC applications, where Microsoft Windows is most popular; the Linux ABI-level compatibility is for server applications, where Linux servers with x86 processors are most popular. Figure 2 shows the Godson-3 binary translation system's software architecture. The process-level and system-level virtual machine monitors (VMMs) are implemented on Linux, which is improved to provide x86-compatible system calls.

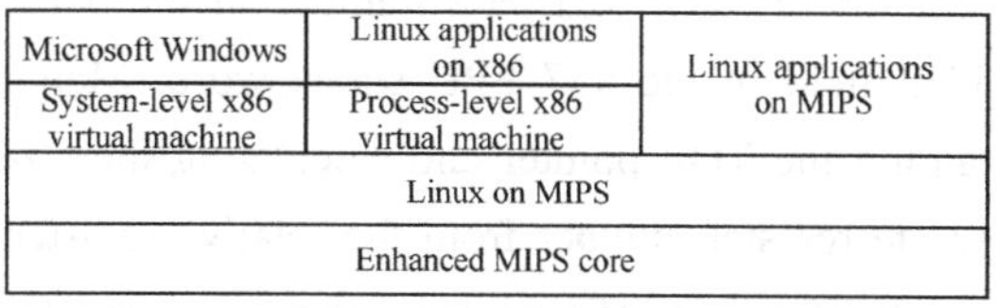

<table>
<tr><td>Microsoft Windows</td><td>Linux applications on x86</td><td rowspan="2">Linux applications on MIPS</td></tr>
<tr><td>System-level x86 virtual machine</td><td>Process-level x86 virtual machine</td></tr>
<tr><td colspan="3">Linux on MIPS</td></tr>
<tr><td colspan="3">Enhanced MIPS core</td></tr>
</table>

Figure 2. The GS464 virtual machine's software architecture. The x86 operating systems and applications are built on MIPS Linux system through virtual machine monitor

## 2.1 Hardware support fo EFlag of x86

A major difference between the x86 and MIPS ISAs is that the x86 ISA uses EFlags. Most x86 fixed-point arithmetic instructions generate 6 bit EFlags as by-products of the arithmetic calculation, and the branch directions of branch instructions are determined according to the EFlag values. MIPS fixed-point arithmetic instructions don't generate EFlags, and MIPS branch instructions decide branch directions according to the general-purpose registers' values.

Software simulation of the 6 bit EFlags with MIPS instructions requires tens of MIPS instructions. As the x86 program segment in Figure 3(a) shows, we need about 40 MIPS instructions to

| Number of instructions | | Instruction | | | | Comment |
|---|---|---|---|---|---|---|
| 0 | | SUB | ECX | EDX | | |
| 1 | | JE | x86_target | | | |

(a)

| | | | | | | |
|---|---|---|---|---|---|---|
| 0.00 | | SUBU | Result | Recx | Redx | |
| 0.01 | | SRL | Rsf | Result | 31 | /*SF=Result[31]*/ |
| 0.02 | | BEQ | Result | R0 | L1 | |
| 0.03 | | ADD | Rzf | R0 | R0 | /*ZF=0*/ |
| 0.04 | | B | L2 | | | |
| 0.05 | | NOP | | | | |
| 0.06 | L1: | ADDI | Rzf | R0 | 1 | /*ZF=1*/ |
| ⋮ | ⋮ | ⋮ | ⋮ | ⋮ | ⋮ | ⋮ |
| 0.35 | | B | LB | | | |
| 0.36 | | NOP | | | | |
| 0.37 | L7: | ADDI | Rcf | R0 | 1 | /*CF=1*/ |
| 0.38 | L8: | ADD | Recx | Result | R0 | |
| 1.00 | | BNE | Rzf | R0 | MIPS_target | |
| 1.01 | | NOP | | | | |

(b)

| | | | | | | |
|---|---|---|---|---|---|---|
| 0.0 | | SUBU | Result | Recx | Redx | /*Generating Sub result*/ |
| 0.1 | | SETFLAG | | | | |
| 0.2 | | SUBU | Reflag | Recx | Redx | /*Generating EFLAGS*/ |
| 1.0 | | x86JE | Reglag | MIPS_target | | /*Branch on EFLAGS*/ |

(c)

| | | | | | | |
|---|---|---|---|---|---|---|
| 0.0 | | SUB | Result | Recx | Redx | /*Generating Sub result*/ |
| 0.1 | | x86SUB | Reflag | Recx | Redx | /*Generating EFLAGS*/ |
| 1.0 | | x86JE | Reglag | MIPS_target | | /*Branch on EFLAGS*/ |

(d)

Figure 3. Example of EFlag translation: Original x86 program (a); the program translated with standard MIPS code(b). We can reduce the number of instructions by adding a SetFlag prefix, which turns the SUB instruction into its EFlag counterpart (c), or by adding a new instruction to define new EFlag counterpart instructions(d). The instructions in boldface type are new instructions for x86 emulation

simulate the "SUB ecx, edx" instruction to produce the subtraction result and the four most commonly used bits of EFlag (SF, ZF, OF, and CF), shown in Figure 3(b).

To reduce the cost of generating the x86 EFlags with MIPS instructions, GS464 provides an EFlag counterpart instruction for each fixed-point arithmetic instruction: either by adding a SetFlag prefix to the original instruction, or, for frequently used instructions, by defining the EFlag counterpart instruction. Using new instructions can significantly reduce the number of translated instructions. For example, adding a SetFlag prefix to the SUB Rl, R2, R3 instruction turns the SUB instruction into its EFlag counterpart, which performs the same calculation as the original SUB instruction but generates x86 EFlags instead of the difference of R2 and R3 (as Figure 3(c) shows). We can also define the new instruction x86SUB for generating EFlags of the subtraction, as Figure 3(d) shows.

An instruction's EFlag counterpart can reuse most of the original instruction's data paths, such as register renaming logic, reorder logic, issue logic, and write-back logic. We only need to slightly adjust the decode logic and execution unit.

GS464 also defines a set of branch instructions corresponding to x86 EFlag-based branch instructions, such as x86JE.

## 2.2 Hardware support for the floating-point format and register

The x87 FPU differs from the FPU of RISC processors in that it is accessed in a stack-based way and supports 80 bit floating-point numbers.

The x87 FPU instructions (math-related instructions for the x86 architecture) treat the eight x87 FPU data registers as the register stack. It addresses the data registers relative to the register on the top of the stack, and it stores the current top-of-stack register's number in the 3 bit TOP field in the x87 FPU status word. Maintaining the TOP pointer and calculating the absolute register number from the relative register number through software at runtime is costly. To solve this problem, GS464 maintains a TOP pointer dynamically. GS464 adds the TOP value to the floating-point register number in the decode stage. It uses the new register number as a logical register number to look up the physical roister number in the register renaming stage. We define some instructions to modify the TOP pointer. GS464 uses a hardware flag in the MIPS floating-point control register to indicate whether a register number for floating-point instructions is relative to TOP. The TOP pointer only affects MIPS instructions translated from x86 instructions. With the support of the GS464 floating-point register, the TOP pointer reduces more than 10 instructions in each x86 floating-point instruction translation. It also reduces the number of switches between the translator and the generated code.

The x87 FPU supports 80 bit floating-point numbers, whereas MIPS supports 64 bit floating-point numbers. Transferring between 80 bit and 64 bit floating-point numbers requires more than 40 integer instructions. GS464 defines one instruction to transfer an 80 bit floating-point number stored in two 64 bit registers to a 64 bit number kept in one register, and two instructions to transfer a 64 bit floating-point number to an 80 bit floating-point number that occupies two registers. Figure 4 shows an example of floating-point format translation. Without any hardware support, about 40 MIPS instructions in Figure 4(b) are needed to simulate the three x86 instructions in Figure 4(a), which GS464 can emulate with four instructions.

The x87 FPU has a 16 bit tag word to indicate the contents of each of the eight registers in the x87 FPU data-register stack (one 2 bit tag per register). The tag codes indicate whether a register contains a valid number, zero, a special floating-point number, or is empty. The x87 FPU uses the tag values to detect stack overflow and underflow. GS464 provides dedicated instructions to simulate the x87 tag with general-purpose registers and defines a new exception to reflect stack overflow or underflow exceptions in tag simulation. Except when turned on explicitly, the binary translator can ignore the tag simulation because correct programs never raise the stack overflow or underflow exception.

## 2.3 Hardware support for x86 multimedia instructions

The x86 ISA defines powerful multimedia instruction sets such as MMX, streaming SIMD extensions (SSE), and SSE2, which differ from the MIPS digital media extension (MDMX) in application-specific extensions of MIPS. GS464 emulates x86 multimedia instructions using its

| Number of instructions | Instruction | | Comment |
|---|---|---|---|
| 0 | FLD | *%R10 | |
| 1 | FMUL | *16(%R10) | |
| 2 | FSTP | *%R10 | |

(a)

| | | | | | |
|---|---|---|---|---|---|
| 0.00 | LD | Rtmp1 | 12(R8) | | /*convert 1st operand*/ |
| 0.01 | LD | Rtmp2 | 4(R8) | | |
| 0.02 | ANDI | Rsign | Rtmp1 | | /*get sign bit and sign bit of exp*/ |
| 0.03 | DSLL32 | Rsign | Rsign | 16 | /*get biased exponent |
| ⋮ | ⋮ | ⋮ | ⋮ | ⋮ | ⋮ |
| 0.23 | DMTC1 | F8 | Rtp2 | | |
| 1.00 | MUL.d | F9 | F7 | F8 | /*64 bit multiply*/ |
| 2.00 | DMFC1 | Rres | F9 | | |
| 2.01 | DSRL32 | Rsign | Rres | 31 | /*get sign bit*/ |
| ⋮ | ⋮ | ⋮ | ⋮ | ⋮ | ⋮ |
| 2.12 | SD | Rres1 | 12(R8) | | /*write back result*/ |
| 2.13 | SD | Rres2 | 4(R8) | | |

(b)

| | | | | | |
|---|---|---|---|---|---|
| 0.0 | GSLQC1 | F4 | 4(R8) | F5 | /*128 bit load to F4 and F5*/ |
| 0.1 | CVT.d.Id | F7 | F4 | | /*80 bit to 64 bit convert*/ |
| 0.2 | GSLQC1 | F2 | 20(R8) | F3 | /*128 bit load to F2 and F3*/ |
| 0.3 | CVT.d.Id | F8 | F2 | | /*80 bit to 64 bit convert*/ |
| 1.0 | MUL.d | F9 | F7 | F8 | /*64 bit multiplication*/ |
| 2.0 | CVT.ud.d | F7 | F9 | | /*64 bit to high part of 80 bit*/ |
| 2.1 | CVT.Id.d | F8 | F9 | | /*64 bit to low part of 80 bit*/ |
| 2.2 | GSSQC1 | F7 | 4(RB) | | /*128 bit store*/ |

(c)

Figure 4. Example of 80 bit floating-point operation translation.The original x86 program (a), translated with standard MIPS code (b), and translated with GS464 floating-point conversion instructions (c). The instructions in boldface are new instructions for x86 emulation

own multimedia instructions with little additional hardware cost.

In GS464, multimedia instructions have similar functions to those in x86 SSE2. GS464 defines and implements its multimedia instructions by extending the *fmt* field in the floating-point operations. Like MIPS floating-point instructions, GS464 internal floating-point operations use a 5 bit *fmt* field to specify data types. The *fmt* values of 16, 17, 20, 21, and 22 represent single-precision floating point, double-precision floating point, fixed-point word, fixed-point long, and paired single-precision floating point, respectively. GS464 SIMD multimedia operations extend the *fmt* fields in floating-point operations to define eight 8 bit or four 16 bit fixed-point data units in the 64 bit floating-point data path. For example, the ADD.fmt operation represents ADD.single, ADD.double, or ADD.PS in MIPS. GS464 extends the operation to represent ADD.8×8 and ADD.4×16. To further facilitate emulating x86 multimedia instructions, GS464 extends MIPS-style unaligned memory-access instructions to floating-point registers.

## 2.4 Other hardware support for x86 binary translation

In addition to support for x86 EFlags, floating-point instructions, and multimedia instructions, GS464 adopts many other techniques to further facilitate x86 emulation.

New addressing mode. The x86 ISA has more flexible addressing modes than MIPS. It supports the SIB addressing mode in the form of "(base) + (index) × scale + disp", whereas MIPS only supports the "(base)+disp" for both fixed-point and floating-point load and store instructions, and the "(base)+(index)" addressing mode for floating-point load and store instructions. To ease the translation of x86 addressing modes, GS464 supports the "(base)+(index) + disp8" addressing mode for both fixed-point and floating-point load and store instructions.

Bounded load and store. GS464 supports bounded load and store instructions, which read the bound register as the memory-access boundary in addition to the normal base roister and value register. The bounded load and store instructions have the same behavior as normal load and store instructions, except an address exception is raised if the memory-access address of a load and store instruction exceeds the boundary address. The bounded load and store instructions help ease the translation of segment address mode instructions in x86.

Fixed-point multiplication and division. MIPS fixed-point multiplication and division instructions use the special Hi/Lo registers as destination registers, and MIPS provides instructions to move data between Hi/Lo registers and general-purpose registers. GS464 implements fixed-point multiplication and division instructions, which use general-purpose registers as destination registers to ease the translation of fixed-point multiplication and division x86 instructions.

Byte insertion and extraction. The x86 ISA supports 8 bit, 16 bit, 32 bit, and 64 bit operations, whereas RISC processors normally support 32 bit and 64 bit operations. To close this gap, GS464 implements flexible byte insertion instructions that can insert a byte, half word, or word from any location of a register to any location of another register; and byte extraction instructions that extract a byte, half word, or word from any register location and store the result to another register after zero or sign extension.

## 2.5 Hardware supports for binary translation mechanism

Binary translation dynamically generates binary codes on the target machine. The binary codes generated during runtime are the data results of the binary translator. Because the binary translator stores these codes in the data cache, executing them requires flushing them from the data cache and loading them into the instruction cache. However, flushing the data cache through software to keep coherence between the data and instruction caches is time consuming. Therefore, GS464 keeps coherence between the data and instruction caches, as well as the L2 cache, through hardware.

Translation of indirect branch instructions is costly because the binary translator must look up the MIPS branch target dynamically according to the x86 branch target from some mapping mechanism, such as a hash table. GS464 implements a 64-entry content-associated memory (CAM) to speed up the translation and execution of indirect branch instructions. Each CAM entry includes a process ID field, an address field, and a data field. GS464 provides instructions to read, write, and probe the CAM. Figure 5 shows an indirect branch target translation. GS464 CAM instructions greatly reduce the complexity of translation results. Although we've added more codes to handle the CAM miss situation, they're rarely executed.

## 2.6 Binary translation system

The Godson-3 binary translator is built on top of the Linux operating system. We improved the Linux operating system based on Godson-3's MIPS ISA to provide a system call compatible with x86 Linux and increase binary translator efficiency. We implemented a process-level binary translator on Linux to run x86 applications. We can also implement a system-level binary translator to

| Number of instructions | Instruction | | | Comment |
|---|---|---|---|---|
| 0 | MOV | %RAX | %R11 | |
| 1 | JMPQ | %*R11 | | |

(a)

| | | | | |
|---|---|---|---|---|
| 0 | MOVE | Rr11 | Rrax | |
| 1.0 | CAMPV | Rtmp | Rr11 | /*Look up the first level indirect jump address*/ |
| 1.1 | CAMPV | Rtgt | Rtmp | /*Look up the final jump address*/ |
| 1.2 | JR | Rtgt | | |

(b)

Figure 5. Example of indirect branch target translation: The original x86 program (a), and the program translated with Godson-3 content-associated memory (CAM) in structions (b). The boldface text indicates new instructions for x86 emulation

achieve x86 compatibility at the ISA level.

Godson-3's binary translation system is an improvement upon the open-source binary translator QEMU[12]. In addition to implementing a new interpreter and translator under the QEMU framework, we redesigned the intermediate representation to allow optimizations, such as fixed register allocation and lazy conditional code evaluation. Like other traditional binary translators, the Godson-3 VMM initially interprets x86 instructions on the MIPS processor and monitors the behavior during interpretation. When the VMM finds hot spots, it translates their x86 codes to MIPS codes for execution and optimizes them at different levels according to the hot degree. The multicore Godson-3 can also perform parallel optimization for very hot spots. The Godson-3 hardware support for x86 instruction translation makes the translation process straightforward, not only improving the translated code's efficiency, but also simplifying the binary translator's implementation.

System optimization also improves the Godson-3 binary translator's performance. To fully use the 64 bit processor resources, such as registers, we implemented the N32/N64 tool chain. The Godson-3 system uses GNU binutils, the GNU compiler collection (GCC), and the GNU C library (GLIBC) as the basic compilation environment. We modified the binutils' GNU Assembler (Gas) to help GCC recognize all the new instructions. We added new instruction templates and pipeline descriptions to GCC to improve the generated code quality. We also implemented auto-vectorization for the 128 bit memory-access instructions and multimedia instructions to fully utilize the Godson-3 processor resources. We profiled GLIBC and rewrote the frequently executed routines.

## 2.7 Preliminary performance results

The first four-core Godson-3 design was taped out in 2008. Before the chip returned from fabrication, we carried out the Godson-3 performance analysis on two platforms: a register-transfer-level (RTL) simulation platform and a field-programmable gate array (FPGA) prototyping platform. In the RTL simulation environment, we set the core clock frequency to 1 GHz, the DDR2/DDR3 clock frequency to 333 MHz, and the HyperTransport clock frequency to 800 MHz. To speed up the simulation, we used Cadence's Xtreme-3 simulation accelerator, which can achieve a speed of 200 000 to 400 000 cycles per second. Because of the difficulty of building a full-scale Godson-3 FPGA prototype system, we built a partial-scale prototype to evaluate a single processor core's performance. The prototype system includes one processor core, a 1 MB L2 cache, one DDR2/DDR3 controller, and one HyperTransport controller. FPGA prototyping speed is 50 MHz, which is much faster than RTL simulation. Because the ratios between the FPGA prototype's core and I/O clocks differ from those in a real system, we carefully adjusted the FPGA prototyping system's I/O latency to obtain accurate performance results.

Table 1 shows the benchmarks we tested on the Xtreme-3 and FPGA platforms. We selected nine typical kernels or full applications to evaluate the hardware improvements and software translation efficiency. We ran all benchmarks in three modes:

• native MIPS mode, in which benchmarks are directly compiled into MIPS binary and run on MIPS hardware;

• basic translator mode, in which benchmarks are compiled into x86 binary and run on MIPS hardware using the basic QEMU binary translator;

• improved translator mode, in which benchmarks are compiled into x86 binary and run on MIPS hardware using the improved QEMU binary translator with x86 binary translation acceleration on RISC processors (XBar) hardware support.

We compiled all programs with the GCC-O3 flag.

Figure 6 shows the relative performance of basic and improved translator modes compared to native MIPS mode. Godson-3's x86 emulation hardware support significantly accelerates binary translation from x86 to MIPS, and on average achieves performance that is nearly 70 percent of the ideal mode.

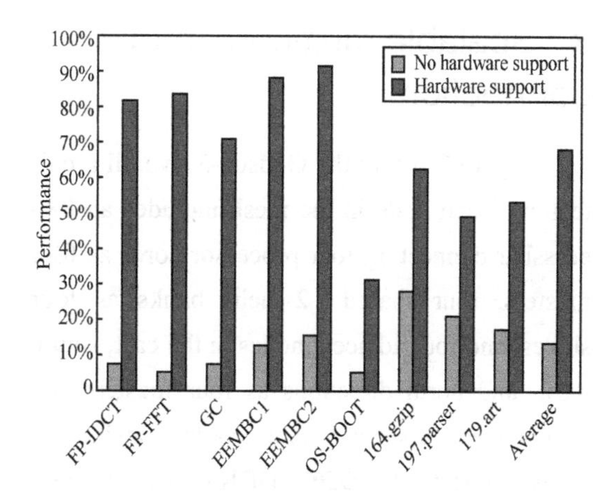


Figure 6. Experimental results for the nine benchmarks. 100 percent performance represents the execution of native MIPS code. Columns represent the execution of x86 emulation with and without hardware support. Higher is better

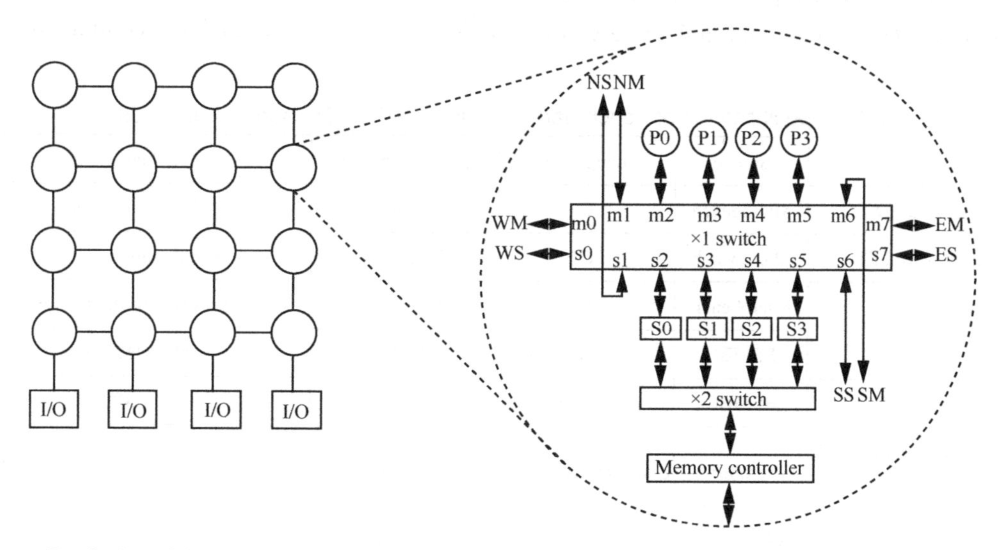


Figure 7. Godson-3 in terconnection topology. A 2D mesh connects nodes, and each node connects cores and L2 modules through a crossbar; memory controllers are connected in nodes, and I/O controllers are connected in the 2D mesh boundary

The performance results in Figure 6 are still preliminary. Much more work must be done to further improve the binary translator.

## 3 Scalable multicore interconnection

Figure 7 shows the Godson-3 overall architecture. Each node in the mesh includes an 8×8 crossbar connecting four processor cores as four masters, four shared L2-cache banks as four slaves, and four adjacent nodes in the east, south, west, and north directions as four masters and four slaves. A second-level crossbar inside the node connects the DDR2/DDR3 memory controllers to L2-cache banks. The Godson-3 HyperTransport I/O controller is connected to the free crossbar ports of boundary nodes.

The four-core, eight-core, and 16-core Godson-3 chips take one, two, and four nodes in the mesh, respectively. Figure 8 shows the architecture of the four-core and eight-core Godson-3. Both have two DDR2/ DDR3 controllers, two HyperTransport controllers, and other necessary peripheral interfaces. The four-core and eight-core Godson-3 have the same I/O package.

The Godson-3 interconnection network takes the 128 bit AXI standard interface, which is simple, efficient, and open. We extended the LI crossbar AXI interface to support cache coherence in Godson-3.

Figure 9 shows the crossbar's architecture. An 8×8 multiplexing matrix connects eight AXI master link (AML) modules and eight AXI slave link (ASL) modules. Each link has five channels according to the AXI protocol:

- write address channel (AW),
- write data channel (W),
- write response channel (B),
- read address channel (AR), and
- read data channel (R).

The AML routes read and write requests (AW, W, and AR channels) according to the

Table 1. Specifications of benchmark kernels tested on Godson-3 prototypes

| Name | Source | Language | Experiment platform | Note |
|---|---|---|---|---|
| Floating-point IDCT | Microbench | C | Xtreme-3 | x86 floating-point |
| Floating-point FFT | Microbench | C | Xtreme-3 | x86 floating-point |
| General control | Microbench | C | Xtreme-3 | x86 control complicated programs |
| Fixed-point IDCT | EEMBC | x86 assembly | FPGA | x86 SIMD |
| Fixed-point FFT | EEMBC | x86 assembly | FPGA | x86 SIMD |
| Operating system startup code Microbench | Microbench | C and x86 assembly | Xtreme-3/FPGA | |
| 164.gzip (train) | SPEC 2000 | C | FPGA | |
| 197.parser (train) | SPEC 2000 | C | FPGA | |
| 179 .art (train) | SPEC 2000 | C | FPGA | |

access address. Each AW or AR channel has eight address windows, each of which assigns a destination ASL port to a matching request. The ASL routes read and write replies (R and B channels) according to the corresponding request's AML port number. Both AML and ASL have two pipeline stages-that is, the crossbar has a latency of four hops. Two buffers for each pipeline stage prevent the pipeline stalling signal from propagating across stages.

## 4 Reconfigurable architecture

Godson-3's reconfigurability involves three features:

- reconfigurable processor cores,
- dynamic L2 cache migration,
- reconfigurable DMA engine.

Because Godson-3's interconnection network adopts the AXI protocol, we can insert any processor cores complying with the AXI protocol

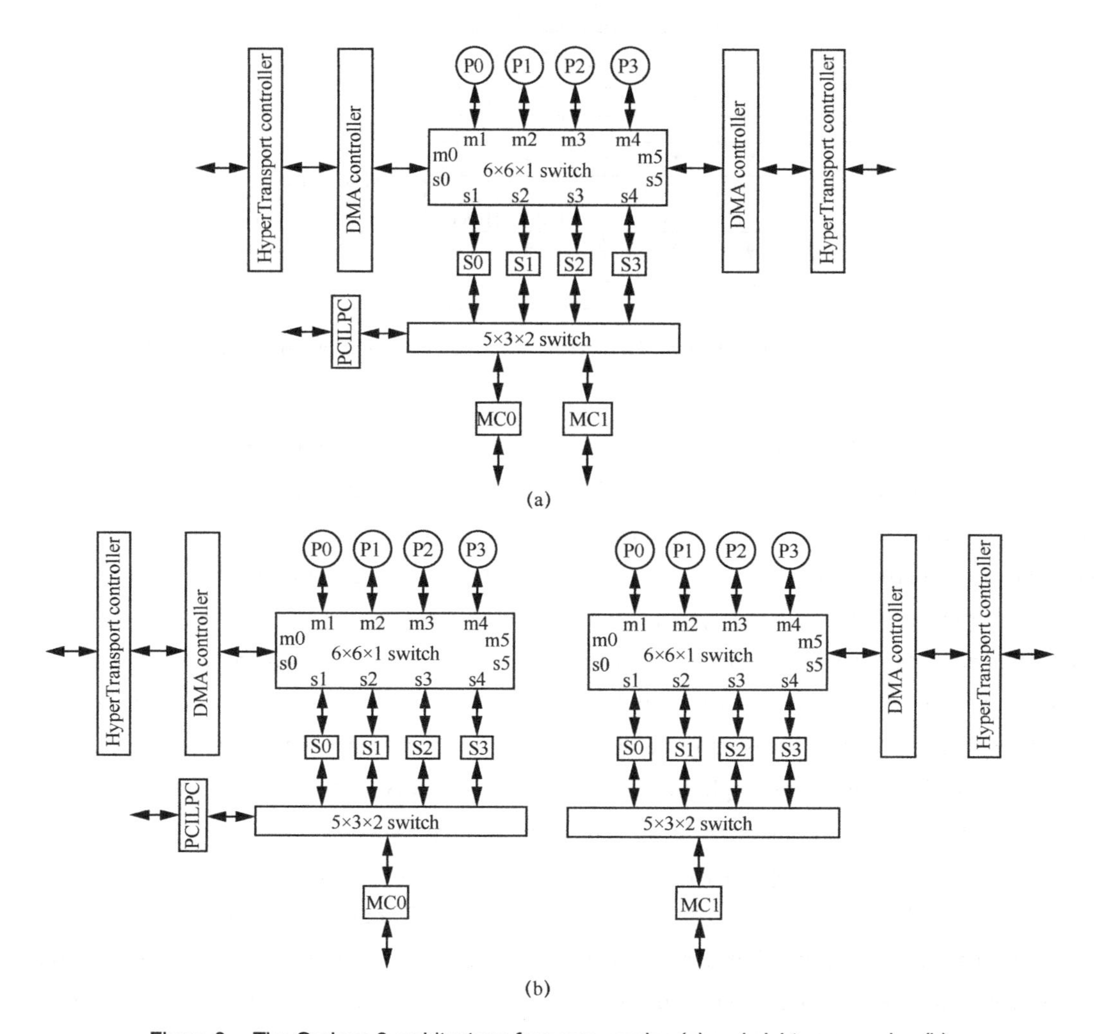


Figure 8. The Godson-3 architecture: four-core version (a) and eight-core version (b)

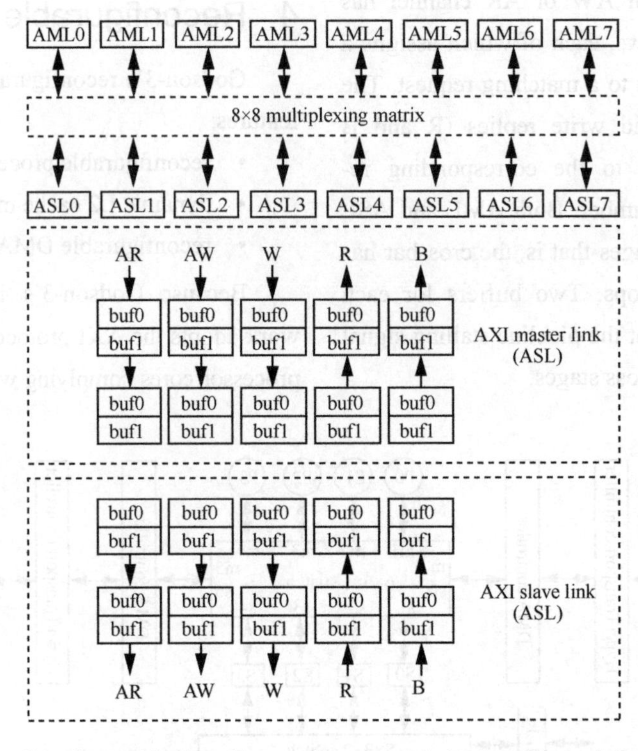


Figure 9. Crossbar architecture. Four cycles are needed to travel through the crossbar

into the network slots. Depending on the application, we can configure the Godson-3 chips' AXI ports to contain either general-purpose GS464 cores or special-purpose cores.

A distributed shared memory system's performance depends heavily on memory-access locality. In a nonuniform cache access (NUCA) system such as Godson-3, accesses to nonlocal L2 cache blocks suffer from high L2 access latency. Godson-3 can increase the locality of L2 cache access by migrating shared L2 cache blocks. Each AW or AR channel in the AXI crossbar has eight reconfigurable address windows, which let the software dynamically bind memory addresses to cache block locations. With the reconfigurable address windows, the software can migrate home blocks across cache bank, or set cache blocks as private or shared blocks according to the locality of memory accesses.

We can also configure Godson-3's DMA engine to achieve high performance. Software can decide whether the DMA data is to or from main memory or to or from the L2 cache directly. The DMA engine maintains cache coherence automatically when transferring data between I/O and memory. In addition, we can configure the DMA engine to prefetch data from memory to L2 cache, or to transpose a matrix in the memory without intervention from the processor core.

## 5 The four-core physical implementation

We built the four-core Godson-3 using the seven-metal 65 nm CMOS process. We designed it using call-based flow with some custom-designed macros and manual placement and routing. Full-custom macros include a four write port and four read port (4w4r) 64×64 register file, a 64×64 CAM for TLB, a phase-locked loop (PLL), and a HyperTransport link. To reduce clock cycle time, we manually mapped specific data path modules or modules with replicated structure to the cell library and placed them in a bit-sliced structure. The clock tree takes an H-tree structure and is mainly manually placed and routed. We use a clock skew technique for the critical-path pipeline stage to borrow time from adjacent pipeline stages.

Figure 10 shows the layout of the four-core Godson-3. The chip includes 425 million transistors and the die size measures 14 240 micrometers by 12 205 micrometers. The chip's highest frequency is 1.0 GHz, and its power dissipation ranges from 5 to 10 W depending on the application.

Our recent work includes an eight-core Godson-3 chip design with improved frequency. Moreover, research on teraflops-scale many-core chips is also in progress.

## Acknowledgments

The work presented here is supported by the National High-Tech Research and Development 863 Program of China under Grant No. 2008AA110901, the National Basic Research 973 Program of China under Grant No. 2005CB321600, and the National Natural Science Foundation of China under Grant No. 60736012. We thank Ling-yi Huang, Huan-dong Wang, Dan Tang, Meng-hao Su, Xiao-yu Li, Song-song Cai, Peng-yu Wang, Ge Zhang, Hai-hua Shen, Dan-dan Huan, Bao-xia Fan, Liang Yang,

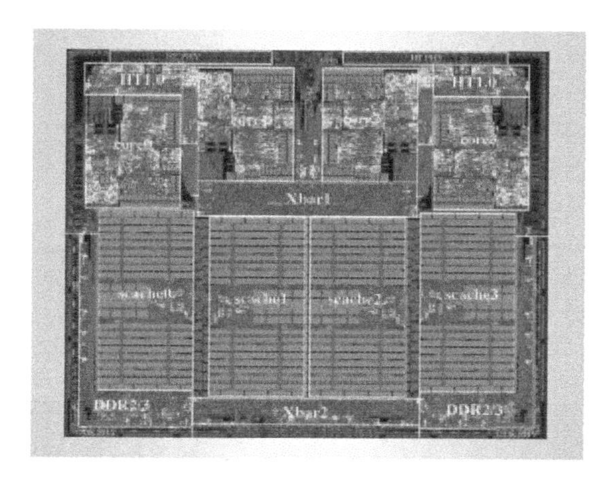


Figure 10. Layout of the four-core Gods on-3. The HyperTransport links are at the top, the CPU cores and L2 caches are in the center, and the DDR2/DDR3 controllers and physical layer are at the bottom. Two crossbars connect these components

Yan-ping Gao, Jiang-mei Wang, Ru Wang, Ying Xu, Bing Xiao, Xu Yang, Shi-qiang Zhong, Feng Zhang, Ying Zhao, Zong-ren Yang, Li-qiong Yang, Lu Zhang, Hao Cui, Hang Yu, Zhuo Gao, Xiang-ku Li, Zi-chu Qi, Cheng Qian, Wei-dan Wu, Jin-zhao He, and Jin Zhang for their contributions to the Godson-3 project.

## References:


[1] KESSLER R. The Alpha 21264 microprocessor[J]. IEEE Micro, 1999, 19(2): 2436.

[2] YEAGER K C. The MIPS R10000 superscalar microprocessor[J]. IEEE Micro, 1996, 16(2): 28-41.

[3] HOREL T, LAUTERBACH G. UntraSparc-lll: designing third-generation 64 bit performance[J]. IEEE Micro, 1999, 19(3): 73-85.

[4] KUMAR A. The HP PA-8000 RISC CPU[J]. EEE Micro, 1997, 17(2): 27-32.

[5] KALLA R, SINHAROY B, TENDLER J M. IBM POWER5 chip: a dual-core multithreaded processor[J]. IEEE Micro, 2004, 24(2): 40-47.

[6] KONGETIRA P, AINGARAN K, OLUKOTUN K. Niagara: a 32-way multithreaded SPARC processor[J]. IEEE Micro, 2005, 25(2): 21-29.

[7] SANKARALINGAM K, NAGARAJAN R, LIU H M, et al. Exploiting ILP, TLP, and DLP with the polymorphous TRIPS architecture[J]. IEEE Micro, 2003, 23(6): 46-51.

[8] HU W W, ZHANG F X, LI Z S. Microarchitecture of the Godson-2 processor[J]. Journal of Computer Science and Technology, 2005, 20(2): 243-249.

[9] KLAIBER A. The technology behind Crusoe processors[Z]. White Paper, Transmeta Corp., 2000.

[10] CHERNOFF A, HERDEG M, HOOKWAY R, et al. FX!32: a profile-directed binary translator[J]. IEEE Micro, 1998, 18(2): 56-64.

[11] EBCIOGLU K, ALTMAN E, GSCHWIND M, et al. Dynamic binary translation and optimization[J]. IEEE Transactions on Computers, 2001, 50(6): 529-548.

[12] BELLARD F. QEMU, a fast and portable dynamic translator[C]//The USENIX Annual Technical Conference. Berkeley: USENIX Association, 2005: 41-46.

发表于 *Journal of Computer Science and Technology*，2011 年第 4 期

# New methodologies for parallel architecture

Dong-rui FAN, Xiao-wei LI, Guo-jie LI

Key Laboratory of Computer System and Architecture, Institute of Computing Technology,

Chinese Academy of Sciences, Beijing 100190, China

**Abstract:** Moore's law continues to grant computer architects ever more transistors in the foreseeable future, and parallelism is the key to continued performance scaling in modern microprocessors. In this paper, the achievements in our research project, which is supported by the National Basic Research 973 Program of China, on parallel architecture, are systematically presented. The innovative approaches and techniques to solve the significant problems in parallel axchitecture design were summarized, including architecture level optimization, compiler and language-supported technologies, reliability, power-performance efficient design, test and verification challenges, and platform building. Two prototype chips, a multi-heavy-core Godson-3 and a many-light-core Godson-T, are described to demonstrate the highly scalable and reconfigurable parallel architecture designs. We also present some of our achievements appearing in ISCA, MICRO, ISSCC, HPCA, PLDI, PACT, IJCAI, Hot Chips, DATE, IEEE Trans. VLSI, IEEE Micro, IEEE Trans. Computers, etc.

**Key words:** architecture, multi-core, many-core, parallelism

## 1 Introduction

With the development of large scale integrated circuit, it has become difficult to improve the performance by increasing the frequency of the clock, referred to as Power Wall. Meanwhile, aggressive ILP (instruction level parallelism) exploitation is also becoming very complex, so-called ILP Wall, and diminishing returns appear. People in the computer industry has widely consensus that the performance can be improved greatly by increasing the number of processing cores on a die. Aggressive superscalar design is being replaced by on-chip multi-core/many-core processor design. Many commercial and academic designs have been introduced in recent few years, such as Intel's 8-core Xeon, IBM's CELL and Cyclops-64, NVIDIA's Tesla and Fermi GPUs.

Parallel architecture is one of the promising choices for the future processor designs. However, there is still a grand challenge to express and exploit the parallelism correctly and efficiently. Generally speaking, parallel architecture can be organized into many light cores, known as many-core architecture, and less heavy cores, known as multi-core architecture. There are several issues need to be addressed for both of these

This work was in part supported by the National Basic Research 973 Program of China under Grant No.2011CB302500, No.2005CB321600, and the National Natural Science Foundation of China under Grant No.60921002.

two parallel architectures:

- scalable performance;
- programmability;
- reliability;
- power efficiency;
- test and verification.

First of all, performance is one of the challenges. High performance can be achieved through both hardware and software solutions. Hardware solutions include circuit and micro-architecture optimizations, while software solutions include compiler and language-supported techniques. Secondly, reliability needs to be concerned for parallel architecture. Due to various technology-related issues, cores in a multi-core or many-core processor will exhibit higher failure rates than a single-core processor. Thus the reliability is required to be enhanced. Thirdly, power efficiency is also an involved issue. Since a significant and increasing fraction of the total cost of ownership is due to operational costs, power efficiency has become an important design goal for parallel system design. Finally, it is still difficult to overcome the test and verification challenges for modern parallel architecture. Along with technology scaling, and rising number of cores, many factors and parameters — such as temperature, voltage fluctuation, and process variation — have caused a significant impact on chip design, test and verification.

In this paper we systematically present the innovative approaches and techniques that we proposed in our 973 project. This work is based on two characteristic architectures, which are multi-heavy-core Godson-3 and many-light-core Godson-T. The detailed achievements include micro-architecture with scalable performance, compiler and language-supported technologies, reliability enhanced methods, power-performance efficient design, test and verification techniques, and platform building. In the rest of the paper, Section 2 describes our approaches in details, Section 3 gives out the two prototype chips, and finally the conclusion and future work are presented in Section 4.

## 2 Innovative approaches for parallelism

Moore's law will grant computer architects ever more transistors for the foreseeable future. It is well known that parallelism is the key to continued performance scaling in modern microprocessors. To overcome the increasing serious problems such as power wall and ILP wall, many kinds of approaches to efficient parallel computing have been achieved at different levels. Here, parallel architecture design and supported technologies are concerned about this issue.

### 2.1 Scalable micro-architecture design

Moore's law suggests that the number of processing cores on a single chip increases exponentially. However, based on current technologies, integrating multi heavy cores or many lightweight cores on a single die for building parallel architectures are two choices. In this research, we build both of these two kinds of parallel architectures: multi-core processor and many-core processor, aiming at scalability and reconfigurable of future processor architectures.

#### 2.1.1 Multi-core architecture

For multi-core architecture, a scalable mesh

of crossbar (SMOC) on-chip network topology is adopted, with multi-level scalable and reconfigurable of processor cores, on-chip memory system and on-chip network in computing unit level, processor cores level and chip level.

Scalable and distributed on-chip network connects processor cores and globally addressed level-two (L2) cache modules are designed on multi-core processors[1]. A directory-based cache-coherence protocol maintains multiple level-one (LI) copies of the same L2 block. Four-core processor based on 65 nm CMOS technology has been designed and fabricated. The 8-core and 16-core chips are still in physical implementation.

### 2.1.2 Many-core architecture

Also, a many-core architecture is proposed to overcome the challenge of power wall and memory wall, with high performance and preferable programmability.

64 homogeneous, in-order and dual-issue processing cores were integrated in Godson-T, a many-core processor[2]. Each processing core has a 16 KB 2-way set-associative private instruction cache and a 32 KB local memory. A data transfer agent (DTA) is built in each core for fast data communication. There are 16 address- interleaved L2 cache banks (256 KB each) distributed along four sides of the chip. The L2 cache is shared by all processing cores and can serve up to 64 cache accessing requests in total. In addition, a dedicated synchronization manager provides architectural support for efficient mutual exclusion and barrier synchronization.

Simulators are one most important tool for computer architecture researchers. While many-core architecture brings big challenges to computer architects on the development of fast corresponding simulators.

A parallel discrete event simulation (PDES) method is proposed to speedup a cycle-accurate event-driven many-core processor simulator in [3]. By applying PDES to a conventional sequential simulator, a speedup of 10.9×(up to 13.6×) on a 16-core host machine while remaining cycle counter differences less than 0.1%, thus providing a fast and accurate way to simulate many-core processors.

### 2.1.3 Architecture level optimization

Based on but not limited to these two kinds of platforms, we explore many architecture level techniques for building efficient parallel computing platforms.

Both advances in bandwidth increasement and latency reduction for I/O buses and devices, I/O data move in/out of memory is becoming more and more important. Observing from different characteristics of I/O and CPU memory reference behavior, the potential benefits of separating I/O data from CPU data are found. Tang et al proposed a DMA cache technique to store I/O data in dedicated on-chip storage and presented two DMA cache designs, decoupled DMA cache (DDC), which adopts additional on-chip storage as the DMA cache to buffer I/O data, and partition-based DMA cache (PBDC), which does not require additional on-chip storage, but can dynamically use some ways of the processor's last level cache (LLC) as the DMA cache[4].

Parallelism is the key to continued performance scaling in modern microprocessors. This parallelism can often contain a surprising amount of instruction redundancy. To exploit this redundancy to improve performance and decrease energy consumption, Long et al proposed a multi-threading micro-architecture, minimal multi-threading (MMT) method in [5]. It leverages register renaming and the instruction window to combine the fetch and execution of identical instructions between threads in SPMD applications.

Debugging parallel program is a well-known difficult problem. A promising method to facilitate debugging parallel program is using hardware support to achieve deterministic replay. Based on Godson-3, Chen and Hu et al proposed a novel and succinct hardware-assisted deterministic replay scheme named LReplay, which has a small log size, as well as low design cost, to be feasible for adopting by industrial processors[6]. The innovation of LReplay is that instead of recording the logical time orders between instructions or instruction blocks as previous investigations, LReplay is built upon recording the pending period information which relies on global clock. For system-on-chip (SOC) without a global clock, Su and Chen et al proposed a synchronization FIFO technique to force the behavior of the chip deterministic[7].

Design space exploration (DSE) is a critical step for the design of a microprocessor. A common approach for DSE is to predict the performance of an architecture configuration through learning from the known performances/powers of several architecture configurations. Given limited learned cases, it is difficult to achieve high prediction accuracy. Guo and Chen et al proposed the COMT approach which can exploit unlabeled design configurations to improve the prediction accuracy[8]. Experiments show that COMT outperforms the state-of-the-art DSE techniques through reducing the mean squared error by 30% to even 84%.

## 2.2 Compiler and language-supported technology

Achieving high performance on multi-core and many-core processors is a huge challenge. To utilize the modern processors more efficiently, parallelism and locality are two key co-related issues. In our research, we tackle the problem from all aspects of the programming system, language extension, compiler analysis and optimizations, iterative compiler, language runtime and general runtime support. Many techniques are developed for parallelism exploitation and sequential performance improvement.

### 2.2.1 Parallel optimization

The bandwidth-aware job scheduling is an effective and relatively easy-to-implement way to relieve the bandwidth contention for shared-memory multiprocessors. In [9], Xu and Wu et al quantified the impact of bandwidth contention on overall performance, and proposed a new workload scheduling policy. Its basic idea is that interference due to bandwidth contention could be minimized when bandwidth utilization is maintained at the level of average bandwidth requirement of the workload.

Unified parallel C (UPC) is designed for high performance computing on large-scale parallel machines. With graphics processing unit (GPU), which is becoming an increasingly important high performance computing platform, Chen et al proposed new language extensions to UPC to take advantage of GPU clusters[10]. They extended UPC with hierarchical data distribution, revised the execution model of UPC to mix

SPMD with fork-join execution model, and modified the semantics of upc_forall to reflect the data-thread affinity on a thread hierarchy.

Work-stealing is a key technique in many multi-threading programming languages to get load balancing. It has a high implementation overhead in some applications, and cannot handle many applications without definitive working sets. To handle these problems, Wang et al proposed a new adaptive task creation strategy[11]. Its basic idea is that except when some thread becomes idle, each busy thread would avoid creating more tasks.

Most previous studies on tiling are focused on the division of iteration space. However, on distributed memory parallel systems, in order to attain load balancing and minimize data migration, the decomposition of computation and the distribution of data must be handled at the same time. In [12], Liu et al presented a global tiling method to select the optimal iteration space tiling and data space tiling, and thus improving parallelism and minimizing the volume of communication.

### 2.2.2 Compiler optimization

Though iterative optimization has become a popular compiler optimization approach, it is based on a premise which has never been truly evaluated: that it is possible to learn the best compiler optimizations across datasets. Chen and Wu et al truly put iterative compilation to the test for the first time by evaluating its effectiveness across a large number of datasets[13]. They composed KDataSets, a dataset suite with 1 000 datasets for 32 programs, which have been released to the public. And they demonstrated that it is possible to derive a robust iterative optimization strategy across datasets.

Pointer analysis is the basis of most other static program analyses and many compiler optimizations. However, none of the existing flow- and context- sensitive (FSCS) pointer analysis algorithms can scale to large real programs. To analyze million lines of code in minutes, Yu et al proposed a practical and scalable method for FSCS pointer analysis for C programs[14]. It analyzes the pointers in a program, level by level in terms of their points-to levels, allowing the points-to relations of the pointers at a particular level to be discovered based on the points-to relations of the pointers at this level and higher levels.

To improve the locality of the heap objects, Wang et al proposed a lightweight dynamic optimizer, which aims to exploit the affinity of the allocated heap objects and improve their layout at runtime[15]. It uses heuristics based on affinity analysis to allocate related objects in the same memory pool even if those objects are allocated at different call sites.

Binary translation has been an important approach to migrate application software across instruction set architectures. To improve the performance of handling misaligned data accesses (MDA), Li and Wu et al proposed a new mechanism based on exception handling[16].

A given program may use one type of registers much more frequently than other types. This creates an opportunity to employ the infrequently used registers as spill destinations for the more frequently used register types. Lyu et al presented a code optimization method named idle register exploitation (IRE) to exploit such opportunities[17]. A model called IRE is developed to determine the static performance gains of IRE versus spilling to the stack.

## 2.3 Reliability technology

Effective reliability techniques are essential for complex integrated circuits, such as multi/many-core processors, to tolerate various faults, extend MTTF (mean time to failures), and improve production yield. This subsection will briefly introduce parts of our research achievements.

A defect tolerance method is proposed by employing redundancy at the core-level instead of at the microarchitecture level[18]. When faulty cores existing, how to reconfigure the topology with redundant cores as well as to balance the system performance and transparency for programmers is a well known NP-complete problem. The concept of virtual topology is borrowed to provide a unified topology which is isomorphic with the target reference topology regardless of various possible underlying physical topologies. An effective heuristic, namely Row Rippling Column Stealing-Guided Simulated Annealing algorithm is then presented to solve the topology reconfiguration problem.

With technology scaling down, process, voltage, and temperature (PVT) variations can significantly degrade the performance benefits expected from the next generation of designs. When multiple programs are running in a multi-core processor, variations create spatial and temporal unbalance across the processing cores. Most prior schemes are dedicated on tolerating PVT variations individually for a single core, but ignore the opportunity of leveraging the complementary effects between variations and the intrinsic variation unbalance among individual cores. Yan et al found out that cores with mild variations can share the violent workloads from cores suffering large variations[19]. If operated correctly, variations on different cores can help mitigating each other and result in a variation-mild runtime environment. They proposed a Timing Emergency Aware Thread Migration (TEA-TM), a delay sensor-based scheme to reduce system timing emergencies caused by PVT variations. From the experimental results, TEA-TM can help save up to 24% throughput loss on average, at the same time improve the system fairness by 85%.

Future shipped microprocessors will be increasingly vulnerable to intermittent faults. Quantitatively characterizing the vulnerability of microprocessor structures to intermittent faults is significantly helpful to balance reliability and performance. Pan et al proposed a metric intermittent vulnerability factor (IVF) to achieve this goal[20]. A structure's IVF is the probability that an intermittent fault in the structure causes an external visible error. They compute IVF for two structures: reorder buffer and register file. Experimental results show that IVF varies significantly across different structures and workloads, which imply partial protection to the most vulnerable structures to enhance system reliability while with minimal performance overhead.

## 2.4 Power efficient design

Power management in high performance processors has been an involved issue. In order to increase power effective with little performance reduction, energy-delay product has become a widely-used metric to scale power-performance gains.

At a hardware point, dynamic voltage/frequency scaling (DVFS) plays a critical role since 1990s. Another way to low-power high-speed computing is scheduling applications with variation effects. Therefore several algorithms for application scheduling with DVFS have been developed to maximize throughput under a given hardware power budget. These scheduling algorithms are also employed to handle core-to-core process variation. The power-performance tradeoffs cannot be easily explained without considering the application behavior.

Since DVFS is adopted in CMP, different clock domains could run at different frequencies. Therefore the asynchronous communication is necessary between clock domains. In order to transmit signals correctly, several cycles are required to synchronize signals, and this overhead will affect the performance, the circuit does not work when the frequency is changing. Some researches on DVFS have proved that changing frequency while cache miss and memory access can reduce some performance reduction.

Although commercial designs have traditionally used full-chip DVFS, per-core frequency scaling has been assumed and significant improvements in energy-efficiency are enabled by fine-grained frequency. Clock domains of different frequencies can communicate synchronously, and also, fine-grained frequency changes can be made without pause. As the work in [21], coarse-grained and fine-grained clock gating are both used in this chip. The first level is clock gating for IPs and other blocks and the second level is fine-grained clock gating for registers within block. Clock gating and DVFS working under operating system can also be used for temperature controlling.

Besides DVFS, at an architecture point, slowing down processors in the uncritical path to save power may not affect the overall performance. On the contrary, slowing down processors in the critical path will negatively impact performance due to longer execution time. In the following, we will mention at her power-efficient design cases which achieve both high speed and low power, assumed in microprocessor architecture.

Our vision in this paper lies in power efficient performance management problems in future CMP. As future CMP scaling to higher core counts, greater complexity, and increased heterogeneity, we expect that the prospective mentioned above will have great effects on both hardware and software.

## 2.5 Test and verification technology

Along with technology scaling, and rising functional frequency and density, delay variability caused by crosstalk and process variations poses a formidable challenge on chip test and verification.

Zhang and Li et al proposed the precise crosstalk-induced path delay fault (PCPDF) model to convert the timing constraints of coupling lines into logic constraints, and proposed path delay test generation methods toward activation of worst case crosstalk effects, which improves test coverage on complex delay faults and decreases the test escape of delay testing[22]. Low cost and low power design-for-testability (DFT) techniques are adopted to face huge test data volume challenges. Han et al proposed X-Config stimulus compression and X-Tolerant response compaction methods to optimize test cost and test power with the limited test resource[23]. More techniques, such as hybrid scan compression structure, built-in self-repair, and at-speed test generation using on-chip PLL clock, are implemented in our prototype chips[24].

To verify an execution against memory consistency is known as NP-hard problem. A fast memory consistency verification method is proposed by identifying a new natural partial time order[25]. Chen et al proposed that in multiprocessor systems with store atomicity, a time order restriction exists between two operations whose pending periods are disjoint: the former operation in time order must be observed by the latter operation. Based on the time order restriction, memory consistency verification is localized, thus its time complexity can be reduced to $O(n^2)$. In [26], Hu and Chen et al further reduced the time complexity of memory consistency verification with global clock to $O(n)$. In [27], Li et al proposed a test program regularization method to reduce the time complexity of memory consistency verification to $O(n)$ even without a global clock.

The essential purpose of verification is to expose as many bugs as possible. In [28], Guo et al proposed a novel verification methodology to leverage the early bug prediction of a DUV. The methodology utilizes the artificial neural networks (ANNs) based predictive models, thus it can model the relationship between the high-level attributes of a design and its associated bug information. Such bug prediction technique can effectively guide the verification flow of a processor.

Semi-formal methods are widely explored to combine the power of simulation-based and formal verification techniques. Zhang and Lyu et al proposed an abstraction-guided simulation approach aiming to cover hard-to-reach states in functional verification of microprocessors[29]. They constructed a Markov model to integrate vector correlations from instruction set architecture (ISA). Effective strategies about state evaluation and dynamical adjustment to the Markov model are proposed, which greatly help the simulation converge to target states more efficiently in comparison with other intelligent engines.

## 3 Prototype chips

Previous efforts improve single-thread performance through faster clock rate or aggressive ILP exploitation. But these techniques have finally run out because of power limits and diminishing returns. In this section, we bring up two solutions, multi-heavy-core, Godson-3, and many-light-core, Godson-T. We describe the two architectures, and evaluate some of their features by analyzing the experimental results.

### 3.1 Multi-heavy-core, Godson-3

Godson-3 is a multi-heavy-core processor, whose target applications include high throughput computing, high performance scientific computing and high end embedded applications[1]. In this subsection, Godson-3 architecture is introduced and some of its features are evaluated.

#### 3.1.1 Godson-3 architecture

As a multi-heavy-core processor, Godson-3 CPU core features include out-of-order execution, aggressive cache design, two fixed-point and two floating-point functional units, 64-bit virtual addresses, 48-bit physical addresses, 64 entries fully associative TLB, EJTAG, hardware support for x86 emulation, and others.

Figure 1 shows the Godson-3 overall architecture, which is a 2D mesh topology. There are 16 nodes in the mesh, and each includes a crossbar connecting 4 processor cores (masters), 4 shared L2-cache banks (slaves), and routers connecting 4 adjacent nodes. The memory controllers are connected to the L2 cache banks through a second-level crossbar, and the I/O controller is connected to the free crossbar ports of boundary nodes[30-31].

Godson-3 is a reconfigurable architecture involving three features: 1) Either general-purpose cores or special-purpose cores can be inserted into the net work slots; 2) With the reconfigurable address windows, the software can migrate home blocks across cache banks, or set cache blocks as private or shared blocks according to the locality of memory accesses; and 3) Godson-3's DMA engine can also be configured. Software can decide the destination (memory or L2 cache) of the DMA data. And data in the memory can be prefetched to the L2 cache by the DMA engine.

To improve the performance-energy efficiency, Godson-3B adopts a processor core with SIMD (single-instruction multi-data) functional units, which is named GS464V[32]. GS464V integrates 2 256 bit SIMD functional units (VPUs). Hence, GS464V can complete 8 double- precision floating-point MADD, or 16 single- precision floating-point MADD per clock cycle. Furthermore, GS464V integrates a programmable memory access coprocessor to both tackle the performance bottleneck of memory access and reduce power consumption.

### 3.1.2 Experimental results and analysis

Experiments were carried out to analyse the Godson-3 performance on two platforms: RTL simulation platform and FPGA prototyping plat-

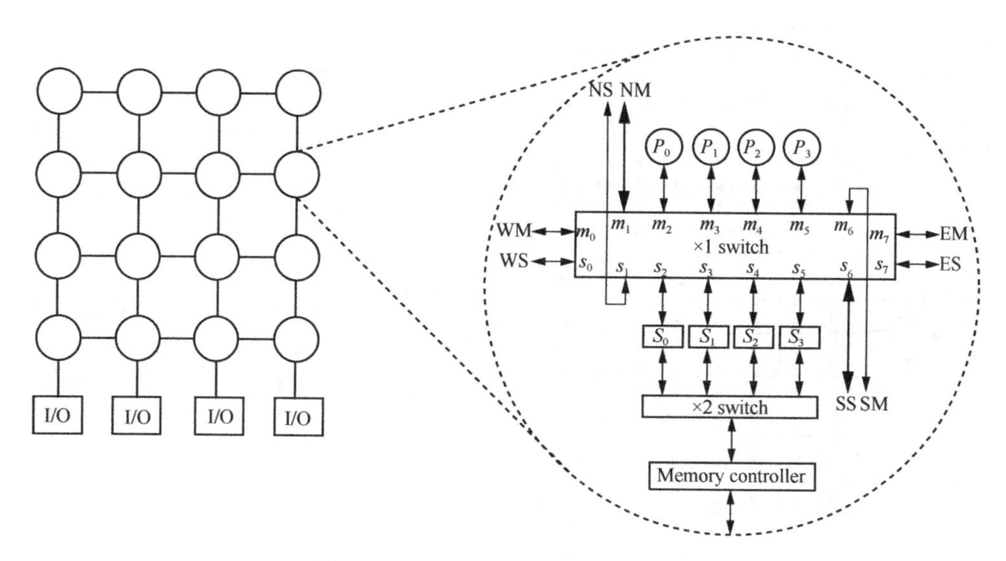


Figure 1. Godson-3 overall architecture

form. In the RTL simulation environment, the core clock frequency was set to 1 GHz, the DDR2/DDR3 clock frequency to 333 MHz, and the HyperTransport clock frequency to 800 MHz. A partial-scale FPGA prototype, whose speed was 50 MHz, was used to evaluate a single processor core's performance, which included one processor core, one 1 Mbyte L2 cache, one DDR2/DDR3 controller, and one HyperTransport controller.

Nine typical kernels or full applications were selected to evaluate hardware support for x86 emulation. It is found that hardware support can significantly accelerate binary translation from x86 to MIPS, and on average achieves performance that is nearly 70% of the ideal mode.

The LMbench was used to measure the dependent-load latency, and it was found that the load-to-use latency was lower than 6 ns from L1 cache and no more than 90 ns from L2 cache to LI cache. The STREAM benchmark was used to evaluate the sustainable memory bandwidth, and the experimental results showed that Godson-3 provides nearly-linear bandwidth scaling as the core number increases.

Godson-3 was also evaluated with SPEC CPU2000. The experimental results showed that performance superiority of Godson-3 is obvious, compared with Intel Pentium series processors.

## 3.2 Many-light-core, Godson-T

Godson-T is a processor prototype of many-core architecture, which targets highly parallelizable applications which require high computational throughput. Figure 2 gives the overview of the Godson-T processor architecture[2].

### 3.2.1 Godson-T architecture

Godson-T has 64 homogeneous, in-order and dual-issue processing cores. The target frequency of each core is 1 GHz. The RISC-like

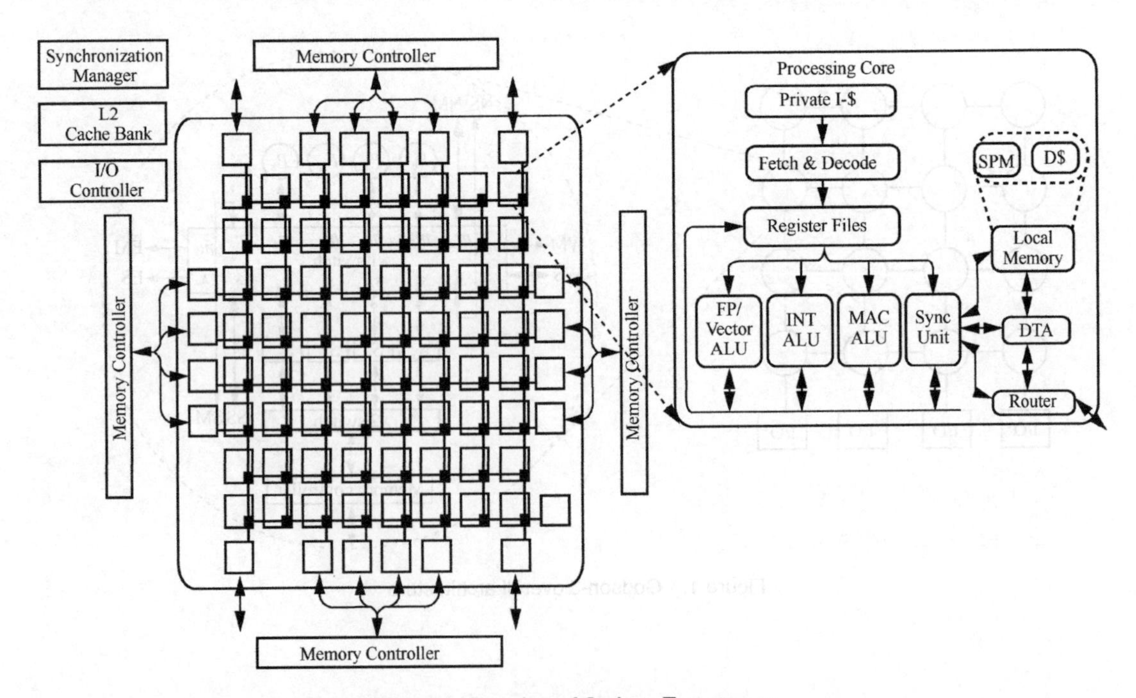


Figure 2. Overview of Godson-T processor

processing core supports 32 bit MIPS ISA (user-level) with synchronization instruction extensions. One floating-point arithmetic operation and one floating-point multiply-add operation can be issued to corresponding fully-pipelined function units in a cycle. Each processing core has a 16 KB 2-way set-associative private instruction cache and a 32 KB local memory. The local memory functions as a 32 KB 4-way set-associative private data cache in default. It can also be configured as an explicitly-controlled scratchpad memory (SPM), or a hybrid of cache and SPM. A data transfer agent (DTA) is built in each core for fast data communication. When the processing core is doing calculations, DTA can be programmed to manage data communication asynchronously and transparently. There are 16 address-interleaved L2 cache banks (256 KB each) distributed along four sides of the chip. The L2 cache is shared by all processing cores and can serve up to 64 cache accessing requests in total. 4 L2 cache banks in the same side of the chip share a memory controller.

A dedicated synchronization manager provides architectural support for efficient mutual exclusion and barrier synchronization. The 8×8, 128-bit-width and packet-switching 2-D mesh network connects all on-chip units. The mesh network employs deterministic X-Y routing policy and provides total 2 TB/s on-chip bandwidth among 64 processing cores.

### 3.2.2 Experimental results and analysis

Experiments are conducted on the cycle-accurate simulator for Godson-T architecture.

Two bioinformatics applications and four kernels of SPLASH-2 are used to evaluate the cache consistent mechanism. All of the programs achieve good speedups. The efficiency and scalability of the proposed region-based cache consistency protocol has been solidly proved for conventional multithreaded programs with cache hierarchy.

A series of micro-benchmarks and methodology are adopted to evaluate the synchronization overhead on Godson-T. It is shown that the hardware support of synchronization is hundreds or even thousands of times faster than software synchronization.

Two important kernels including SGEMM (single-precision general matrix multiplication) and FFT are composed to exploit the potential of high-performance computing of Godson-T architecture. Intensive optimizations are done on the kernels, using SPM and DTA. The efficiency of Godson-T is comparable or superior to other processors because of its explicit but simple memory hierarchy abstraction with fast and versatile data communication mechanism (e.g., horizontal DTA operation).

Two micro-benchmarks are used to evaluate the efficiency of producer-consumer synchronization supported by full/empty-bit mechanism of Godson-T: the 2-D wave-front; and the Loop 6 from the Livermore Loop benchmarks. Both applications achieve good speedups, benefiting from the fine-grained synchronization.

1-D Jacobi and Graph Traverse are used to demonstrate the flexibility and efficiency of the configurable SPM/Cache memory organization. The experiments demonstrate that the performance of the program can benefit from hybrid memory partitions with minor modification of the program previously written with full-cache configuration.

#### 3.2.3 Platform building

A light software runtime system, GodRunner[3] is proposed for many-core architecture, such as Godson-T. It manages hardware resources, and focuses on efficiently abstracting a large number of hardware threading units and dynamic load-balancing. Since continual context-switching incurs save-restore overhead and cache thrashing, GodRunner does not support preemptive execution, but permits programmers to create more tasks than hardware thread units, and transparently maps them to hardware thread units at runtime. It is responsible for swapping the completed task out and scheduling a new one in a thread unit. Two well-known task scheduling algorithms are implemented for efficiency dynamic load balancing: work-stealing and conditional spawning. Different task scheduling algorithms can be incorporated into it flexibly.

Accordingly, a cross compiler toolchain is exploited correlatively. It employs the latest MIPS gcc4.3.2 cross compiler based on x86/Linux platform. It supports series of ABI, such as O32, N32, N64. It equips with Binutils tool sets, C library and mathematics library. In addition, the assembler is modified to support newly added instructions for Godson-T.

As well as the compiler toolchain, an FPGA emulation platform is assembled for Godson-T. The platform composes of six Vertex-5 LX330 chips to validate large scale design which has more than 64 processing cores. A reusable and scalable validating scheme is proposed for logical circuits. The scheme enables the FPGA platform to validate larger many-core architecture design going over the top capacity of the 6 FPGA chips.

## 4 Conclusion and future work

Many challenges have to be considered when designing a high efficient parallel architecture. In this paper, we address these challenges and propose several innovative approaches and techniques, including both hardware and software solutions. Our evaluation results, based on the two prototype chips in different architectures, show the possibility to design highly scalable and reconfigurable parallel architecture.

In addition to high performance computing, high throughput computing (HTC) is another important aspect for parallel architecture. Due to the different features of high-throughput applications, parallel architecture for HTC will face many new challenges. Fortunately, the design methodology used in this paper will contribute to the high-throughput parallel architecture designs in future.

## Acknowledgements

The authors thank Wei-wu Hu, Hua-wei Li, Xiao-bing Feng, Cheng-yong Wu, Yu Hu, Cheng-gang Wu, Hui-wei Lyu, Dan Tang, Guo-ping Long, Yun-ji Chen, Meng-hao Su, Qi Guo, Di Xu, Li Chen, Yang Chen, Lei Wang, Lei Liu, Hong-tao Yu, Zhen-liang Wang, Jian-jun Li, Fang Lyu, Lei Zhang, Gui-hai Yan, Song-jun Pan, Min-jin Zhang, Yin-he Han, Da Wang, Lei Li, Tao Zhang, and Xiang Gao for their contributions to our 973 project and this work. The reference papers were supported by the 973 project, and the major contributors are members of the 973 project.

## References:

[1] HU W, WANG J, GAO X, et al. Godson-3: a scalable

multi-core RISC processor with x86 emulation support[J]. IEEE Micro, 2009, 29(2): 17-29.

[2] FAN D R, YUAN N, ZHANG J C, et al. Godson-T: an efficient many-core architecture for parallel program executions[J]. Journal of Computer Science and Technology, 2009, 24(6): 1061-1073.

[3] LYU H, CHENG Y, BAI L, et al. P-GAS: parallelizing a cycle-accurate event-driven many-core processor simulator using parallel discrete event simulation[C]// The Workshop on Principle of Advanced and Distributed Simulation. Piscataway: IEEE Press, 2010: 1-8.

[4] TANG D, BAO Y, HU W, et al. DMA cache: using on-chip storage to architecturally separate I/O data from CPU data for improving I/O performance[C]// The International Conference on High-Performance Computer Architecture. Piscataway: IEEE Press, 2010: 1-12.

[5] LONG G, FRANKLIN D, BISWAS S, et al. Minimal multi-threading: finding and removing redundant instructions in multi-threaded processors[C]// The IEEE/ACM International Symposium on Micro architecture. Piscataway: IEEE Press, 2010: 337-348.

[6] CHEN Y, HU W, CHEN T, et al. L Replay: a pending period based deterministic replay scheme[C]// The International Symposium on Computer Architecture. New York: ACM Press, 2010: 187-197.

[7] SU M, CHEN Y, GAO X. A general method to make multiclock system deterministic[C]// The Conference on Design, Automation and Test in Europe. Piscataway: IEEE Press, 2010: 1480-1485.

[8] GUO Q, CHEN T, CHEN Y, et al. Effective and efficient microprocessor design space exploration using unlabeled design configurations[C]// The International Joint Conference on Artificial Intelligence. New York: ACM Press, 2011.

[9] XU D, WU C, YEW P C. On mitigating memory bandwidth contention through bandwidth-aware scheduling[C]// The International Conference on Parallel Architectures and Compilation Techniques. Piscataway: IEEE Press, 2010: 237-247.

[10] CHEN L, LIU L, TANG S, et al. Unified parallel C for GPU clusters: language extensions and compiler implementation[C]// The 23rd International Workshop on Languages and Compilers for Parallel Computing. Heidelberg: Springer, 2010: 151-165.

[11] WANG L, CUI H, DUAN Y, et al. An adaptive task creation strategy for work-stealing scheduling[C]// The International Conference on Code Generation and Optimization. New York: ACM Press, 2010: 266-277.

[12] LIU L, CHEN L, WU C Y, et al. Global tiling for communication minimal parallelization on distributed memory systems[C]// The 14th International Euro-Par Conference on Parallel Processing. New York: ACM Press, 2008: 382-391.

[13] CHEN Y, HUANG Y, EECKHOUT L, et al. Evaluating iterative optimization across 1000 data sets[C]// The Conference on Programming Language Design and Implementation. New York: ACM Press, 2010: 448-459.

[14] YU T, XUE J, HUO W, et al. Level by level: making flow- and context-sensitive pointer analysis scalable for millions of lines of code[C]// The International Conference on Code Generation and Optimization. New York: ACM Press, 2010: 218-229.

[15] WANG Z, WU C, YEW P C. On improving heap memory layout by dynamic pool allocation[C]// The International Conference on Code Generation and Optimization. New York: ACM Press, 2010: 92-100.

[16] LI J, WU C, HSU W C. An evaluation of misaligned data access handling mechanisms in dynamic binary translation systems[C]// The International Conference on Code Generation and Optimization. Piscataway: IEEE Press, 2009: 180-189.

[17] LYU F, WANG L, FENG X, et al. Exploiting idle register classes for fast spill destination[C]// The International Conference on Supercomputing. New York: ACM Press, 2008: 319-326.

[18] ZHANG L, HAN Y, XU Q, et al. On topology reconfiguration for defect-tolerant NoC-based homogeneous manycore systems[J]. IEEE Transactions on VLSI Systems, 2009, 17(9): 1173-1186.

[19] YAN G, LIANG X, HAN Y, et al. Leveraging the core-level complementary effects of PVT variations to reduce timing emergencies in multi-core processors[C]// The International Symposium on Computer Architecture. New York: ACM Press, 2010: 485-496.

[20] PAN S, HU Y, LI X. IVF: characterizing the vulnerability of microprocessor structures to intermittent faults[C]// The Conference on Design. Piscataway: IEEE Press, 2010: 238-243.

[21] HU W, WANG R, CHEN Y, et al. Godson-3B: a 1 GHz 40 W 8-Core 128 GFlops processor in 65 nm CMOS[C]// The International Solid-State Circuits Conference. Piscataway: IEEE Press, 2011.

[22] ZHANG M, LI H, LI X. Path delay test generation toward activation of worst case coupling effects[J].

IEEE Transactions on Very Large Scale Integration Systems, 2010, 18(12): 1-14.

[23] HAN Y, HU Y, LI X, et al. Embedded test decompressor to reduce the required channels and vector memory of tester for complex processor circuit[J]. IEEE Transactions on Very Large Scale Integration Systems, 2007, 5(15): 531-540.

[24] WANG D, HU Y, LI H, et al. The design-for-testability features and test implementation of a giga hertz general purpose microprocessor[J]. Journal of Computer Science and Technology, 2008, 23(6): 1037-1046.

[25] CHEN Y, LV Y, HU W, et al. Fast complete memory consistency verification[C]// The International Symposium on High-Performance Computer Architecture. Piscataway: IEEE Press, 2009: 381-392.

[26] HU W, CHEN Y, CHEN T, et al. Linear time memory consistency verification[J]. IEEE Transactions on Computers, 2011, 61(4): 502-516.

[27] LI L, CHEN T, CHEN Y, et al. Brief announcement: program regularization in verifying memory consistency[C]// The Symposium on Parallelism in Algorithms and Architectures. New York: ACM Press, 2011.

[28] GUO Q, CHEN T, SHEN H, et al. Empirical design bugs prediction for verification[C]// The Conference on Design, Automation and Test in Europe. Piscataway: IEEE Press, 2011: 1-6.

[29] ZHANG T, LYU T, LI X. An abstraction-guided simulation approach using Markov models for microprocessor verification[C]// The Conference on Design, Automation and Test in Europe. Piscataway: IEEE Press, 2010: 484-489.

[30] HU W, WANG J, GAO X, et al. Micro-architecture of Godson-3 multi-core processor[C]// The Symposium on High Performance Chips. Piscataway: IEEE Press, 2008.

[31] GAO X, CHEN Y J, WANG H D, et al. System architecture of Godson-3 multi-core processors[J]. Journal of Computer Science and Technology, 2010, 25(2): 181-191.

[32] HU W, CHEN Y. GS464V: a high-performance low-power XPU with 512-bit vector extension[C]// The Symposium on High Performance Chips. [S.l.:s.n.], 2010.

# Godson-T: an efficient many-core processor exploring thread-level parallelism

Dong-rui FAN, Hao ZHANG, Da WANG, Xiao-chun YE,
Feng-long SONG, Guo-jie LI, Ning-hui SUN

Institute of Computing Technology, Chinese Academy of Sciences

**Abstract:** Godson-T is a research many-core processor designed for parallel scientific computing that delivers efficient performance and flexible programmability simultaneously. It also has many features to achieve high efficiency for on-chip resource utilization, such as a region-based cache coherence protocol, data transfer agents, and hardware-supported synchronization mechanisms. Finally, it also features a highly efficient runtime system, a pthreads-like programming model, and versatile parallel libraries, which make this many-core design flexibly programmable.

Although various many-core processors, such as Tilera's TILE64[1], IBM's Power7[2], AMD's Opteron[3], and the SPARC64[4], provide tremendous computational capability, programmers still face the grand challenge of expressing and exploiting parallelism correctly and efficiently. Parallelization always requires significant programming efforts. Even when a parallel program works correctly, performance tuning can be daunting. Parallel programming brings many complex problems that are highly related to performance issues: managing conflicts in accessing shared resources, synchronizing disparate threads, and so on. In many cases, significant programming efforts can't be transformed into performance gain. This problem is unlikely to ever be perfectly resolved by software programming infrastructure alone.

Therefore, the Institute of Computing Technology (ICT) of the Chinese Academy of Sciences (CAS) has developed a many-core processor called Godson-T, which provides a widely used programming paradigm and highly efficient architectural support for multithreaded programs. Thus, it frees programmers from concentrating on efficient parallel execution and lets them focus on expressing parallelism. The Godson-T architecture supports two fundamental multithreading operations: data communication and thread synchronization. At the same time, it provides a multithread programming environment and a runtime system.

## 1 Godson-T architecture overview

As Figure 1 shows, the Godson-T processor

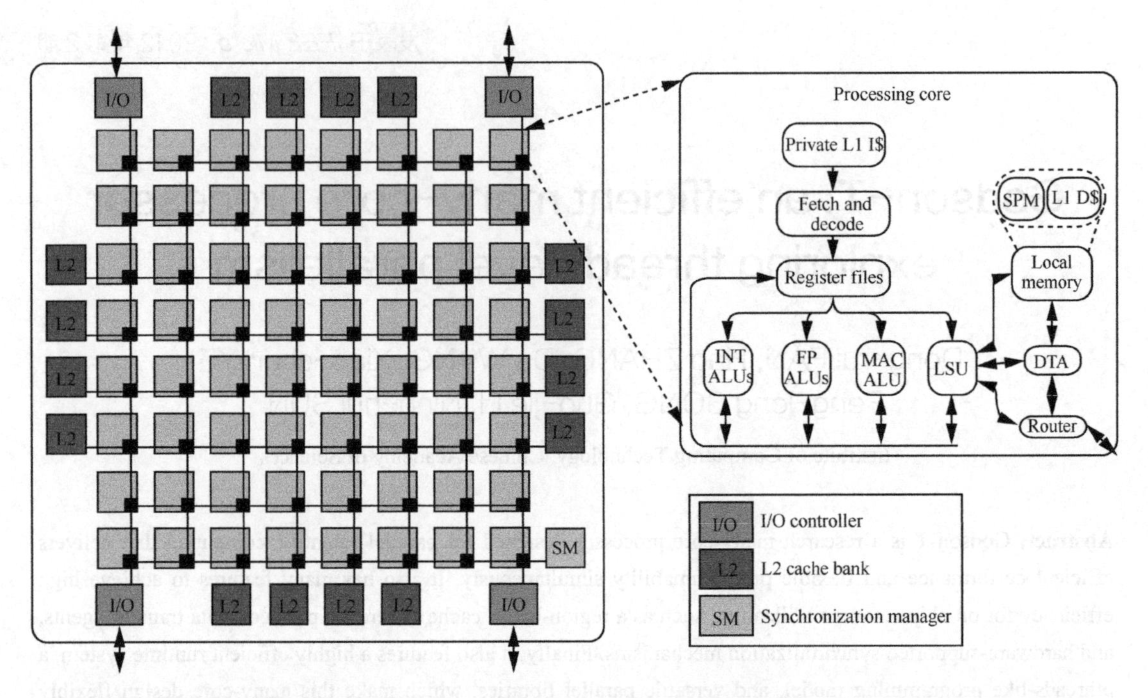


Figure 1. High-level block diagram of Godson-T. (ALU: arithmetic logic unit; D$: data cache; DTA: data transfer agent; I$: instruction cache; INT: integer; FP: floating point; LSU: load-store unit; L1: Level 1; L2: Level 2; MAC: multiply accumulate; SM: synchronization manager; SPM: scratch-pad memory)

is a many-core processor with 64 homogeneous, in-order, and dual-issue processing cores running at 1 GHz. The taped-out prototype chip is a set of 16 cores, with the same architecture and key techniques to verify the proposed solutions. The 64-core chip is now well underway. The following descriptions are based on the 64-core architecture.

Each eight-stage pipeline processing core supports the million instructions per second (MIPS) instruction set architecture and the synchronization instruction extensions. Two floating-point operations, including a multiply-accumulation, can be issued to fully pipelined function units in one cycle.

Godson-T's on-chip memory hierarchy has two levels. The first is the memory attached to each core, which contains a 16 KB two-way set-associative private instruction cache and a 16 KB data cache. The data cache can also be configured as scratch-pad memory (SPM) or a hybrid structure of SPM and a data cache.

The second level is the global L2 cache shared by all the cores. There are 16 address-interleaved L2 cache banks (128 KB each) distributed along four sides of the chip. The L2 cache can serve up to 64 cache-accessing requests. Four L2 cache banks on the same side of the chip share a memory controller (see Figure 1). Each L2 cache bank is connected with a router.

This two-level cache uses a region-based cache coherence (RCC) protocol. Each core has a data transfer agent (DTA) for fast data communication. A dedicated on-chip synchronization manager (SM) provides architectural support for efficient mutual exclusion, barrier, and signal/wait synchronization.

A 128-bit-width packet-switching 2D mesh network connects all on-chip units. The mesh network employs deterministic X-Y routing and a wormhole switching policy and provides 4 terabytes per second (TB/s) of on-chip bisection bandwidth among 8×8 processing cores working at 1 GHz.

## 2 Micro architecture features

Figure 2 shows a Godson-T processing core's floorplan. It includes nine main functional regions: the instruction-fetch unit, floating-point unit, multiply-accumulate unit, integer unit, load-store unit, instruction issue unit, data transfer unit, local memory unit, and router.

### 2.1 Pipeline

In a given cycle, each Godson-T core will fetch two instructions from the instruction cache, and then decode and issue them. There are nine execution units within each core: one double-precision floating-point, multiple-add unit; three floating-point units; two branch execution units; two integer arithmetic logic units; and one load-store unit. The double-precision, floating-point multiple-add unit can also execute two single-precision floating-point operations simultaneously.

The Godson-T core has a combined 64×64 bit register file that's easy to reuse as fixed-point or floating-point registers. This improves the available number of fixed-point registers, as well as floating-point registers, which is more convenient for program optimization. The load-store unit also provides the ability of 128 bit dual-double-precision floating-point load and store in a single cycle.

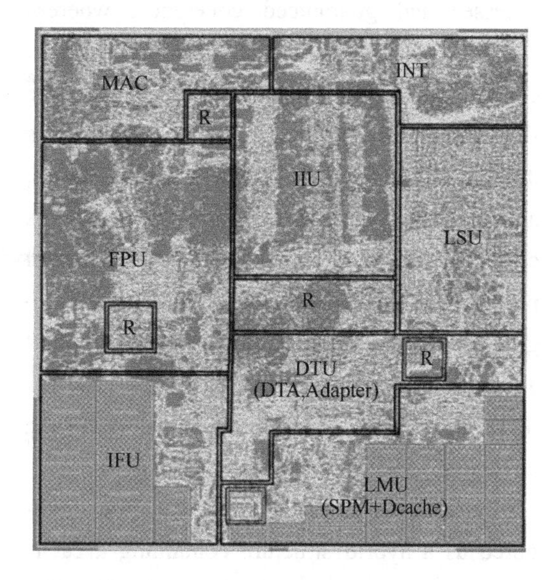


Figure 2. Godson-T processing-core floorplan with nine main regions: instruction-fetch unit (IFU), floating-point unit (FPU), multiply-accumulate unit (MAC), integer unit (INT), load-store unit (LSU), instruction issue unit (IIU), data transfer unit (DTU), local memory unit (LMU), and router (R). The DTU includes a data transfer agent (DTA) and an adapter. The IFU contains a 16 KB instruction cache. The LMU is composed of a 16 KB data cache and control logic

### 2.2 Cache hierarchy

To ease large-scale parallelism, we prefer an RCC protocol to traditional ones[5]. The principle of RCC is that the coherence of data should be lazily guaranteed just upon request. The shared access and private access are expected to be distinguished. The cache hierarchy benefits from differentiating types of accesses and taking appropriate consistency operations to maximize the benefits. In our terminology, a region is a code sequence marked by an open-region primitive and a closed-region primitive. Memory accesses inside the region are regarded as shared accesses and guaranteed coherence, whereas accesses outside the region aren't guaranteed coherence. Ideally, regions embrace only the shared accesses. The RCC's flexibility lets users declare either coarse- or fine-grained regions, and makes it possible to ensure the program's correctness at first, and then make incremental performance improvements by reducing the regions' sizes. It's also convenient to correctly port a parallel program to Godson-T by just inserting these primitives at synchronization points, such as mutually exclusive locks.

In Godson-T, the L1 data cache is configured as a hybrid structure containing a cache and an explicitly controlled and globally addressed SPM to speed up data access. Setting a border register defines the SPM and data cache capacity. Each core can configure its SPM separately. The local SPM's latency is only one cycle, and the remote SPMs' latencies depend on the distance and network congestion. The SPM is globally addressed for all cores. Each core's SPM maps to an interleaving address space of more than 3 Gbytes; thus, the SPM won't clash with data or instruction memory. Processing cores can access any SPM through its mapped address, whether it is local or remote.

L2 cache banks are placed symmetrically around the edge of the core array. Each L2 bank supports four requests and two DTA operations. To meet the need of lock-free programming, this L2 cache supports two extra atomic operations included in the four requests: fetch and add (FAA) and test and set (TAS).

## 2.3 Scratch-pad memory

One of SPM's most important features is that it provides eight full/empty bits for each 32 byte cache line to synchronize fine-grained threads. The Godson-T design extends synchronization instructions to support full/empty bit operations.

When an on-chip private memory is configured as SPM, each cache line's 8 bit write mask works as a set of full/empty bits (see Figure 3). Once the processing core issues a memory access operation, its memory address is used to choose a target from local and remote SPMs.

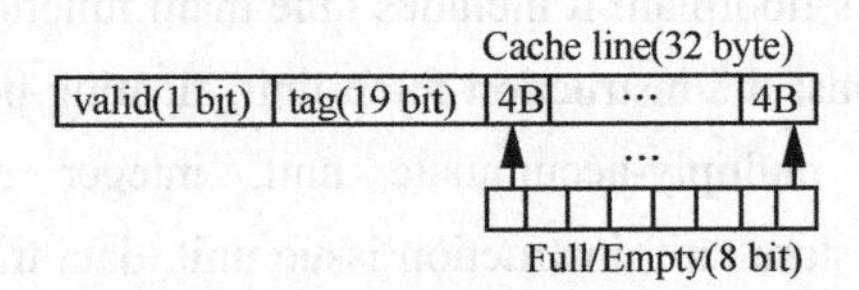


Figure 3. SPM full/empty bit configuration. When an on-chip private memory is configured as SPM, each cache line's 8 bit write mask works as a set of full/empty bits

Data coherence should be maintained by software via new instructions, such as sync_load, sync_store, load_future, and store_future. The sync_load instruction waits until the full/empty bit is set to full in order to read, and then sets it to empty. Conversely, sync_store waits until the full/empty bit is set to empty in order to write, and then sets it to full. The load_future instruction waits until the full/empty bit is set to full in order to read, and then leaves this bit untouched. Similarly, store_future waits until the full/empty bit is set to full in order to write, and then leaves this bit untouched. These two instructions can be used to handle the situation for multiple producers and multiple consumers, together with sync_ load and sync_store.

Whether a sync_store instruction stores a new value depends on the relevant full/empty bit's value. When its full/empty bit's value is 0, it stores this new value and sets the full/empty bit to 1. Otherwise, if the full/empty bit's value is 1, the current sync_store should wait for a sync_load instruction to set it to 0 by loading the data, and then the sync_store can be executed. In an analogy to a sync_store instruction, a sync_load instruction can load the requested data when the full/empty bit is 1, and then sets it to 0. If the requested data's full/empty bit is 0, the current sync_load instruction should be stalled in the pipeline until a sync_store sets it to 1 by storing the data. Once the full/empty bit is set to 1, the load_future and store_future instructions can load or store to the corresponding blocks directly and leave their respective full/empty bits to 1. Otherwise, they should be stalled in the pipeline until a sync_store instruction sets the full/ empty bit to 1. Figure 4 shows a state machine of full/empty bits for explanation.

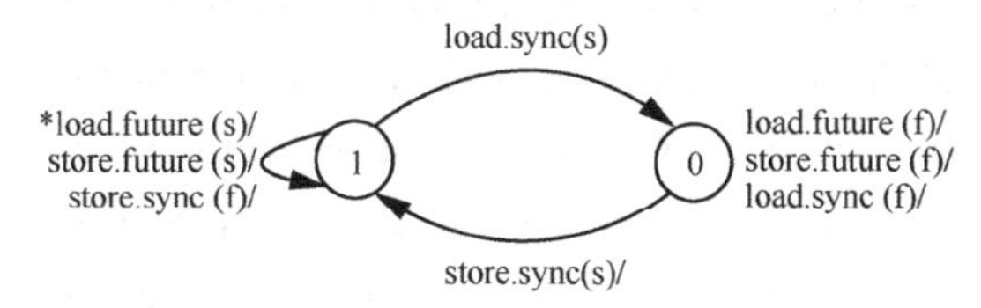


Figure 4. State machine of full/empty bits. Via value transformation of full/empty bits, a synchronization of multiple producers and multiple consumers can be implemented by a sync_store instruction with some store_future instructions and a sync_load instruction with some load_future instructions

Efficient fine-grained synchronization is implemented by a tagged memory or register, in which data can be carried to transfer synchronization information. Compared to the traditional full/empty bit mechanism, the proposed synchronization implements a counter; therefore, various synchronization schemes can be implemented on the basis of it, such as in a multiple-producer, single-consumer situation. We use two micro-benchmarks to evaluate the efficiency of Godson-T's full/empty-bit mechanism: the 2D Wavefront and Livermore Loop 6. The result of fine-grained Livermore Loop 6 gains 70 percent speedup over the coarse-grained synchronization version[5].

## 2.4 Orchestrating data movement

As an execution core coprocessor, the DTA is the main component of our data management framework. Through the DTA, we can transfer data blocks with various strides among SPMs and the L2 caches. Figure 5 shows the overview of the DTA block with five modules. The DTA instruction interpreter receives the instructions issued from the core and translates DTA instructions to operating actions. The request buffer has three entries, and each can keep one DTA request until completion. Therefore, a DTA can deal with three requests concurrently. The DTA request buffer contains the entire network information.

The DTA can assemble the flits to be transferred to local SPM, remote L2 cache banks, and remote SPMs. If the request is to put local data to a remote L2 cache bank or an SPM, the DTA request buffer sends assembled operation instructions to the SPM I/O and then sends data to the DTA I/O module. If the request is to fetch remote data to the local SPM, the DTA request buffer first sends data requests to the DTA I/O. Then, it puts the return data into the local SPM. A flow controller limits and balances the traffic load of the network-on-chip (NoC) traffic load. A flow controller records the number of issued and received flits. When the difference is above some threshold, the controller sends a stall signal to the request buffer to stop successive requests.

Figure 6 shows the data block pattern that the DTA supports. The shadowed area indicates the packets that must be moved. The block is constructed by packets that are smaller than 16 bytes. Figure 6(a) indicates the pattern in which blocks and packets are both continuous; Figure 6(b) indicates the pattern in which packets are continuous but blocks are noncontinuous; and Figure 6(c) indicates the pattern in which both packets and blocks are noncontinuous.

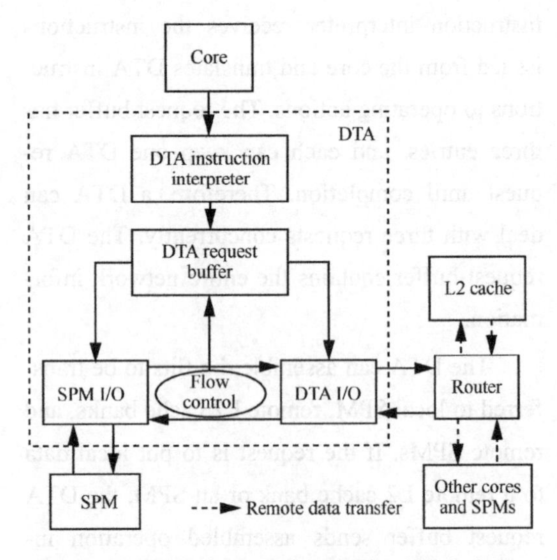


Figure 5. Block diagram of the DTA, which includes five modules: the DTA instruction interpreter, DTA request buffer, SPM I/O, flow controller, and DTA I/O

Godson-T provides DTA-related APIs that are constructed with reduced-instruction-set computing (RISC)-like DTA instructions to implement explicit data movement, which is critical to improving on-chip memory utilization and on-chip network bandwidth efficiency. Our evaluation results show that the performance can be dramatically accelerated by 2 to 3 times when using on-chip DTA operations.

## 2.5 Synchronization manager

There is a hardware SM on chip, which can handle both mutual-exclusion and barrier synchronization. The SM implements fast hardware synchronization to replace a pure software synchronization method. The evaluation results show that the overhead of mutual-exclusion and barrier synchronization on Godson-T is far lower than software-based methods, ranging from tens to hundreds of times lower.

As Figure 7 shows, the SM is a 128-entry and four-way set-associative table in which each entry records a 57-bit synchronization request. Core_ID represents the core, which sends the synchronization requests. State represents the state of the synchronization, such as lock_acquired or lock_waiting. Sync_ID is a unique identifier for a synchronization operation. Bar_Count is used only for barrier synchronization and records the number of threads that haven't arrived at the barrier, so this field differs in each linked entry. Next is a pointer to the next entry, which represents another core waiting for the same lock or barrier.

The mutual-exclusion (lock) requests are organized in the form of queues in the SM. Entries belonging to the same queue are linked with Next pointers. When a lock request is sent to the

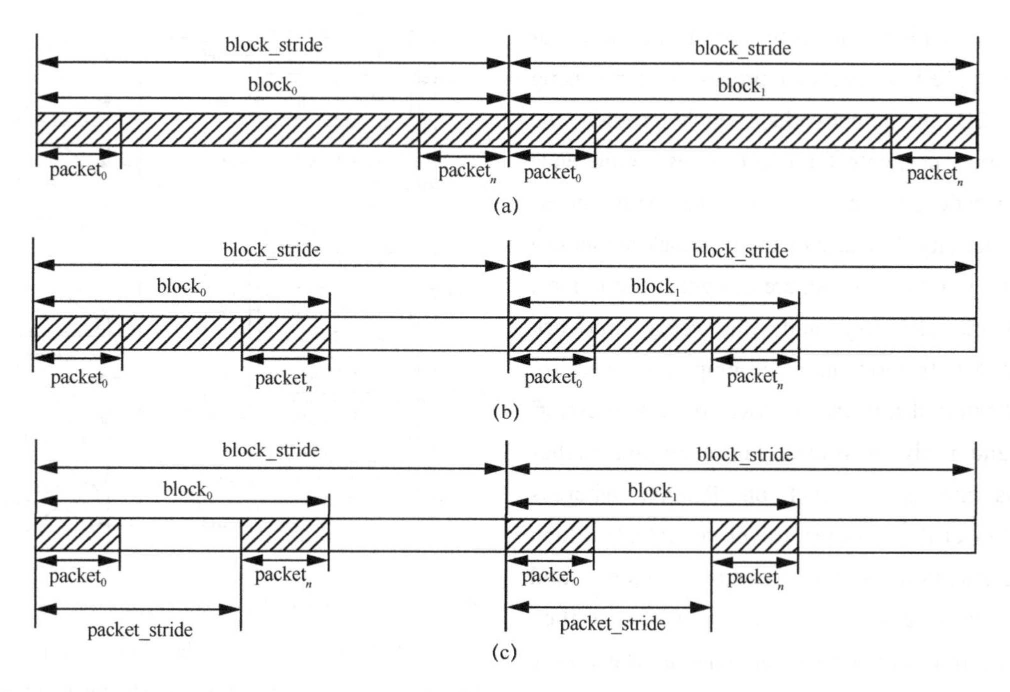


Figure 6. Bulk data patterns supported by the DTA. The figure indicates three patterns: one in which packets and blocks are both continuous (a), one in which packets are continuous but blocks are noncontinuous (b), and one in which packets and blocks are both noncontinuous (c)

SM through the on-chip network, the SM checks the state of entries with the same Sync_ID to decide whether the request will be satisfied or should be put in the waiting queue. Queuing avoids sending too many messages on the network, which is a major disadvantage of traditional test and set lock.

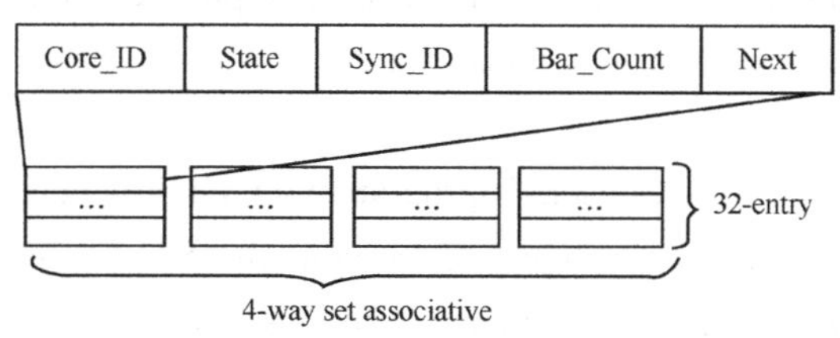


Figure7. The SM structure. A centrie on-chip synchronization manager helps to implement faster synchronization without accessing off-chip shared memory

The barrier synchronization can be implemented similarly: Barrier requests with the same Sync_ID are organized as a queue. When all the barrier requests have arrived at the SM (Bar_Count changes to 0), the ACK messages are sent to all threads participating in the barrier synchronization.

To reduce the search overhead, all 128 entries are organized into a four-way set-associative table. Both Core_ID and Sync_ID are used to index the entry number. A software handler is triggered if the SM table overflows.

## 2.6 Network on chip

Godson-T's on-chip interconnect network is called gMesh, which is a departure from the traditional bus-based on-chip interconnect structure shown in Figure 1. Instead of using buses or rings, gMesh connects the processing cores using two 2D mesh networks, which provide the transport channels for cache access, coherence information, I/O access, and other communication activity. Because of the deadlock avoidance strategy, two networks are designed: the request network (REQN) and the reply network (RPLN). In Godson-T, on-chip messages are orchestrated into two classes: request messages and reply messages. For example, cache-miss messages travel on REQN, whereas cache-refill messages travel on RPLN. Such classification isolates memory request and reply messages into different networks, so they won't interfere with each other at the core's memory port or the L2 caches.

For each processing core connected to the NoC, there's a dedicated router, as shown in Figure 8. In the router, five ports are connected by a fully connected crossbar, which allows all-to-all five-way communication. Each router port contains a two-entry input buffer for each link. Having the two mesh networks leverages the on-chip wiring resources to provide massive on-chip communication bandwidth. When Godson-T works at 1 GHz, the gMesh networks can afford a 4 Tbyte/s bisection bandwidth for an 8×8 mesh network. By using mesh networks, the architecture can support any number of cores with few modifications on the fabric. In fact, the router resource can remain unchanged, even for a fairly large scale. Growing the mesh size provides more connecting bandwidth. The design provides the essential scalability for the Godson-T many-core architecture.

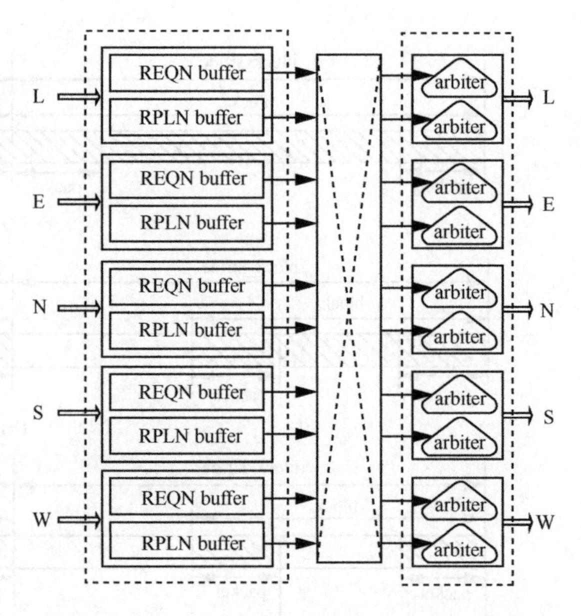


Figure 8. Overview of a wormhole router used in two independent on-chip networks. Each router connects in five directions: east, west, north, south, and to the local core. Each link contains two 128 bit unidirectional links. (REQN: request network; RPLN: reply network; L: local core; E: east; N: north; S: south; W: west; X: crossbar)

## 3 Software environments

Given a many-core processor like Godson-T, one challenge is efficiently utilizing the large on-chip computing capability. Programming on Godson-T is based on multithreading. Providing a simple and reasonable programming interface is undoubtedly critical to programmability.

The Godson-T programming environment adopts conventional C programming and the necessary tool chain. We also provide a Pthreads-like C library for task management. Therefore, a large amount of parallel program sources written with Pthreads can be conveniently ported onto Godson-T. The library provides a rich set of APIs for task management, including batch thread management, which can significantly reduce tasking overhead by grouping identical operations in batches. The programmer is responsible for creating, terminating, and synchronizing tasks by using appropriate Pthreads-like APIs.

## 3.1 GodRunner

We developed a software runtime system named GodRunner to provide efficient abstraction of a large number of hardware thread units and dynamic load balancing. The GodRunner task model (in our terminology, a task resides in software, whereas a thread resides in hardware) adopts a create-join method inspired by Pthreads. GodRunner tasks don't support preemptive execution, because frequent context-switching incurs save-restore overhead and cache thrashing. Therefore, a task will keep executing on a thread unit until it terminates. Nonpreemptive execution can make task initialization simple and fast: for example, the stack is statically allocated to each thread unit, which avoids a time-consuming dynamic allocation. GodRunner lets programmers create more tasks than hardware thread units, and transparently maps them to hardware thread units at runtime. GodRunner swaps the completed task out and schedules a new one into a thread unit:.

GodRunner can flexibly incorporate different task-scheduling algorithms. So far, we implemented two well-known task-scheduling algorithms for efficient dynamic load balancing: work-stealing[6] and conditional spawning[7]. Godson-T supports fine-grained threads very well. For example, GodRunner can create a new thread in just tens of cycles. Figure 9 illustrates a GodRunner task-scheduling example.

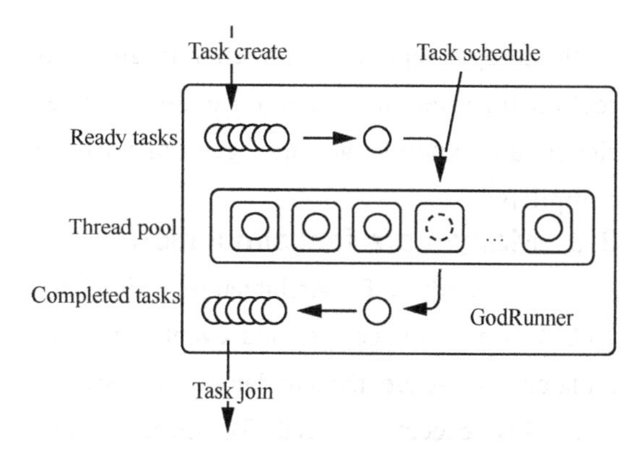


Figure 9. GodRunner lets programmers create more tasks than hardware thread units, and transparently maps them to hardware thread units at runtime. It also swaps the completed task out and schedules a new one in a thread unit

## 3.2 Compiler optimization

Computer architects and system software designers face a unique opportunity to bring together new architectural features as well as corresponding compiler technology, so as to achieve a strong impact on system performance. Using the Godson-T platform, we performed some compiler technology studies. Using a pattern-making methodology, we encapsulated algorithm-specific optimizations into optimization patterns expressed in terms of preprocessor directives so that simple annotations could result in significant performance improvements. To validate this new methodology, we developed a framework named EPOD (Extendable Pattern-Oriented Optimization Directives) to map such directives to the underlying optimization schemes. Our experimental results show that a pattern-guided compiler can outperform the state-of-the-art compilers and even achieve performance as competitive as hand-tuned code. Thus, such a pattern-making methodology represents an encouraging direction for integrating domain expert's experience and knowledge into general-purpose compilers[8].

### 3.3 Many-core fast simulation

The Godson-T Architecture Simulator (GAS) is a cycle-accurate and event-driven simulator. As a sequential simulator, it also suffers from slow execution speed. To speed up this many-core processor simulator, we introduced a parallel discrete event simulation (PDES) method. Our goal was to speed up the sequential simulator with multi-threading, while maintaining its cycle-level accuracy. The basic idea is to divide the global event queue into separate queues for each logic process, while maintaining time synchronization between these queues using PDES. Evaluation against the sequential version shows that the parallelized simulator achieves an average speedup of 10.9× (up to 13.6×) when running the Splash-2 kernel on four quad-core AMD Opteron 8347HE processors running at 1.9 GHz with 64 Gbytes of DRAM.

## 4 Physical implementation

Figure 10 shows the floorplan of the Godson-T 16-core prototype chip. It passes all standard tests and achieves the corresponding performance discussed in previous work[5]. Considering cost and performance tradeoffs, we reduced the 16 data cache banks to four in this 16-core prototype chip. The 64-core prototype chip will be manufactured in 65 nm CMOS technology, targeting an operating frequency of 1 GHz, and with an approximate area of 200 mm$^2$.

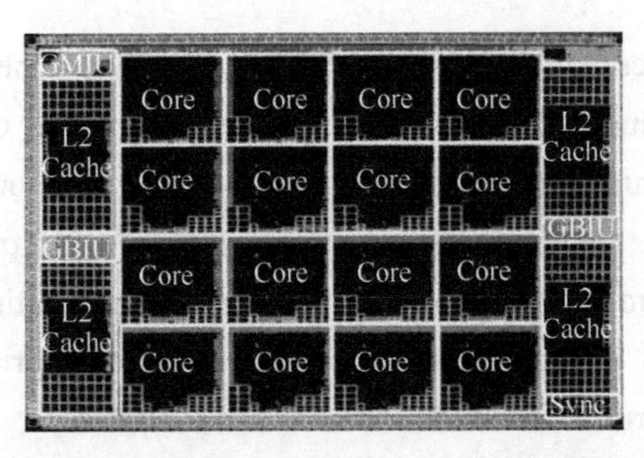


Figure 10. Floorplan of Godson-T 16-core prototype chip. Taped out with SMIC 130 nm, 1P8M CMOS process, working at 300 MHz, with an area of 230 mm$^2$(GBIU: Godson-T bus interface unit; GMIU: Godson-T memory interface unit; Sync: synchronization manager)

Godson-T will form the basis for ICT's future high-performance computing system. In the future, we will continue to investigate architectural innovations for high-performance and high-throughput computing. Once integrated processing cores increase to above 1 000, the scalability of on-chip interconnect networks and memory subsystems becomes an emergent and open problem. One of our future works will be finding a way to make many-core architecture more extendable. Another challenge is how to face the dark silicon problem. We will devote efforts to energy-efficient architecture design with the support of the state-of-art manufacturing technologies.

## Acknowledgments

Godson-T represents the combined efforts of many diligent ICT researchers, engineers, and students across multiple teams in the State Key Laboratory of Computer Architecture of China. We thank professor Paolo Ienne for his Godson-T related research. This work is supported by the National Basic Research 973 Program of

China (under grant 2011CB302501), the National Natural Science Foundation of China (under grants 60921002, 60925009, 61173007, 61100013, and 61100015), and the National Major Science & Technology Program of China (under grant 2011ZX01028-001-002).

## References:

[1] BELL S, EDWARDS B, AMANN J, et al. TILE64 processor: a 64-core SoC with mesh interconnect[C]// IEEE International Solid-State Circuits Conference. Piscataway: IEEE Press, 2008.

[2] KALLA R, SINHAROY B, STARKE W J, et al. Power7: IBM's next-generation server processor[J]. IEEE Micro, 2010, 30(2): 7-15.

[3] CONWAY P, KALYANASUNDHARAM N, DONLEY G, et al. Cache hierarchy and memory subsystem of the AMD Opteron processor[J]. IEEE Micro, 2010, 30(2): 16-29.

[4] MARUYAMA T, YOSHIDA T, KAN R, et al. Sparc64 VIIIfx: a new-generation octocore processor for Petascale computing[J]. IEEE Micro, 2010, 30(2): 30-40.

[5] FAN D R, YUAN N, ZHANG J C, et al. Godson-T: an efficient many-core architecture for parallel program executions[J]. Journal of Computer Science and Technology, 2009, 24(6): 1061-1073.

[6] BLUMOFE R D, LEISERSON C E. Scheduling multithreaded computations by work stealing[J]. Journal of the ACM, 1999, 46(5): 720-748.

[7] PALATIN P, LHUILLIER Y, TEMAM O. CAP-SULE: hardware-assisted parallel execution of component-based programs[C]// 2006 39th Annual IEEE/ACM International Symposium on Microarchitecture. Piscataway: IEEE Press, 2006: 247-258.

[8] CUI H M, XUE J L, WANG L, et al. Extendable pattern-oriented optimized directives[C]// ACM Transactions on Architecture and Code Optimization. Piscataway: IEEE Press, 2011: 107-118.

发表于 IEEE International Parallel and Distributed Processing Symposium 论文集，2012 年

# Building billion-threads computer and elastic processor

Guo-jie LI

Institute for Computing Technology, Chinese Academy of Sciences

**Abstract:** The characteristics of IT applications in the future decades are computing for the masses. What datacenters will deal with is a high number of active users, a high number of applications, a high number of parallel requests, massive amount of data, etc. This challenge is especially serious for China, since there is a huge population. The emergence of Intemet-of-Thing makes the class of applications ever more and more, and the ossified computer architecture cannot be suitable for the various niche applications. To address these big issues, Chinese Academy of Sciences (CAS) has started up the Future Information Technology (FIT) Initiative, a 10-year frontier research project for targeting applications and markets of 2020-2030. The State Key Lab on Computer Architecture (CARCH), which is located at the Institute of Computing Technology (ICT) and is the unique SKL in the area of computer architecture in China, is one of major undertakings of the FIT project. The research directions of CARCH include building billion-threads computer, elastic processor, cloud-sea computing, etc. In this talk, we will survey the motivations and basic ideas of these projects. Moreover, we will briefly introduce another foresighted research going on ICT: service-oriented future Internet architecture.

第 8 章

# 下一代网络与通信

# 走计算与通信技术融合之路

国家智能计算机研究开发中心是我国较早接触互联网的科研单位之一。在全国正式接入国际互联网之前，1993 年我就通过中国科学院高能物理研究所的专线上网，直接登录我在美国伊利诺伊大学厄巴纳-香槟分校（UIUC）使用过的计算机。1994 年中国科学院率先接通国际互联网以后，国家智能计算机研究开发中心立即开通了国内最早的 BBS 网站，还在国内率先开展“天联”“天罗”等互联网搜索产品研制。

2001 年，中国科学院计算技术研究所（以下简称计算所）启动了“IPv6 网络关键技术研究和城域网示范系统”知识创新工程重大项目，在国内率先研发和推广 IPv6 技术，我是此项目的负责人。计算所与中国网通合作，建立了重庆网通信息港 IPv6 实验网。这是世界上首次基于 IPv6 的大规模城域网试运营，还在北京建立了无线移动 IPv6 城域示范网。通过此项目研究，计算所在 IPv6 网络的关键技术及网络运营方面取得了可喜的成绩。当时美国国防部对 IPv6 网络进行测试时，也要使用基于我们的专利技术生产的测试仪器。

我国 863 计划信息领域有计算机主题和通信主题，但没有网络主题，网络技术两不管导致我国的网络技术研究长期投入不足，与国外的差距拉大。2005 年做国家中长期规划战略研究时，我牵头的信息领域专家组对众多前沿技术进行凝练，将“未来网络技术”作为信息领域的主攻方向，提议将建设“8 亿龙网”作为国家未来 15 年的“重大专项”之一，突破网络升级换代的关键技术。2005 年中国的上网人数不到 1 亿人，当时我国科学技术部、信息产业部预测 2020 年我国的网民最多达到 6 亿人，保守的预测只有 3 亿～4 亿人。我们预测可突破 8 亿人，所以称为“8 亿龙网”。截至 2019 年 6 月，我国网民规模已达 8.54 亿，说明我们当时的预测是靠谱的。后来为了集中力量突破 4G 和 5G 技术，将网络技术聚焦为“新一代宽带无线移动通信网”，设立了 03 专项。

制定互联网 IPv6 的初衷是解决 IPv4 地址不足的问题，将互联网地址从 32 位扩展到 128 位，可解决所有人上网以及未来物联网需要的网络地址。但是，20 世纪 70 年代设计互联网时并没有想到它会成为全世界的信息基础设施，因此在可扩展性、服务质量保证（尤其是实时性）、安全性等方面先天不足。21 世纪以来学术界和公司都在探索新的网络基础设施。2010 年，我担任国家重大科技基础设施建设中长期规划工程技术专家组组长，我提出的“未来网络试验设施（CENI）”被列为国家“十二五”期间的重大项目之一，由以计算所为主要参与单位的江苏省未来网络创新研究院牵头实施，刘韵洁院士为项目负责人。国家重大科技基础设施（以前称为“大科学工程”）立项评审过程很长，“十二五”的项目到“十三五”才启动。2019 年 5 月 CENI 正式开通运行，首批骨干节点连接北京、南京等 12 个大城市，边缘节点覆盖全国 40 个城市的 133 个试验节点，已开展服务定制网络、网络可编程及功能虚拟化等新型网络架构试验。

计算所自成立以来一直围绕计算机做科研，没有专门开展通信技术研究。中国科学院也没有主攻通信技术的研究所，通信技术成了其短板。近几十年，我国计算机和通信各自成为独立的“一级学科”。中国的通信（电信）行业很强势，强化了通信与计算机学科的分离。互联网为计算机与通信技术的融合提供了广阔的平台。1995 年我与邬江兴、陈俊亮合作承担“九五”国家科技攻关项目“CIN-02 高级智能网系统”，这是计算与通信技术融合的一次尝试。我担任计算所所长以后，更主动地推动实现计算与通信技术的融合。2001 年计算所从北京邮电大学引进石晶林副教授，由他牵头在计算所拉起一支做无线通信的科研队伍，在 DSP 芯片、卫星通信、超级无线基站等方向做出了国际一流的科研成果。后来成立了移动计算与新型终端北京市重点实验室，还创建了北京中科晶上科技股份有限公司和以移动通信为主要方向的计算所洛阳分所，促进洛阳的中国一拖集团有限公司实现产业转型，在通信领域大展风采。在北京中科晶上科技股份有限公司的赞助下，2014 年开始每年举办一次“通信网络与计算科学融合”香山科学会议，邀请全球知名华人科学家参加，在学术界已产生较大影响。

发表于《计算机研究与发展》2001 年第 7 期

# 一种可扩展大型E-MAIL服务系统的研究与实现

陈明宇，李国杰，赵延波，张晓霞

中国科学院计算技术研究所，北京 100080

**摘　要**：大型e-mail服务系统的用户规模已经开始向千万级发展，从而对系统的可扩展性、可用性和可管理性都提出了新的挑战。本文提出了一种基于机群和分布式存储的层次型大型e-mail服务器系统结构。这种结构通过三级负载平衡机制实现了请求的均匀分配，并通过对各功能单元进行并行化，较好地解决了系统扩展中的瓶颈问题，提高了系统的可用性。文中还介绍了这种结构在曙光2000-II超级服务器上的实现——FreeMail及性能测试。实验表明，这种系统结构具有良好的性能和可扩展性。

**关键词**：e-mail；大型系统；机群；可扩展性；分布式存储；负载平衡

# Study and implementation of a scalable large-scale E-MAIL server system

Ming-yu CHEN, Guo-jie LI, Yan-bo ZHAO, Xiao-xia ZHANG

Institute of Computing Technology, Chinese Academy of Sciences, Beijing 100080

**Abstract:** The user capacity in a large-scale e-mail server system is going up to the level of ten million, challenging the system scalability, availability and manageability. In this article, a hierarchical large-scale e-mail server architecture based on cluster and distributed storage is presented. This architecture can distribute incoming requests evenly through a three-level load balancing mechanism, and can solve the bottleneck problem when scale up through the parallelization of each functional unit. Finally, the implementation of this architecture on the Dawning 2000-II super server – FreeMail and the performance test are introduced. Test result shows this system has high throughput and scalability.

**Key words:** e-mail, large-scale, cluster, scalability, distributed storage, load balance

## 1 引言

近年来随着 Internet 用户量的飞速增长，提供个人邮件服务的大型 ISP/ICP 已经需要同时支持几十万到上百万数量级的 e-mail 用户，并且用户量还有进一步向千万级增长的趋势。

本课题得到国家 863 计划项目资助。

目前大型 e-mail 服务系统多采用机群系统+共享存储设备的方式来实现。早期的系统多基于 NFS/NIS[1-2]，现在已经发展到了使用 SAN、NAS 等专用大容量共享存储设备。但是这种集中式存储系统的固有缺点是系统的扩展性、性能和可靠性完全依赖于一个或几个专用设备上，而这些专用设备成本较高，并且其本身也很容易成为系统的瓶颈或单一故障点。于是人们开始研究利用分布式存储技术实现更大规模的可扩展、高可用 e-mail 服务系统。

实现分布式存储大型 e-mail 系统的关键问题是如何实现邮箱的合理分布和大量用户同时访问时的负载平衡。美国 Washington 大学研制的 Porcupine 系统[3]采用了一种基于动态邮箱迁移的完全对称式的结构。Porcupine 强调每个节点都能服务所有的用户，因而涉及大量内部节点通信和状态信息的维护，使系统实现起来非常复杂，而且没有很好地解决负载平衡问题。

本文提出并实现了一种层次型分布式存储大型 e-mail 服务系统的结构。这种结构中不同层次的节点完成不同的功能。通过基于三级单一登录点的负载平衡机制，保证了邮件请求到达的均匀性；通过集中的索引服务器实现用户邮箱的合理分布。此外，所有的功能节点都可以针对其功能特点进行优化，并可通过简单扩展提高处理能力，从而避免成为整个系统升级的瓶颈。按照这种结构我们在曙光 2000-II 超级服务器上实现了 FreeMail 系统，并进行了性能测试，实验表明系统具有良好的可扩展性和性能。

## 2 分布式存储大型 e-mail 服务器系统中的主要技术问题

### 2.1 E-mail 服务的访问特性

图 1 给出了一封邮件经历的全过程。e-mail 服务器参与了邮件递交、邮件发送、邮件接收、邮件提取 4 个步骤。在大型 e-mail 系统中，一般还需要提供基于浏览器的 web 访问方式，因此还需要包含部分邮件编辑、邮件显示等功能。其中涉及的网络协议有 DNS（发送）、SMTP（递交、发送、接收）、POP3/IMAP4（提取）、HTTP（递交、提取、编辑、显示）等[4]。

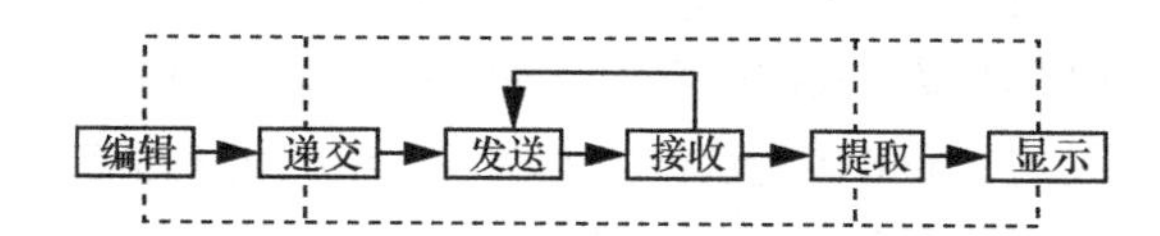


图 1 邮件服务器在邮件传输中的作用

分析可知 e-mail 访问具有以下特性。

- 独立性：每个用户都只访问自己的私有数据——邮箱，而无权访问其他用户的私有数据。用户之间基本不存在共享信息。
- 可靠性：用户邮件必须保证完整可靠送交对方。如果对方的邮件服务器暂时不可用，则发送邮件服务器应该自动在一定时间后重试。如果最终邮件仍不能送达，则应通知发送方发送已失败。
- 非实时性：用户邮件传送不需要很强的实时性，一般邮件在几天之内送达都是允许的，而且允许到达的顺序不一致。
- 均匀性：一般每个用户一天的收发的邮件数有限，并且每次用户的访问时间不会持续很长。总体来看每个用户产生的负载是比较均匀的。
- I/O 密集性：每次邮件访问无论是提取还是发送一般都需要进行磁盘 I/O 操作，而且由于访问之间独立性的原因，磁盘缓冲基本不起作用。

可以看出 e-mail 服务与 web 服务有显著的不同。即使是 web mail 方式，其处理也是以磁盘 I/O 和动态页面生成为主。在大型 e-mail 系统结构设计时必须考虑到这些特性。

### 2.2 分布式存储大型 e-mail 服务器访问负载能力分析

以每个用户在服务器上的数据存储量 1 MB 以上计算，5 万用户的数据存储需求量就在 1 MB×5 万=50 GB 量级。这是一台 I/O 能力较强的服务器可以承受的。

根据 e-mail 访问的均匀性特点，我们认为 e-mail 请求的到达满足泊松分布，并可以采用 M/G/1 模型来描述系统。假设平均每个用户每天的 e-mail 访问次数为 5 次，则 5 万用户的访问到达率 $\lambda$=50 000×5/86 400=2.89。又设每次访问的平均完成时间为 $W$=10 s，则易知系统平均同时服务用户数为 $L=\lambda W$=28.9 个。即使考虑到广域网的延迟和访问的波动性等因素，则 $L$ 也在 $10^2$ 量级，同样一台普通服务器完全可以承受。

通常 e-mail 的数据以短信息为主。设平均每次 e-mail 访问平均交互数据量为 10 KB，则 5 万用户产生的平均网络带宽只有 30 KB/s 左右。一般服务器的网络能力完全可以承受。

如果我们假设每个机群节点可以服务 5 万量级的邮件用户，那么实现百万级以上的大型 e-mail 服务器就需要几十个节点的规模的机群。

虽然理论上每增加一定量用户时，只要增加一个节点，则系统的存储容量和处理能力都得到加强，系统性能也能线性的增长。但是当系统节点规模较大时，到达各节点的请求个数将变得不均匀。很容易形成访问的局部热点，导致系统整体性能下降。因此在请求达到服务节点之前必须首先进行有效的负载平衡。

此外，如果用户分布不均匀，则会导致节点存储空间利用率不一致。很容易形成某个节点存储空间已满，而系统整体存储空间并未得到有效利用的情况。这就要求系统不能采用简单的静态用户分布策略，而是采用动态的分布策略。

## 3 FreeMail：一种层次型分布式大型 e-mail 系统结构

### 3.1 系统结构

如图 2 所示，整个系统由多个不同的功能模块组成。其中 Extranet 是指可以直接向外部 Internet 发送网络包的网络。系统内部节点通信则使用内部高速互连网。

系统可以分为 5 个功能模块：DNS、PowerRouter、Proxy、Index、MServer。

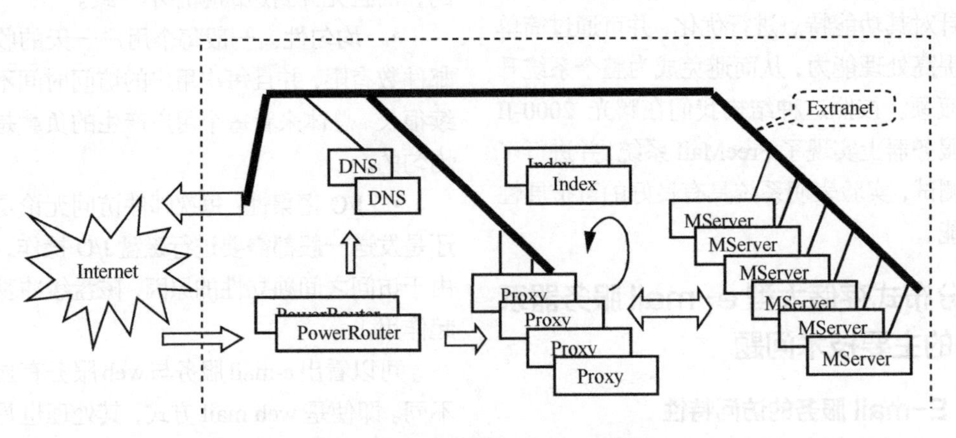


图 2 层次型分布式大型 e-mail 系统

DNS 服务器提供域名查询服务。将用户对机群系统单一域名的查询转化为对多个 PowerRouterIP 的请求。

PowerRouter[5]是一种工作在系统核心的高速包转发软件。它可以把对同一服务地址（IP：Port）的请求，根据后台服务器的实时负载情况分发给不同的 Proxy 节点。PowerRouter 类似于现在的一些第四层交换设备，不同之处是所有从服务器端返回的数据包不再通过 PowerRouter 而是直接返回，因而效率更高。

Proxy 则是工作在用户空间的转发软件，可以实现基于目标（Locality-Aware）的请求分发。Proxy 节点首先完成协议的前期处理，包括身份认证、安全检查和目标服务器定位等，然后进入转发状态，在客户端和负责服务该客户的目标 MServer 之间建立连接直至服务结束。Proxy 服务器分为 SMTP、POP3、HTTP 3 种。DNS、PowerRouter、Proxy 一起构成了系统的三级单一登录点。

Index 服务器用于存放用户认证信息及每个用户对应的 MServer 位置。当用户请求到达 Proxy 后，Proxy 首先向索引服务器发出查询请求，并一次性获得用户身份和位置信息。Index 服务器由高性能数据库服务器构成。很容易实现海量用户信息的存储、管理和查询。

MServer 即邮件存储和处理服务器。每一个 MServer 都运行独立的 SMTP、POP3 和 HTTP 服务器，并配有大容量硬盘存放用户邮件，基本等同于一台可容纳几万用户的独立的邮件服务器。每个用户的邮件由几台固定的 MServer 进行服务。所有的用户则均匀分布在各 MServer 之中。当需要增加用户容量时，只需增加 MServer 节点即可。此外，MServer 集数据存储和处理为一体，也使服务的效率得到保证。

每次客户邮件请求开始时，首先通过 DNS 查询得到机群系统的一个 IP，这个 IP 实际上对应了某个 PowerRouter。客户端向 PowerRouter 发出连接请求，PowerRouter 根据当前各 Proxy 节点的负载情况，将请求转发给一个负载较轻的 Proxy 节点。Proxy 首先得到用户的认证信息，并通过 Index 服务器查询得到用户邮件所在的 MServer 位置，随后 Proxy 将后续用户请求转发给 MServer。MServer 则对用户的请求进行服务。

## 3.2 基于三级单一登录点的负载平衡机制

单一登录点是指整个机群系统在网络上对客户端具有的唯一的系统名，客户端利用此系统名可以访问到机群系统提供的服务，而不需要指明特定机群内部的节点。FreeMail 设计中采用了基于三级单一登录点的负载平衡机制。

第一级单一登录点是轮换式 DNS。DNS 服务器可以将对一个域名的查询映射为不同的 IP 地址，实现部分负载平衡功能。但由于 DNS 信息可能被其他 DNS 服务器缓冲，造成来自某一区域的客户集中访问一个 IP，无法实现实时的负载平衡，更无法进行及时的故障屏蔽。因此 DNS 只适合于广域网和长期的负载平衡[6]。

第二级单一登录点是 PowerRouter。PowerRouter 可以进行实时的细粒度负载平衡，即使来自同一客户端两次相继的 TCP 连接也可能被转发到不同的节点上。PowerRouter 直接在核心网络处理层进行转发工作，因此其效率很高，一个 PowerRouter 就可以带几十个节点。

PowerRouter 可以根据后台 Proxy 节点的性能和当前负载按比例的分发新到来的请求。保证每个后台 Proxy 得到均匀的负载。

结合 DNS 的多 IP 映射，我们还可以利用多个 PowerRouer 同时工作，防止 PowerRouter

自身成为瓶颈。同时，我们将 DNS 服务器也放在了 PowerRouter 后面，使得 DNS 服务器本身也可以并行化。

第三级单一登录点是 Proxy 阵列。Proxy 是真正实现“登录”功能的节点，主要完成定位和转发工作。因为需要对应用层协议数据进行分析并提取用户信息，所以这些工作无法由 PowerRouter 进行，而只能由 Proxy 在用户空间完成。Proxy 的效率要低于 PowerRouter。但 Proxy 节点几乎不进行磁盘 I/O 操作，主要功能是进行数据分析和转发，因此也可以同时支持多个 MServer。

Proxy 节点也进行部分负载平衡工作。当一个用户同时由几个 MServer 提供服务时，Proxy 可以选择其中负载较轻的一个。此外利用 e-mail 传送非实时性的特点，在接收邮件时如果目标 MServer 负载过重或发生故障，Proxy 节点也可以选择一个负载轻的无关 MServer 节点接收，该节点会选择在适当时机自动进行内部转发。

经过了三级单一登录点后，最终到达每个 MServer 的只有与该 MServer 相关的用户请求。这样 MServer 的负载与一个独立的邮件服务器基本相同，而与整个系统的规模无关。从而保证了系统的可扩展性。

### 3.3 基于集中式索引的动态用户分布机制

分布式存储的 e-mail 系统中用户邮箱分布信息的维护是一个关键性的问题。如果采用静态的 hash 分布算法，则可以加快每次访问定位的速度，但不能保证用户邮箱在各节点间的均匀分布。采用动态的索引，为每个用户建立一个映射，则通过修改索引就可以动态移动用户的邮箱位置。但是，在大用户量下，索引本身数据量也很大，其维护开销也不容忽视。如在 Porcupine 系统中，每个节点上都保持了一份用户索引，光用户索引就要占几十 MB 的内存，而且维护多份复制的一致性更需要复杂的算法。

我们在 FreeMail 设计中采用了集中的索引服务器来存放用户索引信息。这样每次 Proxy 接收请求后都要增加一次查询索引服务器操作，从而增加了部分延迟。但因为每个用户信息较少，可以保证在一次往返就得到所需信息，而在数据库服务器进行单记录查询的开销也很小。如果我们对数据库服务器采用高性能的配置，如配备大容量内存将所有索引信息加载在内存中，并配备高性能处理器，则一次查询索引操作的延迟可能小于一次本地磁盘 I/O 所花费的时间。

集中式的索引使得系统的管理变得非常方便，可以通过数据库的 SQL 接口方便地进行查询、统计和大批量用户的调整，同时省去了复杂的一致性维护开销。

### 3.4 可扩展性分析

因为采用分布式存储技术，系统容量的可扩展性很容易通过增加 MServer 得到。

系统的吞吐量与整个系统各部分的处理能力有关。而 FreeMail 的设计中所有的功能层次都可以由多个节点共同完成。这就保证了各部分处理能力的可扩展性。当任何部分成为系统瓶颈时，只需加强相应的处理节点个数或提高性能即可。

FreeMail 系统的节点功能划分清晰，使得我们可以根据不同的功能需求采用不同优化配置的处理节点。表 1 给出了 5 种节点的不同需求特点。

### 3.5 可管理性和可用性

由于采用了 Index 数据库使得系统的可管理性大大增强。所有的用户分布和注册信息等可以通过标准 SQL 语句进行查询、统计、调整等操作。特别是当使用多个 Index 节点时，可以通过并行数据库本身的功能进行一致性

表 1　FreeMail 中的不节点对不同处理性能的需求

| 节点类型 | PowerRouter | DNS | Index | Proxy | MServer |
|---|---|---|---|---|---|
| 节点规模 | 少量 | 少量 | 少量 | 适量 | 大量 |
| 功能特点 | 核心网络包高速转发 | DNS 数据查询、服务 | 对大用户量数据库单记录、查询 | 协议前期处理、用户态网络包转发 | 协议后期处理、大量磁盘 I/O 操作 |
| 较高性能要求部分 | 网络吞吐量 | CPU | 内在量、CPU、磁盘 I/O 速度 | CPU、网络速度 | 磁盘容量、I/O 速度 |
| 较低性能要求部分 | CPU、内存量、磁盘容量、速度 | 内存、硬盘容量、速度 | 磁盘容量 | 磁盘容量/速度 | CPU |

数据维护，减少了复杂的一致性协议设计和实现，同时提高了可靠性。

FreeMail 的分布式结构设计首先保证了任何一个节点故障都不会造成整个系统不可用。例如某个 MServer 故障，只会短时期影响这个 Server 上的用户的提取请求(发送请求可以利用上面提到的内部转发机制屏蔽故障）。其他 Mserver 上的用户则不受任何影响。采用高可用软件[7]通过 Heartbeat 技术及时检测到故障，并自动进行屏蔽，可以使系统功能受到的影响减到最小。

当对用户数据的可用性需求较高时，可以采用邮箱复制或小型共享存储设备。这时只要在 Index 中增加每个用户的备份节点信息即可。由于系统并没有限制每个用户只能使用一台 Mserver，实际上基于 SAN 的共享存储设备同样可以使用在本系统中。

### 3.6 FreeMail 与 Porcupine 系统的比较

Porcupine[3]的系统结构相当于将 Proxy、Index、MServer 3 种节点功能结合在一起，而没有进行针对请求到达的负载平衡，因此不能避免局部热点的问题。此外，Porcupine 不含对 webmail 的支持。

Porcupine 在设计中强调了每个节点都可以完成全部的功能，因此其每个节点的功能比较复杂，要求节点的性能比较全面。而 FreeMail 的中具有节点专用化的特点，每个节点可以针对其功能进行优化配置。

Porcupine 系统的邮件处理和存储是分离的，每个用户的邮箱还可能被分成几个部分，使得每个用户请求的处理时间变长。FreeMail 系统中的处理和存储是一体的，数据读写都通过本地 I/O，特别是对邮箱的锁操作只需在本地进行，因此效率更高。

Porcupine 的状态信息是分布在各节点内存中的，因此在定位时速度快。但这些状态信息本身要耗费相当的资源，管理和一致性维护需要复杂的协议。FreeMail 的集中式索引虽然增加了一次查询操作，但相比与磁盘 I/O 和广域网延迟几乎可以忽略，而同时却得到了集中管理便于维护的好处。

Porcupine 采用了全对称的结构，系统的可用性较高。FreeMail 通过功能节点的并行化同样可以实现高可用性。

Porcupine 通过动态邮箱迁移可以实现自动用户分布调整。FreeMail 中改变用户分布则只能由管理员通过调整数据库索引实现。但根据 e-mail 访问均匀性的特点，实际上每个用户产生的负载变化不大。所以我们认为只有在节点规模变化时才有必要迁移用户，而这种情况并不经常发生。

最后，Porcupine 为实现全对称的结构，使用了自己设计的专用邮件服务程序，因此其功能还很不完全。FreeMail 设计中，MServer 是独立的，可以采用标准的功能完善的邮件服务器，更具有灵活性。

## 4 系统实现和性能测试

### 4.1 系统的实现

FreeMail 大型 e-mail 系统是国家 863 计划曙光 2000-II 超级服务器项目中非科学计算类可扩展应用研制的一部分，目前系统已开发完成并通过了鉴定。

曙光 2000-II 超级服务器是一种典型的机群系统结构，共由 82 个 SMP 节点组成，其中节点分为基于 Power3 的 64 位高性能节点和基于 PowerPC604e 的 32 位普通节点两种。节点间互连采用高速交换式以太网。每个节点运行独立的 IBMAIX 操作系统，并配有 SCSI 硬盘。FreeMail 是由以曙光 2000-II 为平台开发的一系列服务软件构成。

PowerRouter 是工作在系统核心的驱动程序。并且可以工作在任意 PowerPC/AIX 节点上，不需要配备专门的设备。

IndexServer 实现中我们采用了 PostgresSQL。我们编写了专门的连接池程序，并实现了 Proxy 与具体数据库访问接口分离。

针对 SMTP、POP3 和 HTTP 3 种协议我们开发了 3 个 Proxy 程序。

在 MServer 实现中，我们对通用的 Sendmail（SMTP 服务）和 qpopper（POP3 服务）软件做了少量改动以支持大用户数。我们采用了标准 apacheweb 服务器并通过 Module/CGI 方式实现了对邮件的 web 访问。

MServer 是一个相对独立的单机 Mail 服务器。我们没有针对邮件服务器进行专门优化，而是采用了最常用的 free 软件。这对我们得到的最终性能有一定影响。如果可能我们完全可以直接采用其他商用的高性能单机大容量邮件服务软件。

### 4.2 可扩展性测试

对大型 e-mail 系统的测试目前为止还没有标准的测试程序。因此我们编写了简单的测试程序。测试主要包含了 POP3 取信和 SMTP 接收信件这两个主要处理功能。测试的主要目的是验证系统的可扩展性。

我们的测试采用了一种饱和性测试方法：通过一定数量的客户端线程持续发出请求，并记录一定时间内系统完成的邮件请求总量。考虑到 e-mail 以短信息为主，测试中统一采用长度为 1.5 KB 的邮件。其中 POP3 和 SMTP 操作比例为 2:1，以模拟 POP3 访问时而没有邮件的情况。我们只记录 POP3 完成过程在 30 s 内，SMTP 接收在 60 s 内的操作数量。并把平均一次邮件读取和 0.5 次邮件接收定义为一次邮件访问。

首先，我们采用了 128 个客户端线程，对曙光 2000-II 的两种不同节点分别进行单 Proxy 测试。从图 3 可以看出 64 位高性能节点可以带动 5 个以上的 MServer 节点，而 32 位普通节点不到 4 个就饱和了。说明了 CPU 和内存访问性能对 Proxy 节点的性能起决定作用。

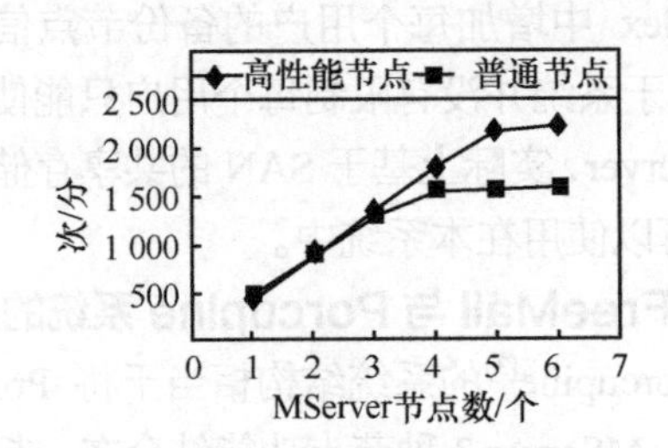


图 3 单 Proxy 性能测试

我们还进行了吞吐量测试，即不限客户端个数，测试一定数量 MServer 在单位时间内能完成的最大邮件访问数。我们测试了最多 16 个 MServer 节点的系统，此时共使用了 4 个 Proxy

节点和 292 个客户端线程。从图 4 可以看出，系统的加速比基本是线性的。

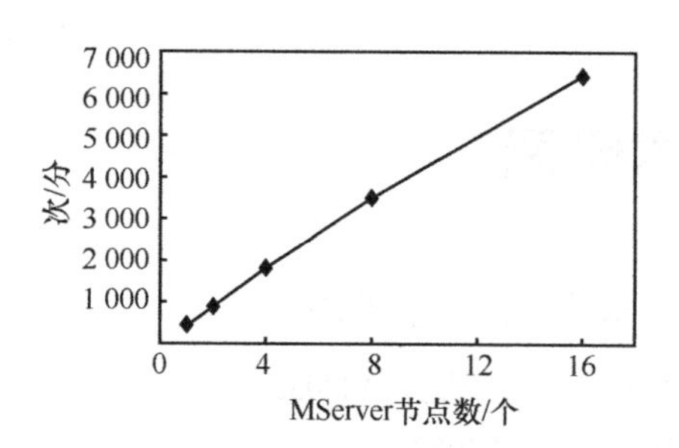


**图 4　系统最大吞吐量测试**

实验中我们还比较了两种节点作为 MServer 的性能，发现两种节点的性能相差不多。说明对于 MServer 来说，磁盘 I/O 能力是主要的，而 CPU 性能并不重要。

测试中 POP3 操作平均完成时间在 2 s 以内，SMTP 操作在 15 s 以内。其中经 Proxy 带来的额外延迟约为 0.1~0.5 s。经 PowerRouter 的延迟在毫秒级，基本可以忽略不计。

在我们最大 16 个 MServer 系统配置中，PowerRouter、DNS、Index 服务器都只使用了一台，且未达到饱和。从上述试验结果可以看出 FreeMail 系统的性能可扩展性非常好。

如果将 16 个 MServer 节点的测试结果转化为以天为单位，那么每天可完成短邮件访问 930 万次以上。

## 5　进一步的工作

目前 FreeMail 中的 Proxy 性能还可以进一步优化，以支持更多的节点。我们还计划在 Proxy 层加入实时邮件过滤系统和安全网络协议，提高系统安全性，并增加 IMAP4 功能。目前的测试工作还比较原始，有待于对实际的 e-mail 访问记录进行分析，以便采用更合适的模型对系统进行测试。进一步我们将考虑通过广域机群使系统规模可以进一步扩大。此外，虽然此系统是针对 e-mail 服务设计的，但同样的结构也可适用于其他大用户量、具有独立性的 I/O 密集型服务（如电子商务等）的系统设计。

## 参考文献：

[1] CHRISTENSON N, BOSSERMAN T, BECKEMEYER D. A highly scalable electronic mail service using open systems[C]// USENIX Symposium on Internet Technologies and Systems. [S.l.:s.n.], 1997.

[2] GRUB M. How to get there from here: scaling the enterprise-wide mail infrastructure[C]// The 10th USENIX Systems Administration Conference. [S.l.:s.n.], 1996.

[3] SAITO Y, BERSHAD B, LEVY H. Manageability, availability and performance in Porcupine: a highly scalable cluster-based mail[C]// The 17th ACM Symposium on Operating System Principles(SOSP'99). New York: ACM Press, 1999.

[4] HUGHES L. Internet e-mail: protocols, standards, and implementation[M]. Boston: Artech House, 1998.

[5] LIU X. PowerRouter-the single entry point of cluster[D]. Beijing: ICT, CAS, 1998.

[6] CARDELLINI V, COLAJANNI M, YU P S. Dynamic load balancing on web-server systems[J]. IEEE Internet Computing, 1999, 3(3): 28-39.

[7] CHEN M Y, GAO W, ZHANG W S, et al. The design of high availability in the dawning server consolidation system[C]// IEEE 4th International Conference on High Performance Computing in Asia-Pacific Region. Piscataway: IEEE Press, 2000.

发表于《计算机科学》2005年第1期

# IPv6业务技术研究

张云勇 [1,2]，张智江 [1]，刘韵洁 [1]，李国杰 [2]，李忠诚 [2]

1. 中国联合通信有限公司技术部博士后科研工作站，北京 100032；

2. 中国科学院计算技术研究所，北京 100080

**摘　要**：IPv6业务技术在以IPv6为核心的下一代互联网中是关键因素之一，本文着重阐述了IPv6业务相关的关键技术、过渡期IPv6业务相关的服务器、IPv6特色业务。对存在的问题进行了分析，并对业务相关的发展给予了展望。

**关键词**：IPv6；服务质量；流标签；AAA认证

# Research of IPv6 service technology

Yun-yong ZHANG[1,2], Zhi-jiang ZHANG[1], Yun-jie LIU[1], Guo-jie LI[2], Zhong-cheng LI[2]

1. Postdoctoral Programme, Technology Department, China Unicom, Beijing 100032

2. Institute of Computing Technology, Chinese Academy of Sciences, Beijing 100080

**Abstract:** IPv6 service technology is one key factor in IPv6 based next generation Internet (NGI). In this paper, IPv6 service-related core technology, transition of IPv6 related server and IPv6 killer applications are mainly focused. Some problems are analyzed and expectation of IPv6 service technology are given out.

**Key words:** IPv6, quality of service, flow label, AAA

## 1 引言

下一代 IPv6 网络是一个基于业务驱动的网络，业务是下一代 IPv6 网络的重点[1-2]。未来的网络带宽将不断提高，如果没有相应的业务支持，将造成网络资源的浪费。相反更多的业务，将会使最终用户直接受益，进而能够更好地留住老客户，吸引新用户，以望繁荣网络市场、推动网络的发展。为此 IPv6 业务技术的研究尤其重要。

## 2 IPv6 业务相关的核心技术

### 2.1 IPv6 服务质量技术

#### 2.1.1 流标签的语义及使用

一个流是指从一个特定的源地址到特定的目的地址（单播或组播地址）发送的一组数

据包，并且源节点希望中间的路由器对流进行特殊的处理。特殊处理的属性可通过控制协议传到路由器，如资源预留协议，或者通过流数据包本身所携带的信息传到路由器，如逐跳选项。流标签由流的源节点指定，必须从1到0xFFFFF（十六进制）范围内（伪）随机并且唯一地选择，以使流标签的任何1比特都能作为路由器中哈希表的键值，用于查找流对应的状态。

所有属于同一个流的数据包发送时必须具有相同的源地址、目的地址和流标签。沿着流路径建立的流处理状态的最大生存周期必须在状态建立机制中说明，例如资源预留协议或建立流的逐跳选项。源节点不允许在任何流处理状态的最大生存周期内把该流标签分配给新的流，因为在使用流标签前，可能状态已经建立起来了。

当一个节点重启时（例如死机后的恢复运行），必须小心使用流标签，因为该流标签有可能在前面的仍处于最大生存周期内的流中使用。这可以通过在静态存储上记录流标签的使用情况来实现，从而在死机恢复后仍然保存该信息，或者避免在任何先前可能建立的流的最大生存周期过期之前使用任何流标签。如果节点的最小重启时间已知，实际重启时间可以从等待分配流标签所需的时间中推算得到。

#### 2.1.2 流标签相关问题

改变流标签带来的后果有时比无QoS保证更为严重，如造成窃取服务攻击和拒绝服务攻击等，所以安全非常重要[3-4]。

由于安全会加密某些字段（如端口、协议类型等），在IPSec隧道模式下还会加密流标签，给流的分类带来了困难。有时还会有防火墙过滤的问题出现。

至于流标签的具体使用，从减少内存机制方面出发，需要使用5元组或者6元组的M-F分类，这样可以减少分类规则的类型及数量；从提高分类速度的角度出发，则需要尽量减少对IPv6头的检查，在draft-conta-ipv6-flow-label-02.txt中提供了一些方法。

在socket接口方面，也需要提供对流标签、业务类型的访问函数，在draft- itojun-ipv6-flowlabel-api-01.txt中定义了IPV6_AUTOFLOWLABEL属性，用于setsockopt函数。在draft-itojun-ipv6-tclass-api-03.txt中则定义了IPV6_TCLA-SS。

另外如果在资源预留协议（Resource Reservation Protocol，RSVP）中利用IPv6流标签，必须重新定义FILTER_SPEC和SENDER_TEMPLATE对象，以让其既包含IPv6流标签，又包含源端口。

### 2.2 移动IPv6及安全

#### 2.2.1 移动IPv6的认证

移动IPv6[5]业务的使用需要Internet提供支持移动IP的AAA服务，即移动用户的认证、授权和计费服务。当移动节点（MN）移动到外地网络时，需要对外地代理或接入设备进行认证，以确定对方的有效性，外地代理也需要对MN进行身份认证，以防止其非法攻击。授权和计费主要涉及MN在外地网络上的资源的使用权和使用情况。目前因特网工程任务组（The Internet Engineering Task Force，IETF）已出台协议草案来支持移动IP的AAA服务（RFC 2977和draft-ietf-aaa-diameter-mobileip-08.txt）。

AAA是英文Authentication、Authorization和Accounting的缩写，分别表示鉴别、授权和记账。鉴别的任务是验明主体的身份；授权是决定是否给予访问权限或级别的过程；记账的功能是记录资源的使用情况。AAA在网络管理和网络安全中起着非常重要的作用；它简化了网络用户的鉴别和授权管理；加强了网络安全；并且通过记账确保网络资源和服务被合理使用。

逐跳的安全关联需要通信路径上每个节点均支持 AAA 服务，实现代价较大，所以实际应用中一般采用基于代理的 AAA 体系，代理服务器可以充当归属和外地两个管理域的代理角色，这两个域与代理均有安全关联，代理能够安全地为两个域中继 AAA 消息。当移动节点在外地网络时，向本地 AAA 服务器请求 AAA 服务，本地 AAA 服务器通过 AAA 代理与归属网络中的 AAA 服务器建立联系，为移动节点提供 AAA 服务。

2.2.2 基于 Diameter 的认证及计费

IETF 的 AAA 工作组致力于网络访问中鉴定、认证和计费的需求开发，需求来自 NASREQ、移动 IP、ROAMOPS 工作组和 TIA45.6。这个 AAA 工作组主要是在 Diameter 提议的基础上，开发 IETF 的标准协议。

在众多 AAA 协议中，目前最普遍采用的是远程认证拨号用户服务（Remote Authentication Dial-In User Service，RADIUS）协议（RFC2138）。但是近年来，随着网络技术的发展，各种新的需求不断出现以及网络设备变得越来越复杂，RADIUS 等旧的 AAA 协议已经不能满足要求了，这些要求主要是：（1）故障恢复；（2）传送级安全：IPSec、TLS；（3）支持可靠性：超时或丢失的重传机制等；（4）支持代理转发消息；（5）服务器端发起的消息；（6）审计；（7）支持协议转换：在不同的 AAA 协议间转换。

Diameter 协议（draft-ietf-aaa-diameter）是在 RADIUS 协议之后发展起来的，主要弥补 RADIUS 的不足，并且可以与 NASREQ、ROAMOPS 和移动 IPv4/IPv6 一起应用。Diameter 协议不但支持以上要求的功能，另外它还支持以下特性。

（1）功能协商机制：与对方协商共同支持的功能。

（2）对方发现和配置：通过域名系统（Domain Name System，DNS）等找到对方实体，并进行配置。

（3）支持漫游：Diameter 在协议说明中明确有对漫游的支持。

Diameter 框架结构包括一个基本协议和一些扩展协议（如安全性、NASREQ、移动 IPv6 和计费等）。服务的通用功能在基本协议中实现，而与应用有关的功能在扩展机制中实现。在针对 MIPv6 扩展了的 Diameter 协议中，Diameter 服务器可以对移动节点上的移动 IPv6 服务进行鉴别、授权和收集计费信息。

移动 IP 工作组最近已经改变了工作重点，重点研究管理域间的移动性，这解决了应用移动性协议的蜂窝发送者的需求；并扩展了 Diameter 基本协议，Diameter 服务器可以对移动节点上的移动 IP 服务进行鉴定、认证和收集计费信息。结合基本协议的域间功能，这个扩展允许移动节点从其他服务提供者获得服务。Diameter 计费扩展用在外地和归属代理上，向 Diameter 服务器传输有用信息。

2.2.3 DHCPv6 与 AAA 的结合

IPv6 节点（客户端）可向归属 AAA 提交认证材料，以便对归属网络进行访问。然而在 IPv4 中，路由器和动态主机配置协议（Dynamic Host Configuration Protocol，DHCP）不一定能处理这种功能，如果由 IPv6 路由器进行这些访问控制，效率将会更加合理，也可能作为执行 DHCPv6 延迟功能的一部分。

DHCPv6 服务器和路由器可以与 AAA 服务器联合使用，决定客户端的认证材料是否有效。通过邻居缓存中的控制条目的设备，路由器能执行这样的网络访问控制。如果没有有效的认证材料，路由器就不会把数据包转发到这个设备，而且这样的设备也不能访问（或者只是有限访问）相邻的其他网络链路。只有通过这种方法，新设备在认证完成之前，才不会滥用网络连接。

## 2.3 IPv6开发接口技术

USAGI提供IPv6开发接口，通过使用getaddrinfo解析网络地址，通过使用getnameinfo解析主机地址，可使程序不依赖于特定的协议。

WinSock2.2支持IPv6，编制IPv6程序时，需包含tpipv6.h头文件。PF_UNSPEC可以接受IPv4和IPv6报文，而PF_INET6只支持IPv6。

Java1.4.1中实现了RFC2373（IPv6 Addressing Architecture）、RFC 2553（BasicSocket Interface Extensions for IPv6）、RFC 2732（Format for Literal IPv6 Addresses in URLs）规范。在双栈情况下，缺省优先支持IPv6，但可以通过设置java.net.preferIPv4Stack=true来使得优先支持IPv4。DNS查询缺省优先支持IPv4，但可通过设置java.net.preferIPv6Addresses= true来使得优先支持IPv6。另外，Java中还增加了InetAddress的两个子类：Inet4Address和Inet6Address，分别代表V4和V6地址。为了支持IPv6多播，增加了以下类和函数：Network Interface getNetworkInterface( )、setNetwork Interface (NetworkInterface netIf)、joinGroup (SocketAddress mcastaddr, NetworkInterface net If)、leave Group (SocketAddress mcastaddr, NetworkInterface netIf)。

目前，某些厂商（如Sun）的RPC也支持IPv6，以后还将支持基于IPv6的TLI和CORBA。

# 3 过渡期IPv6业务相关的服务器

## 3.1 IPv6域名系统

虽然IPv6将取代IPv4的互联网协议，但有许多部分仍然继承了现行IPv4的优点，如域名系统DNS。IPv6网络中的DNS同样非常重要，一些IPv6的新特性和DNS的支持密不可分。

IPv6网络中的DNS与IPv4的在体系结构上是一致的，都是采用树型结构的域名空间。虽然IPv4与IPv6存在着相当大的区别，但这并不意味着需要单独两套DNS体系，相反在DNS的体系和域名空间上两者是一致的，IPv4和IPv6共同拥有统一的域名空间。在IPv4到IPv6的过渡阶段，域名可以同时对应于多个IPv4和IPv6的地址。随着IPv6网络的普及，IPv6地址将逐渐取代IPv4地址。

可聚集全局单播地址[6]是目前主要应用的IPv6地址，因IPv6可聚集全局单播地址是在全局范围内使用的地址，必须进行层次划分及地址聚集。IPv6地址的层次性在DNS中通过地址链技术可以得到很好的支持。

### 3.1.1 正向解析

IPv4的地址正向解析的资源记录是“A”，而IPv6地址的正向解析目前有两种资源记录，即“AAAA”和“A6”记录。其中“AAAA”较早提出，它是对IPv4“A”录的简单扩展，由于IP地址由32位扩展到128位，扩大了4倍，所以资源记录由“A”扩大成4个“A”。但“AAAA”用来表示域名和地址的对应关系，并不支持地址的层次性。“A6”是在RFC2874基础上提出，A6记录类型为38，它是把一个IPv6地址与多个“A6”记录建立联系，每个“A6”记录都只包含了IPv6地址的一部分，结合后拼装成一个完整的IPv6地址。“A6”记录支持一些“AAAA”所不具备的新特性，如地址聚集、地址重编号等。

### 3.1.2 反向解析

IPv6反向解析的记录和IPv4一样，是“PTR”，但地址表示形式有两种。一种是用

“.”分隔的半字节 16 进制数字格式（Nibble Format），低位地址在前，高位地址在后，域后缀是“IP6.INT”。另一种是二进制串（Bit-string）格式，以“\[”开头，16 进制地址（无分隔符，高位在前，低位在后）居中，地址后加“]”，域后缀是“IP6.ARPA”。半字节 16 进制数字格式与“AAAA”对应，是对 IPv4 的简单扩展。二进制串格式与“A6”记录对应，地址也像“A6”一样，可以分成多级地址链表示，每一级的授权用“DNAME”记录。和“A6”一样，二进制串格式也支持地址层次特性。

反向解析也称为指针检索，根据 IP 地址来确定主机名。为了给反向解析创建名字空间，在 IP6.INT 域中，IPv6 地址中所有的 32 位十六进制数字都逆序分隔表示。

以地址链形式表示的 IPv6 地址体现了地址的层次性，支持地址聚集和地址更改。但是，由于一次完整的地址解析要分成多个步骤进行，需要按照地址的分配层次关系到不同的 DNS 服务器进行查询，并且所有的查询都成功才能得到完整的解析结果。这势必会延长解析时间，出错的机会也增加。因此，在技术方面 IPv6 需要进一步改进 DNS 地址链功能，提高域名解析的速度才能为用户提供理想的服务。

#### 3.1.3 IPv6 中的即插即用与 DNS

IPv6 支持地址自动配置，这是一种即插即用的机制，在没有任何人工干预的情况下，IPv6 网络接口可以获得链路局部地址、站点局部地址和全局地址等，并且可以防止地址重复。IPv6 节点通过地址自动配置得到 IPv6 地址和网关地址。但是，地址自动配置中不包括 DNS 服务器的自动配置。如何自动发现提供解析服务的 DNS 服务器也是一个需要解决的问题。DNS 服务器自动发现的解决方法可以分为“无状态”和“有状态”两类。

在无状态的方式下，需要为子网内部的 DNS 服务器配置站点范围内的任播地址。要进行自动配置的节点以该任播地址为目的地址发送服务器发现请求，询问 DNS 服务器地址、域名和搜索路径等 DNS 信息。这个请求到达距离最近的 DNS 服务器，服务器根据请求，回答 DNS 服务器单播地址、域名和搜索路径等 DNS 信息。节点根据服务器的应答配置本机 DNS 信息，以后的 DNS 请求就直接用单播地址发送给 DNS 服务器。

另外，也可以不用站点范围内的任播地址，而采用站点范围内的多播地址或链路多播地址等。还可以一直用站点范围内的任播地址作为 DNS 服务器的地址，所有的 DNS 解析请求都发送给这个任播地址。距离最近的 DNS 服务器负责解析这个请求，得到解析结果后把结果返回请求节点，而不像上述做法是把 DNS 服务器单播地址、域名和搜索路径等 DNS 信息告诉节点。从网络扩展性、安全性、实用性等多方面综合考虑，第一种采用站点范围内的任播地址作为 DNS 服务器地址的方式相对较好。

在有状态的 DNS 服务器发现方式下，是通过类似 DHCP 这样的服务器把 DNS 服务器地址、域名和搜索路径等 DNS 信息告诉节点的。当然，这样做需要额外的服务器。

#### 3.1.4 IPv6 过渡阶段的 DNS

在 IPv4 到 IPv6 的过渡过程中，作为 Internet 基础架构的 DNS 服务也要支持这种网络协议的升级和转换，当然这也要有一个过渡方法，而不是一下子全面改成 IPv6 的 DNS。IPv4 和 IPv6 的 DNS 记录格式等方面有所不同，为了实现 IPv4 网络和 IPv6 网络之间的 DNS 查询和响应，可以采用应用层网关 DNS-ALG 结合 NAT-PT 的方法，在 IPv4 和 IPv6 网络之间起到一个翻译的作用。例如，

IPv4 的地址域名映射使用“A”记录，而使 IPv6 使用“AAAA”或“A6”记录。那么，IPv4 的节点发送到 IPv6 网络的 DNS 查询请求是“A”记录，DNS-ALG 就把“A”改写成“AAAA”，并发送给 IPv6 网络中的 DNS 服务器。当服务器的回答到达 DNS-ALG 时，DNS-ALG 修改回答，把“AAAA”改为“A”，把 IPv6 地址改成 DNS-ALG 地址池中的 IPv4 转换地址，把这个 IPv4 转换地址和 IPv6 地址之间的映射关系通知 NAT-PT，并把这个 IPv4 转换地址作为解析结果返回 IPv4 主机。IPv4 主机就以这个 IPv4 转换地址作为目的地址与实际的 IPv6 主机通过 NAT-PT 通信。

对于采用双协议栈方式的过渡方法，在 DNS 服务器中同时存在“A”记录和“AAAA”（或“A6”）记录。由于节点既可以处理 IPv4，也可以处理 IPv6，因此无需类似 DNSALG 的转换设备。无论 DNS 服务器回答“A”记录还是“AAAA”记录，都可以进行通信。

## 3.2 IPv6 网络管理系统

### 3.2.1 IPv6 相关 MIB

目前 IPv6MIB 包含 6 个表项。

- ipv6IfTable，为 IPv6 接口表，包含设备所有的 IPv6 接口情况。
- ipv6IfStatsTable，包含 IPv6 接口的流量统计情况。
- ipv6AddrPrefixTable，为 IPv6 地址前缀表项，反映设备接口相关的地址前缀。
- ipv6AddrTable，含有关 IPv6 接口的地址信息。
- ipv6RouteTable，为 IPv6 路由表项，IPv6 单播传送地址决定网络包的转发。
- ipv6NetToMediaTable，为 IPv6 地址翻译表项，用来将网络地址翻译成物理地址。

另外传输控制协议（Transmission Control Protocol，TCP）、用户数据报协议（User Datagram Protocol，UDP）、因特网控制消息协议（Internet Control Message Protocol，ICMP）等都增加相应的 IPv6MIB。

### 3.2.2 IPv6 网络管理系统

SNMP 没有定义传输 SNMPPDU 的方式，并且 SNMPPDU 与网络无关，因此在 IPv6 环境下实现 SNMP 是完全可能的。

但对 IPv6 网的管理和测试与对 IPv4 网的管理和测试有很大区别，IPv4 网着重网络的生存能力和信息传送的公平性，IPv6 网也关心网络的生存能力和信息传送的公平性，但更看重网络的可管理性、可维护性、网络提供的端到端 QoS 和承载多业务能力。另外 IPv6 环境下，某些网络管理功能的实现发生了很大变化，如使用邻居发现代替 ARP 功能，使用扩展的 ICMP 来运载邻居发现报文等。IPv6 地址自动配置特性对网络管理也提出了要求，再有 IPv6 也为 SNMP 提供了新的安全保障，IPv6 设备也需要安全管理，因而，它比 IPv4 网络的要求也更复杂和困难。

IPv6 网络综合管理的范围包含所有网络路由设备、交换设备；根据 ITU-T 和 TMF 等相关标准以及电信行业规范，管理的层次包括基本的网元管理、网络管理、业务管理以及事务管理等不同方面。

网络管理模式大体上分为 3 种：远端应急管理系统、带外管理、带内管理。在进行网管平台的建设时，充分考虑到扩展性。目前国内外大规模网络运营商趋向于在网络建设中尽量把网管数据与业务数据分开，为网管数据建立独立通道，以保证重要的管理数据、敏感的统计信息、敏感和占用带宽的实时计费信息的安全、可靠传输。有助于提高网管的效率与可靠性，也有利于提高网管数据的安全性。

IPv6 环境下，SNMP 性能也发生了变化，由于 IPv6 报头长度的变化，SNMP 报文数据量增加，特别是在取单个 MIB 值时，增长达27%，随着每次所取值的增加，两者差别减小。

# 4 IPv6 特色业务

IPv6 网络形成以后，要想迅速促进其发展，必须在其网络之上形成大量业务，一方面可以将 IPv4 网络业务引入进来，另一方面还应研究 IPv6 的特色业务，如视频业务、移动 IP 业务、Peer-to-Peer（P2P）业务等。

## 4.1 杀手应用——IP 电信网

IP 电信网的需求只有 IPv6 的承载才能完全满足，IPv6 具有优异的端到端服务能力；IPv6 有充分的地址空间，多级分层结构，地址规划便利；IPv6 有充分的流标识空间，有利用发展更好的服务质量机制；IP 电信网是一个增值业务网络，营运方式有一定差异，相比传统的数据网，采用 IPv6 在维护性和收益方面要好很多。

## 4.2 杀手应用——在线游戏

游戏业是一个很大的产业，仅美国的游戏市场就达到了 100 亿美元。在线游戏又是游戏业的一个明显的发展趋势。

在线游戏需要把分散在不同地域的用户连接起来，并保证安全、隐私和计费的需要。由于缺少足够的 IP 地址，IPv4 的网络无法满足在线游戏 P2P 的需求。

在线游戏必须支持固定和移动两种网络接入方式。采用基于 IPv6 的游戏终端主要是和游戏服务器进行交互，几乎不需要访问原来大量的 IPv4 的服务器，这也非常符合 IPv6 网络早期的“相互连接的孤岛”的架构。

## 4.3 IPv6 在 3G 中的应用

IPv4 地址空间不足，限制了网络和用户数目的发展；网络地址转换（Network Address Translation，NAT）技术破坏了网络端到端的特性，缺少固定地址、永远在线机制，限制了移动 IP、IP 电话、Push 业务的发展；移动网络向 IPv6 承载过渡是必然趋势。

3G 网络的部署不仅为用户提供了更加高速的无线连接，也为用户接入互联网提供了更为丰富的接入手段。IPv6 的采用，不仅满足了未来移动设备对 IP 地址空间的需求，也让移动终端更易于配置管理（自动配置）；而用户对基于 IP 的应用业务的使用也更为安全方便。

早在 1997 年，IPv6 就被引入 GSM/GPRS 标准中。2001 年 5 月，在 3GPPRelease5 中，IPv6 则被确立为 3G 多媒体业务子系统（IMS）中唯一支持的 IP 协议版本。

另外，IETF 作为互联网协议的标准化组织，也积极推动 IPv6 在 3GPP 中的应用。与 IPv6 相关的标准化研究工作在 IETF 的多个工作组展开，如 AAA 工作组、IPv6 工作组、ROHC 工作组等。

移动用户对基于 IP 灵活多样的、低成本的应用业务的需求，推动了传统电路交换网络运营商着手部署基于 IP 技术的包交换网络，从而拉开了两种网络融合的序幕。

### 4.3.1 基于 IPv6 的 3G 数据传输网络

3G 网络由用户终端（UE）、接入网（UTRAN）和骨干网构成，骨干网内主要有服务 GPRS 节点（SGSN）和网关 GPRS 节点（GGSN）两种。SGSN 承担认证授权以及移动性管理等功能；GGSN 则连接其他网络与内部网络，并承担着收集计费信息的功能。

这样，接入的移动终端与 GGSN 之间建立了基于 IPv4 或 IPv6 的数据包协议上下文（PDPContext），而从移动终端发出的 IP 数据包，由 GGSN 路由，经由用户指定的 GGSN 上的接入点（APN），访问目标网络。对于用户使用其他终端设备，经由移动终端访问目标网络。

### 4.3.2 3G 网络中对移动终端 IPv6 地址的分配

与一般的 IPv6 节点地址配置一样，在 3G

网络中，移动节点的地址配置也有两种方式：有状态的和无状态的自动配置。在 3G 网络中，GGSN 的每一个接入点（APN）都可以单独指定地址配置方式。其中无状态的地址自动配置方式与一般的 IPv6 节点不同。地址配置的协议过程如下：

（1）移动终端发起激活上下文请求，指定 PDP 类型为 IPv6；

（2）SGSN 接收到请求，发起建立 PDP 上下文请求；

（3）GGSN 收到请求，为移动终端分配接口标识以及地址前缀，然后将信息封装在建立 PDP 上下文的响应中，发回给 SGSN；

（4）SGSN 将响应信息封装在激活 PDP 上下文认可中，发回给移动终端；

（5）移动终端得到地址配置信息，然后可以按照配置发起路由器请求；

（6）GGSN 发送路由器广播给移动终端。

经过以上过程，移动终端利用路由器广播信息，与先前收到的接口标识组成 IPv6 地址。而且，由移动终端发出的 IPv6 数据包，则被 3G 网络中的节点直接转发到 GGSN，再由 GGSN 路由到目标网络。

### 4.4 智能终端和家庭网络

IPv6 非常适合拥有巨大数量各种细小设备的网络，将在连接由各种简单装置的超大型网络中运行良好，这些简单设备不仅仅是手机和掌上电脑/PDA，还可是标签机、家用电器、信用卡、水电表等。

IPv6 的特长是可以无限制地为网络终端提供上网所必不可缺的地址，使家用游戏机及电视等接入网络。这样一来，同一个家庭内不光个人电脑而且电子微波炉、空调等所有家电都可拥有一个地址接入因特网。

E3global 目前推出了上网家电及 PDA 的 IPv6 协议堆栈（ProtocolStack）“e3-IPv6”。另外 μITRON4.0 版和 VxWorks 版 IPv6 堆栈协议的开发也基本完成。此外还有公司在开发面向嵌入式终端的 Linux 版本。

### 4.5 SIP 和 IPv6 的结合

会话起始协议（Session Initiation Protocol，SIP）是 IETF 提出的在 IP 网络上进行多媒体通信的应用层控制协议，可用于建立、修改、终结多媒体会话和呼叫。SIP 采用基于文本格式的客户-服务器方式，以文本的形式表示消息的语法、语义和编码，客户机发起请求，服务器进行响应。SIP 独立于低层协议——TCP 或 UDP，而采用自己的应用层可靠性机制来保证消息的可靠传送。有关 SIP 的详细内容可参见 IETF RFC3372。该规范已经定义了对 IPv6 的支持。SIP 与 IPv6 的结合一方面消除了 NAT 带来的效率低下问题，另一方面也提高了安全性。SIP 与 IPv6 的结合，将为下一代网络的融合奠定基础。

### 4.6 杀手应用——组播业务

组播是一种允许一个或多个发送者（组播源）发送同一报文到多个接收者的技术。组播源将一份报文发送到特定组播地址，组播地址不同于单播地址，它并不特定属于某单个主机，而是属于一组主机。一个组播地址表示一个群组，需要接收组播报文者加入这个群组。这样，无论有多少个组播报文接收者，整个网络中任何一条链路只传送单一的报文，大大节省了带宽。未来可开展基于组播的业务，如视频点播等。

### 4.7 具有服务质量保证的语音及其视频业务

如何利用 IPv6 新的服务质量机制，开展一些语音及视频业务，也是研究的一个重点。

### 4.8 安全电子交易业务

在 IPv6 中，IPSec 是必须得到支持，这就为开展一些对安全性要求较高的业务提供了保障。

### 4.9 其他相关应用

如 IPv6 在软交换、ENUM 中的应用等。

## 5 小结

目前在日本，已开展了智能运输系统（Internet Car）应用，Sony 还在进行一些实验，如网络HDD VCR、远程医疗诊断和IPv6 Town等。目前韩国-ETRI 已经开发了 IPv6 多播视频会议、视频流业务，NCA 已经开发了 VoIPv6，另外开展了一些宽带 Internet 业务，如在线网络游戏、网络银行、网络教学、实时 VOD 等。6TALK 目前正在开发 IPv4/IPv6 过渡技术，6ANTS 在开发网络自动配置技术，6NEAT 在开发 IPv6 应用。目前台湾 HiNet 开展了大量的 IPv6 多媒体业务，如 VOD、TV、娱乐等。另外还进行一些实验，如在 IPv6 网络上进行远程科学 Grid 的建立，ITRI 也基于 IPv6 进行 SIP/Enum 的实验。这些都为 IPv6 业务的探索奠定了基础。

## 参考文献:


[1] HUITEMA C. IPv6: the new internet protocol (second edition)[M]. Upper Saddle River: PrenticeHall, 1998.

[2] DEERING S, HINDEN R. Internet protocol, version 6 (IPv6) specification: RFC 2460[S]. [S.l.:s.n.], 1998.

[3] HUSTON G. Next steps for the IP QoS architecture: RFC 2990[S]. [S.l.:s.n.], 2000.

[4] ENGELS M, ERREYGERS J. IPv6 flow label classifier[S]. [S.l.:s.n.], 2003.

[5] JOHNSON D, PERKINS C. Mobility support in IPv6[S]. [S.l.:s.n.], 2000.

[6] HINDEN R, O'DELL M, DEERING S. An IPv6 aggregatable global unicast address format: RFC 2374[S]. [S.l.:s.n.], 1998.

发表于 *Physical Review E*, 2007 年

# Enhancing the transmission efficiency by edge deletion in scale-free networks

Guo-qing ZHANG[1,2], Di WANG[1,2], Guo-jie LI[1]

1. Institute of Computing Technology, Chinese Academy of Sciences, Beijing 100080, China

2. Graduate University of Chinese Academy of Sciences, Beijing 100080, China

**Abstract:** How to improve the transmission efficiency of Internet-like packet switching networks is one of the most important problems in complex networks as well as for the Internet research community. In this paper we propose a convenient method to enhance the transmission efficiency of scale-free networks dramatically by kicking out the edges linking to nodes with large betweenness, which we called the "black sheep." The advantages of our method are of facility and practical importance. Since the black sheep edges are very costly due to their large bandwidth, our method could decrease the cost as well as gain higher throughput of networks. Moreover, we analyze the curve of the largest betweenness on deleting more and more black sheep edges and find that there is a sharp transition at the critical point where the average degree of the nodes $\langle k \rangle \to 2$.

## 1 Introduction

In recent years, network structure and dynamics have attracted tremendous interest [1-5]. One of the most important motivations is that large communication networks, such as the Internet, wireless networks, and so on, have become the most influential media in modern society. The explosive growth of the number of users and hence the amount of traffic has posed a number of problems, which are not only important in practice, e.g., avoiding undesired congestion, but also of theoretical interest as an interdisciplinary topic[6-7]. Prompted by other disciplines like statistical physics, biology, and sociology, such interest has flourished in the broader framework of network science [8].

In transportation systems, e.g., the Internet, the underlying networks play a significant role in the traffic dynamics taking place upon them [9-12]. To the best of our knowledge, recent evidence has shown that the Autonomous System (AS) level topology of the Internet is scale-free, i.e., the degree distribution follows a power law [13-14]. Thus, how to improve the efficiency of the systems, taking account of the topological structures, is a crucial problem. Some recent studies in this direction can be roughly classified into two categories. One involves optimizing the underlying networks [15-16] and the other developing better routing strategies [17-21]. The former is maybe useful to P2P (peer-to-peer) networks [22] as its logical structure is demonstrated artificially by the controlling software, while the latter is practical in packet transmission for some applications on the Internet, of which the topological structure is not expected to be changed. However, compared with greatly redesigning or modifying the topology of the Internet, it is actually easy to delete some edges. In this paper, we propose a method to remarkably improve the transmission efficiency of Internet-like networks by deleting a few edges linking to nodes with relatively large betweenness [23-26].

The paper is organized as follows. In Sec. 2, the prototype traffic model, which was used in many former studies, is introduced. In Sec.3, we present our method, the simulation results, and the corresponding discussions. The conclusion is given in Sec.4

## 2 The model and notation

Several models have been proposed to simulate packet traffic dynamics on complex networks by introducing random generation of packets in each time step and various routing strategies. We here adopt one widely used before.

In each time step, each node generates $R/N$ packets, where $N$ is the size of the network and $R$ is a parameter tuning the generation rate, i.e., there are R packets generated in the network at each step. When $R/N$ is not an integer, we create Int($R/N$) packets determinately and create a packet simultaneously with probability $p= R/N-$ Int($R/N$), where Int($R/N$) is the integral part of R/N and thus $p$ is the fractional part. The packets are initialized with random destinations. Moreover, we set a transmission capacity $C_i$ to the node $i$, $i$=1,2,$\cdots$,$N$, which means that, at each time step, the maximal number of packets transferred by the node $i$ to the next node according to the routing table is $C_i$. When the node cannot transfer all the packets accumulated in its queue, it deals with them following the first-in-first-out rule. Hence, the routing strategy also plays an important role in the traffic dynamics. We see that, in the Internet, the routing strategy of within domains is the shortest path algorithm and, between domains, i.e., for the AS level, the border gateway protocol causes the packets to be transmitted along almost the shortest path. Therefore, in the paper, we adopt the shortest path routing strategy. When a packet reaches its destination through the shortest path routing, the packet will be deleted from the system.

In order to analyze the transition from free flow to a congested state, we use the order parameter presented in previous studies [27].

$$\mu(R)=\lim_{t\to\infty}\frac{\langle \Delta W\rangle}{R\Delta t} \tag{1}$$

where $W(t)$ is defined as the number of packets on the network at time step $t$, and $\Delta W=W(t+\Delta t)-W(t)$, with$\langle\cdots\rangle$ indicating averaging over time windows of width $\Delta t$. In other words, the order parameter $\mu$ represents the ratio between the existing flow and the inflow of packets calculated through a long enough time period. Obviously, in the free flow state, i.e., where there is no congestion in the network, the system can deal with the generated packets and thus the existing flow is close to zero. Otherwise, in the congested state, the number of generated packets is too large to be transmitted and the flow existing in the network will also be large, which causes the order parameter to approach 1.

As mentioned above, we adopt the shortest path routing in this paper, and hence another characteristic quantity, the betweenness of a node, is of utmost importance in traffic dynamics. The betweenness of node v is defined as [23]

$$g_v = \sum_{s \neq t} \frac{\sigma_{st}(v)}{\sigma_{st}} \tag{2}$$

where $\sigma_{st}$ is the number of shortest paths going from node $s$ to node $t$ and $\sigma_{st}(v)$ is the number of shortest paths going from $s$ to $t$ and passing through the node $v$. In general, the nodes with large betweenness (hub nodes) are part of shortest paths than other peripheral nodes. Obviously, for a traffic system networked according to the shortest path routing, these hub nodes will be congested more easily than others. Therefore, we can improve the transmission efficiency of the traffic network by controlling or detouring properly around the nodes with large betweenness. Our method described below follows this idea.

## 3 Method, results, and discussions

In this section we will give our method to improve the transmission efficiency of the traffic network, show the simulation results, and then discuss them extensively.

As analyzed before, the nodes with large betweenness will be congested more easily than the others. When a few hub nodes are congested, the others can still deal with packets successfully. Because of the edges connecting the hub nodes, the hub nodes will become more and more congested. In contrast, if there are fewer edges linking the hub nodes, the packets will make a detour round the hub nodes, and hence the capacity of the networked system will be enhanced.

First, we calculate the betweenness of each node according to the definition mentioned above. Next the edge adjoining two nodes $i$ and $j$ is given a weight $w_{ij}=g_i g_j$, where $g_i$ and $g_j$ are the betweennesses of the two nodes, respectively, and the value of $w_{ij}$ will not be changed in the following deletion process. Then we sort the edges with their weights from large to small. Finally, a fraction $f_d$ of the edges ranked first are deleted. However, we should keep the connectedness of the network which means that, if deleting an edge will cause some nodes to be disconnected, we will not delete it, and go to deal with the edge ranked next.

In order to test our method, we should generate Internet-like topological networks. There exist several models of Internet topology (see, e.g., the positive-feedback preference() model [28] and the Locality-Driven(LD) model). Here, for simplicity, we select the Barabasi-Albert (BA) model [2] to generate the topological network, the degree distribution of which is a power law $p(k) \propto k^{-3.0}$ where $p(k)$is the ratio of the nodes with degree $k$ to the number of all nodes in the network. The model could represent the heterogeneous node degree of many real-world networks, including the Internet AS level topology, the logical topology of unstructured P2P distributed systems (e.g., Gnutella), and so on. In the simulation, we set the capacity of each node mentioned in Sec. 2 to a constant, i.e., $C_i=C$, $i=1,2,\cdots,N$. All the simulations in this paper are performed with $C=1$ and the mean degree of the network $\langle k \rangle=4$. It is worth pointing out that these values of C and $\langle k \rangle$ involve no loss of generality and do not change the analysis and conclusion of this paper.

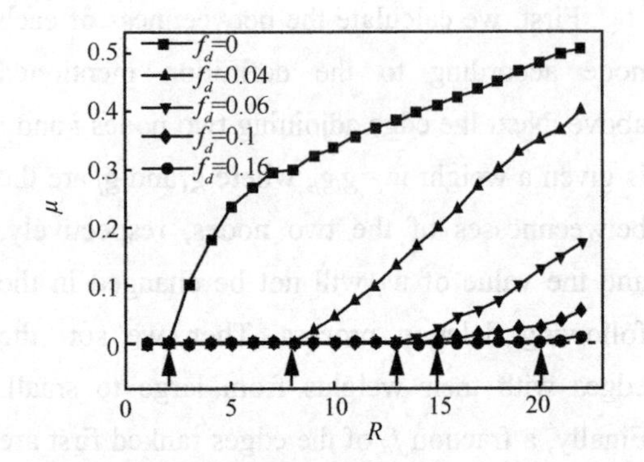


Figure 1. Order parameter $\mu$ versus rate $R$ of generating packets with different values of $f_d$: 0, 0.04, 0.06, 0.1, 0.16 and $C$=1. The network parameters are $N$=1 225, $\langle k\rangle=4$. The arrows point to critical values $R_c$ for different $f_d$. The dashed line is $\mu$=0 as a reference

Figure 1 displays the order parameter $\mu$ versus the packet generation rate $R$ for different values of the fraction $f_d$ of edges deleted. One can see that there is a critical value $R_c$ of the generation rate, above which the order parameter is not zero and grows with increasing $R$, while below which $\mu$ is always equal to zero. Thus, the critical value $R_c$ can be used to represent the transmission capacity of the traffic network system. That is to say, the larger the value of $R_c$, the better the efficiency of the traffic network system. It is shown in this figure that the value of $R_c$ increases with the growth of $f_d$. For example, $R_c$ =2 when $f_d$ =0, which means that scale-free networks have poor efficiency of transmission, while for $f_d$ =0.04, 0.06, 0.1, and 0.16, the values of $R_c$ are 8, 13, 15, and 20, respectively. But this is not the whole story. In Figure 2(c) we show the value of $R_c$ and the average shortest path length $L$ versus the deleted fraction $f_d$. The curve of $R_c$ presents an inverse $U$ shape, and the curve of $L$ goes up very slowly when $f_d \leqslant 0.45$ and rises sharply when $f_d \to 0.5$. This shows that the efficiency of the network will become very much higher on expurgating a small fraction of edges with large weight, whereas the efficiency will be lower when deleting many edges. Figure 2(a) displays the degree distribution of networks with various values of $f_d$. Here we emphasize the remarkable increase of $R_c$ after deleting just a few edges, while the topology structure of the network is still nearly scale-free. For example, when $f_d$ =0.08 the value of $R_c$ can be improved from 2 to 13 [see Figure 2(c)] while the degree distribution is nearly a power law [see Figure 2(a)]. Moreover, the average shortest path length $L$ increases slowly in the range $f_d$ =0.25–0.4 in which the efficiency decreases very fast. We recall previous studies of the relation between $R_c$ and the largest betweenness $g_{\max}$ of the nodes in a network [15].

$$R_c \propto \frac{1}{g_{\max}} \tag{3}$$

This indicates that the value of $R_c$ is approximately inversely proportional to the value of the maximal betweenness $g_{\max}$. We display the curve of $g_{\max}$ versus $f_d$ in Figure 2(b). It is shown that the value of $g_{\max}$ goes up very quickly when $f_d$ approaches 0.5. This can be explained as follows. Since the mean degree $<k>$=4, when $f_d \to 0.5$ the number of edges is almost equal to the number of nodes. This means that, if we delete a few more edges, the network will be disconnected. When $f_d$ approaches the critical point $f_d$ =0.5, the network will be modified to a treelike topology that causes a few crossroad nodes;

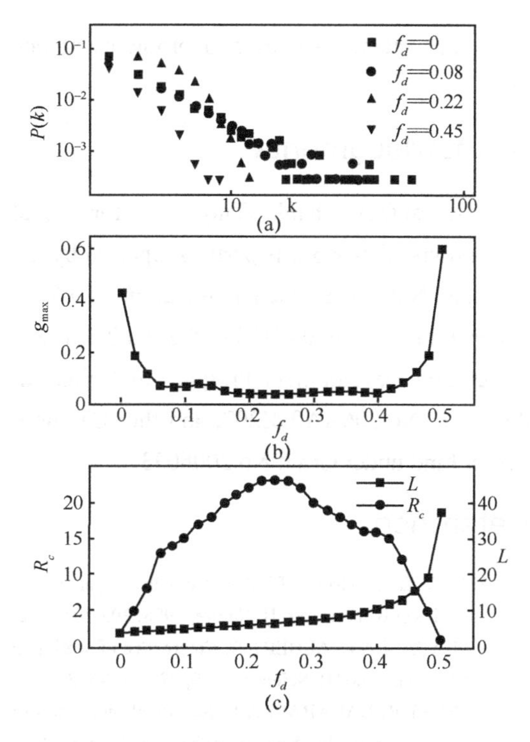


Figure 2. (a) Degree distribution of networks with $f_d$=0, 0.08, 0.22, 0.45. (b) Maximal betweenness in network versus $f_d$. (c) $R_c$ and the average shortest path length $L$ versus $f_d$. The initialized simulation parameters are the same as the previous ones

hence the large betweenness of such nodes. The network topology with three typical values of $f_d$=0, 0.24, 0.5, is displayed in Figure 3, and confirms our analysis. It is worthy of note that the remaining treelike topology, after many links are deleted, looks like but is different from the gradient networks[29] studied by Toroczkai et al. In [29], the authors studied traffic congestion in directed gradient networks and have shown the interesting result that the jamming is limited in scale-free systems, i.e., even though the network is congested, some nodes can also process some packets. In contrast, we study in this paper the traffic congestion in undirected networks and the remaining treelike topology has low efficiency of transmission.

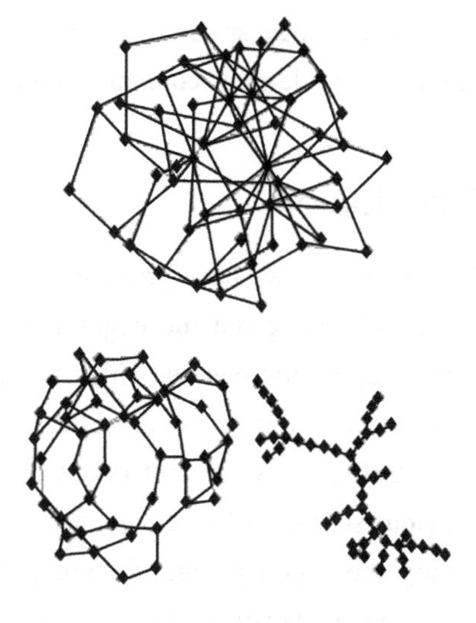

Figure 3. Network topologies with three typical values of $f_d$=0, 0.24, 0.5 (from top to bottom, left to right). The initial network parameters are $N$=50 and $\langle k \rangle$=4

In addition, we should point out that $R_c$=2, in Figure 1, is true just for scale-free networks generated by the BA model with average degree $\langle k \rangle = 4$ and no links removed. When $\langle k \rangle$ increases, the average shortest path length decreases, and thus the maximal betweenness decreases. According to Eq.(3), $R_c$ will increase. However, our method will still be effective.

As shown above one can see that deleting just a few edges with large weights can improve remarkably the transmission efficiency and simultaneously reduce the cost of network construction and maintenance, while the average shortest path length increases very slowly. Thus we call those edges “black sheep” and kick them out. It is noteworthy that the method in Ref. [21]

is different from ours. In the former, the authors proposed an interesting and effective routing scheme inspired by the extreme optimization method. In contrast, our method in this paper is to adjust the topology structure to enhance the throughput of scale-free communication networks.

## 4 Conclusion

To summarize, we have proposed a convenient method, kicking out the edges linking to nodes with large betweenness, which we called the black sheep, to improve the efficiency of scale-free traffic network systems. Compared with previous studies in this direction, including redesigning the topology and designing better routing strategies, its advantages are of convenience and practical importance. In this paper we set the weight of the edge adjoining node $i$ and $j$ as $w_{ij}=g_i g_j$, where $g_i$ and $g_j$ are the betweennesses of nodes $i$ and $j$, respectively. The calculation of the betweenness needs overall information about the network. An alternative choice is to set $w_{ij}=k_i k_j$, where $k_i$ and $k_j$ are the degrees of nodes iand j. Actually, we have implemented the alternative method and found that its performance is just a little worse. Thus, for very large networks, one can choose the alternative method. In addition, we have performed our method on the more realistic PFP, LDPFP model [28] and the actual AS level Chinese Internet topology [14]; The conclusion is almost the same as on the BA model.

Furthermore, it is well known that the black sheep edges are costly lines with large bandwidth in practice. Therefore, our method can improve the efficiency of communication network systems as well as decrease the cost. In short, our method solves two problems at one time.

## Acknowledgments

The authors thank Gang Yan for useful discussions. This work is partly supported by the National Natural Science Foundation of China under Grant No.660673168, the Hi-Tech Research and Development Program of China under Grant No.206AA01Z207, and the ICT innovation fund under Grant No.2006033.

## Reference:

[1] WATTS D J, STROGATZ S H. Collective dynamics of “small-world” networks[J]. Nature, 1998, 393: 440-442.

[2] BARABÁSI A L, ALBERT R. Emergence of scaling in random networks[J]. Science, 1999, 286: 509-512.

[3] ALBERT R, BARABÁSI A L. Statistical mechanics of complex networks[J]. Reviews of Modern Physics, 2002, 74: 47-97.

[4] NEWMAN M E J. The structure and function of complex networks[J]. SIAM Review. 2003, 45: 167-256.

[5] BOCCALETTI S, LATORA V, MORENO Y, et al. Complex networks: structure and dynamics[J]. Physics Reports, 2006, 424, 175-308.

[6] DOROGOVTSEV S N, MENDES J F F. Evolution of networks: from biological nets to the Internet and WWW[M]. New York: Oxford University Press, 2003.

[7] PASTOR-SATORRAS R, VESPIGNANI A. Evolution and structure of the Internet: a statistical physics approach[M]. Cambridge: Cambridge University Press, 2004.

[8] NEWMAN M E J, BARABÁSI A L, WATTS D J. The structure and dynamics of complex networks[M]. Princeton: Princeton University Press, 2005.

[9] TADIĆ B, RODGERS G J. Packet transport on scale-free networks[J]. Advances in Complex Systems, 2002, 5(4): 445-456.

[10] TADIĆ B. Modeling traffic of information packets on graphs with complex topology[M]. Berlin: Springer, 2003.

[11] TADIĆ B, THURNER S, RODGERS G J. Traffic on complex networks: towards understanding global statistical properties from microscopic density fluctuations[J].

Physical Review E, 2004, 69: 036102.

[12] ZHAO L, LAI Y C, PARK K, et al. Onset of traffic congestion in complex networks[J]. Physical Review E, 2005, 71: 026125.

[13] FALOUTSOS M, FALOUTSOS P, FALOUTSOS C. On power-law relationships of the Internet topology[J]. ACM SIGCOMM Computer Communication Review, 1999, 29(4): 251-262.

[14] ZHOU S, ZHANG G Q, ZHANG G Q. Chinese Internet AS-level topology[J]. IET Communication, 2007, 1(2): 209-214.

[15] GUIMERÀ R, DÍAZ-GUILERA A, VEGA-REDONDO F, et al. Optimal network topologies local search congestion[J]. Physics Review Letters, 2002, 89(24): 248701.

[16] CHOLVI V, LADERAS V, LÓPEZ L, et al. Self-adapting network topologies in congested scenarios[J]. Physical Review E, 2005, 71: 035103.

[17] ECHENIQUE P, GÓMEZ-GARDEÑES J, MORENO Y. Improved routing strategies for Internet traffic delivery[J]. Physical Review E, 2004, 70: 056105.

[18] YAN G, ZHOU T, HU B, et al. Efficient routing on complex networks[J]. Physical Review E, 2006, 73: 046108.

[19] WANG W X, YIN C Y, YAN G, et al. Integrating local static and dynamic information for routing traffic[J]. Physical Review E, 2006, 74: 016101.

[20] SREENIVASAN S, COHEN R, LÓPEZ E, et al. Structural bottlenecks for communication in networks[J]. Physical Review E, 2007, 75: 036105.

[21] DANILA B, YU Y, MARSH J A, et al. Optimal transport on complex networks[J]. Physical Review E, 2006, 74: 046106.

[22] ORAM A. Peer-to-peer: harnessing the power of disruptive technologies[M]. Sebastopol: O'Reilly Associates, 2001.

[23] FREEMAN L. A set of measures of centrality based on betweenness[J]. Sociometry, 1977, 40: 35-41.

[24] GOH K I, KAHNG B, KIM D. Universal behavior of load distribution in scale-free networks[J]. Physical Review Letters, 2001, 87: 278701.

[25] NEWMAN M E J. Scientific collaboration networks. I. network construction and fundamental results[J]. Physical Review E, 2001, 64: 016131.

[26] BARTHÉLEMY M. Betweenness centrality in large complex networks[J]. The European Physical Journal B, 2004, 38: 163-168.

[27] ARENAS A, DÍAZ-GUILERA A, GUIMERÀ R. Communication in networks with hierarchical branching[J]. Physical Review Letters, 2001, 86: 3196.

[28] ZHOU S, MONDRAGÓN R J. Accurately modeling the internet topology[J]. Physical Review E, 2004, 70: 066108.

[29] TOROCZKAI Z, BASSLER K E. Jamming is limited in scale-free systems[J]. Nature, 2004, 428: 716.

发表于《电信科学》2018 年第 3 期

# 未来移动通信系统中的通信与计算融合

周一青[1,2,3]，李国杰[1,2,3]

1. 中国科学院大学，北京 100190；2. 中国科学院计算技术研究所，北京 100190；

3. 移动计算与新型终端北京市重点实验室，北京 100190

**摘　要：**本文旨在从不同层面梳理、总结通信与计算融合，明确其未来发展的重点。首先介绍了早期的通信与计算融合。作为信息科学的核心技术，通信与计算是天然融合的。在目前的移动通信系统中，从单个设备和单个技术层面而言，通信与计算已有较好的融合，但要突破传统移动通信系统的瓶颈，更需要从系统的角度开展通信与计算融合的研究与应用。已有研究初步表明，从系统层面融合通信与计算，有望降低对移动通信容量的需求，提升系统支撑业务的能力。面向未来移动通信系统，可从定义系统服务能力（基础理论），构建可扩展、可重塑、透明化、模块化的系统架构以及研究业务感知的跨层信息交流机制等多方面，深入探讨通信与计算融合，推动移动通信系统的可持续发展。

**关键词：**移动通信；通信与计算融合；系统思维；信息交流；服务能力

# Convergence of communication and computing in future mobile communication systems

Yi-qing ZHOU[1,2,3], Guo-jie LI[1,2,3]

1. University of Chinese Academy of Sciences, Beijing 100190, China

2. Institute of Computing Technology, Chinese Academy of Sciences, Beijing 100190, China

3. Beijing Key Laboratory of Mobile Computing and Pervasive Device, Beijing 100190, China

**Abstract:** An overview and summary were provided on the convergence of communication and computing and the future research directions were clarified. As two key technologies of information science, communication and computing were born to be converged. In current mobile communication systems, from a single device or technique point of view, communication and computing had been well converged. However, to break the bottleneck of traditional mobile communication systems, it was necessary to converge communication and computing in a system level. Existing research demonstrates that a systematic convergence of communication and

**基金项目：**北京市自然科学基金—海淀原始创新联合基金重点项目（No.L1792049）

computing will reduce the channel capacity requirement on the mobile system and thus enhancing the system capability to support more services. Facing future mobile communication system, communication and computing convergence should be further studied in-depth from defining system service capabilities (basic theory), building a scalable, remodelable, transparent, modular system architecture and investigating service-aware cross-layer information exchange mechanisms, thus the sustainable development of mobile communication system could be promoted.

**Key words:** mobile communication, convergence of communication and computing, systematic view, information exchanging, serving capability

## 1 引言

移动通信与计算是信息领域的两大核心技术，移动通信产业与计算相关产业的发展一直密切关联、互相促进。移动通信系统基本遵循“十年一代”的发展规律，经过 2G 至 4G 高速发展的 30 年，增速明显趋缓。目前系统构建成本日益高涨，性能逐渐向理论极限趋近，而收益则趋于平坦化，移动通信系统的可持续发展面临巨大的挑战[1]。另外，在计算领域，云计算、大数据、智能计算等新概念、新技术层出不穷，展现出极强的生命力。移动通信业界和学术界正积极地从计算领域借鉴先进的理念和技术，期望借助计算的力量，通过通信与计算的融合，突破传统移动通信系统的限制，促进整个系统的可持续发展。

通信与计算的融合一直在发展。笔者认为，作为信息科学的核心技术，通信关注信息传输，计算关注信息处理，两者都是为信息交流服务。因此，本质上两者就是融合的。经过多年的发展，在通信系统的单个设备（如数字程控交换机）以及单个技术（如信道编解码）层面，通信与计算已呈现很好的融合与协同。但在系统层面，移动通信与计算机两个系统的设计思维、架构、研究特征等不尽相同，通信与计算融合存在很大的空间。已有研究表明，如果能够融合、协同移动通信系统中的计算存储与通信能力，就可以降低对系统传输容量的需求，提升业务支撑能力。因此，为了突破传统移动通信发展瓶颈，系统层面的通信与计算融合是一个有潜力的发展方向。面向未来移动通信系统中的通信与计算融合，首先在基础理论方面，系统发展的核心应从传统通信的信息传输转为信息交流（融合通信、计算与存储能力），衡量系统能力的指标也应从传输容量（通信能力）转为服务能力（融合通信、计算与存储能力）；其次，在系统架构方面，移动通信系统应融合计算机系统设计理念，不断提升系统的可扩展性、可塑性、透明性和模块性；另外，未来移动通信系统的发展应以结构化设计为指导思想；最后，在信息交流机制方面，应融合通信、计算与存储能力，研究业务感知的跨层信息交流机制，提高移动通信系统总体的信息交流能力。

## 2 早期的通信与计算融合[2]

20 世纪 40—50 年代，在信息科学发展的初级阶段，图灵、维纳、香农等大师在计算、控制和通信领域分别提出了奠基性的理论，通信、计算和控制是天然融合的。计算机可以被看作“另一种形式的通信设备”，而自动机器就像人一样，是一个控制和通信融合的系统。其后，随着计算机和通信技术的发展完善，各自成为独立的学科，分离发展，但通信与计算融合的努力一直在进行。20 世纪 90 年代中期，由国家智能计算机研究院开发中心研制成功的“曙光 1000”大规模并行计算机系统就融合了重要的通信技术。“曙光 1000”由 36 个节点机构成，各节点机之间需要进行高速的数据通信来实现并行计算，因此采用了二维 mesh 通信网络架构，并创新地采用了基于蛀洞（Wormhole）机制的异步路由器（国际上当时只实现了同步蛀洞路由），实现了高速、可靠的数据通信。再如，由解放军信息工程大学研制成功的 HJD04 数字程控交换机（存储程序控制交换机）也是计算与通信融合的典范。HJD04 是面向以电路交换提供语音电话业务的通信系统中的交换设备，它将电话的控制、接续等写成程序，通过计算机程序控制电话通信过程，自动完成电话交换，具备大容量交换的能力。如果说“曙光 1000”和 HJD04 体现的是单个设备层面上通信与计算的融合，互联网的发展则为通信与计算融合提供了广阔的平台。互联网本身即通信与计算融合的产物，其网络节点具备丰富的计算、存储能力，节点之间的互联则依赖通信技术。将多个节点互联起来可以形成具有超大计算、存储能力的云，而要应用云计算、云存储能力又离不开依赖通信技术实现的网络。因此云和网之间已经没有明显的界线，云网融合的根本也正是计算与通信的融合。

## 3 移动通信网络中的通信与计算融合

前述通信与计算的融合中，涉及的主要是有线通信系统。近年来迅猛发展的移动通信网络可以分为以有线通信为主的核心网和以无线通信为主的接入网。

### 3.1 核心网中的通信与计算融合

目前移动通信网络的核心网面临的主要挑战是网络部署、维护、升级等成本的日益高涨。与提供尽力而为服务的互联网不同，通信网从一开始就是一个以商业经营为目的的网络，必须确保通信服务质量和服务安全，对网络的部署、维护升级、电信设备的性能等都有严格的要求。通信网最初提供的核心服务就是语音通信，为了满足语音通信的实时性等要求，必须使用基于专用集成电路的专用设备。通信专用设备能够保证服务质量，但成本高昂。经过近几十年有线、移动通信的迅猛发展，各种通信服务需求层出不穷，按照传统通信网专用硬件对应专用服务的思路，通信网络已经从一个简单、低负荷、易运维的网络演变成为一个多网共存、高负荷、高能耗、高运维成本的复杂网络，运营商的收益率逐年下降，迫切地需要改变传统思路，降低通信网络的建设成本、运维成本和能耗成本。

近年来运营商积极研发和应用计算机领域的虚拟化技术[3]。该技术是将计算机的实体资源包括计算资源、存储资源等，通过抽象和转换后呈现出来，能够透明化底层实体资源，打破实体资源不可切割的障碍，不受实体资源地域等限制，最大化实体资源的利用率，降低网络部署成本。目前通信与计算机领域都在热切研讨及应用的虚拟化技术包括软件定义网络（Software Defined Network，SDN）和网络功能虚拟化（Network Function Virtualization，NFV）。SDN 是对网络本身的虚拟化，关注的是网络节点之间的连接。传统互联网中的专用设备如路由器，其数据和控制是混合在一起处理的，由于任一路由器只能获取周围局部的网络信息，造成了网络控制的局部性，降低了数据传输的效率。SDN 将数据面和控制面分离开来，并将控制面从专用设备上提取出来，集中放置，使得网络的控制对网络的数据有一个全面的了解，提高数据传输效率。NFV 则关注各种网络节点功能的虚拟化，基于通用的服务器，通过软件定义的方式，虚拟化地实现网络实体的功能，部署成本低，并能快速适应网络需求变化。将 SDN 和 NFV 联合起来，就可以将原本昂贵的专用通信设备用低成本的通用设备加软件实现。由于数据的控制面集中起来就可以实现标准化，未来网络应用的变革只要改变控制面，无需大量设备的硬件升级，大幅降低了网络运行维护以及升级的成本。

## 3.2 无线接入网中的通信与计算融合

无论是有线通信还是无线移动通信，点到点的传输容量 $C$ 都是有限的，可用香农公式表示如下：

$$C = B \cdot \mathrm{lb}(1 + \mathrm{SNR}) \quad (1)$$

其中，$B$ 代表信道带宽，SNR 代表信噪比。有线通信的信道条件好，信噪比高，对通信速率的限制少，系统设计的重点在于通信协议。相比有线通信，无线移动通信可实现在任何时间、任何地点与任何人、事、物的通信，能够更加灵活、方便、快速地实现信息交流，但是移动信道的条件较差，信噪比低；此外由于无线频谱资源天然的有限性，信道带宽 $B$ 也极其有限。这些都对移动通信的速率提出了较大的限制。因此除了通信协议，移动通信系统设计的重点是如何提升频谱利用率、传输速率。

移动通信无线传输的研究一直致力于提升频谱利用率，逼近香农容量理论极限。例如，信道编码是提升无线信道可靠性的重要技术。它通过编码计算，将冗余度有逻辑地引入有效数字信息中，形成一个码字。码字在无线信道中传输，很可能被信道畸变，造成误码。在接收端通过解码计算，利用编码引入的有逻辑的冗余，可从带有误码的码字中正确地恢复有效数字信息。香农指出，如果采用足够长的随机编码，就能逼近香农信道容量。传统的信道编码都有规则的代数结构，跟“随机”相距甚远；同时，出于译码复杂度的考虑，码长也不可能太长。所以传统的信道编码性能与信道容量之间都有较大的差距。1993 年法国科学家 Claude Berrou[4]提出的 Turbo 码是一个长码，它采用多次迭代伪随机译码达到优越的纠错性能，是第一个能够逼近香农容量的信道编码，但其代价是译码算法的计算复杂度非常高。由于信道编解码所需的存储和计算复杂度与编码长度成正比，可见香农容量公式隐含的一个意义是：这是点对点通信，在计算与存储资源不受限制时，所能达到的最大传输速率，此时该最大速率仅由通信资源，即带宽和发送功率（SNR）所决定。如果实际编解码对计算与存储资源有较大的限制，如传统的分组码，那么编码性能与信道容量之间必然有较大的差距。因此，从信道编码这个单一的传输技术看，计算与存储能力的提高能够带来通信能力的增强。但目前这个方向的努力已经接近极限，Turbo 码、LDPC 码以及 Polar 码都可以逼近香农容量，能够继续改进的空间非常小。

正如信道编码一样，任何一个具体的移动通信无线传输技术，都离不开计算，而且计算的作用越来越重要。从这个角度看，通信与计算在无线传输中已有很好的协同与融合。但通信与计算的融合不应限于单个技术层面的融合，更要从系统的高度，合理地优化计算与通信，满足用户日益增长和变化的需求。系统思维是计算机领域研究的一个主要特征，指的是对计算机系统不同层次的抽象和归纳，对整机系统的性能分析和优化等[2]。而移动通信由于整个系统庞大复杂，包括移动终端、无线传输、基站、核心网等，很难对整个移动通信系统的性能进行分析和优化。已有研究大部分局限于某个技术点，比如信道编码、多天线、干扰管控等。随着移动通信发展瓶颈的显现，对未来移动通信的研究应突破传统的局部思维，借鉴计算机领域的系统思维，在研究设计未来通信机制的时候，更多地考虑系统级的因素。如果能够利用、融合整个通信系统中的计算与存储能力，有可能提升整个系统的通信能力。这是因为，计算存储和数据通信都能起到信息交流的作用[2]，比如，内容分发网络（Content Delivery Network，CDN）在网络各处放置大容量服务器存储内容，可以将用户请求导向离用户最近的服务点，从而降低对远程通信的要求。

目前，移动通信领域已经在这个方向上做了初步的探索。参考文献[5]研究了引入移动终端的存储能力以及针对存储内容特别设计的编解码计算能力后，移动通信系统的传输容量增益极限。考虑有 $K$ 个终端通过移动通信系统连接到一个基站，该基站存放了 $N$（$N \geqslant K$）个大小相同的文件可供终端下载。在传统的移动通信系统中，终端与基站的信息交流仅依赖移动通信，通信容量是衡量系统服务能力的唯一指标。现考虑每个终端有存储 $M$（$M<N$）个文件的能力。那么终端与基站的信息交流将分为两步进行：第一步是内容放置，即在非繁忙时段，将终端有可能需要的内容从基站推送给终端。第一步信息交流的限制因素不在移动通信能力，而是终端的存储能力。第二步是信息分发，终端向基站提出内容请求，基站传输相关内容给终端。这一步的限制因素才是移动通信能力。假定 $K$ 个终端各自向基站发出下载一个文件的需求。在传统的移动通信系统中，基站需要同时传送 $K$ 个文件给 $K$ 个终端，对通信传输容量的需求是 $K$。当每个终端有存储 $M$ 个文件的能力后，在内容放置阶段基站可将每个文件的 $M/N$ 部分推送给终端，共 $N$ 个文件，占满终端的存储空间，这样在信息分发阶段，无论终端需要哪个文件，基站只要向每个终端传输（$1-M/N$）个文件，对通信容量的要求是 $K(1-M/N)$。可见，存储能力给移动通信传输容量带来的增益（即对传输容量需求的下降比例）为（$1-M/N$）。该增益只与单个终端的存储能力相关，故称为局部增益。当终端的存储能力与基站的存储内容量可比拟时，局部增益是相当可观的。

但一般而言，基站端的内容量会远远大于终端存储能力，此时的局部增益就非常小。因此，参考文献[5]进一步提出了编码多播的概念，其主要思想是，在第一步内容放置阶段精细地设计放在每个终端的内容，为第二步信息分发阶段构造多播的机会。这样第二步信息分发时，基站可采用编码多播的机制进一步降低传输数据量，带来的传输容量的增益可达 $1/(1+KM/N)$。该增益与所有终端的存储能力 $KM$ 相关，因此称为全局增益。如图 1 所示，基站端有 $N=2$ 个文件，有 $K=2$ 个终端，每个终端的存储能力 $M=1$。将基站端的两个文件二等分，标记为 $\{A_1,A_2\}$ 以及 $\{B_1,B_2\}$。在第一步内容放置阶段，基站将 $A_1$ 和 $B_1$ 推送给终端 1，将 $A_2$ 和 $B_2$ 推送给终端 2。在第二步信息分发阶段，终端 1 和 2 分别请求文件 A 和 B。此时基站可将 $B_1 \oplus A_2$，即 $B_1$ 和 $A_2$ 按比特异或的结果广播给终端，终端 1 和终端 2 分别根据已存的内容 $B_1$ 和 $A_2$，恢复出 $A_2$ 和 $B_1$。如果不采用编码多播，第二阶段需要传输 $K(1-M/N)=1$ 个文件，采用编码多播后，第二阶段需要传输 $K(1-M/N)/(1+KM/N)=0.5$ 个文件，进一步降低了对传输容量的需求。

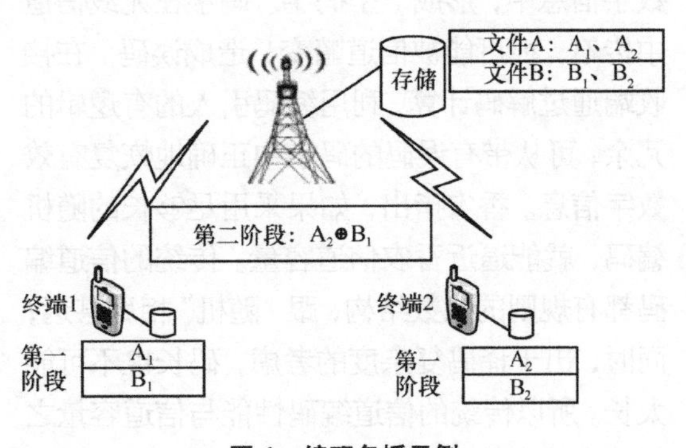


图 1　编码多播示例

参考文献[6]探讨了通信计算存储融合的移动通信网络的一些基础理论问题，试图给出计算和存储资源在理论上的度量方式以及在考虑通信计算存储三维资源时给出移动通信网络的容量。研究认为，通信维度的操作例如编码和调制，只是保护了信息流，并未改变信息流，其度量单位为 bit/(s·Hz)；存储维度的操作则是引入了非因果性，即时域的可逆性，能够增强系统在更长的时间范围内传输信息的能力，其度量单位为 byte；而与计算领域以 Flops 来定义计算性能不同，该研究将移动通信网络中融合的计算认为是信息流之间的代数或者逻辑操作，它会改变信息流，因此在接收端也需要额外的代数或者逻辑操作恢复出原信息流。这种融合计算的度量单位被定义为参与计算的移动通信网络节点数（例如协作多点传输[7]中的节点数）或者信息流数（例如网络编码[8]中涉及的信息流数）。从上述定义可见，通信的度量是一个实数，而存储与计算的度量都是整数，这样的定义能够将存储和计算与通信资源分离开。以编码多播[5]为例，该研究给出了在上述定义下，存储和计算能够给移动通信网络容量带来的增益，该增益随存储和计算资源线性增长。

总结而言，面向移动通信系统不同的业务需求，如果能够从系统层面融合通信、存储与计算，利用计算存储的信息交流能力，有可能突破传统通信系统的容量限制，将移动通信系统的核心功能从传统的信息传输能力转变为信息交流能力，提升系统的业务支撑能力。

## 4 未来的通信与计算融合

面向未来的移动通信系统，可从多个方面进一步深入探讨通信与计算融合。

基础理论：未来移动通信系统服务能力的度量（Metric）。如前所述，传统移动通信系统主要关注通信能力，其核心功能是信息传输。在融合了计算存储以后，未来移动通信系统应利用计算存储的信息交流能力，协同原有的无线传输能力，将系统的核心功能转为信息交流。与此同时，应提出新的系统能力的定义与度量方法。在早期的移动通信网络如 GSM 中，网络只提供单一的实时语音通信业务，此时通信即网络的核心功能也是核心业务，通信的能力（或容量）等同于网络的服务能力；随着移动通信网络的发展，网络提供的业务越来越多样化，通信只能作为网络的核心功能之一，不再等同于核心业务[6]，通信的能力也不再等同于网络的服务能力。移动通信网络提供的业务对通信、计算和存储能力的需求不尽相同。例如实时语音服务主要依赖通信能力，而非实时视频流下载主要依赖存储能力，未来 5G 的杀手级应用——移动 AR/VR 则对网络的边缘计算[9]能力提出很高的要求。可见，面向不同的业务需求，融合通信、计算和存储资源，未来移动通信系统的服务能力远大于通信能力。如何定义服务能力，构建一套新的基础理论体系，是亟须解决的根本与核心的问题。

系统架构：融合计算机系统设计的思维，加强系统可扩展性（Scalable）、可重塑性（Elastic）、透明性和模块性[2]。可扩展性指的是扩大系统规模时，性能应接近线性提高。但受限于通信开销太大，往往难以实现。目前移动通信系统已采用将内容或服务尽量拉近终端侧，降低通信开销的方法，来提升系统的可扩展性[10]。比如移动边缘计算[9]，只有在确实需要的时候移动用户才会接入远端强大的云计算资源帮助其计算，一般情况都可以将其计算卸载在近端的移动边缘计算节点。另外欧洲的 TROPIC 项目计划在小基站上增加云的功能（云增强小基站），这样移动用户可以在近端通过无线接入云服务。由于无线通信和计算资源都被拉到用户近端，降低了通信开销，云增强小基站可同时提升无线通信和计算的可扩展性。可重塑性指的是能够快速重组现有资源，构建新的架构适应新的负载需求；透明性指的是屏蔽系统底层架构的变化，用户只需关注上层的业务逻辑；模块性指的是像堆积木一样构建系统，与可重塑性相辅相成。移动通信领域目前也已经在这几个方面努力推进。正如前文提到的在移动通信核心网应用的 SDN 和 NFV 技术，增加了系统的透明性和可重塑性；此外，软件无线电（Software Radio）技术与 NFV 的思想类似，希望构建一个具有开放性、标准化、模块化的通用硬件平台，通过加载不同的软件来实现不同的移动通信功能，如设置工作频段、通信协议等；网络切片技术则是根据时延、带宽、可靠性、安全性等不同的服务需求，将电信运营商的物理网络划分为多个虚拟网络，每个虚拟网络应对不同的应用场景，满足不同的需求。总体而言，未来移动通信系统的可扩展性、可重塑性、透明性和模块性越强，其服务能力越强。

结构化：未来移动通信系统发展的方向。20 世纪 70 年代 Dijikstra E W 等计算机科学家深入研究了"结构化程序设计"，为后来的软件工程奠定了基础，成为软件技术发展的一个里程碑。目前的移动通信系统与当年的软件情况类似，需要以结构化设计为指导思想。未来通信网络的出路在"结构化"，应深入研究分布式系统的"结构"。移动通信网络像自然界一样，也在不断地进步演化，因此所谓"结构化"不像简单的机械装备一样，先有固定的蓝图，需要在动态演变中逐步提高"结构化"程度。

信息交流机制：融合通信、计算和存储的业务感知信息交流机制。传统移动通信系统的设计目标是作为一个信息传输的管道，不管传送什么信息，力求更快更可靠，信息交流等同于物理层信息传输。随着移动通信系统承载的业务越来越多样化，移动数据流量指数级增长，但物理层信息传输能力受限于有限的频谱资源和信道容量，将无法支撑上述的业务发展。因此必须突破传统思维，打通物理层信息传输管道和上层业务应用之间的联系，感知业务内容的需求与特性，一方面融合计算存储，增强信息交流能力，另一方面，与物理层信息传输能力相适配，提高移动通信系统总体的信息交流能力。

## 5 结束语

移动通信系统经过近几十年的迅猛发展，通信容量趋于饱和，可持续发展能力堪忧。突破传统移动通信发展瓶颈的一种思路是融合计算机领域的思想和技术。事实上，通信与计算融合的努力可追溯到信息科学发展的早期，而目前移动通信系统在核心网和无线接入网方向都开展了通信与计算融合的研究，初步表明通信与计算融合能有效降低网络部署与维护成本，降低对系统的容量需求，从而提升系统支撑业务的能力。未来可在基础理论、系统架构、结构化设计、信息交流机制等多方面进一步深入探讨。期望通过通信与计算的融合，推动移动通信系统可持续发展。

## 参考文献：

[1] China Mobile Research Institute. C-RAN: the road towards green RAN[R]. 2014.

[2] 李国杰. 对通信与计算机技术融合的初浅认识[R]. 2014.

[3] 翟振辉, 邱巍, 吴丽华, 等. NFV 基本架构及部署方式[J]. 电信科学, 2017, 33(6): 179-185.

[4] BERROU C, GLAVIEUX A. Near optimum error correcting coding and decoding: Turbo-codes[J]. IEEE Transactions on Communications, 1996, 44(10): 1261-1271.

[5] MADDAH-ALI M, NIESEN U. Fundamental limits of

caching[J]. IEEE Transactions on Information Theory, 2014, 60(5): 2856-2867.

[6] LIU H, CHEN Z, QIAN L. The three primary colors of mobile systems[J]. IEEE Communications Magazine, 2016, 54(9): 15-21.

[7] GARCIA V, ZHOU Y, SHI J L. Coordinated multipoint transmission in dense cellular networks with user-centric adaptive clustering[J]. IEEE Transactions on Wireless Communications, 2014, 13(8): 4297-4308.

[8] AHLSWEDER, CAI N, LI S Y R, et al. Network information flow[J]. IEEE Transactions on Information Theory, 2000, 46(4): 1204-1216.

[9] ETSI. Mobile edge computing: a key technology towards 5G[S]. 2015.

[10] BARBAROSSA S, SARDELLITTI S, LORENZO P D. Communication while computing: distributed mobile cloud computing over 5G heterogeneous networks[J]. IEEE Signal Processing Magazine, 2014, 31(6): 45-55.

# 第9章

# 大数据系统与算法

# 大数据研究的“顶天立地”

人工智能的复兴有标志性的事件，Google 子公司研制的 Alpha Go 围棋机器人下赢人类围棋冠军事件再次掀起人工智能的高潮。大数据的兴起是自然而然的事，没有标志性的事件。21 世纪初还没有人讲“大数据”，我指导的学生已经在做数据的分类、聚类等理论研究，这是数据处理的关键技术，也是人工智能研究的主要内容之一。本章收录的早期论文实质上就是大数据的研究成果。计算机理论和计算机系统研究、人工智能特别是人工神经网络的研究，已为大数据的兴起做了长时间的技术储备。

2007—2009 年，我在牵头做中国科学院组织的中国至 2050 年信息科技发展路线图战略研究时，强调 21 世纪计算机科学技术的主要发展方向是普惠计算。这次战略研究的成果不但以中英文图书正式出版，而且在 *Communications of the ACM* 期刊上发表了前瞻性文章 *Computing for the masses*（此文收录在本章），明确指出普惠计算是 e-People，而不是 e-Business、e-Science、e-Government。这里也包含了大数据的理念，因为大数据一定是来源于海量用户，又服务于海量用户。如果只是少数科学家采集数据，就不可能形成今天的大数据热潮。

2012 年 5 月，我和成思危、姚期智、华云生、程学旗等共同发起举办了题为“网络数据科学与工程”的第 424 次香山科学会议。此次会议受到美国 *Science* 杂志重视，2012 年 6 月该杂志发表了会议期间采访我们的文章。也许是巧合，也许是香山科学会议起到“布谷鸟”报春的作用，从“百度搜索指数”可以查到，从 2012 年 6 月开始，媒体和网络上“大数据”一词的出现频次明显上升。2019 年 11 月，我们又召开了第 667 次香山科学会议，主题是“数据科学与计算智能”，会议执行主席增加了梅宏教授和赵伟教授。这两次香山科学会议都重点讨论了什么是“数据科学”和如何发展“数据科学”，提出了一些独到的看法和见解。

为了促进我国大数据技术的研究和发展，2012 年 11 月中国计算机学会成立了“大数据专家委员会”。中国计算机学会的基层组织一般称为“专业委员会”，但大数据领域的专家来自人工智能、数据库、计算机系统等各个专业，因此破格称为“专家委员会”。大概是因为众口难调，我作为最年长的学者被大家选为首任主任。在大数据领域，几年前中国科学院计算技术研究所就与国外企业合作，每年组织一次有关 Hadoop 的技术大会，请国内外第一线专家介绍大数据技术的最新进展，因为讲的都是“干货”，会议越开越红火。大数据专家委员会成立以后，这个技术大会改为中国大数据技术大会，尽管一张门票要几千元，仍然人满为患，一票难求。大数据专家委员会还举办了全球范围的大数据竞赛，由企业出题，全球各大学参赛，每年有几万人参加。大数据专家委员会虽然成立较晚，但现在已是中国计算机学会非常活跃的专委会之一。

2003 年我在《科学美国人》杂志上读到巴拉巴西（Albert-Laszlo Barabasi）的一篇科普文章《无尺度网络》，高深的网络理论写得通俗易懂，引人入胜。我多次在《中国计算机学会通讯》上呼吁中国的学者向他学习，写出像他那样深入浅出的好文章。复杂网络理论与大数据有密切联系，因为大数据的背后就是复杂的数据网络。巴拉巴西是全球复杂网络研究的开创者之一，现在是美国东北大学的教授，从 2003 年以后我一直关注他的研究进展，读过他写的《爆发：大数据时代预见未来的新思维》《链接》等著作。2015 年我指导的博士生黄俊铭毕业后，就去了美国东北大学，去巴拉巴西门下做博士后。探索信息在网络上传播的规律是大数据研究的重要内容，他在国内写的论文 *Temporal scaling in information propagation* 发表在 *Nature* 子刊 *Scientific Reports* 上，被收录在本选集中。

为本选集压轴的是一篇重磅论文，是由程学旗研究员牵头写的《大数据系统和分析技术综述》，该论文 2014 年在《软件学报》上发表后被大量引用。程学旗是我的在职博士，现任中国科学院计算技术研究所副所长。大数据专家委员会成立以来，他一直担任秘书长，在大数据学术圈做了许多组织工作。他还创办了一家大数据公司——中科天玑数据科技股份有限公司，在舆情和情报分析方面为国家安全部门做了大量工作。如果说召开香山科学会议探索“数据科学”是“顶天”的工作，创办和发展“中科天玑数据科技股份有限公司”就是“立地”的努力。在大数据领域，中国科学院计算技术研究所也继承了一贯坚持的“顶天立地”传统，体现了“科研为国分忧，创新与民造福”的价值追求。

发表于《计算机研究与发展》2001年第2期

# 基于关系结构的轻量级工作流引擎

何清法，李国杰，焦丽梅，刘力力

中国科学院计算技术研究所，北京 100080

**摘　要：**针对关键业务应用的开发离不开工作流技术的支持，通过对关键业务的实际开发需求的分析，在传统的关系数据库基础上，提出了一个适用于关键业务开发的基于关系结构的轻量级工作流引擎的框架结构。此工作流模型由机构模型、信息模型和控制模型3部分组成。深入讨论了采用关系结构和轻量级理念来设计工作流引擎的原因，并详细地给出了相关的机构模型、信息模型和控制模型的设计原理以及具体的表示和实现方法。其原型已经应用到实际的应用系统中。实践证明，利用此工作流引擎可以显著地缩短关键业务的开发周期。

**关键词：**关系；轻量级；工作流引擎；关键业务；活动

# Relation-based lightweight workflow engine

Qing-fa HE, Guo-jie LI, Li-mei JIAO, Li-li LIU

Institute of Computing Technology, Chinese Academy of Sciences, Beijing 100080

**Abstract:** Workflow technology plays an important role in the development of critical business applications. According to the analysis of the requirements to develop critical business applications, a framework of a relationship-based lightweight workflow engine is presented, which is built over the conventional relational database system. It consists of three components, namely: organization model, information model and control model. The reasons that the concepts of relationship and lightweight are adopted to design the workflow engine are thoroughly discussed. The principle to design, the representation of and the way to implement the organization, information and control model of the workflow engine are described respectively. A prototype of this workflow engine has been applied to a project about the automation of trademark registration and administration, which shows the application of the workflow engine can markedly shorten the cycle of the development critical business applications.

**Key words:** relationship, lightweight, workflow engine, critical business, activity

## 1 引言

目前，针对企业或者部门的计算机应用已不仅仅停留在诸如文档处理、公文流转以及信息发布等这些简单的业务层面上。越来越多的企业或部门要求将信息技术的应用扩展到关键业务中。关键业务的普遍特征是：

（1）企业或部门赖以生存的；

（2）业务过程往往由许多业务活动组成，业务逻辑和业务规则复杂；

（3）业务的完成依赖于其中众多业务活动之间的交互和众多的业务人员的协作参与；

（4）涉及的数据量经常是海量数据；

（5）如果能将信息技术恰当地应用到这些关键业务中，不仅能够提高工作效率，还可以减少出错的可能性。

例如，产品的设计和制造过程，银行的借贷和划账业务，还有商标的申请、审查和注册业务等，都属于相应企业或部门的关键业务。工作流技术所具有的协调本质决定了其在关键业务的信息化过程中将扮演重要的角色。

正如文献[1-2]所述，工作流是业务过程的计算模型，即将相应的业务逻辑和业务规则在计算机中以恰当的模型进行表示并对其实施计算。业务过程是若干业务活动的集合，这些业务活动按照一定的规则前后链接在一起，相互协作，以便达到一个共同的目标。业务活动则是能够完成特定的功能的一个实际环节，它在信息系统中通常针对具体的应用逻辑。为了对工作流管理系统的开发起到一个指导作用，工作流管理联盟（WfMC）给出了工作流系统的一个通用框架——工作流参考模型[2]。在工作流参考模型中，工作流引擎是工作流管理系统的核心。工作流引擎是为工作流管理系统在定义提供支持、同时在运行时提供解释和执行服务的一组数据模型和软件。根据文献[3]中对工作流引擎体系结构的讨论，我们认为工作流引擎主要包括机构模型、信息模型和控制模型 3 种模型，前两者合称为工作流引擎的数据模型。

本文以国家智能计算机研究开发中心所承担的一个具体的应用项目——国家商标局的商标注册与管理信息化系统为实例，同时分析了其他不同的企业和部门的关键业务的基本特征，针对关键业务的开发需求，在传统的关系 DBMS 的基础上，讨论一个基于关系结构的轻量级工作流引擎的具体的设计原理与实现方法。它充分考虑了关键业务开发过程中对工作流功能的需求，利用此工作流引擎，可以使用传统的开发工具构造出具有工作流特征的大型信息系统（注：如果没有特别说明，本文所述及的应用、应用系统以及信息系统等概念都是针对关键业务而言的）。

本文的第 2 节讨论为什么要采用关系结构和轻量级这两个概念来设计工作流引擎；第 3 节给出相关的数据模型及其表示；在第 4 节中将描述工作流引擎的控制模型；最后还将给出具体的应用实例。

## 2 与关系结构和轻量级相关的一些讨论

目前，软件开发的一个普遍现象是软件产品的规模和功能越来越向大型化和复杂化的方向发展，而本文所提出的基于关系结构的轻量级工作流引擎却强调其小型化的特征。许多工作流产品从数据存储到运行环境往往都有自己的一整套独特的体系结构，它们除了具备工作流的基本功能外，还同时宣称可以任意集成第三方的应用，甚至有的还嵌入了程度不等的应用开发的功能。但是，开发人员真正需要的可能并非这些复杂特性。

## 2.1 工作流的设计中心

所谓工作流的设计中心指的是由谁来定义和开发工作流的应用。那么，工作流的设计中心到底应该在哪里？Patricia Seybold Group 的 RonniMarshak 曾经对这个问题进行过一些有益的论述。

（1）完全由实际的业务人员来负责工作流应用的定义和开发，一些工作流产品也大力提倡这一点。的确，实际的业务人员对自己的业务规则最为熟悉，但他们对计算机技术了解不多。因此，这只适合简单的工作流应用，一旦业务逻辑比较复杂尤其是面对关键业务时，要他们将业务逻辑转换为工作流并且自己定制相应的应用逻辑则非常困难。

（2）业务人员和专业技术人员相结合。工作流产品提供图形化的界面供业务人员（或者结合专业人员）定义业务逻辑和规则，具体的应用逻辑则由专业开发人员完成。应用的开发可以利用工作流所提供的集成开发工具，也可以利用第 3 方的开发工具。可能存在的问题是如果使用工作流产品所集成的开发工具，则其所提供的开发应用的功能是否足够；此外，如果使用第 3 方的开发工具，又该如何实现工作流机制的集成。

（3）还有一种观点认为，要建立真正复杂、灵活而且可扩展的应用系统，必须将工作流的开发融合到信息系统的开发过程中，从整个信息系统的角度来定义工作流中的业务规则、任务流转以及相关角色。甚至有一种极端的观点认为应该把业务规则硬编码到具体的应用中，如果业务规则比较稳定，这种方法可以得到非常紧凑的应用系统，其缺点是系统的重构和复用非常困难。

我们认为，对于关键业务，第（2）种或者是第（3）种模式是可行的，但必须有一个恰当的工作流引擎的支持，否则会显著地增加实际开发的难度和工作量。

## 2.2 为什么要基于关系结构

所谓基于关系的工作流引擎指的是工作流引擎中的数据模型（即机构模型和信息模型）全部通过关系结构来表达；控制工作流引擎运作的各种程序逻辑（即控制模型）也是通过常规关系数据库管理系统中所提供的存储过程、包以及触发器等机制来实现；同时，事务的并发控制也通过数据库系统所提供的机制来实现。

从技术角度来说，使用关系结构来表达工作流引擎中的数据模型可以降低工作流引擎开发过程中的技术难度和工作量。具体表现在：

（1）与工作流引擎相关的各种控制数据（包括业务活动的状态数据）可以存储在数据库系统中；

（2）与此相关的数据的完整性可以由数据库管理系统来维护；

（3）利用关系结构可以方便地定义工作流引擎中的各种数据格式和数据结构；

（4）可以方便地利用数据库管理系统提供的各种 DML 语句来操纵工作流引擎所需的各种数据。

从开发应用系统的角度来看，在同一数据库环境下为开发者提供一个基于关系结构的工作流引擎，并且如果这个工作流引擎所提供的功能可以方便地嵌入应用的开发环境中，则可以降低开发应用的难度。这是因为：

（1）针对关键业务的应用系统通常会采用一个常规的关系数据库系统作为后台的支撑；

（2）应用系统的开发者往往会采用一种他们所熟悉的并且适合此数据库系统的前端开发工具来开发具体应用，这些前端开发工具一个显著特征是开发功能强大，但一般不具备工作流机制。因此，采用基于关系结构的工作流引擎很容易与应用的开发环境做到无缝集成。

## 2.3 为什么要采用轻量级

轻量级的工作流引擎指的是从够用、灵活和低成本的设计原则出发，不追求工作流引擎的功能的完备和复杂，只是实现其中必不可少的功能和特征。

在设计工作流引擎时主要考虑对其数据模型的定义和解释、活动之间的协调以及任务的分配和控制等功能提供支持，而不支持诸如提供内建（built-in）的应用开发工具、对应用数据的定义和完整性维护、完善的异常处理以及长事务控制等功能。

我们之所以采用"轻量级"这一特征来刻画工作流引擎主要出于如下考虑。

（1）许多现有的工作流产品都在不同程度上提供了对外部工具的集成功能，部分产品还提供了基于表单的应用逻辑的定制和开发环境。但是，外部工具的多样性和复杂性决定了对外部工具的集成难以做到无缝；而工作流产品内建的开发工具除了与流行的开发工具不兼容外，其开发功能往往都比较简单。因此，对于简单的应用（例如公文流转、订单的审批等），这些产品是合适的。但是，如果是开发关键业务的应用系统（特别是行业应用系统），现有工作流产品所能提供的开发功能是远远不够的。

（2）许多针对 DBMS 的开发工具提供了极强的应用开发手段，但是这些开发工具往往不具备对工作流机制的支持，而现有的工作流产品由于其出发点不同，很难与其他开发环境有机地融合在一起。因此开发人员往往苦于找不到一套合适的工作流支撑系统来开发具有工作流特征的应用。

（3）具有工作流特征的应用的形态千变万化，要想在工作流系统中对不同的应用（包括应用数据）进行统一的表示往往不遂人意。利用这种所谓灵活的工作流系统开发出来的应用在实际运作过程中反而表现不灵活。因此，另外一种趋势是，应用的逻辑仍旧由专用的应用开发工具去完成，工作流引擎只管理相关的控制数据，对应用数据只提供必要的关联手段将其与控制数据链接在一起。

（4）目前已经有许多中间件产品（各种应用服务器、TP 等）提供了对应用事务包括长事务的控制能力，对事务控制有特殊需求的应用系统可以使用这些产品。

新西兰 Massey 大学的 Tagg 等学者[4]对工作流引擎的描述曾经使用过"轻量级"这一术语，但其侧重点在于如何构造一个"瘦客户端"。我们的侧重点则是设计一个充分支持工作流特征的小型内核，它可以无缝地嵌入传统的应用开发环境中。

综上所述，基于关系的轻量级工作流引擎是这样一种产品：它可以在传统的关系数据库基础之上定义工作流数据模型；它利用 DBMS 内嵌的编程语言来实现工作流引擎的控制逻辑；它提供了一系列比较完备的 APIs，应用的开发者可以将这些 APIs 嵌入自己的应用系统中，从而实现具有工作流性质的信息系统。基于关系的轻量级工作流引擎的适用对象并非应用系统的最终用户，而是利用专用开发工具构造相应应用系统的专业开发人员。它为开发人员提供了驱动工作流机制的支持，从而构造出各种灵活的具有工作流特征的应用系统。其具体表现形式为一套表结构、一组建模工具和一系列供实际应用调用的 APIs。

# 3 数据模型

基于关系结构的轻量级工作流引擎的数据模型包括机构模型和信息模型两部分。机构模型描述的是企业或者部门的组织机构关系，信息模型则定义工作流引擎中所用到的各种控制数据。通过数据模型，可以方便地描述关键业务的业务规则、活动的依赖关系以及任务的指派等特征。它们都通过统一的关系结构来定义。图 1 给出了基于关系结构的轻量级工作流引擎的数据模型的 ER 图（限于篇幅只给出核心表结构）。

### 3.1 机构模型

与机构模型相关的表主要有 STAFF、DEPART-MENT、TEAM、STAFF- TEAM、ROLE 和 S TAFF-IN-ROLE，表之间的关系已经在图中通过不同含义的连线标出。下面将有重点地对其中的一些含义做出一些解释。

DEPARTMENT 和 TEAM 分别表示部门和团队，部门通常表示纵向的行政隶属关系，而团队通常表示横向的合作关系。DEPARTMENT 和 TEAM 分别通过相应的 SUPER-DEPT-ID 和 SUPER-TEAM-ID 关联使得在部门之间和团队之间分别形成树状的上下级关系。

STAFF 记录与人员相关的个体信息，其中的 LOGGING-ON 指示相应的人员目前是否已经登录到系统，ON-LEAVE 表示该人员是否正处于休假期，这两种信息在工作流引擎中

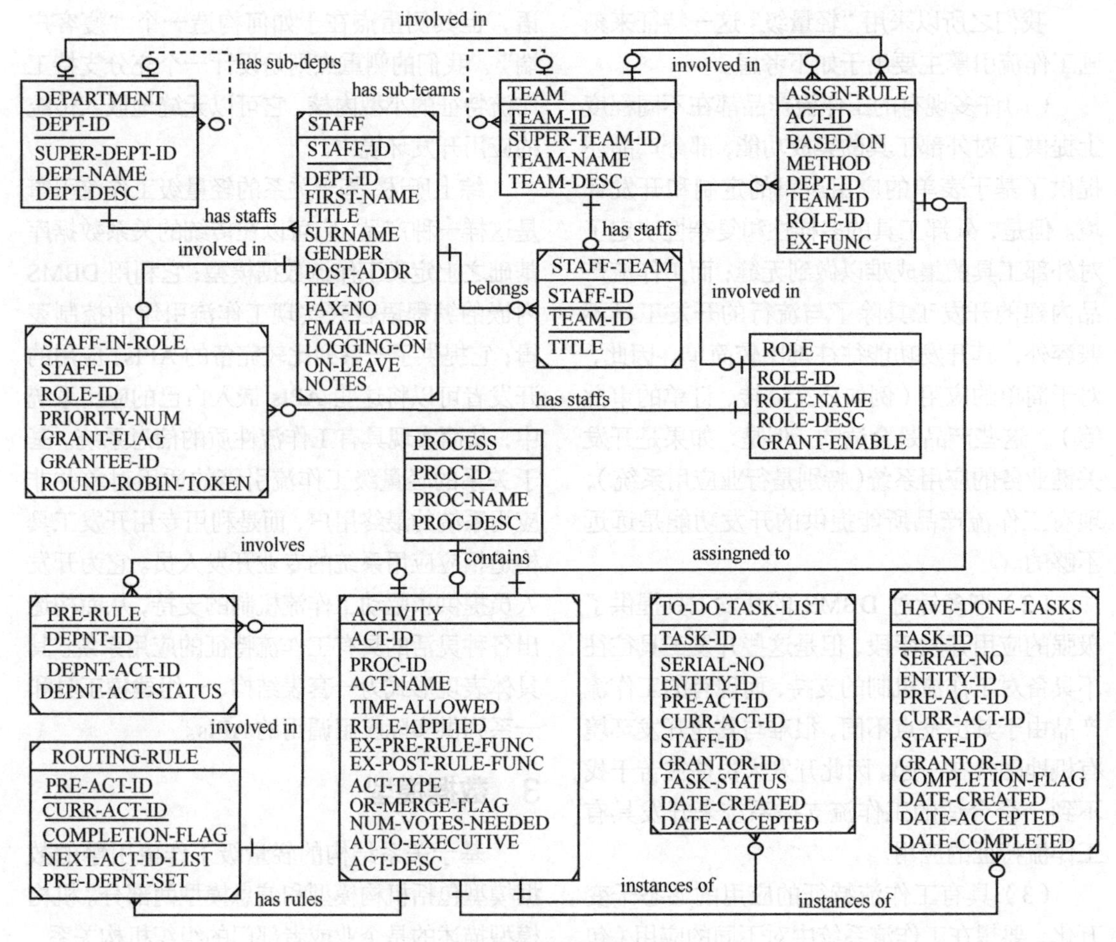


图 1 数据模型 ER 图

将被作为任务分配的指示信息。STAFF 中的外码 DEPT-ID 指明该人员所隶属的部门。关系 STAFF-TEAM 指定了人员与团队之间的隶属关系。

机构模型中的部门、团队、人员以及相互间的关系为大型企业尤其是从事技术工作的企业的机构建模提供了有力的支持，同时也为现代企业流行的管理模式——“矩阵管理”提供了支持。当然，对于小型机构而言，完全可以考虑只定义 DEPARTMENT 或者 TEAM 其中之一。由于 DEPARTMENT 和 TEAM 之间在 ER 图中并无联系，因此缺其一并不会破坏机构模型的完整性。

ROLE 进一步扩展了机构模型的建模能力，关系 STAFF-IN-ROLE 在 STAFF 和 ROLE 之间建立起来关联。在机构模型中引入角色这一概念主要是为了增强任务指派的能力，在本文的后续内容中将对角色中的有关概念作进一步的解释。

## 3.2 信息模型

信息模型的核心是业务活动表（以下简称活动）ACTIVITY，其他相关的表结构主要有业务过程 PROCESS、业务规则（活动流转规则）ROUTING-RULE、活动前依赖规则 PRE-RUL E、任务指派规则 ASSGN-RULE、任务列表 TO-DO-TASK-LIST 以及已完成的任务列表 HAVE-DONE-TASKS。从图 1 中可以看出，ACTIVITY 与其他表之间都存在联系。

### 3.2.1 活动类型

每个业务过程由若干业务活动组成，不同的业务活动通过各不相同的 ACT-ID 来唯一标识，ACT-TYPE 则指明相应活动的类型。同一个业务活动在工作流运行时可能具有多个实例。我们将活动的实例称为任务（注：为了叙述的方便，我们有时可能不一定严格区分活动、活动的实例以及任务这 3 个概念，在某些情况下，活动指的是活动的实例，这可以通过上下文区分出来），将属于同一业务过程的任务称为属于同一批次的任务。有的业务活动可能针对具体的业务环节，即在前台（后台）对应实际的应用逻辑；有的业务活动则不针对具体的业务环节。活动类型可以进行如下分类。

（1）INITIAL，初始化活动，业务过程的第 1 个活动，不针对具体业务环节。

（2）INTERACTION，常规交互活动，INTER-ACTION 活动对应实际的业务环节，在前台对应实际的应用逻辑，完成此活动需要实际人员的参与。在所有活动类型中，只有 INTERACTION 活动才需要与实际人员交互。

（3）AUTOMATION，常规自动活动，同样对应实际的业务环节，但是实际的应用逻辑位于后台，由工作流引擎自动调用完成。AUTO-EXECUTIVE 指明相应应用逻辑的执行体。

（4）AND-BRANCH，与分支活动，不针对具体业务环节，此活动将同时派生出若干后继活动。

（5）AND-MERGE，与汇聚活动，是一同步活动，不针对具体业务环节，流经此处的任务将进行与汇聚同步。此活动将进行活动的前依赖规则检查，只有所有的前依赖规则均被满足，才可流向后继活动。

（6）OR-MERGE，或汇聚活动，是一同步活动，不针对具体业务环节，流经此处的任务将进行或汇聚同步。它同样将进行活动的前依赖规则检查，但是在前依赖规则只要存在一条满足指定条件的，就可以流向后继活动。OR-MERGE-FLAG 用于指定或汇聚条件。

（7）VOTE-MERGE，投票汇聚活动，是一同步活动，不针对具体业务环节，同一批次的任务只有达到 NUM-VOTES-NEEDED 所指定的票数才可流向后继活动。

（8）DUMMY，哑活动，不针对具体业务环节，它可以作为某些活动的虚拟后继活动，还可以使用它来构造更为复杂的业务规则。若哑活动有后继活动，则可以立即流向后继活动。

（9）COMPLETION，终结活动，表明相应业务过程的终结，不针对具体业务环节。

3.2.2 业务规则的表示

在工作流引擎中，业务规则可以分解成活动的前依赖规则和活动的后转发规则。活动的前依赖规则指明相应活动的启动条件，启动条件是通过相应活动的直接前趋活动以及相应的状态标志来表示的，前依赖规则包含顺序、与汇聚、或汇聚和投票汇聚 4 种规则。活动的后转发规则指的是当前活动所对应的任务结束后该启动哪些后继活动，后转发规则包含顺序、或分支和与分支 3 种规则。图 1 中的 PRE-RULE 表、ROUTING-RULE 表以及 ACTIVITY 表中的 ACT-TYPE 和 RULE-APPLIED 等字段联合表示活动的前依赖规则和后转发规则。

由于我们将各种汇聚活动单独抽取出来，因此可用很简洁的关系结构来表达活动的前依赖和后转发规则。首先 ACTIVITY 表中的 RULE-APPLIED 字段指示相应活动应采用何种规则判断准则，它可以有 4 种取值：DEFAULT、USER-DEFINED-PRE-RULE、USER-DEFINED-POST-ROUTING-RULE 和 USER-DEFINED-BOTH-RULE。DEFAULT 表示由工作流引擎自动根据 PRE-RULE 表和 ROUTING-RULE 表来进行规则检查。考虑到业务规则的多样性，本文提供了自定义方式来表达那些无法用缺省规则表示的特殊业务规则，ACTIVITY 表中的 EX-PRE-RULE-FUNC 和 EX-POST-RUL E-FUNC 分别指定了前依赖和后转发规则的自定义调用接口。自定义业务规则的行为完全由相应的程序确定。一般情况下，大多数业务规则都可以直接通过 DEFAULT 方式表达。接下来将讨论在 DEFAULT 方式下前依赖规则和后转发规则的表示。

活动的后转发规则主要通过表 ROUTING-RULE 表示，后转发规则可以用如下四元组来表达：Post-Routing-Rule= (PRE-ACT-ID, CURR-ACT-ID, COMPLETION-FLAG, NEXT-ACT-ID-LIST)，其含义是：在当前活动的 ACT-ID 为 CURR-ACT-ID 的情况下，如果当前活动的前趋活动的 ACT-ID 为 PRE-ACT-ID，并且当前活动的结束标记为 COMPLETION-FLAG 的话，工作流将流向由 NEXT-ACT-ID-LIST 所指明的后继活动。

前依赖规则需联合 PRE-RULE、ROUTING-RULE 和 ACTIVITY 共同表示，前依赖规则可以用一个三元组来表达，即 Pre-Dependency-Rule=(PRE-ACT-ID,CURR-ACT-ID,PRE-DEPNT-SET)，PRE-DEPNT-SET 为前依赖活动集，其中的每一个元素又可以用另外一个三元组来表示：Element-Pre-Depnt-Set=(DEPNT-ID, DEPNT-ACT-ID, DEPNT-ACT-STATUS)，Pre- Dependency-Rule 的含义是：由前趋活动 PRE-ACT-ID 流转过来的当前活动 CURR-ACT-ID 能否启动取决于前依赖活动集 PRE-DEPNT-SET 中所包含的那些活动是否已经到达各自应该到达的结束状态 DEPNT-ACT-STATUS。可以看出，只有在前依赖活动集中出现的那些前趋活动才可以联合构成对当前活动的约束关系，如果某个前依赖规则三元组中的 PRE-DEPNT-SET 为空集，则表明由此前趋活动流到当前活动的流转过程跟其他前趋活动没有任何关系，与此相应的当前活动可以立即启动。

#### 3.2.3 任务队列和已完成任务队列

一个活动可以同时具有多个实例，即任务，这些实例可以是属于同一批次的，也可能属于不同的批次，流水号 SERIAL-NO 用来标识任务所属的批次，所有属于同一批次的任务具有相同的流水号；不同的任务之间则通过唯一的 TASK-ID 进行标识。

本文所讨论的轻量级工作流引擎并不涉及对具体的应用逻辑和应用数据的管理，但是，在工作流引擎中必须提供一种手段将任务与应用实体有机地关联起来，否则，单独的任务将不具有任何实际意义。实体标识 ENTITY-ID 便起到了这种桥梁作用，其取值的真实含义完全取决于应用逻辑的自身解释，例如它可以是某个编号，也可以是某个文件的名字。ENTITY-ID 在业务初始化的时候设置，其值也可以在业务流转过程中被随时改变。

任务队列 TO-DO-TASK-LIST 用于记录那些已经创建但尚未完成的任务，位于任务队列中的任务具有 4 种状态：

（1）PENDING，任务正处于“与汇聚”同步状态，即正在等待其他相关的前趋任务的结束；

（2）WAITING，任务已经就绪，处于“等待处理”的状态；

（3）PROCESSING，任务处于“正在处理”的状态；

（4）PAUSING，任务处于“暂停”的状态。

已完成任务队列 HAVE-DONE-TASKS 用于记录那些已经正常结束的任务，COMPLETION-FLAG 表示相应任务的结束标记。

STAFF-ID 表示执行此任务的实际人员，GRANTOR-ID 若不为空，则表示此次任务的执行过程是经过授权的，GRANTOR-ID 指明相应的授权人员。

### 3.3 任务指派

任务指派指的是依照何种准则将任务分配给具体人员来执行。只有常规交互活动才涉及任务指派的问题；其他活动要么在前台不具备实际的应用逻辑，要么由工作流引擎自动调用，因此与任务指派没有关系。

在前文中已经提到了机构模型和信息模型中的许多表和字段将联合用于工作流引擎的任务指派，其核心表结构为 ASSGN-RULE，每一个常规交互活动在 ASSGN-RULE 表都对应一条记录。BASED-ON 指明任务指派的基准，它可以取如下 4 个值之一：

（1）DEPT-BASED，基于部门进行任务指派，DEPTID 用于指定执行此活动的部门；

（2）TEAM-BASED，基于团队进行任务指派，TEAM-ID 用于指定执行此活动的团队；

（3）ROLE-BASED，基于角色进行任务指派，ROLE-ID 用于指定执行此活动的角色；

（4）USER-DEFINED，基于自定义的方式进行任务指派，EX-FUNC 指明相应的自定义执行程序。

任务指派的基准确定了可以执行相应任务的群体，具体指派到哪些实际人员还取决于任务指派方法 METHOD，METHOD 可以取如下值。

（1）ALL，任务将分配给由 BASED-ON 指定的群体中的所有人员。

（2）LEASTWORKINGLIST，任务将分配给指定群体中的工作量最少的人员，工作量的多少可以通过 TO-DO-TASK-LIST 的统计数据得到。

（3）FCFA，先来先分配（First Coming First Assigning），即将任务队列中最早创建的任务分配给相应群体中最先提出执行任务请求的个体，任务的创建时间由 DATE-CREATED 指示。

（4）PRIORITY，基于优先数分配，只适合于 ROLE-BASED 任务指派基准。在表 STAFF-IN-ROLE 中有个字段 PRIORITY-NUM 用于指定相应人员的优先数，此方法将把任务分配给优先数最大的人员。

（5）ROUNDROBIN，轮转法，只适合于 ROLE-BASED 任务指派基准。ROUND-ROBIN-TOKEN 为轮转令牌，任务将分配给携有轮转令牌的人员。

# 4 工作流引擎的控制模型

控制模型将机构模型和信息模型有机地结合在一起，它根据其中定义的业务规则对业务过程中的各项业务活动的流转以及任务指派等工作进行控制和协调。控制模型是工作流引擎的控制中心。

## 4.1 应用框架

图 2 给出了使用本文所提出的工作流引擎的应用系统的框架结构。

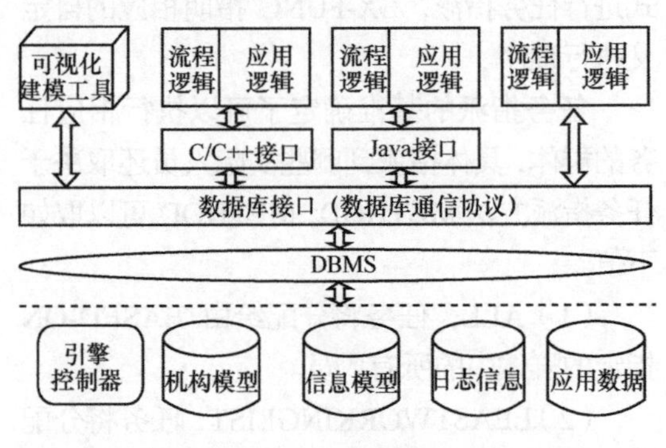


图 2 应用框架

如图 2 所示，“可视化建模工具”即采用一套恰当的图示化的工具来对业务过程进行描述，然后将其转换成如机构模型和信息模型中所述及的关系结构，从而建立起工作流引擎的数据模型。因此，“可视化建模工具”是工作流引擎在构造时的定义中心，而“引擎控制器”则是工作流引擎在运行时的控制中心，它负责工作流引擎在运行时的协调、调度和控制功能。根据具体应用的开发环境的不同，工作流引擎在应用框架中为不同类型的应用提供了不同的接口，例如 C/C++接口、Java 接口以及直接基于数据库通信协议的接口，从而为不同类型的应用与工作流引擎的交互提供了方便。应用框架中的“应用数据”则由具体的应用逻辑自行管理，工作流引擎并不关心这部分的数据格式。

## 4.2 引擎控制器

引擎控制器是工作流引擎在运行时的控制中心，图3给出了引擎控制器的控制结构图。

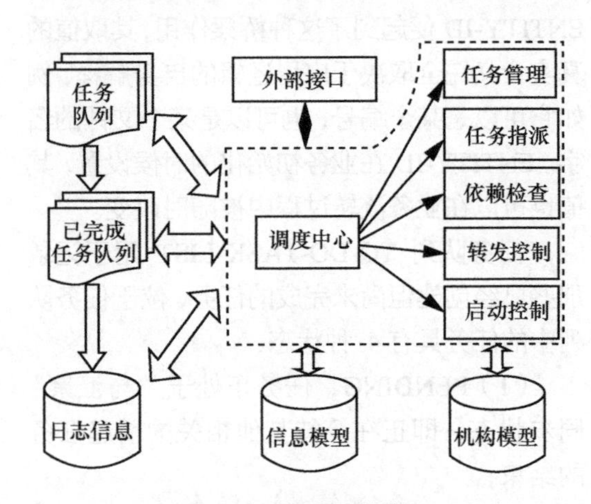


图 3 引擎控制器结构图

（1）调度中心

调度中心接受从外部接口发送过来有关流程控制的请求（如业务初始化、获取任务以及结束任务等），然后根据不同的请求类型调用相应的处理模块完成与本次请求相关的操作并将结果返回. 由于是在 DBMS 内部实现工作流引擎的控制模型，因此有关请求的并发处理等问题完全可以交给数据库管理系统来完成，也不需要诸如请求队列等形式的数据结构。因此，事实上可以将调度中心看成一个多线程的并发服务器，它可以对多个外部请求提供并发服务。对外部请求的处理过程中肯定会涉及对内部数据结构（即工作流引擎的数据模型）中有关数据的读写和更改操作，这些数据的完整性和互斥操作则可以通过 DBMS 提供的各种加锁机制来实现，从而实现了多个外部请求之间的独立性。

（2）任务管理

任务管理主要根据调度中心的指示完成诸如任务创建、任务状态的转换以及相关数据的维护等工作。每次“结束任务”的外部请求将触发调度中心调用“任务管理”为后继活动（如果存在的话）创建新的实例，其状态为“Pending”；同时，其他不同的外部请求也将触发“任务管理”实施任务状态的切换。任务状态转换图如图 4 所示。

（3）任务指派

任务指派处理只是针对常规交互活动，通常情况下，在任务状态由“Pending”切换到“Waiting”过程中完成任务的指派工作，即处于就绪状态的任务在通常情况下都确定了其执行者（FCFA 除外）。任务指派过程首先根据任务指派基准确定可以执行此任务的群体人员，通常情况下这是一个包含多个人员的集合；然后根据任务指派方法确定由这个群体中的哪些个体来执行任务，执行任务的个体标识记录在相应任务记录的 STAFF-ID 字段中。在第 3.4 节中已经对任务指派方法进行了解释，这里有两点需要特别强调。

一是如果任务指派方法是“ALL”，将对当前的任务记录进行拷贝，即保证每一执行任务的个体在 TO-DO-TASK-LIST 中都有一条对应的记录。

二是如果任务指派方法是“FCFA”，事实上在任务指派阶段不作任何工作，即相应任务记录的 STAFF-ID 字段为空。此时任务指派工作自动隐含在获取任务的请求中，即谁先发出获取任务的请求，就自动将此类型的任务分配给谁。

（4）依赖检查

依赖检查指的是活动的前依赖规则的检查，调度中心在将任务切换到就绪状态之前将进行相关的前依赖规则检查，只有满足检查条件的任务才可以进行状态的切换。前文已经描述了前依赖规则在数据模型中的表示方法，这里主要讨论在控制模型中是如何对各种前依赖规则进行处理的。

对于顺序前依赖规则，很显然，从前趋活动流转到当前活动跟其他前趋活动没有关系，PR E-DEPNT-SET 为空集，当前活动的启动没有其他约束条件，相应任务可以立即由“Pending”状态转换到“Waiting”状态。

对于与汇聚前依赖规则，PRE-DEPNT-SET 中指明所有参与与汇聚的其他前趋活动，只有所有相关的前趋活动均到达各自指定的结束状态 DEPNT-ACT-STATUS，当前活动方可启动。

对于或汇聚前依赖规则，PRE-DEPNT-

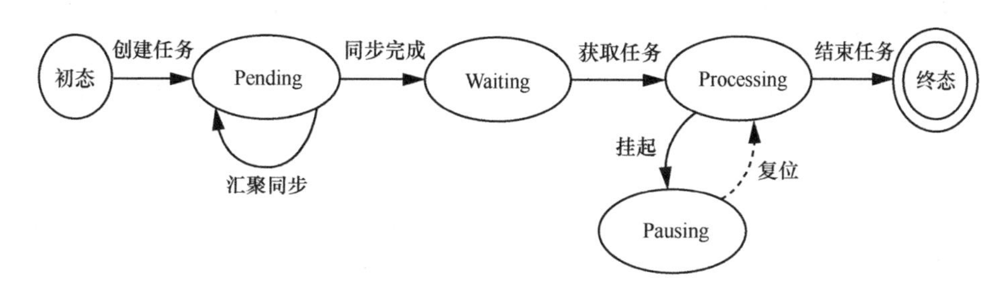


图 4 任务状态转换图

SET 为空集，此规则的检查将涉及 ACTIVITY 表中的 OR-MERGE-FLAG、OR-MERGE-FLAG 的取值可以是所有相关的前趋活动的结束标记之一或者是一个特殊的标记“ANY”。如果 OR-MERGE-FLAG 的值不是“ANY”，则将检查相应前趋活动的结束标记 COMPLETION-FLAG 是否与 OR-MERGE-FLAG 相同，若相同，则启动当前活动，若不相同，则不作任何处理；否则，如果 OR-MERGE-FLAG 的值为“ANY”，则首先结束的前趋活动将启动当前活动，后结束的活动将被丢弃。

对于投票汇聚活动，PRE-DEPNT-SET 同样为空集，当前活动要等到属于同一批次任务数目达到 NUM-VOTES-NEEDED 的要求方可启动。属于同一批次的任务数目可以通过对 TO-DO-TASK-LIST 按照 ACT-ID 和 SERIAL-NO 进行统计得到。

（5）转发控制

当应用发出“结束任务”的外部请求时，该请求将触发调度中心启动“转发控制”。转发控制的主要依据为工作流数据模型中定义的后转发规则，后转发规则定义了当前活动与其后继活动之间的关系。转发控制的处理过程是根据“结束任务”请求中所携带的“任务结束标记”以及相应前趋活动和当前活动的活动标识匹配 ROUTING-RULE 表中的记录，从而得到相应的后继活动列表 NEXT-ACT-ID-LIST；然后由调度中心根据后继活动列表启动“任务管理”为相应的后继活动新建任务。

对于顺序转发以及或分支转发规则，NEXT-ACT-ID-LIST 只包含一个活动；对于与分支转发规则，NEXT-ACT-ID-LIST 中将包含多个活动。

（6）启动控制

启动控制负责常规自动活动所对应的自动执行体的启动并对其活动进行监控。

## 5 应用实例

本文所讨论的基于关系结构的轻量级工作流引擎已经有一个具体的应用对象，即国家商标局的商标注册与管理信息化系统。这是一个大型的信息系统，总共涉及商标局各个业务处大约 18 个商标处理业务，如商标新申请业务、商标变更业务以及商标转让业务等，所有这些业务都属于商标局的关键业务；同时，这也是一个具有海量数据的信息系统，在实施我们的项目之前，商标局已经有一个早期系统存储所有已经注册的商标的信息，总共有约 100 万条商标，超过 10GB 的历史数据，以后随着商标业务的发展以及其他商标业务的电子化，数据量估计还将增加一个数量级。

在所有的商标业务中，大部分业务过程都比较复杂，例如商标新申请业务，其中主要的业务活动就超过 15 个，这些业务活动发生既有顺序关系也有并行关系，大部分都包含往复关系，相互间的依赖关系也比较复杂。通过调查，我们发现现有的工作流产品在流程控制、数据处理以及应用开发上都难以满足系统的需求。这迫使我们自己去设计和实现一个如本文所述的工作流引擎。我们采用 Oracle 公司的 Developer/2000 开发商标业务的应用逻辑，然后在应用逻辑中嵌入对工作流引擎的调用从而实现流程的控制。

目前，该系统已经全部开发完毕并完成测试工作，马上要投入正式运行。实际的应用实例表明，由于我们有一个基于关系结构的工作流引擎的支持，使得我们可以将注意力集中于应用逻辑的开发。基于关系结构的轻量级工作流引擎在关键业务的开发过程中体现出它的价值。

## 参考文献：

[1] HOLLINGSWORTH D. Workflow management coalition: the workflow reference model: WfMC-TC00-1003[S]. [S.l.:s.n.], 1994.

[2] WfMC. Workflow management coalition specification: ter-minology & glossary: WfMC-TC-1011[S]. [S.l.:s.n.], 1996.

[3] LEUNG K R P H,CHUNG J M L. The liaison workflow engine architecture[C]// The 32nd Hawaii International Conference on System Sciences. [S.l.:s.n.], 1999.

[4] TAGG R, et al. Preliminary design of a lightweight workflow server[C]// The 8th Australian Conference on Information Systems. [S.l.:s.n.], 1997.

发表于《计算机学报》2002年第8期

# 聚类/分类中的粒度原理

卜东波，白硕，李国杰

中国科学院计算技术研究所，北京 100080

**摘　要：**从信息粒度的角度来剖析聚类和分类技术，试图使用信息粒度原理的框架来统一聚类和分类。从信息粒度的观点来看，聚类是在一个统一的粒度下进行计算，而分类却是在不同的粒度下进行计算。本文还根据粒度原理设计了一种崭新的分类算法，在大规模中文文本分类的应用实践表明这种分类算法有较强的泛化能力。

**关键词：**信息粒度；聚类；分类；粗集

# Principle of granularity in clustering and classification

Dong-bo BU, Shuo BAI, Guo-jie LI

Institute of Computing Technology, Chinese Academy of Sciences, Beijing 100080

**Abstract:** From the view of information granularity, it is clear that there exists a tight link between clustering and classification. Clustering is a computation under the same granularity, while classification is under different ones. This paper proposes a new classification algorithm based on theory of information granularity, and it shows better generalization ability in experiments.

**Key words:** information granularity, clustering, classification, rough sets

## 1　引言

从表面上看，聚类和分类有着很大的差异——聚类是指无导师的学习，而分类是有导师的学习。更进一步说，聚类的目的是发现样本点之间最本质的“抱团”性质，在选定了表示样本的特征之后，样本点就表示为特征空间中一个点，如果再选定样本点之间的相似性测度函数，那么聚类的结果就应该是确定的。总之，聚类是样本点“抱团”性质的一种客观反映。

**基金项目：**国家自然科学基金资助项目（No.69773008）；国家重点基础研究发展计划（973计划）基金资助项目（No.G1998030510）

而分类在这一点上却大不相同，分类需要一个训练样本集，由领域专家指明哪些样本属于一类，哪些样本属于另一类，但是分类的这种先验知识却常常是纯粹主观的。有些领域专家认为某些样本点应该属于同一类，而另外一些专家从另外的角度出发，可能认为这些样本点应该分属不同的类别，总之，你无法预先限定别人的奇思异想。从特征空间的角度来看，聚类是把那些相似性测度较大（比如欧氏距离最小）的样本点归为一类，而先验知识可能规定距离非常遥远的两个样本点属于同一类别。换句话说，先验知识极有可能和特征以及相似性测度函数不协调。

然而，如果从信息粒度的角度看的话，就会发现聚类和分类有很大的相通之处：聚类操作实际上是在一个统一的粒度下进行计算，而分类操作是在不同的粒度之下进行计算。依据信息粒度原理，我们设计了一种新的分类算法，充分利用了分类是在不同粒度之下进行计算这一特性。实验表明这种分类算法能够有效地提高泛化能力。

## 2 信息粒度原理

粒度（Granularity）本来是一个物理学概念，意指“微粒大小的平均度量”，在这里则被借用做“信息粗细的平均度量”。物理粒度涉及对物理对象的细化划分，而信息粒度则是对信息和知识细化的不同层次的度量[1]。

张钹和张铃提出了信息粒度的概念[2]，并且做出了非常精辟和透彻的论述。之所以提出信息粒度的概念，是因为人工智能和认知科学研究者们观察到人类智能的一个公认特点，那就是在认知和处理现实世界的问题时，常常采用从不同层次观察问题的策略，往往从极不相同的粒度上观察和分析同一问题。

但是，无论是状态空间法，还是问题归约法，都难以把上述现象描述清楚。而信息粒度的概念则有助于把人类的这种能力形式化。

### 2.1 信息粒度的形式化描述

文献[2]使用一个三元组$(X,F,\Gamma)$来描述一个问题，其中，$X$表示问题的论域，也就是我们要考虑的基本元素的集合。并设$F$是属性函数，定义为$F$：$X\rightarrow Y$，$Y$表述基本元素的属性集合。$\Gamma$表示论域的结构，定义为论域中各个基本元素之间的关系。

从一个较“粗”的角度看问题，实际上是对$X$进行简化，把性质相近的元素看成等价的，把它们归入一类，整体作为一个新元素，这样就形成一个粒度较大的论域$[X]$，从而把原问题$(X,F,\Gamma)$转化成新层次上的问题$([X],[Y],[\Gamma])$。

粒度和等价关系有着非常密切的联系。实际上，上面所说的简化过程和商集的概念完全相同。

### 2.2 不同粒度世界的关系

对于一个问题，有时需要同时在粗细不同的粒度世界中进行问题求解，因此有必要研究不同粒度世界之间的关系。

设$R$表示由$X$上一切等价关系所组成的集合，可以如下定义等价关系，也就是粒度的“粗”和“细”。

**定义 1** 设 $R_1,R_2\in R$，如果对于任意的$x,y\in X$，都有 $xR_1y \Rightarrow xR_2y$，那么就称 $R_1$ 比 $R_2$ 细，记为 $R_1\leqslant R_2$。

可以证明如下的定理[2]。

**定理 1** $R$ 在如上定义的“$\leqslant$”关系下形成一个完备半序格。

这是一个非常深刻的定理。它揭示了有关粒度的核心性质，其他性质都以此为基础。有关它的详细证明请参阅文献[2]。

根据这个定理，我们可以得到如下的序列：

$$R_n\leqslant R_{n-1}\leqslant\cdots\leqslant R_1\leqslant R_0$$

直观地看，如上操作得到的序列和一棵 $n$ 层的树是相对应的。设 $T$ 是一棵 $n$ 层的树，所有叶节点构成集合 $X$，那么每一层节点都对应着 $X$ 的一个划分。而聚类操作得到的聚类谱系图恰好也是一棵 $n$ 层树，因此必定存在一个等价关系序列与之对应，这也就是聚类和粒度之所以相通的原因。

## 3 聚类/分类中的粒度原理

### 3.1 聚类中的粒度原理

聚类算法的结果一般都使用聚类谱系图来表示。比如对于图 1 中的 4 个样本点，相应的聚类谱系图如图 2 所示。

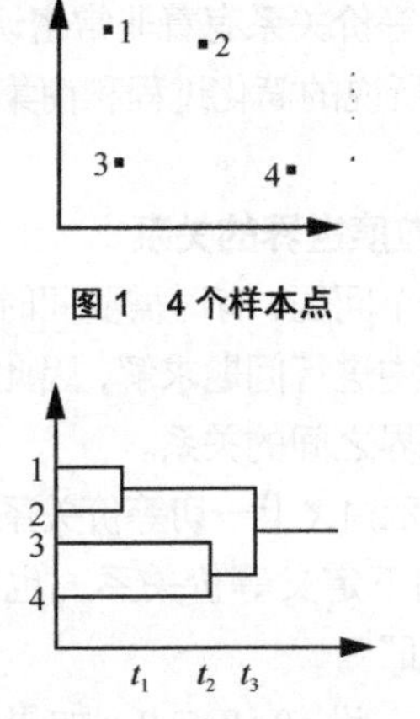


图 1 4 个样本点

图 2 4 个样本点的聚类谱系图

从聚类谱系图中可以看出，如果我们选取的分类阈值 $t$ 足够大的话（$t \geqslant t_3$），那么所有的样本点都被归为一类；如果 $t_2<t<t_3$，那么所有的样本点被分为两类，样本点{3, 4}归为一类，剩余的样本点归为另一类。随着 $t$ 的减小，类数越来越多，直到所有的样本点自成一类。

聚类操作实质上是在样本点之间定义一种等价关系。属于同一类的任意两个样本点被看作等价的，可以认为它们具有相近的性质，在当前的阈值尺度下是没有区别的。一个等价关系就定义了样本点集合的一个划分，它把样本点划分成一些子集，一个子集就对应着聚类形成的一个类。

和由大到小的一系列阈值相对应，会形成由粗到细的一族等价关系。采用较大的阈值时，展现在我们面前的是样本点集比较“粗”的轮廓，一些细枝末节被忽略掉了；而采用较小的阈值时，就能够比较精细地刻画样本点之间一些细微差别。进一步说，这一族粗细不等的等价关系之间又存在一种奇妙的关系，“细”等价关系继承了“粗”等价关系的部分性质，用数学语言来说，这一族等价关系形成一个偏序格结构[2]。

### 3.2 分类中的粒度原理

和聚类试图尽可能忠实地反映样本点之间“抱团”性质的目标不同，分类实际上是一个学习过程。对于给定的一堆样本点，先由领域专家确定分成几类，每一类都包含哪些样本点，分类算法的目标就是探索每一类样本点的规律。不妨从着色的角度来更加直观形象地理解分类的本质：选定特征之后，样本集就被表示成特征空间中一群点，我们可以按照先验知识来给样本点着色，使得同一类样本点有同样的颜色，而不同类的样本点的颜色也不同，研究目标就是摸清楚：红色点有什么规律？绿色点又有什么规则可循？再来一个新的样本点，当投射到同一个特征空间之后，我们判断它应该染成红色、绿色还是其他某种颜色呢？

#### 3.2.1 聚类结果和先验知识的不协调

最理想的先验知识自然是以下的情形：在特征空间中，异类样本点之间有明显的界限，相似性测度较小；而同类样本点则聚集成一团，相似性测度较大。或者从聚类的角度说，先验知识中规定的某一类中的样本点，依照选定的特征空间和相似性测度，也应当聚成一类。然而不幸的是，在绝大多数情况下，都不会达到这种理想境界，常常遇到情况的却是领域专家认为应该归为一类的点，往往在特征空间中距离特别远，或者说相似性测度特别小；而那些被认为分属不同类的样本点，却距离非常近，相似性测度特别大。换言之，聚类结果和先验知识之间往往存在某种不协调性。XOR 问题可以作为这种不协调性的一个典型代表。

假设有如图 3 所示的 4 个样本点 A、B、C、D，选定的特征 X、Y，采用欧氏距离作为相似性测度，这 4 个样本点在特征空间中的位置如图所示。领域专家从某个角度出发认为分两类比较合适，其中 A、C 构成一类，B、D 构成另一类，或者从着色的角度说，A、C 被染成红色，B、D 被染成绿色。然而在选定的特征空间中，A 和 B 离得比较近，和 C 离得反而较远，B 的情况和 A 类似，如果要求对这 4 个样本点聚成两类的话，应该是 A、B 一类，C、D 一类。

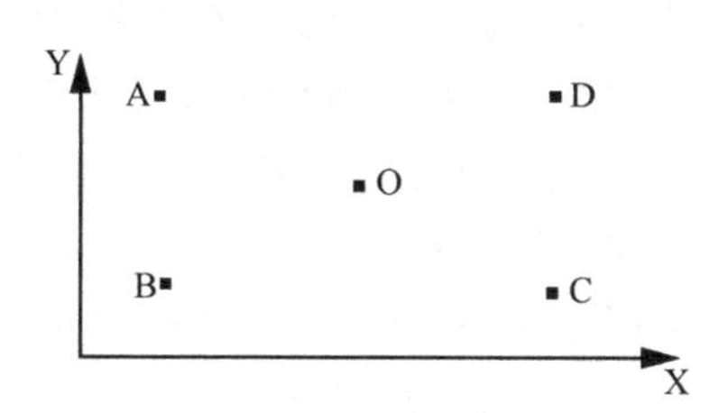


**图 3 4 个样本点在特征空间中的位置**

这种不协调性增加了分类的难度。因为如果同一类的样本点在特征空间中分布得很开，甚至和其他类的样本点混杂在一块的话，我们就很难找到隐藏在表象之后的、之所以将这些样本点如此分类的原因。再比如对于距离函数分类算法，它使用同一类样本点的重心作为此类的代表，新样本点只需和各类的重心比较，归入距离较近的那一类。然而对于上面的例子来说，事情就变得复杂了：A、C 的重心 $O_1$ 和 B、D 的重心 $O_2$ 是重合的，在图 3 中表示为点 O。任何一个新样本点到两个类中心的距离都是相同的，新样本点将无所适从。

造成这种不协调性有两个方面的原因：首先，聚类是一个试图客观忠实反映样本点之间的抱团性质的过程，当选定了特征空间和相似性度量函数之后，那么聚类结果就确定了；其次，分类的先验知识却是由人来主观决定的，对于同一个样本集合，不同人会有不同的、甚至是迥异的分类方案。这种主观和客观的差异是不可避免的，反映在分类过程中，就是聚类结果和先验知识之间的不协调性。

### 3.2.2 从粒度的角度理解不协调性

Rough Sets 理论是 1991 年由波兰数学家 Pawlak 提出的、对不确定知识进行表示的理论。Rough Sets 对于知识表示是一种非常有效的理论，它和粒度是密不可分的，在这里我们借用 Rough Sets 的体系和术语来描述聚类结果和先验知识的不协调性[3-5]。

Rough Sets 理论认为知识就体现在分类上。设 $U$ 是非空有限集合，称为论域。$U$ 中每个元素称为对象，并设有定义在 $U$ 上的等价关系 $R$，商集 $U/R$ 表示根据 $R$ 对 $U$ 中对象的划分：$U/R=\{X_1,X_2,\cdots,X_n\}$，每一个等价类 $X_1$ 称为一个范畴。将二元组$(U,R)$称为一个知识库。从这种知识表示体系出发，Rough Sets 理论定义了一个重要的不确定性度量：粗糙度。

设 $R$ 是论域 $U$ 上的等价关系，$U$ 的任意一个子集 $X$ 都可以由以下两个集合来近似描述，或者说使用现有的知识体系来近似描述：

$$R(X)=\bigcup\{X_i|X_i\subseteq X,X_i\in U/R\}$$

$$R(X)=\bigcup\{X_i|X_i\cap X\neq\varnothing,X_i\in U/R\}$$

$$\text{Boundary}(X,R)=R(X)-R(X)$$

这两个集合分别称为 $X$ 的 $R$ 下近似（R-Lower Approximation）$R$ 上近似（R-Upper Approximation），下近似表示 $X$ 中可以完全使用现有知识表示的对象，上近似则表示所有和

$X$有关的范畴。如果上近似和下近似不同，就说明待表达的集合 $X$ 不能由现有的知识体系精确地反映，就说 $X$ 是粗糙的。而上近似和下近似的差异，即边界 Boundary($X$,$R$)的大小就能够定量地说明$X$在现有的知识体系$R$之下的粗糙程度。

采用 Rough Sets 框架可以很容易地表示聚类和先验知识之间不协调性。从第 3.1 节可以看出，聚类谱系图实际上是定义了一个粒度逐渐变细的等价关系序列，选择一个阈值实际上就是选定一个等价关系 $R$，进而可以得到商集，也就是知识体系 $U/R$。如果由先验知识规定的类 $X$ 能够使用现有的知识体系精确表达，就表示聚类结果和先验知识是协调的；而如果上近似和下近似不相同的话，就说明聚类结果和先验知识是不协调的，这种不协调性的程度可以使用 $X$ 的 $R$ 边界 Boundary($X$,$R$)的大小来定量表示，如图 4 所示。

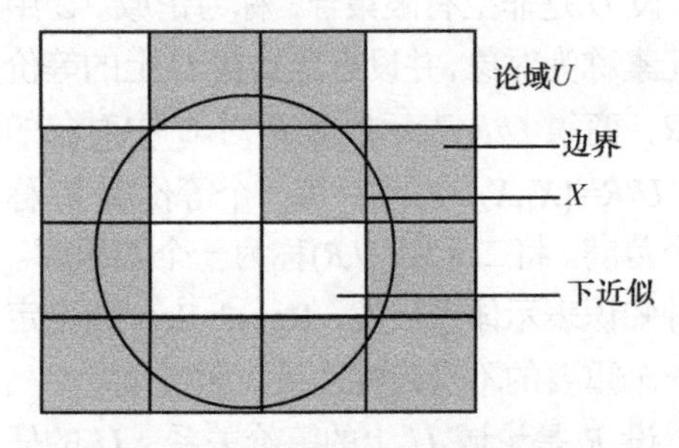


**图 4　使用 $X$ 的 $R$ 边界 Boundary($X$,$R$)定量表示聚类结果和先验知识的不协调性程度示意**

## 3.3　统一在粒度框架下的聚类和分类

有两种策略来消除这种不协调性。一种方法是变换特征空间。虽然在当前的特征空间中不容易区分各个类，但是变换一下特征空间就有可能使得各类之间的区别暴露出来。比如 Vapnik 提出了支持向量机理论（Supported Vector Machine，SVM）[6]，其核心思想就是用简单的线性分类器划分样本空间，对于那些非线性可分的模式，则采用一个核函数将样本空间映射到适当的高维空间，在这个高维空间下是线性可分的；张铃和张钹教授提出的球面投影算法和 SVM 有异曲同工之妙，其基本思想是将 $n$ 维的样本点投影至（$n$+1）维的球面上，使用超平面对其进行划分。然而，在有些应用场合下，不希望引入新的特征，只能继续使用原有的特征空间。我们以下的讨论即是沿着这条思路。

其实，如果都从粒度的角度看的话，就会发现聚类和分类之间存在某些相通之处。仍然就上节中的例子继续讨论，对于由先验知识规定的类$X$，如果在一个粒度比较粗的 $R$ 决定的知识体系下，只能得到比较粗糙的表示，Boundary($X$,$R$)也比较大；而在一个粒度比较细的 $R$ 决定的知识体系下，就能得到比较精细的表示，Boundary($X$,$R$)也比较小。

考虑一种极端的情况，假如我们选择粒度最细的等价关系 $R$，也就是每个对象自成一个等价类，那么在这种 $R$ 决定的知识体系之下，类 $X$ 能够得到最精细的表达，此时的边界 Boundary($X$,$R$)等于 0，换句话说，在这种粒度的表示下粗糙度等于 0。然而粗糙度的降低是有代价的，因为采用如此细的粒度来表达$X$实际上只是对$X$中元素的简单枚举，我们并没有挖掘出构成类$X$元素的任何规律。

上面所说的极端情况给我们以很大的启发：我们的目标是寻找一个知识体系，这种知识体系一方面能够精确地表达先验知识规定的类$X$,同时又能表示构成类$X$诸元素的规律。因此我们放弃统一或者均匀粒度的设想，而采用一种非均匀粒度的知识体系。具体的方案描述如下。

对于类$X$，我们首先选择一个比较粗的粒度 $R_1$，计算$X$的 $R_1$下近似和 $R_1$边界，因为在当前的知识体系之下，$X$的 $R_1$下近似已经能够精细表达，无须进一步的操作，我们只需考虑尚不能精确表达的 $X$ 的 $R_1$ 边界。对于边界中诸元素，我们可以适当进入细节，采用更细的粒度 $R_2$，再计算 $X$ 的 $R_2$ 下近似和 $R_2$ 边界，扣除掉能够由 $R_2$ 精细表达的 $R_2$ 下近似，继续进行这种操作直到边界为 0。采用非均匀粒度表示 $X$ 的详细情况，如图 5 所示。

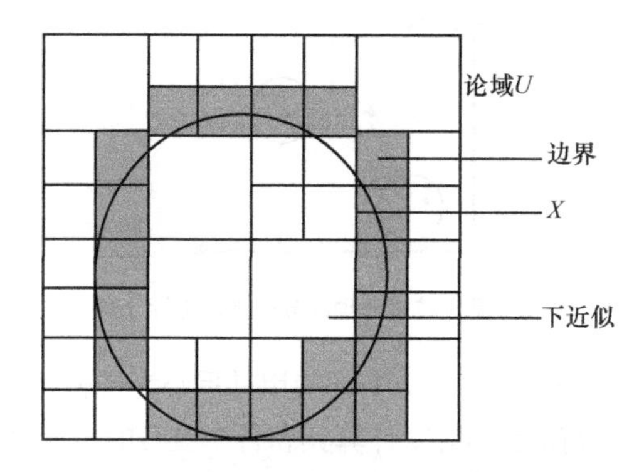


**图 5　采用非均匀粒度表示 $X$ 的详细情况**

采用非均匀粒度方案的结果，就是把类 $X$ 分解成一些子类的并集 $X=X_1\cup X_2\cup \quad \cup X_n$，每一个子类都采用不同的粒度，是在相应粒度之下能够精细表达的最大子集。而连接符号“∪”的数目，恰恰能够定量地反映出当前选择的特征空间和相似性测度函数与先验知识之间的协调程度，如果“∪”的数目很少，就表示当前选择的特征空间和相似性测度函数比较支持这种先验知识，否则就表示其中存在着较大的不协调性。

概括地说，我们首先尽可能地使用粗的粒度来表示，因为如果能用粗粒度的、简单的区别手续体现出类内元素的构成规律，使聚类结果与先验知识协调起来，就没有必要把问题复杂化；而对于那些边边角角中的元素，由于在大粒度世界里不容易看出它们之间的区别，易于造成和其他类元素的混淆，适当地采用较小一些的粒度，才能够更准确地对它们加以区分。

至此，我们可以用一句话来总结聚类和分类：聚类是在一个均匀的、统一的粒度下来描述样本集，而分类是在非均匀粒度下来描述样本集上的先验知识。

图 6 中左图表示聚类情形，在某个确定的阈值处只“砍”一刀，将样本点分成两类；图 6 中右图表示分类情形，在不同的阈值处“砍”多次，使得每一刀“砍”出的分支都是同颜色的样本点。这里的阈值即表示信息粒度。

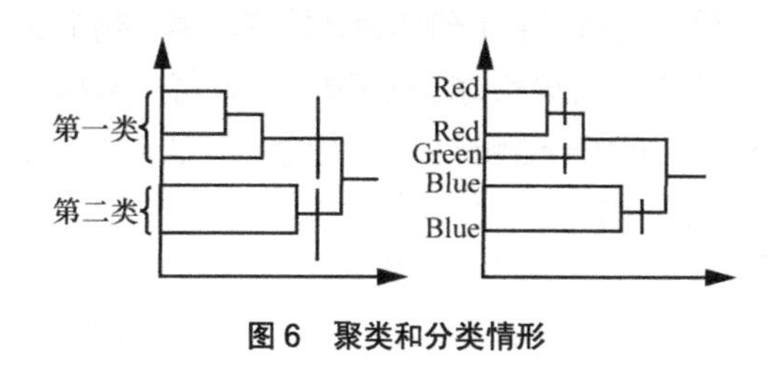


**图 6　聚类和分类情形**

## 4　基于信息粒度原理的分类算法

第 3 节中表示 $X$ 的非均匀粒度方案实际上隐含了一种崭新的分类算法，在这种分类算法中，每一类被拆分成一些小类，每一个小类在特征空间都呈现出很强的“抱团”性质。我们把这种表示方案形式化地表示成如下的算法：Procedure Classify($U$)。

**步骤 1**　在规定的特征空间和相似性度量函数下，对 $U$ 执行聚类操作；得到聚类谱系图 $G$，初始阈值 $T$ 设置为最大阈值。

**步骤 2**　在阈值 $T$ 处切聚类谱系图 $G$，得到一些分支，每一分支都构成一类。

**步骤 3**　逐个考察每个分枝。

如果此分支的每个叶节点都有相同的颜色，也就是属于先验知识规定的同一类，则表示已经能够精确表达，不再讨论此分支；否则，降低阈值 $T$，跳转至步骤 2。

## 5　实验结果

我们在一个具有 2 830 篇中文文本的语料

库上测试上述基于粒度原理的分类算法，并和直接使用类重心作为类代表元的分类算法[7]做了比较。

语料库中的文档都是新闻电讯稿，绝大部分采自新华社，还有200余篇采自中国新闻社和人民日报。所有的新闻稿都由领域专家事先进行分类，按照中图分类法分成政治、经济、军事等共38类。我们依照文献[7]的方法选择训练集和测试集，即将这些分好类的语料平均分成10份，选择其中1份作为开放测试集，剩余的9份作为训练集和封闭测试集。这样每一份都依次轮流作为开放测试集，运行分类算法，共执行10次分类操作，计算其平均值。附录中列出我们对分类正确率的测试数据。

从附录中可以看出，采用基于粒度原理的分类算法，在封闭测试中其分类正确率可以大大提高，达到99%~100%；而在开放测试中，除了对军事类、心理类、电子类和服务类的分类正确率低于距离函数法之外，其他34类的分类正确率都有不同程度的提高。因此，采用基于粒度原理的分类算法能够有效地提高分类正确率。

例如图7中所示的3篇训练文档，其中文档1和文档2同属于经济类，文档3属于采矿类。文档1主要讲述有关金融方面的知识，文档2是有关银行方面的文章，新的待测试文档也属于采矿类，使用“*”来表示。如果采用距离函数法，即使用1和2的重心O来代表经济类，那么就会发现待测试文档距离重心O比文档3更近一些，因此会错分至经济类；而在基于粒度原理的分类算法中，经济类被分为两个子类，文档1和文档2各成一个子类，重新计算待测试文档到各个子类中心的距离，就会把它正确地归入采矿类中，避免了上述的错误分类，如图8所示。

以上采用的基于粒度原理的分类算法还有改进的余地。在上述算法中，一个类被分解成若干子类，每个子类使用其重心为代表元。对于待测试样本，计算其和各个重心的夹角余弦，并归入夹角余弦最大的类中。然而这样简单的处理并没有考虑到子类内样本数目的多少，比如：一个经济子类中拥有100个样本，而另一个军事子类中只有1个样本，不考虑子类中的样本数，只是简单地使用重心作代表，会使信息挖掘得不够充分。这也是军事等4类的分类正确率不如距离函数法的原因之所在。

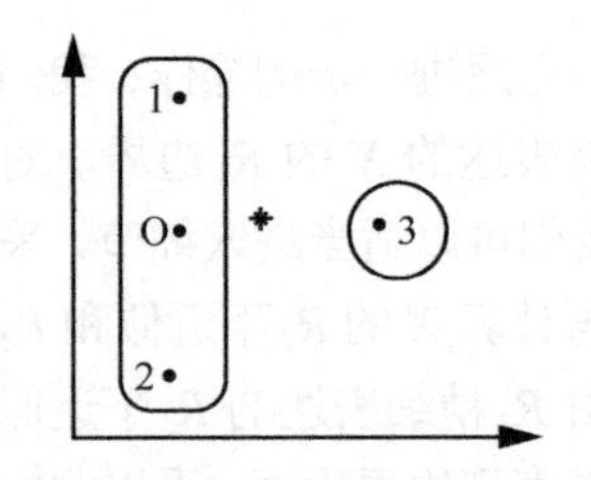

图7　采用距离函数法进行分类

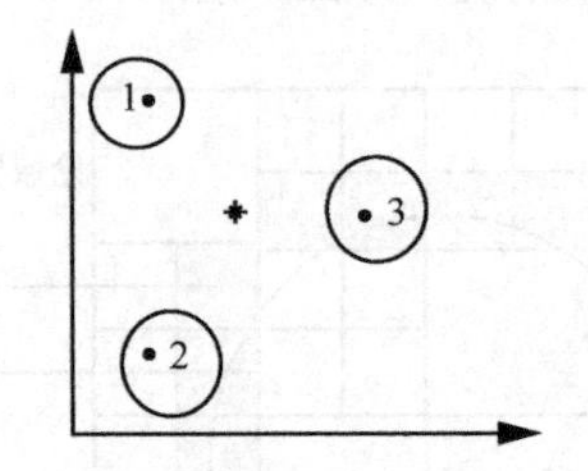

图8　基于粒度原理的分类算法进行分类

## 6　结束语

从表面上看，聚类和分类有很大的区别。然而从信息粒度的角度，我们能够更清楚地认识聚类和分类之间的联系。聚类是在一个统一的粒度之下看问题，而分类是在不同的粒度层次之下进行计算。根据这种认识，我们提出了一种新的分类算法，它把先验知识中的一个类拆分成不同粒度之下的一些子类，每个子类在各自的粒度之下显示出更加明显的“抱团”性质，采用不同的粒度表示有助于消除分类先验知识和特征选取之间的不协调性。实验结果表明了这种算法能够有效地提高泛化能力。

## 附录

| 类别 | 类内文档数 | 距离函数法 | | 基于粒度原理的分类法 | |
|---|---|---|---|---|---|
| | | 封闭测试 | 开放测试 | 封闭测试 | 开放测试 |
| 经济类 | 135 | 0.76 | 0.67 | 1.00 | 0.81 |
| 政治类 | 510 | 0.91 | 0.90 | 1.00 | 0.91 |
| 法律类 | 93 | 0.63 | 0.63 | 1.00 | 0.77 |
| 军事类 | 117 | 0.83 | 0.78 | 1.00 | 0.76 |
| 语言类 | 47 | 0.80 | 0.77 | 1.00 | 0.79 |
| 文学类 | 41 | 0.92 | 0.88 | 1.00 | 0.93 |
| 艺术类 | 104 | 0.76 | 0.71 | 1.00 | 0.80 |
| 文化文明 | 77 | 0.78 | 0.71 | 1.00 | 0.74 |
| 教育科技 | 104 | 0.70 | 0.67 | 1.00 | 0.88 |
| 体育类 | 218 | 0.89 | 0.88 | 1.00 | 0.93 |
| 心理类 | 19 | 0.95 | 0.89 | 1.00 | 0.79 |
| 历史类 | 50 | 0.84 | 0.76 | 1.00 | 0.80 |
| 数学类 | 16 | 0.99 | 0.94 | 0.99 | 0.94 |
| 物理学 | 48 | 0.83 | 0.88 | 1.00 | 0.81 |
| 化学类 | 12 | 0.73 | 0.50 | 1.00 | 0.83 |
| 空间天文 | 44 | 0.95 | 0.93 | 1.00 | 0.95 |
| 地理地质类 | 64 | 0.78 | 0.77 | 0.99 | 0.94 |
| 生物类 | 36 | 0.89 | 0.86 | 1.00 | 0.89 |
| 哲学类 | 62 | 0.66 | 0.61 | 1.00 | 0.77 |
| 医学类 | 72 | 0.80 | 0.75 | 1.00 | 0.82 |
| 服务类 | 58 | 0.87 | 0.81 | 1.00 | 0.79 |
| 农牧业 | 109 | 0.85 | 0.82 | 0.99 | 0.85 |
| 能源核电 | 39 | 0.84 | 0.82 | 1.00 | 0.90 |
| 电子类 | 61 | 0.81 | 0.77 | 1.00 | 0.79 |
| 通信类 | 38 | 0.86 | 0.82 | 1.00 | 0.87 |
| 自动化 | 11 | 1.00 | 1.00 | 1.00 | 1.00 |
| 计算机类 | 74 | 0.79 | 0.72 | 0.99 | 0.81 |
| 轻工业 | 35 | 0.85 | 0.54 | 1.00 | 0.60 |
| 化工类 | 34 | 0.99 | 0.88 | 1.00 | 0.91 |
| 石油油气 | 41 | 0.95 | 0.90 | 1.00 | 0.93 |
| 采矿矿山 | 45 | 0.90 | 0.78 | 1.00 | 0.82 |
| 冶金 | 29 | 0.94 | 0.79 | 1.00 | 0.90 |
| 机械制造 | 36 | 0.82 | 0.72 | 1.00 | 0.72 |
| 建筑建设 | 58 | 0.81 | 0.76 | 1.00 | 0.84 |
| 水利 | 64 | 0.93 | 0.91 | 1.00 | 0.94 |
| 环保污染 | 98 | 0.71 | 0.63 | 0.99 | 0.79 |
| 交通运输 | 74 | 0.83 | 0.80 | 1.00 | 0.85 |
| 航空航天 | 49 | 0.92 | 0.82 | 1.00 | 0.90 |

## 参考文献：


[1] SHAO J. Information granularity computing based on rough sets [D]. Beijing: Institute of Automatics, Chinese Academy of Sciences, 2000.

[2] ZHANG B, ZHANG L. Theory of problem solving and its application[M]. Beijing: Tsinghua University Press, 1990.

[3] MIAO D Q. Research on application of rough sets in machine learning[D]. Beijing: Institute of Automatics, Chinese Academy of Sciences, 1997.

[4] WANG J, et al. Review on theory and application of rough sets[J]. Pattern Recognition and Artificial Intelligence, 1996, 9(4): 292-299.

[5] WANG J, et al. Data condense based on theory of rough sets[J]. Chinese Journal of Computers, 1998, 21(5): 393-400.

[6] VAPNIK V N. The essence of statistical learning[M]. Beijing: Tsinghua University Press, 2000.

[7] HUANG X J. Retrieval, classification and summarization of large-scale Chinese text [D]. Shanghai: Fudan University, 1998.

发表于 *Communications of the ACM*, 2011 年第 10 期

# Computing for the masses

Zhi-wei XU, Guo-jie LI

Institute of Computing Technology, Chinese Academy of Sciences

The fields of computer science and engineering have witnessed amazing progress over the last 60 years. As we journey through the second decade of the 21st century, however, it becomes increasingly clear that our profession faces some serious challenges. We can no longer solely rely on incremental and inertial advances. Fundamental opportunities and alternative paths must be examined.

In 2007, the Chinese Academy of Sciences (CAS) sponsored a two-year study on the challenges, requirements, and potential roadmaps for information technology advances into year 2050[1]. We present a perspective on a key finding of this study: a new paradigm, named computing for the masses, is needed to cope with the challenges facing IT in the coming decades.

Computing for the masses is much more than offering a cheap personal computer and Internet connection to the low-income population. It means providing essential computing value for all people, tailored to their individual needs. It demands paradigm-shifting research and discipline rejuvenation in computer science, to create augmented value (*V*), affordability (*A*) and sustainability (*S*) through ternary computing (*T*). In other words, computing for the masses is VAST computing.

The CAS study focuses on China's needs. However, the issues investigated are of interest to the worldwide computing community. For instance, when considering the drivers of future computing profession, it is critical not to underestimate the requirements and demands from the new generations of digital native population. As of July 2010, 59% of China's 420 million Internet users are between the ages of 6–29 years old. The time frame of 2010–2050 is not too distant a future for them. These digital natives could drive a ten-fold expansion of IT use.

## 1 Challenges in the coming decades

The first challenge to address is the sobering fact that IT market growth appears to have reached a point of stagnation. The IT market size is measured by the total expenditure on computer and network hardware, software, and services. According to the Organization for Economic Co-operation and Development (OECD)[2], the worldwide IT spending in 2008 grew only 4.49% to $1 540 billion. The picture is even bleaker if we look at the market's long-term history, as illustrated in Table 1. The Compound Annual Growth Rate (CAGR) of IT spending declined from double digits before 1980 to low single digits today. These numbers are based on nominal U.S. dollar values. Taking inflation into account, the IT market has seen barely any growth.

Table 1. IT market growth declines to single digit (in compound annual growth rate)

| | | | | |
|---|---|---|---|---|
| World | Before 1980 | 1980-2000 | 2000-2008 | 2010-2030 |
| | double digits | near 10% | 5.5% | 1%~3%? |
| China | 1981-1990 | 1990-2000 | 2000-2008 | 2010-2030 |
| | 30% | 44% | 17% | 5%~10%? |

Compiled by the authors based on data from IDC, OECD and CCID. The China numbers are calculated based on data from China Center of Computer and micro-electronics Industry Development (CCID). The worldwide numbers are calculated based on data from IDC (before 1980 and 1980-2000) and OECD(2000-2008). See also calculations by Shane Greenstein of Northwestern University ("Are the Glory Days" Long Gone for IT?" New York Times, Ang. 8, 2009.)

China is a relative new market for computing. Its IT market has shown higher growth rates than the worldwide market in the past three decades. Still, we do see signs of slowing down. By estimation from International Data Corporation (IDC), China's IT market growth in 2008–2013 would drop to an annual rate of only 10.8%, in a context of 3.3% worldwide.

Realistic questions must be asked. Will the IT market both in China and worldwide, shrink (in real terms) between 2010–2030? Will it shrink further in the 2030–2050 timeframe? If IT becomes a shrinking sector, what will be the impact to the IT workforce, to computer science education, and to computing research?

The second challenge is that inertial and incremental technology progress faces limitations. The International Technology Roadmap for Semiconductors (ITRS) is the authoritative report on the 15-year assessment of future technology requirements for the semiconductor industry. Its 2009 edition states: "ITRS is entering a new era as the industry begins to address the theoretical limits of CMOS scaling".

The current Internet architecture and protocols have inherent limitations in scalability, manageability, security, and quality of service. A number of "clean slate" future Internet research programs have started up, such as the NSF Future Internet Architecture (FIA) program, to explore radically new approaches. Supercomputers face issues such as power and system complexity. Zettaflops supercomputers cannot be built without revolutionary devices and architecture.

Figure 1 depicts parameters of representative computer systems built by the Institute of Computing Technology, Chinese Academy of Sciences (ICT-CAS) over the past 50 years. They range from the vacuum tube computer (Model 103) to the latest parallel computer Dawning Nebulae. The figure reveals two critical concerns: power and system software complexity. For 45 years (before 2000), power needs per system never exceeded 100 KW. But in the last 18 years, power requirements continuously grew, reaching 3 MW in 2010. The system software complexity also grew rapidly in this time. If these trends continue, we may have an exaflops system by 2020, but the power requirement will exceed 100 MW, and the system software will grow to over 100 million lines of source code. Can we overcome industrial inertia and fundamentally address power, systems complex-

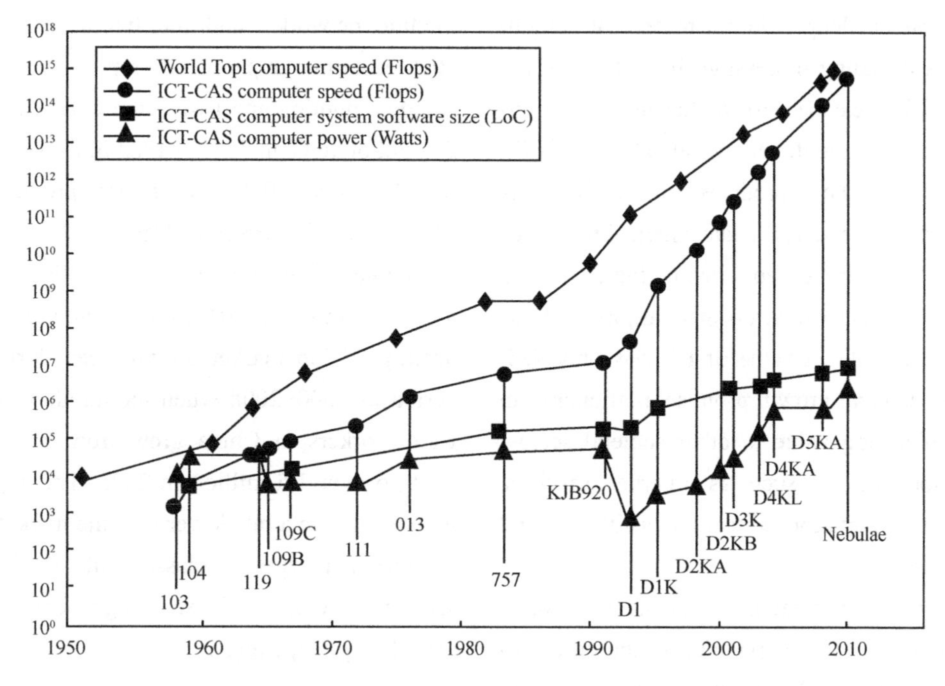


Figure 1. Speed, power, system software complexity trends of ICT-CAS computers in the past 50 years, compare to the world frontier speed

ity, concurrency, cost, and reliability issues to reach realistic zettascale ($10^{21}$ Flops) systems?

Computer science as an academic discipline also faces challenges. For the past five decades, computing has always been a highly regarded profession in China. However, we now see a disturbing phenomenon: Parents of university freshmen are advising their children to study business or law to make a good living. Moreover, parents urge students interested in pursuing a Ph.D. to choose a major in mathematics or physics. In China's Ph.D.- granting universities, computer science as an elected major has dropped from the top spot to out of the top five[3].

Our study reveals a primary reason for this trend is that young people today see computer science as a tool. That is, computing is a universal tool, thus offers good employment. However, computing is regarded as just a tool in many professions, not a primary value driver; computing requires long working hours doing boring tasks such as documentation, programming, and testing; and computing is a rapidly changing field, quickly turning skills learned in college obsolete.

In contrast, a medical school education provides life-long benefits. Medicine is not a boring tool, but a rewarding profession providing essential value to society. The coming generations of Chinese youth are manifestly more self-aware and socially aware, caring about environment and community values. How do we convince perspective students that computing is an academically beautiful, intellectually stimulating, and socially rewarding profession?

These challenges do not mean that computing is a declining profession. In fact, its future potential is great because of the need to continually increase the direct and added value of IT, especially for those products and services targeting mass users. As a latecomer, China's IT use lags by several decades behind the developed countries. Its per-capita IT spending in 2008, as a measure of direct value of IT, was only $85. Still, it has great growth promise, particularly in terms of adding value to other societal sectors throughout China, such as Internet services, energy, transportation, education, health care and entertainment.

China's urbanization in the next 30 years will move 1% of its population to the cities annually. In the next decade alone, China needs to build 80 million new homes to house the expanding urban population. Over 200 million smart meters will be installed in urban homes to support a smart grid and to save energy. More than 10 million cars were sold in China in 2009. All these modern automobiles, homes, smart meters need increasing IT use to add value.

China's Internet services sector ($15 billion revenue in 2009) serves hundreds of millions of users but is not counted in the IT market. It continues to grow on a double-digit scale. Taobao, China's top online shopping Web site, predicts its number of page views per day will grow 80% annually for the next five years, to reach 33 billion page views per day. The Internet services sector is being augmented by cloud computing. China education television is using cloud computing to create a China education open mall, to provide lifelong education to 800 million people via telecommunication, digital television, two-way satellite networks, and the Internet. The IDC 2009 economic impact study report predicts that cloud computing and associated clients will add $80 billion to China's IT market in 2014.

The same IDC report also predicts that China's IT job market will grow 7.2% annually for the next five years, from 4.5 million in 2008 to 6.4 million in 2013, thus maintaining the healthy IT job market growth seen during the period of 2000–2008 when the number of software workers in China grew from 30 000 in 2000 to over 1.5 million in 2008. The expected growth may be a bit slower in some markets, but IT areas that target the masses—like the Internet services sector—will see high double-digit growth in job openings.

## 2 Call for a new paradigm

The challenges and opportunities we have noted in this article suggest that computing must transform itself, yet again, from the social perception of a high-tech tools discipline to the manifestation of its authentic core of providing essential informational value to all people; that is, computing for the masses.

Computing for the masses should include the following four specific features:

• Value-augmenting mass adoption: By 2050, China's IT users should cover 80% of the population, with the per capita IT expenditure increasing 13 times over.

• Affordable computing: Each mass user's total cost of ownership of IT use should go down significantly, when experiencing the same or increasing IT value.

• Sustainable computing: The order of magnitude growth in IT use should not see the order of magnitude growths in energy consumption and emission.

• Ternary computing: These features can only be achieved with transformative innovations. A major source of opportunities for such innovations is the coupling and interaction of the human society, the cyberspace, and the physical world.

Value-augmenting mass adoption is the overarching goal of computing for the masses. Affordability and sustainability are two important constraints for the 21st century. Ternary computing represents the transformative innovations needed to achieve the goal under the two constraints, utilizing trends in computer science research. Let's look at these four features in more detail.

## 3 Value augmenting mass adoption

An obvious goal of computing for the masses is to expand the user base to 80% of the population. For China, this translates to 1.2 billion users by 2050. To achieve this goal, computer use should go far beyond institutions such as companies, governments, or science labs. It should target the masses as personal or community users. This implies that future computing work-loads should change significantly. According to IDC's data, the worldwide servers market reached its peak of $65.5 billion in 1997, where over 95% ran institutional workloads, but only 4.3% ran personal workloads. Computing for the masses advocates that by 2050, personal workloads may expand many times to use over 50% of servers worldwide.

Furthermore, computing should offer value services personalized to the masses' individual requirements, thus enlarge the per-capita IT value. We are far from really understanding what IT value is for the masses. IT expenditure only partially reflects IT value. IT provides added value to other sectors which is not counted as IT expenditure. IT also provides personal and societal value that is not measured in economic numbers. The French economist Yann Moulier Boutang recently likened IT to bees, where the added value (pollination) is much larger than the direct value (honey).

Studies in other fields could give us some hints on defining IT value. Lazo, et al[4]. surveyed U.S. households to measure how much dollar value households place on weather forecasts, to show that the total value of weather services ($31.5 billion) is much larger than their total cost ($5.1 billion). Nordhaus[5] used Lumen-Hour to measure the value of lighting, to show that advances in lighting technology improved value/price by five orders of magnitude in 1800–2000. The CAS study uses the annual IT expenditure per user as an approximation of IT value. Five classes of IT value are defined:

The IT poverty line measures the minimal value that should be provided to any citizen. Today such value manifests in an inexpensive personal computer enabling basic applications and Internet access. The commodity value, in addition, offers commodity IT products and services targeted to a wide range of users. The ubiquity value adds mobile or "anywhere" IT services to the commodity value. The expertise value provides extra IT value for a specific field of expertise, such as an animation system for a cartoon artist. In 2008, we estimate the IT poverty line value in China to be $150, the commodity value to be $300–500, the ubiquity value to be $500–1 000, and the expertise value to be $1 000–10 000, respectively. The personalized value refers to the user-centric situation, where IT hardware, software, and services are customized to an individual user's needs. For example, the Loongson (Godson) CPU research team at the authors' institute uses a highly tuned computing environment in designing microprocessor chips. The EDA software providers and the downstream manufacturer provide onsite, customer specific services. The team's annual IT spending is over $20 000 per user.

Table 2 contrasts two growth paths for China, both extending IT users to 1.2 billion people from 270 million in 2008. The poverty-line growth scenario offers the 930 million new users only poverty line IT value. This is a common conception for reaching the masses, but an unlikely scenario. It ignores the fact that by 2050, the majority of China's population will enter the middle class and most of the Chinese population will be digital natives. They will demand more than the IT poverty line value. The value-augmenting scenario makes two assumptions: by 2050, China's per-capita annual IT spending should approach that of the U.S. in 2000, which was about $1 400; and the digital divide will not worsen; that is, the user distribution among the value classes stays the same.

The value-augmenting projection is not overly optimistic. Even when the projection is achieved, China's information welfare in 2050 will still be 50 years behind that of developed countries. Another fundamental support lies in China's education effort for the masses. In 2008, seven million students graduated from colleges. They all took one-semester to one-year's worth of computer literacy courses. This trend is likely to continue for decades. The implication is that by 2050, China will have educated over 300 million new computing literate college graduates. They will require at least ubiquity value in IT use.

## 4 Affordable computing

Computing for the masses is affordable computing for everyone's value creation and

Table 2. Two scenarios of IT growth in China 2008-2050

| Historic data and projections | IT market ($ trillion) | IT market (CAGR) | IT users (million) | IT spending per capita |
|---|---|---|---|---|
| 2000 (Actual data) | 0.026 | 25.0% | 22.5 | s21 |
| 2008 (Actual data) | 0.11 | 12.7% | 270 | S85 |
| 2050 (Poverty Line growth) | 0.25 | 2.0% | 1.200 | S190 |
| 2050 (Value-augmenting growth) | 2.0 | 7.1% | 1.200 | s1.321 |

CAGR: Compound Annual Growth Rate

consumption. For instance, future high school students in China may learn protein folding in a biology course based on computational thinking, by accessing petaflops computer simulation service for a week, with a lab cost of $10 per student. However, four barriers to such value affordability exist today. They are:

- Cognition barrier. Computing must offer readily cognizable value, for example, a learning experience in protein folding, not just a petaflops tool. Showing value instead of ware to the society is often difficult. Computing may also be associated with negative value. A 2009 study by Beijing Women Association showed that 17% of parents strongly oppose children's access to the Internet, and 66% allow supervised access. The top concerns are Internet games (cited by 45% of parents) and online porn (40% of parents). How to make the cognizable value of computing outweigh its negative effect is a challenge.
- Cost barrier. Petaflops capability can be rented via cloud computing services today, but the price must be reduced by six orders of magnitude.
- Control barrier. A user's computing activities today are often tied to a few platforms or vendors. User creativity is hindered by platform or vendor control. Switching a computing provider or platform is much more difficult than switching a TV channel.
- Usability barrier. Computing needs to provide value-level user interface, and hide low-level details such as low-level coding, deployment, configuration, maintenance, monitoring, debugging, optimization, and adjustments to technical changes.

Affordable computing implies that we should aim for reducing the total cost of ownership for the masses, including purchasing cost, learning cost, use cost, and platform switching cost. An affordable computing product or service not only comes with a low price, but also provides readily identifiable value, freedom from platform control, and ease of use.

## 5 Sustainable computing

An important constraint for computing in the 21st century is sustainability requirements. From 1980 to 2008, China's energy consumption per dollar GDP decreased 208%. A more desirable scenario is to maintain a reasonable GDP growth while achieving zero growth in both fossil fuel consumption and $CO_2$ emission. A projection based on this scenario is illustrated in Figure 2, utilizing data from a CAS energy roadmap study[6]. China's per-capita IT expenditure will reach $1 300 by 2050, an increase 13 times over the 2010 value. But growths in fossil fuel consumption and $CO_2$ emission will significantly slow down, so the per-capita numbers in 2050 will return to the 2010 levels.

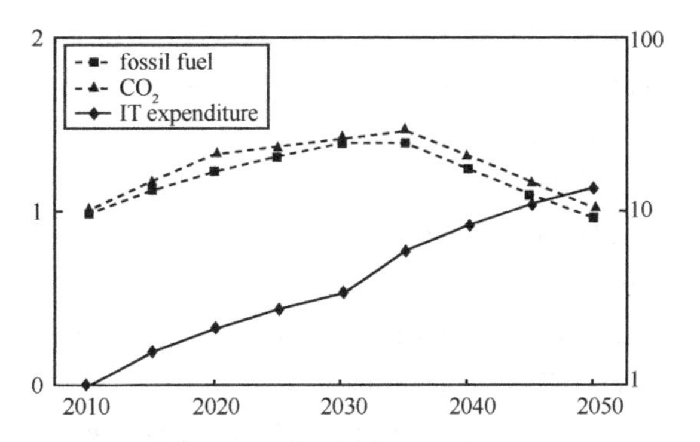


Figure 2. Predictions of China's fossil fuel consumption and $CO_2$ emission (left axis), compared to IT expenditure (right axis). All are per-capita data normalized to year 2010 velues

Sustainability is a difficult constraint that computing for the masses must address. Koomey[7] calculated that electricity use for datacenters doubled worldwide from 2000 to 2005, and the Asia Pacific region grew faster than the worldwide annual rate of 16.7%. IDC predicted that from 1998 to 2012, the number of servers installed worldwide will grow at an average annual rate of 11.6% to reach 42 million servers, with power and cooling costs reaching $40 billion[8]. In 2008, China's telecommunication carriers consumed 24 billion KWH of electricity, growing 14% annually. Such growth trends of IT energy consumption are contrary to the objectives of Figure 2, and cannot be sustained. When computing users in China grow 3.4 times to 1.2 billion in 2050 with intensified IT use, current practices will require 150 billion KWH just to power China's datacenters.

In the past 40 years, energy and environment impact for computing research and IT use in China was not a main issue, because the IT user base was only a small part of the population, and the IT sector is more efficient and clean than the rest of the economy. In designing future IT systems for the masses, saving energy, reducing pollution, and using recyclable material will become hard design objectives, as important as functionality, cost, and speed. Innovations are needed to achieve these objectives while maintain healthy growth of the IT market. Utilizing the diversity factor in the workloads of the mass users offers an opportunity.

Computing can also help sustainability in other sectors that involve mass users. IDC predicts that technologies such as smart meters, energy management systems for buildings, intelligent building design, and teleworking, all involving IT and affecting mass users, can help reduce China's $CO_2$ emission by 532 million tons in 2020[8].

## 6 Ternary computing

Computing for the masses is not just a lofty ideal to conquer the digital divide. It is based on and required by an economical and social trend in China and in the world: people are increasingly living in an intermingling ternary universe of the physical world, the human society, and the cyberspace. Computing is becoming a common fabric of the ternary universe that relates to value creation through information transformation[9-11]. People expect computing services and computational thinking to enrich their physical and societal lives, through IT devices, tangible interfaces, and intangible interfaces.

Thus, computing for the masses calls for ternary computing—transformative innovations in computing that utilize the ternary universe trend. We focus here on five requirements pertinent to ternary computing for the masses, ranging from computer science foundation, ease of personalized access, efficiency in use, effectiveness in value creation and consumption, and ways to measure and regulate. From the academic angle, these form the five pillars of ternary computing for the masses.

**Pillar 1: Computer science for the ternary universe.** When our traditional computer science was established, we assumed a man-machine symbiosis system consisting of a human user interacting with a computer[12]. In the 21st century, the masses will live in a ternary universe of the cyberspace, the human society, and the physical world, clustered into various value communities. This calls for an augmented computer science that provides a foundation to investigate the computational processes and phenomena in the ternary universe. The main target of computer science study will be the Net, that is, the common computational fabric of the ternary universe, which is a Net of people, bits, and things. Here, we suggest five research areas:

- Algorithm networks. We need to enhance traditional computability and algorithmic theories to account for time and space bounds of networks of interacting algorithms. New complexity metrics, such as energy complexity and effort complexity, are required to measure and study energy consumption and human labor needs.
- Ternary systems modularity. We need to discover modularity rules in ternary computing systems, similar to the Liskov substitution principle[13] in object-oriented systems. Such rules involve human-cyber-physical resources and should enable seamless substitution of a Net service with a better one.
- Fundamental impossibility. We need more impossibility results like Brewer's CAP theorem (the impossibility of simultaneously satisfying Consistency, Availability, and Partition tolerance)[14] that rigorously relate basic properties of Net services. We especially need results that bridge qualitative issues (for example, privacy and safety) to quantitative issues (for example, scalability and energy saving), that are important in the ternary universe.
- Emergence and locality. We need to positively embrace complex systems, by learning to utilize their emergent properties as an advantage, not a hurdle. An example is to map various network phenomena to new locality principles[15-16] to enable the design of better Net services and to facilitate the emergence of desired ternary communities.
- Computing in Nature. We need to understand Les Valiant's evolution problem[17]. How did Nature "compute" the Homo sapiens so efficiently, generating a genome with $3\times10^9$ base pairs in only $3\times10^9$ years? The answer could help us learn how Nature computes. The principles learned can be used in building good ternary systems.

**Pillar 2: A universal compute account for everyone.** Each of the 1.2 billion users in China will have a tetherless, seamless, life-time Universal Compute Account (UCA) to access the Net. The UCA is not tied to any device, location, network, resource, or vendor. A user will "log on" to his or her UCA. The UCA is not merely a user's identifier, but a personal information environment, a uniform grip to the local and Net resources that the user is entitled to use. In effect, it offers a user his personal Net, where resources are used on demand and the services are charged by usage. The UCA could also behave as an entry to a personal server, enabling the user to contribute and share. Over one billion such UCAs can generate great community value, leading to the $2^W$ network effect[18].

For computing to reach the masses, IT must offer some fundamental stability and constancy. The UCA is such a fundamental invariant as it could help deliver ever occurring new values while making technical idiosyncrasies and upgrades invisible. It provides a stable focal point for continued innovation and value delivery, improvement in ease of use, and personality accumulation. We have seen similar precedents, but targeting resources instead of users, in URI and RESTful services[19].

**Pillar 3: Lean system platforms for the masses.** Once a user logs onto the Net, he or she should see a lean system, where most resources are used for generating application value. We do not yet have good metrics for measuring applications value delivered per watt. For datacenters, we can define an approximation called the Performance-Energy-Efficiency (PEE) measure, where the percentages of energy spent on application speed is used as a value indicator: as shown in Figure 3.

$$\text{PEE} = \text{Efficiency} \times \text{Utilization}/\text{PUE} = \frac{\text{App speed}}{\text{Peak speed}} \times \frac{\text{Time for App}}{\text{Total time}} \Big/ \frac{\text{Total power}}{\text{IT power}}$$

Figure 3. PEE

The Power Usage Effectiveness (PUE) is defined as the amount of total power entering a datacenter divided by the power consumed by the IT hardware (servers, storage, networks, among others). Assume a total life time of four years for datacenter IT hardware. Utilization is defined as the percentage of this total time when the IT hardware is busy running applications, instead of running system tasks or staying idle. Efficiency is defined as the application speed achieved divided by the peak speed of the IT hardware, when the hardware is running applications.

Data generated from the CAS field studies are summarized in Table 3. The most efficient scenario refers to an efficient supercomputing center running optimized Linpack-like applications. For every watt spent, 0.4 watts are used for applications. The efficiencies are much lower for typical usage scenarios. The potential for improvement could grow two to three orders of magnitude.

We are far from fully understanding existing platform inefficiencies. We need new micro and macro benchmarks and metrics that characterize workloads for the masses, to guide the design of parallel, distributed, and decentralized computer architectures, especially the cooperation and divide of labor among application software, compiler, execution model, and hardware. We lack a design formula for the servers and datacenters that can relate value, resource, and energy in a precise way, as Hennessey and Patterson's formula[20] did for the CPU microarchitecture designs by precisely relating performance to the number of instructions, cycle-per- instruction and clock frequency.

There exist much space for enhancing efficiency in the entire Net, including networks, client devices, and sensors. New internetwork architectures supporting network virtualization must be investigated. Client machines should

Table 3. PEE values for different datacenter usage scenarios

| Usage Scenario | PEE | Efficiency | Utilization | PUE |
| --- | --- | --- | --- | --- |
| Most Efficient | 40% | 0.75 | 0.70 | 1.3 |
| Typical | 0.12%-1.2% | 0.05-0.10 | 0.07-0.25 | 2.1-3.0 |

support more intuitive interactions via three-dimensional, multimodal and semantic interfaces. When comparing today's PC systems to the Xerox PARC personal computer, Alan Kay estimated that "approximately a factor of 1000 in efficiency has been lost"[21].

**Pillar 4: A science of the Net ecosystem.** IT growth in the past 30 years benefited significantly from a rich IT ecosystem, manifesting as the whole of guiding principles, infrastructure interfaces, technical standards, and the interactions of users, academia, industry, and volunteer communities. However, we do not yet have a science of IT ecology. Computing for the masses will lead to mass creation and innovations utilizing network effects. This calls for a science of the Net ecosystem to study the effectiveness of mass creation and consumption in the ternary universe.

Two research topics are important to start establishing such a science. The first is to identify the basic objective attributes of the Net ecosystem. For creators, we must inherit the Internet openness and neutrality principle[22] to enable mass creation and to prevent monopoly. For users, we need a safe Net, with trust, security, and privacy. Both the creators and users should not feel isolated, but free to share and contribute to community assets, such as shared data, software, hardware, groups, and processes. A Net ecosystem with these six attributes is called a harmonious ecosystem.

The second topic is to develop a rigorous computing model that can precisely define and classify these six attributes, and relate them to new technology. We can start with simple ternary computing systems, where component elements can be superimposed. For instance, the studied system can be a centralized cloud service for exchanging household best practices of saving energy. Model checking techniques can be developed to verify the service has sound mechanisms, such that adding a community asset will not negatively affect household privacy.

**Pillar 5: National information accounts.** As IT use becomes an indispensable part of people's life and a primary value driver of national economy, it demands meticulous and scientific accounts, in a way similar to how economists created the national economic accounts system[23] now used worldwide. This national accounts system for the Net, supported by countrywide information meters, can give us objective pictures of the stocks and flows of information materials, goods, services, and users. It helps identify and understand IT's use, cost, value, bottlenecks and opportunities, as well as societal and environment impacts. It serves as a foundation for objectively assessing individual IT projects and national IT policies, by scientifically linking IT resources to IT values.

Such a national information accounts system is also good for rejuvenating the computing discipline. It could add to a basis for a pervasive yet unified computing discipline based on computational thinking, by scientifically organizing massive data from the ternary universe. Micro and macro informatics insights could be gained from such organized and linked data, as our discipline evolves from numerical computing, symbolic computing, process computing to data computing[24-25].

## 7 Conclusion

Over the decades our profession did invent many transformative technologies benefiting the masses, such as the personal computer and the Web. For instance, the PC significantly reduced computing cost, enlarged the user base, and augmented the IT market. It achieved all these benefits not by dumbing down a mainframe computer, but through transformative innovations such as interactive computing, desktop-based graphical user interface technology, object-oriented programming, and the Ethernet. In the 21st century, we must study and understand such successful endeavor and do it again, but in a more fundamental and systematic way. In the next 40 years, the rich demands, workloads, and experiences of the mass users will generate unprecedented economic scale and intellectual stimulus, expanding computing's width and depth in ways that we are now barely able to see.

## References:


[1] LI G J. Information science and technology in China: a roadmap to 2050[M]. Beijing: Science Press, 2010.

[2] OECD. The OECD information technology outlook 2008[R]. 2009.

[3] ZHANG M, LO V. Undergraduate computer science education in China[C]// ACM SIGCSE 2010. New York: ACM Press, 2010.

[4] LAZO J K, MORSS R E, DEMUTH J L. 300 billion served: sources, perceptions, uses, and values of weather forecasts[J]. Bulletin of American Meteorological Society, 2009: 785-798.

[5] NORDHAUS W D. Do real output and real wage measures capture reality? the history of lighting suggests not[M]// BRESNAHAN T F, GORDON R J. The economics of new goods. Chicago: The University of Chicago Press, 1997.

[6] CHEN Y. Energy science and technology in China: a roadmap to 2050[M]. Beijing: Science Press, 2010.

[7] KOOMEY J G. Worldwide electricity used in data centers[J]. Environmental Research Letters, 2008, 3: 034008.

[8] TURNER V, BIGLIANI R, INGLE C. Reducing greenhouse gases through intense use of information and communication technology[R]. 2009.

[9] KARP R. Understanding science through the computational lens[J]. Journal of Computer Science and Technology, 2011, 26(4): 569-577.

[10] SNIR M. Computer & information science & engineering-what's all this[C]// The 2nd NSF-NSFC Sino-USA Computer Science Summit. [S.l.:s.n.], 2008.

[11] WING J. Computational thinking[J]. Communications of the ACM, 2006, 49(3): 33-35.

[12] LICKLIDER J. Man-computer symbiosis[J]. IRE Transaction on Human Factors in Electronics, 1960, HFE-1(1): 4-11.

[13] LISKOV B, WING J. A behavioral notion of subtyping[J]. ACM Transactions on Programming Languages and Systems, 1994, 16(6): 1811-1841.

[14] GILBERT S, LYNCH N. Brewer's conjecture and the feasibility of consistent, available, partition-tolerant web services[J]. ACM SIGACT News, 2002, 33(2): 51-59.

[15] EASLEY D, KLEINBERG J. Networks, crowds, and markets: reasoning about a highly connected world[M]. Cambridge: Cambridge University Press, 2010.

[16] LEE R, XU Z. Exploiting stream request locality to improve query throughput of a data integration system[J]. IEEE Transactions on Computers, 2009, 58(10): 1356-1368.

[17] VALIANT L. Examples of computational models for neuroscience and evolution[C]// Speech at the Princeton Workshop on the Computational Worldview and the Sciences. [S.l.:s.n.], 2006.

[18] RAMAN T V. Toward $2^W$, beyond web 2.0[J]. Communications of the ACM, 2009, 52(2): 52-59.

[19] FIELDING R T, TAYLOR R N. Principled design of the modern web architecture[J]. ACM Transactions on Internet Technology, 2002, 2(2): 115-150.

[20] O'HANLON C. A conversation with John Hennessy and David Patterson[J]. ACM Queue, 2006-2007, 4(10): 14-22.

[21] FELDMAN S. A conversation with Alan Kay[J]. ACM Queue, 2004, 2(9): 20-30.

[22] CERF V. The open Internet[J]. Telecommunications Journal of Australia 2009, 59(2): 18.1-18.10.
[23] STONE R. Nobel memorial lecture 1984. The accounts of society[J]. Journal of Applied Econometrics, 1986, 1: 5-28.
[24] HENDLER J, SHADBOLT N, HALL W, et al. Web science: an interdisciplinary approach to understanding the web[J]. Communications of the ACM, 2008, 51(7): 60-69.
[25] HEY T, TANSLEY S, TOLLE K. The fourth paradigm: data-intensive scientific discovery[R]. 2009.

发表于*Scientific Reports*, 2014年6月

# Temporal scaling in information propagation

Jun-ming HUANG[1], Chao LI[1,2], Wen-qiang Wang[2], Hua-wei Shen[1], Guo-jie LI[1], Xue-qi CHENG[1]

1. Institute of Computing Technology, Chinese Academy of Sciences, Beijing, China;

2. Web Sciences Center, School of Computer Science and Engineering, University of Electronic Science and Technology of China, Chengdu, China

**Abstract:** For the study of information propagation, one fundamental problem is uncovering universal laws governing the dynamics of information propagation. This problem, from the microscopic perspective, is formulated as estimating the propagation probability that a piece of information propagates from one individual to another. Such a propagation probability generally depends on two major classes of factors: the intrinsic attractiveness of information and the interactions between individuals. Despite the fact that the temporal effect of attractiveness is widely studied, temporal laws underlying individual interactions remain unclear, causing inaccurate prediction of information propagation on evolving social networks. In this report, we empirically study the dynamics of information propagation, using the dataset from a population-scale social media website. We discover a temporal scaling in information propagation: the probability a message propagates between two individuals decays with the length of time latency since their latest interaction, obeying a power-law rule. Leveraging the scaling law, we further propose a temporal model to estimate future propagation probabilities between individuals, reducing the error rate of information propagation prediction from 6.7% to 2.6% and improving viral marketing with 9.7% incremental customers.

## 1 Introduction

In recent years, information propagation on social networks has been attracting much attention from academia and industry[1-9]. Understanding the mechanisms of information propagation, with or without exogenous and endogenous factors, is a fundamental task to uncover the universal laws governing the process of information propagation, which is important for better explaining the dynamics of information propagation[10], predicting information popularity[11], and initiating viral marketing campaign[12-16]. This task, from the microscopic perspective, is formulated as inferring and estimating the propagation probability that a piece of information propagates from one individual to another along social links connecting them.

The difficulty of estimating propagation probability lies in the complex interaction pattern between individuals and the co-existence of various confounding factors, such as the interplay between social selection and social influence. Previous studies empirically identified two classes of factors that drive information propagation: the attractiveness of information and the interactions between individuals. Existing studies on the first class mainly discussed three fundamental mechanisms with respect to message attractiveness[17]: the time-invariant intrinsic attractiveness or fitness[18-19], the Matthew effect in the popularity accumulation[17], and the freshness of messages decaying in a power-law[20], exponential[21-22], Rayleigh[23-24], or log-normal[17] manner with respect to the time span since the message is posted[25]. In contrast, most conventional studies on the second class were limited to static or quasi-static scenarios, assuming time-invariant interactions between any pair of individuals. Researchers estimated a propagation probability by indifferently aggregating recent and long-ago interactions[21,26], or by learning a probability function with static features including structural characteristics of the underlying network[11,27-29], demographic features[30], and topical and contextual features[31-33]. Few studies explored the possibility that individual interactions change with time. A recent study modeled social influence as a Markovian chain on temporally sliced snapshots of a social network, but did not reveal the intrinsic temporal scaling how social influence evolved[34].

Actually, most real-world social networks are far from static. On evolving social networks, whether a piece of information will be propagated is more related to instant frequency of individual interactions rather than average frequency indifferently aggregated over recent and long-ago interactions. Hence, it is problematic to neglect the dynamic nature of individual interactions and its crucial role at information propagation, leading to inaccurate predictions. A possible solution is working only on recent interactions based on temporally sliced snapshots of interactions. However, it is hard to determine the appropriate temporal scale of snapshots since the frequency of interactions is scale-free[35]. Therefore, we lack a full understanding about the temporal scaling of information propagation, which is crucial to grasp the propagation dynamics of information.

In this report, we study whether and how individual interactions vary temporally and their role at predicting the instant propagation probability. Intuitively, a high frequency of recent communication implies strong instant interaction and a high propagation probability. As the delegate of recency, latency is defined as the idle time since the latest communication between two individuals. A long latency generally reflects a low tendency of future interaction. Thus, analyzing the interdependence between the latency and the trend of a propagation probability provides us a peculiar delegate for understanding the temporal effect of information propagation. With this delegate, we study on a population-scale social media dataset and conduct an empirical validation for the intuition that a longer latency indicates a relatively lower instant propagation probability.

To focus on analyzing the temporal scaling of propagation probabilities from the perspective of individual interactions, in this report we do not consider the factors of information attractiveness, and instead calculate a propagation probability between two individuals as the ratio of retweeted and neglected messages that are propagated from one to another. This methodology is reasonable when the number of messages is sufficient to largely average out information attractiveness. In this way the temporal scaling of information propagation fully reflects the temporal scaling of individual interactions.

## 2 Results

The studies are based on a publicly available dataset (WISE 2012 Challenge) collected from Sina Weibo, the largest Chinese micro blogging website, like Twitter. In the dataset with some simple preprocessing (see Section S1), half a million users created 1.2 million following relations among them, providing channels for propagation of 8 million messages. We denote with an edge ($v_i$, $v_j$) the relation that a user $v_j$ (called the follower) follows another user $v_i$ (called the followee). Each time $v_j$ sees a message $k$ posted or retweeted by $v_i$ that $v_j$ has not retweeted before, we say $\delta_{i,j,k}$ =1 if $v_j$ retweets $k$, forming a positive example indicating $v_i$ successfully activates $v_j$ to retweet $k$; otherwise $\delta_{i,j,k}$ =0 for a negative example if $v_j$ neglects $k$. For each positive/negative example, we measure the latency $\tau_{i,j,k}$ as the time span since the latest time $v_j$ retweets a message from $v_i$.

We start to explore the temporal scaling of information propagation by examining time stamps of positive examples on two randomly selected edges, a followee and two of his followers. Figure 1(a) and Figure 1(b) reveal a non-uniform density of positive examples that the followers frequently retweet messages from the followee in several short time periods, separated by long idle periods. This implies a burst phenomenon on individual interactions: short time frames of intense interactions are separated by long idle periods[35]. To provide a solid evidence for the existence of burst in retweeting behaviors, we depict in Figure 1(e) the distribution of latency of all positive examples. The power-law distribution of latency, reflecting the emergence of bursty retweeting behaviors, exhibits the temporal nature of individual interactions. Note that static individual interactions lead to a time-invariant propagation probability on each edge in this scenario, which views retweeting behaviors as a homogeneous Poisson process, resulting in an exponential distribution of latency.

The temporal nature of individual interactions results in a necessity to assign a unique propagation probability to every retweeting/neglecting behavior even occurred on the same edge, reflecting the instant tendency that a follower retweets a followee's message at the time that message arrives. To uncover the temporal scaling of instant propagation probabilities, we investigate the interdependence between the propagation probability behind every retweeting/neglecting behavior and the latency associated with it. The interdependence is suggested by the distribution of retweeting/neglecting behaviors on those two edges against associated latency, where most retweeting behaviors occur with short latency (Figure 1(c) and 1(d)). We calculate the ratio of retweeting and neglecting behaviors over all edges to estimate the invisible instant propagation probability given certain latency. The propagation probability decreases with the latency in a power-law manner (Figure 1(f)). Fitting the log-log curve in Figure 1(f) produces a consistently decaying speed of -0.71 slope,

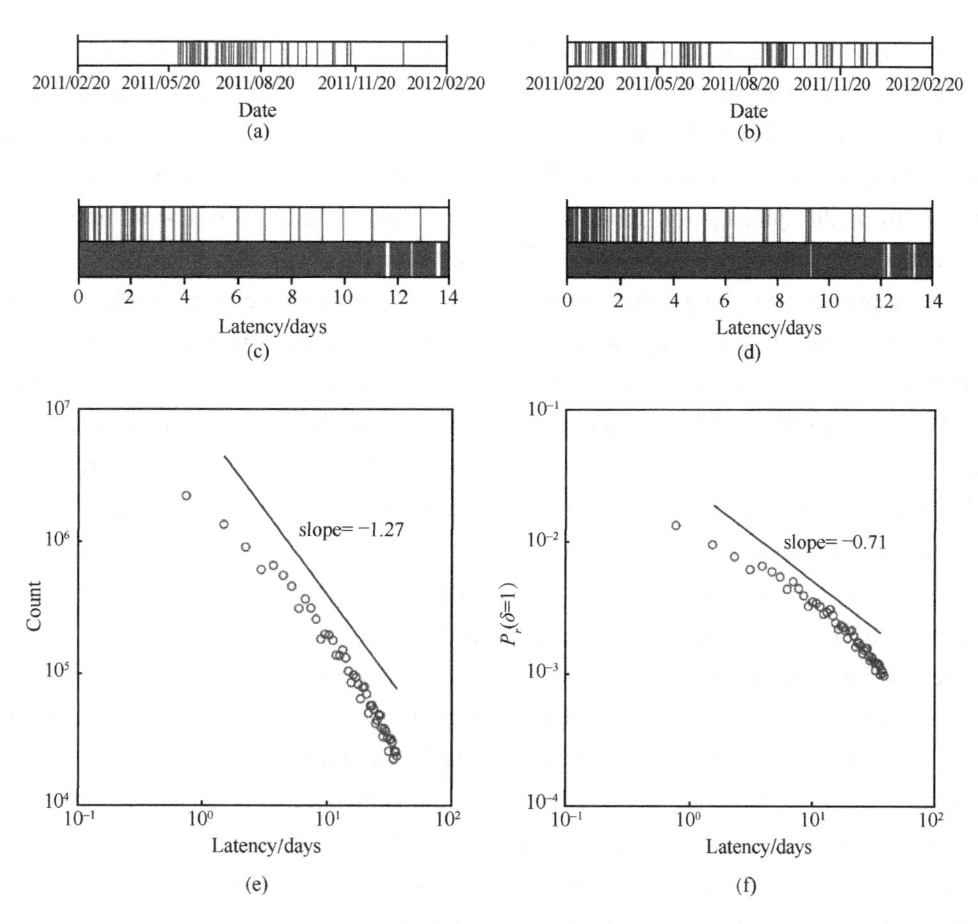


Figure 1. Characterizing propagation probabilities. (a),(b) Time stamps of positive examples (retweeting behaviors) on two random edges. Each vertical line represents a retweeting behaviors occurring with the time stamp marked on the horizontal axis. (c),(d) Positive (retweeting) and negative (neglecting) examples on those two edges. Vertical lines in upper half represent positive examples, while those in lower half represent negative ones. It shows an obvious tendency that most positive examples are concentrated on the left zone, i.e., most retweeting behaviors occur with short latency. The tendency is stronger on (c) than that on (d). (e) Distribution of latency of retweeting behaviors over all edges. (f) Ratio of positive examples upon all examples on all edges with respect to the associated latency, demonstrating the power-law interdependence between the propagation probability and the latency.

suggesting the temporal scaling between a propagation probability $Pr(\delta=1)$ behind a retweeting/neglecting behavior and its associated latency $\tau$ as follows,

$$Pr\left(\delta_{i,j,k}=1\right) \propto \tau_{i,j,k}^{-0.71} \tag{1}$$

We further study whether retweeting behaviors on different edges share the same power exponent, governing the temporal scaling. As shown in Figure 1(a–d), although the retweeting behaviors on the two edges both obey the power-law temporal scaling, the power exponents are quite different. Therefore, we need to assign an edge-specific exponent on each edge, in order to model the temporal scaling of information propagation on various edges of social networks.

Motivated by the observed temporal scaling, we propose a temporal model, namely Decay model, to predict propagation probability. We evaluate the performance of the model by applying it to predict retweeting behaviors and to launch a viral marketing strategy, compared with four mainstream baselines, namely MLE, EM[26], Static Bernoulli[21], and Static PC Bernoulli[21].

The first evaluation experiment measures the probability a model correctly predicts whether or not an individual will retweet an incoming message. Figure 2(a) reports AUC, the area under the Receiver Operating Characteristic (ROC) curve, equivalent to the probability that a classifier correctly distinguishes a positive example from a negative one. The Decay model outperforms all baselines, raising AUC from 93.3% to 97.4%. Intuitively speaking, when facing a randomly selected pair of a retweeting behavior and a neglecting behavior, the error rate to incorrectly distinguish them is reduced by a half by the Decay model over the best baseline. We then report the perplexity on the testing set against the training set ratio to obtain the probability that a model, trained with incomplete observations, correctly generates the testing examples. As shown in Figure 2(b), the Decay model achieves the lowest (best) perplexity among all tested models. The priority of the Decay model is consistent in all examined training set ratios, with a more significant improvement on a relatively smaller training set. We also evaluate the Decay model with ROC curve, which is a metric appropriate for extremely imbalanced datasets such as the one we use in this report (as well as most real-world social media) where positive examples occupy less than 1%. ROC, measuring the sensitivity (true positive rate) against specificity (one minus false positive rate), is insensitive to the ratio between positive and negative examples. Figure 2(c) reports the ROC curves of the Decay model and baselines with 90% examples held out as the training set. Results of other training set ratios are similar. The figure shows that the Decay model achieves the best capability at distinguishing retweeting behaviors from neglecting behaviors with a significant improvement upon all baselines.

The second evaluation measures the accuracy a model predicts propagation probabilities. Intuitively, predictions that are more accurate would help select a better initial seed set, triggering a larger fraction of individuals. We split all examples into 4 groups in a chronological order with respect to example time stamps. Each group contains examples in 30 weeks (see Section S6 for details). The Decay model and baselines train on examples in the earlier 205 days (training phase) and predict the propagation probabilities in the last 5 days (evaluation phase). Based on those predictions, a state-of-the-art influence maximization algorithm (CELF++[15]) is used to select an initial seed set maximizing the expected eventual influence spread. We then estimate the pseudo actual spread of such a seed set as the number of nodes reachable from the seed set on a propagation network, which is a subgraph of the social network consisting of edges with at least one actual retweeting behavior in the last 5 days.

As reported in Figure 2(d) (one group shown only), the largest pseudo actual spread comes from the seed set selected on propagation probabilities predicted by the Decay model, which eventually reaches 2 590 nodes, achieving a 9.7% increase upon what is reached by the best baseline, i.e., Static PC Bernoulli which reaches 2 361 nodes. The increase in pseudo actual spread demonstrates the advantage that the Decay model more accurately predicts the propagation probabilities, confirming our finding that individual interactions decay with latency.

## 3 Discussion

In this report, we uncovered the temporal scaling in information propagation from the perspective of individual interactions: a propagation probability decays slowly in a power-law manner with the latency since their latest interaction. Such a dynamic nature was demonstrated by empirical studies on a large-scale public social media dataset, showing the power-law interdependence between a propagation probability and latency.

With the observed temporal scaling, a Decay model was proposed to predict future propagation probability among individuals, incorporating a time-invariant base probability and a time-decaying exponent on each edge. The model is applicable in scenarios where an underlying social network and tractable information propagation with time stamps are observed, such as micro blogging (Twitter and Sina Weibo), blog sites, book sharing sites and email promotion networks. Empirical evaluations supported that the Decay model outperformed mainstream baselines in predicting retweeting behaviors, significantly reducing by a half the expected error rate of incorrectly identifying a retweeting behavior.

From the perspective of machine learning, the discovered temporal scaling provides an additional feature to estimate propagation probability. While traditional models assume static propagation probability, the proposed Decay model additionally explores the temporal effect of a propagation probability, explaining the increased accuracy. Generally speaking, a model with more features requiring more data for training suffers severe over-fitting problem on sparse data. This partly explains why traditional models do not consider temporal features. In order to reduce the pain of sparsity, the Decay model introduces a prior distribution of the decaying exponent $p(\alpha)$, suggested by the global decaying exponent in empirical study results. The prior distribution successfully reduces the pain of sparsity: the improvement of the Decay model upon baselines is even more significant with a relatively smaller training set (Figure 2(a) and 2(b)). Note that typically only several retweeting behaviors are observed on an edge in a real-world scenario, the outstanding performance of the Decay model on sparse data is of great importance in practice.

It is worth noting that the viral marketing evaluation is not conducted using Monte Carlo simulations, as done in most influence maximization studies. That is because what we compare is the configurations of propagation probabilities estimated with various model, and thus it is unfair to run Monte Carlo simulations with any estimated configuration, otherwise estimating all

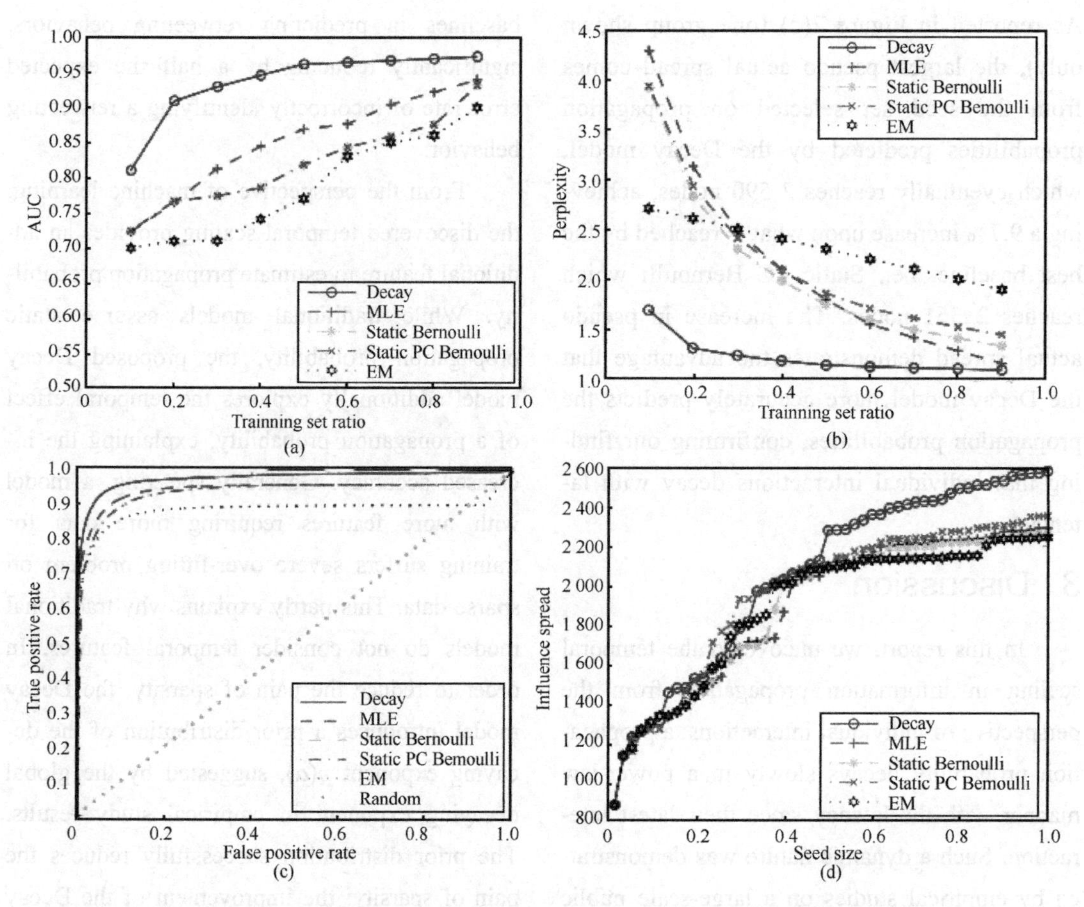


Figure 2. Model evaluation. (a) AUC of the Decay model and baselines. AUC measures the area under the ROC curves, and thus is equivalent to the probability thata trained model correctly distinguish a randomly selected positive example from another randomly selected negative example. (b) Perplexity of the Decay model and baselines when predicting retweeting behaviors, against the training set ratio. A lower perplexity indicates a better prediction accuracy, meaning less extent a testing example surprises a trained model. (c) ROC curves with a training set of 90% examples. (d) Influence spreads of an initial seed set selected on propagation probabilities predicted by the Decay model and baselines

probabilities equal to one will surely win. Instead, we estimate the propagation spread in a pseudo-actual way. We build a propagation network, a subgraph of the social network, with edges where at least one retweeting behavior occurs in the 5-day evaluation phase. Therefore, the reachability of a node on the propagation network measures its pseudo actual influence spread during that 5 days. It is equivalent to one Monte Carlo simulation that is produced from the (unknown) actual individuals and observed by actual retweeting behaviors. The estimated propagation spread is deterministic without any random deviation.

In the Decay model, the base probability $q$ is considered as a free variable whose value is fully determined by maximum-a-posteriori inference with a prior distribution. In fact, the Decay model can certainly incorporate any endogenous or exogenous factors through rewriting $q$ as a function of those factors, such as demographical, structural, content and context features. Parameters of such a function could also be estimated in maximum-a-posteriori inference.

In the first evaluation experiment, the Decay model is tested with only one testing example on each edge, for the ease of calculating latency. When facing multiple testing examples (e.g., predicting whether an individual will retweet a series of messages in a month), one should predict those examples one by one in a chronological order and calculate the expected latency of a later example over the joint probability distribution of predicted results of all previous testing examples.

Choosing the latency as a delegate of recency is equivalent to approximating the information propagation occurrences as a first order Markov process, i.e., only the idle time since the latest interaction, instead of all historical interactions, affects the current decision. Such an approximation, effectively avoiding expensive calculation with a nondeterministic number of parameters required to build a complicated function defined on all historical interactions, succeeds in revealing strong evidence of interdependence between propagation probabilities and latency and in building an outperforming prediction model. That supports the important role that the temporal scaling plays in characterizing a propagation probability.

As an open question in future, it would be attractive to characterizing influential nodes identified with high propagation probabilities estimated by the Decay model, and to demonstrate the evolving distribution of instant influential nodes on a social network.

## 4 Methods

The proposed Decay model describes the propagation probability $P(\delta_{i,j,k}=1)$, that an individual $v_i$ will successfully activate another individual $v_j$ to retweet a message $k$, which is believed to be determined by two factors:

- $q_{i,j} \in [0,1]$: the base probability associated with the edge $(v_i, v_j)$;
- $\tau_{i,j,k} \in [1,+\infty)$: latency, the time span since the latest time $v_i$ activated $v_j$, i.e., $\tau_{i,j,k} = t_{k,i} - t_{k',j}$, where $t_{k,i}$ is the time stamp when $v_i$ posts or retweets $k$, and $k'$ is the latest message before $k$ that $v_i$ activates $v_j$ to retweet.

Specifically, the propagation probability is as follows,

$$P(\delta_{i,j,k}=1) = q_{i,j}\tau_{i,j,k}^{-\alpha_{i,j}} \tag{2}$$

where $\alpha_{i,j} > 0$ is a decaying exponent associated with the edge $(v_i, v_j)$. The decaying exponent is edge-specific, with a prior distribution $p(\alpha)$ reflecting the global decaying exponent. Traditional models without temporal scaling of propagation probabilities can be viewed as special cases of the Decay model with constant $\alpha$=0.

Latency is required to be bounded, i.e., $\tau \geqslant 1$, to guarantee $P(\delta_{i,j,k}=1) \in [0,1]$. Specifically, $\tau_{i,j,k}$=1 results in that $q_{i,j}\tau_{i,j,k}^{-\alpha_{i,j}} = q_{i,j}$, revealing the intuitive meaning of the base probability that $q_{i,j}$ equals to the probability $v_i$ successfully activate $v_j$ to retweet a message $k$ which arrives immediately after a previous successful activation.

The hidden parameters $q$ and $\alpha$ are inferred with a maximum-a-posteriori estimate with prior distributions $p(q)$ and $p(\alpha)$. See Section S3 for details.

To demonstrate the performance of the Decay model, four mainstream baselines are implemented to estimate and predict propagation probabilities on all edges, including MLE, EM[26], Static Bernoulli[21], and Static PC Bernoulli[21] (see Section S4). Some other widely used models are not compared because those models require user profiles or message content that are absent in this scenario.

In the retweeting prediction experiment, we apply a next-one strategy to split a training set and a testing set. On each edge, we sort all examples in a chronological order, take the earliest $N\%$ examples as the training set, and leave the next one example as the testing set. Thus, the size of the training set increases with $N\%$, the training set ratio, while the size of the testing set is a constant equal to the number of edges. With parameters trained on the training set, the Decay model predicts the label $\delta$ of examples in the testing set.

The evaluation metrics include perplexity, ROC curve and AUC. The perplexity measures how the testing examples surprise a trained model. A lower perplexity demonstrates better prediction ability.

$$\text{perplexity} = e^{\frac{\sum_{\{v_i,v_j,k\}\in D_{\text{test}}} \delta_{i,j,k} \ln \tilde{P}(\delta_{i,j,k}=1) + (1-\delta_{i,j,k}) \ln\left(1-\tilde{P}(\delta_{i,j,k}=1)\right)}{|D_{\text{test}}|}} \quad (3)$$

where $D_{\text{test}}$ represents the testing set, and $\tilde{P}(\delta_{i,j,k}=1)$ is the estimated propagation probability. The ROC curve plots sensitivity (true positive rate) against specificity (one minus false positive rate). AUC measures the area under the ROC curve, which is equivalent to the probability that a model correctly distinguishes a randomly selected positive example from a randomly selected negative example. A higher AUC indicates a better distinguish ability. See Section S5 for details.

## Acknowledgments

We thank Tao Zhou, Yan-yan Lan, Su-qi Cheng and Ming Tang for valuable discussions. This work was funded by the National Basic Research Program of China (973 Program) under grant number 2014CB3430401, The National High-tech R&D Program of China (863 Program) under grant number 2014AA015 103, and the National Natural Science Foundation of China under grant number 61232010, 61202215, 61272536.

## Reference:

[1] LESKOVEC J, MCGLOHON M, FALOUTSOS C, et al. Patterns of cascading behavior in large blog graphs[C]// The 7th SIAM International Conference on Data Mining. Philadelphia: SIAM, 2007.

[2] BAKSHY E, HOFMAN J M, MASON W A, et al. Everyone's an influencer: quantifying influence on twitter[C]// The 4th ACM International Conference on Web Search and Data Mining. New York: ACM Press, 2011.

[3] HUANG J, CHENG X Q, GUO J, et al. Social recommendation with interpersonal influence[C]// The 19th European Conference on Artificial Intelligence Lisbon. The Netherlands: IOS Press, 2010.

[4] HUANG J, CHENG X Q, SHEN H W, et al. Exploring social influence via posterior effect of word-of-mouth recommendations[C]// The 5th ACM International Conference on Web Search And Data Mining. New York: ACM Press, 2012.

[5] WATTS D J, DODDS P S. Influentials, networks, and public opinion formation[J]. Journal of Consumer Research, 2007, 34: 441-458.

[6] TANG J, SUN J, WANG C, et al. Social influence

analysis in large-scale networks[C]// The 15th ACM SIGKDD International Conference on Knowledge Discovery and Data Mining. New York: ACM Press, 2009.

[7] MUCHNIK L, ARAL S, TAYLOR S J. Social influence bias: a randomized experiment[J]. Science, 2014, 341: 647-651.

[8] ARAL S, MUCHNIK L, SUNDARARAJAN A. Distinguishing influence-based contagion from homophily-driven diffusion in dynamic networks[J]. Proceedings of the National Academy of Sciences, 2009, 106: 21544-21549.

[9] CENTOLA D. An experimental study of homophily in the adoption of health behavior[J]. Science, 2011, 334: 1269-1272.

[10] LESKOVEC J, ADAMIC L A, HUBERMAN B A. The dynamics of viral marketing[J]. ACM Transactions on the Web, 2007, 1: 5.

[11] BAO P, SHEN H W, HUANG J, et al. Popularity prediction in microblogging network: a case study on Sina Weibo[C]// The 22nd International Conference on World Wide Web. New York: ACM Press, 2013.

[12] KEMPE D, KLEINBERG J, TARDOS E. Maximizing the spread of influence through a social network[C]// The 9th ACM SIGKDD International Conference on Knowledge Discovery and Data Mining. New York: ACM Press, 2003.

[13] LESKOVEC J, KRAUSE A, GUESTRIN C, et al. Cost-effective outbreak detection in networks[C]// The 13th ACM SIGKDD International Conference on Knowledge Discovery and Data Mining. New York: ACM Press, 2007.

[14] CHEN W, WANG Y, YANG S. Efficient influence maximization in social networks[C]// The 15th ACM SIGKDD International Conference on Knowledge Discovery and Data Mining. New York: ACM Press, 2009.

[15] GOYAL A, LU W, LAKSHMANAN L V. Celf11: optimizing the greedy algorithm for influence maximization in social networks[C]// The 20th International Conference Companion on World Wide Web. New York: ACM Press, 2011.

[16] CHENG S, SHEN H W, HUANG J, et al. StaticGreedy: solving the scalability-accuracy dilemma in influence maximization[C]// The 22nd ACM International Conference on Information and Knowledge Management. New York: ACM Press, 2013.

[17] WANG D, SONG C, BARABÁSI A L. Quantifying long-term scientific impact[J]. Science, 2013, 342: 127-132.

[18] LERMAN K, HOGG T. Using a model of social dynamics to predict popularity of news[C]// The 19th International Conference on World Wide Web. New York: ACM Press, 2010.

[19] YE M, LIU X, LEE W C. Exploring social influence for recommendation: a generative model approach[C]// The 35th International ACM SIGIR Conference on Research and Development in Information Retrieval. New York: ACM Press, 2012.

[20] GOMEZ-RODRIGUEZ M, LESKOVEC J, KRAUSE A. Inferring networks of diffusion and influence[C]// The 16th ACM SIGKDD International Conference on Knowledge Discovery and Data Mining. New York: ACM Press, 2010.

[21] GOYAL A, BONCHI F, LAKSHMANAN L V. Learning influence probabilities in social networks[C]// The 3rd ACM International Conference on Web Search and Data Mining. New York: ACM Press, 2010.

[22] SONG X, CHI Y, HINO K, et al. Information flow modeling based on diffusion rate for prediction and ranking[C]// The 16th International Conference on World Wide Web Banff. New York: ACM Press, 2007.

[23] GOMEZ-RODRIGUEZ M, BALDUZZI D, SCHÖLKOPF B. Uncovering the temporal dynamics of diffusion networks[C]// The 28th International Conference on Machine Learning. [S.l.:s.n.], 2011.

[24] MANUEL GOMEZ R, JURE L, BERNHARD S, et al. Structure and dynamics of information pathways in online media[C]// The 6th ACM International Conference on Web Search and Data Mining. New York: ACM Press, 2013.

[25] WU F, HUBERMAN B A. Novelty and collective attention[J]. Proceedings of the National Academy of Sciences, 2007, 104: 17599-17601.

[26] SAITO K, NAKANO R, KIMURA M. Prediction of information diffusion probabilities for independent cascade model[J]. Lecture Notes in Computer Science, 2008, 5179: 67-75.

[27] ZHANG J, LIU B, TANG J, et al. Social influence locality for modeling retweeting behaviors[C]// The 23th International Joint Conference on Artificial Intelligence. Menlo Park: AAAI Press, 2013.

[28] LÜ L, CHEN D B, ZHOU T. The small world yields the most effective information spreading[J]. New Journal of Physics, 2011, 13(12): 123005.

[29] FANG X, HU P J H, LI Z, et al. Predicting adoption

probabilities in social networks[J]. Information Systems Research, 2013, 24: 128-145.
[30] ARAL S, WALKER D. Identifying influential and susceptible members of social networks[J]. Science, 2012, 337: 337-341.
[31] SUH B, HONG L, PIROLLI P, et al. Want to be retweeted? large scale analytics on factors impacting retweet in twitter network[C]// IEEE 2nd International Conference on Social Computing. New York: Institute of Electrical and Electronics Engineers, 2010.
[32] FENG W, WANG J. Retweet or not?: personalized tweet re-ranking[C]// The 6th ACM International Conference on Web Search and Data Mining. New York: ACM Press, 2013.
[33] YANG Z, GUO J Y, CAI K K, et al. Understanding retweeting behaviors in social networks[C]// The 19th ACM International Conference on Information and Knowledge Management. New York: ACM Press, 2010.
[34] WANG C, TANG J, SUN J, et al. Dynamic social influence analysis through time-dependent factor graphs[C]// 2011 IEEE/ACM International Conference on Advances in Social Network Analysis and Mining. Piscataway: IEEE Press, 2011: 239-246.
[35] BARABÁSI A L. The origin of bursts and heavy tails in human dynamics[J]. Nature, 2005, 435: 207-211.

发表于《软件学报》2014年第9期

# 大数据系统和分析技术综述

程学旗，靳小龙，王元卓，郭嘉丰，张铁赢，李国杰
中国科学院计算技术研究所网络数据科学与技术重点实验室，北京 100190

**摘　要**：首先根据处理形式的不同，介绍了不同形式数据的特征和各自的典型应用场景以及相应的代表性处理系统，总结了大数据处理系统的三大发展趋势；随后，对系统支撑下的大数据分析技术和应用（包括深度学习、知识计算、社会计算与可视化等）进行了简要综述，总结了各种技术在大数据分析理解过程中的关键作用；最后梳理了大数据处理和分析面临的数据复杂性、计算复杂性和系统复杂性挑战，并逐一提出了可能的应对之策。
**关键词**：大数据；数据分析；深度学习；知识计算；社会计算；可视化

# Survey on big data system and analytic technology

Xue-qi CHENG, Xiao-long JIN, Yuan-zhuo WANG, Jia-feng GUO, Tie-ying ZHANG, Guo-jie LI
Key Laboratory of Network Data Science and Technology, Institute of Computing Technology, Chinese Academy of Sciences, Beijing 100190, China

**Abstract:** This paper first introduces the key features of big data in different processing modes and their typical application scenarios, as well as corresponding representative processing systems. It then summarizes three development trends of big data processing systems. Next, the paper gives a brief survey on system supported analytic technologies and applications (including deep learning, know ledge computing, social computing, and visualization), and summarizes the key roles of individual technologies in big data analysis and understanding. Finally, the paper lays out three grand challenges of big data processing and analysis, i.e., data complexity, computation complexity, and system complexity. Potential ways for dealing with each complexity are also discussed.
**Key words:** big data, data analysis, deep learning, knowledge computing, social computing, visualization

**基金项目**：国家重点基础研究开发计划（973 计划）基金资助项目（No.2014CB340401，No.2012CB316303）；国家自然科学基金资助项目（No.61232010，No.61100175，No.61173008，No.61202214）；北京市科技新星计划（No.Z121101002512063）

## 1 引言

近几年，大数据迅速发展成为科技界和企业界甚至世界各国政府关注的热点。*Nature* 和 *Science* 等相继出版专刊专门探讨大数据带来的机遇和挑战。著名管理咨询公司麦肯锡称：“数据已经渗透到当今每一个行业和业务职能领域，成为重要的生产因素。人们对于大数据的挖掘和运用，预示着新一波生产力增长和消费盈余浪潮的到来”[1]。美国政府认为大数据是“未来的新石油”，一个国家拥有数据的规模和运用数据的能力将成为综合国力的重要组成部分，对数据的占有和控制将成为国家间和企业间新的争夺焦点。大数据已成为社会各界关注的新焦点，“大数据时代”已然来临。

什么是大数据，迄今并没有公认的定义。从宏观世界角度来讲，大数据是融合物理世界（Physical World）、信息空间和人类社会（Human Society）三元世界的纽带，因为物理世界通过互联网、物联网等技术有了在信息空间中的大数据反映，而人类社会则借助人机界面、脑机界面、移动互联等手段在信息空间中产生自己的大数据映像[2-3]。从信息产业角度来讲，大数据还是新一代信息技术产业的强劲推动力。所谓新一代信息技术产业本质上是构建在第三代平台上的信息产业，主要是指大数据、云计算、移动互联网（社交网络）等。IDC 预测，到 2020 年第三代信息技术平台的市场规模将达到 5.3 万亿美元，而 2013—2020 年，IT 产业 90%的增长将由第三代信息技术平台驱动。从社会经济角度来讲，大数据是第二经济（Second Economy[4]）的核心内涵和关键支撑。第二经济的概念是由美国经济学家 Auther 在 2011 年提出的。他指出由处理器、链接器、传感器、执行器以及运行在其上的经济活动形成了人们熟知的物理经济（第一经济）之外的第二经济（不是虚拟经济）。第二经济的本质是为第一经济附着一个“神经层”，使国民经济活动能够变得智能化，这是 100 年前电气化以来最大的变化。Auther 还估算了第二经济的规模，他认为到 2030 年，第二经济的规模将逼近第一经济。而第二经济的主要支撑是大数据，因为大数据是永不枯竭并不断丰富的资源产业。借助于大数据，未来第二经济下的竞争将不再是劳动生产率而是知识生产率的竞争。

相较于传统的数据，人们将大数据的特征总结为 5 个 V，即体量大（Volume）、速度快（Velocity）、模态多（Variety）、难辨识（Veracity）和价值大密度低（Value）。但大数据的主要难点并不在于数据量大，因为通过对计算机系统的扩展可以在一定程度上缓解数据量大带来的挑战。其实，大数据真正难以对付的挑战来自于数据类型多样（Variety）、要求及时响应（Velocity）和数据的不确定性（Veracity）。因为数据类型多样使得一个应用往往既要处理结构化数据，同时还要处理文本、视频、语音等非结构化数据，这对现有数据库系统来说难以应付；在快速响应方面，在许多应用中时间就是利益。在不确定性方面，数据真伪难辨是大数据应用的最大挑战。追求高数据质量是对大数据的一项重要要求，最好的数据清理方法也难以消除某些数据固有的不可预测性。

为了应对大数据带来的上述困难和挑战，以 Google、Facebook、Linkedin、Microsoft 等为代表的互联网企业近几年推出了各种不同类型的大数据处理系统。借助于新型的处理系统，深度学习、知识计算、可视化等大数据分析技术也得已迅速发展，已逐渐被广泛应用于不同的行业和领域。本文从系统支撑下的大数据分析角度入手，介绍了不同的大数据处理模式与代表性的处理系统，并对深度学习、知识计算等重要的大数据分析技术进行综述，最后指出大数据处理和分析所面临的 3 个核心挑战，并提出可能的解决思路。

## 2 大数据处理与系统

大数据中蕴含的宝贵价值成为人们存储和处理大数据的驱动力。在文献[5]中，Mayer-Schönberger 指出了大数据时代处理数据理念的三大转变，即要全体不要抽样，要效率不要绝对精确，要相关不要因果。因此，海量数据的处理对于当前存在的技术来说是一种极大的挑战。目前，人们对大数据的处理形式主要是对静态数据的批量处理，对在线数据的实时处理[6]，以及对图数据的综合处理。其中，在线数据的实时处理又包括对流式数据的处理和实时交互计算两种。本节将详细阐述上述4种数据形式的特征和各自的典型应用以及相应的代表性系统。

### 2.1 批量数据处理系统

利用批量数据挖掘合适的模式，得出具体的含义，制定明智的决策，最终做出有效的应对措施实现业务目标是大数据批处理的首要任务。大数据的批量处理系统适用于先存储后计算，实时性要求不高，同时数据的准确性和全面性更为重要的场景。

#### 2.1.1 批量数据的特征与典型应用

（1）批量数据的特征

批量数据通常具有 3 个特征。第一，数据体量巨大。数据从 TB 级别跃升到 PB 级别。数据是以静态的形式存储在硬盘中，很少进行更新，存储时间长，可以重复利用，然而这样大批量的数据不容易对其进行移动和备份。第二，数据精确度高。批量数据往往是从应用中沉淀下来的数据，因此精度相对较高，是企业资产的一部分宝贵财富。第三，数据价值密度低。以视频批量数据为例，在连续不断的监控过程中，可能有用的数据仅仅有一两秒。因此，需要通过合理的算法才能从批量的数据中抽取有用的价值。此外，批量数据处理往往比较耗时，而且不提供用户与系统的交互手段，所以当发现处理结果和预期或与以往的结果有很大差别时，会浪费很多时间。因此，批量数据处理适合大型的相对比较成熟的作业。

（2）典型应用

物联网、云计算、互联网以及车联网等无一不是大数据的重要来源，当前批量数据处理可以解决前述领域的诸多决策问题并发现新的洞察。因此，批量数据处理可以适用于较多的应用场景。本节主要选择互联网领域的应用、安全领域的应用以及公共服务领域的应用这 3 个典型应用场景加以介绍[7-12]。

在互联网领域中，批量数据处理的典型应用场景主要包括 3 个。

（a）社交网络：Facebook、新浪微博、微信等以人为核心的社交网络产生了大量的文本、图片、音视频等不同形式的数据。对这些数据的批量处理可以对社交网络进行分析，发现人与人之间隐含的关系或者他们中存在的社区，推荐朋友或者相关的主题，提升用户的体验。

（b）电子商务：电子商务中产生大量的购买历史记录、商品评论、商品网页的访问次数和驻留时间等数据，通过批量分析这些数据，每个商铺可以精准地选择其热卖商品，从而提升商品销量；这些数据还能够分析出用户的消费行为，为客户推荐相关商品，以提升优质客户数量。

（c）搜索引擎：Google 等大型互联网搜索引擎与 Yahoo!的专门广告分析系统，通过对广告相关数据的批量处理用来改善广告的投放效果，以提高用户的点击量。

在安全领域中，批量数据主要用于欺诈检测和IT安全。在金融服务机构和情报机构中，欺诈检测一直都是关注的重点。通过对批量数据的处理，可对客户交易和现货异常进行判断，从而对可能存在欺诈行为提前预警。另一方面，企业通过处理机器产生的数据，识别恶意软件和网络攻击模式，从而使其他安全产品判断是否接受来自这些来源的通信。

在公共服务领域，批量数据处理的典型应用场景主要包括2个。

（a）能源：例如，对来自海洋深处地震时产生的数据进行批量的排序和整理，可能发现海底石油的储量；通过对用户能源数据、气象与人口方面的公共及私人数据、历史信息、地理数据等的批量处理，可以提升电力服务，尽量为用户节省在资源方面的投入。

（b）医疗保健：通过对患者以往的生活方式与医疗记录进行批量处理分析，提供语义分析服务，对病人的健康提供医生、护士及其他相关人士的回答，并协助医生更好的为患者进行诊断。当然，大数据的批量处理不只应用到这些领域，还有移动数据分析、图像处理以及基础设施管理等领域。随着人们对数据中蕴含价值的认识，会有更多的领域通过对数据的批量处理挖掘其中的价值来支持决策和发现新的洞察。

#### 2.1.2 代表性的处理系统

由Google公司2003年研发的Google文件系统GFS[13]和2004年研发的MapReduce编程模型[14]以其Web环境下批量处理大规模海量数据的特有魅力，在学术界和工业界引起了很大反响。虽然Google没有开源这两项技术的源码，但是基于这两篇开源文档，2006年Nutch项目子项目之一的Hadoop实现了两个强有力的开源产品：HDFS和MapReduce。Hadoop成为了典型的大数据批量处理架构，由HDFS负责静态数据的存储，并通过MapReduce将计算逻辑分配到各数据节点进行数据计算和价值发现。Hadoop顺应了现代主流IT公司的一致需求，之后以HDFS和MapReduce为基础建立了很多项目，形成了Hadoop生态圈。

MapReduce编程模型之所以受到欢迎并迅速得到应用，在技术上主要有3方面的原因[15-16]。首先，MapReduce采用无共享大规模集群系统。集群系统具有良好的性价比和可伸缩性，这一优势为MapReduce成为大规模海量数据平台的首选创造了条件。其次，MapReduce模型简单、易于理解、易于使用。它不仅用于处理大规模数据，而且能将很多繁琐的细节隐藏起来（比如自动并行化、负载均衡和灾备管理等），极大地简化了程序员的开发工作。而且，大量数据处理问题，包括很多机器学习和数据挖掘算法，都可以使用MapReduce实现。最后，虽然基本的MapReduce模型只提供一个过程性的编程接口，但在海量数据环境、需要保证可伸缩性的前提下，通过使用合适的查询优化和索引技术，MapReduce仍能够提供很好的数据处理性能。

### 2.2 流式数据处理系统

Google于2010年推出了Dremel，引领业界向实时数据处理迈进。实时数据处理是针对批量数据处理的性能问题提出的，可分为流式数据处理和交互式数据处理两种模式。在大数据背景下，流式数据处理源于服务器日志的实时采集，交互式数据处理的目标是将PB级数据的处理时间缩短到秒级。

#### 2.2.1 流式数据的特征及典型应用

（1）流式数据的特征

通俗而言，流式数据是一个无穷的数据序列，序列中的每一个元素来源各异，格式复杂，序列往往包含时序特性，或者有其他的有序标签（如 IP 报文中的序号）。从数据库的角度而言，每一个元素可以看作一个元组，而元素的特性则类比于元组的属性。流式数据在不同的场景下往往体现出不同的特征，如流速大小、元素特性数量、数据格式等，但大部分流式数据都含有共同的特征，这些特征便可用来设计通用的流式数据处理系统。下面简要介绍流式数据共有的特征[17]。

首先，流式数据的元组通常带有时间标签或其余含序属性。因此，同一流式数据往往是被按序处理的。然而数据的到达顺序是不可预知的，由于时间和环境的动态变化，无法保证重放数据流与之前数据流中数据元素顺序的一致性。这就导致了数据的物理顺序与逻辑顺序不一致。而且，数据源不受接收系统的控制，数据的产生是实时的、不可预知的。此外，数据的流速往往有较大的波动，因此需要系统具有很好的可伸缩性，能够动态适应不确定流入的数据流，具有很强的系统计算能力和大数据流量动态匹配的能力。其次，数据流中的数据格式可以是结构化的、半结构化的甚至是无结构化的。数据流中往往含有错误元素、垃圾信息等。因此流式数据的处理系统要有很好的容错性与异构数据分析能力，能够完成数据的动态清洗、格式处理等。最后，流式数据是活动的（用完即弃），随着时间的推移不断增长，这与传统的数据处理模型（存储→查询）不同，要求系统能够根据局部数据进行计算，保存数据流的动态属性。流式处理系统针对该特性，应当提供流式查询接口，即提交动态的 SQL 语句，实时地返回当前结果。

（2）典型应用

流式计算的应用场景较多，典型的有两类[6]。

（a）数据采集应用：数据采集应用通过主动获取海量的实时数据，及时地挖掘出有价值的信息。当前数据采集应用有日志采集、传感器采集、Web 数据采集等。日志采集系统是针对各类平台不断产生的大量日志信息量身订做的处理系统，通过流式挖掘日志信息，达到动态提醒与预警功能。传感器采集系统（物联网）通过采集传感器的信息（通常包含时间、位置、环境和行为等内容），实时分析提供动态的信息展示，目前主要应用于智能交通、环境监控、灾难预警等。Web 数据采集系统是利用网络爬虫程序抓取万维网上的内容，通过清洗、归类、分析并挖掘其数据价值。

（b）金融银行业的应用：在金融银行领域的日常运营过程中会产生大量数据，这些数据的时效性往往较短，不仅有结构化数据，也会有半结构化和非结构化数据。通过对这些大数据的流式计算，发现隐含于其中的内在特征，可帮助金融银行进行实时决策。这与传统的商业智能（BI）分析不同，BI 要求数据是静态的，通过数据挖掘技术，获得数据的价值。然而在瞬息万变的场景下，诸如股票期货市场，数据挖掘技术不能及时地响应需求，就需要借助流式数据处理的帮助。

总之，流式数据的特点是，数据连续不断、来源众多、格式复杂、物理顺序不一、数据的价值密度低。而对应的处理工具则需具备高性能、实时、可扩展等特性。

#### 2.2.2 代表性的处理系统

流式数据处理已经在业界得到广泛的应用，典型的有 Twitter 的 Storm、Facebook 的 Scribe、Linkedin 的 Samza、Cloudera 的 Flume、Apache 的 Nutch。

（1）Twitter 的 Storm 系统

Storm 是一套分布式、可靠、可容错的用于处理流式数据的系统。其流式处理作业被分发至不同类型的组件，每个组件负责一项简单的、特定的处理任务。Storm 集群的输入流由名为 Spout 的组件负责。Spout 将数据传递给名为 Bolt 的组件，后者将以指定的方式处理这些数据，如持久化或者处理并转发给另外的 Bolt。Storm 集群可以看成一条由 Bolt 组件组成的链（称为一个 Topology）。每个 Bolt 对 Spout 产生出来的数据做某种方式的处理。

Storm 可用来实时处理新数据和更新数据库，兼具容错性和扩展性。Storm 也可被用于连续计算，对数据流做连续查询，在计算时将结果以流的形式输出给用户。它还可被用于分布式 RPC，以并行的方式运行复杂运算。一个 Storm 集群分为 3 类节点：（a）Nimbus 节点，负责提交任务，分发执行代码，为每个工作结点指派任务和监控失败的任务；（b）Zookeeper 节点，负责 Storm 集群的协同操作；（c）Supervisor 节点，负责启动多个 Worker 进程，执行 Topology 的一部分，这个过程是通过 Zookeeper 节点与 Nimbus 节点通信完成的。因为 Storm 将所有的集群状态在保存在 Zookeeper 或者本地磁盘上，Supervisor 节点是无状态的，因此其失败或者重启不会引起全局的重新计算。

Storm 的主要特点如下。

（a）简单的编程模型：Storm 提供类似于 MapReduce 的操作，降低了并行批处理与实时处理的复杂性。一个 Storm 作业只需实现一个 Topology 及其所包含的 Spout 与 Bolt。通过指定它们的连接方式，Topology 可以胜任大多数的流式作业需求。

（b）容错性：Storm 利用 Zookeeper 管理工作进程和节点的故障。在工作过程中，如果出现异常，Topology 会失败。但 Storm 将以一致的状态重新启动处理，这样它可以正确地恢复。

（c）水平扩展：Storm 拥有良好的水平扩展能力，其流式计算过程是在多个线程、进程和服务器之间并行进行的。Nimbus 节点将大量的协同工作都交由 Zookeeper 节点负责，使得水平扩展不会产生瓶颈。

（d）快速可靠的消息处理：Storm 利用 ZeroMQ 作为消息队列，极大提高了消息传递的速度，系统的设计也保证了消息能得到快速处理。Storm 保证每个消息至少能得到一次完整处理。任务失败时，它会负责从消息源重试消息。

（2）Linkedin 的 Samza 系统

Linkedin 早期开发了一款名叫 Kafka[18]的消息队列，广受业界的好评，许多流式数据处理系统都使用了 Kafka 作为底层的消息处理模块。Kafka 的工作过程简要分为 4 个步骤，即生产者将消息发往中介（Broker），消息被抽象为 Key-Value 对，Broker 将消息按 Topic 划分，消费者向 Broker 拉取感兴趣的 Topic。2013 年，Linkedin 基于 Kafka 和 YARN 开发了自己的流式处理框架——Samza。Samza 与 Kafka 的关系可以类比 MapReduce 与 HDFS 的关系。Samza 系统由 3 个层次组成，包括流式数据层（Kafka）、执行层（YARN）、处理层（SamzaAPI）。一个 Samza 任务的输入与输出均是流。Samza 系统对流的模型有很严格的定义，它并不只是一个消息交换的机制。流在 Samza 的系统中是一系列划分了的、可重现的、可多播的、无状态的消息序列，每一个划分都是有序的。流不仅是 Samza 系统的输入与输出，它还充当系统中的缓冲区，能够隔离相互之间的处理过程。Samza 利用 YARN 与 Kafka 提供了分步处理与划分流的框架。Samza 客户端向 Yarn 的资源管理器提交流作业，生成多个 TaskRunner 进程，这些进程执行用户编写的 Stream Tasks 代码。该系统的输入与输出来自于 Kafka 的 Broker 进程。

Samza 的主要特性如下。

（a）高容错：如果服务器或者处理器出现故

障，Samza 将与 YARN 一起重新启动流处理器。

（b）高可靠性：Samza 使用 Kafka 来保证所有消息都会按照写入分区的顺序进行处理，绝对不会丢失任何消息。

（c）可扩展性：Samza 在各个等级进行分割和分布；Kafka 提供一个有序、可分割、可重部署、高容错的系统；YARN 提供了一个分布式环境供 Samza 容器运行。

## 2.3 交互式数据处理

### 2.3.1 交互式数据处理的特征与典型应用

（1）交互式数据处理的特征

与非交互式数据处理相比，交互式数据处理灵活、直观、便于控制。系统与操作人员以人机对话的方式一问一答——操作人员提出请求，数据以对话的方式输入，系统便提供相应的数据或提示信息，引导操作人员逐步完成所需的操作，直至获得最后处理结果。采用这种方式，存储在系统中的数据文件能够被及时处理修改，同时处理结果可以立刻被使用。交互式数据处理具备的这些特征能够保证输入的信息得到及时处理，使交互方式继续进行下去。

（2）典型应用

在大数据环境下，数据量的急剧膨胀是交互式数据处理系统面临的首要问题。下面主要选择信息处理系统领域和互联网领域作为典型应用场景进行介绍。

在信息处理系统领域中，主要体现了人机间的交互。传统的交互式数据处理系统主要以关系型数据库管理系统（DBMS）为主，面向两类应用，即联机事务处理（OLTP）和联机分析处理（OLAP）。OLTP 基于关系型数据库管理系统，广泛用于政府、医疗以及对操作序列有严格要求的工业控制领域；OLAP 基于数据仓库（Data Warehouse）系统广泛用于数据分析、商业智能（BI）等。最具代表性的处理是数据钻取，如在 BI 中，可以对于数据进行切片和多粒度的聚合，从而通过多维分析技术实现数据的钻取。目前，基于开源体系架构下的数据仓库系统发展十分迅速，以 Hive、Pig 等为代表的分布式数据仓库能够支持上千台服务器的规模。

互联网领域。在互联网领域中，主要体现了人际间的交互。随着互联网技术的发展，传统的简单按需响应的人机互动已不能满足用户的需求，用户之间也需要交互，这种需求诞生了互联网中交互式数据处理的各种平台，如搜索引擎、电子邮件、即时通讯工具、社交网络、微博、博客以及电子商务等，用户可以在这些平台上获取或分享各种信息。此外，各种交互式问答平台，如百度的知道、新浪的爱问以及 Yahoo!的知识堂等。由此可见，用户与平台之间的交互变得越来越容易，越来越频繁。这些平台中数据类型的多样性，使得传统的关系数据库不能满足交互式数据处理的实时性需求。目前，各大平台主要使用 NoSQL 类型的数据库系统来处理交互式的数据，如 HBase 采用多维有续表的列式存储方式；MongoDB 采用 JSON 格式的数据嵌套存储方式。大多 NoSQL 数据库不提供 Join 等关系数据库的操作模式，以增加数据操作的实时性。

### 2.3.2 代表性的处理系统

交互式数据处理系统的典型代表系统是 Berkeley 的 Spark 系统和 Google 的 Dremel 系统。

（1）Berkeley 的 Spark 系统

Spark[19]是一个基于内存计算的可扩展的开源集群计算系统。针对 MapReduce 的不足，即大量的网络传输和磁盘 I/O 使得效率低效，Spark 使用内存进行数据计算以便快速处理查询，实时返回分析结果。Spark 提供比 Hadoop 更高层的 API，同样的算法在 Spark 中的运行速度比 Hadoop 快 10 倍~100 倍[19]。Spark 在技术层面兼容 Hadoop 存储层 API，可访问 HDFS、HBASE、SequenceFile 等。Spark-Shell 可以开启交互式 Spark 命令环境，能够提供交互式查询。

Spark 是为集群计算中的特定类型的工作负载而设计，即在并行操作之间重用工作数据集（比如机器学习算法）的工作负载。Spark 的计算架构具有 3 个特点。

（a）Spark 拥有轻量级的集群计算框架。Spark 将 Scala 应用于他的程序架构，而 Scala 这种多范式的编程语言具有并发性、可扩展性以及支持编程范式的特征，与 Spark 紧密结合，能够轻松地操作分布式数据集，并且可以轻易地添加新的语言结构。

（b）Spark 包含了大数据领域的数据流计算和交互式计算。Spark 可以与 HDFS 交互取得里面的数据文件，同时 Spark 的迭代、内存计算以及交互式计算为数据挖掘和机器学习提供了很好的框架。

（c）Spark 有很好的容错机制。Spark 使用了弹性分布数据集（RDD），RDD 被表示为 Scala 对象分布在一组节点中的只读对象集中，这些集合是弹性的，保证了如果有一部数据集丢失时，可以对丢失的数据集进行重建。

Spark 高效处理分布数据集的特征使其有很好的应用前景，现在四大 Hadoop 发行商 Cloudera、Pivotal、MapR 以及 Hortonworks 都提供了对 Spark 的支持。

（2）Google 的 Dremel 系统

Dremel[20]是 Google 研发的交互式数据分析系统，专注于只读嵌套数据的分析。Dremel 可以组建成规模上千的服务器集群，处理 PB 级数据。传统的 MapReduce 完成一项处理任务，最短需要分钟级的时间，而 Dremel 可以将处理时间缩短到秒级。Dremel 是 MapReduce 的有力补充，可以通过 MapReduce 将数据导入 Dremel 中，使用 Dremel 来开发数据分析模型，最后在 MapReduce 中运行数据分析模型。

Dremel 作为大数据的交互式处理系统可以与传统的数据分析或商业智能工具在速度和精度上相媲美。Dremel 系统主要有以下 5 个特点。

（a）Dremel 是一个大规模系统。在 PB 级数据集上要将任务缩短到秒级，需要进行大规模的并发处理，而磁盘的顺序读速度在 100 MB/S 上下，因此在 1 s 内处理 1TB 数据就意味着至少需要有 1 万个磁盘的并发读，但是机器越多，出问题概率越大，如此大的集群规模，需要有足够的容错考虑，才能够保证整个分析的速度不被集群中的个别慢（坏）节点影响。

（b）Dremel 是对 MapReduce 交互式查询能力不足的有力补充。Dremel 利用文件系统 GFS 作为存储层，常常用它来处理 MapReduce 的结果集或建立分析原型。

（c）Dremel 的数据模型是嵌套的。Dremel 类似于 Json，支持一个嵌套的数据模型。对于处理大规模数据，有大量的 Join 操作，而传统的关系模型显得力不从心，Dremel 却可以很好地处理相关的查询操作。

（d）Dremel 中的数据是用列式存储的。使用列式存储，在进行数据分析的时候，可以只扫描所需要的那部分数据，从而减少 CPU 和磁盘的访问量。同时，列式存储是压缩友好的，通过压缩可以综合 CPU 和磁盘从而发挥最大的效能。

（e）Dremel 结合了 Web 搜索和并行 DBMS 的技术。首先，它借鉴了 Web 搜索中查询树的概念，将一个相对巨大复杂的查询，分割成较小、较简单的查询，分配到并发的大量节点上。其次，与并行 DBMS 类似，Dremel 可以提供了一个 SQL-like 的接口。

## 2.4 图数据处理系统

图由于自身的结构特征，可以很好地表示事物之间的关系，在近几年已成为各学科研究的热点。图中点和边的强关联性，需要图数据处理系统对图数据进行一系列的操作，包括图数据的存储、图查询、最短路径查询、关键字查询、图模式挖掘以及图数据的分类、聚类等。随着图中节点和边数的增多（达到几千万甚至上亿数），图数据处理的复杂性给图数据处理系统提出了严峻的挑战。下面主要阐述图数据的特征和典型应用以及代表性的图数据处理系统。

### 2.4.1 图数据的特征及典型应用

（1）图数据的特征

图数据中主要包括图中的节点以及连接节点的边，通常具有 3 个特征。第一，节点之间的关联性。图中边的数量是节点数量的指数倍，因此，节点和关系信息同等重要，图结构的差异也是由于对边做了限制，在图中，顶点和边实例化构成各种类型的图，如标签图、属性图、语义图以及特征图等。第二，图数据的种类繁多。在许多领域中，使用图来表示该邻域的数据，如生物、化学、计算机视觉、模式识别、信息检索、社会网络、知识发现、动态网络交通、语义网、情报分析等。每个领域对图数据的处理需求不同，因此，没有一个通用的图数据处理系统满足所有领域的需求。第三，图数据计算的强耦合性。在图中，数据之间是相互关联的，因此，对图数据的计算也是相互关联的。这种数据耦合的特性对图的规模日益增大达到上百万甚至上亿节点的大图数据计算提出了巨大的挑战。大图数据是无法使用单台机器进行处理的，但如果对大图数据进行并行处理，对于每一个顶点之间都是连通的图来讲，难以分割成若干完全独立的子图进行独立的并行处理；即使可以分割，也会面临并行机器的协同处理，以及将最后的处理结果进行合并等一系列问题。这需要图数据处理系统选取合适的图分割以及图计算模型来迎接挑战并解决问题。

（2）典型应用

图能很好地表示各实体之间的关系，因此，在各个领域得到了广泛的应用，如计算机领域、自然科学领域以及交通领域。

（a）互联网领域的应用。随着信息技术和网络技术的发展，以 Web 2.0 技术为基础的社交网络（如 Facebook、人人网）、微博（如 Twitter、新浪微博、腾讯微博）等新兴服务中建立了大量的在线社会网络关系，用图表示人与人之间的关系。在社交网络中，基于图研究社区发现等问题；在微博中，通过图研究信息传播与影响力最大化等问题。除此之外，用图表示如 E-mail 中的人与人之间的通信关系，从而可以研究社会群体关系等问题；在搜索引擎中，可以用图表示网页之间相互的超链接关系，从而计算一个网页的 PageRank 得分等。

（b）自然科学领域的应用。图可以用来在化学分子式中查找分子，在蛋白质网络中查找化合物，在 DNA 中查找特定序列等。

（c）交通领域的应用。图可用来在动态网络交通中查找最短路径，在邮政快递领域进行邮路规划等。当然，图还有一些其他的应用，如疾病爆发路径的预测与科技文献的引用关系等。图数据虽然结构复杂，处理困难，但是它有很好的表现力，因此得到了各领域的广泛应用。随着图数据处理中所面临的各种挑战被不断地解决，图数据处理将有更好的应用前景。

### 2.4.2 代表性图数据处理系统

现今主要的图数据库有 GraphLab、Giraph（基于 Pregel 克隆）、Neo4j、HyperGraphDB、

InfiniteGraph、Cassovary、Trinity 以及 Grappa 等。下面介绍 3 个典型的图数据处理系统，包括 Google 的 Pregel 系统，Neo4j 系统和微软的 Trinity 系统。

（1）Google 的 Pregel 系统

Pregel[21]是 Google 提出的基于批量同步并行（Bulk Synchronous Parallel，BSP）模型的分布式图计算框架，主要用于图遍历（BFS）、最短路径（SSSP）、PageRank 计算等。BSP 模型是并行计算模型中的经典模型，采用的是“计算-通信-同步”的模式。它将计算分成一系列超步（Superstep）的迭代。从纵向上看，它是一个串行模式，而从横向上看，它是一个并行的模式，每两个超步之间设置一个栅栏，即整体同步点，确定所有并行的计算都完成后再启动下一轮超步。Pregel 的设计思路是以节点为中心计算，节点有两种状态：活跃和不活跃。初始时每个节点都处于活跃状态，完成计算后每个节点主动“VotetoHalt”进入不活跃状态。如果接收到信息，则激活。没有活跃节点和消息时，整个算法结束。

Pregel 架构有 3 个主要特征。一是采用主/从（Master/Slave）结构来实现整体功能。一个节点为 Master，负责对整个图结构的任务进行切分，根据节点的 ID 进行散列计算分配到 Slave 机器，Slave 机器进行独立的超步计算，并将结果返回给 Master。二是有很好的容错机制。Pregel 通过 Checkpoint 机制实行容错，节点向 Master 汇报心跳维持状态，节点间采用异步消息传输。三是使用 GFS 或 BigTable 作为持久性的存储。Apache 根据 Google 于 2010 年发表的 Pregel 论文开发了高可扩展的迭代的图处理系统 Giraph，现在已经被 Facebook 用于分析社会网络中用户间的关系图中。

（2）Neo4j 系统

Neo4j 一个高性能的、完全兼容 ACID 特性的、鲁棒的图数据库。它基于 Java 语言开发，包括社区版和企业版，适用于社会网络和动态网络等场景。Neo4j 在处理复杂的网络数据时表现出很好的性能。数据以一种针对图形网络进行过优化的格式保存在磁盘上。Neo4j 重点解决了拥有大量连接的查询问题，提供了非常快的图算法、推荐系统以及 OLAP 风格的分析，满足了企业的应用、健壮性以及性能的需求，得到了很好的应用。

Neo4j 系统具有以下 5 个特性。

（a）支持数据库的所有特性：Neo4j 的内核是一种极快的图形引擎，支持事物的 ACID 特性、两阶段提交、符合分布式事务以及恢复等。

（b）高可用性：Neo4j 通过联机备份实现它的高可用性。

（c）可扩展性：Neo4j 提供了大规模可扩展性，可以在一台机器上处理数十亿节点/关系/属性的图，也可以扩展到多台机器上并行运行。

（d）灵活性：Neo4j 拥有灵活的数据结构，可以通过 Java-API 直接与图模型进行交互。对于 JRuby/Ruby、Scala、Python 以及 Clojure 等其他语言，也开发了相应的绑定库。

（e）高速遍历：Neo4j 中图遍历执行的速度是常数，与图的规模大小无关。它的读性能可以实现每毫秒遍历 2 000 关系，而且完全是事务性的。Neo4j 以一种延迟风格遍历图，即节点和关系只有在结果迭代器需要访问它们的时候才会被遍历并返回，支持深度搜索和广度搜索两种遍历方式。

（3）微软的 Trinity 系统

Trinity[22]是 Microsoft 推出的一款建立在分布式云存储上的计算平台，可以提供高度并行查询处理、事务记录、一致性控制等功能。Trinity 主要使用内存存储，磁盘仅作为备份存储。

Trinity 有以下 4 个特点。

（a）数据模型是超图：超图中，一条边可以连接任意数目的图顶点。此模型中图的边称为超边。基于这种特点，超图比简单图的适用性更强，保留的信息更多。

（b）并发性：Trinity 可以配置在一台或上百台计算机上。Trinity 提供了一个图分割机制，由一个 64 位的唯一标识 UID 确定各结点的位置，利用散列方式映射到相应的机器上，以尽量减少延迟。Trinity 可以并发执行 PageRank、最短路径查询、频繁子图挖掘以及随机游走等操作。

（c）具有数据库的一些特点：Trinity 是一个基于内存的图数据库，有丰富的数据库特点，如：在线高度并行查询处理、ACI 交易支持、并发控制以及一致性维护等。

（d）支持批处理：Trinity 支持大型在线查询和离线批处理，并且支持同步和不同步批处理计算。相比之下，Pregel 只支持在线查询处理，批处理必须是严格的同步计算。

微软现在使用 Trinity 作为 Probase 的基础架构，可以从网上自动获得大规模的知识库。Trinity 主要作用是分类建设、数据集成以及查询 Probase。Trinity 也被用于其他的项目中，如 Aether 项目，其功能也在不断的增加中。

### 2.5 小结

面对大数据，各种处理系统层出不穷，各有特色。总体来说，我们可以总结出 3 种发展趋势。

（1）数据处理引擎专用化：为了降低成本，提高能效，大数据系统需要摆脱传统的通用体系，趋向专用化架构技术。为此，国内外的互联网龙头企业都在基于开源系统开发面向典型应用的大规模、高通量、低成本、强扩展的专用化系统。

（2）数据处理平台多样化：自 2008 年以来克隆了 Google 的 GFS 和 MapReduce 的 ApacheHadoop 逐渐被互联网企业所广泛接纳，并成为大数据处理领域的事实标准。但在全面兼容 Hadoop 的基础上，Spark 通过更多的利用内存处理大幅提高系统性能。而 Scribe、Flume、Kafka、Storm、Drill、Impala、TEZ/Stinger、Presto、Spark/Shark 等的出现并不是取代 Hadoop，而是扩大了大数据技术的生态环境，促使生态环境向良性化和完整化发展。

（3）数据计算实时化：在大数据背景下，作为批量计算的补充，旨在将 PB 级数据的处理时间缩短到秒级的实时计算受到越来越多的关注。

## 3 大数据分析

要挖掘大数据的大价值必然要对大数据进行内容上的分析与计算。深度学习和知识计算是大数据分析的基础，而可视化既是数据分析的关键技术也是数据分析结果呈现的关键技术。本节主要介绍深度学习、知识计算和可视化等大数据分析的关键技术，同时也对大数据的典型应用包括社会媒体计算等进行简要综述。

### 3.1 深度学习

大数据分析的一个核心问题是如何对数据进行有效表达、解释和学习，无论是对图像、声音还是文本数据。传统的研究也有很多数据表达的模型和方法，但通常都是较为简单或浅层的模型，模型的能力有限，而且依赖于数据的表达，不能获得很好的学习效果。大数据的出现提供了使用更加复杂的模型来更有效地表征数据、解释数据的机会。深度学习就是利用层次化的架构学习出对象在不同层次上的表达，这种层次化的表达可以帮助解决更加复杂抽象的问题。在层次化中，高层的概念通常是通过低层的概念来定义的。深度学习通常使用人工神经网络，常见的具有多个隐层的多层感知机（MLP）就是典型的深度架构。

深度学习的起源要追溯到神经网络，20 世纪 80 年代，后向传播（BP）算法的提出使得人们开始尝试训练深层次的神经网络。然而，BP 算法在训练深层网络的时候表现不够好，以至于深层感知机的效果还不如浅层感知机。于是很多人放弃使用神经网络，转而使用凸的更容易得到全局最优解的浅层模型，提出诸如支持向量机、boosting 等浅层方法，以至于此前大部分的机器学习技术都使用浅层架构。转机出现在 2006 年，多伦多大学的 Hinton 等人[23]使用无监督的逐层贪婪的预训练（Greedy Layer-Wise Pre-Train）方法成功减轻了深度模型优化困难的问题，从而掀起了深度学习的浪潮。Hinton 引入了深度产生式模型 DBN，并提出高效的逐层贪婪的学习算法，使用 DBN 初始化一个深度神经网络（DNN）再对 DNN 进行精调，通常能够产生更好的结果。Bengio 等人[24]基于自动编码器（Auto-Encoder）提出了非概率的无监督深度学习模型，也取得了类似的效果。

近几年，深度学习在语音、图像以及自然语言理解等应用领域取得一系列重大进展。从 2009 年开始，微软研究院的 Dahl 等人[25]率先在语音处理中使用深度神经网络（DNN），将语音识别的错误率显著降低，从而使得语音处理成为成功应用深度学习的第 1 个领域。在图像领域，2012 年，Hinton 等人[26]使用深层次的卷积神经网络（CNN）在 ImageNet 评测上取得巨大突破，将错误率从 26%降低到 15%，重要的是，这个模型中并没有任何手工构造特征的过程，网络的输入就是图像的原始像素值。在此之后，采用类似的模型，通过使用更多的参数和训练数据，ImageNet 评测的结果得到进一步改善，错误率下降至 2013 年的 11.2%。Facebook 人工智能实验室的 Taigman 等人[27]使用了与文献[26]中类似的神经网络在人脸识别上也取得了很好的效果，将人脸识别的正确率提升至接近人类水平。此外，图像领域还有一些基于无监督的深度学习研究，比如在 GoogleBrain 项目中，Le 等人[28]尝试使用完全无标注的图像训练得到人脸特征检测器，使用这些学习到的特征可以在图像分类中取得非常好的效果；Google 的深度学习系统（DistBelief）在获取数百万 YouTube 视频数据后，能够精准地识别出这些视频中的关键元素——猫。在自然语言领域，从 2003 年开始，Bengio 等人[29]使用神经网络并结合分布式表达（Distributed Representation）的思想训练语言模型并取得很好的效果，不过当时还没有使用到更深层次的模型。2008 年，Collobert 等人[30]训练了包含一个卷积层的深度神经网络，利用学习得到的中间表达同时解决多个 NLP 问题。尽管这些工作没有取得像图像和语音处理领域如此重大的进展，但也都接近或超过了已有的最好方法。近年来，斯坦福大学的 Socher 等人[31]的一系列工作也值得关注。他们使用递归神经网络（Recursive Neural Network，RNN）在情感分析等问题上取得一系列进展，将现有的准确率从 80%提升到 85%。在国内，2011 年科大讯飞首次将 DNN 技术运用到语音云平台，并提供给开发者使用，并在讯飞语音输入法和讯飞口讯等产品中得到应用。百度成立了深度学习研究院（IDL），专门研究深度学习算法，目前已有多项深度学习技术在百度产品上线。深度学习对百度影响深远，在语音识别、OCR 识别、人脸识别、图像搜索等应用上取得了突出效果。此外，国内其他公司如搜狗、云知声等纷纷开始在产品中使用深度学习技术。

## 3.2 知识计算

基于大数据的知识计算是大数据分析的基础。知识计算是国内外工业界开发和学术界研究的一个热点。要对数据进行高端分析，就需要从大数据中先抽取出有价值的知识，并把它构建成可支持查询、分析和计算知识库。目前，世界各国各个组织建立的知识库多达 50 余种，相关的应用系统更是达到了上百种。其中，代表性的知识库或应用系统有 KnowItAll[32-33]、TextRunner[34]、NELL[35]、Probase[36]、Satori[37]、PROSPERA[38]、SOFIE[39] 以及一些基于维基百科等在线百科知识构建的知识库，如 DBpedia[40]、YAGO[41-43]、Omega[44]和 WikiTaxonomy[45-46]。除此之外，一些著名的商业网站、公司和政府也发布了类似的知识搜索和计算平台，如 Evi 公司的 TrueKnowledge 知识搜索平台，美国官方政府网站 Data.gov，Wolfram 的知识计算平台 wolframalpha，Google 的知识图谱（Knowledge Graph）、Facebook 推出的类似的实体搜索服务 GraphSearch 等。在国内，中文知识图谱的构建与知识计算也有大量的研究和开发工作。代表性工作有中国科学院计算技术研究所的 OpenKN，中国科学院数学研究院陆汝钤院士提出的知件（Knowware），上海交通大学最早构建的中文知识图谱平台 zhishi.me，百度推出了中文知识图谱搜索，搜狗推出的知立方平台，复旦大学 GDM 实验室推出的中文知识图谱展示平台等。

支持知识计算的基础是构建知识库，这包括 3 个部分，即知识库的构建、多源知识的融合与知识库的更新。知识库的构建就是要构建几个基本的构成要素，包括抽取概念、实例、属性和关系。从构建方式上，可以分为手工构建和自动构建。手工构建是依靠专家知识编写一定的规则，从不同的来源收集相关的知识信息，构建知识的体系结构[47]。比较典型的例子是知网（Hownet）[48]、同义词词林[49]、概念层次网络（HNC）[50]和中文概念词典（CCD）[51]、OpenCyc[52]等。自动构建是基于知识工程、机器学习、人工智能等理论自动从互联网上采集并抽取概念、实例、属性和关系[53-54]。比较著名的例子是 Probase[36]、YAGO[41-43]等。手工构建知识库，需要构建者对知识的领域有一定的了解，才能编写出合适的规则，开发过程中也需要投入大量的人力物力。相反地，自动构建的方法依靠系统自动的学习经过标注的语料来获取规则的，如属性抽取规则、关系抽取规则等，在一定程度上可以减少人工构建的工作量。随着大数据时代的到来，面对大规模网页信息中蕴含的知识，自动构建知识库的方法越来越受到人们的重视和青睐。自动构建知识库的方法主要分为有监督的构建方法和半监督的构建方法两种。有监督的构建方法是指系统通过学习训练数据，获取抽取规则，然后根据这些规则，提取同一类型的网页中的概念、实例、属性和关系。这类方法的缺点是规则缺乏普适性。而且，由于规则是针对特定网页的，当训练网页发生变化，需要重新进行训练来获取规则。半监督的构建方法是系统预先定义一些规则作为种子，然后通过机器学习算法，从标注语料中抽取相应的概念、实例、属性和关系。进一步地，系统根据抽取的结果，发现新的规则，再用来指导抽取相应的概念、实例、属性和关系，从而使抽取过程能够迭代的进行。

多源知识的融合是为了解决知识的复用问题。如前文所述，构建一个知识库的代价是非常大的，为了避免从头开始，需要考虑知识的复用和共享，这就需要对多个来源的知识进行融合，即需要对概念、实例、属性和关系的冲突，重复冗余，不一致进行数据的清理工作，包括对概念、实例进行映射、消歧，对关系进行合并等。这其中概念间关系或分类体系的融合是很关键一部分。按融合方式可以分为手动融合和自动融合。对于规模较小的知识库，手动融合是可行的，但这是一种非常费时而且容易出错的融合方式。相比于手动融合方式，建立在机器学习、人工智能和本体工程等算法上的融合方式具有更好的可扩展性，相关工作包括 YAGO[41-43]、Probase[36]等。YAGO 知识库将维基百科，WordNet 和 GeoNames 等数据源的知识整合在知识库中。其中，将维基百科的分类体系和 WordNet 的分类体系进行融合是 YAGO 的重要的工作之一。维基百科的分类是一个有向无环图生成的层次结构[41]，这种结构由于仅能反映主题信息，所以容易出错。Probase 提出了一种基于概率化的实体消解（Entity Resolution）的知识整合技术[35]，将现有结构化数据，如 Freebase、IMDB、Amazon 等整合到 Probase 当中。对多源知识的融合，除了分类体系的融合外，还包括对实体和概念的消解问题，实体和概念的消歧问题等。面对海量知识库时，建立若干个针对不同领域，不同需求的有效的知识融合算法，快速进行多元知识的融合，是亟待进一步解决的问题之一。

大数据时代数据的不断发展与变化带给知识库构建的一个巨大的挑战是知识库的更新问题。知识库的更新分为两个层面，一是新知识的加入；二是已有知识的更改。目前专门针对开放网络知识库的更新工作较少，很多都是从数据库的更新角度展开的，如对数据库数据的增加、删除和修改工作的介绍。虽然对开放网络知识库的更新，与数据库的更新有很多相似之处，但是其本身对更新的实时性要求较高。目前这方面的工作，从更新方式来讲分为两类：一是基于知识库构建人员的更新；二是基于知识库存储的时空信息的更新。前者准确性较高，但是对人力的消耗较大。后者多由知识库自身更新，需要人工干预的较少，但是存在准确率不高的问题。总体上讲，对知识库的更新仍然没有很有效的方法。尤其在面对用户对知识的实时更新需求方面，远远达不到用户的要求。在更新数据的自动化感知方面，缺乏有效的办法自动识别知识的变化，也没有能够动态响应这些变化的更新机制。

## 3.3 社会计算

以 Facebook、Twitter、新浪微博、微信等为代表的在线社交网络和社会媒体正深刻改变着人们传播信息和获取信息的方式，人和人之间结成的关系网络承载着网络信息的传播，人的互联成为信息互联的载体和信息传播的媒介，社会媒体的强交互性、时效性等特点使其在信息的产生、消费和传播过程中发挥着越来越重要的作用，成为一类重要信息载体。正因如此，当前在线社会计算无论在学术圈和工业界都备受重视，大家关注的问题包括了对在线社会网络结构、信息传播以及信息内容的分析、建模与挖掘等一系列问题。

### 3.3.1 在线社会网络的结构分析

在线社会网络在微观层面上具有随机化无序的现象，在宏观层面上往往呈现出规则化、有序的现象，为了理清网络具有的这种看似矛盾的不同尺度的结构特性，探索和分析连接微观和宏观的网络中观结构（也称社区结构）成为了本领域一个重要的研究方向。一般意义上讲，社区结构是指网络节点按照连接关系的紧密程度不同而自然分成若干个内部连接紧密、与外部连接稀疏的节点组，每个节点组相应地被称为社区[55]。社区分析研究目前主要包括社区的定义和度量、社区结构发现和社区结构演化性分析等基本问题[56]。

社区定义或度量大体上分为 4 类，基于节点的社区定义、基于节点组（社区）的社区定义、基于网络整体的社区定义、基于层次结构的社区定义。目前，社区结构的研究主要集中在基于某种给定社区定义或度量的社区发现上。最具代表性的社区发现算法包括密歇根大学 Newman 等人提出的模块度（Modularity）优化方法[57]、匈牙利科学院 Palla 等人提出的完全子图渗流（Clique Percolation）方法[58]、华盛顿大学的 Rosvall 等人提出的基于网络最短编码的 InfoMap 方法[59]、Airoldi 等人[60]提出的 Mixed Membership Stochastic Block（MMSB）模型，这些社区发现方法在人工构造的测试网络和一些小规模的真实网络上取得了很好的效果。真实世界在线社交网络中的社区结构具有多尺度、重叠等特点，近几年逐步引起研究人员的关注，成为一个研究热点[61-65]。

网络社区的演化性是信息网络的一个基本特性，也是促使大规模信息网络的内容与结构涌现现象及信息大规模传播的基本原因[66]。近几年，在前述社区发现研究的基础上，人们开始研究社区随时间演化的规律[67]。例如，Palla 等人[68]基于完全子图渗流社区发现方法研究社区演化，得到一个有趣结论，小社区的稳定性是保证它存在的前提，大社区的动态性是它存在的基础。Song 等人[69]考虑了网络结构变化的时间因素，并认为网络演化过程是平滑的，他们使用扩展了的动态贝叶斯网络来建模网络的演化过程，取得了很好的效果。Xing 等人[70]将网络演化的观点引入结点的角色分析中。在 MMSB 模型中加入时间因素，他们认为两个相邻的时间片内角色选择方式和角色之间的关系具有一阶马尔可夫性质。此外，社区结构被用于预测网络中潜在存在的边，对于网络演化具有重要意义[71]。

### 3.3.2 在线社会网络的信息传播模型

在信息传播模型的研究中，最广泛深入研究的是传染病模型[72-73]，除了传染病模型，随机游走模型也是信息传播的基本模型之一[74]，作为基本的动力学过程之一，随机游走与网络上的许多其他动力学过程（如反应–扩散过程、社团挖掘、路由选择、目标搜索）紧密相关。

近几年，研究人员开始注意到信息传播和传染病传播具有显著不同的特性[75]，包括信息传播的记忆性、社会增强效应、不同传播者的角色不同、消息内容的影响等。Romero 等人[76]提出了 Stickiness 和 Persistence 两个重要概念，分析不同领域内的 Hashtag 在 Twitter 上的传播过程。Wu 等人[77]分析名人、机构、草根等不同群体之间的消息流向，并分析了不同类型的消息被转发的情况及其生命周期。Lerman 等人[78]从网络动力学角度，通过实际数据分析了 Twitter 中消息传播的特性。Castillo 等人[79]通过特征提取，利用机器学习中分类的方法，对 Twitter 中消息的可信度建模，并预测其中消息的可信性。Phelan 等人[80]提出了一种 Twitter 消息新颖度的度量，并建立了向用户实时推荐新消息的系统。Lerman 等人[81]利用概率方法和先验知识，对 Digg 中的消息建模，预测消息的流行度。当前，对在线社交网络中信息传播的研究主要集中在实证分析和统计建模，对于信息传播机理仍然缺乏深入的理解和有效的建模。

### 3.3.3 社会媒体中信息检索与数据挖掘

社会媒体的出现对信息检索与数据挖掘的研究提出了新的挑战。不同于传统的 Web 数据，社会媒体中的数据呈现出一些新的特征：信息碎片化现象明显，文本内容特征越发稀疏；信息互联被人的互联所取代，社会媒体用户形成的社会关系网络的搜索和挖掘过程中的重要组成部分；社会媒体的易参与性使得人人具有媒体的特征，呈现出自媒体现象，个人影响力、情感与倾向性掺杂其中。针对这些特点，研究人员在传统信息检索与数据挖掘技术基础上提出了一系列的新模型[82-83]。

鉴于用户所创造的信息往往具有很强的时效性，Yang 等人[84]提出了一种时间序列聚类的方法，从 Twitter 数据中挖掘热门话题发展趋势的规律。因为用户的状态和评论中包含了大众的观点和态度，所以 Bollen 等人[85]通过对 Twitter 中用户的信息进行情感分析，将大众情绪的变化表示为 7 种不同的情绪时间序列，进而发现这些序列能够预测股票市场的走势。此外，基于用户在协作平台上所贡献的内容和标签等信息往往蕴含有丰富的大众知识和智慧这一现象，Hu 等人[86]利用 Wikipedia 中的文章和类别信息来确定用户的查询意图，进而辅助信息检索。社会媒体的检索与挖掘研究在国内也受到了越来越多的重视，包括北京大学、清华大学、哈尔滨工业大学、上海交通大学、浙江大学、复旦大学、中国科学院、微软亚洲研究院等大学和研究机构已经取得了一定的进展，涉及的研究内容包括社会化标签系统中的标签学习和排序[87-88]、信息抽取和分类[89]、社会化多媒体检索[90]、协作搜索和推荐[91-92]等。

### 3.4 可视化

大数据引领着新一波的技术革命，对大数据查询和分析的实用性和实效性对于人们能否及时获得决策信息非常重要，决定着大数据应用的成败。但产业界面对大数据常常显得束手无策。一是因为数据容量巨大，类型多样，数据分析工具面临性能瓶颈。另一原因在于，数据分析工具通常仅为 IT 部门熟练使用，缺少简单易用、让业务人员也能轻松上手实现自助自主分析即时获取商业洞察的工具。因此，数据可视化技术正逐步成为大数据时代的显学。对大数据进行分析以后，为了方便用户理解也需要有效的可视化技术，这其中交互式的展示和超大图的动态化展示值得重点关注。

大数据可视化，不同于传统的信息可视化，面临最大的一个挑战就是规模，如何提出新的可视化方法能够帮助人们分析大规模、高维度、多来源、动态演化的信息，并辅助作出实时的决策，成为了这个领域最大的挑战。为了解决这个问题，我们可以依赖的主要手段是两种，即数据转换和视觉转换。现有研究工作主要聚焦在 4 个方面。

（1）通过对信息流进行压缩或者删除数据中的冗余信息对数据进行简化。其中很多工作主要解决曲面的可视化，使用基本的数据转换方法来对数据进行简化。例如，文献[93-94]提出通过删除节点以及包含这个节点的三角形进行网格的简化，而 Hoppe 等人[95-96]则提出了一种渐进网格表达方法，通过有效地删除边及其所属的三角形实现。一些研究人员把上述的这些曲面算法进一步扩展到四面体上[97-98]。上述这些工作存在主要的不同之处在于基本的数据转换步骤中使用不同的错误近似方法。

（2）通过设计多尺度、多层次的方法实现信息在不同的解析度上的展示，从而使用户可自主控制展示解析度。很多已有多尺度算法集中在对地形类数据的渲染上。例如，一些使用固定网格方法的系统建立在直角三角形的层次结构之上[99-100]，而一些不规则三角形网[101-102]则是通过不把三角形限制在固定网格上的方式来解决这个问题，这两类方法各有利弊。Cignoni 等人[103-104]则通过利用四叉树纹理层次以及利用三角片面二叉树显示几何形，展示了实时显示大型地形数据集中自适应的几何形和纹理的能力。

（3）利用创新的方法把数据存储在外存，并让用户可以通过交互手段方便地获取相关数据，这类研究也成为核外算法（Out-of-Core Algorithm）。为了应对大规模数据结构无法在内存中存放，而外存访问时间又极大地依赖于外部存储单元的位置，人们设计了一些新的算法与分析工具来解决几何算法[105-106]以及可视化方法。这类工作重点解决两个问题：（a）对算法进行分析得到数据访问的模式，从而重新设计数据结构来最大化访问的局部性；（b）在二级存储设备中的数据需要有与算法访问模式匹配的存放布局。Pascucci 和 Frank[107]引入了一种新的静态索引体系使得在层次化遍历 ***n*** 维规则网格时候，数据的存放布局能够满足上述两个要求。

（4）提出新的视觉隐喻方法以全新的方式展示数据。其中，一类典型的方法是“焦点+上下文”方法，它重点对焦点数据进行细节展示，对不重要数据的则简化表示，例如鱼眼视图[108]。Plaisant 提出了空间树（Space Tree）[109]，一种树形浏览器通过动态调整树枝的尺寸来使其最好地适配显示区域。分层平行坐标方法，作为平行坐标方法的多尺度版本，通过在不同的细节层次使用多的视图来对大规模数据进行表达。

对大数据进行探索和可视化仍然还处在初始阶段，特别是对于动态多维度大数据流的可视化技术还非常匮乏，非常需要扩展现有的可视化算法，研究新的数据转换方法以便能够应对复杂的信息流数据。也需要设计创新的交互方式来对大数据进行可视化交互和辅助决策。

### 3.5 小结

大数据处理和分析的终极目标是借助对数据的理解辅助人们在各类应用中做出合理的决策。在此过程中，深度学习、知识计算、社会计算和可视化起到了相辅相成的作用。

（1）深度学习提高精度：如前所述，要挖掘大数据的大价值必然要对大数据进行内容上的分析与计算，而传统的数据表达模型和方法通常是简单的浅层模型学习，效果不尽人意。深度学习可以对人类难以理解的底层数据特征进行层层抽象，凝练具有物理意义的特征，从而提高数据学习的精度。因此，深度学习是大数据分析的核心技术。

（2）知识计算挖掘深度：每一种数据来源都有一定的局限性和片面性，只有对各种来源的原始数据进行融合才能反映事物的全貌，事物的本质和规律往往隐藏在各种原始数据的相互关联之中。而借助知识计算可以将碎片化的多源数据整合成反映事物全貌的完整数据，从而增加数据挖掘的深度。因此，基于大数据的知识计算是大数据分析的基础。如何基于大数据实现新知识的感知，知识的增量式演化和自适应学习是其中的重大挑战。

（3）社会计算促进认知：IT 技术的发展使得社会媒体成了一类重要的信息载体，承载着对事物的客观或主观描述信息。因此，通过基于社会媒体数据的社会计算可以促进人们对事物的认知。但是，社会媒体大数据往往蕴含着一个体量庞大、关系异质、结构多尺度和动态演化的网络，对它的分析既要有效地计算方法，更需要支持大规模网络结构的图数据存储和管理结构，以及高性能的图计算系统结构和算法。

（4）强可视化辅助决策：对大数据查询和分析的实用性和实效性对于人们能否及时获得决策信息非常重要。而强大的可视化技术，不仅可以对数据分析结果进行更有效的展示，而且可以在大数据分析过程中发挥重要作用。

## 4 大数据计算面临的挑战与应对之策

尽管大数据是社会各界都高度关注的话题，但时下大数据从底层的处理系统到高层的分析手段都存在许多问题，也面临一系列挑战。这其中有大数据自身的特征导致的，也有当前大数据分析模型与方法引起的，还有大数据处理系统所隐含的。本节对这些问题与挑战进行梳理。

## 4.1 数据复杂性带来的挑战

大数据的涌现使人们处理计算问题时获得了前所未有的大规模样本，但同时也不得不面对更加复杂的数据对象，如前所述，其典型的特性是类型和模式多样、关联关系繁杂、质量良莠不齐。大数据内在的复杂性（包括类型的复杂、结构的复杂和模式的复杂）使得数据的感知、表达、理解和计算等多个环节面临着巨大的挑战，导致了传统全量数据计算模式下时空维度上计算复杂度的激增，传统的数据分析与挖掘任务如检索、主题发现、语义和情感分析等变得异常困难。然而目前，人们对大数据复杂性的内在机理及其背后的物理意义缺乏理解，对大数据的分布与协作关联等规律认识不足，对大数据的复杂性和计算复杂性的内在联系缺乏深刻理解，加上缺少面向领域的大数据处理知识，极大地制约了人们对大数据高效计算模型和方法的设计能力。

因此，如何形式化或定量化地描述大数据复杂性的本质特征及其外在度量指标，进而研究数据复杂性的内在机理是个根本问题。通过对大数据复杂性规律的研究有助于理解大数据复杂模式的本质特征和生成机理，简化大数据的表征，获取更好的知识抽象，指导大数据计算模型和算法的设计。为此，需要建立多模态关联关系下的数据分布理论和模型，理清数据复杂度和时空计算复杂度之间的内在联系，通过对数据复杂性内在机理的建模和解析，阐明大数据按需约简、降低复杂度的原理与机制，使其成为大数据计算的理论基石。

## 4.2 计算复杂性带来的挑战

大数据多源异构、规模巨大、快速多变等特性使得传统的机器学习、信息检索、数据挖掘等计算方法不能有效支持大数据的处理、分析和计算。特别地，大数据计算不能像小样本数据集那样依赖于对全局数据的统计分析和迭代计算，需要突破传统计算对数据的独立同分布和采样充分性的假设。在求解大数据的问题时，需要重新审视和研究它的可计算性、计算复杂性和求解算法。因此，研究面向大数据的新型高效计算范式，改变人们对数据计算的本质看法，提供处理和分析大数据的基本方法，支持价值驱动的特定领域应用，是大数据计算的核心问题。而大数据样本量充分，内在关联关系密切而复杂，价值密度分布极不均衡，这些特征对研究大数据的可计算性及建立新型计算范式提供了机遇，同时也提出了挑战。

因此，需要着眼于大数据的全生命周期，基于大数据复杂性的基本特征及其量化指标，研究大数据下以数据为中心的计算模式，突破传统的数据围绕机器式计算，构建以数据为中心的推送式计算模式，探索弱 CAP 约束的系统架构模型及其代数计算理论，研究分布化、流式计算算法，形成通信、存储、计算融合优化的大数据计算框架；研究适应大数据的非确定性算法理论，突破传统统计学习中的独立同分布假设；也需要探索从足够多（Large Enough）的数据，到刚刚好（Just Enough）的数据，再到有价值（Valuable Enough）的数据的按需约简方法，研究基于自举和采样的局部计算和近似方法，提出不依赖于全量数据的新型算法理论基础。

## 4.3 系统复杂性带来的挑战

针对不同数据类型与应用的大数据处理系统是支持大数据科学研究的基础平台。对于规模巨大、结构复杂、价值稀疏的大数据，其处理亦面临计算复杂度高、任务周期长、实时性要求强等难题。大数据及其处理的这些难点不仅对大数据处理系统的系统架构、计算框架、处理方法提出了新的挑战，更对大数据处理系统的运行效率及单位能耗提出了苛刻要求，要求大数据处理系统必须具有高效能的特点。对于以高效能为目标的大数据处理系统的系统架构设计、计算框架设计、处理方法设计和测试基准设计研究，其基础是大数据处理系统的效能评价与优化问题研究。这些问题的解决可奠定大数据处理系统设计、实现、测试与优化的基本准则，是构建能效优化的分布式存储和处理的硬件及软件系统架构的重要依据和基础，因此是大数据分析处理所必须解决的关键问题。

大数据处理系统的效能评价与优化问题具有极大的研究挑战性，其解决不但要求理清大数据的复杂性、可计算性与系统处理效率、能耗间的关系，还要综合度量系统中如系统吞吐率、并行处理能力、作业计算精度、作业单位能耗等多种效能因素，更涉及实际负载情况及资源分散重复情况的考虑。因此，为了解决系统复杂性带来的挑战，人们需要结合大数据的价值稀疏性和访问弱局部性的特点，针对能效优化的大数据分布存储和处理的系统架构，以大数据感知、存储与计算融合为大数据的计算准则，在性能评价体系、分布式系统架构、流式数据计算框架、在线数据处理方法等方面展开基础性研究，并对作为重要验证工具的基准测试程序及系统性能预测方法进行研究，通过设计、实现与验证的迭代完善，最终实现大数据计算系统的数据获取高吞吐、数据存储低能耗和数据计算高效率。

## 5 结束语

互联网、物联网、云计算技术的快速发展，各类应用的层出不穷引发了数据规模的爆炸式增长，使数据渗透到了当今每一个行业和业务领域，成为重要的生产因素。大数据因此成为社会各界关注的新焦点，大数据时代已然来临。为了应对不同的业务需求，以 Google、Facebook、Linkedin、Microsoft 等为代表的互联网企业近几年推出了各种大数据处理系统，深度学习、知识计算、可视化等大数据分析技术也得到迅速发展，已被广泛应用于不同的行业和领域。本文根据处理形式的不同，介绍了批量处理数据、流式处理数据、交互处理数据和图数据4种不同形式数据的突出特征和各自的典型应用场景以及相应的代表性处理系统，并总结出引擎专用化、平台多样化、计算实时化是当前大数据处理系统的三大发展趋势。随后，对系统支撑下的深度学习、知识计算、社会计算与可视化4类大数据分析技术和应用进行了简要综述，总结了各种技术在大数据分析理解过程中的关键作用，即深度学习提高精度，知识计算挖掘深度，社会计算促进认知，强可视化辅助决策。本文最后梳理了大数据处理和分析面临的 3 个核心挑战，包括数据复杂性、计算复杂性和系统复杂性，并提出了可能的应对之策。

## 参考文献：


[1] MANYIKA J, CHUI M, BROWN B, et al. Big data: the next frontier for innovation, competition, and productivity[R]. 2011.

[2] LI G J, CHENG X Q. Research status and scientific thinking of big data[J]. Bulletin of the Chinese Academy of Sciences, 2012, 27(6): 647-657.

[3] WANG Y Z, JIN X L, CHENG X Q. Network big data: present and future[J]. Chinese Journal of Computers, 2013, 36(6): 1125-1138.

[4] ARTHUR W B. The second economy[R]. 2011.

[5] MAYER-SCHÖNBERGER V, CUKIER K. Big data: a revolution that will transform how we live, work, and think[M]. Boston: Houghton Mifflin Harcourt, 2013.

[6] SUN D W, ZHANG G Y, ZHENG W M. Big data stream computing: technologies and instances[J]. Journal of Software, 2014, 25(4): 839-862.

[7] TSOURAKAKIS C E. Fast counting of triangles in

large real networks without counting: algorithms and laws[C]// The 8th IEEE International Conference on Data Mining. Piscataway: IEEE Press, 2008.

[8] CHEN Y, ALSPAUGH S, KATZ R. Interactive analytical processing in big data systems: a cross-industry study of MapReduce workloads[J]. Proceedings of the VLDB Endowment, 2012, 5(12): 1802-1813.

[9] STUPAR A, MICHEL S, SCHENKEL R. Rank reduce-processing k-nearest neighbor queries on top of MapReduce[J]. Large-Scale Distributed Systems for Information Retrieval, 2010: 13-18.

[10] ZHOU M Q, ZHANG R, XIE W, et al. Security and privacy in cloud computing: a survey[C]// The 6th International Conference on Semantics, Knowledge and Grids. Piscataway: IEEE Press, 2010.

[11] FEBLOWITZ J. Analytics in oil and gas: the big deal about big data[C]// The SPE Digital Energy Conference. [S.l.:s.n.], 2013.

[12] YU H, WANG D. Research and implementation of massive health care data management and analysis based on Hadoop[C]// IEEE 4th International Conference on Computational and Information Sciences. Piscataway: IEEE Press, 2012.

[13] GHEMAWAT S, GOBIOFF H, LEUNG S T. The Google file system[J]. ACM SIGOPS Operating Systems Review, 2003, 37(5): 29-43.

[14] DEAN J, GHEMAWAT S. MapReduce: simplified data processing on large clusters[J]. Communications of the ACM, 2008, 51(1): 107-113.

[15] DEAN J, GHEMAWAT S. MapReduce: a flexible data processing tool[J]. Communications of the ACM, 2010, 53(1): 72-77.

[16] WHITE T. Hadoop: the definitive guide[M]. Sebastopol: O'Reilly Media, Inc., 2012.

[17] CHAKRAVARTHY S, JIANG Q. Stream data processing: a quality of service perspective-modeling, scheduling, load shedding, and complex event processing[M]. Berlin: Springer-Verlag, 2009.

[18] GOODHOPE K, KOSHY J, KREPS J, et al. Building LinkedIn's real-time activity data pipeline[J]. IEEE Data Engineering Bulletin, 2012, 35(2): 33-45.

[19] ZAHARIA M, CHOWDHURY M, FRANKLIN M, et al. Spark: cluster computing with working sets[C]// HotCloud 2010. [S.l.:s.n.], 2010.

[20] MELNIK S, GUBAREV A, LONG J J, et al. Dremel: interactive analysis of web-scale datasets[J]. Proceedings of the VLDB Endowment, 2010, 3(1-2): 330-339.

[21] MALEWICZ G, AUSTERN M H, BIK A J, et al. Pregel: a system for large-scale graph processing[C]// The 2010 ACM SIGMOD International Conference on Management of Data. New York: ACM Press, 2010.

[22] SHAO B, WANG H, LI Y. Trinity: a distributed graph engine on a memory cloud[C]// The 2013 ACM SIGMOD International Conference on Management of Data. New York: ACM Press, 2013.

[23] HINTON G, OSINDERO S, TEH Y W. A fast learning algorithm for deep belief nets[J]. Neural Computation, 2006, 18(7): 1527-1554.

[24] BENGIO Y, LAMBLIN P, POPOVICI D, et al. Greedy layer-wise training of deep networks[C]// Conference and Workshop on Neural Information Processing Systems. [S.l.:s.n.], 2007.

[25] DAHL G E, YU D, DENG L, et al. Context-dependent pre-trained deep neural networks for large-vocabulary speech recognition[J]. IEEE Transactions on Audio, Speech, and Language Processing, 2012, 20(1): 30-42.

[26] KRIZHEVSKY A, SUTSKEVER I, HINTON G E. Imagenet classification with deep convolutional neural networks[C]// Conference and Workshop on Neural Information Processing Systems. [S.l.:s.n.], 2012.

[27] TAIGMAN Y, YANG M, RANZATO M, et al. DeepFace: closing the gap to human-level performance in face verification[C]// 2014 IEEE Conference on Computer Vision and Pattern Recognition. Piscataway: IEEE Press, 2014.

[28] LE QV. Building high-level features using large scale unsupervised learning[C]// 2013 IEEE International Conference on Acoustics, Speech and Signal Processing. Piscataway: IEEE Press, 2013.

[29] BENGIO Y, SCHWENK H, SENÉCAL J S, et al. Neural probabilistic language models[M]// Innovations in Machine Learning. Berlin: Springer-Verlag, 2006: 137-186.

[30] COLLOBERT R, WESTON J. A unified architecture for natural language processing: deep neural networks with multitask learning[C]// The 25th International Conference on Machine Learning. New York: ACM Press, 2008.

[31] SOCHER R, PERELYGIN A, WU J Y, et al. Recursive deep models for semantic compositionality over a sentiment treebank[C]// The Conference on Empirical Methods in Natural Language Processing. [S.l.:s.n.], 2013.

[32] ETZIONI O, CAFARELLA M, DOWNEY D, et al. Unsupervised named-entity extraction from the web: an

experimental study[J]. Artificial Intelligence, 2005, 165(1): 91-134.

[33] ETZIONI O, CAFARELLA M, DOWNEY D, et al. Web-scale information extraction in knowitall: (preliminary results)[C]// The 13th International Conference on World Wide Web. New York: ACM Press, 2004.

[34] BANKO M, CAFARELLA M J, SODERLAND S, et al. Open information extraction for the web[C]// 2007 IJCAI. [S.l.:s.n.], 2007.

[35] CARLSON A, BETTERIDGE J, KISIEL B, et al. Toward an architecture for never-ending language learning[C]// The 24th AAAI Conference on Artificial Intelligence. Menlo Park: AAAI Press, 2010.

[36] WU W, LI H, WANG H, et al. Probase: a probabilistic taxonomy for text understanding[C]//The 2012 ACM SIGMOD International Conference on Management of Data. New York: ACM Press, 2012.

[37] GALLAGHER S. How Google and Microsoft taught search to understand the web[R]. 2012.

[38] NAKASHOLE N, THEOBALD M, WEIKUM G. Scalable knowledge harvesting with high precision and high recall[C]// The 4th ACM International Conference on Web Search and Data Mining. New York: ACM Press, 2011.

[39] SUCHANEK F M, SOZIO M, WEIKUM G. SOFIE: a self-organizing framework for information extraction[C]// The 18th International Conference on World Wide Web. New York: ACM Press, 2009.

[40] AUER S, BIZER C, KOBILAROV G, et al. DBpedia: a nucleus for a web of open data[C]// The 6th International the Semantic Web and the 2nd Asian Conference on Asian Semantic Web Conference. Piscataway: IEEE Press, 2007

[41] BIEGA J, KUZEY E, SUCHANEK F M. Inside YAGO2s: a transparent information extraction architecture[C]// The 22th International Conference on World Wide Web. New York: ACM Press, 2013.

[42] HFFART J, SUCHANEK F, BERBERICH K, et al. YAGO2: a spatially and temporally enhanced knowledge base from Wikipedia[J]. Artificial Intelligence Journal, 2013, 194(4): 28-61.

[43] SUCHANEK F, KASNECI G, WEIKUM G. YAGO: a core of semantic knowledge[C]// The 16th International Conference on World Wide Web. New York: ACM Press, 2007.

[44] PHILPOT A, HOVY E H, PANTEL P. Ontology and the Lexicon[M]// The omega ontology. Cambridge: Cambridge University Press, 2008: 35-78.

[45] PONZETTO S, NAVIGLI R. Large-scale taxonomy mapping for restructuring and integrating Wikipedia[C]// The 21st International Joint Conference on Artificial Intelligence. San Francisco: Morgan Kaufmann Publishers, 2009.

[46] PONZETTO S, STRUBE M. Taxonomy induction based on a collaboratively built knowledge repository[J]. Artificial Intelligence, 2011, 175(9-10): 1737-1756.

[47] SHI Z Z. Knowledge discovery[M]. Beijing: Tsinghua University Press, 2002.

[48] DONG Z D, DONG Q, HAO C L. Theoretical findings of HowNet[J]. Journal of Chinese Information Processing, 2007, 21(4): 3-9.

[49] MEI L J, ZHOU Q, CANG L, et al. Merge information in HowNet and TongYiCi CiLin[J]. Journal of Chinese Information Processing, 2005, 19(1): 63-70.

[50]黄曾阳. HNC 理论概要[J]. 中文信息学报, 1997, 11(4): 11-20.

[51] YU J S, YU S W. The structure of Chinese concept dictionary[J]. Journal of Chinese Information Processing, 2002, 16(4): 12-20.

[52] XU W Y, LIU S Y. Logic for knowledgebase systems[J]. Chinese Journal of Computers, 2009, 32(11): 2123-2129.

[53] ZHONG X Q, LIU Z, DONG P P. Construction of knowledge base on hybrid reasoning and its application[J]. Chinese Journal of Computers, 2012, 35(4): 761-766.

[54] CHEN L W, FENG Y S, ZHAO D Y. Extracting relations from the web via weakly supervised learning[J]. Journal of Computer Research and Development, 2013, 50(9): 1825-1835.

[55] GIRVAN M, NEWMAN M E J. Community structure in social and biological networks[J]. Proceedings of the National Academy of Sciences, 2002, 99(12): 7821-7826.

[56] FORTUNATO S. Community detection in graphs[J]. Physics Reports, 2010, 486(3): 75-174.

[57] NEWMAN M E J. Finding community structure in networks using the eigenvectors of matrices[J]. Physical Review E, 2006, 74(3): 36-104.

[58] PALLA G, DERÉNYI I, FARKAS I, et al. Uncovering the overlapping community structure of complex networks in nature and society[J]. Nature, 2005, 435(7043): 814-818.

[59] ROSVALL M, BERGSTROM C T. Maps of information flow reveal community structure in complex networks[J]. PNAS, 2008, 105(4): 1118-1123.

[60] AIROLDI E M, BLEI D M, FIENBERG S E, et al. Mixed membership stochastic blockmodel[J]. Journal of Machine Learning Research, 2008, 9: 1981-2014.
[61] FORTUNATO S, BARTHELEMY M. Resolution limit in community detection[J]. Proceedings of the National Academy of Sciences, 2007, 104(1): 36-41.
[62] SALES-PARDO M, GUIMERA R, MOREIRA A A, et al. Extracting the hierarchical organization of complex systems[J]. Proceedings of the National Academy of Sciences, 2007, 104(39): 15224-15229.
[63] MUCHA P J, RICHARDSON T, MACON K, et al. Community structure in time-dependent, multiscale, and multiplex networks[J]. Science, 2010, 328(5980): 876-878.
[64] DELVENNE J C, YALIRAKI S N, BARAHONA M. Stability of graph communities across time scales[J]. Proceedings of the National Academy of Sciences, 2010, 107(29): 12755-12760.
[65] AHN Y Y, BAGROW J P, LEHMANN S. Link communities reveal multiscale complexity in networks[J]. Nature, 2010, 466(7307): 761-764.
[66] PASTOR-SATORRAS R, VESPIGNANI A. Epidemic spreading in scale-free networks[J]. Physical Review Letters, 2001.
[67] HOPCROFT J, KHAN O, KULIS B, et al. Tracking evolving communities in large linked networks[J]. Proceedings of the National Academy of Sciences, 2004, 101: 5249-5253.
[68] PALLA G, BARABÁSI A L, VICSEK T. Quantifying social group evolution[J]. Nature, 2007, 446(7136): 664-667.
[69] SONG L, KOLAR M, XING E P. Time-varying dynamic Bayesian networks[C]// The 23rd Neural Information Processing Systems. [S.l.:s.n.], 2009.
[70] XING E P, FU W, SONG L. A state-space mixed membership blockmodel for dynamic network tomography[J]. Annals of Applied Statistics, 2010, 4(2): 535-566.
[71] CLAUSET A, MOORE C, NEWMAN MEJ. Hierarchical structure and the prediction of missing links in networks[J]. Nature, 2008, 453: 98-101.
[72] ANDERSON R M, MAY R M C. Infectious diseases of humans: dynamics and control[M]. Oxford: Oxford Science Publications, 1992.
[73] HETHCOTE H W, VAN DEN DRIESSCHE P. Two SIS epidemiologic models with delays[J]. Journal of Mathematical Biology, 2000, 40: 3-26.
[74] NOH J D, RIEGER H. Random walks on complex networks[J]. Physical Review Letters, 2004, 92(11): 118701.
[75] LÜ L, CHEN D B, ZHOU T. The small world yields the most effective information spreading[J]. New Journal of Physics, 2011, 13: 123005.
[76] ROMERO D M, MEEDER B, KLEINBERG J. Differences in the mechanics of information diffusion across topics: idioms, political hashtags, and complex contagion on Twitter[C]// The 20th International Conference on World Wide Web. New York: ACM Press, 2011.
[77] WU S, HOFMAN J M, MASON W A, et al. Who says what to whom on Twitter[C]// The 20th International Conference on World Wide Web. New York: ACM Press, 2011.
[78] LERMAN K, GHOSH R. Information contagion: an empirical study of the spread of news on Digg and Twitter social network[C]// The 4th International Conference on Weblogs and Social Media. [S.l.:s.n.], 2010.
[79] CASTILLO C, MENDOZA M, POBLETE B. Information credibility on Twitter[C]// The 20th International Conference on World Wide Web. New York: ACM Press, 2011.
[80] PHELAN O, MCCARTHY K, SMYTH B. Using Twitter to recommend real-time topical news[C]// The 3rd ACM Conference on Recommender Systems. New York: ACM Press, 2009.
[81] LERMAN K, HOGG T. Using a model of social dynamics to predict popularity of news[C]// The 19th International Conference on World Wide Web. New York: ACM Press, 2010.
[82] CHENG X Q, GUO J F, JIN X L. A retrospective of web information retrieval and mining[J]. Journal of Chinese Information Processing, 2011, 25(6): 111-117.
[83] 沈华伟，靳小龙，任福新，等. 面向社会媒体的舆情分析[J]. 中国计算机学会通讯, 2012, 8(4): 32-36.
[84] YANG J, LESKOVEC J. Patterns of temporal variation in online media[C]// The 4th ACM International Conference on Web Search and Data Mining. New York: ACM Press, 2011.
[85] BOLLEN J, MAO H, ZENG X. Twitter mood predicts the stock market[J]. Journal of Computational Science, 2011, 2(1): 1–8.
[86] HU J, WANG G, LOCHOVSKY F, et al. Understanding user's query intent with Wikipedia[C]// The 18th International Conference on World Wide Web. New York: ACM Press, 2009.
[87] WU L, YANG L, YU N, et al. Learning to tag[C]// The

18th International Conference on World Wide Web. New York: ACM Press, 2009.

[88] LIU D, HUA X S, YANG L, et al. Tag ranking[C]// The 18th International Conference on World Wide Web. New York: ACM Press, 2009.

[89] LUO P, LIN F, XIONG Y, et al. Towards combining web classification and web information extraction: a case study[C]// The 15th ACM SIGKDD International Conference on Knowledge Discovery and Data Mining. New York: ACM Press, 2009.

[90] QI G J, HUA X S, ZHANG H J. Learning semantic distance from community-tagged media collection[C]// ACM Multimedia. New York: ACM Press, 2009.

[91] XUE G R, HAN J, YU Y, et al. User language model for collaborative personalized search[J]. ACM Transactions on Information Systems, 2009, 27(2): 1-28.

[92] LI B, YANG Q, XUE X. Transfer learning for collaborative filtering via a rating-matrix generative model[C]// ICML. [S.l.:s.n.], 2009.

[93] SCHROEDER W J, ZARGE J A, LORENSEN W E. Decimation of triangle meshes[J]. Computer Graphics, 1992, 26(2): 65-70.

[94] RENZE K J, OLIVER J H. Generalized unstructured decimation[J]. IEEE Computer Graphics and Applications, 1996, 16(6): 24-32.

[95] HOPPE H. Progressive meshes[C]// SIGGRAPH. New York: ACM Press, 1996.

[96] HOPPE H. View-dependent refinement of progressive meshes[C]// SIGGRAPH. New York: ACM Press, 1997.

[97] CHOPRA P, MEYER J. Tetfusion: an algorithm for rapid tetrahedral mesh simplification[C]// The Visualization 2002. Washington: IEEE Computer Society, 2012.

[98] STAADT O G, GROSS M H. Progressive tetrahedralizations[C]// The Visualization'98. Los Alamitos: IEEE Computer Society, 1998.

[99] EVANS W, KIRKPATRICK D, TOWNSEND G. Right triangular irregular networks[R]. 1997.

[100] MIRANTE A, WEINGARTEN N. The radial sweep algorithm for constructing triangulated irregular networks[J]. IEEE Computer Graphics and Applications, 1982, 2(3): 11-13, 15-21.

[101] FOWLER R J, LITTLE J J. Automatic extraction of irregular network digital terrain models[J]. Computer Graphics, 1979, 13(2): 199-207.

[102] SILVA C T, MITCHELL J S B, KAUFMAN A E. Automatic generation of triangular irregular networks using greedy cuts[C]// The Visualization'95. Los Alamitos: IEEE Computer Society, 1995.

[103] CIGNONI P, GANOVELLI F, GOBBETTI E, et al. BDAM: batched dynamic adaptive meshes for high performance terrain visualization[C]// The 24th Annual Conference of the European Association for Computer Graphics. Blackwell: IEEE Computer Society, 2003.

[104] CIGNONI P, GANOVELLI F, GOBBETTI E, et al. Interactive out-of-core visualization of very large landscapes on commodity graphics platforms[C]// ICVS 2003. New York: Springer-Verlag, 2003.

[105] GOODRICH M T, TSAY J J, VENGROFF D E, et al. External memory computational geometry[C]// The 34th Annual IEEE Symposium on Foundations of Computer Science. Piscataway: IEEE Press, 1993.

[106] MATIAS Y, SEGAL E, VITTER J S. Efficient bundle sorting[C]// The 11th Annual ACM-SIAM Symposium on Discrete Algorithms. New York: ACM Press, 2000.

[107] PASCUCCI V, FRANK R J. Global static indexing for real-time exploration of very large regular grids[C]// The 2001 ACM/IEEE Conference on Supercomputing (CDROM). New York: ACM Press, 2001.

[108] PLAISANT C, CARR D, SHNEIDERMAN B. Image-browser taxonomy and guidelines for designers[J]. IEEE Software, 1995, 12(2): 21-32.

[109] PLAISANT C, GROSJEAN J, BEDERSON B B. Spacetree: supporting exploration in large node link tree, design evolution and empirical evaluation[C]// The IEEE Symposium on Information Visualization. Washington: IEEE Computer Society, 2002.